Ill-Posed Problems
In Natural Sciences

Ill-Posed Problems in Natural Sciences

Proceedings of the International Conference
Held in Moscow - August 19-25, 1991

Editor-in-Chief
Andrei N. Tikhonov

Editors
Leonov A.S.
Prilepko A.I.
Vasin I.A.
Vatutin V.A.
Yagola A.G.

///VSP///
UTRECHT, THE NETHERLANDS
TOKYO, JAPAN

TVP
SCIENCE PUBLISHERS
MOSCOW, RUSSIA

VSP BV
P.O. 346
3700 AH Zeist
The Netherlands

TVP Sci. Publ.
Vavilov st. 42
117966 Moscow GSP-1
Russia

First published in 1992
ISBN 90-6764-141-3

CIP-DATA KONINKLIJKE BIBLIOTHEEK, DEN HAAG

Ill-posed problems in natural sciences: Proceedings of the International Conference / eds.: A.N. Tikhonov et al. - Utrecht: VSP / Moscow: TVP Sci. Publ.
ISBN 90-6764-141-3 bound
NUGI 8011
Subject heading: natural sciences

Printed in Russia by Contact, Moscow

CONTENTS

Part 2. INVERSE PROBLEMS IN MATHEMATICAL PHYSICS

PREFACE

This volume contains the Proceedings of the First International Conference "Ill-Posed Problems in Natural Sciences", which was held in Moscow during the week of August 19–25, 1991. The conference was managed by Moscow State University and Keldysh Institute of Applied Mathematics and sponsored by several institutes from Moscow, Novosibirsk, Ekaterinburg, Krasnoyarsk.
A great interest to ill-posed problems and their applications leads to sufficiently large amount of participants: USSR — 210, Japan — 8, USA — 4, Italy — 3, Germany — 2, Australia — 1, Austria — 1, Canada— 1, Finland — 1, France — 1, Sweden — 1, Czechoslovakia — 1.

The Conference had five plenary sessions with following plenary lectures:

Monday, August 19
A.N. Tikhonov (USSR). "Problems with approximate information". *M.M. Lavrentiev* (USSR). "Mathematical problems of tomography". *A.A. Samarsky, P.N. Vabishchevich* (USSR). "Regularization of difference schemes".

Tuesday, August 20
V.G. Romanov (USSR). "Stability in inverse problems of geoelectrics". *A.V. Goncharsky* (USSR). "Computational diagnostics and image processing". *A.G. Ramm* (USA). "Property C and inverse scattering problems".

Wednesday, August 21
A.S. Leonov, A.G. Yagola (USSR). "Tikhonov's approach for constructing regularizing algorithms". *H.W. Engl, A. Neubauer* (Austria). "Convergence rates for Tikhonov regularization of linear and nonlinear ill-posed problems". *A.B. Bakushinsky* (USSR). "Iterative methods for solving nonlinear ill-posed problems and applications". *A.V. Goncharsky, A.M. Cherepashchuk, A.G. Yagola* (USSR). "Inverse problems of astrophysics".

Thursday, August 22
G. Vainikko (Estonia). "Discretization and regularization of ill-posed problems with non-compact operators". *A.I. Prilepko* (USSR). "Inverse nonlocal problems and predictioncontrollable systems of equations in mathematical physics". *G. Anger* (Germany). "A measure theoretic approach for studying inverse problems relative to wave equation and the heat equation". *A. Lorenzi* (Italy). "Identification problems for integrodifferential equations".

Friday, August 23
M.Z. Nashed (USA). "Ill-posed problems and signal processing". *A.M. Fedotov* (USSR). "Constructing the numerical algorithms for ill-posed problems". *V.V. Vasin* (USSR). "Ill-posed problems and iterative approximation of fixed points of pseudocontractive mappings".

In addition to plenary sessions 14 sections were organized:

Section 1. Inverse problems in mathematical physics. (Co-chairmen: *M.M. Lavrentiev, A.I. Prilepko*).

Section 2. Inverse problems for differential equations. (Co-chairmen: *Yu.E. Anikonov, V.G. Romanov*).

Section 3. Inverse spectral problems and related topics. (Co-chairmen: *A.M. Denisov, V.G. Cherednichenko*).

Section 4. Statistical methods. (Co-chairmen: *A.V. Krianev, A.M. Fedotov*).

Section 5. Iterative methods. (Co-chairmen: *A.B. Bakushinsky, G.M. Vainikko*).

Section 6. Variational methods. (Co-chairmen: *J.T. King, V.P. Tanana, F.P. Vasiliev*).

Section 7. Numerical methods. (Co-chairmen: *L. Elden, A.S. Leonov, A.A. Samarsky*).

Section 8. Special problems of the regularization theory. (Co-chairmen: *A.D. Iskenderov, V.A. Morozov, M.Z. Nashed*).

Section 9. Inverse problems in the control theory. (Co-chairmen: *Yu.S. Osipov, V.V. Vasin*).

Section 10. Tomography and computational diagnostics. (Co-chairmen: *A.V. Goncharsky, V.P. Palamodov*).

Section 11. Inverse problems of electro-, hydro-, gazo-dynamics. (Co-chairmen: *E.V. Zakharov, A.G. Sveshnikov*).

Section 12. Inverse problems in geophysics and astro-physics. (Co-chairmen: *V.I. Dmitriev, V.N. Strakhov, A.M. Cherepashchuk*).

Section 13. Inverse heat transfer problems. (Co-chairmen: *O.M. Alifanov, E.A. Artyukhin*).

Section 14. Inverse problems in optics and spectroscopy. (Co-chairmen: *A.V. Tikhonravov, A.G. Yagola*).

Some lectures and talks presented to the conference are included in this book. Regretfully not all plenary lecturers prepared their lectures for publication and some limitations did not allow to all authors of communications to take part in this volume.

We wish to acknowledge with thanks all contributors and participants, who made the conference a success, and to TVP Science Publishers for the possibilities to publish this volume. We greatly appreciated the presence of so many distinguished scientists in Moscow. We hope that such conferences will become regular.

Professor A.N. Tikhonov,
Chairman of the National Organizing Committee

Professor A.G. Yagola,
Secretary of the National Organizing Committee

Part One

THEORY AND METHODS OF SOLVING ILL-POSED PROBLEMS

Ill-Posed Problems in Natural Sciences, pp. 3 – 12
A. Tikhonov (Ed.)
1992 VSP/TVP

GENERALIZED MAXIMUM LIKELIHOOD METHOD AND ITS APPLICATION FOR SOLVING ILL-POSED PROBLEMS

V.Ya. ARSENIN and A.V. KRIANEV
Moscow Engineering Physics Institute, Kashirskoe shosse, 31,
115409 Moscow, Russia

ABSTRACT

When solving applied problems one can often use certain prior information on the desired solution. Two classical, commonly used methods allowing to employ additional prior information are the minimax method and the Bayes method.
In this paper, one more method for using prior information is proposed which is a generalization of the maximum likelihood method. This method, named the "generalized maximum likelihood method" (GMLM), enables one to make use of prior information on the solution in stochastic and deterministic forms. The GMLM can be applied for the estimation of solutions of systems of algebraic equations (linear and nonlinear).
Additional prior information on the desired vector is supposed to be present except for the case of the system of main equations. This additional prior information can be given in various forms.
The main problem to be solved with the aid of the GMLM is to find optimal estimators of the desired vector taking into account additional prior information.
A great many of applied problems are reducible to systems of algebraic equations, among them problems of processing and interpretation of complicated physical experiments. Some of these problems, for example, inverse problems of mathematical physics, are ill-posed and therefore unstable even with respect to small errors in measurable quantity. Effective methods and algorithms based on the GMLM are proposed which yield optimal approximate solutions of input systems; these solutions are stable with respect to deviations of input data and prior assumptions.
The efficiency of the GMLM is illustrated with the examples of certain applied problems. A robust method for solving unstable problems including computerized tomography problems, based on the GMLM, is proposed.

1. INTRODUCTION

Approximate solution of many applied problems, including the problems of mathematical physics and the interpretation of results of a physical experiment, reduces to systems of equations of the form

$$Au = f_T, \tag{1}$$

where A is a linear or nonlinear operator acting from a Euclidean space $\mathbf{E}^m$ into another Euclidean space $\mathbf{E}^n$ and u is an unknown vector.

When solving certain problems, a measurable vector is known instead of a coexact vector $f = f_T + \xi$, where ξ is an error vector and the problem is to find an approximate solution of system (1).

In many problems the vector ξ is random and its distribution law or some characteristics of this law are known. Under these conditions it is natural to use the stochastic approach for finding an approximate solution of the system. This stochastic approach takes into account the existing information on the distribution law of the error vector and enables one to get an approximate solution with minimum possible deviation from the desired exact solution.

The stochastic approach for solving ill-posed problems and, among them, system (1), was used in (Tikhonov and Arsenin, 1986; Lavrentiev and Vasiliev, 1966; Turchin et al., 1970; Jukovsky, 1972; Muravieva, 1973; Meleshko, 1980; Kuks and Olman, 1972; Lavrentiev and Fedotov, 1982; Fedotov, 1982; Fedotov, 1988; Hocking et al., 1976; Hoerl, 1962; Hoerl and Kennard, 1970; Marquardt, 1970; Rao, 1976) and many other papers.

Different methods for solving approximately ill-posed problems by using the stochastic approach employ different kinds of prior information about the desired solution. Basically, there are two methods taking into account the prior information when solving ill-posed problems using the stochastic approach. These are the Bayes method and the minimax method.

The Bayes method treats the desired solution u as a random vector, and the prior information is represented by the probability density function of a random vector u. As an approximate solution of problem (1) (the Bayes estimator $\hat{u}_B$), one of the characteristics of posterior distribution (mean, mode, etc.) is given.

If the system (1) is linear (A is an $n \times m$ matrix) and both the distribution ξ and the prior distribution are normal ($\mathcal{N}(0, \sigma_f^2 K_f)$ and $\mathcal{N}(u_0, \sigma_a^2 K_a)$, respectively), then the Bayes estimator is linear (as the mode of posterior distribution is given):

$$\hat{u}_B = \left(\frac{1}{\sigma_f^2} A^T K_f^{-1} A + \frac{1}{\sigma_a^2} K_a^{-1}\right)^{-1} \left(\frac{1}{\sigma_f^2} A^T K_f^{-1} f + \frac{1}{\sigma_a^2} K_a^{-1} u_0\right). \tag{2}$$

Thus, the Bayes method makes use of the prior information about the desired stochastic solution and, moreover, interprets this solution as a random vector. To our mind, this restricts the applicability of the Bayes method to solving applied problems. The Bayes method has been used for solving ill-posed problems in (Lavrentiev and Vasiliev, 1966; Turchin et al., 1970; Jukovsky, 1972; Muravieva, 1973; Meleshko, 1980).

The minimax method allows one to utilize prior information in a deterministic form which is given as a feature of the desired solution with respect to a prior set $u \in R_a$.

The minimax estimator $\hat{u}_M$ is given by

$$\hat{u}_M = \arg\min_{\hat{u} \in W,} \max_{u \in R_a} \mathbf{E}\|\hat{u} - u\|, \tag{3}$$

where **E** is the expectation and W is a given class of estimations (usually the class of linear estimations is chosen).
The "narrower" the prior set, the "more extensive" the prior information and the "closer" the minimax estimator $\hat{u}_M$ to the desired solution u.
The minimax method gives the best estimation class W under the assumption that the assumption in prior set R_a has the worst possible order u (when the norm on the right-hand side of (3) is fixed).
The minimax method has been used for solving ill-posed problems in (Kuks and Olman, 1972; Lavrentiev and Fedotov, 1982; Fedotov, 1982; Fedotov, 1988).

2. STOCHASTIC FORM OF PRIOR INFORMATION

Now we will consider another method of taking into account the prior information. This method, based on the stochastic approach, is the generalized maximum likelihood method (GMLM); it has been developed in (Krianev, 1989; Arsenin and Krianev, 1991). The GMLM allows one to exploit prior information in both the stochastic and the deterministic form.
First we will study a general scheme of the GMLM which takes into account the prior information in the stochastic form.
Assume that, in addition to system (1), we are given a realization

$$Y_1, \ldots, Y_r \tag{4}$$

of a sequence of random variables $\eta_1, \ldots, \eta_r$ which do not depend on the random variable ξ. The joint probability density $p_a(y_1, \ldots, y_r; u)$ of the sequence $\eta_1, \ldots, \eta_r$ is supposed to depend on the solution u we seek for.
A realization is called a prior sample. For solving practical problems a prior sample can be obtained as a result of additional direct or indirect measurements with respect to the desired solution u.
Let $p(x, y_1, \ldots, y_r; u)$ denote the joint probability density of the random variables $\xi, \eta_1, \ldots, \eta_r$. From what was said above it follows that

$$p(x, y_1, \ldots, y_r; u) = p_\xi(x) p_a(y_1, \ldots, y_r; u), \tag{5}$$

where $p_\xi(x)$ is the probability density of the random variable ξ.
The functions $L_g(u) = p(f - Au, Y_1, \ldots, Y_r; u)$, $L(u) = p_\xi(f - Au)$, $L_a(u) = p_a(Y_1, \ldots, Y_r; u)$ are called, respectively, the generalized input function, the prior function, and the likelihood function. Assuming that the functions $L(u)$ and $L_a(u)$ are twice continuously differentiable, we introduce matrices $I_g(u), I(u), I_a(u)$ having the elements

$$(I_g(u))_{ij} = -\mathbf{E}\frac{\partial^2 \ln L_g(u)}{\partial u_i\, \partial u_j},$$

$$(I(u))_{ij} = -\mathbf{E}\frac{\partial^2 \ln L(u)}{\partial u_i\, \partial u_j},$$

$$(I_a(u))_{ij} = -\mathbf{E}\frac{\partial^2 \ln L_a(u)}{\partial u_i\, \partial u_j}, \quad i, j = 1, \ldots, m.$$

$I_g(u), I(u), I_a(u)$ are called, respectively, the generalized input matrix, the prior matrix, and the informative matrix. Each of the matrices is symmetric and non-negative.
From (5) we obtain

$$I_g(u) = I(u) + I_a(u). \tag{6}$$

From (6) it follows that the generalized informative matrix unites the information about the desired solution u contained in the input system (1) and in the prior sample (4).
According to the GMLM, we choose the estimator

$$\hat{u}_g = \arg\max_u \; L_g(u) \tag{7}$$

as an approximate solution of system (1) on the base of prior information about the desired solution determined by the prior sample (4).
It is often more convenient to determine estimator (7) from the equality

$$\hat{u}_g = \arg\min_u \left(-\ln L(u) - \ln L_a(u)\right). \tag{8}$$

If the matrix $I_g(u)$ is nonsingular, then under some additional generalized conditions for the regularity of (7) and (8) such estimators exist and are unique (see (Krianev, 1989)).
Nonsingularity of the matrix $I_g(u)$ means that the input system (1) and the prior sample (4) contain sufficient amount of information about the desired solution to get its estimator having a finite covariance matrix. Moreover, under the same conditions, for any asymptotically unbiased estimator $\hat{u}$ found from system (1) (with the measured value f) and the prior sample (4), the generalized Kramer–Rao inequality

$$K_{\hat{u}} \geqslant I_g^{-1}, \tag{9}$$

holds, where $K_{\hat{u}}$ is the covariance matrix of the estimator $\hat{u}$ and I_g^{-1} is the inverse matrix for $I_g(u)$.
If we have equality in (9), then estimators $\hat{u}$ are said to be jointly effective.
Let system (1) be linear, $\xi \sim \mathcal{N}(0, \sigma_f^2 K_f)$, Y_i be independent realizations of the same random variable $\eta \sim \mathcal{N}(u, \sigma_a^2 K_a)$. Then estimator (7) exists, is unique and linear, and is given by the equality

$$\hat{u}_g = \left(\frac{1}{\sigma_f^2} A^T K_f^{-1} A + \frac{r}{\sigma_a^2} K_a^{-1}\right)^{-1} \left(\frac{1}{\sigma_f^2} A^T K_f^{-1} f + \frac{1}{\sigma_a^2} K_a^{-1} \sum_{i=1}^{r} Y_i\right). \tag{10}$$

Estimator (10) is unbiased and, due to the relations

$$I(u) = \frac{1}{\sigma_f^2} A^T K_f^{-1} A, \qquad I_a(u) = \frac{r}{\sigma_a^2} K_a^{-1},$$

the generalized Kramer–Rao inequality with the covariance matrix $K_{\hat{u}} = \left(\frac{1}{\sigma_f^2} A^T K_f^{-1} A + \frac{r}{\sigma_a^2} K_a^{-1}\right)^{-1}$ turns into equality.

Therefore, estimator (10) is jointly effective. It is worth mentioning that the informative matrices here do not depend on the desired vector u.
If $r = 1$, then estimator (10) is analogous to the Bayes estimator (2), where the role of the random realization Y_1 is played by u_0, the centre of the prior distribution. For estimator (10), the prior informative matrix is nonsingular, while the input informative matrix can be nonsingular as well as singular. The latter circumstance means that input system (1) contains no sufficient information about the desired solution to find its unbiased estimator (based on system (1) only) with the finite covariance matrix.
When solving practical problems, one often encounters the situation where, together with the desired solution u, the statement of the input problem contains some other unknown parameters v. As a rule, such unknown additional parameters v are characteristics of the distribution of the random variables $\xi, \eta_1, \ldots, \eta_r$. In these cases the likelihood function L_g depends not only on u, but also on v. Instead of estimator (7), we define the estimator of the generalized maximum likelihood method by

$$\hat{w}_g = \arg\max_{w} \; L_g(w), \tag{11}$$

where $w = (u, v)$ is comprised of the desired solution u and the vector of the unknown parameters v.
Assume that $\xi \sim \mathcal{N}(0, \sigma_f^2 K_f)$, $\eta \sim \mathcal{N}(u, \sigma_a^2 K_a)$, the elements of the prior sample (4) are independent realizations, and the "noise levels" σ_f^2 or σ_a^2 are unknown. Then, instead of estimator (10), we obtain, according to (11), the GMLM estimators

$$\hat{u}_g = \left(A^T K_f^{-1} A + \hat{\alpha} K_a^{-1}\right)^{-1} \left(A^T K_f^{-1} f + \hat{\alpha} K_a^{-1} \overline{Y}\right),$$

$$\hat{\alpha} = r\hat{\sigma}_f^2 / \hat{\sigma}_a^2, \qquad \hat{\sigma}_f^2 = \frac{1}{n}(f - A\hat{u}_g)^T K_f^{-1}(f - A\hat{u}_g), \tag{12}$$

$$\hat{\sigma}_a^2 = \frac{1}{rm} \sum_{i=1}^{r} (\hat{u}_g - Y_i)^T K_a^{-1} (\hat{u}_g - Y_i), \qquad \overline{Y} = \frac{1}{r} \sum_{i=1}^{r} Y_i.$$

Estimators (12), same as estimator (1), are effective. Comparing estimators (12) with the approximate solution obtained by the regularization method with the quadratic stabilizer $\Omega(u) = (u - \overline{Y})^T K_a^{-1} (u - \overline{Y})$, we conclude that the quantity $\hat{\alpha}$ from (12) can be interpreted as the optimal value of the "regularization parameter".
Under some regularity conditions estimators (7) are consistent, in particular, under certain conditions they converge to the desired solution u. Estimator (10) converges to u in mean square as $r \to \infty$; if $\sigma_f \to 0$, then estimator (10) converges to a solution of the system $Au = f_T$ in mean square. In particular, if the input informative matrix is not singular, then under the condition $\sigma_f \to 0$, $\hat{u}_g$ converges to u in mean square. All the results stated above have been proved in (Krianev, 1989).

3. DETERMINISTIC FORM OF PRIOR INFORMATION

Now we will consider an application of the GMLM scheme allowing one to use the deterministic form of prior information.

Let the prior information about the desired solution be given as a characteristic of u determined by a prior set R_a (the same situation is considered usually when one uses the minimax method).

The GMLM estimator $\hat{u}_g$ is determined from

$$\hat{u}_g = \arg\max_{u \in R_a} L(u). \tag{13}$$

Let (1) be a linear system, $\xi \sim \mathcal{N}(0, \sigma_f^2 K_f)$, and let the prior set R_a be a multidimensional ellipsoid $R_a = \{u\colon (u-u_0)^T B(u-u_0) \leqslant 1\}$, where B is a symmetric positive definite matrix. Then estimator (13) exists, is unique and is given by the formula

$$\hat{u}_g = \begin{cases} u^*, & \text{if } u^* \in R_a \\ (A^T K_f^{-1} A + \alpha^* B)^{-1}(A^T K_f^{-1} f + \alpha^* B u_0), & \text{if } u^* \overline{\in} R_a \\ & \text{or } u^* \text{ does not exist,} \end{cases} \tag{14}$$

where $u^* = (A^T K_f^{-1} A)^{-1} A^T K_f^{-1} f$ is the LSM-estimator, α^* can be found from the equality $(\hat{u}_g - u_0)^T B(\hat{u}_g - u_0) = 1$. The family of estimators $\hat{u}_g$ of the form (14) converges in mean square to a solution of system (1), as $\sigma_f \to 0$.

In particular, if $I(u)$ is a nonsingular matrix, then the family $\hat{u}_g$ converges to u as $\sigma_f \to 0$. The above statements have been proved in (Krianev, 1989).

It is worth mentioning that if $u_0 = 0$ and K_f is the identity matrix, then estimator (14) is obtained by the regularization method as an approximate solution. In (Pergament and Telkovskaya, 1979) a connection between the maximum likelihood method and the regularization method is established in the case where the distribution of the error vector ξ is normal.

The GMLM makes it possible to use simultaneously the prior information about the desired solution in stochastic and deterministic forms. Let the prior information concerning the prior sample (4) and the relationship $u \in R_a$ be given in stochastic and deterministic forms. Then the GMLM estimator $\hat{u}_g$ is given by

$$\hat{u}_g = \arg\max_{u \in R_a} L_g(u). \tag{15}$$

Assume that (1) is a linear system, $\xi \sim \mathcal{N}(0, \sigma_f^2 K_f)$, $Y_i \sim \mathcal{N}(u, \sigma_a^2 K_a)$, $i = 1, \ldots, r$; Y_i is independent, and R_a is a multidimensional ellipsoid. Then estimator (15) exists, is unique and has the form

$$\hat{u}_g = \begin{cases} u^*, & \text{if } u^* \in R_a \\ (A^T K_f^{-1} A + \frac{r\sigma_f^2}{\sigma_a^2} K_a^{-1} + \alpha^* B)^{-1} & \\ \times (A^T K_f^{-1} f + \frac{\sigma_f^2}{\sigma_a^2} K_a^{-1} \sum_{i=1}^r Y_i + \alpha^* B u_0), & \text{if } u^* \overline{\in} R_a, \end{cases}$$

where $u^* = (A^T K_f^{-1} A + \frac{r\sigma_f^2}{\sigma_a^2} K_a^{-1})^{-1}(A^T K_f^{-1} f + \frac{\sigma_f^2}{\sigma_a^2} K_a^{-1} \sum_{i=1}^r Y_i)$ and α^* is found from the equality $(\hat{u}_g - u_0)^T B(\hat{u}_g - u_0) = 1$.

4. ROBUST APPROXIMATE SOLUTIONS

Now we will examine the case where there are s normally distributed realizations $f_j \in \mathbf{E}^n$, $j = 1,\dots,s$, of the right-hand side of equation (1) and $\sigma_f^2 K_f = \text{diad}\,(\sigma_1^2,\dots,\sigma_n^2)$. In this case, according to what was said above, the GMLM estimator of equation solution (1) is given by

$$\hat{u}_g = \arg\max_u \prod_{k=1}^{n}\prod_{j=1}^{s} p_{\mathcal{N}}\left(\frac{f_{jk}-(Au)_k}{\sigma_k}\right)\prod_{i=1}^{r} p_\eta(Y_i;u), \tag{16}$$

where $p_{\mathcal{N}}(x)$ is the density of the normal distribution and f_{jk} is the j-realization of the kth component of the vector f.

In practice, when neither true error values nor their norms are known, a part of component measurements of the vector f, can have dispersions which exceed considerably the given values σ_k^2. Such error measurements are usually referred to as "large ejections" or "blockages". A large ejection can happen as a result of misfunctioning of the measurement equipment or, for example, as a result of some unforeseen effect. The GMLM estimator (16) has no optimal properties if a large ejection occurs.

Theoretical and numerical experiments show that traditional methods of solving the ill-conditioned system (1), which do not take into account the large ejections, give improper estimators of a solution. In view of that it is desirable to propose robust (in particular, stable to blockage) methods which would provide acceptable estimators of a solution, by excluding or diminishing effects of the large ejections. According to (Arsenin et al., 1989) we will examine, instead of (16), the GMLM estimator

$$\hat{u}_r = \arg\max_u \prod_{k=1}^{n}\prod_{j=1}^{s} p_r\left(\frac{f_{jk}-(Au)_k}{\sigma_k}\right)\prod_{i=1}^{r} p_\eta(Y_i;u), \tag{17}$$

where $p_r(x)$ is the robust Huber density of probability:

$$p_r(x) = \begin{cases} C\,\exp\{-x^2/2\}, & |x| \leqslant K \\ C\,\exp\{-K|x| + K^2/2\}, & |x| > K, \end{cases}$$

and the Huber parameter K depends on the blockage value and on the error of the blocked part under consideration (see (Huber, 1981; Arsenin et al., 1989)).

To clarify the problem we will examine now the special case $\sigma_k = \sigma$, $k = 1,\dots,n$. Let $\eta \sim \mathcal{N}(u, \sigma_a^2 K_a)$ be given and let $\alpha B = (rK_a^{-1})/\sigma_a^2$. Then estimator (17) takes thc form

$$\hat{u}_r = \arg\min_u \left\{\sum_{k=1}^{n}\sum_{j=1}^{s} \rho(f_{jk}-(Au)_k) + \alpha(u-\overline{Y})^T B(u-\overline{Y})\right\}, \tag{18}$$

where

$$\rho(x) = \begin{cases} x^2/2K, & |x| \leqslant K, \\ |x| - K/2, & |x| > K, \end{cases}$$

is the Huber function, α is the regularization parameter, and $\overline{Y} = \frac{1}{r}\sum_{i=1}^{r} Y_i$. Suppose that more than a half of s realizations of each component of the vector f are normally distributed random variables with dispersion σ^2 and the remaining realizations of each component may have distributions with arbitrary large dispersions.

Let u be the unique solution of equation (1). Then if the error of measuring the unblocked part σ^2 tends to zero and the condition $\lim_{\sigma\to 0} \sigma/\alpha = 0$ is satisfied, the sequence of estimators (18) converges to u in probability:

$$p \lim_{\sigma \to 0} \|\hat{u}_r - u\| = 0. \tag{19}$$

If it is known a priori that $u \in R_a$, where R_a is a compact set in $\mathbf{E}^m$, then the following can be used instead of (18) as a robust regularized solution estimator:

$$\hat{u}_r = \arg\min_{u \in R_a} \sum_{k=1}^{n} \sum_{j=1}^{s} \rho(f_{jk} - (Au)_k). \tag{20}$$

If the above conditions hold, then formula (19) is valid for the sequence of estimators (20).

In order for the above statements about the convergence of a sequence of robust regularized solutions to the exact solution u to be valid it is necessary that the relation $0 < K < C\sigma$ hold, where C is a positive constant not depending on σ.

Asymptotical properties of robust regularized solutions can be investigated not only in the statistical form of the problem under consideration but also in the deterministic form.

Assume that more then a half of s realizations of the k-component of vector f lay in σ-neighborhood of a point f_{Tk}, while the remaining realizations have larger errors. The parameter K will be chosen to satisfy the relation $0 < K < C\delta$, where C does not depend on δ.

If u is the unique solution of equation (1), then the sequence of estimators (18) converges to u: $\lim \|\hat{u}_r - u\| = 0$, $\delta \to 0$, $\alpha \to 0$, $\delta/\alpha \to 0$.

If equation (1) has more then one solution and the operator A is linear, then $\|\hat{u}_r - u_B\| \to 0$, $\delta \to 0$, $\alpha \to 0$, $\delta/\alpha \to 0$, where $u_B = \arg\min_{u:\, Au = f_T} (u - \overline{Y})^T B (u - \overline{Y})$.

In the case where $u \in R_a$, R_a being a compact set, the relation $\lim \|\hat{u}_r - u\| = 0$, $\delta \to 0$, remains valid for the regularized solution (20).

It should be noted that, if large ejections occur, then none of the properties concerning the convergence of robust regularized solutions to the exact ones are valid for the traditional nonrobust smoothing functional for any choice of dependence of parameter α on the error of the unblocked part of a measurement.

All the statements of this section were proved in (Arsenin et al., 1989).

We note that in (Arsenin et al., 1989) an example of a robust smoothing functional for equation (1) is given, u being an element of a Hilbert space $\mathbf{H}$, $f \in C(a,b)$, A being an operator from $\mathbf{H}$ into $C(a,b)$, and some properties of the relative regularized solutions are investigated.

5. APPLICATION OF GMLM TO SOLVING SOME APPLIED PROBLEMS

The GMLM developed in this paper is used in various fields of science for solving many applied problems of experiment processing and interpretation.

Approximate solutions found by GMLM and given by formulas (10), (12), (14) were used for solving inverse problems of the theory of reactors, in particular, for correction of calculating the neutron distributions and temperature in the active zone of a reactor in the case where one takes into account the reading of inside supervision transducer (Krianev and Lisov, 1984).

To find the nonlinear robust approximate solutions given by formulas (17), (18), (20), efficient iterative procedures and numerical algorithms were developed in (Krianev et al., 1985; Krianev, 1989; Arsenin et al., 1989; Krianev and Tsupko-Sitnikov, 1989). Using these algorithms, problems of data interpretation of neutron-physical experiments and reduction of fissionable nuclides characteristics (Volkov et al., 1986)), problems of reduction of beam characteristics of charged particles according to experimental data (Krianev and Tsupko-Sitnikov, 1989) and problems of emission tomography were solved.

The results of solving the problems described above by robust regularization methods show high effeciency and stability of the methods.

REFERENCES

Arsenin, V.Ya. and Krianev, A.V. (1989). Statistical methods of solving systems of algebraic equation and their applications for processing and interpretation of results. In: *Mathematical Processing and Interpreting of Physical Experiments Results*, Energoatomizdat, Moscow, pp. 3–15 (in Russian).

Arsenin, V.Ya. and Krianev, A.V. (1991). Generalized maximum likelihood method for solving finite dimensional ill-posed problems. *J. Comp. Math. and Math. Phys.* **31**, 643–653 (in Russian).

Arsenin, V.Ya., Krianev, A.V., and Tsupko-Sitnikov, M.V. (1989). Applications of robust methods for solving ill-posed problems. *J. Comp. Math. and Math. Phys.* **29**, 653–661 (in Russian).

Fedotov, A.M. (1982). *Linear Ill-posed Problems with Random Error in Data.* Nauka, Novosibirsk (in Russian).

Fedotov, A.M. (1988). Designing of optimal decisive procedures for operator equations. *J. Comp. Math. and Math. Phys.* **28**, 963–970 (in Russian).

Hocking, R.R., Speed, F.M., and Lynn, M.J. (1076). A class of based estimators in linear regression. *Technometrics* **18**, 425–437.

Hoerl, A.E. (1962). Application of ridge analysis to regression problems. *Chemical Eng. Prog.* **58**, 23–35.

Hoerl, A.E. and Kennard, R.W. (1970). Ridge regression: Biased estimation for nonortogonal problems. *Technometrics* **12**, 55–67.

Huber, P.J. (1981). *Robust Statistics.* John Wiley, New York.

Jukovsky, E.L. (1972). Statistical regularization of algebraic equations systems. *J. Comp. Math. and Math. Phys.* **12**, 185–191 (in Russian).

Krianev, A.V. (1986). Statistical form of least square regularized method by A.N. Tikhonov. *Dokl. Akad. Nauk SSSR* **291**, 780–785 (in Russian).

Krianev, A.V. (1989). *Analitical and numerical methods for solving correctness problems of mathematical theory of reactors.* Doctor Thesis, United Inst. Nucl. Res. Dubna (in Russian).

Krianev, A.V. and Lisov, D.A. (1984). Correction of calcularity field according to transducer system read-out. In: *Physical Researches Automatization,* Energoatomizdat, Moscow, pp. 169–172 (in Russian).

Krianev, A.V., Lisov, D.A., and Tsupko-Sitnikov, M.V. (1985). Nonlinear stochastic regularization. In: Studies of all-Union school-seminar *Theory and Methods for Ill-Posed Problems Solving and its Application.* Ed. by A.N. Tikhonov. Saratov Univ. Press., Saratov, pp. 88–89 (in Russian).

Kuks, J. and Olman, V. (1972). Minimax linear estimation of regression coefficients. *Izv. Akad. Nauk EstSSR,* Ser. Math. and Phys. **21**, 66–72 (in Russian).

Lavrentiev, M.M. and Vasiliev, V.G. (1966). About setting some ill-posed problems of mathematical physics. *Siberian Math. J.* **7**, 358–364 (in Russian).

Lavrentiav, M.M. and Fedotov, A.M. (1982). About setting some problems of mathematical physics with random output data. *J. Comp. Math. and Math. Phys.* **22**, 133–143 (in Russian).

Marquardt, D.W. (1970). Generalised inverses, ridge regression, biased linear estimation and nonlinear estimation. *Technometrics* **12**, 591–612.

Meleshko, V.M. (1980). Solution of ill-posed stochastic algebraic equations. *J. Comp. Math. and Math. Phys.* **20**, 11–25 (in Russian).

Muravieva, M.V. (1973). About optimal and limited properties of Bayes solution of a linear system of algebraic equations. *Comp. Math. and Math. Phys.* **13**, 819–828 (in Russian).

Pergament, A.H. and Telkovskaia, O.V. (1979). *Regularization method and maximum likelihood method solving.* Preprint No 111. Appl. Math. Inst. Academy Sci. USSR.

Rao, C.R. (1976). Estimation of parametres in a linear model. *Ann. Statist.* **4**, 1023–1037.

Tikhonov, A.N. and Arsenin, V.Ya. (1977). *Solution of Ill-Posed Problems.* John Wiley, New York.

Turchin, V.F., Kozlov, V.P., and Malkevich, M.C. (1970). The usage of mathematical statistics methods for solving ill-posed problems with random errors in data. *Uspekhi Phys. Nauk* **102**, 345–386 (in Russian).

Volkov, N.G., Dorogov, V.I., Konshin, V.A., and Krianev, A.V. (1986). *Complex Programme of Regression Restoration and Estimation* σ_f, σ_t, α, $\bar{\nu}$ *Corresponding to* ^{235}U. Preprint No 1. Nuclear Energie Inst., Minsk (in Russian).

Ill-Posed Problems in Natural Sciences, pp. 13 – 17
A. Tikhonov (Ed.)
1992 VSP/TVP

ITERATIVE METHOD FOR THE SOLUTION OF NONLINEAR ILL-POSED PROBLEMS AND THEIR APPLICATIONS

A.B. BAKUSHINSKII
VNIISI, Prospekt 60-letiya Oktyabrya, 9, 117312 Moscow, Russia

ABSTRACT

Modern iterative methods are considered.

1. A mathematical model of some system, phenomenon or process induces a mapping G associated with this model.
The mapping connects the "input" and the "output" of the model and is usually known implicitly. Moreover, in many cases the input of the mapping is only given approximately (in the sense of a certain metric).
To explain this let us consider a typical mathematical model represented in the form of operator equation

$$Ax = f, \tag{1}$$

where A is a "given" linear or nonlinear operator, f is the input of the model, and x is its output. Here, x and f are elements of some metric spaces (Banach or Hilbert ones). The mapping G for model (1) is the inverse operator A^{-1}:

$$x = A^{-1}f.$$

The input data may be included in the mathematical model not only via the right-hand side of equation (1) but also directly via the operator A.
In this case the mathematical model is represented in the form

$$F(x) = 0. \tag{2}$$

The mapping G connects these input data with the output data x.
Models of optimization, game theory, etc. may be reduced to problems of solving the variational inequality:
to find x such that, for any $z \in Q \subseteq H$,

$$(F(x), x - z) \leq 0. \tag{3}$$

In (3) Q is a closed convex set in a Hilbert space H, F is an operator (linear or nonlinear) $H \to H$ and $(\cdot, \cdot)$ is the inner product in H.

In this case, the above mentioned mapping G is the one that connects some input data with the output data. The mapping G connected with models (1)–(3) may be continuous or not (it is assumed that the input/output data lie in some metric spaces). This is very important for the construction of approximate methods for finding the values of G. If G is not continuous, model (1)–(3) is said to be ill-posed. We consider in what follows the problem of construction of abstract numerical methods for finding the values of G in the case where G is not continuous (correspondingly, model (1)–(3) is ill-posed).

Any numerical method for (1)–(3), no matter ill-posed or well-posed, is based on some scheme of approximation of the mapping G. This scheme is based on the construction of a sequence of continuous mappings $R_n, \lim n \to \infty\ R_n x = G(x)$. The evaluation of $R_n x$ requires the computation of $F(\cdot)$ and $A(\cdot)$ in (1)–(3), or their approximations.

Unfortunately, for the ill-posed case this approximation can not always be realized. In case this family can be constructed, it is necessary to add "a stopping rule" connecting the admittable choice of n with the error of input data (Bakushinskii and Goncharskii, 1989).

To solve the well-posed problem (1)–(3), some iterations are often used. Principles of their construction are well-known to specialists. Iterated approximations of the solutions of ill-posed problems (1)–(3) are investigated and used also.

The linear case (A in (1) is a linear operator in a Banach or Hilbert space) has been investigated in detail. The main result of this investigation is that any linear "standard" iteration (the operator of a step is a linear operator) in a Hilbert space generates a sequence R_n such that $\lim n \to \infty\ R_n f = A^{-1} f$. A set of stopping rules for these iterations was offered too (see (Bakushinskii and Goncharskii, 1989)). Many results of this kind may be extended to linear problems in Banach spaces.

For nonlinear ill-posed problems (1)–(3), the standard ways to obtain iterative methods (Newton, Gauss–Newton, etc.) are not in use. In this case iterative methods are constructed using "the principle of iterative regularization" (PIR). The idea of this principle is to combine standard iterative methods with a modification of problem (1)–(3) on each iteration.

2. The PIR was thoroughly investigated in the case where the operator F or A in (1)–(3) is a monotone operator in a Hilbert space.

The PIR is based on the fact that the unique solution x_α of the equation

$$F(x) + \alpha x = 0, \quad \alpha > 0 \tag{4}$$

or the inequality

$$(F(x) + \alpha x, x - z) \leq 0, \quad \alpha > 0 \tag{5}$$

under the condition of monotonicity of F and some other conditions (smoothness, growth, etc.) exists and there holds $\lim \alpha \to 0\ x_\alpha = x^*$, where x^* is a solution of (2) or (3).

Generally, if $\alpha > 0$, then equation (4) or inequality (5) can be solved by any common iterative method (without using F', or Newton's method, etc.).

An iterative procedure for solving (1)–(3) ($\alpha = 0$) can be devised in the following way. We select some iterative method for solving (4), (5) (a base method) and choose α_0, x_0. A step of the base method for equation (4) or inequalities (5) with

$\alpha = \alpha_0$ and the initial point x_0 generates a point x_1. If we choose α_1, then it is possible to make a step of the base iterative method with $\alpha = \alpha_1$ and the initial point x_1. It generates a point x_2, etc. For the set of base methods it is possible to indicate a priori a lot of sequences $\{\alpha_n\}$ such that $\lim n \to \infty\; x_n = x^*$, where x^* is a solution of (2) or (3). For general results, see (Bakushinskii and Goncharskii, 1989). Below is an example of the results of this kind. Let the base method be "the gradient projection method":

$$x_{n+1} = \mathcal{P}_Q(x_n - \gamma_n F(x_n)), \quad x_0 \in H, \tag{6}$$

where $\mathcal{P}_Q$ is the projector on Q.
If we apply PIR to (6) then we obtain the iteration

$$x_{n+1} = \mathcal{P}_Q(x_n - \gamma_n(F(x_n) + \alpha_n x_n)). \tag{7}$$

THEOREM 1. *Let the set of solutions of (2) or (3) be nonempty and the operator F be a monotone operator and let its Freche derivative F' satisfy the condition*

$$\sup_{z \in Q} \|F'(z)\| \leq L < \infty. \tag{8}$$

If sequences α_n, γ_n satisfy the conditions

$$0 < \alpha_n \leq 1, \quad \lim_{n \to \infty} \alpha_n = 0, \quad \sum_1^{\infty} \alpha_n^2 = \infty$$

$$\begin{gathered} \gamma_n = \theta \alpha_n, \quad 0 < \theta \leq 1/(L^2 + 1) \\ \lim_{n \to \infty} |(\alpha_{n+1} - \alpha_n)/\alpha_n^4| = 0, \end{gathered} \tag{9}$$

then $\lim n \to \infty\; x_n = x^*$, where x^* is a solution of (2) or (3) and its norm is minimal. The behavior of sequence (7) under the perturbation of the operator F can be demonstrated putting $F = A - f$ and substituting $\tilde{f}$ for f: $\|f - \tilde{f}\| \leq \delta$.

THEOREM 2. *Let α_n, γ_n satisfy conditions (9), $n(\delta)$ be an integer function of δ and let*

$$\begin{gathered} \lim_{\delta \to 0} n(\delta) = \infty, \\ \lim_{\delta \to 0} \frac{\delta}{\alpha_{n(\delta)}} = 0. \end{gathered} \tag{10}$$

Then $\lim \delta \in 0 \;\sup\; \|f - \tilde{f}\| \leq \delta\; \|x_{n(\delta)}(\tilde{f}) - x^*\| = 0$, where $x_{n(\delta)}(\tilde{f})$ is the $n(\delta)$-th term of sequence (7). Condition (10) gives the "stopping rule" for approximation (7) as wellas for many other algorithms generated by

PIR (see (Bakushinskii and Goncharskii, 1989)).
The PIR can partly be extended to equations or inequalities (2)–(3) in Banach spaces, if F is a monotone operator.
3. The condition of monotonicity of F in (1)–(3) is very restrictive. Recently some results on the iterative were obtained that make no use of this condition.

Let us consider equation (2), where F is a nonlinear operator acting from H_1 to H_2 and H_1, H_2 are Hilbert spaces. In this case for constructing iterative methods for solving (2) the same method of minimization of the functional

$$\Phi(x) = \|F(x)\|^2 \tag{11}$$

is used. A typical method for this minimization is the Gauss–Newton method:

$$x_{n+1} = x_n - [F'^*(x_n)F'(x_n)]^{-1}F'^*(x_n)F(x_n). \tag{12}$$

However, if problem (2) is ill-posed, then the linear operator F'^*F' is noninvertible and iterations (12) are not defined. There exist many formal modifications of (12) allowing to overcome this difficulty. The PIR makes use of one of them:

$$x_{n+1} = x_n - (F'^*(x_n)F'(x_n) + \alpha_n E)^{-1}(F'^*(x_n)F(x_n) + \alpha_n(x_n - \xi)). \tag{13}$$

The principle of studying the PIR formulated above gives no possibility to analyse process (3) in the general case.
New approaches to the analysis of this process are given in Theorems 3 and 4.

THEOREM 3. *Let x^* be a solution of equation (2) $\Phi(x^*) = 0$, and let the following conditions be satisfied*

$$\|F'(x)\| \le N_1, \quad \|F''(x)\| \le N_2, \tag{14}$$

$$x^* - \xi = F'^*(x^*)F'(x^*)v, \tag{15}$$

$$\alpha_n = \alpha_0 q^n, \quad 0 < q < 1, \tag{16}$$

$$\|x^* - x_0\| \le \epsilon_0, \tag{17}$$

$$\frac{N_1 N_2}{2}\frac{\epsilon_0}{\alpha_0} + \frac{\alpha_0}{\epsilon_0}\|v\| + 2N_1N_2\|v\| \le q. \tag{18}$$

Then

$$\|x_n - x^*\| \le q^n \epsilon_0, \quad n = 1, 2, \ldots. \tag{19}$$

Condition (15) plays a key role. It extends the condition of "sourcewise representability" of the solution in the linear case (Bakushinskii and Goncharskii, 1989). If problem (2) is well-posed, then condition (15) is satisfied automatically. Condition (15) may be extended are follows:

$$x^* - \xi = \varphi(F'^*(x^*)F'(x^*))v, \quad \varphi \in C^\infty[0, N_1^2]$$

This extended condition generates a number of conditions on x^* and ξ that have natural interpretation.

THEOREM 4. *Let $\Phi(x^*) = 0$ and condition (14) be satisfied. Let*

$$\|x_\alpha - x^*\| = D(\xi)\alpha + 0(\alpha),$$

$$D(\xi) < C, \tag{20}$$

$$x_\alpha = \arg\min_{x \in H}(\Phi(x) + \alpha\|x - \xi\|^2)$$

and

$$\overline{\lim}_{n\to\infty}\alpha_n = 0,$$

$$|\alpha_n/\alpha_{n+1}| = O(1), \tag{21}$$

$$\overline{\lim}_{n\to\infty}|(\alpha_n - \alpha_{n+1})/\alpha_n| \le \delta.$$

If $\|x_0 - x^*\|$ *and* $\|\xi - x^*\|$ *are sufficiently small, then*

$$\|x_n - x^*\| = O(\alpha_n).$$

If we fix $\{\alpha_n\}$ in (13), say by conditions (16) or (21), then it is possible to investigate $\{x_n\}$ under perturbations of different kinds (for instance, under the perturbation of ξ, which can be treated as an approximate variant of condition (15)). Thus a "stopping rule" for sequence (14) can be obtained.
By using an approximate variant of condition (15), it is possible to choose $\{\alpha_n\}$ in (13).
More precisely, let

$$x^* - \xi = F'^*(x^*)F'(x^*)v + \eta, \quad \|\eta\| \le \Delta.$$

There exists a sequence $\{\alpha_n\}$ such that

$$\overline{\lim}_{n\to\infty}\|x_n - x^*\| = O(\Delta).$$

4. Applications of the PIR to the condition of monotonicity of $F(x)$ in (2), (3) are given in (Bakushinskii and Goncharskii, 1989).
Iteration (13) is convenient for numerical solution of "inverse problems" (in acoustics, heat conduction, etc.). There are many examples of its practical application.

REFERENCES

Bakushinskii, A.B. and Goncharskii, A.V. (1989). *Iterative Methods for the Solution of Ill-Posed Problems.* Nauka, Moscow (in Russian).

Ill-Posed Problems in Natural Sciences, pp. 18 – 23
A. Tikhonov (Ed.)
1992 VSP/TVP

UNIFORM REGULARIZATION OF THE FIRST KIND OPERATOR EQUATIONS IN HILBERT SPACE

A.V. CHEREPAKHIN
Institute of Control Sciences, Profsojuznaya str., 65, 117806 Moscow, Russia

ABSTRACT

The complete description of convex correctness sets (in Tikhonov's sense) for the first kind operator equations in a Hilbert space is given. Our main theorem shows the necessary and sufficient conditions under which a subset of a Hilbert space can be used as a correctness set for the uniform regularization of a linear operator equation. Being reformulated in terms of L_2 space it allows us to find new noncompact correctness sets for solving the convolution type operator equations numerically.

1. INTRODUCTION

We consider as an example of improperly posed inverse problem the first kind operator equation

$$Ax = y_\delta, \tag{1}$$

$$\|y_\delta - \bar{y}\| \leq \delta, \tag{2}$$

where A: $H \to H$ is a bounded linear operator in a Hilbert space H, $\ker A = 0$, and A^{-1} is an operator with the dense domain $D(A^{-1})$. It is assumed that there exists the true solution $\bar{x}$ of problem (1), while the element $\bar{y}$, $\bar{y} = A\bar{x}$, is unknown and only the value y_δ defined by (2) is available. Since the operator A^{-1} may be not bounded, this problem is ill-posed, and in order to find an approximate solution it is necessary usually to attract an additional prior information about $\bar{x}$. In accordance with Tikhonov's theory we will assume that $\bar{x}$ belongs to a set $M \subset H$ which is called a correctness set.
To make a uniform regularization of problem (1) means to point a set M and a one-parameter family of operators R_δ: $H \to H$ such that for an arbitrary $\varepsilon > 0$ there exists $\delta > 0$ such that for all $x \in M$, $y_\delta \in H$, the inequality

$$\|y_\delta - Ax\| \leq \delta$$

implies the relation

$$\|R_\delta y_\delta - x\| \leq \varepsilon.$$

The aggregate of correctness sets is described completely only for the case when the operator A is completely continuous. It was shown by V.K.Ivanov (Ivanov,

1966), that under this condition a correctness set M can always be represented as the direct sum of a compact set and a finite-dimensional subspace of H. In the general case to obtain a uniform regularization of problem (1) one uses, as a rule, only either, compact sets or images of a sphere under the mappings determined by the monotonous functions of the modulus of the operator A. So the aim of this paper is to get as complete as possible description of this class of sets.

2. CORRECTNESS CLASSES FOR THE FIRST KIND OPERATOR EQUATIONS

As it was shown in (Ivanov, 1966), problem (1) can be regularized uniformly only when A^{-1} is uniformly continuous on AM. We denote the class of sets possessing this property by $UC(A)$.

THEOREM 1. *A convex set $M \subset H$ belongs to $UC(A)$ if and only if for any $\varepsilon > 0$ there exist $\alpha > 0$ and $\eta > 0$ such that for any $x_1 \in M$, $x_2 \in M$ satisfying the inequality*

$$\left\|(I - E_\alpha)(x_1 - x_2)\right\| \le \eta, \tag{3}$$

the inequality

$$\left\|(x_1 - x_2)\right\| \le \varepsilon \tag{4}$$

is valid also.
Here I is the identity operator and E_α is the spectral representation of the positive self-adjoint operator $(A^*A)^{1/2}$.

Proof. Since $\|Ax\| = \left\|(A^*A)^{1/2}x\right\|$ for any $x \in H$, it is easy to see that $UC(A) = UC\left((A^*A)^{1/2}\right)$.
Thus, without lost of generality, we can assume that A is a positive self-adjoint operator. Let us prove the "sufficiency" part.

Sufficiency. Let us assume that M satisfies conditions (3)–(4). In this case for any $\varepsilon > 0$ we can choose $\alpha(\varepsilon)$, $\eta(\varepsilon)$ such that for an arbitrary $x_1 \in M$, $x_2 \in M$ inequality (3) implies the inequality

$$\left\|E_\alpha(x_1 - x_2)\right\| \le \varepsilon/\sqrt{2}.$$

Let $\delta = \min\left\{\eta(\varepsilon),\, \alpha(\varepsilon)\varepsilon\sqrt{2}\right\}$. Then if

$$\|Ax_1 - Ax_2\| \le \delta, \tag{5}$$

then the inequality

$$\left\|(I - E_\alpha)(x_1 - x_2)\right\| \le \varepsilon/\sqrt{2} \tag{6}$$

is valid also since

$$\left\|(I - E_\alpha)(x_1 - x_2)\right\| \le \frac{1}{\alpha}\|Ax_1 - Ax_2\|.$$

We note also that

$$\left\|(I - E_\alpha)(x_1 - x_2)\right\| \le \eta(\varepsilon),$$

which, in its turn, implies the estimate

$$\|E_\alpha(x_1 - x_2)\| \le \varepsilon/\sqrt{2}. \tag{7}$$

Summing (6) and (7) we find that if inequality (5) holds true then

$$\|x_1 - x_2\| \le \varepsilon.$$

Thus A^{-1} is uniformly continuous on AM.

Necessity. Let us show that if Theorem 1 is valid for the sets having centre of symmetry, then it is valid also for all convex sets. Let M be a convex set. We denote by $\operatorname{abs\,conv}(M)$ the intersection of all convex sets containing M and satisfying the property: if an element Z belongs to such a set then the element $-Z$ belongs to it also.

Let us show that if the operator A^{-1} is uniformly continuous on AM, then A^{-1} is uniformly continuous on $\operatorname{abs\,conv}(AM)$. Note that any element y of $\operatorname{abs\,conv}(AM)$ can be represented as

$$y = y_1 - y_2; \quad y_1 \in AM, \quad y_2 \in AM. \tag{8}$$

Since A^{-1} is uniformly continuous on AM then for any $\varepsilon > 0$ we can find $\delta > 0$ such that if

$$\|y_1 - y_2\| \le \delta,$$

then

$$\|A^{-1}(y_1 - y_2)\| \le \varepsilon.$$

Let $y \in \operatorname{abs\,conv}(AM)$ and $\|y\| \le \delta$. Then in view of (8)

$$y = y_1 - y_2, \quad y_1 \in AM, \quad y_2 \in AM$$

and $\|y_1 - y_2\| \le \delta$.

Consequently,

$$\|A^{-1}y\| = \|A^{-1}(y_1 - y_2)\| \le \varepsilon.$$

Thus the operator A^{-1} is continuous on the set $\operatorname{abs\,conv}(AM)$ at point 0.

We consider now two elements $y_1 \in \operatorname{abs\,conv}(AM)$ and $y_2 \in \operatorname{abs\,conv}(AM)$ which satisfy the inequality

$$\|(y_1 - y_2)/2\| \le \delta. \tag{9}$$

It is easy to see that the element $(y_1 - y_2)/2$ belongs to $\operatorname{abs\,conv}(AM)$. The continuity of the operator A^{-1} at point 0. Together with (9) implies

$$\|A^{-1}(y_1 - y_2)\| \le 2\varepsilon.$$

Thus the operator A^{-1} is uniformly continuous on $\operatorname{abs\,conv}(AM)$. So we see that any convex set $M \subset UC(A)$ belongs to some convex set with central symmetry which also belongs to $UC(A)$. Hence if Theorem 1 is valid for central-symmetric sets, it is also valid for any convex set.

Let us prove the necessity part of Theorem 1. Let M be a central-symmetric set, and let us assume that A^{-1} is uniformly continuous on AM, but the conditions of Theorem 1 are not fulfilled, i.e., there exists $R > 0$ such that for any m there exist elements $x_1^m \in M$ and $x_2^m \in M$ satisfying the inequalities

$$\|(I - E_\alpha)(x_1^m - x_2^m)\| \leq 1/m, \quad \alpha = 1/m, \tag{10}$$

and

$$\|E_\alpha(x_1^m - x_2^m)\| \geq R, \quad \alpha = 1/m. \tag{11}$$

In view of the central symmetry of M the elements x_1^m and x_2^m can be chosen in such a way for to satisfy the inequality

$$\|E_\alpha(x_1^m - x_2^m)\| = R, \quad \alpha = 1/m,$$

for any m. Hence

$$\|AE_\alpha(x_1^m - x_2^m)\| \leq R/m, \quad \alpha = 1/m, \tag{12}$$

and

$$\|A(I - E_\alpha)(x_1^m - x_2^m)\| \geq \|A\|/m, \quad \alpha = 1/m. \tag{13}$$

Considering expressions (10)–(13) together we see that, for any m, both the inequalities

$$\|x_1^m - x_2^m\| \geq R$$

and

$$\|A(x_1^m - x_2^m)\| \leq (R + \|A\|)/m$$

hold true and the operator A^{-1} is not uniformly continuous on AM.
The proof of Theorem 1 is completed.

3. CORRECTNESS CLASSES FOR THE EQUATIONS OF CONVOLUTION TYPE

Consider the integral equation

$$\int_{-\infty}^{+\infty} K(t - \tau)x(\tau)\, d\tau = y(t).$$

We assume that $K(t)$, $x(t)$, $y(t)$ belong to L_2 space. We assume also that the Fourier transform of the function $K(t)$ is bounded almost everywhere, which makes the operator A of this equation continuous in L_2 space.
Let $\widetilde{K}(\omega)$ be the Fourier transform of $K(t)$. Consider the class B of convolution type equations, for which the function $1/\widetilde{K}(\omega)$ is locally bounded, i.e., that $1/\widetilde{K}(\omega)$ is bounded on each finite interval. For example, this class contains all the equations for which the function $1/\widetilde{K}(\omega)$ is continuous.
Theorem 1 allows to obtain for the equations of this class a very useful sufficient condition for some function class $M \subset L_2$ to belong to $UC(A)$.

THEOREM 2. *A convex set $M \subset L_2$ belongs to $UC(A)$ if for any $\varepsilon > 0$ there exist $a > 0$, $\eta > 0$ such as for any $x_1 \in M$, $x_2 \in M$ satisfying the inequality*

$$\left\| I_a(\tilde{x}_1(\omega) - \tilde{x}_2(\omega)) \right\| \leq \eta$$

the relation

$$\left\| (I - I_a)(\tilde{x}_1(\omega) - \tilde{x}_2(\omega)) \right\| \leq \varepsilon$$

is valid also.
Here $\tilde{x}_1(\omega)$, $\tilde{x}_2(\omega)$ are the Fourier transforms of $x_1(t)$ and $x_2(t)$, and I_a is defined by

$$I_a(\omega) = \begin{cases} 1 \text{ if } |\omega| \leq a, \\ 0 \text{ if } |\omega| > a. \end{cases}$$

Proof. Let us show that M satisfies the conditions of Theorem 1. Denote by $1/\alpha$ the maximum of the function $1/\widetilde{K}(\omega)$ on the interval $(-a, a)$, where a is chosen to satisfy the inequality

$$\left\| (1 - I_a)(\tilde{x}_1(\omega) - \tilde{x}_2(\omega)) \right\| \leq \varepsilon \tag{14}$$

if

$$\left\| I_a(\tilde{x}_1(\omega) - \tilde{x}_2(\omega)) \right\| \leq \eta \tag{15}$$

holds true on M. It is easy to see that when the inequality

$$\left\| (I - E_\alpha)(x_1 - x_2) \right\| \leq \eta \tag{16}$$

is fulfilled, then

$$\left\| I_a(\tilde{x}_1(\omega) - \tilde{x}_2(\omega)) \right\| \leq \eta$$

also holds true, and, in its turn, implies (14), and, consequently,

$$\left\| E_\alpha(x_1 - x_2) \right\| \leq \varepsilon.$$

Thus for any ε, we can choose $\eta(\varepsilon)$, $\alpha(\varepsilon)$ in correspondence with Theorem 1, and $M \subset UC(A)$.
The theorem proved allows to find new noncompact correctness sets for the equations under consideration. For example, one can show that the subspace of stepwise functions from L_2 which are constants on the intervals $(-d/2, d/2)$, $(\pm d - d/2, \pm d + d/2)$, $(\pm 2d - d/2, \pm 2d + d/2)$, ... belongs to $UC(B)$.
As it was shown by (Akhmetzyanov and Cherepakhin, 1986) and by (Meerov et al., 1988), the a priori restriction of the true solution set to this class allows to calculate numerically the approximate solutions of the ill-posed problems for the linear equations in partial derivatives, which can be reduced to the convolution type operator equations.

REFERENCES

Akhmetzyanov, A.V. and Cherepakhin, A.V. (1986). The simulation of nonstationary gas flow in pipelines. In: *Transaction of 11th IMACS World Congress, Vol.* 3, *Modelling and Simulation in Engineering.* Ed. by B. Wahlström and K. Leiviskä, North–Holland/Elsevier, Amsterdam, pp. 215–223.

Ivanov,V.K. (1966). On a uniform regularization of unstable problems. *Siberian Math. J.* **7**, 546–558 (in Russian).

Meerov, M.V., Akhmetzyanov, A.V., and Cherepakhin, A.V. (1988). Modelling of nonstationary flows in closed-loop systems of oil and gas pipelines *Control-Theory Appl. Technol.* **4**, 137–152.

Ill-Posed Problems in Natural Sciences, pp. 24 – 28
A. Tikhonov (Ed.)
1992 VSP/TVP

ON THE DISCRETIZATION ERROR IN REGULARIZED PROJECTION METHODS WITH PARAMETER CHOICE BY DISCREPANCY PRINCIPLE

U. HÄMARIK
Tartu University, Liivi 2-415, 202400 Tartu, Estonia

ABSTRACT

Regularized projection methods for solving linear ill-posed problems are considered. The regularization parameter is chosen by the (modified) discrepancy principle. Our error estimate has the optimal order with respect to the data errors and our estimate of discretization error is better than those given in earlier papers.

1. REGULARIZED PROJECTION METHODS

Consider the equation

$$Au = f, \quad f \in \mathcal{R}(A) \neq \overline{\mathcal{R}(A)}, \tag{1}$$

where $A \in \mathcal{L}(H, F)$, H, F are Hilbert spaces. Assume that only $f_\delta \in F$ is available with $\|f_\delta - f\| \leq \delta$. Let $h > 0$ be the discretization step and P_h, Q_h be the orthoprojectors in H, F, respectively, with $P_h \to I$, $Q_h \to I$ pointwise in H, F respectively, $\|A(I-P_h)\| \to 0$, $\|(I-Q_h)A\| \to 0$ $(h \to 0)$. The projection method for (1) has the form $A_h u_h = Q_h f_\delta$, $A_h = Q_h A P_h$, $u_h \in \mathcal{R}(P_h)$. We regularize this equation using a Borel measurable function g_r: $[0,a] \to \mathbf{R}$ $(r \geq 0)$ with the following properties for $r > 0$:

1) $|g_r(\lambda)| \leq \gamma r$ $(0 \leq \lambda \leq a,\ \gamma = \text{const})$,

2) $\lambda^p|1 - \lambda g_r(\lambda)| \leq \gamma_p r^{-p}$ $(0 \leq \lambda \leq a,\ 0 \leq p \leq p_0,\ p_0 > 0,\ \gamma_p = \text{const})$,

3) $r \to g_r(\lambda)$ is continuous and there holds $\left.\frac{\partial(g_s(\lambda))}{\partial s}\right|_{s=r} \leq \gamma' \beta_r(\lambda)(1 - \lambda g_r(\lambda))$, $\gamma' = \text{const}$, where $\beta_r(\lambda) = 1$ for $p_0 = \infty$, $\beta_r(\lambda) = (1-\lambda g_r(\lambda))^{1/p_0}$ for $p_0 < \infty$,

4) $r \to |1 - \lambda g_r(\lambda)|$ is decreasing for any $\lambda \geq 0$.

Let $u_0 \in H$ and let u_* be the solution of (1), nearest to u_0. We find

$$u_{h,r} = \left(I - g_r(A_h^* A_h) A_h^* A_h\right) P_h u_0 + g_r(A_h^* A_h) A_h^* f_\delta, \tag{2}$$

assuming that $\|A\|^2 \leq a$. In the special case

$$F = H, \quad A = A^* \geq 0, \quad Q_h = P_h \tag{3}$$

we assume that $\|A\| \leq a$ and use the regularized Ritz–Galerkin method

$$u_{h,r} = \big(I - g_r(A_h)A_h\big)P_h u_0 + g_r(A_h)P_h f_\delta. \tag{4}$$

Examples of regularization methods of the form (2), (4) with the corresponding functions g_r, satisfying **1)–4)** are the ordinary Lavrentiev and Tikhonov methods ($p_0 = 1$), their iterated versions of order m ($p_0 = m$), explicit and implicit iteration methods ($p_0 = \infty$) (see (Plato and Vainikko, 1989, 1990; Gfrerer, 1987)).

2. RULES FOR CHOOSING r

Let $B_{h,r} = B'_{h,r} = I$ if $p_0 = \infty$, $B_{h,r} = \big(I - A_h A_h^* g_r(A_h A_h^*)\big)^{1/(2p_0)}$ if $p_0 < \infty$. In the case (3) define $B'_{h,r} = \big(I - g_r(A_h)A_h\big)^{1/p_0}$ if $p_0 < \infty$.

RULE 1. *Let $b_2 \geq b_1 \geq 1$, $0 < \Theta \leq 1$. If $\|B_{h,r}(A_h u_0 - Q_h f_\delta)\| \leq b_1\delta$, choose $r = 0$. Otherwise choose $0 < r \leq \bar{r} := \sup_{q>0}\{(2q)^{1/q}\|(I - P_h)|A|^q\|^{-2/q}\}$ ($|A| = (A^*A)^{1/2}$) such that*

$$\big\|B_{h,r}(A_h u_{h,r} - Q_h f_\delta)\big\| \leq b_2\delta, \quad \exists s \in [\Theta r, r]\colon \big\|B_{h,s}(A_h u_{h,s} - Q_h f_\delta)\big\| \geq b_1\delta. \tag{5}$$

If no $r \leq \bar{r}$ exists such that (5) holds, choose $r = \bar{r}$ or $r = \mathrm{int}(\bar{r}) + 1$.

Rule 2 = Rule 1, where $B_{h,r}$ is replaced by I.
If (3) holds, then we choose in r approximation (4) according to Rules 1′, 2′ which we obtain from Rules 1, 2 by using the substitutions $B_{h,r} \to B'_{h,r}$, $\bar{r} \to (\bar{r})^{1/2}$, $Q_h \to P_h$.

3. CONVERGENCE AND RATE OF CONVERGENCE

THEOREM 1. **a)** *Let r in (2) be chosen by Rule 1. Then $u_{h,r} \to u_*$ ($\delta \to 0$, $h \to 0$). If*

$$u_* = |A|^p z, \quad \|z\| \leq \rho, \quad u_* - u_0 = |A|^p v, \quad \|v\| \leq \rho, \tag{6}$$

with $p \leq 2p_0$, then

$$\|u_{h,r} - u_*\| \leq c\Big\{\rho^{1/(p+1)}\delta^{p/(p+1)} + \rho\big\|(I - P_h)|A|^p\big\| + \rho e(Q_h)\Big\}, \tag{7}$$

where

$$e(Q_h) = \begin{cases} O\Big(\big\|(I - Q_h)|A^*|^\mu\big\|^{\min\{p/\mu,2\}}\Big)\ \forall\mu,\ 0 < \mu \leq 1,\ \mu \neq p/2 & \text{for } p \leq 2, \\ O\Big(\Big(1 + \log\big\|(I - Q_h)|A^*|^{p/2}\big\|\Big)\ \big\|(I - Q_h)|A^*|^{p/2}\big\|^2\Big) & \text{for } p \leq 2, \\ O\Big(\big\|(I - Q_h)A\big\|\ \big\|(I - Q_h)|A^*|^{p-1}\big\|\Big) & \text{for } p \geq 2. \end{cases}$$

b) *Let $p_0 > 1/2$ and r in (2) be chosen by Rule 2. Then $u_{h,r} \to u_*$ $(\delta \to 0,\ h \to 0)$. If (6) holds with $p \le 2p_0 - 1$, then (7) holds.*

THEOREM 2. *Let (3) hold.* **a)** *Let r in (4) be chosen by Rule 1'. Then $u_{h,r} \to u_*$ $(\delta \to 0,\ h \to 0)$. If (6) holds with $p \le p_0$, then (7) holds with the term $\rho e(Q_h)$ omitted.*
b) *Let r in (4) be chosen by Rule 2'. Then $u_{h,r} \to u_*$ $(\delta \to 0,\ h \to 0)$. If (6) holds with $p \le p_0 - 1$, then (7) holds with the term $\rho e(Q_h)$ omitted.*

The proof of Theorem 2 can be found in (Hämarik, 1991).

4. CHOICE OF h

A reasonable choice of h in (2) is determined by the rule

$$\left\|(I-P_h)|A|^\lambda\right\|^{1/\lambda} + \left\|(I-Q_h)|A^*|^\mu\right\|^{1/\mu} \approx \delta, \tag{8}$$

and in (4) by the rule $\|(I-P_h)A^\lambda\|^{1/\lambda} \approx \delta$ (provided that (3) holds). Table 1 presents the values of λ and μ depending on p_0 and the rule for choosing r. The choice of h guarantees that if (6) holds (with an unknown p) then $\|u_{h,r} - u_*\| \le c\delta^{p/(p+1)}$ for $p \le \bar{p}$, where the values of $\bar{p}$ are given in the last row of Table 1.

Table 1. Values of λ and μ in rules for choice of h

approx.	(4) (provided(3))			(2)		
p_0	∞	$p_0 < \infty$	$1 < p_0 < \infty$	∞	$p_0 < \infty$	$1/2 < p_0 < \infty$
r by Rule	2'	1'	2'	2	1	2
λ	1	$\frac{p_0}{p_0+1}$	$\frac{p_0-1}{p_0}$	1	$\frac{2p_0}{2p_0+1}$	$\frac{2p_0-1}{2p_0}$
μ				$\frac{1}{2}$	$\frac{p_0}{2p_0+1}$	$\frac{2p_0-1}{4p_0}$
$\bar{p}$	∞	p_0	$p_0 - 1$	∞	$2p_0$	$2p_0 - 1$

5. APPLICATION TO INTEGRAL EQUATIONS

Let $Au(t) = \int_0^1 \mathcal{K}(t,s)u(s)\,ds$,

$$\int_0^1\int_0^1 \left|\frac{\partial^{l_1}\mathcal{K}(t,s)}{\partial s^{l_1}}\right|^2 dt\,ds < \infty, \qquad \int_0^1\int_0^1 \left|\frac{\partial^{l_2}\mathcal{K}(t,s)}{\partial t^{l_2}}\right|^2 dt\,ds < \infty,$$

A: $\mathbf{L}_2[0,1] \to \mathbf{L}_2[0,1]$. Let $\mathcal{R}(P_{h_1})$, $\mathcal{R}(Q_{h_2})$ be spline spaces of degrees $k_1 - 1$, $k_2 - 1$ and discretization step sizes h_1, h_2, respectively. Then

$$\|(I-P_{h_1})|A|^p\| = O(h_1^{\min\{pl_1,k_1\}}), \tag{9}$$

$$\|(I - Q_{h_2})|A^*|^p\| = O(h_2^{\min\{pl_2,k_2\}}), \tag{10}$$

if $p \leq 1$. Hence, (8) means $h_1^{\min\{l_1,k_1/\lambda\}} + h_2^{\min\{l_2,k_2/\mu\}} \approx \delta$ and in § 2 the value $h_1^{-2l_1}$ may be used instead of $\bar{r}$.

6. COMPARISON WITH EARLIER RESULTS

In (Plato and Vainikko, 1989, 1990), for (2) in the case (6), the error estimate

$$\begin{aligned} \|u_{h,r} - u_*\| \leq c\Big\{ & \rho^{1/(p+1)}\delta^{p/(p+1)} \\ & + \rho\|A(I - P_h)\|^{\min\{p,1\}} + \rho\|(I - Q_h)A\|^{\min\{p,2\}}\Big\} \end{aligned} \tag{11}$$

was obtained and, for (4), estimate (11) was stated without the last term under the assumption that (3) and (6) hold. Note that estimates (7) and (11) for the regularized least squares method (the case $Q_hAP_h = AP_h$) hold without the last term and for the regularized least error method (the case $Q_hAP_h = Q_hA$) without the second term. Corresponding estimate (11) for the first of these methods was earlier given in (Groetsch et al., 1982; Groetsch, 1984; Gfrerer, 1987) and for the second method in (King and Neubauer, 1988; Neubauer, 1988). Note also that estimate (7) is always not worse that (11), but there are examples when

$$\|(I - P_h)|A|^p\| = O(h^{\max\{p,1\}}), \qquad \|A(I - P_h)\|^{\min\{p,1\}} = O(h^{\min\{p,1\}}),$$
$$e(Q_h) = O(h^{\max\{p,2\}}), \qquad \|(I - Q_h)A\|^{\min\{p,2\}} = O(h^{\min\{p,2\}}).$$

In the case $p < 1$ such estimates follow from (9) and (10) in view of the example of § 5 with $k_i = 1$, $l_i > i/p$, $(i = 1, 2)$. It is worth noting that rule (8) with $\lambda = \mu = 1$ (independently from p_0) was proposed for choosing h in (Plato and Vainikko, 1990).

REFERENCES

Gfrerer, H. (1987). An a posteriori parameter choice for ordinary and iterated Tikhonov regularization of ill-posed problems leading to optimal convergence rates. *Math. Comput.* **49**, 507–522.

Groetsch, C.W., King, J.T., and Murio, D. (1982). Asymptotic analysis of a finite element method for Fredholm equations of the first kind. In: *Treatment of Integral Equations by Numerical Methods.* Ed. by C.T.H. Baker and G.F. Miller. Academic Press, London, pp. 1–11.

Groetsch, C.W. (1982). *The Theory of Tikhonov Regularization for Fredholm Equations of the First Kind.* Pitman, Boston.

Hämarik, U. (1991). Quasioptimal error estimate for the regularized Ritz–Galerkin method with the a posteriori choice of the parameter. *Acta et Comment. Univ. Tartuensis* **937**, 63–76.

King J.T. and Neubauer, A. (1988). A variant of finite-dimensional Tikhonov regularization with a posteriori parameter choice. *Computing* **40**, 91–109.

Neubauer, A. (1988). An a posteriori parameter choice for Tikhonov regularization in the presence of modeling error. *Appl. Numer. Math.* **4**, 507–519.

Plato, R. and Vainikko, G. (1990). On the regularization of projection methods for solving ill-posed problems. *Numer. Math.* **57**, 63–79.

Plato, R. and Vainikko, G. (1989). On the regularization of the Ritz–Galerkin method for solving ill-posed problems. *Uch. Zap. Tartu Univ.* **863**, 3–17.

Ill-Posed Problems in Natural Sciences, pp. 29 – 32
A. Tikhonov (Ed.)
1992 VSP/TVP

REGULARIZATION AND UNIQUENESS OF SOLUTIONS OF SYSTEMS OF VOLTERRA NONLINEAR INTEGRAL EQUATIONS OF THE FIRST KIND WITH TWO ARGUMENTS

M.I. IMANALIEV and A.A. ASANOV
Institute of Mathematics, Kirgizstan Academy of Sciences,
Leninskii Prospekt 265a, Bishkek, Kirgizstan

ABSTRACT

Systems of nonlinear integral equations with a nondifferentiable vector kernel are considered. Volterra regularization operators are constructed in various spaces and theorems on uniqueness are proved.
For linear systems an existence theorem is proved.

Consider a system of nonlinear integral equations of the form

$$\int_{t_0}^{t}\int_{x_0}^{x} K\big(t,x,s,y,u(s,y)\big)\,dy\,ds = f(t,x), \quad (t,x)\in\overline{G}, \tag{1}$$

where K, f, and u are n-dimensional vector functions, K and f being known and u unknown, and $\overline{G}$ is the closure of the set $G=(t_0,T)\times(x_0,l)$.
Throughout we assume that

$$K(t,x,s,y,u) = Q(t,x,s,y)u + K_1(t,x,s,y,u), \tag{2}$$

$$Q(t,x,t,x) = A(t)B(x)K_0(t,x), \tag{3}$$

where Q, A, and B are $n\times n$ matrix valued measurable functions, K_1 is an n-dimensional measurable vector function one, K_0 is a $n\times n$ matrix valued continuous function in $\overline{G}$ and $\det K_0(t,x)\neq 0$ in $\overline{G}$. Without loss of generality we assume that $K_0(t,x)$ is the unit $n\times n$ matrix E_n every where in $\overline{G}$.
Systems of one-dimensional and two-dimensional Volterra integral equations of the first kind were considered in (Asanov, 1980; Imanaliev and Asanov, 1988; Imanaliev and Asanov, 1989; Ten Men Jan, 1975).
Here we do not need to assume the vector-functions K and f to be smooth with respect to t and x. Moreover, $\det[A(t)B(x)]$ may vanish in $\overline{G}$.

Remark . In the sequel we use the following notation.
1) For any $n \times n$ matrix A and any n-dimensional vector u respectively, $\|A\|$ and $\|u\|$ mean, respectively, their norms.
2) $C^{\gamma_1,\gamma_2}_{\varphi_1,\varphi_2}(\overline{G})$, $0 < \gamma_i \leq 1$, $i = 1, 2$, is the set of n-dimensiond vector functions $u(t, x)$ satisfying the inequalities

$$\|u(t,x) - u(s,x) - u(t,y) + u(s,y)\| \leq c|\varphi_1(t) - \varphi_1(s)|^{\gamma_1} |\varphi_2(x) - \varphi_2(y)|^{\gamma_2},$$

$$\|u(t,x) - u(s,x)\| \leq c|\varphi_1(t) - \varphi_1(s)|^{\gamma_1}, \quad \|u(t,x) - u(t,y)\| \leq c|\varphi_2(x) - \varphi_2(y)|^{\gamma_2},$$

where C depends solely on u.
3) $C_n(\overline{G})$ and $L^p_n(G)$ are the spaces of n-dimensional vector functions with components in $C(\overline{G})$ and $L^p(G)$ respectively.
Along with system (1)–(3), let us consider the system of integral equations of the second kind:

$$\varepsilon^2 v(t,x,\varepsilon) + \varepsilon \int_{t_0}^{t} A(s) v(s,x,\varepsilon)\, ds + \varepsilon \int_{x_0}^{x} B(y) v(t,y,\varepsilon)\, dy$$

$$+ \int_{t_0}^{t}\int_{x_0}^{x} K(t,x,s,y,v(s,y,\varepsilon))\, dy\, ds = f(t,x) + \varepsilon^2 \big[u(t_0,x) + u(t,x_0)$$

$$-u(t_0,x_0)\big] + \varepsilon \Big[\int_{t_0}^{t} A(s) u(s,x_0)\, ds + \int_{x_0}^{x} B(y) u(t_0,y)\, dy \Big], \tag{4}$$

where ε is a small positive parameter.
Let $\lambda_i(t)$, $i = 1, \ldots, n$, be the eigenvalues of the matrix $(1/2)[A(t) + A^*(t)]$, where $*$ denotes conjugation,

$$\lambda(t) = \min_i \lambda_i(t),$$

$\mu_i(x), i = 1, \ldots, n,$ be the eigenvalues of the matrix $1/2[B(x) + B^*(x)]$,

$$\mu(x) = \min_i \mu_i(x).$$

We say that condition (M) is fulfilled if:
I. For any fixed $t \in (t_0, T]$ and $x \in (x_0, l]$, we have

$$\|Q(t,x,s,y)\| \in L^q\big([t_0,t] \times [x_0,x]\big), \quad q > 1,$$

$$\|Q(t,x,t,x)\| \in L^1(G), \qquad \|A(t)\| \leq N_0 \lambda(t)$$

and $\lambda(t) \geq 0$ for $t \in [t_0, T]$, $N_0 > 0$, $\lambda(t) \in L^q(t_0, T)$, $\|B(x)\| \leq N_0 \mu(x)$ and $\mu(x) \geq 0$ for $x \in [x_0, l]$, $\mu(x) \in L^q(x_0, l)$;
II. For $t \geq \tau$ and $x \geq \nu$ and for every $(t,x,s,y), (t,\nu,s,y)$ (τ,x,s,y), $(\tau,\nu,s,y) \in \overline{\Omega} = \{(t,x,s,y)\colon\ t_0 \leq s \leq t \leq T,\ x_0 \leq y \leq x \leq l\}$ the following inequalities are fulfilled:

$$\|Q(t,x,s,y) - Q(\tau,x,s,y)\| \leq C_0 \lambda(s) \mu(y) \Big(\int_{\tau}^{t} \lambda(s)\, ds \Big),$$

$$\|Q(t,x,s,y)-Q(t,\nu,s,y)\| \le C_1\lambda(s)\mu(y)\Big(\int_\nu^x \mu(\nu)\,d\nu\Big),$$

$$\|Q(t,x,s,y)-Q(\tau,x,s,y)-Q(t,\nu,s,y)+Q(\tau,\nu,s,y)\|$$

$$\le C_2\lambda(s)\mu(y)\Big(\int_\tau^t \lambda(\tau)\,d\tau\Big)\Big(\int_\nu^x \mu(\nu)\,d\nu\Big),$$

where C_0, C_1, and C_2 are non-negative constants;
III. For $t>\tau$ and $x>\nu$ and for every (t,x,s,y,u_1), (t,ν,s,y,u_1), (t,ν,s,y,u_2), (t,x,s,y,u_2), (τ,x,s,y,u_1), (τ,x,s,y,u_2), (τ,ν,s,y,u_1), $(\tau,\nu,s,y,u_2) \in \overline{\Omega}\times \mathbf{R}^n$ the following inequalities are fulfilled:

$$\|K_1(t,x,s,y,u_2)-K_1(t,\nu,s,y,u_2)-K_1(t,x,s,y,u_1)$$

$$+K_1(t,\nu,s,y,u_1)\| \le C_0'\lambda(s)\mu(y)\Big(\int_\nu^x \mu(\nu)\,d\nu\Big)\|u_1-u_2\|,$$

$$\|K_1(t,x,s,y,u_2)-K_1(\tau,x,s,y,u_2)-K_1(t,x,s,y,u_1)$$

$$+K_1(\tau,x,s,y,u_1)\| \le C_1'\lambda(s)\mu(y)\Big(\int_\tau^t \lambda(s)\,ds\Big)\|u_1-u_2\|,$$

$$\|K_1(t,x,s,y,u_2)-K_1(t,x,s,y,u_1)-K_1(\tau,x,s,y,u_2)$$
$$+K_1(\tau,x,s,y,u_1)-K_1(t,\nu,s,y,u_2)+K_1(t,\nu,s,y,u_1)$$
$$+K_1(\tau,\nu,s,y,u_2)-K_1(\tau,\nu,s,y,u_1)\|$$

$$\le C_2'\lambda(s)\mu(y)\Big(\int_\tau^t \lambda(\tau)\,d\tau\Big)\Big(\int_\nu^x \mu(\nu)\,d\nu\Big)\|u_1-u_2\|,$$

where C_0', C_1' and C_2' are non-negative constants;
IV. $K_1(t,x,t,x,u)\equiv 0$ for $(t,x,u)\in G\times\mathbf{R}^n$ and $K_1(t,x,s,y,0)\equiv 0$, $(t,x,s,y)\in \overline{\Omega}$.

THEOREM 1. *Let condition* (M) *be fulfilled. Then*

1) *if system (1) has a solution* $u(t,x)\in C^{\gamma_1,\gamma_2}_{\varphi_1,\varphi_2}(\overline{G})$, $0<\gamma_i\le 1$, $i=1,2$, *where* $\varphi_1(t)=\int_{t_0}^t \lambda(s)\,ds$ *for* $t\in[t_0,T]$ *and* $\varphi_2(x)=\int_{x_0}^x \mu(\nu)\,d\nu$ *for* $x\in[x_0,l]$, *then the solution* $v(t,x,\varepsilon)$ *of the system (4) converges in norm* $\|\cdot\|_C$ *to* $u(t,x)$ *as* $\varepsilon\to 0$. *Moreover,*
$$\|v(t,x,\varepsilon)-u(t,x)\|_C \le C_3\varepsilon^{\gamma_0},$$
where $\gamma_0=\min\{\gamma_1,\gamma_2\}$, C_3 *is a known constant not depending on* ε;

2) *if* $\lambda(t)>0$ *for almost all* $t\in[t_0,T]$, $\mu(x)>0$ *for almost all* $x\in[x_0,l]$ *and system (1) has a solution* $u(t,x)\in C_n(\overline{G})$, *then the solution* $v(t,x,\varepsilon)$ *of system (4) converges in norm* $\|\cdot\|_C$ *to* $u(t,x)$ *by* $\varepsilon\to 0$. *Moreover,*

$$\|v(t,x,\varepsilon)-u(t,x)\|_C \le C_4\omega_{\bar u}(\sqrt{2}\varepsilon^\beta)+C_5\exp\Big\{-\frac{1}{\varepsilon^{1-\beta}}\Big\},$$

where $0 < \beta < 1$, C_4 and C_5 are non-negative constants not depending on ε,

$$\omega_{\bar{u}}(\delta) = \sup_{|z'-z|\leq\delta} \|u(\varphi_1^{-1}(z_1), \varphi_2^{-1}(z_2)) - u(\varphi_1^{-1}(z_1'), \varphi_2^{-1}(z_2'))\|,$$

$\varphi_1^{-1}(z_1)$ and $\varphi_2^{-1}(z_2)$ are the corresponding inverse functions, $z \in [0, \varphi_1(T)] \times [0, \varphi_2(l)]$.

Scheme of the proof. We use the following substitution in system (4):

$$v(t,x,\varepsilon) = u(t,x) + \xi(t,x,\varepsilon),$$

and invert the operator $A_0 = (\varepsilon E_n + A_1)(\varepsilon E_n + B_1)$, $A_1\xi = \int_{t_0}^{t} A(s)\xi(s,x,\varepsilon)ds$, $B_1\xi = \int_{x_0}^{x} B(\nu)\xi(t,\nu,\varepsilon)\,d\nu$. Then we estimate $\xi(t,x,\varepsilon)$.

COROLLARY. *Let condition* (M) *be fulfilled,* $\|K(t,x,s,y,v)\|$ *be a continuous function in* $\overline{\Omega} \times \mathbf{R}^n$, $\lambda(t_0) \neq 0$, $\mu(x_0) \neq 0$. *Then*

(*i*) *the solution of system* (1) *is unique in* $C^{\gamma_1,\gamma_2}_{\varphi_1,\varphi_2}(\overline{G})$, $0 < \gamma_i \leq 1$, $i = 1, 2$;

(*ii*) *if* $\lambda(t) > 0$ *for almost all* $t \in [t_0, T]$ *and* $\mu(x) > 0$ *for almost all* $x \in [x_0, l]$, *then the solution of system* (1) *is unique in* $C_n(\overline{G})$.

If $K_1(t,x,s,y,u) \equiv 0$ in $\Omega \times \mathbf{R}^n$ then (1) and (4) then reduce to systems of linear equations. In this case the following result is proved.

THEOREM 2. *Let condition* (M) *be fulfilled,* $q > 1$, $f(t,x) \in C^{1,1}_{\varphi_1,\varphi_2}(\overline{G})$, $f(t_0,x) \equiv 0$ *for* $x \in [x_0, l]$ *and* $f(t,x_0) \equiv 0$ *for* $t \in [t_0, T]$. *Then system* (1) *has a solution in* $L^p_n(G)$, $1/p + 1/q = 1$.

The proof of Theorem 2 is based on the reflexivity of the space $L^p_n(G)$.

REFERENCES

Asanov, A.A. (1980). Regularization and uniqueness of solutions of Volterra linear integral equations of the first kind with two arguments. *Research Int.-Different. Equat.* **13**, 207–214 (in Russian).

Imanaliev, M.I. and Asanov, A.A. (1988). Regularization, uniqueness and existence of solutions of Volterra integral equations of the first kind. *Research Int.-Different. Equat.* **21**, 3–38 (in Russian).

Imanaliev, M.I. and Asanov, A.A. (1989). Of solutions of systems of nonlinear Volterra integral equations of the first kind. *Soviet Math. Dokl.* **309**, 1052–1055 (in Russian).

Ten Men Jan (1975). The cubature method to solve two-dimensional integral Volterra equations of the first kind. In: *Modeling and Optimization in Large Energy Systems*, Irkutsk Univ. Press, Irkutsk, pp. 107–116.

Ill-Posed Problems in Natural Sciences, pp. 33 – 43
A. Tikhonov (Ed.)
1992 VSP/TVP

POINTWISE RESIDUAL METHOD FOR SOLVING SYSTEMS OF LINEAR ALGEBRAIC EQUATIONS AND INEQUALITIES

A.Yu. IVANITSKII, F.P. VASIL'EV, and V.A. MOROZOV
Physics and Mathematics Department, Chuvash State University
428015 Cheboksary, Russia
Department of Mathematics and Cybernetics, Moscow State University,
119899 Moscow, Russia
Scientific Research Computer Center, Moscow State University,
119899 Moscow, Russia

ABSTRACT

A stable method for the solution of systems of linear algebraic equations in general form and of systems of inequalities with approximate data given in interval form is considered. The estimates (of speed of convergence for the method) of the same order as the order of the input data error are obtained.

1. STATEMENT OF THE MAIN PROBLEM

Let us consider the system of linear algebraic equations and inequalities:

$$Au = f, \quad Bu \leq d, \tag{1}$$

where $u = (u_1, u_2, \ldots, u_n)^T \in \mathbf{R}^n$ is the unknown vector,

$$A = \{a_{ij}\} \in \mathbf{R}^{m\times n}, \quad B = \{b_{lj}\} \in \mathbf{R}^{p\times n},$$

$$f = (f_1, f_2, \ldots, f_m)^T \in \mathbf{R}^m, \quad d = (d_1, d_2, \ldots, d_p)^T \in \mathbf{R}^p,$$

is the input data, $\mathbf{R}^s$ is a real linear s-dimensional space, $\mathbf{R}^{k\times n}$ is a linear space of real $k \times n$ matrices, a symbol T means transposition. We assume that a solution of system (1) exists but may be not unique. We shall denote by U the set of its solutions. The cases of absence of inequality ($B = 0$, $d = 0$) nor equality ($A = 0$, $f = 0$) in (1) are not excluded. Here and elsewhere we shall denote by 0 the matrix or the vector with zero components of corresponding dimension.
Let instead of exact data A, f, B, d be given their approximations $\tilde{A} = \{\tilde{a}_{ij}\} \in \mathbf{R}^{m\times n}$, $\tilde{f} = (\tilde{f}_1, \tilde{f}_2, \ldots, \tilde{f}_m)^T \in \mathbf{R}^m$, $\tilde{B} = \{\tilde{b}_{lj}\} \in \mathbf{R}^{p\times n}$, $\tilde{d} = (\tilde{d}_1, \tilde{d}_2, \ldots, \tilde{d}_p)^T \in \mathbf{R}^p$, satisfying the following relations:

$$|\tilde{a}_{ij} - a_{ij}| \leq \Delta_{ij}, \quad |\tilde{f}_i - f_i| \leq \delta_i, \quad i = 1, \ldots, m; \quad j = 1, \ldots, n; \tag{2a}$$

$$|\tilde{b}_{lj} - b_{lj}| \leq \Lambda_{lj}, \quad |\tilde{d}_l - d_l| \leq \lambda_l, \quad l = 1, \ldots, p; \quad j = 1, \ldots, n; \tag{2b}$$

where Δ_{ij}, δ_i, Λ_{lj}, λ_l are some given pointwise levels of input data errors. It is easy to verify by simple examples that the following system of equations and inequalities

$$\tilde{A}u = \tilde{f}, \quad \tilde{B}u \leq \tilde{d} \tag{3}$$

can be unsolvable in the classic sense, and that its solutions in the case of solvability can be unstable. Therefore elsewhere we shall consider instead of (3) the following systems of inequalities

$$|\sum_{j=1}^{n} \tilde{a}_{ij}u_j - \tilde{f}_i| \leq \sum_{j=1}^{n} \Delta_{ij}|u_j| + \delta_i, \quad i = 1, \ldots, m, \tag{4a}$$

$$\sum_{j=1}^{n} \tilde{b}_{lj}u_j - \tilde{d}_l \leq \sum_{j=1}^{n} \Lambda_{lj}|u_j| + \lambda_l, \quad l = 1, \ldots, p. \tag{4b}$$

It can be verified easily that each solution $u = (u_1, u_2, \ldots, u_n)^T$ of system (1) satisfies relations (4a) and (4b). Consequently, the set $\tilde{U}$ of vectors satisfying inequalities (4a) and (4b), $\tilde{U}$ is nonempty. However an arbitrary element of the set $\tilde{U}$ can not be chosen as an approximate solution of system (1), because the closeness of this element to the set of exact solutions U can not be guaranteed. It is necessary to have a selection principle for stable approximate solutions one of which is considered bellow. This principle is based on the transition from the initial problem (1) to the following minimizing problem

$$\|Lu\|_I \to \inf, \quad u \in \tilde{U}, \tag{5}$$

where $L \in \mathbf{R}^{q \times n}$ is some matrix, and $\|\cdot\|_I$ is a norm in the $\mathbf{R}^q$-space.
Many applied problems have additional a prove information about the solution of problem (1) (for example, the non-negativeness of some variables: $u_j \geq 0$; the monotony of these variables: $u_j \leq u_{j+1}$ etc.). We shall denote D the set, taking in to account the a priori information about the solutions of system (1). Then it is naturally to consider instead of (5) the following problem:

$$\|Lu\|_I \to \inf, \quad u \in \tilde{U}_D = \tilde{U} \cap D. \tag{6}$$

If $D = \mathbf{R}^n$, then this problem coincides with (5). Here in (6) and everywhere below we assume that the set D is known precisely, so we assume that the input data determining this set, are given precisely.
Below we will show that any solution of problem (6) can be taken as an "approximation" to a solution of (1) with a suitable choice of the matrix L. Moreover, the is a set $U_*(\Delta)$ of solutions of problem $\sigma = (\{\Delta_{ij}\}, \{\Delta_{lj}\}, \{\delta_i\}, \{\lambda_l\}) \to 0$; $U_*(\Delta)$ will converge in some sense to U_*, where is the set U_* of solutions of the following problem:

$$\|Lu\|_I \to \inf, \quad u \in U_D = U \cap D. \tag{7}$$

In particular the unit matrix $E \in \mathbf{R}^{n\times n}$ can be always taken as L. Then the set U_* consists of the so called normal solutions of problem (1), which have the minimal norm. Other possibilities of the choice of matrix L will be considered below.
We shall call the suggested selection algorithm (5) or (6) of the approximate solutions of problem (1) under the pointwise representation of errors (2a), (2b) a pointwise residual method. It is an analogue of the well known residual method for the solving of the unstable equations (see (Morozov, 1984b; Tikhonov and Arsenin, 1986)). The pointwise residual method for systems of linear equations and some particular classes of inequalities was also investigated in (Morozov, 1984b; Ivanitskii, 1989).

2. WEIERSTRASS THEOREM. CONDITION OF ADDITIONALITY

We investigate problem (7), assuming that $U_D \neq 0$ and the following condition, of additionality is fulfilled there exist such constants $\alpha_0 > 0$; $\alpha_i \geq 0$; $i = 1,2,3$; $\alpha_1 + \alpha_2 + \alpha_3 > 0$, that

$$\alpha_0\|u\| \leq \alpha_1\|Lu\|_{\mathrm{I}} + \alpha_2\|Au\|_{\mathrm{II}} + \alpha_3\|(Bu)^+\|_{\mathrm{III}}, \quad \forall u \in D, \tag{8}$$

where $\|\cdot\|, \|\cdot\|_{\mathrm{I}}, \|\cdot\|_{\mathrm{II}}, \|\cdot\|_{\mathrm{III}}$ are some norms in the respective spaces, $z^+ = (z_1^+, z_2^+, \dots, z_p^+)^T$, $z_i^+ = \max\{0, z_i\}$, if $z = (z_1, z_2, \dots, z_p)^T$. Note that $L = E$ the condition of additionality is satisfied for arbitrary matrices $A \in \mathbf{R}^{m\times n}$, $B \in \mathbf{R}^{p\times n}$ and for any nonempty set $D \subseteq \mathbf{R}^n$ with any $\alpha_2 \geq 0$, $\alpha_3 \geq 0$ and $\alpha_0 > 0$, $\alpha_1 > 0$, determined from the condition equivalence of the norms $\|\cdot\|, \|\cdot\|_{\mathrm{I}}$ in the finite-dimensional space $\mathbf{R}^n$. A more detailed discussion of condition (8) will be presented in Theorem 2.
The same argument as in the proof of Theorem 1.1.2 from (Ivanitskii, 1988) shows that the following theorem is valid.

THEOREM 1. *Let D be a nonempty closed subset of $\mathbf{R}^N$ such that $U_D = U \cap D \neq \varnothing$ and let the condition of additionality* (8) *be fulfiled. Then the set U_* of solutions of problem* (7) *is nonempty, compact and any minimizing sequence $\{u^k\}$ for the function $\|Lu\|_{\mathrm{I}}$ on the set U_D converges to U_*, i.e. $\lim_{k\to\infty} \rho(u^k, U_*) = 0$, where $\rho(u, U_*) = \inf_{v\in U_*} \|u - v\|$ is the deviation of the vector U from the set U_*.*

Let us analyse the condition of additionality (8) in more detail. Note that if $B = 0$, $d = 0$ than this condition (8) is the known condition of additionality, for the first time introduced by V.A. Morozov (Morozov, 1984a) for the construction of regularization methods for operator equations.
We shall formulate necessary and sufficient conditions for the fulfilment of the condition of additionality (8). Let ker$A$ and ker$L$ be the kernels of matrices A and L respectively, $K_D = \{u \in \mathbf{R}^n\colon\ u = \lambda v, \lambda \geq 0, v \in D\}$, $\overline{K}_D$ be of closure of K_D. Analogomly to the proof of Theorem 1 in (Ivanitskii and Fedotov, 1989) one can prove

THEOREM 2. *The condition of additionality* (8) *and the following condition*

$$\ker A \cap \ker L \cap \{u \in \mathbf{R}^n\colon\ Bu \leq 0\} \cap \overline{K}_D = \{0\} \tag{9}$$

are equivalent.

We remark that in (Ivanitskii and Fedotov, 1989) Theorem 2 was proved for the infinite dimensional spaces when $B = 0,\ d = 0$.
It follows from Theorem 2 that if $\ker L = \{0\}$ for the matrix L, then the condition of additionality (8) is fulfilled for arbitrary matrices $A \in \mathbf{R}^{m\times n}$, $B \in \mathbf{R}^{p\times n}$ and for any closed sets $D \subseteq \mathbf{R}^n$. In this case set $\mathcal{L}_y = \{u \in \mathbf{R}^n \colon \|Lu\|_{\mathrm{I}} \le y\}$ is compact for any $y \ge 0$ and the function $\|Lu\|_{\mathrm{I}}$ is a stabilizer (Tikhonov and Arsenin, 1986; Vasil'ev, 1988a) for problems (1), (7). We give an example showing that condition (8) can be fulfilled in the case when $\|Lu\|_{\mathrm{I}}$ is not a stabilizer.

Example. Let $\|v\|_I = \|v\|_2 = \left(\sum_{j=1}^{3} v_j^2\right)^{1/2}$ be the Euclidean norm at $\mathbf{R}^3$,

$$D = \{u \in \mathbf{R}^3 \colon u_3 = 0\}, \quad Au = (u_1, 0, 0)^T, \quad Lu = (0, u_2, 0)^T \quad \text{for} \quad \forall u \in D^3.$$

It can be verified easily that condition (9) and also condition (8) are satisfied with any matrix Lebeg's $B \in \mathbf{R}^{p\times 3}$. However set $\mathcal{L}_y = \{u \in \mathbf{R}^3 \colon \|Lu\|_2 = |u_2| \le y\}$ is unbounded here, and hence $\|Lu\|_2$ is not a stabilizer.
Thus the condition of additionality (8) simultaneously universely connects the four main objects of problem (7), namely: the matrices A, B, L and the set D, and it contains the idea of compensation of "bad" properties of some objects by "good" properties of the others. It allows to consider sufficiently wide class of unstable problems and methods of their regularization with relaxed requirements on matrices A, B, L and on the set D as compared with the traditional cases, where a priory $\|Lu\|_{\mathrm{I}}$ is considered as a stabilizer.

3. CONVERGENCE OF THE POINTWISE RESIDUAL METHOD

THEOREM 3. *Let the conditions of Theorem* 1, (2a), (2b) *be fulfilled. Then* $\tilde{U}_* = \{u \in \tilde{U}_D \colon \|Lu\|_{\mathrm{I}} = \tilde{\mu}_* = \inf_{u\in\tilde{U}_D} \|Lu\|_{\mathrm{I}}\}$ *is the set solutions of problem* (6) *well be nonempty and compact if* $\{\Delta_{ij}\}$, $\{\Lambda_{lj}\}$ *are sufficiently small and any minimizing sequence* $\{u^k\}$ *for the function* $\|Lu\|_{\mathrm{I}}$ *on the set* $\tilde{U}_D$ *converges to* U_*.

The proof of Theorem 3 is similar to the proof of Theorem 1.

Remark. If D is bounded or $\alpha_2 = \alpha_3 = 0$, then the condition that $\{\Delta_{ij}\}$ and $\{\Lambda_{ij}\}$ are small is superfluous (Ivanitskii, 1988).
It is not necessary to solve problem (6) exactly in the numerical realizing of the pointwise residual method. It is sufficient to determine the vector $u = u(\varepsilon, \sigma)$ from the condition:

$$u \in \tilde{U}_D \colon \|Lu\|_{\mathrm{I}} \le \tilde{\mu}_* + \varepsilon, \quad \varepsilon \ge 0. \tag{10}$$

We shall denote by $U_*(\delta, \varepsilon)$ the set of vectors sutisfying conditions (10).
The convergence of the method under consideration is established by

THEOREM 4. *Let the conditions of Theorem 3 be fulfilled. Then* $\beta(U_*(\sigma, \varepsilon), U_*) \to 0$ *as* $\delta \to 0$ *and* $\varepsilon \to 0$, *where*

$$\beta(V, W) = \sup_{v\in V} \inf_{w\in W} \|v - w\|.$$

Proof. We consider arbitrary sequences $\{\sigma^k\} = \{\Delta^k, \Lambda^k, \delta^k, \lambda^k\} \to 0$, $\{\varepsilon^k\} \to 0$ as $k \to \infty$. According to the definition of the upper bound there exists such $u^k = u(\sigma^k, \varepsilon^k) \in U_*(\sigma^k, \varepsilon^k)$ that

$$\rho(u^k, U_*) = \inf_{w \in U_*} \|u^k - w\| \geq \beta(U_*(\sigma^k, \varepsilon^k), U_*) - \frac{1}{k}, \quad k = 1, 2, \ldots \tag{11}$$

Since $U_D \subseteq \tilde{U}_D$, we have $\tilde{\mu}_* \leq \mu_*$, and hence by (10)

$$\|Lu^k\|_I \leq \mu_* + \varepsilon^k, \quad k = 1, 2, \ldots. \tag{12}$$

Taking into account (8), (12) we obtain

$$\alpha_0 \|u^k\| \leq \alpha_1(\mu_* + \varepsilon^k) + \alpha_2 M_3 \{\|f\|_\infty + 2(\|\Delta^k\|_{IV} \cdot \|u^k\|_1 + \|\delta^k\|_\infty)\}$$

$$+ \alpha_3 M_2 \{\|d^+\|_\infty + 2(\|\Lambda^k\|_{IV} \|u^k\|_1 + \|\lambda^k\|_\infty)\}, \tag{13}$$

where $\|\Delta^k\|_{IV} = \max_{i,j,} \Delta^k_{ij}$, $\|u^h\|_1 = |u^k_1| + |u^k_2| + \ldots + |u^k_n|$, $\|z\|_\infty = \max_i |z_i|$.

This implies the boundedness of the sequence $\{u^k\}$: $\|u^k\| \leq C_0$, $C_0 = \text{const}$ for sufficiently large. We can assume that $\{u^k\} \to v_*$ as $k \to \infty$, selecting a subsequence if necessary. Let us prove that $v_* \in U_*$. It is not difficult to show that

$$\|Au^k - f\|_\infty \leq 2(\|\Delta^k\|_{\mathrm{IV}} C_0 + \|\delta^k\|_\infty), \tag{14a}$$

$$\|(Bu^k - d)^+\|_\infty \leq 2(\|\Lambda^k\|_{\mathrm{IV}} C_0 + \|\lambda^k\|_\infty), \tag{14b}$$

when k is sufficiently large using relations from correlations (2a), (2b), (4a), (4b), Theorem 3 and the inequality $\|u^+\|_\infty \leq \|(u - v)^+\|_\infty + \|v^+\|_\infty$ for $\forall u, v \in \mathbf{R}^n$. If $k \to \infty$, then we deduce from this

$$\|Av_* - f\|_\infty = 0, \quad \|(Bv_* - d)^+\|_\infty = 0,$$

i.e. the vector v_* is a solution of system (1). Since D is closed, $v_* \in D$. Hence, $v_* \in U_D$. On the other hand we have from (12) $\|Lu_*\|_I \leq \mu_*$. Thus $v_* \in U_*$. Therefore

$$\lim_{k \to \infty} \rho(u^k, U_*) = \rho(v_*, U_*) = 0.$$

It follows from this hence and from (11) that $\lim_{k \to \infty} \beta(U_*(\sigma^k, \varepsilon^k), U_*) = 0$. Then we get due to the arbitrariness of the choice of the sequences $\{\sigma^k\} \to 0$, $\{\varepsilon^k\} \to 0$ that $\beta(U_*(\sigma, \varepsilon), U_*) \to 0$ as $\sigma \to 0$, $\varepsilon \to 0$. The Theorem is proved.

4. ESTIMATION OF THE SPEED OF CONVERGENCE

Some additional restrictions on the set D and a special choice of the norm $\|\cdot\|_{\mathrm{I}}$ allow to get estimates of the speed of convergence for the pointwise residual method. In particular the following theorem is valid.

THEOREM 5. *Let in problem* (7) *the norm* $\|\cdot\|_I = \|\cdot\|_\infty$, *the set* $D = \{u \in \mathbf{R}^n\colon\ A^1u = f^1,\ B^1u \leq d^1,\ A^1 \in \mathbf{R}^{m_1 \times n},\ B^1 \in \mathbf{R}^{p_1 \times n},\ f^1 \in \mathbf{R}^{m_1},\ d^1 \in \mathbf{R}^{p_1}\}$

and the conditions of Theorem 3 are fulfilled. Then the following estimate is true if $\{\Delta_{ij}\}, \{\Lambda_{lj}\}$ *are sufficiently small*

$$\sup_{u\in U_*(\sigma,\varepsilon)} \rho(u,U_*) = O(\|\Delta\|_{IV} + \|\Lambda\|_{IV} + \|\delta\|_\infty + \|\lambda\|_\infty + \varepsilon), \tag{15}$$

where $\rho(u,U_*) = \inf_{w\in U_*} \|u-w\|_2$, $\|z\|_2 = \left(\sum_{j=1}^{n} z_j^2\right)^{1/2}$, $z = (z_1, z_2, \dots, z_n)^T \in \mathbf{R}^n$.

Proof. The set of solutions U_* of problem (7) can be presented in our case as

$$U_* = \{u \in \mathbf{R}^n\colon\ \|Lu\|_\infty - \mu_* \le 0,\ Au = f,\ Bu \le d,\ A^1u = f^1,\ B^1u \le d^1\}$$

or as

$$U_* = \{u \in \mathbf{R}^n\colon\ \ (Lu)_s - \mu_* \le 0,\ -(Lu)_s - \mu_s \le 0,\ s = 1,\dots,q,$$
$$Au = f,\ Bu \le d,\ A^1u = f^1,\ B^1u \le d^1\},$$

where $(Lu)_s$ is the component of the vector Lu. This set is polihedral. Then, as it follows from the results of (Hoffman, 1952), see also (Vasil'ev, 1988a, Lemma 3, p.375), there exists such a constant $M > 0$ depending only on the elements of the matrices L, A, B, A^1 and B^1 that

$$\rho(u,U_*) \le M \max\{\max_{1\le s\le q}((Lu)_s - \mu_*)^+, \max_{1\le s\le q}(-(Lu)_s - \mu_*)^+,$$
$$\|Au-f\|_\infty, \|(Bu-d)^+\|_\infty, \|A^1u - f^1\|_\infty, \|(B^1u-d^1)^+\|_\infty\},\quad \forall u \in \mathbf{R}^n. \tag{16}$$

In particular this inequality is correct for all the points $u(\sigma,\varepsilon) \in U_*\ (\sigma,\varepsilon) \subseteq D \subseteq \mathbf{R}^n$. Let us estimate the quantities from the right side of (16) when $u = u(\sigma,\varepsilon)$. We have analogously to (12) $\|Lu\|_\infty \le \mu_* + \varepsilon$, $\forall u \in U_*(\sigma,\varepsilon)$. Hence we obtain

$$\max_{1\le s\le q}((Lu)_s - \mu_*)^+ \le \varepsilon,\quad \max_{1\le s\le q}((-Lu)_s - \mu_*)^+ \le \varepsilon,\quad \forall u \in U_*(\sigma,\varepsilon), \tag{17}$$

If $\{\Delta_{ij}\}$ and $\{\Lambda_{lj}\}$ are sufficiently small, as in (14a) and (14b) we have:

$$\|Au-f\|_\infty \le 2(\|\Delta\|_{IV}C_1 + \|\delta\|_\infty),\quad \forall u \in U_*(\sigma,\varepsilon), \tag{18a}$$

$$\|(Bu-d)^+\|_\infty \le 2(\|\Lambda\|_{IV}C_1 + \|\lambda\|_\infty),\quad \forall u \in U_*(\sigma,\varepsilon), \tag{18b}$$

where C_1 is a constant, limiting the compact set $U_*(\sigma,\varepsilon)$. Obviously, $\|A^1u - f^1\|_\infty = 0$, $\|(B^1u-d^1)^+\|_\infty = 0$ for $\forall u \in D$. Now we obtain from (16), (17), (18a), (18b)

$$\rho(u,U_*) \le M\{2[(\|\Delta\|_{IV} + \|\Lambda\|_{IV})C_1 + \|\delta\|_\infty + \|\lambda\|_\infty] + \varepsilon\},\quad \forall u \in U_*(\sigma,\varepsilon).$$

The statement of Theorem 5 follows from this.

Thus as it follows from the last estimation, the solution of problem (7) can be found with required accuracy $\{\Delta_{ij}\}$ and $\{\Lambda_{ij}\}$ are sufficiently small.
We remark that in (Gilyazov and Morozov, 1987) and (Morozov et al., 1987) was obtained the estimate

$$\sup_{u\in U_*(\sigma,\epsilon)} \|u-u_*\| = O(\|\Delta\|_V^{1/2} + \|\delta\|_\infty^{1/2}),$$

where $\|\Delta\|_V = \max_{1\le i\le m}(\sum_{j=1}^n \Delta_{ij}^2)^{1/2}$, for the case $B=0,\ d=0,\ \|\cdot\|_I = \|\cdot\|_2$, and in (Gilyazov and Morozov, 1987) it was also assumed that $D = \mathbf{R}_+^n = \{u \in \mathbf{R}^n\colon u \ge 0\}$, and in (Morozov et al., 1987) D was a convex and closed set.
Estimates like (15) were obtained in (Ivanitskii et al., 1990) separatly for the cases: 1)$\ker L = \{0\},\ A=0,\ f=0$; 2)$B=0,\ d=0$.
Other approaches to obtain stable solutions of problem (1) with inexact input data were discussed in (Morozov, 1988a; Tikhonov and Arsenin, 1986; Morozov et al., 1989; Vasil'ev, 1988a; Molodtsov, 1987; Vasil'ev et al., 1990).

5. STABLE METHOD FOR DETERMINATION OF PSEUDOSOLUTIONS

We consider a system of linear algebraic equations

$$Au = f,\ A = \{a_{ij}\} \in \mathbf{R}^{m\times n},\ f = (f_1, f_2, \dots, f_m)^T \tag{19}$$

without assuming its solvability. We shall define the pseudosolutions of system (19), determined from the following conditions, as a solution of the system:

$$\|Au - f\|_2 \to \inf,\ u \in \mathbf{R}^n, \tag{20}$$

where $\|\cdot\|_2$ is the Euclidean norm. It is known that the set U (U is of solutions of problem (20)) is nonempty for any A and f.
In the general case problem (20) is unstable with respect to disturbance of the input data $\{A, f\}$. To construct a stable approximation to the set U one can also apply here the pointwise residual method (6) after a certain modification. We consider the following problem by analogy with (7):

$$\|Lu\|_I \to \inf,\ u \in U, \tag{21}$$

where $L \in \mathbf{R}^{q\times n}$ is a matrix, $\|\cdot\|_I$ is a norm in $\mathbf{R}^q$ to describe this method. We denote by U_* the set of solutions of problem (21) and we shall call them L are pseudosolutions of system (19). We assume that the following condition of additionality is fulfilled: there exist such constants $\alpha_0 > 0,\ \alpha_1 \ge 0,\ \alpha_2 \ge 0,\ \alpha_1 + \alpha_2 > 0$, that

$$\alpha_0\|u\| \le \alpha_1\|Lu\|_I + \alpha_2\|Au\|_{II},\quad \forall u \in \mathbf{R}^n. \tag{22}$$

According to Theorem 2 the condition (22) is equivalent to the relation

$$\ker A \cap \ker L = \{0\} \tag{23}$$

We shall represent the set U_* in a more suitable form. As it is known (Morozov, 1984b; Vasil'ev, 1988a) $u \in U$ if and only if $A^T(Au - f) = 0$. Putting $\xi = f - Au$, we can reduce this system can be deduced to the system of equations

$$Au + \xi = f, \quad A^T\xi = 0 \tag{24}$$

with respect to the variable $v = (u|\xi)^T \in \mathbf{R}^s$, $s = n + m$. Let V be the set of solutions of system (24): if $v = (u|\xi)^T \in V$, then $u \in U$ and $\|\xi\|_2 = \inf_{u \in \mathbf{R}^n} \|Au - f\|_2$. Let

$$\overline{L} = [L|0] \in \mathbf{R}^{q \times s}, \quad \overline{A} = \begin{pmatrix} A & E \\ 0 & A^T \end{pmatrix} \in \mathbf{R}^{s \times s}, \quad \bar{f} = (f|0)^T \in \mathbf{R}^s,$$

where $s = n + m$, and E is the unit matrix. Then problem (21) can be reduced to the problem:

$$\|\overline{L}v\|_{\text{I}} \to \inf, \quad v \in V = \{v \in \mathbf{R}^s \colon \overline{A}v = \bar{f}\} \tag{25}$$

in the space of variables $u_* = (u_*|\xi_*)^T \in V_*$. We denote by V_* the set of solutions for this problem. It is clear that if $v_* = (u_*|\xi_*)^T \in V_*$, then $u_* \in U_*$ and conversly, if $u_* \in U_*$, then $v_* = (u_*|\xi_*)^T \in V_*$, where $\xi_* = Au_* - f$.

Problem (25) is of the same type as problem (7) if $B = 0$, $d = 0$, $D \subseteq \mathbf{R}^n$. We shall verify the correctness of the condition of additionality for problem (25) to use the theory developed in §1–4. We shall use the following Theorem.

THEOREM 6. *If condition of additionality (22) is fulfilled for matrices A and L, then an analogous condition is fulfilled for matrices $\overline{A}$ and $\overline{L}$, i.e.*

$$\alpha_0' \|v\| \le \alpha_1' \|\overline{L}v\|_{\text{I}} + \alpha_2' \|\overline{A}v\|_{\text{II}}, \quad \forall v \in \mathbf{R}^s, \tag{26}$$

where constants $\alpha_0' > 0$, $\alpha_1' \ge 0$, $\alpha_2' \ge 0$, $\alpha_1' + \alpha_2' > 0$.

The proof of Theorem 6 may be found in (Ivanitskii, 1988).

Thus Theorem 1 is correct for the problem (25) if conditions (22) are fulfilled, and therefore the set V_* of its solutions is nonempty and compact. Then the set U_* of solutions of problem (21) is also nonempty and compact.

Let instead of the exact data $\{A, f\}$ we know their approximations

$$\{\tilde{A}, \tilde{f}\} \colon \tilde{A} = \{\tilde{a}_{ij}\} \in \mathbf{R}^{m \times n}, \quad \tilde{f} = (\tilde{f}_1, \tilde{f}_2, \ldots, \tilde{f}_m)^T,$$

connected by relations (2a). It can be shown that stable approximations to the solutions of problem (25) are the solutions of the following variational problem:

$$\|\overline{L}v\|_{\text{I}} \to \inf, \quad v \in \tilde{V}, \tag{27}$$

where

$$\tilde{V} = \{v \in \mathbf{R}^s \colon \big|\sum_{j=1}^{s} \hat{a}_{ij} v_j - \hat{f}_i\big| \le \sum_{j=1}^{s} \hat{\Delta}_{ij} |v_j| + \delta_i, \ i = 1, \ldots, s\},$$

$$\hat{A} = \{\hat{a}_{ij}\} = \begin{pmatrix} \tilde{A} & E \\ 0 & \tilde{A}^T \end{pmatrix}, \quad \hat{f} = (\tilde{f}, 0) \in \mathbf{R}^s,$$

$$\hat{\Delta} = \{\hat{\Delta}_{ij}\} = \begin{pmatrix} \Delta & 0 \\ 0 & \Delta^T \end{pmatrix}, \quad \hat{\delta} = (\delta|0)^T \in \mathbf{R}^s.$$

Problem (27) can be solved approximately, i.e., it is sufficient to determine a point v from the condition

$$v \in \tilde{V}\colon \|\overline{L}v\|_{\mathrm{I}} \le \tilde{\mu}_* + \varepsilon, \quad \tilde{\mu}_* = \inf_{v \in \tilde{V}} \|\overline{L}v\|, \quad \varepsilon \ge 0. \tag{28}$$

Let $V_*(\sigma, \varepsilon)$ be the set of points v satisfying (28), $U_*(\sigma, \varepsilon)$ be the set of all points $\tilde{u}_*$ for which $\tilde{v}_* = (\tilde{u}_*|\tilde{\varepsilon}_*)^T \in V_*(\sigma, \varepsilon)$. An estimate of the speed of the pointwise residual method for the solution of problem (21) with inexact data is contained in the following theorem.

THEOREM 7. *Let the condition of additionality* (22), *and relations* (2a) *be fulfilled and let* $\|\cdot\|_I = \|\cdot\|_\infty$. *Then the following estimates hold true if sufficiently small* $\{\Delta_{ij}\}$ *are:*

$$\sup_{v \in V_*(\sigma,\varepsilon)} \rho(v, V_*) = O(\|\Delta\|_{\mathrm{IV}} + \|\delta\|_\infty) + \varepsilon, \tag{29}$$

$$\sup_{u \in U_*(\sigma,\varepsilon)} \rho(u, U_*) = O(\|\Delta\|_{\mathrm{IV}} + \|\delta\|_\infty) + \varepsilon. \tag{30}$$

Proof. In our case the set V_* is polihedral. Hence reasoning in just the same way as in the proof of Theorem 5, we obtain estimate (29).
Let us prove (30). We take a point $\tilde{v}_* = (\tilde{u}_*|\tilde{\varepsilon}_*)^T \in V_*(\sigma, \varepsilon)$. Let $w_* = (u_*|\xi_*)^T$ be the projection of $\tilde{v}_*$ on the set V_*. Then, as it was noted $u_* \in U_*$. Therefore

$$\rho(\tilde{u}_*, U_*) \le \|\tilde{u}_* - u_*\|_2 \le \|\tilde{v}_* - w_*\|_2 = \rho(\tilde{v}_*, V_*).$$

Using now (29), we deduce (30). The Theorem is proved.
Let us make several final remarks. The auxiliary matrix L is of a great importance for all the above developed versions of the pointwise residual method. The matrix L together with the matrices A, B and the set D is required only to satisfy that condition of additionality of type (8). As it is obvious from Theorem 2 there exists sufficiently large freedom arbitrariness is the choice of this matrix, which can be used to get stable approximate solutions of the considered above problems with additional properties. For example if it is necessary to get a solution with minimal first coordinate, then one can try to take such L that $\|Lu\|_{\mathrm{I}} = |u_1|$, i.e.

$$L = \begin{pmatrix} 1 & 0 \\ 0 & 0 \end{pmatrix} \in \mathbf{R}^{q \times n}.$$

If a condition of type (8) is fulfilled for such matrix L, then the pointwise residual method permits to get a stable solution with the demanded property. Some other interesting examples of the choice of L have been considered in (Morozov, 1984b; Morozov et al., 1989).

Everywhere above it was assumed that the matrix L is known exactly. But all the developed versions of the pointwise residual method can be easly modified for the case when instead of the exact matrix L only its approximation is known.

REFERENCES

Bereznev, V.A., Carmanov, V.G., and Tretyakov, A.A. (1987). *Stable Methods for Solving Extremum Problems with Approximate Data. Preprint.* Scient. Soc. on Complex Problem "Cybernetics", USSR Academy of Sciences, Moscow (in Russian).

Eremin, I.I. (1988). *Contradicting Models of Optimal Planning.* Nauka, Moscow (in Russian).

Gilyazov, S.F. and Morozov, V.A. (1987). On estimate of error for the solution of linear algebraic equations systems. *USSR Comput. Math. and Math. Phys.* **27**, 1416–1418 (in Russian).

Hoffman, A.J. (1952). On approximate solutions of systems of linear inequalities. *J.Res.Nat. Bur. Standards.* **49**, 263–265.

Ivanitskii, A.Yu. (1988). *Stable Methods for Solving Systems of Linear Equations and Inequalities with Interval Coefficient.* Candidate Thesis. Moscow Univ. Press, Moscow (in Russian).

Ivanitskii, A.Yu. (1989). On the Solving of systems of linear inequalities with approximate data. In: *Methods and Algorithms of Numerical Analysis and Applications.* Moscow Univ. Press, Moscow, pp. 42–49 (in Russian).

Ivanitskii, A.Yu. and Fedotov, V.P. (1989). On V.A.Morozov's condition of additionality. In: *Methods and Algorithms of Numerical Analysis and Applications.* Moscow Univ. Press, Moscow, pp. 50–55 (in Russian).

Ivanitskii, A.Yu., Vasil'ev, F.P., and Morozov, V.A. (1990). Estimation of speed of convergence for the pointwise residual method for solving linear algebra problems. In: *On Cooperated Scientific Researches of Comput. Centre of Moscow State Univ. and Comp. Center of Budapest Univ.*, Moscow Univ. Press, Moscow, pp. 24–31.

Molodtsov, D.A. (1987). *Stability of Optimality Principles.* Nauka, Moscow (in Russian).

Morozov, V.A. (1984a). *Methods for Solving Incorrectly Posed Problems.* Springer-Verlag, New-York, Berlin, Heidelberg, Tokyo.

Morozov, V.A. (1984b). On stable numerical methods for solving compatible linear algebraic equations systems. *USSR Comput. Math. and Math. Phys.* **24**, 179–186 (in Russian).

Morozov, V.A. and Ivanitskii, A.Yu. (1987). Stable method for solving one class of linear programming problems. In: *Numerical Analysis: Methods, Algorithms, Programs.* Moscow Univ. Press, Moscow, pp. 3–17 (in Russian).

Morozov, V.A., Gilyazov, S.F. and Ivanitskii, A.Yu. (1987). On error of the solution of systems of linear algebraic equations with approximate data. In: *Numerical Analysis: Methods and Algorithms.* Moscow Univ. Press, Moscow, pp. 3–12 (in Russian).

Morozov, V.A., Medvedev, N.V. and Ivanitskii, A.Yu. (1989). *Regularization of Algebra and Analysis Problems.* Moscow Univ. Press, Moscow (in Russian).

Tikhonov, A.N. and Arsenin, V.Ya. (1986). *Methods for Solving Incorrect Problems.* Nauka, Moscow (in Russian).

Vasil'ev, F.P. (1988a). *Numerical Methods for Solving Extremum Problems.* Nauka, Moscow (in Russian).

Vasil'ev, F.P. (1980). *Methods for Solving Extremum Problems.* Nauka, Moscow (in Russian).

Vasil'ev, F.P., Morozov, V.V., and Yachimovich,M. (1989). Estimation of speed of convergence of regularisation method for linear programming problems. *USSR Comput. Math. and Math. Phys.* **29**, 631–635 (in Russian).

Vasil'ev, F.P., Ivanitskii, A.Yu., and Morozov,V.A. (1990). Estimation of speed of convergence for the residual method for the problems of linear programming with approximate data. *USSR Comput. Math. and Math. Phys.* **30**, 1257–1262 (in Russian).

Ill-Posed Problems in Natural Sciences, pp. 44 – 47
A. Tikhonov (Ed.)
1992 VSP/TVP

OPTIMALITY REGION OF THE m-STEP LAVRENTIEV METHOD

T. KIHO
"Akadeemia", Küütri str. 1, 202400 Tartu, Estonia

ABSTRACT

This paper deals with the problem of a priori parameter choice in the iterated versions of Lavrentiev's method for solving ill-posed problems in classes of sourse-like solutions. Operator equations in Hilbert spaces are considered. The problem is to determine the best possible estimation for approximate solutions and to indicate the corresponding choice of the regularization parameter that leads to this estimation, when the iterative versions of Lavrentiev's method is in use and it is known a priori that the exact solution is a source-like element.

In fact, the principal form of the optimal estimation has already been known including its main point, namely, the exact power of the initial error bound. Thus, for us in our aspiration to find an unimprovable estimation of the approximation and the corresponding parameter choice, the only way is to use the constants of the estimation.

Our aim is to indicate, for each iterated version of Lavrentiev's method, conditions for the optimal parameter choice. If the optimal parameter choice is not possible, then we are looking for quasi-optimality, i.e., for a choice of the regularization parameter that gives an order-optimal estimation of the approximate solution.

In this paper, the exact optimal and quasi-optimal estimations along with the corresponding parameter choices in the case of the ordinary Lavrentiev method are indicated. For the iterated versions conditions for optimality are introduced (together with the corresponding estimation and parameter choice) as well as conditions for quasi-optimality (supplied with the desired estimation and parameter choice nowhere but in the limit cases).

1. The situation we deal with is as follows:

a) we consider an ill-posed problem, an operator equation $Au = f$ in a Hilbert space H with self-conjugate non-negative linear bounded operator $A = A^* \geq 0$, the range of the operator $R(A) \subset H$ is assumed to be nonclosed;

b) we assume that the right-hand side of the equation is known approximately, i.e., instead of f we deal with another element $f_\delta \in H$ such that $\|f - f_\delta\| \leq \delta$, $\delta > 0$;

c) we know the exact operator but keep in mind that the potential pollution of the operator would affect merely the details, not the essence of the matter;

d) it is known a priori that the exact solution is a source-like element, i.e.,

$$u \in M_{p\rho u_0} = \{v \in H\colon v - u_0 = A^p w,\ \|w\| \leq \rho\},$$

where $p > 0,\ \rho > 0,\ u_0 \in H$;
e) we use the m-step iterated Lavrentiev's method to find an approximate solution:

$$u_\alpha = u_{m,\alpha};$$

$$u_{j,\alpha} = (\alpha I + A)^{-1}(\alpha u_{j-1,\alpha} + f_\delta), \quad j = 1, ..., m;$$

$$u_{0,\alpha} = u_0,$$

where the regularization parameter $\alpha > 0$.
Recall that if it is possible to choose the parameter α so that

$$\sup_{u \in M_{p\rho u_0}, f_\delta \in H:\ \|Au - f_\delta\| \le \delta} \|u - u_\alpha\| \le c_{pm} \rho^{\frac{1}{p+1}} \delta^{\frac{p}{p+1}} \tag{1}$$

(p and m are fixed), then the method is quasi-optimal (=order-optimal), and if $c_{pm} = 1$, then it is optimal on $M_{p\rho u_0}$. The relation $p \notin (0, m]$ leads to the loss of both optimality and quasi-optimality, see (Vainikko, 1987) and (Vainikko et al., 1986).
2. We need the following.

Definition. A set $\Omega_m (\Omega_m \subset (0, m])$ is called the optimality region of the m-step Lavrentiev's method if, for each $p \in \Omega_m$, it is possible to choose α, the parameter of regularization, so that the method is optimal on the source set $M_{p\rho u_0}$.
3. Let us pose two problems:
For each natural $m \geq 1$,
I° find the optimality region Ω_m and indicate (for each $p \in \Omega_m$) the corresponding parameter choice satisfying (1) with $c_{pm} = 1$;
II° for each $p \in (0, m] \backslash \Omega_m$, find the unimprovable quasioptimal estimation, i.e., indicate the parameter choice minimizing the constant c_{pm} in estimation (1).
4. The following assertions are true.
I° The optimality region of the m-step Lavrentiev's method is the half-interval $\Omega_m = (0, p_m]$, where p_m is the only solution of the following equation

$$\frac{p+1}{p}\left[\sqrt[m]{p+1} - 1\right]^2 m^2 - 1 = 0, \quad (0 < p \le m) \tag{2}$$

the corresponding parameter choice is

$$\alpha = \frac{1}{\sqrt[m]{p+1} - 1}\left(\frac{\delta}{\rho}\right)^{\frac{1}{p+1}}.$$

The first solutions (for $m = 1, 2$), expressed analytically, are: $p_1 = (\sqrt{5} - 1)/2 \approx 0.6180$; $p_2 = (a^4 + 4a^3 - 18a^2 + 28a + 49)/36a^2$, where $a = \sqrt[3]{44 + 3\sqrt{177}}$, i.e., $p_2 \approx 0.7675$. The subsequent values of p_m (found using a computer) are listed below, see Table 1.
Table 1. *Solutions of equation* (2).

m	1	2	3	4	5	10	∞
p_m	0.62	0.77	0.84	0.89	0.91	0.97	1.04

The increasing sequence of the solutions $(p_m)_{m=1}^{\infty}$ converges to the solution of the limit equation

$$\frac{p+1}{p}\log^2(p+1)-1=0,$$

where $p_\infty = 1.043$, which determines the asymptotic optimality region $\Omega_\infty = (0, p_\infty]$.

For proofs, consult (Kiho, 1988a). Note also that the asymptotic optimality has been studied in (Vainikko et al., 1986).

II° For $m = 1$, the unimprovable quasioptimal estimation has the form (1) with

$$c_{p1} = \left[\frac{(2p-1)^{2p-1}(1-p)^{2(1-p)}(1+p)^{1+p}}{p^p}\right]^{\frac{1}{2(1+p)}},$$

the corresponding parameter choice is

$$\alpha = \left[\frac{1}{p(2p-1)^{2p-1}(1-p)^{2(1-p)}}\right]^{\frac{1}{2(1+p)}}\left(\frac{\delta}{\rho}\right)^{\frac{1}{p+1}}.$$

The proof can be found in (Kiho, 1987).

Table 2. *Values of* c_{p1}.

p	$\frac{1}{2}(\sqrt{5}-1)=0.62$	0.70	0.80	0.90	1
c_{p1}	1	1.02	1.08	1.19	$\sqrt{2}=1.41$

Note that the result contains the unimprovable, smallest value of the constant $c_{1,1} = \sqrt{2}$ which is of much importance. Thus, the estimations of the type (1) which one can meet in literature with greater constants, see, e.g., (Ivanov et al., 1978), $c_{1,1} = 2$ can be slightly improved.

For $m \geq 2$, we are able only to recommend the easy choice

$$\alpha = \left(\sqrt{m}\frac{\delta}{\rho}\right)^{\frac{1}{m+1}},$$

which provides quasioptimality if $p = m$. Then estimation (1) holds with

$$c_{mm} = \sqrt{m^{\frac{m}{m+1}}(m+1)}.$$

See (Kiho, 1988b).

Table 3. *Values of* c_{mm}.

m	1	2	3	5	10	100	1000
c_{mm}	1.41	2.18	3.02	4.79	9.45	98.23	997.05

5. We can recapitulate that for the m-step Lavrentiev method the best, unimprovable estimation of the approximate solution and the corresponding parameter choice have been found for the small values of p (from the optimality region) and for the greatest possible value ($p = m$). The exact error estimate and parameter choice in the ordinary, noniterated Lavrentiev method have been indicated for all p.

REFERENCES

Ivanov, V.K., Vasin, V.V., and Tanana, V.P. (1978). *Theory of Linear Ill-Posed Problems and Its Applications.* Nauka, Moscow (in Russian).

Kiho, T. (1987). Optimal choice of the parameter in the Lavrentiev method. *Acta et commentationes Universitatis Tartuensis* **762**, 31–39.

Kiho, T. (1988a). Optimal choice of the parameter in the iterated versions of the Lavrentiev method on the source classes of solutions. *Acta et commentationes Universitatis Tartuensis* **833**, 97–106.

Kiho, T. (1988b). Error estimation in iterated version of Lavrentiev's method. In: *Abstracts of Conference "Methods of Algebra and Analysis"*, Tartu, pp. 163–165, (in Russian).

Vainikko, G. (1982). *Methods for Solving Linear Ill-Posed Problems in Hilbert Spaces.* Tartu.

Vainikko, G.M. and Veretennikov, A.J. (1986). *Iterative Procedures in Ill-Posed Problems.* Nauka, Moscow (in Russian).

Ill-Posed Problems in Natural Sciences, pp. 48 – 56
A. Tikhonov (Ed.)
1992 VSP/TVP

MULTILEVEL ITERATIVE METHODS FOR ILL-POSED PROBLEMS

J.T. KING
Department of Mathematical Sciences, University of Cincinnati, Cincinnati, Ohio 45221-0025, USA

ABSTRACT

In this paper we consider the use of multilevel operators to precondition an iterative method for the solution of first kind operator equations of the form $K\ u = g$. We restrict our attention to a particular method of Tikhonov regularization that is to be solved by an iterative method such as Landweber-Fridman iteration or the method of conjugate gradients. We present theoretical bounds for the error reduction factors and estimate these quantities for some example problems. Our emphasis here is on the algebraic problem rather than the selection of regularization parameters as a function of the noise level, δ, in the data. Numerical experiments are given that compare the iterative methods to solution by elimination.

INTRODUCTION

Let H_1 and H_2 be real separable Hilbert spaces and suppose $K \in \mathcal{B}(H_1, H_2)$ is a compact operator whose adjoint, K^*, has trivial nullspace. We denote the inner products on H_1 and H_2 by $(\cdot\,,\cdot)$ and the corresponding norms by $\|\cdot\|_1$ and $\|\cdot\|_2$ respectively. We consider the first kind operator equation

$$Ku = g, \tag{1}$$

where $g \in R(K)$ is imprecisely known, say g^δ is given with $\|g - g^\delta\|_2 \leq \delta$, where $\delta > 0$ is fixed.
To approximate $u =: K'g$, where K' denotes the Moore-Penrose generalized inverse of K, we consider a particular finite dimensional variant of (Tikhonov, 1963) regularization introduced by (King and Neubauer, 1988). Let $W_1 \subset W_2 \subset \ldots \subset W_m$ be a nested sequence of finite dimensional subspaces of H_2 and denote by Q_j: $H_2 \to W_j$ the orthogonal projection onto W_j. Set $\gamma_j =: \|K^*(I - Q_j)\|$ and we assume $\lim_{m\to\infty} \gamma_m = 0$. Our approximation $u^\delta_{m,\lambda}$ to u is given by $u^\delta_{m,\lambda} =: K^* w^\delta_{m,\lambda}$, where

$$(K^* w^\delta_{m,\lambda}, K^*\phi) + \lambda(w^\delta_{m,\lambda}, \phi) = (g^\delta, \phi), \quad \phi \in W_m = (Q_m g^\delta, \phi) \tag{2}$$

and $\lambda \geq 0$ is fixed. We do not address the selection of the regularization parameter λ here, but see (King and Neubauer, 1988) for an optimal a posteriori choice. Rather we want to consider the solution of (2) for a given λ by preconditioned iterative methods using a multilevel approach. We point out that $u^{\delta}_{m,\lambda}$ minimizes, the Tikhonov functional over $V_m =: K^*(W_m)$

$$F(v) = \|Kv - g^{\delta}\|_2^2 + \lambda\|v\|_1^2.$$

MULTILEVEL OPERATORS

To describe the multilevel operators and iterative methods we find it convenient to introduce the symmetric bilinear form

$$a(w, z) = (K^*w, K^*z) + \lambda(w, z), \quad w, z \in H_2$$

and the operators $A_{j,\lambda}$: $W_j \to W_j$ given by

$$a(w, \phi) = (A_{j,\lambda}w, \phi), \quad \phi \in W_j$$

Then (2) is equivalent to

$$A_{m,\lambda}w^{\delta}_{m,\lambda} = Q_m g^{\delta}. \tag{3}$$

For later reference we denote the minimal eigenvalue of $A_{j,0}$ by μ_j and remark that $\mu_j \leq \gamma^2_{j-1}$.

The multilevel iterative methods for the solution of (3) are described as follows. Let B_j: $W_j \to W_j$ denote a symmetric positive definite operator that is (in a sense to be made clear later) an approximate inverse of $A_{j,\lambda}$. Let n_j denote the number of iterations on level j , that is on W_j. On W_1 set $B_1 = A^{-1}_{1,\lambda}$ and define $g^{\delta}_j = Q_j g^{\delta}$ for each j, $1 \leq j \leq$ m.

Full Multilevel Algorithm. Define $w_1^{n_1} = B_1 g_1^{\delta}$.
Suppose $w_{j-1}^{n_{j-1}} \in W_{j-1}$ has been defined and define $w_j^k \in W_j$ by

$$w_j^k = w_j^{k-1} + B_j(g_j^{\delta} - A_{j,\lambda}w_j^{k-1}), \tag{4}$$

where $w_j^0 = w_{j-1}^{n_{j-1}}$ and with n_j determined by the least index k such that

$$\|w_j^k - w_j^{k-1}\|_2 \leq \text{ TOL } \|w_j^k\|_2.$$

The sequence is terminated at level j when

$$\|A_{j,0}w_j^k - g_j^{\delta}\|_2 \leq \text{RTOL}.$$

The tolerances, TOL and RTOL, are specified so that one does not iterate too often per level and so that the residual is comparable to the (Morozov, 1966) discrepancy principle, say RTOL $= C\delta$, for some constant C.

We note that (4) is simply a preconditioned version of Landweber-Fridman iteration applied to $A_{j,\lambda}w^{\delta}_{j,\lambda} = Q_j g^{\delta}$. In the context of finite differences or finite elements for partial differential equations this sort of iteration is called the full multigrid algorithm (see (McCormick, 1987)). To completely specify the procedure described above we need to define the operators B_j.

One can also use the multilevel operators to precondition the iterative solution of $A_{j,0}w_{j,0}^{\delta} = Q_j g^{\delta}$. In this case (4) is replaced by

$$w_j^k = w_j^{k-1} + B_j(g_j^{\delta} - A_{j,0}w_j^{k-1}). \tag{5}$$

Remark. In this paper we have restricted our attention to the preconditioning of Landweber-Fridman iteration but the operators B_j can be used to precondition conjugate gradients (see (King, 1989 and King, 1991)) or steepest descent, etc. For example, to precondition the conjugate gradient method for (3) one applies conjugate gradients to the equivalent problem

$$B_m A_{m,\lambda} w_{m,\lambda}^{\delta} = B_m Q_m g^{\delta}$$

with respect to the inner product $(B_m^{-1}\cdot,\cdot)$. Of course this presumes B_m is symmetric positive definite on W_m.
To motivate the definition of B_j consider the operator $T_j = I - \alpha_j(I - Q_{j-1})A_{j,\lambda}$ defined on W_j. Then T_j is a symmetric positive definite operator and is a contraction on the orthogonal complement of W_{j-1} for suitable α_j .

LEMMA 1. *For* $0 < \alpha_j < (\lambda + \gamma_{j-1}^2)^{-1}, T_j$ *is symmetric in* $a(\cdot,\cdot)$ *and the spectrum of* $T_j, \sigma(T_j)$*, satisfies* $\sigma(T_j) \subset (0,1]$. *Moreover, if* $w \in W_{j-1}^{\perp} = \{w \in W_j \colon a(w,\phi) = 0, \phi \in W_{j-1}\}$*, then*

$$a(T_j w, w) \leq \delta_j a(w,w),$$

where $\delta_j = 1 - (\mu_j + \lambda)\alpha_j < 1$.

Proof. We only demonstrate that T_j is a contraction. Let P_{j-1} denote the orthogonal projection onto W_{j-1} with respect to the $a(\cdot,\cdot)$ inner product. Then $P_{j-1}Q_{j-1} = Q_{j-1}$ and it follows that $(I-P_{j-1})T_j = (I-P_{j-1})(I-\alpha_j A_{j,\lambda})$. Thus, for any $w \in W_j$,

$$a(T_j(I-P_{j-1})w,(I-P_{j-1})w) = a((I-\alpha_j A_{j,\lambda})(I-P_{j-1})w,(I-P_{j-1})w)$$

$$\leq [1-(\lambda+\mu_j)\alpha_j]\,a((I-P_{j-1})w,(I-P_{j-1})w)$$

by the definition of μ_j.

Remark. We assume throughout that $0 < \alpha_j < (\lambda + \gamma_{j-1}^2)^{-1}$.
A direct consequence of Lemma 1 is the following. Consider one iteration of

$$z_j^k = z_j^{k-1} + \alpha_j(I - Q_{j-1})(f - A_{j,\lambda}z_j^{k-1})$$

Then for $z_{j,\lambda}$ defined by $A_{j,\lambda}z_{j,\lambda} = f \in W_j$ we have

$$z_{j,\lambda} - z_j^1 = T_j(z_{j,\lambda} - z_j^0)$$

and hence Lemma 1 says the "high frequency components" of the initial error are reduced by the factor δ_j in the norm $a^{1/2}(\cdot,\cdot)$. Thus we may think of T_j as

a smoothing operator. To further reduce the error we need to reduce the "low frequency components" by restriction to the space W_{j-1}.
We define the symmetric multilevel operators B_j: $W_j \to W_j$ by induction. Set $B_1 = A_{1,\lambda}^{-1}$ and suppose B_k has been defined for $k \le j-1$. For $f \in W_j$ define $B_j f = z_2 + z_3$ where:

(i) $z_1 = B_{j-1}Q_{j-1}f$;

(ii) $z_2 = z_1 + \alpha_j(I - Q_{j-1})(f - A_{j,\lambda}z_1)$;

(iii) $z_3 = B_{j-1}Q_{j-1}(f - A_{j,\lambda}z_2)$

and α_j is specified in Lemma 1.
These multilevel operators are a generalization of the two-level operators introduced in (King, 1990). The operator B_j is linear and requires the calculation of three orthogonal projections and two applications of $A_{j,\lambda}$. Note, for example, that the application of B_4 requires two applications of B_3 and hence four applications of B_2 and finally eight applications of $B_1 = A_{1,\lambda}^{-1}$. Thus a single application of B_4 requires eight inversions (using, e.g., Cholesky) of the same low dimensional symmetric matrix. Typically the dimension of W_4 is at least eight times that of W_1. From the definition of B_j it is not difficult to establish the identity

$$I - B_j A_{j,\lambda} = S_j T_j S_j, \tag{6}$$

where $S_j = I - P_{j-1} + (I - B_{j-1}A_{j-1,\lambda})P_{j-1}$ and P_{j-1}: $H_2 \to W_{j-1}$ is the orthogonal projection in the inner product $a(\cdot,\cdot)$. This identity gives a recurrence relation for the multilevel operators that can be used to show that B_j is symmetric in the H_2 inner product. The identity (6) is also a key factor in establishing the following result (see (King, 1991)).

Theorem . *There exists $\epsilon_j \in (0,1)$ such that for all $w \in W_j$*

$$0 \le a((I - B_j A_{j,\lambda})w, w) \le \epsilon_j a(w,w),$$

where $\epsilon_j = \epsilon_{j-1}^2 + \delta_j(1 - \epsilon_{j-1}^2)$ and δ_j is given in Lemma 1.

One can obtain a bound for the reduction factor that depends only on $\hat{\delta} =:$ $\max\{\delta_j\colon 2 \le j \le m\}$. We find for $1 \le j \le m$

$$\hat{\delta} \le \epsilon_j \le 1 - (1-\hat{\delta})(1-\hat{\delta}^2)^{j-2}$$

and hence $\hat{\epsilon} =: \max\{\epsilon_j\} \le 1 - (1-\hat{\delta})(1-\hat{\delta}^2)^{m-2}$. The theory predicts the reduction factor ϵ_j degenerates as $j \to \infty$. However, for ill-posed problems it is not advantageous to let the gridsize tend to zero (i.e. $j \to \infty$) when solving a problem whose data is in error by some fixed amount δ.

Remark. An immediate consequence of Theorem 2 is the inequality

$$(1-\epsilon_j)a(w,w) \le a(B_j A_{j,\lambda}w, w) \le a(w,w)$$

and hence by the definition of $A_{j,\lambda}$ and an obvious change of variables we obtain

$$(1-\epsilon_j)(A_{j,\lambda}^{-1}w, w) \le (B_j w, w) \le (A_{j,\lambda}^{-1}w, w)$$

which shows that B_j is an approximation to $A_{j,\lambda}^{-1}$ in the sense of spectral equivalence. Note also that B_j is invertible.
Another consequence of this result justifies the use of B_j to precondition the iterative solution of $A_{j,0}w_{j,0}^\delta = Q_j g^\delta$. Define on H_2 the form $\mathcal{A}(w,z) = (K^*w, K^*z)$; then we have the following result (King, 1991).

COROLLARY. *The operator $I - B_jA_{j,0}$ is symmetric in $\mathcal{A}(\cdot,\cdot)$ and, for all $w \in W_j$*

$$0 \le \mathcal{A}((I - B_jA_{j,0})w, w) \le (1 - \tau(1 - \hat{\epsilon}))\mathcal{A}(w,w),$$

where $\tau = \min\{1, \mu_m\lambda^{-1}\}$.

From the Corollary it follows that the iterative method

$$w_j^k = w_j^{k-1} + B_j(g_j^\delta - A_{j,0}w_j^{k-1})$$

satisfies for $w_j^0 = 0$

$$\|u_{j,0}^\delta - K^*w_j^k\|_1 \le \rho^k \|u_{j,0}^\delta\|_1,$$

where $\rho = 1 - \tau(1 - \hat{\epsilon}) < 1$. Note that λ in the definition of B_j can be chosen independently of the choice of λ in $A_{j,\lambda}$.

EXAMPLES AND NUMERICAL EXPERIMENTS

We consider an integral operator of the form

$$Ku(s) = \int_a^b k(s,t)u(t)\,dt$$

defined on $\mathbf{L}_2[a,b]$ with square integrable kernel on $[a,b]\times[a,b]$. We also assume that $\partial^2 k/\partial t^2 \in \mathbf{L}_2[a,b]$. Our choice of finite dimensional subspaces consists of linear splines on a sequence of grids. Let $N > 0$ be an integer and set $h_1 = (b-a)/N$ and $t_i^1 = a + ih_1$ for $0 \le i \le N$. Then W_1 is the set of functions that are linear on each subinterval $[t_i^1, t_{i+1}^1]$ and continuous on $[a,b]$. The spaces W_j are constructed in the same way with $h_j = h_{j-1}/2$ and $t_i^j = a + ih_j$. In this case it can be shown that $\gamma_j = O(h_j^2)$ and a proof is given in (Groetsch et al., 1982). The dimension of W_j is $N_j = 2^{j-1}N + 1$. In our experiments we use $N = 8$ and spaces $W_1, \ldots, W_4$. Thus we solve (2) for $m = 4$ with $N_4 = 65$ by a four level iterative method using the multilevel operators B_j, $1 \le j \le 4$. All of the inner products in $H_1 = \mathbf{L}_2[a,b]$ were computed using a composite 5 point Gaussian quadrature based on $4(N_j - 1) + 1$ points. On $W_j \subset H_2 = \mathbf{L}_2[a,b]$ we used the equivalent discrete inner product

$$(v,w)_j = h_j \sum_{k=1}^{N_j} v(t_k^j)w(t_k^j)$$

in place of the inner product on H_2 in (2). This permits the determination of the projection Q_{j-1} in a simple way that avoids inversion of L_2 Gram matrices. We consider three example problems each of which satisfies $N(K^*) = \{0\}$.

Example 1. Define the kernel

$$k(s,t) = \begin{cases} 0 & \text{if } s \le t, \\ e^{-1/4(s-t)}/\sqrt{\pi(s-t)} & \text{if } s > t. \end{cases}$$

Then it follows that $Ku = g$ on $\mathbf{L}_2[0,1]$ is equivalent to the inverse heat conduction problem:

$$w_t - w_{xx} = 0, \quad 0 < x < \infty,\ 0 < t \le 1,$$

$$w(1,t) = g(t), w(x,t) \text{ bounded as } x \to \infty,$$

$$w(x,0) = 0, \quad w_x(0,t) = -u(t),$$

where t is time and x is the distance to the heated surface. Thus one wants to determine the heat flux at $x = 0$ based on the measured temperature at $x = 1$. For numerical methods for this and the related problem of recovering $w(0,t)$ see (Carasso, 1982), (Elden, 1983), (Engl and Manselli, 1989), (Murio, 1981) and the references contained therein. Using MATLAB we computed the minimal eigenvalue μ_j of $A_{j,0}$ and the theoretical and computed reduction factors for the parameter choices $\lambda = 10^3 h_4^4, \alpha_j = (\lambda + h_j^4)^{-1}$. The results are given in Table 1. The computed reduction factors were determined by selecting 20 random vectors $w_{j,\lambda}$ and the corresponding right hand side by matrix multiplication $g_j =: A_{j,\lambda} w_{j,\lambda}$. Then using the zero initial guess we applied the iteration (4) with the above mentioned choices of λ and α_j. After 40 iterations we determined the average reduction per iteration for each choice of g_j. The arithmetic mean of these 20 reduction factors is given in Table 1. The theoretical ϵ_j are given in Theorem using the computed minimal eigenvalues and $\epsilon_2 = \delta_2 = 1 - (\mu_2 + \lambda)\alpha_2$. The observed reduction factors are smaller than that predicted by the theory.

Table 1. Reduction factors, Example 1

j	μ_j	theoretical ε_j	computed ε_j
2	.58(−12)	.983	.943
3	.71(−17)	.999128	.936
4	.24(−22)	.999952	.945

We applied the full multilevel algorithm for data g chosen such that the solution is given by $u(s) = 1$ if $.25 \le s \le .75$ and $u(s) = 0$ otherwise. In this case it is known that $u \notin R((K^*K)^p)$ for all $p > 0$ and hence there is no convergence rate predicted by the theory (see e.g. Groetsch, 1984) of Tikhonov regularization for $\lambda = \lambda(\delta)$ as $\delta \to 0$. In Table 2 we list the resulting $\mathbf{L}_2$ errors for the multilevel iterative method and direct solution of (3) for several noise levels and for the parameters specified above. The $\mathbf{L}_2$ errors are comparable except for the largest noise level where the iterative method required only one iteration on W_2. We used TOL $= 10^{-3}$ and RTOL $= \delta\|g^\delta\|_2$ in these calculations.

Table 2. $\mathbf{L}_2$ errors for Example 1

δ	n_2	n_3	n_4	iterative error	direct error
.5(−5)	1	42	56	.1601	.158
.5(−4)	1	42	56	.1603	.158
.5(−3)	1	41	56	.1618	.160
.5(−2)	1	0	0	.2168	.179

Example 2. Define the kernel

$$k(s,t) = \begin{cases} s(1-t) & \text{if } s \leq t, \\ t(1-s) & \text{if } s > t. \end{cases}$$

Then $Ku = g$ if and only if $g(0) = g(1) = 0$ and $u = -g''$. As is well known, this problem is much less difficult to solve than the previous example. Again we computed the minimal eigenvalues μ_j and the reduction factors for the parameter choices $\lambda = h_4^4$ and $\alpha_j = (\lambda + (h_j/10)^4)^{-1}$. The results are given in Table 3. For data $g(t) = \sin(\pi t)/\pi^2$ and corresponding solution $u(t) = \sin(\pi t)$ we list in Table 4 the resulting $\mathbf{L}_2$ errors for various noise levels with the parameters specified above and TOL $= 10^{-5}$, RTOL $= .025\delta\|g^\delta\|_2$. In each case the $\mathbf{L}_2$ errors are superior to those for direct solution by Gaussian elimination. We mention that for larger values of TOL and RTOL the iterative process terminates prematurely.

Table 3. Reduction factors, Example 2

j	μ_j	theoretical ε_j	computed ε_j
2	.19(−9)	.879	.855
3	.12(−10)	.963135	.845
4	.76(−12)	.987957	.878

Table 4. $\mathbf{L}_2$ errors for Example 2

δ	n_2	n_3	n_4	iterative error	direct error
.5(−5)	85	48	70	.156(−2)	.17(−1)
.5(−4)	79	0	0	.165(−1)	.29(−1)
.5(−3)	98	94	0	.151(0)	.40(0)
.5(−2)	71	0	0	.131(+1)	.24(+1)

Example 3. Consider the kernel

$$k(s,t) = \frac{\Omega}{\pi} \begin{cases} 1 & \text{if } s = t, \\ \operatorname{sinc} \Omega(s-t) & \text{if } s \neq t. \end{cases}$$

The integral equation

$$\int_{-l}^{l} k(s,t)u(t)\,dt = g(s)$$

arises in optical imaging systems (Groetsch, 1989) and (Ross, 1982) and laser anemometry (Abbiss et. al., 1983). In optics g represents the blurred (due to a limited scattering angle that determines Ω) image of an object u of limited extent $2l$. We have selected $l = 1$ and $\Omega = 10\pi$ for simplicity. In optics the value of Ω can be large.
Using TOL $= .5 \times 10^{-3}$, RTOL $= \delta \|g^\delta\|_2$, and data g chosen such that $u(t) = 1 - t^2$ we computed the direct solution and the multilevel iterative solution of (3) using the parameter choices $\lambda = (10h_4)^4$ and $\alpha_j = (\lambda + h_j^4)^{-1}$. The resulting $\mathbf{L}_2$ errors for several noise levels are given in Table 5. Again we observe that the $\mathbf{L}_2$ errors are comparable. Note also that very few iterations are required.

Table 5. $\mathbf{L}_2$ errors for Example 3

δ	n_2	n_3	n_4	iterative error	direct error
.5(−5)	3	2	1	.105(−1)	.105(−1)
.5(−4)	3	0	0	.124(−1)	.122(−1)
.5(−3)	2	0	0	.162(−1)	.169(−1)
.5(−2)	1	0	0	.234(0)	.257(0)

It should be emphasized that no attempt has been made to choose an "optimal" regularization parameter λ in these calculations.

REFERENCES

Abbiss, J.B., DeMol, C. and Dhadwal, H.S. (1983). Regularized iterative and noniterative procedures for object restoration from experimental data, *Opt. Acta* **30**, 107–124.

Carasso, A. (1982). Determining surface temperatures from interior observations, *SIAM J. Appl. Math.* **42**, 558–574.

Elden, L. (1983). The numerical solution of a noncharacteristic Cauchy problem for a parabolic problem. In: *Numerical Treatment of Inverse Problems in Differential and Integral Equations*. Ed by P. Deuflhard, and E. Hairer, Birkhauser, Basel, pp. 246–268.

Engl, H.W. and Manselli, P. (1989). Stability estimates and regularization for an inverse heat conduction problem in semi-infinite and finite time intervals, *Numer. Funct. Anal. Optimization* **10**, 517–540.

Groetsch, C.W. (1984). *The Theory of Tikhonov Regularization for Fredholm Equations of the First Kind.* Pitman, London.

Groetsch, C.W. (1989). Remarks on some iterative methods for an integral equation in Fourier optics. In: *Transport Theory, Invariant Imbedding, and Integral Equations.* Ed by P. Nelson et.al., Marcel Dekker, New York, pp. 313–324.

Groetsch, C.W., King, J.T. and Murio, D. (1982). Asymptotic analysis of a finite element method for Fredholm integral equations of the first kind. In: *Treatment of Integral Equations by Numerical Methods.* Ed by C.T.H. Baker and G.F. Miller, Academic Press, London. pp. 1–11.

King, J.T. (1989). A minimal error conjugate gradient method for ill-posed problems. *J. Optimization Theory Appl.* **60**, 297–304.

King, J.T. (1990). On the construction of preconditioners by subspace decomposition. *J. Comp. Appl. Math.* **29**, 195–205.

King, J.T. (1991). Multilevel algorithms for ill-posed problems. *Numer. Math.* (to appear).

King, J.T., Neubauer, A. (1988). A variant of Tikhonov regularization with a posteriori parameter choice. *Computing J.* **30**, 91–109.

McCormick, S.F. (1987). *Multigrid Methods.* SIAM, Philadelphia.

Morozov, V.A. (1966). On the solution of functional equations by the method of regularization. *Soviet Math. Dokl.* **7**, 414–417.

Murio, D. (1981). The mollification method and the numerical solution of an inverse heat conduction problem. *SIAM J. Sci. Statist. Comp.* **2**, 17–34.

Ross, G. (1982). Iterative methods in information processing for object restoration. *Opt. Acta* **29**, 1523–1542.

Tikhonov, A.N. (1963). Regularization of incorrectly posed problems. *Soviet Math. Dokl.* **4**, 1624–1627.

Ill-Posed Problems in Natural Sciences, pp. 57 – 62
A. Tikhonov (Ed.)
1992 VSP/TVP

THE METHOD OF MINIMAL PSEUDOINVERSED MATRIX. BASIC STATEMENTS

A.S. LEONOV
Moscow Engineering Physics Institute, Kashirskoe shosse, 31, 115409 Moscow, Russia

ABSTRACT

Basic concepts and statements of the method of minimal pseudoinversed matrix for solving ill-posed linear systems are considered. The main feature of this method is its optimal order of accuracy

1. Let us consider a system of linear algebraic equations

$$\overline{A}z = \bar{u}; \quad z \in \mathbf{R}^n, \quad \bar{u} \in \mathbf{R}^m, \quad \overline{A} \in \mathfrak{A} \equiv \mathfrak{A}_{m\times n}; \quad \overline{A} \neq 0, \quad \bar{u} \neq 0 \tag{1}$$

whose unknowns are represented by the vector z. Here $\mathfrak{A}$ is the normed space of real $m \times n$ matrices having the Euclidian norm. The norms in $\mathbf{R}^n$ and $\mathbf{R}^m$ are also Euclidian. System (1) is not consistent in general. However, there always exists a set of "pseudosolutions" $\widetilde{Z} \neq \varnothing$ of this system, i.e., the set of vectors $\tilde{z} \in \mathbf{R}^n$ such that

$$\|\overline{A}\tilde{z} - \bar{u}\|_{\mathbf{R}^m} = \inf\{\|\overline{A}z - \bar{u}\|_{\mathbf{R}^m} \colon z \in \mathbf{R}^n\}.$$

It is known that this system has also a unique normal pseudosolution $\bar{z} \in \widetilde{Z}$, i.e., the solution of extremal problem

$$\|\bar{z}\|_{\mathbf{R}^n} = \inf\{\|\tilde{z}\|_{\mathbf{R}^n} \colon \tilde{z} \in \widetilde{Z}\}$$

(see (Moore, 1920; Penrose, 1955)). Let us suppose that instead of the precise data $(\overline{A}, \bar{u})$ in system (1) we are given approximate data (A_h, u_δ): $A_h \in \mathfrak{A}$, $\|\overline{A} - A_h\| \leq h$, $u_\delta \in \mathbf{R}^m$, $\|\bar{u} - u_\delta\| \leq \delta$ are given. We assume also that the numerical values h and δ for the accuracy of approximate data are known.
We consider the following basic problem: find a stable approximation of normal pseudosolution of (1) from the perturbed data $(A_h, u_\delta, h, \delta)$, i.e., a vector $z_{h\delta} = z_{h\delta}(A_h, u_\delta, h, \delta)$ such that $\|z_{h\delta} - \bar{z}\| \to 0$ as $h, \delta \to 0$. This problem has been considered in many papers (see, e.g., (Tikhonov, 1965; Gilyazov and Morozov, 1984; Leonov, 1987)). Solving the basic problem by the Tikhonov method of

regularization one meets an interesting phenomenon: the accuracy of the approximate solution we have got depends essentially on whether system (1) is consistent or not. This accuracy is $O(h+\delta)$ for consistent systems and $O(\delta+\sqrt{h})$ for inconsistent systems. In view of this fact the important problem is to construct a method solving the basic problem and giving the same optimal accuracy $O(\delta+h)$ in both cases.
Some approaches to constructing such methods were suggested by Gilyazov and Morozov, Vainikko (Gilyazov and Morozov, 1984; Vainikko, 1986) and others. In particular, the method of minimal pseudoinversed matrix (MPM) was developed by Leonov (Leonov, 1987). We present here some essential aspects of the theory of the MPM method.

2. The MPM method is based on solving the following extremum problem: find a matrix $\widetilde{A}_h \in \mathfrak{A}$ such that

$$\|\widetilde{A}_h^+\|_* = \inf\{\|A^+\|_*\colon\ A \in \mathfrak{A},\ \|A - A_h\| \le h\}. \tag{2}$$

Here A^+ is a pseudoinversed matrix for A (Moore, 1920; Penrose, 1955) and $\|\cdot\|_*$ is the Euclidian norm in the space $\mathfrak{A}^*$ of $n \times m$-matrices. If $\widetilde{A}_h$ is a solution of (2), then we can take the element $z_\eta = \widetilde{A}_h^+ u_\delta$, $\eta = (h, \delta)$ as an approximation to $\bar{z}$. Any solution $\widetilde{A}_h$ of (2) is called a "matrix of the MPM method" and $\widetilde{A}_h^+$ is called a "minimal pseudoinversed matrix".
The following statements hold (Leonov, 1990).

LEMMA 1 (Stability of the rank). *If* $0 \le h < h_0(\bar{A}) = \|\bar{A}^+\|_*^{-1}/2$, *then* $\operatorname{rank}\widetilde{A}_h = \operatorname{rank}\bar{A}$ *for any solution* $\widetilde{A}_h$ *of* (2).

LEMMA 2 (Stability of the minimal pseudoinversed matrix). *If* $0 \le h < h_0(\bar{A})$ *then*

$$\|\widetilde{A}_h^+ - \bar{A}^+\|_* \le 2h\|\bar{A}^+\|_*^2/(1 - 2h\|\bar{A}^+\|_*)^3.$$

Lemma 2 implies the following theorem.

THEOREM 1. *If* $0 \le h < h_0(\bar{A})$, *then the following estimate for the accuracy of the MPM method is valid:*

$$\|z_\eta - \bar{z}\| \le \|\bar{A}^+\|_*[\delta + 2h\|\bar{A}^+\|_*\|\bar{u}\|/(1 - 2h\|\bar{A}^+\|_*)^3]. \tag{3}$$

This estimate can not be improved: the order of involved values δ, h, $\|\bar{A}^+\|_*$ and constant 2 are exact.

3. To implement the MPM method numerically it is sufficient to find any solution of problem (2). We indicate a special class of such solutions which are calculated effectively. Let us introduce a singular value decomposition (SVD) of A_h: $A_h = U_h D_h V_h^T$. Here $D_h = \operatorname{diag}(\tilde{\rho}_1, \dots, \tilde{\rho}_M) \in \mathfrak{A}$ is a rectangular diagonal matrix and the non-negative numbers $\tilde{\rho}_k$ $(k = 1, \dots, M = \min(m, n))$ are the singular values of A_h. The matrices U_h and V_h of size $m \times m$ and $n \times n$ respectively are orthogonal. We introduce the following mathematical programming problem: find numbers $(\hat{\rho}_1,\ldots,\hat{\rho}_M) \in R_0^M = \{(\rho_1,\ldots,\rho_M)\colon \rho_1 \geq \rho_2 \geq \ldots \geq \rho_M \geq 0\}$ such that

$$\sum_{k=1}^{M} \theta(\hat{\rho}_k^2) = \inf\Big\{ \sum_{k=1}^{M} \theta(\rho_k^2)\colon (\rho_1,\ldots,\rho_M) \in R_0^M,\ \sum_{k=1}^{M} (\rho_k - \tilde{\rho}_k)^2 = h^2 \Big\}. \quad (4)$$

Here $\theta(\rho) = \{\rho^{-1}$ for $\rho \neq 0$; 0 for $\rho = 0\}$.

THEOREM 2 (Leonov, 1990). i) *Problem* (4) *is solvable for each* $h \geq 0$. ii) *Under the conditions:* $0 \leq h < h_0(\bar{A})$, $\|A_h\| > h$ *problem* (4) *has a unique solution and this solution is given by*

$$\hat{\rho}_k = \begin{cases} \tilde{\rho}_k x_k[\alpha(h)] & \text{if } 1 \leq k \leq r = \operatorname{rank} \bar{A}; \\ 0 & \text{if } k > r. \end{cases}$$

Here $x_k(\alpha)$ *is the unique solution of the equation*

$$x^4 - x^3 = \alpha \tilde{\rho}_k^{-4}$$

satisfying the inequality $x > 1$, *and* $\alpha(h) > 0$ *is the unique solution of the equation*

$$\varepsilon(\alpha) \equiv \sum_{k=1}^{r} \tilde{\rho}_k^2 [x_k(\alpha) - 1]^2 + \sum_{k=r+1}^{M} \tilde{\rho}_k^2 = h^2$$

with continuous and increasing function $\varepsilon(\alpha)$. *The following two theorems clarify the interconnection between problems* (2) *and* (4).

THEOREM 3. *Let the condition* $\|A_h\| > h$ *hold for a given* h *and let* $(\hat{\rho}_1,\ldots,\hat{\rho}_M)$ *be a solution of* (4). *Then the matrix* $\widehat{A}_h = U_h \widehat{D}_h V_h^T$, *which is defined by the matrix* $\widehat{D}_h = \operatorname{diag}(\hat{\rho}_1,\ldots,\hat{\rho}_M) \in \mathfrak{A}$ *and by the matrices* U_h, V_h *of the SVD for* A_h *is a solution of problem* (2).

THEOREM 4. *If the conditions* $0 \leq h < h_0(\bar{A})$, $\|A_h\| > h$ *are satisfied, then any solution of* (2) *has the same singular values* $\hat{\rho}_1,\ldots,\hat{\rho}_M$.

It is clear from Theorem 3 that the element $z_\eta = V_h \widehat{D}_h^{+} U_h^T u_\delta$ can be taken as an MPM approximation to $\bar{z}$. Here, $\widehat{D}_h^{+} = \operatorname{diag}(\theta(\hat{\rho}_1),\ldots,\theta(\hat{\rho}_M)) \in \mathfrak{A}^*$.

Remark 1. The numerical implementation of the MPM method does not require knowledge of the exact SVD of A_h. It is very important since the algorithms for the calculation of the SVD used in practice give an approximate singular value decomposition which has some accuracy $\varkappa$. I.e., by using A_h, we can calculate the orthogonal matrices $U_{h\varkappa} \in \mathfrak{A}_{m\times m}$, $V_{h\varkappa} \in \mathfrak{A}_{n\times n}$, and $D_{h\varkappa} = \operatorname{diag}(\tilde{\rho}_1^{\varkappa},\ldots,\tilde{\rho}_M^{\varkappa}) \in \mathfrak{A}$, such that

$$\|U_{h\varkappa} D_{h\varkappa} V_{h\varkappa}^T - A_h\| \leq \varkappa.$$

Here, the non-negative numbers $\tilde{\rho}_k^{\varkappa}$ $(k = 1,\ldots,M)$ are the approximate singular values for A_h. This enables us to take the new approximate data $(A_{h\varkappa}, u_\delta, h+\varkappa)$ instead of the initial approximate data (A_h, u_δ, h). The SVD for the matrix $A_{h\varkappa} = U_{h\varkappa} D_{h\varkappa} V_{h\varkappa}^T$ is known exactly and has the accuracy $h+\varkappa$ with respect to $\bar{A}$. Using the MPM method for the data $(A_{h\varkappa}, u_\delta, h+\varkappa)$ in the form, found in Theorem 3, we obtain the element $z_{\eta\varkappa} = V_{h\varkappa} \widehat{D}_{h\varkappa}^{+} U_{h\varkappa}^T u_\delta$. From Theorem 1 the convergence $z_{\eta\varkappa} \to \bar{z}$ follows as $\eta, \varkappa \to 0$.

4. It follows from Theorem 2 that in order to construct a solution of (4) it is necessary to know a stable estimate of $r = \operatorname{rank} \bar{A}$. Now we consider the problem of finding such an estimate. We introduce the Lagrange function for problem (4)

$$M^\lambda[\rho_1, \dots, \rho_M] = \lambda \sum_{k=1}^{M} \theta(\rho_k^2) + \sum_{k=1}^{M} (\rho_k - \tilde{\rho}_k)^2; \quad \lambda \geq 0, \quad \rho_1, \dots, \rho_M \geq 0,$$

and consider the extremum problem: find numberes $\rho_1(\lambda), \dots, \rho_M(\lambda) \geq 0$ such that

$$M^\lambda[\rho_1(\lambda), \dots, \rho_M(\lambda)] = \inf\{M^\lambda[\rho_1, \dots, \rho_M]\colon \rho_1, \dots, \rho_M \geq 0\}. \tag{5}$$

We set $\lambda_p = 27\, \tilde{\rho}_p^4/16 \quad (p = 1, \dots, M)$ and note that $\lambda_1 \geq \lambda_2 \geq \dots \geq \lambda_M \geq 0 = \lambda_{M+1}$.

LEMMA 3. i) *Problem* (5) *is solvable for each* $\lambda \geq 0$. *All solutions of* (5) *are given by the formula*

$$\rho_k(\lambda) = \begin{cases} \tilde{\rho}_k x_k(\lambda) & \text{if } 0 \leq \lambda \leq \lambda_k; \\ 0 & \text{if } \lambda \geq \lambda_k, \quad k = 1, \dots, M. \end{cases}$$

Here $x_k(\lambda)$ *is the solution of the equation* $x^4 - x^3 = \lambda \tilde{\rho}_k^{-4}$ *satisfying the inequality* $1 < x \leq 3/2$.
ii) *Problem* (5) *has an unique solution provided* $\lambda \neq \lambda_p \quad (p = 1, \dots, M)$. *The numbers* $\lambda = \lambda_p$ *are the points of nonuniqueness for the solution of* (5).
iii) *The function*

$$\beta(\lambda) = \sum_{k=1}^{M} [\rho_k(\lambda) - \tilde{\rho}_k]^2, \quad \lambda \geq 0$$

is monotonically increasing and continuous everywhere for $\lambda \geq 0$ *except the points of discontinuity of the first kind* λ_p $(p = 1, \dots, M)$. *Moreover,* $\beta(+0) = 0$, $\beta(+\infty) = \|A_h\|^2$.

We consider an equation

$$\beta(\lambda) = h^2 \tag{6}$$

for a monotonic discontinuous function $\beta(\lambda)$. Using Lemma 3, one can prove that there exists an unique "generalized" solution $\lambda(h) > 0$ of (6). The generalized solution $\lambda(h)$ of the equation with monotonic function $\beta(\lambda)$ is an ordinary solution if $\lambda(h)$ is a point of continuity of β or a point of discontinuity in which the function β "jumps" over the value h^2. We can find the rank of $\bar{A}$ using $\lambda(h)$.

THEOREM 5. *Suppose that* $\|A_h\| > h$ *for each* h, $0 \leq h < h_0(\bar{A})/2$. *Then there exists a number* $p \quad (1 \leq p \leq M)$ *not depending on* h *which can be found from inequality* $\lambda_{p+1} \leq \lambda(h) < \lambda_p$ *and which determines the rank of* $\bar{A}$: $r = \operatorname{rank} \bar{A} = p$.

5. In this section we formulate a version of a numerical algorithm for the method of minimal pseudoinversed matrix.

(a) Find a singular value decomposition of A_h or an approximate SVD as we noted in Remark 1 (see (Forsythe et al., 1977)).

(b) Find a generalized solution $\lambda(h) > 0$ of (6) using a function $\beta(\lambda)$ of the form specified in Lemma 3.

(c) Find a stable estimation p of rank $\bar{A}$ using Theorem 5.

(d) Calculate the singular values $\tilde{\rho}_k$ $(k = 1, \ldots, M)$ of the matrix of the MPM method using Theorem 2 and the obtained number $r = p$.

(e) Find the matrix of the MPM method, the minimal pseudoinversed matrix and the approximate normal pseudosolution of (1) using the results of §3.

It is possible to estimate the number N of operations for the numerical implementation of the algorithm described. Under some assumptions this requires about $(8n^3/3)$ operations (see (Leonov, 1987)) for $n \times n$ - matrices.

6. The MPM method has some optimal properties (Leonov, 1990; 1991).

Definition 1. The mapping P, which puts in to correspondence with any approximate data $(A_h, u_\delta, h, \delta)$ an element $z_\eta = P(A_h, u_\delta, h, \delta) \in \mathbf{R}^n$, is said to be a standard method for finding normal pseudosolution of (1) if z_η has the form $z_\eta = V_h F(D_h, \alpha_\eta) U_h^T u_\delta$. Here $\alpha_\eta \geq 0$ is a regularization parameter chosen in a special way which generally depends on $(A_h, u_\delta, h, \delta)$. The matrix $F(D_h, \alpha_\eta) \in \mathfrak{A}^*$ is diagonal: $F(D_h, \alpha_\eta) = \mathrm{diag}(f_1(\tilde{\rho}_1, \alpha_\eta), \ldots, f_M(\tilde{\rho}_M, \alpha_\eta))$ and the functions $f_k(\rho, \alpha)$ $(k = 1, \ldots, M)$ which are defined for $\rho, \alpha \geq 0$, are measurable in ρ for every fixed $\alpha \geq 0$ and bounded in ρ on the semiintervals $[\varepsilon, +\infty)$, for any $\varepsilon > 0$. Definition 1 admits that there may be no dependence on α_η. For instance, this is the case for the least squares method: $f_k = \theta(\tilde{\rho}_k)$. Other examples of standard methods are represented by the Tikhonov regularization $(f_k = (\alpha_\eta + \tilde{\rho}_k^2)^{-1} \tilde{\rho}_k)$, by its generalization, which was studied by Gilyazov and Morozov, Vainikko (Gilyazov and Morozov, 1984; Vainikko, 1986) $(f_k = (\alpha_\eta + \tilde{\rho}_k^2)^{-2} \tilde{\rho}_k^3)$ and others. We denote the set of all standard methods by S.

THEOREM 6. *The version of the* MPM *method which is based on Theorem* 3 *is a standard algorithm.*

Let R_1, R_2 be positive numbers. We specify the class of exact systems (1) which corresponds to R_1 and R_2:

$$\Sigma = \Sigma(R_1, R_2) = \{(\bar{A}, \bar{u}) \colon \; \|\bar{A}^+\|_* \leq R_1, \; \|\bar{u}\| \leq R_2\}.$$

For every system $(\bar{A}, \bar{u}) \in \Sigma$, let us define a class of permissible approximate data

$$\Sigma_\eta(\bar{A}, \bar{u}) = \{(A_h, u_\delta) \colon \; \|A_h - \bar{A}\| \leq h, \|u_\delta - \bar{u}\| \leq \delta\}$$

corresponding to the accuracy characterizations of data $\eta = (h, \delta)$. We introduce also an accuracy characterization of the method P on the class Σ of exact systems, when the approximate data are taken from the class Σ_η:

$$\begin{aligned} \Delta(P, \eta) &\equiv \Delta(P, \eta; R_1, R_2) \\ &= \sup_{A_h, u_\delta, \bar{A}, \bar{u}} \{\|P(A_h, u_\delta, h, \delta) - \bar{z}\| \colon \; \bar{z} = \bar{A}^+ \bar{u}, \\ &\qquad (\bar{A}, \bar{u}) \in \Sigma(R_1, R_2), \; (A_h, u_\delta) \in \Sigma_\eta(\bar{A}, \bar{u})\}. \end{aligned} \tag{7}$$

Definition 2. The value

$$\Delta_0(\eta) \equiv \Delta_0(\eta; R_1, R_2) = \inf\{\Delta(P, \eta):\ P \in S\}$$

in called the optimal accuracy for approximate solution of system (1) using the standard methods.

Definition 3. The method P, $P \in S$, is said to have optimal order of accuracy, if

$$\limsup_{\eta \to 0} [\Delta(P, \eta; R_1, R_2)/\Delta_0(\eta; R_1, R_2)] \le g.$$

The constant g here does not depend on R_1 and R_2.

LEMMA 4. (Leonov, 1991). *For each* $\eta = (h, \delta)$, $0 \le h < R_1^{-1}/2$, $0 \le \delta < R_2$, *the following estimate holds:*

$$\Delta_0(\eta) \ge R_1(\delta + R_1 R_2 h)\xi(\eta, R_1, R_2).$$

Here the function $\xi(\eta, R_1, R_2)$ *converges to* 1 *as* $\eta \to 0$ *for any positive* R_1, R_2.

Let $\Delta_{MPM}(\eta)$ be the accuracy characterization of the MPM method given by (7). Theorems 1, 6 and Lemma 3 imply the following statement.

THEOREM 7. *The* MPM *method has optimal order of accuracy:*

$$\limsup_{\eta \to 0} [\Delta_{MPM}(\eta)/\Delta_0(\eta)] \le 2.$$

To compare our method with some other methods, recall that the optimality constant g for the method due to (Gilyazov and Morozov, 1984; Vainikko, 1986) is $g = (7 + 2\sqrt{5})^{1/2} \approx 3.387$.

We note in conclusion that the method of minimal pseudoinversed matrix was effectively used in applications (see (Leonov, 1987)).

REFERENCES

Gilyazov, S.F. and Morozov, V.A. (1984). Optimal regularization of ill-posed normally solvable operator equations. *USSR Comput. Math. and Math. Phys.* **24**, 1737–1742 (in Russian).

Forsythe, G.E., Malcolm, M.A., and Moler, C.B. (1977). *Computer Methods for Mathematical Computations.* Prentice-Hall, Englewood Cliffs, NJ.

Leonov, A.S. (1987). The method of minimal pseudoinversed matrix. *USSR Comput. Math. and Math. Phys.* **27**, 1123–1138 (in Russian).

Leonov, A.S. (1990). On the theory of minimal pseudoinversed matrix. *Dokl. Akad. Nauk SSSR.* **314**, 89–93 (in Russian).

Leonov, A.S. (1991). On the problem of accuracy of the method of minimal pseudoinversed matrix. *Math. Notes.* **49**, 81–87 (in Russian).

Moore, E.H. (1920). On the reciprocal of the general algebraic matrix. *Bull. Amer. Math. Soc.* **26**, 394–395.

Penrose, R. (1955). On a generalized inverse for matrices. *Proc. Cambrige. Phil. Soc.* **51**, 406–413.

Tikhonov, A.N. (1965). On the stability of algorithms for solving degenerate systems of linear algebraic equations. *USSR Comput. Math. and Math. Phys.* **5**, 718–722 (in Russian).

Vainikko, G.M. and Veretennikov, A.Yu. (1986). *Iterational Procedures in Ill-Posed Problems.* Nauka, Moscow (in Russian).

Ill-Posed Problems in Natural Sciences, pp. 63 – 70
A. Tikhonov (Ed.)
1992 VSP/TVP

VARIATIONAL ALGORITHMS WITH A POSTERIORI CHOICE OF THE REGULARIZATION PARAMETER FOR SOLVING ILL-POSED EXTREMAL PROBLEMS

A.S. LEONOV
Moscow Engineering Physics Institute, Kashirskoe shosse, 31, 115409 Moscow, Russia

ABSTRACT

The variational algorithms with a posteriori choice of regularization parameter for solving ill-posed extremal problems are considered. Finite-dimensional variants of these algorithms and their numerical implementation are discussed.

1. FORMULATION OF PROBLEM

Let (Z, τ) be a Hausdorff topological space with a topology τ. We assume that a functional $J(z)$ is defined on a subset $D \subset Z$, $D \neq \varnothing$. We consider the conditional extremum problem: find an element $z^* \in D$ such that

$$J(z^*) = \inf \{J(z): \ z \in D\} \equiv J^*. \tag{1}$$

Assume that (1) has a nonempty solution set Z^* and consider the problem of finding Ω-optimal solutions, i.e., elements $\bar{z} \in Z^*$ such that

$$\Omega(\bar{z}) = \inf \{\Omega(z^*): \ z^* \in Z^*\} \equiv \overline{\Omega}. \tag{2}$$

Here $\Omega(z)$ is a bounded below functional on D: $\Omega^* = \inf\{\Omega(z): \ z \in D\} \geq 0$. Suppose that, instead of the exact functional $J(z)$ in (1), we are given some approximate data $(J_\delta, \ D, \ \Psi_0)$. Here, the functional $J_\delta(z)$ treated as a "perturbation" of the functional in (1) belongs to a set $\mathcal{F}$ of permissible approximate functionals bounded below on the domain D and satisfying the approximation condition

$$\left|J(z) - J_\delta(z)\right| \leq \Psi_0(\delta, \Omega(z)) \quad \forall z \in D.$$

We assume that the measure of approximation Ψ_0 and the number $\delta \geq 0$ determining the level of approximation error, are known.

We consider the following Basic problem: find an element $z_\delta = z(J_\delta,\ \delta,\ \Psi_0) \in D$ such that z_δ converges τ-sequentially to the set $\overline{Z}$ of Ω-optimal solutions of (1), (2) as $\delta \to 0$.
To solve this problem one must generally use a computer. Therefore, it is necessary to construct a finite-dimensional approximation of the problem. In many cases the data of the Basic Problem are directly given in a finite-dimensional form. In view of that we formulate the problem of finite-dimensional approximation of (1), (2).
Let Z_N be an N-dimensional normed space with elements $\hat{z}_N$. We introduce operators P_N: $Z \to Z_N$, $\overline{P}_N$: $Z_N \to Z$. The essential properties of these operators are given below. We define a set $\widehat{D}_N = \{\hat{z}_N\colon \overline{P}_N \hat{z}_N \in D\}$. Sets of this kind will represent finite-dimensional analogues of D. In the sequel we sometimes omit the index N when it is fixed (e.g., $\hat{z}_N = \hat{z}$, $\widehat{D}_N = \widehat{D}$, $P_N = P$, and so on).
We assume that, instead of the functional $J_\delta(z)$, its "finite-dimensional approximation" is given, i.e., a finite-dimensional functional $\widehat{J}(\hat{z}) \equiv \widehat{J}_\eta(\hat{z})$, $\eta = (\delta, 1/N)$, with a domain $\widehat{D}_N$, which satisfies an approximation condition

$$\left|\widehat{J}_\eta(P_N z) - J_\delta(z)\right| \leq \widehat{\Psi}(\eta, \Omega(z)) \quad \forall z \in D.$$

Here, the function $\widehat{\Psi}$ is the known measure of the finite-dimensional approximation.
The Basic Finite-Dimensional Problem (BFDP) is to find, using the data $(\widehat{J}_\eta,\ \widehat{\Psi},\ \Psi_0,\ \delta,\ N,\ Z_N)$, an element $\hat{z}_\eta \in \widehat{D}_N$ such that the family $\{z_\eta\}$, $z_\eta = \overline{P}_N \hat{z}_\eta \in D$, converges τ-sequentially to the set $\overline{Z}$ as $\eta \to 0$.
The element $z_\eta = \overline{P}_N \hat{z}_\eta$ found in this way is taken to be an approximate solution of Basic problem.

2. THE MAIN ASSUMPTIONS AND THE SCHEME OF SOLUTION

Let us formulate our main assumptions.
a) The functional J is τ-sequentially lower semicontinuous on D: for any $z_0 \in D$ and any sequence $\{z_n\} \subset D$ such that $z_n \xrightarrow{\tau} z_0$ as $n \to \infty$ the following inequality holds

$$\liminf_{n\to\infty} J(z_n) \geq J(z_0);$$

the same property is assumed to be fulfilled for the functionals J_δ and Ω.
b) The set $\Omega_k = \{z \in D\colon \Omega(z) \leq K\}$ is τ-sequentially compact for any $K \geq \Omega^*$.
c) The linear operators P_N and $\overline{P}_N$ are continuous for any N. They possess the following properties:
c1) $P_N \overline{P}_N = E_N$ for any N, where E_N is the unity operator in Z_N;
c2) the inclusion $\overline{P}_N P_N z \in D$ holds for any $z \in D$ and any N.
d) For any $\bar{z} \in \overline{Z}$ there holds

$$\limsup_{N\to\infty} \Omega(\overline{P}_N P_N \bar{z}) \leq \overline{\Omega}.$$

e) The functional $\widehat{J}_\eta$ is lower semicontinuous on $\widehat{D}_N$ for any η.

f) The measures of approximation $\Psi_0(x,\Omega)$, $\widehat{\Psi}(x,\Omega)$ have the domain $\{x \geq 0\} \times \{\Omega \geq \Omega^*\}$. They are continuous and increase monotonically with respect to Ω for any $x > 0$. The equalities $\Psi_0(0,\Omega) = \widehat{\Psi}(0,\Omega) = 0,\ \forall\Omega \geq \Omega^*$ are also satisfied.
We note that assumptions a), b) guarantee that $\overline{Z} \neq \varnothing$.
Assumption c2) guarantees that $\widehat{D}_N \neq \varnothing$ because the set $\widehat{D}_N$ contains elements of the form $\hat{z}_N = P_N z,\ z \in D$.
Now let us describe a scheme of solving the BFDP. It is based on the Tikhonov regularization (Tikhonov, 1966). For a fixed N, we define the smoothing functional

$$\widehat{M}^\alpha[\bar{z}] = \alpha\widehat{\Omega}(\bar{z}) + \widehat{J}(\hat{z}), \quad \alpha > 0, \quad \hat{z} \in \widehat{D}. \tag{3}$$

Here $\widehat{\Omega}(\hat{z}) \equiv \Omega(\overline{P}\hat{z})$. Next, we introduce the extreme value problem; for a given $\alpha > 0$, find an element $\hat{z}^\alpha \in \widehat{D}$ such that

$$\widehat{M}^\alpha[\hat{z}^\alpha] = \inf\left\{\widehat{M}^\alpha[\hat{z}]\colon\ \hat{z} \in \widehat{D}\right\}. \tag{4}$$

Assumptions a) - c), e) imply that the set of solutions $\widehat{Z}^\alpha$ of this problem is not empty. However, $\widehat{Z}^\alpha$ can contain more than one element. The scheme of solving BFDP consists of two stages.

i) Choosing a regularization parameter $\alpha_\eta = \alpha(\widehat{J},\ \widehat{\Psi},\ \Psi_0,\ \delta,\ N,\ Z_N)$.

ii) Fixing the set $\widehat{Z}^{\alpha_\eta}$ for $\alpha = \alpha_\eta$, and selecting in $\widehat{Z}^{\alpha_\eta}$ an element $\hat{z}_\eta = \hat{z}^{\alpha_\eta}$ by a certain rule.

In this scheme, the element $z_\eta = \overline{P}_N\hat{z}^{\alpha_\eta}$ is taken to be an approximation to $\overline{Z}$. The convergence $z_\eta \xrightarrow{\tau} \overline{Z}$ is guaranteed by a special implementation of i) and ii).
Sometimes it is not necessary to use ii). It is the case for example, when the regularization parameter is chosen a priori, i.e., the dependence $\alpha_\eta = \alpha(\delta, N)$ is given (Tikhonov, 1966; Tikhonov and Vasil'ev, 1976). However, one cannot guarantee in this case that $\widehat{J}(\hat{z}^{\alpha_\eta})$ and $\widehat{\Omega}(\hat{z}^{\alpha_\eta})$ are "near" their desired values known a posteriori. The last demand appears rather often whille solving practical problems. That is why α_η is often chosen a posteriori, i.e., depending not only on δ and N but also on given realizations of $\widehat{J}$, Ψ_0, $\widehat{\Psi}$, etc. For ill-posed extreme value problems, a posteriori choices of α are proposed in (Leonov, 1979; 1982a; 1986; 1990).
At the same time, there are examples (see, e.g., (Leonov, 1982b)) which show the necessity of selecting $\hat{z}^{\alpha_\eta}$ by the a posteriori choice of α.

3. THE FINITE-DIMENSIONAL ALGORITHMS WITH THE A POSTERIORI CHOICE OF THE REGULARIZATION PARAMETER

Let η be fixed and let

$$\hat{\mu}_\eta = \inf\left\{\widehat{J}(\hat{z}) + \Psi\big(\eta, \widehat{\Omega}(\hat{z})\big)\colon\ \hat{z} \in \widehat{D}\right\}. \tag{5}$$

Here, the function $\Psi(\eta,\Omega) = \Psi_0(\delta,\Omega) + \widehat{\Psi}(\eta,\Omega)$ is the joint measure of approximation of the exact functional J by a finite-dimensional functional $\widehat{J}$. The number

$\hat{\mu}_\eta$ is a stable estimate for the unknown value J^* (Leonov, 1982a): $\hat{\mu}_\eta \geq J^*$ for each η and $\hat{\mu}_\eta \to J^*$ as $\eta \to 0$.
In order to formulate the algorithms with the a posteriori choice of α for the BFDP we introduce some auxiliary functions and functionals:

$$\widehat{\gamma}(\alpha) = \widehat{\Omega}(\hat{z}^\alpha), \quad \hat{\beta}(\alpha) = \widehat{J}(\hat{z}^\alpha), \quad \hat{\varphi}(\alpha) = \widehat{M}[\hat{z}^\alpha] \quad \forall \hat{z}^\alpha \in \widehat{Z}^\alpha \tag{6a}$$

and

$$\begin{gathered} \hat{\pi}(\alpha) = \Psi(\eta, \widehat{\gamma}(\alpha)) \equiv \widehat{\Pi}(\hat{z}^\alpha), \\ \hat{\rho}(\alpha) = \hat{\beta}(\alpha) - \hat{\pi}(\alpha) - \hat{\mu}_\eta \equiv \widehat{P}(\hat{z}^\alpha), \\ \varepsilon(\hat{\alpha}) = \hat{\varphi}(\alpha) - \Psi^p(\eta, \widehat{\gamma}(\alpha)) - \hat{\mu}_\eta \equiv \widehat{E}(\hat{z}^\alpha), \\ p = const, \ 0 < p < 1. \end{gathered} \tag{6b}$$

Functions (6a) are analogous to the functions introduced by Morozov, Goncharskii and Leonov (Morozov, 1968; Goncharskii et. al., 1973; Leonov, 1979) for solving operator equations. Functions (6a), (6b) are in the general case multiple-valued since for some $\alpha > 0$ the set $\widehat{Z}^\alpha$ can contain more than one element.

LEMMA 1. *The function $\hat{\varphi}(\alpha)$ is single-valued and continuous for all $\alpha > 0$. The other functions in (6a) and (6b) are single-valued and continuous everywhere for $\alpha > 0$ except, possibly, for not more than a countable number of common points of multiple-valuedness. If $\alpha > 0$ is such a point of multiple-valuedness, then the set $\widehat{Z}^\alpha$ contains at least two elements, $\hat{z}^\alpha_+$ and $\hat{z}^\alpha_-$, such that*

$$\widehat{\gamma}(\alpha \pm 0) = \widehat{\Omega}(\hat{z}^\alpha_\pm), \quad \hat{\rho}(\alpha \pm 0) = \widehat{P}(\hat{z}^\alpha_\pm), \quad \hat{\varepsilon}(\alpha \pm 0) = \widehat{E}(\hat{z}^\alpha_\pm).$$

The functions $\hat{\beta}$, $\hat{\varphi}$, $\hat{\rho}$, $\hat{\varepsilon}$ are monotonically nondecreasing, and $\widehat{\gamma}, \hat{\pi}$ are monotonically nonincreasing.

The algorithm of finite-dimensional generalized discrepancy principle (FDGDP) for solving extremal problems develops the approaches due to (Morozov, 1968; Goncharskii et. al., 1973; Leonov, 1979) and consists in the following (see (Leonov, 1982a; 1986)).

i) The regularization parameter $\alpha_\eta > 0$ is chosen as a generalized solution of an equation with monotonic function:

$$\hat{\rho}(\alpha) = 0. \tag{7}$$

Here and in what follows we say that α_η is a generalized solution of equation $\widehat{\chi}(\alpha) = 0$ with a monotonically nondecreasing function $\widehat{\chi}$, if $\widehat{\chi}(\alpha)(\alpha - \alpha_\eta) \geq 0, \ \forall \alpha \in (0, +\infty)$.

ii) An approximate solution $\hat{z}^{\alpha_\eta}$ is selected by a certain rule. Let $C > 1, \ q > 1$ be given constants of the algorithm, and let $\alpha_1 = \alpha_\eta / q, \ \alpha_2 = \alpha_\eta q$. We consider any two auxiliary extremals of (4), $\hat{z}^{\alpha_1}$ and $\hat{z}^{\alpha_2}$, associated with $\alpha = \alpha_1$ and $\alpha = \alpha_2$, respectively. If

$$\widehat{J}(\hat{z}^{\alpha_2}) \geq C\widehat{\Pi}(\hat{z}^{\alpha_1}) + \hat{\mu}_\eta \tag{8}$$

for the given η, then we take an element $\hat{z}^{\alpha_\eta}$ in $\widehat{Z}^{\alpha_\eta}$ such that $\widehat{P}(\hat{z}^{\alpha_\eta}) \leq 0$ as the desired approximate solution. For instance, we can put $\hat{z}^{\alpha_\eta} = \hat{z}_-^{\alpha_\eta}$. Otherwise, if

$$\widehat{J}(\hat{z}^{\alpha_2}) \leq C\widehat{\Pi}(\hat{z}^{\alpha_1}) + \hat{\mu}_\eta, \tag{9}$$

we select an $\hat{z}^{\alpha_\eta} \in \widehat{Z}^{\alpha_\eta}$ satisfying the condition $\widehat{P}(\hat{z}^{\alpha_\eta}) \geq 0$. For example, we can take $\hat{z}^{\alpha_\eta} = \hat{z}_+^{\alpha_\eta}$.

The algorithm of finite-dimensional generalized quasisolution principle (FDGQP) is a generalization of the methods due to (Ivanov, 1966; Tikhonov and Vasil'ev, 1976). It is possible to use this method if, in addition to the data

$(\widehat{J},\ \widehat{\Psi},\ \Psi_0,\ \delta,\ N,\ Z_N)$, we know an estimate $\widehat{\omega}_\eta$ for $\overline{\Omega}$ possessing the properties: $\widehat{\omega_\eta} \geq \overline{\Omega} \forall \eta$, $\widehat{\omega}_\eta \to \overline{\Omega}$ as $\eta \to 0$.

The algorithm consists of the following steps.

i) Choosing $\alpha_\eta > 0$ as a generalized solution of the equation

$$\hat{\gamma}(\alpha) - \widehat{\omega}_\eta = 0. \tag{10}$$

ii) Choosing an element $\hat{z}^{\alpha_\eta}$ according to the following rule. If (8) holds for a given η, then we select $\hat{z}^{\alpha_\eta} \in \widehat{Z}^{\alpha_\eta}$ such that $\widehat{\Omega}(\hat{z}^{\alpha_\eta}) \geq \widehat{\omega}_\eta$ (e.g., $\hat{z}^{\alpha_\eta} = \hat{z}_-^{\alpha_\eta}$). If (9) holds, then we take $\hat{z}^{\alpha_\eta} \in \widehat{Z}^{\alpha_\eta}$ satisfying the inequality $\widehat{\Omega}(\hat{z}^{\alpha_\eta}) \leq \widehat{\omega}_\eta$ (e.g., $\hat{z}^{\alpha_\eta} = \hat{z}_+^{\alpha_\eta}$).

The algorithm of the finite-dimensional generalized principle of a smoothing functional (FDGSFP) is an advanced version of the algorithm suggested by Morozov and Liskovets (Morozov, 1966; Liskovets, 1976). For general extreme value problems it has the form:

i) Choosing $\alpha_\eta > 0$ as a generalized solution of

$$\hat{\varepsilon}(\alpha) = 0. \tag{11}$$

ii) The selection rule: we take an arbitrary element in $\widehat{Z}^{\alpha_\eta}$ such that $\widehat{E}(\hat{z}^{\alpha_\eta}) \geq 0$.

We note that the second step can be omitted if $\alpha_\eta > 0$ is an ordinary solution if (7), or (10), or (11). In this case any element $\hat{z}^{\alpha_\eta} \in \widehat{Z}^{\alpha_\eta}$ can be taken for constructing the approximate solution $z_\eta = \overline{P}_N \hat{z}^{\alpha_\eta}$. A widely known example of such a situation is the case where problem (4) has a unique solution for each $\alpha > 0$. In this case, functions (6a), (6b) are continuous and equations (7), (10), and (11) have ordinary solutions.

Let α_η, $\hat{z}^{\alpha_\eta}$ and $z_\eta = \overline{P}_N \hat{z}^{\alpha_\eta}$ be the values which are found by one of the algorithms formulated above.

THEOREM 1. *Suppose that the inequality $\Omega(z^*) > \Omega^*$ holds for any solution $z^* \in D$ of problem* (1). *Then:*

a) *equations* (7), (10), *and* (11) *have positive generalized solutions at least for "small" values δ and $1/N$;*

b) *equation* (11) *has a unique generalized solution which converges to* 0 *as $\eta \to 0$;*

c) *equations (7), (10) have also unique generalized solutions if problem (4) has the property:* $\widehat{Z}^{\alpha'} \cap \widehat{Z}^{\alpha''} = \emptyset$ *whenever* $\alpha' > \alpha'' > 0$;
d) *for any sequence* $\eta_n = (\delta_n, 1/N_n)$ *which converges to 0 as* $n \to \infty$ *the corresponding sequence* $z_n = z_{\eta_n}$ *converges in the sense that is such that*

$$z_n \xrightarrow{\tau} \overline{Z}, \qquad \Omega(z_n) \to \overline{\Omega} \quad \text{as} \quad n \to \infty.$$

Thus, the algorithms of generalized principles which have been formulated above solve the Basic finite-dimensional problem.

4. ON THE NUMERICAL TREATMENT OF THE GENERALIZED PRINCIPLES

The algorithms considered are based on solving the extreme value problem (4). Methods for solving such problems are well known (see (Fiacco and McCormick, 1968; Polak, 1976; Vasil'ev, 1980)) and used in many applications (Tikhonov and Arsenin, 1979). We assume in what follows that there is a method which finds one of the solutions of (4) for an arbitrary $\alpha > 0$. Since we usually cannot find all the solutions of (4) for a given α, we must modify the algorithms of §3 in order to use only one extremal $\hat{z}^\alpha$ for every $\alpha > 0$.
The first constituent of each of the algorithms in question is solving an equation of the form $\widehat{\chi}(\alpha) = 0$ (see (7), (10), and (11)). Because of possible discontinuity of the monotonic function $\widehat{\chi}(\alpha)$ we cannot use "fast" methods of the Newton type for solving this equation. However, this equation can be solved by simple bisection method (or by combinations of the bisection method and some "fast" methods) (see (Bakhvalov, 1975)). It is easy to prove the following statement.

THEOREM 2. *Let a function* $\widehat{\chi}(\alpha)$ *be defined for* $\alpha > 0$ *and be monotonically nondecreasing. Suppose that the equation* $\widehat{\chi}(\alpha) = 0$ *has a generalized solution on a given segment* $[a_0, b_0]$, *such that* $a_0 > 0$, *and* $\widehat{\chi}(a_0) < 0$, $\widehat{\chi}(b_0) > 0$. *Then the sequence* α_n, *which is constructed by the bisection method, converges to a generalized solution of this equation.*

It is sufficient to use only one extremal of (4) to implement the bisection method. Now we consider a modification of selection rules (§3), which use also one extremal of (4) for every α. Since the generalized solution $\alpha_\eta > 0$ of (7) (or (10), or (11)) is found approximately, we suppose, that there is a sequence $\alpha_n = \alpha_n(\eta)$ such that: $\alpha_n > \alpha_\eta$ $\forall n$, $\alpha_n \to \alpha_\eta + 0$ as $n \to \infty$. Such a sequence can be determined by the bisection method. Then there exists a number $n(\eta)$ satisfying the condition: $\alpha_{n_\eta}/q < \alpha_\eta$. Choosing such a $n(\eta)$, we fix the following values of the regularization parameter: $a = \alpha_{n(\eta)}$, $a_1 = a/q$, $a_2 = aq$ ($q =$ const > 1), and find arbitrary extremals $\hat{z}^a$, $\hat{z}^{a_1}$, $\hat{z}^{a_2}$ of (4) for these α. Let us formulate analogues of algorithms described in §3 which use these values.

a) The generalized discrepancy principle. If

$$\widehat{J}(\hat{z}^{a_2}) \geq C\widehat{\Pi}(\hat{z}^{a_1}) + \hat{\mu}_\eta \tag{12}$$

holds, then we take $\alpha(\eta) = a_1$ as the regularization parameter and the element $z_\eta = \overline{P}_N \hat{z}^{a_1}$ as the approximate solution of (1), (2). In the opposite case, where

$$\hat{J}(\hat{z}^{a_2}) \leq C\widehat{\Pi}(\hat{z}^{a_1}) + \hat{\mu}_\eta, \tag{13}$$

we put $\alpha(\eta) = a,\ z_\eta = \overline{P}_N \hat{z}^a$.

b) The generalized quasisolution principle is constructed in a similar way, by using (10) instead of (7).

c) The generalized principle of a smoothing functional. We take the number $\alpha(\eta) = a$ as the approximate solution of (11) and $z_\eta = \overline{P}_N \hat{z}^a$ as the approximate solution of (1), (2).

It can be proved that the elements z_η constructed in this way possess the properties established in Theorem 1 d). Note that the condition $\alpha_{n(\eta)}/q < \alpha_\eta$ holds for the bisection method if $n(\eta) > \log_2\{(b_0 - a_0)/[a_0(q-1)]\}$. Here $[a_0, b_0]$, $a_0 > 0$, is an initial localization segment for the solution of (7) (or (10), or (11)).
The algorithms of generalized principles have optimal order of precision (see (Leonov, 1990)). The finite-dimensional algorithms constructed above can be used for finding stable solutions of operator equations by the variational method (Leonov, 1979).
Examples of special formulations of problem (1), (2) in various functional spaces as well as examples of applications of the algorithms are given in (Tikhonov et al., 1992).

REFERENCES

Bakhvalov, N.C. (1975). *Numerical Methods.* Nauka, Moscow (in Russian).

Fiacco, A.V. and McCormick, G.P. (1968). *Nonlinear Programming: Sequential Unconstrained Minimization Techniques.* John Wiley, New York.

Goncharskii, A.V., Leonov, A.S., and Yagola, A.G. (1973). Generalized discrepancy principle. *USSR Comput. Math. and Math. Phys.* **13**, 294–302 (in Russian).

Ivanov, V.K. (1966). On approximate solving of operator equations of the first kind. *USSR Comput. Math. and Math. Phys.* **6**, 1089–1094 (in Russian).

Leonov, A.S. (1979). On algorithms of approximate solving nonlinear problems with operator approximately determined. *Dokl. Akad. Nauk SSSR.* **245**, 300–304 (in Russian).

Leonov, A.S. (1982a). On an application of the generalized residual principle for solving ill-posed extremal problems. *Dokl. Akad. Nauk SSSR.* **262**, 1306–1310 (in Russian).

Leonov, A.S. (1982b). On a relation between the method of generalized discrepancy and generalized discrepancy principle for nonlinear ill-posed problems. *USSR Comput. Math. and Math. Phys.* **22**, 783–790 (in Russian).

Leonov, A.S. (1986). On some algorithms for solving ill-posed extremal problems. *Math. USSR Sbornik.* **129(171)**, 218–231 (in Russian).

Leonov, A.S. (1990). The optimality of precision order for some algorithms of solving ill-posed extremal problems. *Izv. Vyssh. Ucheb. Zaved. Mat.* No **6**, 30–38 (in Russian).

Liscovets, O.A. (1976). A choice of regularization parameter in solving nonlinear ill-posed problems. *Dokl. Akad. Nauk SSSR.* **229**, 292–295 (in Russian).

Morozov, V.A. (1966). On regularization of ill-posed problems and on the choice of regularization parameter. *USSR Comput. Math. and Math. Phys.* **6**, 170–175 (in Russian).

Morozov, V.A. (1968). On discrepancy principle in solving operator equations by regularization method. *USSR Comput. Math. and Math. Phys.* **8**, 295–309 (in Russian).

Polak, E. (1971). *Computational Methods in Optimization. A Unified Approach.* Academy Press, New York, London.

Tikhonov, A.N. (1966). On a stability of the problems of functional's optimization. *USSR Comput. Math. and Math. Phys.* **6**, 631–634 (in Russian).

Tikhonov, A.N. and Vasil'ev, F.P. (1976). The methods for solving ill-posed extremal problems. *Banach Center Publ.* 297–342.

Tikhonov, A.N. and Arsenin, V.Ya. (1979). *Methods of Solving Ill-Posed Problems.* Nauka, Moscow (in Russian).

Tikhonov, A.N., Leonov, A.S., and Yagola, A.G. (1992). *Nonlinear Ill-Posed Problems.* Nauka, Moscow (in Russian).

Vasil'ev, F.P. (1980). *Numerical Methods for Solving Extremal Problems.* Nauka, Moscow (in Russian).

Ill-Posed Problems in Natural Sciences, pp. 71 – 83
A. Tikhonov (Ed.)
1992 VSP/TVP

TIKHONOV'S APPROACH FOR CONSTRUCTING REGULARIZING ALGORITHMS

A.S. LEONOV and A.G. YAGOLA
Moscow Institute of Engineering Physics, Kashirskoe shosse, 31, 115409 Moscow, Russia
Moscow State University, Len. Gory, 119899 Moscow, Russia

ABSTRACT

A general scheme for constructing regularizing algorithms on the base of Tikhonov variational approach is considered. It is used for solving linear and nonlinear ill-posed problems. The algorithms with a priori and a posteriori choice of the regularization parameter are reviewed. The optimality of the order of accuracy is studied for a posteriori parameter choice.

1. FORMULATION OF THE PROBLEM

Let (Z, τ) be a topological space with a topology τ and let D be a nonempty set in Z and U be a metric space with metric ρ. We assume $\mathcal{A}$ to be a class of operators from D into U. Let us fix an operator $A \in \mathcal{A}$ and an element $u \in U$ and let us consider the so-called quasisolution problem for the operator equation

$$Az = u \tag{1}$$

on D: find a $z^* \in D$ for which

$$\rho(Az^*, u) = \inf \left\{ \rho(Az, u) \colon \; z \in D \right\} \equiv \mu_0. \tag{2}$$

In the case $D = Z$ problem (2) gives the pseudosolutions of (1). If the measure of incompatibility μ_0 is equal zero, then the solutions of (2) are solutions of (1) on D.

The quasisolution problem (2) may be ill-posed. Namely, problem (2) is not solvable for some equations of the form (1). The solution of (2) may be not unique and unstable in (Z, τ) with respect to perturbations of the data (A, u).

We assume that to some element $u = \bar{u} \in U$ there corresponds the nonempty set $Z^* \subset D$ of quasisolutions and that Z^* may consist of more than one element. Furthermore, we suppose that a functional $\Omega(z)$ is defined on D and bounded below:

$$\Omega(z) \geq \Omega^* \equiv \inf \left\{ \Omega(z) \colon \; z \in D \right\} \geq 0.$$

The Ω-optimal quasisolution problem for (1) is as follows: find a $\bar{z} \in Z^*$ such that

$$\Omega(\bar{z}) = \inf \{\Omega(z): z \in Z^*\} \equiv \overline{\Omega}. \tag{3}$$

We denote the set of all Ω-optimal quasisolutions of (1) by $\overline{Z}$. If $D = Z$, then $\overline{Z}$ is the set of Ω-optimal pseudosolutions of (1).
We suppose that, instead of the unknown exact data (A, u), we are given approximate data (A_h, u_δ) which satisfy the following conditions

$$u_\delta \in U, \quad \rho(\bar{u}, u_\delta) \leq \delta; \quad A_h \in \mathcal{A}, \quad \rho(Az, A_h z) \leq \psi(h, \Omega(z)) \quad \forall z \in D.$$

Here the function ψ represents the known measure of approximation of the precise operator A by the approximate operator A_h. We are given also numerical characterizations $h, \delta \geq 0$ of the "closeness" of (A_h, u_δ) to (A, u).
The main problem is to construct from the approximate data $(A_h, u_\delta, \psi, h, \delta)$ in (1) an element $z_\eta = z_\eta(A_h, u_\delta, \psi, h, \delta) \in D$ which τ-sequentially converges to the set $\overline{Z}$ of Ω-optimal solutions as $\eta = (h, \delta) \to 0$.
We will consider also the problem of the stable estimation of the measure of incompatibility for (1) on D: using the data $(A_h, u_\delta, \psi, h, \delta)$, find a number μ_η such that $\mu_\eta \to \mu_0$ as $\eta \to 0$.
Let us formulate our basic assumptions.

1. The class $\mathcal{A}$ consists of the operators A from D into U τ-sequentially continuous: for any $z_0 \in D$ and any sequence $\{z_n\} \subset D$ which τ-converges to z_0 the sequence $\{\rho(Az_n, Az_0)\}$ converges to 0 as $n \to \infty$.

2. The functional $\Omega(z)$ is τ-sequentially lower semicontinuous on D:

$$\forall z_0 \in D, \forall \{z_n\} \subset D: z_n \xrightarrow{\tau} z_0 \Rightarrow \lim_{n\to\infty} \inf \Omega(z_n) \geq \Omega(z_0).$$

3. If K is an arbitrary number such that $K \geq \Omega^*$, then the set $\Omega_K = \{z \in D: \Omega(z) \leq K\}$ is τ-sequentially compact in Z.

4. The measure of approximation $\psi(h, \Omega)$ is assumed to be defined for $h \geq 0$, $\Omega \geq \Omega^*$, to depend continuously on all its arguments, to be monotonically increasing with respect to Ω for any $h > 0$, and to satisfy the equality $\psi(0, \Omega) = 0$, $\forall \Omega \geq \Omega^*$.

Conditions 1—3 guarantee that $\overline{Z} \neq \varnothing$.

2. TIKHONOV'S APPROACH

An operator $\mathcal{R}$ is called a regularizing algorithm for problem (1) — (3) if it gives a solution $z_\eta = \mathcal{R}(A_h, u_\delta, \psi, h, \delta) \in D$ of the main problem. This notion was introduced by A.N.Tikhonov in basic papers (Tikhonov, 1963a,b). These papers proposed also an approach for constructing the regularizing algorithms named later as Tikhonov's (or variational) scheme. Preparatory to considering this scheme, we first illustrate its applications by two examples.

Example 1. Let A be a linear continuous operator from a Hilbert space Z into a normed space U. The operator A is also assumed to be injective over Z. Suppose that (1) is compatible for some $u_\delta = \bar{u} \in U$ and suppose that $\bar{z} \in Z$ is the corresponding solution. The operator A is known exactly but instead of $\bar{u}$ we are given an approximation $u_\delta \in U$: $\|\bar{u} - u_\delta\| \le \delta$. The scheme of constructing an approximation to the element $\bar{z}$ from the data $(A,\ u_\delta,\ \delta)$ in (1) is as follows. We introduce the parametrical functional (Tikhonov's, or smoothing, functional)

$$M^\alpha[z] = \alpha\|z\|^2 + \|Az - u_\delta\|^2, \quad z \in Z, \quad \alpha > 0 \tag{4}$$

and consider the variational problem: for a fixed $\alpha > 0$, find an element $z^\alpha \in Z$ such that

$$M^\alpha[z^\alpha] = \inf\{M^\alpha[z]:\ z \in Z\}. \tag{5}$$

Problem (5) has an unique minimizer (extremal) for each $\alpha > 0$. As an approximation to $\bar{z}$, the element $z^{\alpha(\delta)} \in Z$ is used, i.e., the solution of (5) for a specially choosed regularization parameter $\alpha = \alpha(\delta)$. It is the usage of the special choice of regularization parameter that provides the convergence $z^{\alpha(\delta)} \xrightarrow{Z} \bar{z}$ as $\delta \to 0$. Two approaches exist to choosing the regularization parameter: 1) the a priori choice, where α is chosen as a function of the "noise" level δ; 2) the a posteriori choice, in which not only characteristics of data perturbation are used but also the given realizations of data are taken into account. A certain dependence of regularization parameter on the noise level of data is a necessary condition for constructing the regularizing algorithms for (1) in the general case (see (Bakushinskii, 1984)).
The a priory choice conditions were first suggested in (Tikhonov, 1963a, b; 1965): $\alpha(\delta) \to 0,\ \delta^2/\alpha(\delta) \to 0$ as $\delta \to 0$. A widely known a posteriori choice of regularization parameter is the discrepancy principle (Ivanov, 1966; Morozov, 1967), where $\alpha(\delta) > 0$ is chosen such that

$$\beta(\alpha) \equiv \|Az^\alpha - u_\delta\| = \delta \tag{6}$$

holds. Since $\beta(\alpha)$ is continuous and monotonically increasing, the solution $\alpha(\delta)$ of (6) is unique if $\overline{A} \ne 0,\ \bar{u} \ne 0$ and $0 < \delta \le \delta_0 = \text{const}$.
The scheme described in this example fails in the general case, where A is nonlinear operator. On view of this fact we consider

Example 2. (Goncharskii et al., 1974; Leonov, 1982b). Let $Z = U$ be the space of real numbers $\mathbf{R}$ with the norm $\|z\| = |z|$. Fix positive numbers $\alpha_0,\ z_0,\ u_0$ and define the continuous nonlinear operator A_0: $Z \to U$:

$$A_0 z = \begin{cases} \left[\alpha_0(z_0^2 - z^2) + z(z_0 - z)\right]^{1/2} + u_0 & \text{if } 0 \le z \le z_0; \\ \left[\alpha_0(z^2 - z_0^2)\right]^{1/2} + u_0 & \text{if } z > z_0; \\ \alpha_0^{1/2} z_0 + z^2 + u_0 & \text{if } z < 0. \end{cases}$$

Consider equation (1) with the data (A_0, u_0). It has the unique solution $\bar{z} = z_0$. Suppose that $u_\delta = u_0$ for each $\delta \ge 0$. Then problem (5) has the solution $z^\alpha = \{z_0$ for $0 < \alpha < \alpha_0; 0, z_0$ for $\alpha = \alpha_0; 0$ for $\alpha > \alpha_0\}$. Hence, $\beta(\alpha) = \{0$ for $0 < \alpha \le$

$\alpha_0; u_0$ for $\alpha \geq \alpha_0\}$ and equation (6) has no solutions in the ordinary sense for any δ, $0 < \delta < u_0$.

This example shows that the discrepancy principle gives generally no possibility of choosing the value of the regularization parameter for nonlinear problems. Similiar difficulties arise in other strategies of the a posteriori choice. These difficulties are connected with the nonuniqueness of the solution of (5) for the nonlinear case. The nonuniqueness mentioned leads to another difficulty. Let us assume that the regularization parameter $\alpha(\delta) > 0$ is chosen and problem (5) has more than one minimizer for $\alpha = \alpha(\delta)$. The question is whether it is possible to take an arbitrary element $z^{\alpha(\delta)}$ as an approximate solution or a special choice of such element is needed. Example 2 shows that if we choose $\alpha(\delta)$ as a solution of (6) in general sense, i.e., as a "jump" point of a monotonic function $\beta(\alpha)$ over δ, then the "unsuccessful" choice of $z^{\alpha(\delta)}$: $z^{\alpha(\delta)} = 0$, from the set of the extremals corresponding to this $\alpha(\delta)$ can result in that the approximate solutions $z^{\alpha(\delta)}$ do not converge to the precise solution $\bar{z} = z_0$. Thus, the algorithm for Tikhonov's regularization method of solving nonlinear problem (1) should in general contain, along with a rule for choosing the parameter $\alpha(\delta)$ also a rule for selecting the extremal $z^{\alpha(\delta)}$ in the set of solutions of (5) for $\alpha = \alpha(\delta)$.

Let us give a general formulation of Tikhonov's scheme of constructing a regularizing algorithm for solving the main problem. It is based on using the smoothing functional

$$M^{\alpha}[z] = \alpha\Omega(z) + f\big[\rho(A_h z, u_\delta)\big], \quad z \in D, \quad \alpha > 0 \tag{7}$$

in the conditional extremum problem: for a fixed $\alpha > 0$, find an element $z^{\alpha} \in D$ such that

$$M^{\alpha}[z^{\alpha}] = \inf\big\{M^{\alpha}[z]\colon z \in D\big\}. \tag{8}$$

Here $f(x)$ is an auxiliary function. A common choice is $f(x) = x^m$, $m \geq 2$.

We denote the set of extremals of (8) which correspond to a given $\alpha > 0$ by Z^{α}. Conditions 1—3 imply that $Z^{\alpha} \neq \varnothing$. The scheme of constructing an approximation to the set $\overline{Z}$ includes: (i) the choice of the regularization parameter $\alpha_\eta = \alpha_\eta(A_h,\ u_\delta,\ h,\ \delta)$; (ii) the fixation of a set Z^{α_η}, corresponding to α_η, and a special selection of an element z^{α_η} in this set. We take the element z^{α_η} chosen in this way as a solution of main problem. Procedures (i), (ii) must be accomplished so as to guarantee the convergence $z^{\alpha_\eta} \xrightarrow{\tau} \overline{Z}$ as $\eta \to 0$. Thus, the Tikhonov's regularizing algorithms differ from each other by the method of choosing α_η and by the method of selecting z^{α_η}. Sometimes, the latter procedure is not necessary, and one can take an arbitrary element z^{α_η} from Z^{α_η} as an approximate solution. In this case we say that the corresponding Tikhonov's algorithm is simple.

3. TIKHONOV'S ALGORITHMS FOR LINEAR PROBLEMS

Suppose that $\mathcal{A}$ is a set of the linear bounded operators acting from a locally and uniformly convex reflexive Banach space Z into a normed space U. We assume that a set D is convex and strongly closed in Z. As a topology τ we take the topology of weak convergence in Z. Let $\Omega(z) = g(\|z\|)$. Here $g(x) \geq 0$ is a continuous, monotonically increasing and strictly convex function for $x \geq 0$. Suppose that

$f(x)$ is continuous increasing and convex function for $x \geq 0$ satisfying the equality $f(0) = 0$. In the linear case we can put also $\psi(h, \Omega) = h\|z\|$.
Problem (1)—(3) is now the problem of finding the unique normal quasisolution $\bar{z}$ for a linear equation on the set D. The Tikhonov functional for this problem has the form

$$M^\alpha[z] = \alpha g(\|z\|) + f(\|A_h z - u_\delta\|), \quad z \in D. \tag{9}$$

LEMMA 1. *Functional* (9) *is strictly convex on* D *and hence problem* (8) *for this functional has an unique solution for each* $\alpha > 0$.

Thus, the choice of regularization parameter α_η determines uniquely the approximate solution z^{α_η}, i.e., the Tikhonov's regularizing algorithm is simple in our case.

THEOREM 1. *Let* $\alpha = \alpha(\eta)$ *be such the a priori choice of regularization parameter that:* $\alpha(\eta) \to 0$, $(\delta + h)/\alpha(\eta) \to 0$ *as* $\eta \to 0$. *Then* $z^{\alpha_\eta} \xrightarrow{T} \bar{z}, \|z^{\alpha_\eta}\| \to \|\bar{z}\|$ *as* $\eta \to 0$ *for compatible operator equations* (1) *and hence the strong convergence* $z^{\alpha_\eta} \xrightarrow{Z} \bar{z}$ *holds.*

Consider the algorithm with the a posteriori choice of α. They use for their construction the following auxiliary functions

$$\beta(\alpha) = \|A_h z^\alpha - u_\delta\|, \quad \gamma(\alpha) = \|z^\alpha\|, \quad \varphi(\alpha) = M^\alpha[z^\alpha], \quad \alpha > 0. \tag{10}$$

These functions were introduced by Morozov (Morozov, 1968) and were studied later by Goncharskii, Yagola and Leonov (Goncharskii et al., 1973; Yagola, 1979; Leonov, 1979, 1986). From Lemma 1 the continuity of functions (10) follows (see (Leonov, 1982b)). The functions β, φ are monotonically nondecreasing and γ is monotonically nonincreasing.
Let us introduce a generalized measure of incompatibility for (1)

$$\mu_\eta = \inf\left\{\|A_h z - u_\delta\| + \delta + h\|z\|:\ z \in D\right\}.$$

The value μ_η is a stable approximation to μ_0: $\mu_\eta \geq \mu_0$, $\mu_\eta \to \mu_0$ as $\eta \to 0$ (Levin, 1984).
The most widely known method of the a posteriori regularization parameter choice for linear problems is the generalized discrepancy principle. It was introduced in (Goncharskii et al., 1973) as a development of the discrepancy principle for the case of operator equations where the operator is given approximately. Later the generalized discrepancy principle was studied in the linear case in different aspects by Morozov, Ivanov, Yagola, Kochikov, Vainikko and others (Morozov, 1974; Ivanov et al., 1978; Yagola, 1979, 1980; Kochikov et al., 1983, 1987; Vainikko, 1986). According to the generalized discrepancy principle the regularization parameter $\alpha_\eta > 0$ is a solution of the equation

$$\rho(\alpha) \equiv \beta(\alpha) - \delta - h\gamma(\alpha) - \mu_\eta = 0. \tag{11}$$

Another a posteriori choice strategy for α is the generalized principle of a smoothing functional. It consist in choosing $\alpha_\eta > 0$ as a solution of the equation

$$\varepsilon(\alpha) \equiv \varphi(\alpha) - f\left\{\mu_\eta + [\delta + h\gamma(\alpha)]^p\right\}, \quad p = \text{const}, \quad 0 < p < 1. \tag{12}$$

This strategy turns out to be natural generalization of the choice of (Morozov, 1966) in terms of a smoothing functional.
If we have an additional a priori information about the value $\overline{\Omega}$, we can use the generalized quasisolution principle to develope the method of (Ivanov, 1966). When using this choice strategy, the parameter $\alpha_\eta > 0$ is taken to be a solution of the equation

$$\gamma(\alpha) = \omega_\eta. \tag{13}$$

Here, ω_η is an known estimation of $\overline{\Omega}$ for which $\omega_\eta \geq \overline{\Omega}$, $\omega_\eta \to \overline{\Omega}$ as $\eta \to 0$.

THEOREM 2. *Assume that for any quasisolution $z^* \in D$ of the equation* (1) *the nonequality $\|z^*\| > \inf\{\|z\|\colon z \in D\}$ is satisfied. Then:*
(a) *equations* (11)–(13) *have positive solutions for "sufficiently small" values of h and δ;*
(b) *equation* (12) *has an unique solution and this solution converges to* 0 *as $\eta \to 0$;*
(c) *equation* (11), *as well as equation* (13), *has an unique solution provided that $z^{\alpha'} \neq z^{\alpha''}$ for any α', $\alpha'' > 0$, $\alpha' \neq \alpha''$;*
(d) *the element $z^{\alpha_\eta} \in D$, which is taken to be an approximate solution to* (1)–(3), *has the following convergence properties as $\eta \to 0$: $z^{\alpha_\eta} \xrightarrow{\tau} \bar{z}$, $\|z^{\alpha_\eta}\| \xrightarrow{\tau} \|\bar{z}\|$; consequently the convergence $z^{\alpha_\eta} \xrightarrow{Z} \bar{z}$ holds.*

If we know that equation (1) has a solution on D, then we can simplify the choice rules for α_η by omitting the value μ_η in (11), (12). In this case Theorem 2 remains true.
There exist some other a priori and a posteriori methods for choosing the regularization parameter (see e.g., (Tikhonov and Arsenin, 1979; Leonov, 1983; Vainikko, 1986; Gfrerer, 1987)).

4. TIKHONOV'S ALGORITHMS FOR NONLINEAR PROBLEMS

We will consider now a general formulation of the problem in question (see §1) and the scheme of solution introduced in §2. Assume that $f(x)$ has the following properties for some natural m: $f(x) \in \mathbf{C}^m[0,+\infty)$, $f^{(r)}(x) \geq 0$ for $r = 0, 1, \cdots, m-1$, $f^{(m)}(x) \geq \kappa = \text{const} > 0$, $f(0) = 0$. Then the conditions 1—3 guarantee the solvability of minimization problem (8) for Tikhonov's functional for every $\alpha > 0$. However, (8) can have more then one solution (see example 2). Therefore it is necessary not only to choose the regularization parameter α_η, but also to select a special element z^{α_η} in the set Z^{α_η} of solutions of problem (8) with $\alpha = \alpha_\eta$.
It is in this way that the generalized analogues of a posteriori parameter choice strategies mentioned in §3 are used. They were introduced in (Leonov, 1979, 1982a, 1986). We define for their formulations some auxiliary functions and functionals similiar to (10):

$$\gamma(\alpha) = \Omega(z^\alpha), \quad \beta(\alpha) = f\left[\rho(A_h z^\alpha, u_\delta)\right] \equiv I(z^\alpha),$$

$$\varphi(\alpha) = M^\alpha[z^\alpha], \quad \pi(\alpha) = f\left[\psi(h, \gamma(\alpha)) + \delta + \mu_\eta\right] \equiv \Pi(z^\alpha),$$

$$\rho(\alpha) = \beta(\alpha) - \pi(\alpha) \equiv \mathrm{P}(z^\alpha),$$

$$\varepsilon(\alpha) = \varphi(\alpha) - f\left\{\mu_\eta + \left[\psi(h, \gamma(\alpha)) + \delta\right]^p\right\} \equiv \mathrm{E}(z^\alpha), \tag{14}$$

$$p = \text{const}, \quad 0 < p < 1; \qquad \forall z^\alpha \in Z^\alpha; \quad \alpha > 0.$$

Here, $\mu_\eta = \inf\{\rho(A_h z, u_\delta) + \psi(h, \Omega(z)) + \delta\colon z \in D\}$ is a generalized measure of incompatibility for nonlinear problem (11) having the properties: $\mu_\eta \geq \mu_0$, $\mu_\eta \to \mu_0$ as $\eta \to 0$ (Leonov, 1982a). All the functions (14) (except φ) are generally many-valued. They have the following properties.

LEMMA 2. *The functions γ, β, π, ρ, ε are single-valued and continuous everywhere for $\alpha > 0$ except perhaps not more than a countable set of their common points of discontinuity of the first kind, which are points of multiple-valuedness. If α is such a point of multiple-valuedness, then there exist at least two elements z_+^α and z_-^α in the set Z^α such that*

$$\gamma(\alpha \pm 0) = \Omega(z_\pm^\alpha), \quad \rho(\alpha \pm 0) = \mathrm{P}(z_\pm^\alpha), \quad \varepsilon(\alpha \pm 0) = \mathrm{E}(z_\pm^\alpha).$$

The functions β, ρ, ε are monotonically nondecreasing and γ, π are nonincreasing. The function φ is single-valued, continuous and nondecreasing.

We note that function (14) may be discontinuous not only for nonlinear problems but also in the linear case if the Banach space Z is not strictly normed (see(Yagola, 1979, 1980)).

The generalized discrepancy principle for nonlinear problems (GDP) consists of the following steps.

(i) The choice of regularization parameter as a generalized solution $\alpha_\eta > 0$ of the equation

$$\rho(\alpha) = 0. \tag{15}$$

Here and in the sequel we say that a_0 is *a* generalized solution of the equation $\chi(\alpha) = 0$ for a monotonic function χ if α_0 is an ordinary solution or α_0 is a "jump"-point of this function over 0.

(ii) The method of selecting an approximate solution z^{α_η} from the set Z^{α_η} by means of the following selection rule. Let $q > 1$ and $C > 1$ be fixed constants, $\alpha_1 = \alpha_\eta / q$ and $\alpha_2 = \alpha_\eta q$ be auxiliary regularization parameters, and let z^{α_1} and z^{α_2} be any extremals of (8) for $\alpha = \alpha_{1,2}$.

If the inequality

$$I(z^{\alpha_2}) \geq C\Pi(z^{\alpha_1}) - (C-1)f(\mu_\eta) \tag{16}$$

holds for z^{α_1} and z^{α_2}, then any elements $z^{\alpha_\eta} \in Z^{\alpha_\eta}$, subject to the condition $\mathrm{P}(z^{\alpha_\eta}) \leq 0$ can be taken as the approximate solution. For instance, we can take $z^{\alpha_\eta} = z_-^{\alpha_\eta}$. But if

$$I(z^{\alpha_2}) \leq C\Pi(z^{\alpha_1}) - (C-1)f(\mu_\eta), \tag{17}$$

then we choose z^{α_η} so as to have $\mathrm{P}(z^{\alpha_\eta}) \geq 0$, for example, $z^{\alpha_\eta} = z_+^{\alpha_\eta}$.

Note that we do not need any selection rule if α_η is an ordinary solution of equation (15). In this case the equality $\mathrm{P}(z^{\alpha_\eta}) = \rho(\alpha_\eta) = 0$ holds, and an arbitrary element $z^{\alpha_\eta} \in Z^{\alpha_\eta}$ can be taken as the approximate solution of (1)—(2). The same situation is valid also for other algorithms formulated below.

In the generalized quasisolution principle for nonlinear problems (GQP) use is made not only of the data (A_h, u_δ, ψ, h, δ) but also of an estimation ω_η for the unknown value $\overline{\Omega}$. We assume that $\omega_\eta \geq \overline{\Omega}$, $\omega_\eta \to \overline{\Omega}$ as $\eta \to 0$. This method is as follows.

(i) The regularization parameter $\alpha_\eta > 0$ is chosen as a generalized solution of the equation

$$\gamma(\alpha) - \omega_\eta = 0. \tag{18}$$

(ii) The selection rule for z^{α_η} looks as follows. If inequality (16) holds, then we choose any element $z^{\alpha_\eta} \in Z^{\alpha_\eta}$ subject to the condition $\Omega(z^{\alpha_\eta}) \geq \omega_\eta$ as an approximate solution. But if (17) holds, then we select z^{α_η} so as to have $\Omega(z^{\alpha_\eta}) \leq \omega_\eta$.

The generalized principle of a smoothing functional (GSFP)

(i) The regularization parameter is taken as a generalized solution of the equation

$$\varepsilon(\alpha) = 0. \tag{19}$$

(ii) An approximate solution z^{α_η} is selected from Z^{α_η} in such a way that $\mathrm{E}(z^{\alpha_\eta}) \geq 0$. For instance, we put $z^{\alpha_\eta} = z_+^{\alpha_\eta}$.

Let α_η and z^{α_η} be found by using some of the above generalized principles.

THEOREM 3. *Suppose that, for any quasisolution* $z^* \in D$ *of* (1)–(2), *the inequality* $\Omega(z^*) > \Omega^* = \inf\{\Omega(z)\colon z \in D\}$ *holds. Then*
(a) *equations* (15), (18), (19) *have positive generalized solutions for "sufficiently small" values* h *and* δ;
(b) *the generalized solution of* (19) *is unique and converges to* 0 *as* $\eta \to 0$;
(c) *the generalized solution of* (15) *or* (18) *is unique also if the condition* $Z^{\alpha'} \cap Z^{\alpha''} = \varnothing$ *holds for each* α', $\alpha'' > 0$, $\alpha' \neq \alpha''$;
(d) *for any sequence* $\eta_n = (h_n, \delta_n)$ *such that* $\eta_n \to 0$ *as* $n \to \infty$, *the corresponding sequence* $\{z_n\}$ *of approximate solutions, which is found by some of the above generalized principles has the following properties:* $z_n \xrightarrow{\tau} \overline{Z}$, $\Omega(z_n) \to \overline{\Omega}$ *as* $n \to \infty$.

If it is known in addition that equation (1) has a solution on D, then the value μ_η in (14) can be omitted. Theorem 3 is valid in this case. The methods described in this section revert to the usual generalized principles (see §3) if functions (14) are continuous.

In order to construct a simple Tikhonov's algorithm we consider now an a priori choice of the regularization parameter for nonlinear problems. We introduce a generalized Tikhonov's functional of the form

$$N^\alpha[z] = \alpha\Omega(z) + f\Big[\rho(A_h z, u_\delta) + \psi(h, \Omega(z)) + \delta\Big]. \tag{20}$$

Here $f(x) \in \mathbf{C}^1[0, +\infty)$ is a monotonically increasing function having the properties: $f(0) = 0$, $f(+\infty) = +\infty$. Conditions 1—3 (see §1) guarantee the solvability of the following problem for any fixed $\alpha > 0$: find $z^\alpha \in D$ for which

$$N^\alpha[z^\alpha] = \inf\big\{N^\alpha[z]\colon z \in D\big\}. \tag{21}$$

THEOREM 4. *Let $\alpha = \alpha(\eta)$ be an a priori choice of the regularization parameter subject to the conditions: $\alpha(\eta) \to 0, [\delta+\psi(h,\overline{\Omega})]/\alpha(\eta) \to 0$ as $\eta \to 0$. Furthermore, let $\eta_n = (h_n, \delta_n) \to 0$ be an arbitrary sequence and z_n be any solution of (21) corresponding to $\alpha = \alpha(\eta_n)$. Then $z_n \xrightarrow{\tau} \overline{Z}$, $\Omega(z_n) \to \overline{\Omega}$ as $n \to \infty$.*

5. THE OPTIMALITY OF A PRECISION ORDER FOR TIKHONOV'S ALGORITHMS WITH A POSTERIORI CHOICE OF THE REGULARIZATION PARAMETER

In this section we study the precision of Tikhonov's algorithms for nonlinear case. Let Z be a Banach space, D be a closed set in Z and U be a metric space with a metric ρ. A set $\mathcal{A}$ is assumed to consist of continuous injective operators from D into U. We consider the equation

$$A_0 z = u_0, \quad z \in D, \quad A_0 \in \mathcal{A}, \quad u_0 \in U \tag{22}$$

and we assume that it has an unique solution $\bar{z}_0$ on the set D. In addition, we assume that the solution $\bar{z}_0$ has the form $\bar{z}_0 = B\bar{v}$. Here B is linear, injective and completely continuous operator acting from a reflexive Banach space V into Z, and $\bar{v} \in V$, $\bar{v} \neq 0$. For instance, an operator of the compact imbedding of V into Z can be chosen as the operator B.
Furthermore, suppose that instead of (A_0, u_0) we are given approximate data $(A_h,\ u_\delta,\ h,\ \delta)$ and, moreover, $A_h \in \mathcal{A}$ and $\rho(A_h z, A_0 z) \leq \psi(h, \|z\|)$ for every $z \in D$.
An operator P is called a method of solving problem (22), if it puts an "approximate solution" $z_\eta = P(A_h,\ u_\delta,\ h,\ \delta) \in D$ in to correspondence to the data $(A_h,\ u_\delta,\ h,\ \delta)$.
A specific feature of the ill-posed problems is the impossibility of estimating the precision of the approximate solutions without using some essential information concerning the properties of the exact solutions. In our case, we use such a priori information in the form $\bar{z}_0 \in M_R$, where $M_R = \{z \in D\colon\ z = Bv,\ v \in V,\ \|v\| \leq R\}$ and $R \geq \|\bar{v}\|$. Taking it into account, we consider a modified Tikhonov scheme for (22).
We introduce the set $\mathcal{V} = \{v \in V\colon\ Bv \in D\}$ and define on $\mathcal{V}$ Tikhonov's functional as follows:

$$M^\alpha[v] = \alpha g\big(\|v\|\big) + f\big[\rho(A_h Bv, u_\delta)\big].$$

Here the functions $g(x), f(x)$ are non-negative, continuous and increasing in their domain $x \geq 0$ and, in addition, $g(+\infty) = f(+\infty) = +\infty$. Moreover, the function $f(x)$ is assumed to be convex downward. Then the problem: find an element $v^\alpha \in \mathcal{V}$ such that

$$M^\alpha[v^\alpha] = \inf\big\{M^\alpha[v]\colon\ v \in \mathcal{V}\big\} \tag{23}$$

is solvable for each $\alpha > 0$. We denote by $\mathcal{V}^\alpha$ the set of extremals of (23) which correspond to a given α and introduce the following functions and functionals

$$\gamma(\alpha) = g\big(\|v^\alpha\|\big) \equiv \Omega(v^\alpha), \quad \beta(\alpha) = f\big[\rho(A_h Bv^\alpha, u_\delta)\big] \equiv I(v^\alpha),$$

$$\varphi(\alpha) = M^\alpha[v^\alpha], \quad \pi(\alpha) = f\Big[\delta + \psi(h, \|B\|\|v^\alpha\|)\Big] \equiv \Pi(v^\alpha),$$

$$\rho(\alpha) = \beta(\alpha) - \pi(\alpha) \equiv \mathrm{P}(v^\alpha),$$

$$\varepsilon(\alpha) = \varphi(\alpha) - f\Big\{C\Big[\delta + \psi(h, \|B\|\|v^\alpha\|)\Big]\Big\} \equiv \mathrm{E}(v^\alpha) \tag{24}$$

$$\forall v^\alpha \in \mathcal{V}^\alpha; \qquad C = \mathrm{const} > 1.$$

These functions have the same properties as in Lemma 2. Using functions (24) we can reformulate the algorithms of §4. Finding by such algorithms the values $\alpha_\eta > 0,\ v^{\alpha_\eta} \in \mathcal{V}^{\alpha_\eta}$ we obtain the approximate solution of (22) in the form: $z^{\alpha_\eta} = Bv^{\alpha_\eta} \in D$. According to an analogue of Theorem 3, we have the convergence $z^{\alpha_\eta} \overset{Z}{\to} \bar{z}_0$ as $\eta \to 0$.

THEOREM 5 (Leonov, 1988). *For v^{α_η} the following estimates are valid:*

$$\|v^{\alpha_\eta}\| \le R, \quad \rho(A_h Bv^{\alpha_\eta}, u_\delta) \le C\Big[\delta + \psi(h, \|B\|R)\Big] \equiv CH.$$

In this theorem $R = g^{-1}(\Omega_0)$, where g^{-1} is the inverse of the function g, $\Omega_0 = g(\|\bar{v}\|)(C + q - 1)/(C - 1)$ for GDP and GQP, and $\Omega_0 = Cg(\|\bar{v}\|)/(C - 1)$ for GSFP.
In view of the inequalities $\|\bar{v}\| < R,\ \|v^{\alpha_\eta}\| \le R$ the inclusion $\bar{z}_0,\ z^{\alpha_\eta} \in M_R$ is valid.
Along with problem (22) we consider a set of the operator equations

$$Az = u, \quad z \in D, \quad A \in \mathcal{A}, \quad u \in U \tag{25}$$

assumed to have an unique solution $\bar{z}(A, u)$ on D. We define the set $\sum_R$ which corresponds to the number R given by Theorem 5 and to some fixed approximate data $(A_h,\ u_\delta,\ h,\ \delta)$ of problem (25), which consist of the permissible precise data (A, u) for (25):

$$\sum\nolimits_R \equiv \sum\nolimits_R(A_h,\ u_\delta,\ h,\ \delta) = \{(A, u)\colon\ A\bar{z} = u,\ \bar{z} = \bar{z}(A, u) \in M_R,$$

$$\big|\rho(Az, u) - \rho(A_h z, u_\delta)\big| \le CH,\ \forall z \in M_R\Big\}.$$

Let a method P for solving (22), (25) be given. We define the following characterization of the precision of the method P:

$$\Delta(P) \equiv \Delta(P;\ A_h,\ u_\delta,\ h,\ \delta;\ R) = \sup_{A,u}\Big\{\big\|P(A_h,\ u_\delta,\ h,\ \delta) - \bar{z}(A, u)\big\|\colon\ (A, u) \in \sum\nolimits_R\Big\}$$

for the data $(A_h,\ u_\delta,\ h,\ \delta)$ on the set of the permissible problems (25). We introduce also the optimal precision

$$\Delta_0 \equiv \Delta_0(A_h,\ u_\delta,\ h,\ \delta;\ R) = \inf\big\{\Delta(P;\ A_h,\ u_\delta,\ h,\ \delta;\ R)\colon\ P \in \mathcal{P}\big\}$$

over the class $\mathcal{P}$ of all possible methods for solving operator equations. We say that a method P_0 for solving problem (22) has an optimal order of precision if

$$\left\|P_0(A_h,\ u_\delta,\ h,\ \delta) - \bar{z}_0\right\| \le k\Delta_0(A_h,\ u_\delta,\ h,\ \delta;\ R)$$

for arbitrary data $(A_h,\ u_\delta,\ h,\ \delta)$, $0 \le h \le h_0 = \text{const}$, $0 \le \delta \le \delta_0 = \text{const}$, and if the constant $k > 0$ does not depend on $A_h,\ u_\delta,\ h,\ \delta,\ R$.

THEOREM 6 (Leonov, 1990). *The algorithms of the generalized principles (see* §4) *have an optimal order of precision:*

$$\|z^{\alpha_\eta} - \bar{z}_0\| = \|Bv^{\alpha_\eta} - \bar{z}_0\| \le 2\Delta_0.$$

The optimality of the precision order is the characteristic feature of Tikhonov's regularizing algorithms with the a posteriori choice of the regularization parameter. Such an optimality was first studied in (Ivanov et. al., 1978; Tanana, 1981) for the GDP algorithm in the case of linear solvable operator equations. The theory of optimality of the precision order for the general case of nonlinear operator equations and variational problems was developed later in (Leonov, 1988, 1990) for GDP, GQP and GSFP algorithms. There are also other approaches to the problem of optimal precision of Tikhonov's algorithms. Some results in this frield have been presented in (Ivanov et al., 1978; Tikhonov and Arsenin, 1979; Engl and Neubauer, 1988). All these authors use the a priori knowledge about the exact solution of the type $\bar{z} = (A^*A)^\nu w,\ \nu = \text{const}$.
We conclude by noting that the complete theory of Tikhonov regularizing algorithms including finite-dimensional variants of algorithms, their numerical treatment and some applications is described in (Tikhonov et. al., 1992).

REFERENCES

Bakushinskii, A.B. (1984). The remarks on the regularization parameter choice using the quasioptimality and ratio criteria. *USSR Comput. and Math. Phys.* **24**, 1258–1259 (in Russian).

Bakushinckii, A.B. and Goncharski, A.V. (1989). *Ill-Posed Problems. Numerical Methods and Applications.* Moscow Univ. Press, Moscow (in Russian).

Engl, H.W. and Neubauer, A. (1988). A parameter choice for (iterated) Tikhonov regularization of ill-posed problems leading to superconvergence with optimal rates. *Appl. Anal.* **27**, 5–18.

Gfrerer, H. (1987). An a posteriori parameter choice for ordinary and iterated Tikhonov regularization of ill-posed problems leading to optimal convergence rates. *Math. Comp.* **19**, 507–522.

Goncharskii, A.V., Leonov, A.S, and Yagola, A.G. (1973). Generalized discrepancy principle. *USSR Comput. Math. and Math. Phys.* **13**, 294–302 (in Russian).

Goncharskii, A.V., Leonov, A.S., and Yagola, A.G. (1974). On the residual principle for solving nonlinear ill-posed problems. *Soviet Math. Dokl.* **15**, 166–168.

Ivanov, V.K. (1966). On an approximate solving operator equations of the first kind. *USSR Comput. Math. and Math. Phys.* **6**, 1089–1094 (in Russian).

Ivanov, V.K., Vasin, V.V., and Tanana, V.P. (1978). *Theory of Linear Ill-Posed Problems and Its Applications.* Nauka, Moscow (in Russian).

Kochikov, I.V., Matvienko, A.N., and Yagola, A.G. (1983). On some modification of the generalized discrepancy principle. *USSR Comput Math. and Math. Phys.* **23**, 1298–1303 (in Russian).

Kochikov, I.V., Matvienko, A.N., and Yagola, A.G. (1987). Generalized discrepancy principle for the solution of incompatible equations. *USSR Comput. Math. and Math. Phys.* **27**, 1087–1090 (in Russian).

Leonov, A.S. (1979). On the choice of regularization parameter for nonlinear ill-posed problems with operator approximately defined. *USSR Comput. Math. and Math. Phys.* **19**, 1363–1376 (in Russian).

Leonov, A.S. (1982a). On an application of the generalized residual principle for the solution of ill-posed extremal problems. *Soviet Math. Dokl.* **25**, 227–231.

Leonov, A.S. (1982b). On the relation between the generalized discrepancy method and the generalized discrepancy principle for solving nonlinear ill-posed problems. *USSR Comput. Math. and Math. Phys.* **22**, 783–790 (in Russian).

Leonov, A.S. (1983). On solving linear ill-posed problems on the base of a modified quasioptimality criterion. *Math. USSR Sbornik.* **122 (164)**, 405–415 (in Russian).

Leonov, A.S. (1986). On some algorithm for solving ill-posed extremal problems. *Mat. USSR Sbornik.* **129(171)**, 218–231 (in Russian).

Leonov, A.S. (1988). The precision order optimality of the generalized dicrepancy principle and of some different algorithms for solving nonlinear ill-posed problems with approximate data. *Siberian Math. J.* **29**, 85–94 (in Russian).

Leonov, A.S. (1990). The precision order optimality of some algorithms for solving ill-posed extremal problems. *Izv. Vyssh. Ucheb. Zaved. Mat.* No **6**, 30–38 (in Russian).

Levin, A.M. (1984). On the regularization of calculation of the function low bound. *USSR Comput. Math. and Math. Phys.* **24**, 1123–1128 (in Russian).

Morozov, V.A. (1966). On the regularization of ill-posed problems and on the choice of regularization parameter. *USSR Comput. Math. and Math. Phys.* **6**, 170–175 (in Russian).

Morozov, V.A. (1968). On the discrepancy principle for solving operator equations by the method of regularization. *USSR Comput. Math. and Math. Phys.* **8**, 295–309 (in Russian).

Morozov, V.A. (1974). *Regular Methods for Solving Ill-Posed Problems.* Moscow Univ. Press, Moscow (in Russian).

Tanana, V.P. (1981). *The Methods of Solving Operator Equations.* Nauka, Moscow (in Russian).

Tikhonov, A.N. (1963a). On solving ill-posed problems and the regularization method. *Dokl. Akad. Nauk SSSR.* **151**, 501–504 (in Russian).

Tikhonov, A.N. (1963b). On regularization of ill-posed problems. *Dokl. Akad. Nauk SSSR.* **153**, 49–52 (in Russian).

Tikhonov, A.N. (1965). On nonlinear equations of the first kind. *Dokl. Akad. Nauk SSSR.* **161**, 1023–1026 (in Russian).

Tikhonov, A.N. and Arsenin, V.Ya. (1979). *Methods of Solving Ill-Posed Problems.* Nauka, Moscow (in Russian).

Tikhonov, A.N., Goncharskii, A.V., Stepanov, V.V., and Yagola, A.G. (1983). *Regularizing Algorithms and a Priori Information.* Nauka, Moscow (in Russian).

Tikhonov, A.N., Leonov, A.S., and Yagola, A.G. (1992). *Nonlinear Ill-Posed Problems.* Nauka, Moscow (in Russian).

Vainikko, G.M. and Veretennikov, A.Yu. (1986). *Iteration Procedures in Ill-Posed Problems.* Nauka, Moscow (in Russian).

Yagola, A.G. (1979). The generalized discrepancy principle in reflexive spaces. *Dokl. Akad. Nauk SSSR.* **249**, 71–73 (in Russian).

Yagola, A.G. (1980). On the choice of regularization parameter in solving ill-posed problems in reflexive spaces. *USSR Comput. Math. and Math. Phys.* **20**, 586–596 (in Russian).

Ill-Posed Problems in Natural Sciences, pp. 84 – 87
A. Tikhonov (Ed.)
1992 VSP/TVP

REGULARIZATION ALGORITHMS FOR SOLVING ILL-POSED CRITERION PROBLEMS

A.M. LEVIN
Kommunarsk Institute of Mining and Metallurgy, 349104 Kommunarsk, Russia

ABSTRACT

In this paper some results for a class of minimization problems unstable by value are presented. Necessary and sufficient conditions are obtained for the continuity of the measure of incompatibility

$$\mu[A_0, u_0] = \inf_z F_0(z), \quad F_0(z) = \|A_0 z - u_0\|$$

of a linear operator equation $A_0 z = u_0$ in a Banach space. We give a definition of stability by value for the problem of finding the minimizing sequences for the functional $F_0(z)$ in the case when the exact solution of the minimization problem does not exist. We obtain a criterion of stability by value and establish the connection between the argument and the criterion minimization problems.
For the considered ill-posed criterion problems we find the regularizing by value algorithm, named the Method of the Least Estimate of Discrepancy (MLED). If one knows the approximate data A_h, u_δ, $\eta = (h, \delta)$, $\|A_h - A_0\| \leq h$, $\|u_\delta - u_0\| \leq \delta$, then MLED consists of finding the modified measure of incompatibility

$$\mu_\eta[A_h, u_\delta] = \inf_z \left(\|A_h z - u_\delta\| + h\|z\| + \delta \right)$$

and its minimizing element z_η. We show, that MLED gives the unimprovable upper estimate for $\mu[A_0, u_0]$. An effective numerical realization of this method is considered.

1. CRITERION PROBLEMS OF MINIMIZATION

Let Z and U be Banach spaces, $L = L(Z, U)$ be the normed space of linear bounded operators, $A_0 \in L$, $u_0 \in U$. The ill-posed argument problems of solving the operator equation

$$A_0 z = u_0 \tag{1.1}$$

and finding pseudosolutions of equation (1.1)

$$F_0(z) = \|A_0 z - u_0\| \to \min \tag{1.2}$$

are well known (see (Tikhonov and Arsenin, 1977)).
We consider the following criterion problems.

PROBLEM 1 (P1). *Find the measure of incompatibility of equation (1.1)*

$$\mu[A_0, u_0] = \inf_{z \in Z} \|A_0 z - u_0\|. \tag{1.3}$$

PROBLEM 2 (P2). *Find a minimizing sequence $\{z_n\}$ of the functional $F_0(z)$*

$$\|A_0 z_n - u_0\| \to \mu[A_0, u_0], \quad n \to \infty. \tag{1.4}$$

For P1 and P2 we do not assume the existence of solutions and pseudosolutions of (1.1) and the convergence of minimizing sequences $\{z_n\}$. In the general case the function $\mu[A, u]$ is not continuous (see (Morozov, 1973)). It is only upper semicontinuous (see (Levin, 1986)).

Definition 1. P1 is correct for (A_0, u_0), if the function $\mu[A, u]$ is continuous at the point (A_0, u_0).

THEOREM 1. *P1 is correct for (A_0, u_0) if and only if $\mu[A_0, u_0] = 0$ or there is $\lambda > 0$ such that $\|A_0 z\| \geq \lambda \|z\|$ for any $z \in Z$.*

By stability of P2 we mean the possibility of finding minimizing sequences for exact functional (1.2) by solving approximate problems of minimization (see (Levin, 1987)).

Definition 2. We say that P2 is stable by value for (A_0, u_0), if for any sequence of approximate data $\{(A_n, u_n)\}$, $\|A_n - A_0\| \to 0$, $\|u_n - u_0\| \to 0$, $n \to \infty$, any sequence of approximate solutions $\{z_n\}$ of the corresponding minimizing problems

$$\|A_n z_n - u_n\| - \inf_{z \in Z} \|A_n z - u_n\| \leq \varepsilon_n, \quad \varepsilon_n > 0, \quad \varepsilon_n \to 0,$$

satisfies (1.4) with exact data (A_0, u_0).
The same definition of stability by value one can give for the problems with general functionals (see (Levin, 1987)). E.g. unstable criterion problems of this class arise in backward in time heat conduction problems (see (Lattes and Lions, 1969)) and in the method of fundamental solutions for boundary value elliptic problems (see (Vencel et al., 1984)). A criterion of stability in the sense of Definition 2 was found in (Levin, 1987).

THEOREM 2. *P2 is stable by value for (A_0, u_0) if and only if there is $\lambda > 0$ such that*

$$\|A_0 z\| \geq \lambda \|z\| \tag{1.5}$$

for all $z \in Z$.

Condition (1.5) is equivalent to the existence of the bounded operator $A_0^{-1}: \mathrm{Im} A_0 \to Z$. Thus Theorem 2 establishes the connection between stability by value of P2 and the correctness of argument minimization problem (1.2). It implies that one has to use regularization by value algorithms to solve P2 in the case of integral equation of the first kind.

2. METHOD OF THE LEAST ESTIMATE OF DISCREPANCY (MLED)

Let one knows the approximate data $A_h \in L$, $u_\delta \in U$, $\eta = (h, \delta)$, $h \geq 0$, $\delta \geq 0$, such that

$$\|A_h - A_0\| \leq h, \quad \|u_\delta - u_0\| \leq \delta. \tag{2.1}$$

Let Σ_η be the class of equivalent by accuracy input data

$$\Sigma_\eta = \Big\{(A, u) : A \in L,\ \|A - A_h\| \leq h,\ u \in U,\ \|u - u_\delta\| \leq \delta\Big\}.$$

According to MLED one has to find

$$\mu_\eta[A_h, u_\delta] = \inf_{z \in Z} \Phi_\eta(z), \quad \Phi_\eta(z) = \sup_{(A,u) \in \Sigma_\eta} \|Az - u\|. \tag{2.2}$$

LEMMA. *For any $A_h \in L$, $u_\delta \in U$, $h \geq 0$, $\delta \geq 0$, $z \in Z$, there are $(\bar{A}, \bar{u}) \in \Sigma_\eta$ such that*

$$\Phi_\eta(z) = \|\bar{A}z - \bar{u}\| = \|A_h z - u_\delta\| + h\|z\| + \delta. \tag{2.3}$$

THEOREM 3. *If Z is reflexive and $h > 0$, then there is $z_\eta \in Z$ (for Hilbert spaces Z, U and sufficiently small h, δ this element is unique) such that*

$$\Phi_\eta(z_\eta) = \mu_\eta[A_h, u_\delta] = \min_{z \in Z} \Big(\|A_h z - u_\delta\| + h\|z\| + \delta\Big). \tag{2.4}$$

Proof of this theorem see in (Levin, 1986). One can show the convergence of MLED for P1 and P2.

THEOREM 4. *For every (A_h, u_δ) satisfying (2.1) we have*

$$\mu_\eta[A_h, u_\delta] \geq \mu[A_0, u_0], \quad \lim_{\eta \to 0} \mu_\eta[A_h, u_\delta] = \mu[A_0, u_0]$$

and under the conditions of Theorem 3 the sequence $\{z_n\}$, $z_n = z_{\eta_n}$ as $\eta_n \to 0$ is a minimizing sequence for $F_0(z)$.

The value (2.4) is an unimprovable upper estimate of (1.3).

THEOREM 5. *Under the conditions of Theorem 3, if $z_\eta \neq 0$ and $A_h z_\eta \neq u_\delta$, then*

$$\sup_{(A,u) \in \Sigma_\eta} \inf_{z \in Z} \|Az - u\| = \inf_{z \in Z} \sup_{(A,u) \in \Sigma_\eta} \|Az - u\| = \mu_\eta[A_h, u_\delta]$$

For Hilbert spaces Z and U the effective realization of MLED is based on the following theorem (Levin, 1988).

THEOREM 6. *If $h > 0$, then MLED in the form (2.4) is equivalent to the following "Principle of the Least Estimate of Discrepancy":*
1) if $\|A'_h u_\delta\| \leq h\|u_\delta\|$, then $z_\eta = 0$;

2) *if* $\|A'_h u_\delta\| > h\|u_\delta\|$, *then it is necessary to solve the equation*

$$(A'_h A_h + \alpha E)z = A'_h u_\delta, \quad \alpha > 0, \tag{2.5}$$

and if for all $\alpha > 0$ *the inequality* $\alpha\|z_\alpha\| > h\|A_h z_\alpha - u_\delta\|$ *holds, then there is a solution* $\tilde{z}_\eta$ *of equation* $A_h z = u_\delta$ *and* $z_\eta = \tilde{z}_\eta = \lim z_\alpha$ *as* $\alpha \to +0$*; otherwise there is a unique solution* $\alpha_0 > 0$ *of the equation*

$$\alpha\|z_\alpha\| = h\|A_h z_\alpha - u_\delta\| \tag{2.6}$$

and $z_\eta = z_{\alpha_0}$.

For the second part of Theorem 6 one can show that

$$0 \le \alpha_0 \le h\|u_\delta\|\|A_h\|^2\left(\|A'_h u_\delta\| - h\|u_\delta\|\right)^{-1} = \alpha_{\max}$$

and the iterative sequence $\{\alpha_n\}$, $\alpha_1 = \alpha_{\max}$,

$$\alpha_{n+1} = \varphi(\alpha_n) = h\|A_h z_{\alpha_n} - u_\delta\|\|z_{\alpha_n}\|^{-1}$$

converges to a solution α_0 of equation (2.6) (see (Levin, 1991)). Equation (2.5) is the Euler equation for the problem of minimization of Tikhonov's functional

$$M_\alpha(z) = \|A_h z - u_\delta\|^2 + \alpha\|z\|^2.$$

One can use MLED to regularize nonlinear problems.

REFERENCES

Vencel, E.S. and Kobylinsky, V.G. and Levin, A.M. (1984). Using the regularization method for solving the problems of thin elastic plates theory. *USSR Comput. Math. and Math. Phys.* **24**, 323–328.

Lattes, R. and Lions, J.L. (1969). *The Method of Quasi Reversibility.* American Elsevier, New York.

Levin, A.M. (1986). On computing the measure of incompatibility of operator equations of the first kind. *USSR Comput. Math. and Math. Phys.* **26**, 499–507.

Levin, A.M. (1987). On stability of the problem of finding the minimizing sequences of functionals. *USSR Comput. Math. and Math. Phys.* **27**, 483–489.

Levin, A.M. (1988). Modifications of Newton's and secant methods and their application using in regularization algorithms. *USSR Comput. Math. and Math. Phys.* **28**, 1123–1134.

Levin, A.M. (1991). Iterative algorithm for determination of a regularization parameter in the method of the least estimate of discrepancy. *USSR Comput. Math. and Math. Phys.* **31**, 151–153.

Morozov, V.A. (1973). On calculation of lower bounds of functionals by approximate data. *USSR Comput. Math. and Math. Phys.* **13**, 1045–1049.

Tikhonov, A.N. and Arsenin, V.Y. (1977). *Solutions of Ill-Posed Problems.* Winston, Washington.

Ill-Posed Problems in Natural Sciences, pp. 88 – 92
A. Tikhonov (Ed.)
1992 VSP/TVP

ON THE NUMERICAL SOLUTION OF THE ILL-POSED PROBLEMS OF CONSTRUCTION OF APPROXIMATE SOLUTIONS OF SINGULAR INTEGRAL EQUATIONS OF THE SECOND KIND

A.F. MATVEEV
Institute of Theoretical and Experimental Physics, B.Cheremushkinskaja, str. 25, 117259 Moscow, Russia

ABSTRACT

In this note we present a direct method for approximate solution of singular integral equations (SIE). We encounter equations of this kind when solving problems in aerodynamics, elasticity theory, eiectrodynamics, nuclear physics, and other branches of science and technology. SIE belong to the class of ill-posed problems, where any of Hadamard's three conditions for being well-posed may be violated.

0. INTRODUCTION

Let us consider the singular integral eguations (SIE)

$$a(t_0)\varphi(t_0) + \frac{1}{\pi i} \int\limits_L \frac{K(t_0,t)\varphi(t)\,dt}{t-t_0} = f(t_0), \quad t_0 \in L, \tag{0.1}$$

$$a(x_0,z_0)\varphi(x_0,z_0) + \frac{1}{\pi} \int\limits_{-l}^{l} \int\limits_{-d}^{d} \frac{M(x_0,z_0,x,z)\,dx\,dz}{(z_0-z)^2(x_0-x)} = f(x_0,z_0), \tag{0.2}$$

where L is any piecewise smooth path (see (Muskhelishvili, 1968, § 1)), $x_0 \in (-d,d)$, $z_0 \in (-l,l)$. We encounter eguations of this kind when solving problems in aerodynamics, elasticity theory, electrodynamics, nuclear physics, and other branches of science and technology (Belotserkovskii and Lifanov, 1985). SIEs (0.1) and (0.2) belong to the class of ill-posed problems, where any of Hadamard's three conditions for being well-posed my be violated (Tikhonov and Arsenin, 1979).
Thus, the problem is to construct an approach to the regularization in the sense of Tikhonov for SIEs (0.1) and (0.2) and to construct a method of approximate solution of equations (0.1) and (0.2) which can be conveniently programmed on a computer.

1. WELL-POSED PROBLEM OF SOLUTION OF EQUATION (1)

Following (Muskhelisvili, 1968), looking for a solution of (0.1), given a class h, given coefficients $a(t_0)$ and $K(t_0, t_0)$, we define index $\varkappa$ and the canonical function $z(t_0)$ of (0.1) from the class h. Thus the SIE (0.1) can be written either in the form

$$(K^0, w_1 u)(t_0) + (k, M(t_0, \cdot) w_1 u) = f(t_0), \quad t_0 \in L \tag{1.1a}$$

or in the form

$$(R', w_1 u)(t_0) + (k, M(t_0, \cdot) w_1 u) = f(t_0), \quad t_0 \in L, \tag{1.1b}$$

where

$$(K^0, \varphi)(t_0) \equiv a(t_0)\varphi(t_0) + \frac{b(t_0)}{\pi i} \int_L \frac{\varphi(t)\, dt}{t - t_0},$$

$$(R', \varphi)(t_0) \equiv a(t_0)\varphi(t_0) + \frac{1}{\pi i} \int_L \frac{b(t)\varphi(t)\, dt}{t - t_0},$$

$$(k_\nu, g) \equiv \int_L t^\nu g(t) dt, \quad \nu = 0, 1, 2, \ldots,$$

$$(k, g) \equiv (k_0, g), \quad w_1(t) = \frac{z(t)}{a^2(t) - b^2(t)}, \quad b(t_0) = K(t_0, t_0),$$

$a(t_0)$, $b(t_0) \not\equiv 0$, $M(t_0, t)$ and $f(t_0)$ are known functions defined on L, $u(t)$ is an unknown function. We assume that $a(t)$ and $b(t)$ can be discontinuous but have to be from the class H_0 on L, $M(t_0, t)$ is a function from the class H_0 and $f(t)$ is from the class H^*, and everywhere on L we have $a^2(t) - b^2(t) \neq 0$. Definitious of the classes H, H_0 and H^* are given in (Muskhelishvili, 1968, §8)).

We shall use in what follows definitions and notation from (Muskhelishvili, 1968). Notice that under the above assumptions on the coefficients the equations (1.1a) and (1.1b), generally speaking, do not reduce to each other (Muskhelishvili, 1968). Together with the equations (1.1a) and (1.1b) we shall study the corresponding characteristic SIEs

$$(K^0, w_1 u)(t_0) = f(t_0), \tag{1.2a}$$

$$(R', w_1 u)(t_0) = f(t_0). \tag{1.2b}$$

SIEs (1.2a) and (1.2b) are also ill-posed in the sense of Hadamard. In order to get rid of the ill-posed equations (1.2a) and (1.2b) one must consider the equations

$$\begin{gathered} \eta(-\varkappa) b(t_0) \tau_{|\varkappa|-1}(t_0) + (K^0, w_1 u)(t_0) = f(t_0), \\ \eta(\varkappa)(k_\nu, w_1 u) = c_\nu \eta(\varkappa), \quad \nu = 0, \ldots, |\varkappa| - 1, \end{gathered} \tag{1.3a}$$

$$\begin{gathered} \eta(-\varkappa) \tau_{|\varkappa|-1}(t_0) + (R', w_1 u)(t_0) = f(t_0), \\ \eta(\varkappa)(k_\nu, w_1 b u) = c_\nu \eta(\varkappa), \quad \nu = 0, \ldots, |\varkappa| - 1, \end{gathered} \tag{1.3b}$$

where

$$\tau_{|\varkappa|-1}(t_0) \equiv \sum_{k=0}^{|\varkappa|-1} \alpha_k t_0^k, \qquad \eta(\varkappa) = \begin{cases} 1, & \varkappa > 0, \\ 0, & \varkappa \le 0, \end{cases}$$

α_k $(k = 0, \dots, |\varkappa| - 1)$ are regularising unknown constants, c_ν $(\nu = 0, \dots, |\varkappa| - 1)$ are known constauts. Systems (1.3a) and (1.3b) are well-posed in the sense of Hadamard.

When $\varkappa > 0$, the system of equations $(k_\nu, w_1 u) = c_\nu$, $\nu = 0, \dots, |\varkappa| - 1$, is the additional information, whish specifies the unique solution of (1.2a), and the system of equations $(k_\nu, w_1 bu) = c_\nu$, $\nu = 0, \dots, |\varkappa| - 1$, specified the unique solution of (1.2b).

When $\varkappa \le 0$ the regularizing unknown constants α_k make equations (1.3a) and (1.3b) solvable for an arbitrary right-hand side $f(t_0)$.

Finnaly, we note that calculation chemes of numerical solution of SIEs, which we shall consider below, have the property of autoregularization (Tikhonov and Arsenin, 1979). Now return to equations (1.1a) and (1.1b). SIEs (1.1a) and (1.1b) can be reduced to Fredholm integral equations of the second kind (Muskhelishvili, 1968). If these Fredholm integral equations have a unique solution, then Tikhonov's method for ill-posed SIEs (1.1a) and (1.1b) can be constructed similarly to the case of equations (1.2a) and (1.2b).

In this way, in order to get rid of the ill-posed equations (1.1a) and (1.1b) one must consider the systems

$$\begin{aligned} \eta(-\varkappa) b(t_0) \tau_{|\varkappa|-1}(t_0) + (K^0, w_1 u)(t_0) + (k, M(t_0, \cdot) w_1 u) &= f(t_0), \\ \eta(\varkappa)(k_\nu, w_1 u) = c_\nu \eta(\varkappa), \quad \nu = 0, \dots, |\varkappa| - 1&, \end{aligned} \tag{1.4a}$$

and

$$\begin{aligned} \eta(-\varkappa) \tau_{|\varkappa|-1}(t_0) + (R', w_1 u)(t_0) + (k, M(t_0, \cdot) w_1 u) &= f(t_0), \\ \eta(\varkappa)(k_\nu, w_1 bu) = c_\nu \eta(\varkappa), \quad \nu = 0, \dots, |\varkappa| - 1&. \end{aligned} \tag{1.4b}$$

Systems (1.4a) and (1.4b) are well-posed in the sense of Hadamard.

2. APPROXIMATE SOLUTION OF EQUATION (1)

The problem of solution of SIE (1) reduces to the problem of solution of system of equations (1.4a) or (1.4b).

Let us denote by

$$P_{m_0;m}(t) = \sum_{k=0}^{m_0} \mu_k^0 t^k + \sum_{\nu=1}^{s} \sum_{k=-m_\nu}^{-1} \mu_k^{(\nu)} (t - \tilde{c}_\nu)^k, \quad \tilde{c}_\nu \notin L, \quad \nu = 1, \dots, s,$$

the function which we shall call a generalized polynomial of degree $n = m_0 + m$, $m = \sum_{\nu=1}^{s} m_\nu$.

ASSERTION 2.1. *Let the coefficients $a(t_0)$ and $b(t_0)$ of the* SIEs (1.1a) *and* (1.1b) *be from the class H_0 on L, $b(t_0) = b_{r_1}(t_0) b_1(t_0)$, where $b_1(t_0)$ is a polynomial of* *degree $r_1 \geq 0$ and $b_1(t_0)$ is an arbitrary function which can have zeros on L; $\varkappa$ and $z(t_0)$ are the index and the canonical functions respectively of* SIE (1) *from the class h. Then for any generalized polynomial $p_{m_0;m}(t)$ there exist polynomials $q_{m_0-\varkappa;m}(t)$ and $\tilde{q}_{m_0-\varkappa;m}(t)$ such that everywhere on L except (perhaps) knots of the path L, the following relations hold:*

$$(K^0, b_1 w_1 P_{m_0;m})(t_0) = b_1(t_0) q_{m_0-\varkappa;m}(t_0), \tag{2.1a}$$

$$(R', w_1 P_{m_0;m})(t_0) = \tilde{q}_{m_0-\varkappa;m}(t_0), \tag{2.1b}$$

where $m_0 > r_1 + \varkappa$.

Relations (2.1a) and (2.1b) are the basis of the methods of construction of approximate solutions of SIEs (1.1a), (1.1b) and (2). We illustrate the stated above, using the SIE (1.1b) as an example (see also (Afendikova et al., 1987; Matveev, 1989)).

To obtain an approximate solution with a given index $\varkappa$ of SIE (1.1b) of class h and a given integer n we construct two nets $\bar{w}_n = \{t_1, t_2, t_3, \dots, t_n\}$ and $\bar{w}_{0n-\varkappa} = \{t_{01}, t_{02}, \dots, t_{0n-\varkappa}\}$ with pairwise different nodes which differ from the roots of the function $b(t)$.We take as $f(t)$ the approximating generalized polinomial $f_{m_0-\varkappa-1;m}(t)$ of degree $n-\varkappa-1 \quad (n = m_0 + m)$.

An approximate solution of system of equations (1.4b), where $u_{m_0-\varkappa-1,m}(t)$ is an unknown generalized interpolatory polynomial at nodes of $\bar{w}_n$ can be found from the system

$$\eta(-\varkappa)\tau_{|\varkappa|-1}(t_{0j}) = (R', w_1, u_{m_0-1;m})(t_{0j})$$

$$+(k, M(t_{0j}, \cdot) w_1 u_{m_0-1;m}) = f_{m_0-\varkappa-1;m}(t_{0j}), \quad j = 1, \dots, n-\varkappa,$$

$$\eta(\varkappa)(k_\nu, w_1 b u_{m_0-1;m}) = c_\nu \eta(\varkappa), \quad \nu = 0, \dots, |\varkappa| - 1.$$

The latter system can be written in the form (Matveev, 1989)

$$\eta(-\varkappa)\tau_{|\varkappa|-1}(t_{0j}) + \sum_{k=1}^{n} \Big(\frac{B_{k,0s}}{t_{0s} - t_k} + M_{0s,k}\lambda_{k;n}\Big) u_k = f_{0s}, \quad s = 1, \dots, n-\varkappa,$$

$$\eta(\varkappa) \sum_{k=1}^{n} t_k^\nu \lambda_{k;n} u_k = \eta(\varkappa) c_\nu, \quad \nu = 0, \dots, |\varkappa| - 1, \quad t_{0s} \in \bar{w}_{0n-\varkappa}, \quad t_k \in \bar{w}_n.$$

This calculation sheme is accurate for the polinomial of degree n, $n = m_0 + m$, $m_0 > r_1 + \varkappa$.

It is important to note that the problem of solution of equation (2) reduces to the problem of solution of an integral equation, on the left-hand side of which is the product of singular integral operators. The numerical solution of the latter problem is similar to the numerical solution of SIE (1).

REFERENCES

Afendikova, N.G., Lifanof, I.K., and Matveev, A.F. (1987). On the approximate solution of singular integral equations. *Different. Equat.* **23**, 1392–1402.

Belotserkovskii, S.M. and Lifanov I.K. (1985). *Numerical Methods in Singular Integral Equations and Their Application in Aerodynamics, Elasticity Theory, and Electrodynamics.* Nauka, Moscow (in Russian).

Matveev, A.F. (1989). On the construction of approximate solutions of singular integral equations of the second kind. *Dokl. Akad. Nauk SSSR* **307**, 193–197 (in Russian).

Muskhelishvili, N.I. (1968). *Singular Integral Equations.* 3rd ed. Nauka, Moscow (in Russian).

Tikhonov, A.N. and Arsenin, V.Ya. (1977). *Methods of Solving Ill-Posed Problems.* John Wiley, New York.

Ill-Posed Problems in Natural Sciences, pp. 93 – 103
A. Tikhonov (Ed.)
1992 VSP/TVP

REGULARIZATION METHODS FOR ILL-POSED BOUNDARY PROBLEMS

I.V. MEL'NIKOVA
Ural State University, 620151 Ekaterinburg, Russia

ABSTRACT

In this paper the author studies two basic regularization methods for ill-posed differential problems in Banach and Hilbert spaces. They are the quasireversion method and the auxillary boundary conditions method (ABC). Exact by the order error estimates are obtained on characteristic classes of correctness. On the basis of these estimates the comparison of the indicated methods is given for the ill-posed Cauchy problem. It is shower that in the regularization of the Dirichlet problem there is a duality similar to the diality in the questions of well–posedness of the Cauchy and Dirichlet problems.

1. INTRODUCTION

At the present time the theory of operator equations of the first kind is the most advanced area in the theory of ill-posed problems (Tikhonov and Arsenin, 1979; Ivanov et al., 1978; Lavrent'ev, 1973). The reduction of ill-posed problems for differential, integrodifferential and in common case differential-operator equations to the first kind equation usually is not effective. Here we need methods taking into account the specific differential character of these problems. Such methods are the quasireversion method, consisting in the introduction of additional terms with a small parameter into the equation, and the method of auxiliary boundary conditions (ABC), which regularizes by introducing new terms with a small parameter into the boundary conditions.
In this paper the above mentioned methods are investigated for the ill-posed Cauchy problem for the abstract first order equation

$$u'(t) = Au(t), \quad t \geq 0, \quad u(0) = u_0, \tag{1}$$

as wellas for the basic ill-posed problems for the second order equation. The latter are the Cauchy problem for the elliptic case

$$u''(t) = Bu(t), \quad t \geq 0, \quad u(0) = u_0, \quad u'(0) = u_1, \tag{2}$$

and the Dirichlet problem for the hyperbolic case

$$u''(t) = Bu(t), \quad 0 < t < T, \quad u(0) = u_0, \quad u(T) = u_T. \tag{3}$$

Originally the quasireversion method was proposed to solve the control problems for evolution processes in Hilbert space (Lattes and Lions, 1967) and in Banach space (Ivanov, 1972). The fact that the quasireversion method is regularizing for the ill-posed Cauchy problem was proved in (Muzilev, 1975 and Mel'nikova, 1981). The same fact for the ABC method was obtained in (Carasik, 1978; Mel'nikova and Cudrjavtsev, 1983). In (Mel'nikova, 1986, 1988; Freyberg, 1987) the error estimates for the above mentioned methods in the Banach space were obtained. For the quasireversion method it is an unimprovable in the order of δ (the initial values error) logarithmic estimate, and for the ABC method it is a power estimate. However the Banach spaces techniques stipulated the construction obtaining of estimates on different correctness classes. They are $\mathfrak{M}_1 = \{u(T): \|Au(T)\| \leq M_1\}$ for the quasireversion method and $\mathfrak{M}_2 = \{u(T): \|u(T+\tau)\| \leq M_2, \tau > 0\}$ for the ABC method. This fact made it difficult to compare the methods in the first place, and created the impression that the power estimates are inherent to the ABC method and the logarythmic ones are inherent to the quasireversion method in the second place.

In the present paper the both above mentioned methods for the ill-posed Cauchy problem in the Hilbert space are studied. The error estimates for them are obtained on both $\mathfrak{M}_1$ and $\mathfrak{M}_2$ classes. It is shown that the estimates for the quasireversion method are logarythmic and exact by the order both on $\mathfrak{M}_1$ and $\mathfrak{M}_2$. For the ABC method the saturation does not occures: the estimate is logarythmic on $\mathfrak{M}_1$ and power on $\mathfrak{M}_2$.

As for the Dirichlet problem, which is les studied as compared to the Cauchy problem (see, for example, (Liskovets, 1968; Papi, 1979; Mel'nikova and Freyberg, 1985)), the duality was shown to occur in the regularization by the quasireversion and ABC methods with real parameter, which is similar to the duality in the correctness of the Cauchy and the Dirichlet problems.

Besides the possibility to obtain error estimates the technique of Hilbert spaces allowed us to study the equations with irreversible operator at the derivative as well as the connection between the "differential" methods of solving the Cauchy and the Dirichlet problems studied here with the variational methods of solving the corresponding first kind equations. The ABC method was shown to coinside after the regularization parameters were made compatible with the quasisolution, Tichonov's and the discrepancy methods.

2. REGULARIZATION OF ILL-POSED CAUCHY PROBLEMS IN BANACH SPACE

In this section we present the results on regularization of the ill-posed Cauchy problem for the first and second order equations in a Banach space E, which were obtained in above mentioned author's papers.

Consider the Cauchy problem (1) with a linear, densely defined in E operator A satisfying the following conditions

a) $\exists L > 0: \ \forall \lambda \notin G = \{\lambda \in \mathbf{C} : |\arg \lambda| < \beta < \pi/4\}$,

$$\|R_\lambda(A)\| \leq L(1+|\lambda|)^{-1}$$

or

b) $\exists L > 0: \ \forall \lambda \notin G = \{\lambda \in \mathbf{C} : |\operatorname{Im}\lambda| < \alpha < \pi/T, \quad \operatorname{Re}\lambda > \omega, \ \omega \in R^1\}$,

$$\|R_\lambda(A)\| \le L.$$

Denote by E_a and E_b the classes of operators satisfying the conditions a) and b) respectivaly. Let the solution $u(t)$ exist for some exact initial value u_0, but u_0^δ is given with an error δ:

$$\|u_0 - u_0^\delta\| \le \delta.$$

Under these conditions on the operator A the Cauchy problem is ill-posed and therefore its solution may not exist for any u_0^δ and is unstable.
In accordance with the quasireversion method the problem

$$\frac{du_{\varepsilon,\delta}(t)}{dt} = (A - \varepsilon A^2) u_{\varepsilon,\delta}(t), \quad 0 \le t \le T, \quad u_{\varepsilon,\delta}(0) = u_0^\delta, \tag{4}$$

was considered and the operator $R_{\varepsilon,T}$, giving the solution of the problem (4) at $t = T$ on E was shown to be regularizing for the problem (1), i.e.,

1) the operator $R_{\varepsilon,T}$ is defined on all E,

2) there exists a dependence $\varepsilon = \varepsilon(\delta)$ such that

$$R_{\varepsilon(\delta),T} u_0^\delta \longrightarrow u(T), \quad \delta \to 0.$$

The following estimates of the error

$$\Delta(\mathfrak{M}_1, E_a, \delta) = \sup_{u_0:\ \|u_0 - u_0^\delta\| \le \delta,\ u_0 \in \mathfrak{M}_1,\ A \in E_a} \|u(T) - R_{\varepsilon,T} u_0^\delta\|$$

were obtained.

THEOREM 1. *Let $A \in E_a$. Then the solution of problem (4) on $D(A)$ is of the form*

$$u_{\varepsilon,\delta}(t) = -\frac{1}{2\pi i} \int_{\partial G} \exp\left((\mu - \varepsilon\mu^2)T\right) R_\mu(A) u_0^\delta \, d\mu \equiv R_{\varepsilon,T} u_0^\delta,$$

$R_{\varepsilon,T}$ is the regularizing operator for problem (1) at $t = T$ and we have for it the following error estimates

$$\Delta(\mathfrak{M}_1, E_a, \delta) \le C_1 \left(\delta \exp(T/\varepsilon) + \varepsilon\right), \tag{5}$$

$$\Delta(\mathfrak{M}_1, E_a, \delta) \ge C_2 \varepsilon, \quad \Delta(\mathfrak{M}_1, E_a, \delta) \ge C_2 \delta \exp(T/\varepsilon). \tag{6}$$

In accordance with the ABC method the problem

$$\frac{d\tilde{u}_{\varepsilon,\delta}(t)}{dt} = A \tilde{u}_{\varepsilon,\delta}(t), \quad 0 \le t < T + \tau, \quad \tau > 0, \quad \tilde{u}_{\varepsilon,\delta}(0) + \varepsilon \tilde{u}_{\varepsilon,\delta}(T + \tau) = u_0^\delta \tag{7}$$

was considered and the following theorem was proved.

THEOREM 2. *Let $A \in E_b$. Then*

$$\widetilde{u}_{\varepsilon,\delta}(T) = -\frac{1}{2\pi i}\int_{\partial G} \frac{\exp(\lambda T)}{1+\varepsilon\exp(\lambda(T+\tau))} R_\lambda(A)u_0^\delta d\lambda \equiv \widetilde{R}_{\varepsilon,T}u_0^\delta, \qquad (8)$$

$\widetilde{R}_{\varepsilon,T}$ is the regularizing operator for problem (1) and we have for it the following estimates

$$\Delta(\mathfrak{M}_2, E_b, \delta) \le C(\varepsilon^{\gamma\tau/T} + \delta/\varepsilon), \quad 0 < \gamma < 1. \qquad (9)$$

If we narrow the correctness class $\mathfrak{M}_1$ to $\mathfrak{M}_2$ in the quasireversion method, then estimates (5) and (6) remain true with respect to the order. Thus the comparison by estimates of $\Delta(\mathfrak{M}_2, E_a \cap E_b, \delta)$ of the quasireversion and the ABC methods shows us the advantage of the ABC method. It follows from (9) that under the coordination of the parameters ε and δ we have a power (with respect to δ) estimate for the ABC method. From estimates (5) and (6) it follows that the error estimate for the quasireversion method is not better than logarithmic.

For the second order Cauchy problem (2) with B, generating the elliptic case (i.e., $\exists L > 0: \ \forall\lambda < 0, \ \|R_\lambda(B)\| \le L(1+|\lambda|)^{-1}$) we consider the quasireversion method, in which

$$u_{\alpha,\delta}(T) = \frac{1}{2}\Big[x_{\alpha,\delta}(T) + y_\delta(T)\Big] \equiv R_T^\alpha(u_0^\delta, u_1^\delta),$$

$x_{\alpha,\delta}(t), y_\delta(t)$ are the solutions of the system

$$x'_{\alpha,\delta}(t) = (A-\alpha B)x_{\alpha,\delta}(t), \quad x_{\alpha,\delta}(0) = u_0^\delta + A^{-1}u_1^\delta, \quad A = B^{1/2},$$

$$y'_\delta(t) = -Ay_\delta(t), \quad 0 \le t \le T, \quad y_\delta(0) = u_0^\delta - A^{-1}u_1^\delta, \quad \alpha > 0,$$

and the ABC method, in which

$$\widetilde{u}''_{\alpha,\delta}(t) = B\tilde{u}_{\alpha,\delta}(t), \quad 0 \le t < T+\tau,$$

$$\widetilde{u}_{\alpha,\delta}(0) + \alpha\tilde{u}_{\alpha,\delta}(T+\tau) = u_0^\delta, \quad \widetilde{u}'_{\alpha,\delta}(0) = u_1^\delta. \qquad (10)$$

Estimates similar to (5), (6), (9) are obtained.

As for the numerical computation in the ABC method, the solution obtained in the form of Danford integral is not appropriate for this purpose. Such integrals give only the method of proof and of obtaining estimates.

(Al'shanskii et al., 1991) an effective algorithm of reduction of difference schemes for the boundary value problems (7) and (10) to a series of well developed difference schemes for the well-posed Cauchy and Dirichlet problems was suggested.

3. REGULARIZATION OF ILL-POSED CAUCHY PROBLEMS IN HILBERT SPACE.

In order to obtain error estimates on both $\mathfrak{M}_1$ and $\mathfrak{M}_2$ correctness classes we consider here the ill-posed Cauchy problems in a more simple case, – namely with a self-adjoint operator in a Hilbert space $\mathbf{H}$.

Let the self-adjoint operator A generate an orthonormal basis consisting of the eigenvectors $\{x_k\}$ corresponding to the eigenvalues $|\lambda_1| \leq |\lambda_2| \leq \dots$. Let the sequence $\{\lambda_k\}$ be bounded from below. Denote by H_c the class of operators with these properties . Evidently we have $H_c \subset E_a \cap E_b$.
First we consider the ABC method. Using it we solve the well-posed boundary value problem (7).

THEOREM 3. *Let $A \in H_c$. Then the operator solving giving the (7) is the regularizing operator of problem (1). Moveover on the class the exact by the order of δ logarithmic estimate $\mathfrak{M}_1$ holds and on the class $\mathfrak{M}_2$ the exact by the order of δ power estimate holds.*

Proof. If $A \in H_c$ and the solution of the problem (1) with the initial value u_0 exists, it may be expanded into a Fourier series with respect to $\{x_k\}$:

$$u(t) = \sum_{k=1}^{\infty} \exp\Big(\lambda_k t\Big) u_{0k} x_k, \quad u_{0k} = (u_0, x_k),$$

similarly the solution of regularized problem (7) is

$$\widetilde{u}_{\varepsilon,\delta}(t) = \sum_{k=1}^{\infty} \frac{\exp\Big(\lambda_k t\Big)}{1 + \varepsilon \exp\big(\lambda_k (T+\tau)\big)} u_{0k}^{\delta} x_k.$$

Consider the error at $t = T$:

$$\Delta\Big(A,\ u(T),\ \varepsilon, \delta\Big) \equiv \Delta = \sup_{\|u_0 - u_0^{\delta}\| \leq \delta} \|u(T) - \widetilde{u}_{\varepsilon,\delta}(T)\| \leq \Delta_1 + \Delta_2,$$

where

$$\Delta_1 = \|u(T) - \widetilde{u}_{\varepsilon}(T)\|, \qquad \Delta_2 = \sup_{\|u_0 - u_0^{\delta}\| \leq \delta} \|\widetilde{u}_{\varepsilon}(T) - \widetilde{u}_{\varepsilon,\delta}(T)\|,$$

$\widetilde{u}_{\varepsilon}(T) = \widetilde{u}_{\varepsilon,0}(T)$. First we estimate Δ_2. We have

$$\begin{aligned}
\Delta_2 = \sup_{\|u_0 - u_0^{\delta}\| \leq \delta} &\Big\| \sum_{k=1}^{\infty} \frac{\exp\big(\lambda_k T\big)}{1 + \varepsilon \exp\big(\lambda_k (T+\tau)\big)} (u_0 - u_0^{\delta})_k x_k \Big\| \leq \\
&\delta \Big| \frac{\exp(\lambda T)}{1 + \varepsilon \exp\big(\lambda (T+\tau)\big)} \Big|_{\lambda = \lambda_{\text{extr}} = \frac{1}{T+\tau} \log \frac{T}{\varepsilon}} = \\
&= \frac{\delta}{\varepsilon^{T/(T+\tau)}} \frac{(T/\tau)^{T/(T+\tau)}}{(1 + T/\tau)} \leq C_1 \frac{\delta}{\varepsilon}.
\end{aligned} \tag{11}$$

Let us estimate Δ_1. We have $\Delta_1 \leq \Delta_1' + \Delta_1''$, where $\Delta_1' = \| \sum\limits_{k=1}^{N} \varkappa_k \|$, $\Delta_1'' = \| \sum_{k=N+1}^{\infty} \varkappa_k \|$,

$$\varkappa_k = \exp\Big(\lambda_k T\Big) u_{0k} \Big[\frac{\varepsilon \exp\big(\lambda_k (T+\tau)\big)}{1 + \varepsilon \exp\big(\lambda_k (T+\tau)\big)} \Big].$$

Taking into account the fact that if $u(T) \in \mathfrak{M}_i, i = 1,2$, then $\sum_{k=1}^{\infty} \exp(2\lambda_k T)u_{0k}^2 \le M < M_i$, we deduce

$$\Delta_1' \le M \max_{1\le k\le N}\left[\frac{\varepsilon \exp\big(\lambda_k(T+\tau)\big)}{1+\varepsilon\exp\big(\lambda_k(T+\tau)\big)}\right] =$$

$$M\frac{\varepsilon}{\varepsilon+\exp\big(-\lambda_N(T+\tau)\big)}.$$

On the class $\mathfrak{M}_1$ we have

$$\Delta_1'' = \left\{\sum_{k=N+1}^{\infty}\frac{1}{\lambda_k^2}\Big[\lambda_k\exp(\lambda_k T)u_{0k}\Big]^2\Big(\frac{\varepsilon\exp\big(\lambda_k(T+\tau)\big)}{1+\varepsilon\exp\big(\lambda_k(T+\tau)\big)}\Big)^2\right\}^{1/2} \le \frac{M_1}{\lambda_{N+1}}.$$

Now let us get rid of the intermediate parameter N in the dependence ε in such a way that

$$\varepsilon_{N+1} \equiv \exp\big(-\lambda_{N+1}(T+\tau)\big) \le \varepsilon^s \le \exp\big(-\lambda_N(T+\tau)\big) \equiv \varepsilon_N, \quad 0 < s < 1, \tag{12}$$

Then

$$\Delta_1 \le \frac{M\varepsilon}{\varepsilon+\varepsilon_N} + \frac{M_1(T+\tau)}{|\log\varepsilon_{N+1}|} \le \frac{M\varepsilon}{\varepsilon+\varepsilon^s} + \frac{M_1(T+\tau)}{|\log\varepsilon|} = C_2|\log\varepsilon|^{-1}. \tag{13}$$

Hence, since both Δ_1 and Δ_2 do not depend on the particular $u_0 \in \mathfrak{M}_1$ and the operator $A \in H_c$, we obtain the following upper estimate

$$\Delta\Big(\mathfrak{M}_1, H_c, \delta\Big) \le C_3\Big(\frac{\delta}{\varepsilon} + |\log\varepsilon|^{-1}\Big). \tag{14}$$

Let us estimate Δ from below on classes $\mathfrak{M}_1$ and H_c. From the definitions of $\Delta, \Delta_1, \Delta_2$ we have $\Delta \ge \Delta_1$, $\Delta \ge \Delta_2 - \Delta_1$, (i.e.) $\Delta \ge \frac{1}{2}\Delta_2$.
In (Mel'nikova, 1991) it was shown how, to choose sequences of parameters and operators in order to obtain for Δ_1 the estimate from below similar to (13), and the sequences of initial values to obtain for Δ_2 the estimate from below similar to (11). Hence choosing $\varepsilon = \delta^\gamma$, $0 < \gamma < 1$, we have

$$c|\log\delta|^{-1} \le \Delta(\mathfrak{M}_1, H_c, \delta) \le C|\log\delta|^{-1}.$$

Now let us obtain estimates on the class $\mathfrak{M}_2$. Estimates for Δ_2 and Δ_1' do not depend on the correctness class. For Δ_1'' on $\mathfrak{M}_2$, we have

$$\Delta_1 = \left\{\sum_{k=N+1}^{\infty}\frac{1}{\exp(2\lambda_k\tau)}\Big[\exp(\lambda_k(T+\tau))u_{0k}\Big]^2\Big(\frac{\varepsilon\exp\big(\lambda_k(T+\tau)\big)}{1+\varepsilon\exp\big(\lambda_k(T+\tau)\big)}\Big)^2\right\}^{1/2} \le$$

$$\frac{M_2}{\exp(\lambda_{N+1}\tau)}.$$

Choosing here the parameter N according to (12), and taking $\varepsilon = \delta^{\gamma}$, $0 < \gamma < 1$, we have the power error estimate:

$$\Delta\Big(\mathfrak{M}_2,\ H_c,\ \delta\Big) \le C_5\Big(\delta^{1-\gamma} + \delta^{\gamma(1-s)} + \delta^{\gamma s T/(T+\tau)}\Big).$$

In particular if the indeces are equal we have

$$\Delta\Big(\mathfrak{M}_2,\ H_c,\ \delta\Big) \le C_5\delta^{\frac{T}{sT+\tau}}. \tag{15}$$

The exactness of estimate (15) with respect to the order on the classes $\mathfrak{M}_2$ and H_c may be proved analogously to the case of $\mathfrak{M}_1$.
For the quasireversion method we have the following

THEOREM 4. *Let $A \in H_c$. Then the operator, solving problem (4) is the regularizating operator for problem* (1). *Moreover the exact by the order logarythmic error estimate holds*

$$\Delta\Big(\mathfrak{M}_2,\ H_c,\ \delta\Big) < C|\log\delta|^{s-1}, \quad 0 < s < 1.$$

In the paper by (Filinkov, 1989) similar estimates were obtained for the ill-posed second order Cauchy problem (2).

Remark. It is interesting to note the regularization by the ABC method with the help of well-posed boundary value problems expert for the traditional way of obtaining estimates, of Δ_1 and Δ_2 used in Theorem 3, makes it possible to obtain estimates, based on the idea of (Carasso, 1976). For example, regularizing ill-posed Cauchy problem (2) with the help of well-posed problem (10), we have

$$\|u(t) - \widetilde{u}_{\alpha,\delta}(t)\| \le C_1(t,\alpha)\|u(0) - \widetilde{u}_{\alpha,\delta}(0) + \alpha\big(u(T+\tau) - \widetilde{u}_{\alpha,\delta}(T+\tau)\big)\|$$

$$+C_2(t,\alpha)\|u'(0) - \widetilde{u}'_{\alpha,\delta}(0)\| \le C_1(t,\alpha)(\delta + \alpha M_2) + C_2(t,\alpha)\delta.$$

4. REGULARIZATION OF THE ILL-POSED DIRICHLET PROBLEM. COMPARISON OF "DIFFERENTIAL" AND VARIATIONAL METHODS

Consider the Dirichlet problem

$$u''(t) = Bu(t), \quad 0 \le t \le T, \quad u(0) = u_0,\ u(T) = u_T, \tag{16}$$

with an operator $B \in -H_c$ (i.e. $-\ B \in H_c$). Such operator B generates the well-posed Cauchy problem, or the hyperbolic case and a family of cos-functions (Travis and Webb, 1978). The Dirichlet problem is ill-posed on the class $-H_c$. The formal solution of (16) can be expanded into a Fourier-series with respect to x_k: $u(t) = \sum_{k=1}^{\infty} u_k(t)x_k$, where the coefficients $u_k(t)$ satisfy the following relations

$$u_k(t)S\Big(\sqrt{\mu_k}T\Big) = S\Big(\sqrt{\mu_k}(T-t)\Big)u_{0k} + S\Big(\sqrt{\mu_k}t\Big)u_{Tk}, \tag{17}$$

$$S\Big(\sqrt{\mu}t\Big) = \frac{1}{2\sqrt{\mu}}\Big(\exp(\sqrt{\mu}t) - \exp(-\sqrt{\mu}t)\Big); \qquad Bx_k = \mu_k x_k.$$

Hence the cause of ill-posedness of the Dirichlet problem may be the following: $\exists_k\ S(\sqrt{\mu_k}T) = 0$, or $\exists\{k_j\}\ S(\sqrt{\mu_{k_j}}T) \longrightarrow 0$, $j \to \infty$. If for some u_0, u_T the solution of (16) exists and, for any k, $S(\sqrt{\mu_k}T) \neq 0$, i.e., $\mu_k \neq -(\pi k/T)^2$, then the solution is of the form

$$u(t) = \sum_{k=1}^{\infty}\Big[\frac{S(\sqrt{\mu_k}(T-t))}{S(\sqrt{\mu_k}T)}u_{0k} + \frac{S(\sqrt{\mu_k}t)}{S(\sqrt{\mu_k}T)}u_{Tk}\Big]x_k.$$

The attempt to apply the quasireversion method to ill-posed problem (16) leads to the Dirichlet problem for an equation with a small parameter of the following type:

$$u''_{\varepsilon,\delta}(t) = (B + \varepsilon B^2)u_{\varepsilon,\delta}(t), \quad 0 \le t \le T,$$

$$u_{\varepsilon,\delta}(0) = u_0^{\delta}, \qquad u_{\varepsilon,\delta}(T) = u_T^{\delta}. \tag{18}$$

However for the real parameter $\varepsilon \to 0$ we may have $\mu_k - \varepsilon\mu_k^2 = -(\pi k/T)^2$, i.e., $S(\sqrt{\mu_k - \varepsilon\mu_k^2}T) = 0$. Therefore problem (18) in general cannot serve for the regularization of (16).

The attempt to apply the ABC method leads to the same equation as in (16) and the irregular boundary conditions with $d_{12} + d_{34} \neq 0$, where d_{ij} are the minors of the boundary conditions matrix (see(Mel'nikova, 1988)). The simpliest of this problems are

$$u''_{\varepsilon,\delta}(t) = Bu_{\varepsilon,\delta}(t), \quad 0 \le t \le T,$$

$$u_{\varepsilon,\delta}(0) + \varepsilon u'_{\varepsilon,\delta}(T) = u_0^{\delta}, \qquad u_{\varepsilon,\delta}(T) = u_T^{\delta}, \tag{19}$$

or

$$u''_{\varepsilon,\delta}(t) = Bu_{\varepsilon\delta}(t), \qquad 0 \le t \le T, \qquad u_{\varepsilon,\delta}(0) = u_0^{\delta}, \qquad \varepsilon u'_{\varepsilon,\delta}(0) + u_{\varepsilon,\delta}(T) = u_T^{\delta}. \tag{20}$$

However in the general case as $\varepsilon \to 0$ the necessary condition of well-posedness

$$\Lambda \cap \{\mu_k\} = \varnothing$$

is not fulfilled for (19) (see (Mel'nikova,1988)). Here Λ is the set of solutions of the scalar eigenvalue problem, corresponding to (19)

$$y''(t) = \lambda y(t), \qquad y(0) + \varepsilon y'(T) = y(T) = 0.$$

It is equal to the set of solutions of the equation

$$\exp(\sqrt{\lambda}T) - \exp(-\sqrt{\lambda}T) = 2\varepsilon\sqrt{\lambda}.$$

Therefore, problem (19) can not serve for the regularization of (16) too. Similar reasoning can be carried out for (17). Thus in the regularization of the Dirichlet problem with a real parameter the duality with respect to the Cauchy problem takes place. In (Filinkov, 1989) the possibility of the regularization of the ill-posed Dirichlet problem with a complex parameter was shown.

It is interesting to compare the "differential" methods we have used researched with the well studied variational methods of regularization of the first kind equations. The techniques of Hilbert spaces allows to do it for the considered problems.

Conditions on the operator provide the well-posedness of the inverse Cauchy problem. Writing down the solution of the inverse Cauchy problem we obtain problem (1) in the form of the first kind equation:

$$K_1 u \equiv \sum_{k=1}^{\infty} \tilde{u}_k \exp(-\lambda_k T)\tilde{x}_k = f,$$

where $u = u(T)$, $\tilde{u}_k = (u, x_k)$, $f = u_0$.

Similarly using the well-posedness of the Dirichlet problem for the second order equation with an operator $B \in H_c$ we can reduce the ill-posed Cauchy problem for this equation to the operator equation

$$K_2 u \equiv \sum_{k=1}^{\infty} \frac{\tilde{u}_k x_k}{S(\sqrt{\mu}_k T)} = f; \qquad f = \sum_{k=1}^{\infty} f_k x_k = \sum_{k=1}^{\infty} \Big[\frac{c(\sqrt{\mu}_k T)}{S(\sqrt{\mu}_k T)} u_{0k} + u_{1k}\Big] x_k,$$

where $u = u(T)$, $c(\sqrt{\mu}T) = \frac{1}{2}(\exp(\sqrt{\mu}T) + \exp(-\sqrt{\mu}T))$.

For the ill-posed Dirichlet problem with $B \in -H_c$ we obtain similarly the equation

$$K_3 u \equiv \sum_{k=1}^{\infty} \frac{\tilde{u}_k x_k}{S(\sqrt{\mu}_k T)^{-1}} = f; \qquad f = \sum_{k=1}^{\infty} \Big[S(\sqrt{\mu}_k t) u_{T_k} + S(\sqrt{\mu}_k (T-t)) u_{0k}\Big] x_k,$$

where $u = u(T)$, $0 \le t \le T$.

The regularization by the quasisolution method of the equation $Ku = f_\delta$ in H on the compact

$$\mathfrak{M} = \Big\{ u = \sum_{k=1}^{\infty} \tilde{u}_k x_k \in H \colon \sum_{k=1}^{\infty} \gamma_k^2 \tilde{u}_k^2 \le M^2, \quad 0 < \gamma_k \longrightarrow \infty, \quad k \to \infty \Big\}$$

reduces to finding the extremal function of the problem

$$\inf_{u \in \mathfrak{M}} \|Ku - f_\delta\|.$$

If $K = K_1$ and $\gamma_k = \lambda_k$, then the compact $\mathfrak{M}$ coincides with the introduced above class $\mathfrak{M}_1$; if $\gamma_k = \exp(\lambda_k \tau)$, then it coincides with $\mathfrak{M}_2$. Solving this conditional extremum problem by the Lagrange factors method on the set $\mathfrak{M}$ for the operator K_1, we obtain

$$u_{\lambda,\delta}(T) = \sum_{k=1}^{\infty} \exp(\lambda_k T)\Big[1 + \lambda \gamma_k^2 \exp(2\lambda_k T)\Big]^{-1} f_k^\delta x_k; \qquad f_k^\delta = u_{0k}^\delta.$$

For K_2, K_3 see (Mel'nikova, 1991).
From the obtained expressions for solutions we see that the ABC method coincides or is close to the quasisolution method under the coordination of parameters. For example for problem (1) with $\varepsilon = \lambda$, $\gamma_k = \exp(\lambda_k T)$, $\tau = 3T$ these methods coincide. As we know (Ivanov et al, 1978) under the coordination of parameters the quasisolution method coincides with two other variational methods – that of Tichonov and the discrepancy method and is optimal. Therefore the ABC method which coincides with it on the correspondent correctness class is optimal too.

REFERENCES

Al'shanskii, M., Mel'nikova, I. and Freyberg, A. (1991). The numerical realization of the "boundary value problems" method. *Zh. Vychisl. Mat. i Mat. Fiz.* **31**, 929–933 (in Russian).

Carasik, B. (1978). *The Conditionally Well-Posed Cauchy Problem for Differential-Operator Equations and Applications.* Candidate Thesis, Baku (in Russian).

Carasso, A. (1976). Error bounds in the final value problem for the heat equation. *SIAM J. Math. Anal.* **7**, 195–199.

Filinkov, A. (1989). *Well-posedness and regularization of abstract boundary value problems.* Doctor Thesis, Ural State Univ. Press, Sverdlovsk (in Russian).

Freyberg, A. (1987). The regularization of the ill-posed Cauchy problem for the differential-operator equation of the first order. *Deposited at VINITI.* 586–B88 (in Russian).

Lattes, R. and Lions, J.- L. (1967). *Methode de Quasi-Reversibilite et Applications.* Dunod, Paris.

Lavrent'ev, M. (1973). *Conditionally Well-Posed Problems for Differential Quations.* Novosibirsk Univ. Press, Novosibirsk (in Russian).

Liskovets, O. (1968). The numerical solution of some ill-posed problems by quasisolution method. *Differentsial'nye Uravneniya.* **4**, 735–742 (in Russian).

Ivanov, V. (1972). The quasireversion problem for the heat equation in the uniform metrics. *Differentsial'nye Uravneniya.* **8**, 652–658 (in Russian).

Ivanov, V., Vasin, V. and Tanana, V. (1978). *The Theory of Linear Ill-Posed Problems and Applications.* Nauka, Moscow (in Russian).

Mel'nikova, I. (1981). The solution of the inverse Cauchy problem by quasireversion method. *Izv. Vyssh. Uchebn. Zaved. Mat.* **6**, 36–38.

Mel'nikova, I. and Kudrjavtsev, A. (1983). The regularization of the Cauchy problem for the first order equation in a Banach space with the help of the boundary value problems. In: *Theory and Methods of Solving Ill-Posed Problems.* Novosibirsk Univ. Press, Novosibirsk, pp. 260–261 (in Russian).

Mel'nikova, I. and Freyberg, A. (1985). On the regularization of the boundary value problem for the vibration equation. *Zh. Vychisl. Mat. i Mat. Fiz.* **5**, 783–789 (in Russian).

Mel'nikova, I. and Freyberg, A. (1986). The regularization of the illposed Cauchy problem for the second order equation in a Banach space. *Differentsial'nye Uravneniya* **24**, 1332–1338 (in Russian).

Mel'nikova, I. (1988). *Well-Posed and Ill-Posed Problems for Differential-Operator Equations in Banach Spaces.* Doctor Thesis, Novosibirsk Univ. Press, Novosibirsk (in Russian).

Mel'nikova, I. (1991). Regularization of ill-posed differential problems. *Sibirsk. Mat. Zh.* **2**, 125–134 (in Russian).

Papi, F. (1979). On the stability of the Dirichlet problem for the vibrating string equation. *Ann. scuola norm. super. Piza Sci. fis. e mat.* **6**, 719–728.

Travis, C. and Webb, G. (1978). Cosine families and abstract nonlinear second order differential equations. *Acta Math. Acad. Scient. Hung.* **32**, 75–96.

Ill-Posed Problems in Natural Sciences, pp. 104 – 107
A. Tikhonov (Ed.)
1992 VSP/TVP

MODIFIED TIKHONOV REGULARIZATION FOR NONLINEAR ILL-POSED PROBLEMS LEADING TO OPTIMAL CONVERGENCE RATES

A. NEUBAUER
Institut für Mathematik, Universität Linz, A-4040 Linz, Austria

ABSTRACT

We present a variant of Tikhonov regularization for nonlinear ill-posed problems that always leads to optimal convergence rates in Hilbert scales and that yields the rate $O(\delta^{\frac{1}{2}})$ in Hilbert spaces, where δ is the noise level of the data, if the exact solution is smooth enough. The regularization parameter in this variant can be chosen a priori and independently from the smoothness of the exact solution.

We study nonlinear ill-posed problems of the form

$$F(x) = y_0, \tag{1}$$

where F: $D(F) \subset X \to Y$ is a nonlinear operator between Hilbert spaces X and Y. The norms in X and Y will be denoted by $\|\cdot\|_X$ and $\|\cdot\|_Y$, respectively.
By ill-posedness of problem (1) we mean that the solutions do not depend continuously on the data. It was shown in (Engl et al., 1989) that Tikhonov regularization, i.e., the approximation of problem (1) by a solution of the minimization problem

$$\|F(x) - y_\delta\|_Y^2 + \alpha\|x - x^*\|_X^2 \to \text{ min over } D(F), \tag{2}$$

where $\alpha > 0$ is a small parameter and $y_\delta \in Y$ are the available noisy data with

$$\|y_\delta - y_0\|_Y \leq \delta, \tag{3}$$

is a stable method to solve nonlinear ill-posed problems, if F is cotinuous and weakly (sequentially) closed in X. In (Engl et al., 1989; Neubauer, 1988) conditions are given that guarantee for the regularized solutions in X the convergence rates $O(\delta^{\frac{1}{2}})$ and $O(\delta^{\frac{2}{3}})$, respectively, if the regularization parameter $\alpha(\delta)$ is chosen appropriately.
As in the linear case, the a priori choice of this parameter depends on the smoothness of the exact solution. But, in general, this information is not available. Therefore, in practice the problem is, how to choose the regularization parameter,

so that optimal rates are obtained without needing the knowledge of the smoothness of the exact solution. In the linear case, this problem was solved for Tikhonov regularization by certain a posteriori parameter selection methods (Gfrerer, 1987; King and Neubauer, 1988; Neubauer, 1988) and carried over to general regularization methods, based on spectral theory, in (Engl et al., 1989; Groetsch and Neubauer, 1989).

For nonlinear problems, it was shown in (Engl et al., 1989) that the rate $O(\delta^{\frac{1}{2}})$ is obtained, if the necessary conditions on the exact solution are satisfied and if $\alpha(\delta)$ can be chosen according to Morozov's discrepancy principle (Morozov, 1966), i.e., $\alpha = \alpha(\delta)$ solves

$$\|F(x_\alpha^\delta) - y_\delta\|_Y = \delta. \tag{4}$$

Unfortunately, for a nonlinear ill-posed problem, (4) has only a solution under rather restrictive assumptions on the regularized solutions (Kravaris and Seinfeld, 1985). However, even if (4) has a solution, one has to solve two nonlinear problems simultaneously, namely (2) and (4).

So the question is: Is it possible to calculate a regularized solution to problem (1) by solving just one nonlinear minimization problem, so that the optimal rates are obtained without using the knowledge of the smoothness of the exact solution?

The answer is "yes" for regularization in Hilbert scales (Neubauer, 1991).

In Hilbert scales, the regularization term $\|x - x^*\|_X^2$ in (2) is replaced by $\|x - x^*\|_s^2$, where $\|x\|_s := \|L^s x\|_X$, L is a densely defined, self-adjoint, strictly positive operator on X and X_s is the completion of $\cap_{k=0}^{\infty} D(L^k)$ with respect to the norm $\|\cdot\|_s$ (cf. (Krein and Petunin, 1966) for details on Hilbert scales). For linear problems this approach was developed in (Natterer, 1984; Neubauer, 1988).

The significant advantages of this approach are that by replacing the regularizing norm $\|\cdot\|_X$ in (2) by a stronger one better rates with respect to the norm in X can be obtained, and that these rates are obtained by merely requiring smoothness conditions for the exact solution as in the linear case and not smoothness and closeness conditions that are necessary for usual nonlinear Tikhonov regularization (cf. (Engl et al., 1989, Theorem 2.4) and Theorem 2 below).

One can show that under certain conditions the problem

$$\|x - x^*\|_s \to \min \text{ over } M_\delta := \left\{ x \in D(F) \cap X_s \,\middle|\, \|F(x) - y_\delta\|_Y \leq \delta \right\} \tag{5}$$

has a stable solution that converges with optimal rates. Since this problem is still complicated to solve, we concider the Lagrange multiplier approach for this problem, i.e., we look for a solution $\hat{x}_\beta^\delta$ of

$$\left(\|F(x) - y_\delta\|_Y - \delta \right)^2 + \beta \|x - x^*\|_s^2 \to \min \text{ over } D(F) \cap X_s. \tag{6}$$

We now state the main result of (Neubauer, 1991).

THEOREM 1. *Let* (3) *and the following assumptions hold:*

(i) *$D(F)$ is convex and F is continuous and Fréchet-differentiable on X and weakly (sequentialy) closed on X_s for some $s \geq 0$,*

(ii) *a unique* x^* *– MNS* x_0 *exists in* X_s *for data* $y_0 \in Y$, *i.e.,* $x^*, x_0 \in D(F) \cap X_s$, $F(x_0) = y_0$ *and* $\|x_0 - x^*\|_s = \min\left\{\|x - x^*\|_s \mid F(x) = y_0\right\}$ *and assume that* $x^* \neq x_0$,
(iii) *there exist* $\varepsilon > 0$ *and* $\rho > 0$ *such that* $\left\|F'(x_0) - F'(x)\right\|_Y \leq \rho \|x_0 - x\|_X$ *for all* $x \in D(F) \cap U_\varepsilon(x_0)$.
(iv) $\left\|F'(x_0)x\right\|_Y \sim \|x\|_{-a}$ *for all* $x \in X$ *for some* $a > 0$.
(v) $x_0 - x^* \in X_u$ *for some* $u \in [s, a + 2s]$, $s \geq a$.
Then (6) has a stable solution $\hat{x}_\beta^\delta$ *and for the choise* $\beta = O(\delta^2)$ *we obtain*

$$\|\hat{x}_\beta^\delta - x_0\|_X = O(\delta^{u/(a+u)}).$$

Note that now only one minimization problem has to be solved and that the parameter $\beta = \beta(\delta)$ can be chosen independently from the smoothness parameter u and the parameter a, which can be interpreted as a "degree of ill-posedness" of the linearized problem in x_0. Moreover, note that for a very smooth exact solution the rate can be arbitrarily close to $O(\delta)$, if the regularization norm is chosen strong enough, whereas for usual Tikhonov regularization in the Hilbert space X, $O(\delta^{\frac{2}{3}})$ is the best possible rate (Neubauer, 1989). For a numerical realization of this approach (see (Neubauer, 1991)).
The same variant can be used for regularization in Hilbert spaces to obtain the rate $O(\delta^{\frac{2}{3}})$. The proof of the next theorem is just a consequence of the proofs of (Engl et al., 1989, Theorem 2.4) and (Neubauer, 1991, Theorem 2.6).

THEOREM 2. *Let* (3) *and the following assumptions hold:*
(i) *D(F) is convex and F is continuous, weakly (sequentially) closed and Fréchet-differentiable on X,*
(ii) *an* x^* *–MNS* x_0 *exists for the data* $y_0 \in Y$, *i.e.,* $x^*, x_0 \in D(F) \cap X$, $F(x_0) = y_0$ *and* $\|x_0 - x^*\|_X = \min\left\{\|x - x^*\|_X \mid F(x) = y_0\right\}$ *and assume that* $F(x^*) \neq y_0$,
(iii) *there exist* $\varepsilon > 2\left\|x_0 - x^*\right\|_X$ *and* $\rho > 0$ *such that* $\|F'(x_0) - F'(x)\|_Y \leq \rho \|x_0 - x\|_X$ *for all* $x \in D(F) \cap U_\varepsilon(x_0)$,
(iv) $x_0 - x^* = F'(x_0)^* w$ *for some* $w \in Y$,
(v) $\rho \|w\| < 1$.
Then

$$\left(\|F(x) - y_\delta\|_Y - \delta\right)^2 + \beta \|x - x^*\|_X^2 \rightarrow \min \text{ over } D(F) \cap X$$

has a stable solution $\hat{x}_\beta^\delta$ *and for the choice* $\beta = O(\delta^2)$ *we obtain*

$$\|\hat{x}_\beta^\delta - x_0\|_X = O(\delta^{\frac{1}{2}}).$$

Condition **(v)** in Theorem 2 is the closeness assumption, which is not needed in Hilbert scales.
Since in the linear case this method is equivalent to Tikhonov regularization where the regularization parameter is chosen according to a modified discrepancy principle (see Remark 2.7 in (Neubauer, 1991)), and since the discrepancy principle

does not yield the optimal rate $O(\delta^{\frac{2}{3}})$ even if the exact solution is smooth enough (Groetsch, 1983), this approach will not yield the optimal rate $O(\delta^{\frac{2}{3}})$ for regularization in Hilbert spaces.

REFERENCES

Engl, H.W. and Gfrerer, H. (1988). A posteriory parameter choice for general regularization methods for solving linear ill-posed problems. *Appl. Numer. Math.* **4**, 395–417.

Engl, H.W., Kunisch, K., and Neubauer, A. (1989). Convergence rates for Tikhonov regularization of nonlinear ill-posed problems. *Inverse Problems* **5**, 523–540.

Gfrerer, H. (1987). An a posteriori parameter choice for ordinary and iterated Tikhonov regularization of ill-posed problems leading to optimal convergence rates. *Math. Comp.* **49**, 507–522, S5–S12.

Groetsch, C.W. (1983). Comments on Morozov's discrepancy principle. In: *Improperly Posed Problems and Their Numerical Treatment,* Ed. by G. Hämmerlin and K.H. Hoffmann. Birkhäuser, Basel, pp. 97–104.

Groetsch, C.W. and Neubauer, A. (1989). Regularization of ill-posed problems: Optimal parameter choice in finite dimensions. *J. Approx. Theory* **56**, 184–200.

King, J.T. and Neubauer, A. (1988). A variant of finite-dimensional Tikhonov regularization with a-posteriori parameter choice. *Computing J.* **40**, 91–109.

Kravaris, C. and Seinfeld, J.H. (1985). Identification of parameters in distributed parameter systems by regularization. *SIAM J. Control Optimization* **23**, 217–241.

Krein, S.G. and Petunin, J.I. (1966). Scales of Banach spaces. *Russian Math. Surveys* **21**, 85–160.

Morozov, V.A. (1966). On the solution of functional equations by the method of regularization, *Soviet Math. Dokl.* **7**, 414–417.

Natterer, F. (1984). Error bounds for Tikhonov regularization in Hilbert scales. *Appl. Analysis* **18**, 29–37.

Neubauer, A. (1988). An a posteriori parameter choice for Tikhonov–Regularization in Hilbert scales leading to optimal convergence rates. *SIAM J. Numer. Anal.* **25**, 1313–1326.

Neubauer, A. (1989). Tikhonov regularization for nonlinear ill-posed problems: optimal convergence rates and finite-dimensional approximation. *Inverse Problems* **5**, 541–557.

Neubauer, A. (1991). Tikhonov regularization of nonlinear ill-posed problems in Hilbert scales. *Appl. Analysis* (to appear).

Ill-Posed Problems in Natural Sciences, pp. 108 – 117
A. Tikhonov (Ed.)
1992 VSP/TVP

ON DYNAMICAL RESTORATION OF PARAMETERS OF ELLIPTIC SYSTEMS

Yu.S. OSIPOV and A.I. KOROTKII
Institute of Mathematics and Mechanics, Ural Branch of Russia Academy of Sciences, Kovalevskoi str., 16, 620219 Sverdlovsk, Russia

ABSTRACT

The problem under consideration consists in reconstruction (restoration) of parameters of an elliptic system based on the results of approximate measurement of its state. The problem belongs to the class of inverse problems actively studied at present and has many applications (see, e.g., (Tikhonov and Arsenin, 1977; Tikhonov et al., 1990; Ivanov et al., 1978; Lavrent'ev et al., 1980; Latters and Lions, 1967; Aliphanov et al., 1988; Anikonov, 1978; Bukhgeim, 1983; Glasko, 1984; Krutko, 1988; Morozov, 1974; Prilepko, 1973; Kunisch and White, 1987) and bibliography given there). This problem is usually ill-posed (Tikhonov and Arsenin, 1977; Tikhonov et al., 1990; Ivanov et al., 1978; Lavrentev et al., 1980).

To solve the problem, the modification of dynamical approximation method suggested by A.V. Kryazhimskii and Yu.S. Osipov (Osipov and Kryazhimskii, 1983, 1983a) is applied. The method is based on the ideas of the theory of positional control (Krasovskii and Subbotin, 1987; Kurzhanskii, 1977) and of the theory of ill-posed problems (Tikhonov and Arsenin, 1977; Tikhonov et al., 1990; Ivanov et al., 1978; Lavrentev et al., 1980). The essence of modification of the method is the following. An appropriate controlled dynamical system called a model is constructed, the model acts simultaneously with the initial elliptic system whose parameter is to be reconstructed. The model is controlled positionally (by feedback): at each time instant the control is formed on the basis of the state of the model and measurements of the state of the system realized up to this time instant. It turns out that, for a sufficiently wide class of inverse problems, a control low for the model can be chosen in such a way that the integral mean value of the control realization with respect to time variable approximates (reconstructs) the unknown parameter of the system in the desired sense. The suggested parameters reconstruction algorithm is regularizing and stable with respect to the errors of measurements. The algorithm can restor the desired parameters in real time. The model and the control low for the model are imaginary mathematical constructions, but they can be physicaly realized by a computer. It seems therefore that the suggested solution method is constructive. It is shown that the initial inverse problem is reduced to a direct problem of the control theory for a dynamical model.

The present paper continues investigations (Osipov and Kryazhimskii, 1983, 1983a; Kim et al., 1990; Maksimov, 1988).

1. RECONSTRUCTION OF AN OPERATOR VALUE

Let us introduce notation and state the problem. Let V, H be real Hilbert spaces with scalar products $(\cdot,\cdot)_V$, $(\cdot,\cdot)_H$ and corresponding norms $|\cdot|_V$, $|\cdot|_H$. Let V^* and H^* be the spaces dual to V and H respectively. Assume that V is densely and continuously imbedded into H. The spaces H and H^* are identified.
Let an operator $A: V \to V^*$ be given. Assume that A is linear continuous and symmetric. Denote by $a(\cdot,\cdot)$ the bilinear form on V corresponding to the operator A. It is assumed that for some number $\lambda > 0$ the coercivity condition is fulfiled

$$a(w,w) \geq \lambda |w|_V^2, \quad w \in V.$$

Consider an elliptic system (Lions, 1971, pp.52–53)

$$Ay = u, \quad u \in H, \quad y \in V. \tag{1.1}$$

In order to emphasize that y is dependent on u it is also written $y = y_u$ when necessary. The solution y will also be called a state of the system (1.1).
The considered problem is the following. The parameter u in (1.1) is unknown. It is known only that it belongs to A convex bounded closed set $P \subset H$. It is necessary to construct an approximation u_h of the parameter u based on an approximation $\xi \in H$ to the solution y of the system (1.1) with error not larger than h:

$$|\xi - y|_H \leq h. \tag{1.2}$$

The approximation u_h should satisfy the condition

$$|u_h - u|_H \longrightarrow 0 \quad \text{as} \quad h \to 0. \tag{1.3}$$

Condition (1.3) should be valid for all approximations ξ satisfying (1.2). The problem can be formulated now like this.

PROBLEM 1.1. *Construct an operator $D_1 : (0,\infty) \times H \to P$ with the property*

$$\sup_{\xi \in \Xi} |D_1(h,\xi) - u|_H \longrightarrow 0 \quad \text{as} \quad h \to 0, \tag{1.4}$$

where Ξ is the set of all elements $\xi \in H$ satisfying (1.2).

An operator D_1 satisfying property (1.4) will be called a reconstruction algorithm. The element $u_h = D_1(h,\xi)$ will be taken as for an approximation to the unknown parameter u.
Let us construct an algorithm D_1. Fix an $h > 0$ and a $\xi \in H$ which satisfy (1.2). Consider an argument $t \in T$ (imaginery time), $T = [t_0, \vartheta]$, $-\infty < t_0 < \vartheta < +\infty$. Partition the interval T by points t_i:

$$t_0 < t_1 < \ldots < t_m = \vartheta, \quad m = m(h),$$

$$\Delta = \Delta(h) = \max_i (t_{i+1} - t_i) \leq C \cdot h, \quad C = \text{const} > 0,$$

where $C > 0$ is a fixed number. The function $m = m(h)$ is assumed to be fixed.

Consider the control system on T (called a model) which is described by the equation

$$\dot{z}(t) = Az(t) - v_h(t), \quad z(t_0) = \xi, \quad t_0 \le t \le \vartheta. \tag{1.5}$$

Here $v_h : T \to P$ is a piecewise constant control formed according to the rule:

$$v_h(t) = v_i^*, \quad t_i \le t < t_{i+1}, \quad i = 0, \dots, m-1,$$

where v_i^* is an element of P which corresponds to the minimal value of the functional $F(v)$ on the set P,

$$F(v) = 2(\xi - z(t_i), v)_H + \alpha(h)|v|_H^2,$$

$\alpha(h)$ is a non-negative function on $(0, \infty)$ with the property

$$\alpha(h) \longrightarrow 0, \quad \sqrt{h}/\alpha(h) \longrightarrow 0 \quad \text{as} \quad h \to 0$$

(according to the terminology of the ill-posed problems theory (Tikhonov and Arsenin, 1977; Tikhonov et al., 1990; Ivanov et al., 1978) $\alpha(h)$ can be called a regularization parameter). Note that these always exists the single minimizing element v_i^*. Note also that (see, e.g., Lions, 1971, pp. 108–110)

$$z \in \mathbf{L}^2(T; V) \cap \mathbf{C}(T; H), \quad \dot{z} \in \mathbf{L}^2(T; V^*).$$

Now D_1 can be defined as the rule which to every pair (h, ξ) puts into correspondence the element u_h:

$$u_h = \frac{1}{\vartheta - t_0} \int_T v_h(t)\, dt = D_1(h, \xi) \in P. \tag{1.6}$$

THEOREM 1.1. *The operator D_1 solves the problem* 1.1.

Proof. To prove the theorem it suffices to show that, whatever are the sequences $\{h_k\}$ $(h_k > 0, h_k \to 0)$, $\{\xi_k\}$ $(\xi_k \in H, |\xi_k - y|_H \le h_k)$, the following relation holds

$$u_k = D_1(h_k, \xi_k) \longrightarrow u \quad \text{in} \quad H.$$

Fix arbitrary sequences $\{h_k\}$ and $\{\xi_k\}$ which satisfy the above-mentioned properties and denote by $v_k = v_{h_k}$, $z_k(\cdot)$ the solution of the model (1.5) with $h = h_k$, $\xi = \xi_k$.

$$\Lambda_k(t) = |y - z_k(t)|_H^2 + \alpha(h_k) \int_{t_0}^{t} |v_k(\tau)|_H^2\, d\tau - \alpha(h_k) \int_{t_0}^{t} |u|_H^2\, d\tau.$$

Taking into account the rule according to which the function v_k is constructed, one can obtain the following estimate for the functional Λ_k:

$$\max_{t \in T} \Lambda_k(t) \le \gamma \sqrt{h_k},$$

where $\gamma > 0$ is some number which depends on the known data of the problem, but not on the number k. In particular, this estimate implies

$$\max_{t\in T} |y - z_k(t)|_H^2 \leq \gamma\sqrt{h_k} + 2\alpha(h_k)b(\vartheta - t_0), \tag{1.7}$$

$$b = \max_{v\in P} |v|_H^2,$$

$$\int_{t_0}^{\vartheta} |v_k(\tau)|_H^2 \, d\tau \leq \int_{t_0}^{\vartheta} |u|_H^2 \, d\tau + \gamma\sqrt{h_k}/\alpha(h_k). \tag{1.8}$$

Choose a subsequence $\{v_{k_j}\}$ of the sequence $\{v_k\}$, which provides the upper limit $\limsup_{k\to\infty} |u_k - u|$ and weakly converges in $\mathbf{L}^2 = \mathbf{L}^2(T; P)$ to some element $v_* \in \mathbf{L}^2$ (v_k and u_k are connected by (1.6)). From the properties of model (1.5) and estimate (1.7) it follows that for all $t \in T$

$$z_{k_j}(t) \longrightarrow y \text{ in } H, \quad z_{k_j}(t) \longrightarrow z_*(t) \text{ in } H,$$

where $z_*(\cdot)$ is the solution of model (1.5) with $v_h = v_*$, $\xi = y$. Hence, the uniqueness of the limit implies

$$z_*(t) = y, \quad t \in T,$$

and it follows that $v_*(t) = u$ for almost all $t \in T$.
From estimate (1.8) it follows that $\limsup_{j\to\infty} |v_{k_j}|_{\mathbf{L}^2}^2 \leq |u|_{\mathbf{L}^2}^2$. Furthermore from this inequality and the weak lower semicontinuity of the norm $|\cdot|_H$ we obtain

$$|v_{k_j}|_{\mathbf{L}^2} \longrightarrow |u|_{\mathbf{L}^2} = |v_*|_{\mathbf{L}^2},$$

and so

$$v_{k_j} \longrightarrow v_* \quad \text{in} \quad H,$$

$$u_{k_j} = \frac{1}{\vartheta - t_0} \int_T v_{k_j}(t)\, dt \longrightarrow u = \frac{1}{\vartheta - t_0} \int_T v_*(t)\, dt \quad \text{in} \quad H.$$

Consequently, $u_k \to u$ in H. The theorem is proved.

Remark 1.1. If the set M is a compact subset of P, then

$$\sup_{u\in M} \sup_{\xi\in\Xi} |D_1(h,\xi) - u|_H \longrightarrow 0 \quad \text{as} \quad h \to 0.$$

2. RECONSTRUCTION OF AN OPERATOR

Let U be a real Hilbert space with scalar product $(\cdot,\cdot)_U$ and the corresponding norm $|\cdot|_U$. Let an operator $A: V \to V^*$ be given which depends on the parameter u,

$$u \in P \subset U: \; A = A[u].$$

Assume that for every $u \in P$ the operator $A[u]$ is linear continuous and self-conjugate. Denote by $a(u;\cdot,\cdot)$ the bilinear form on V corresponding to the operator $A[u]$. It is assumed that for some numbers $\lambda_1 > 0, \lambda_2 > 0$ and any $u \in P$

$$\lambda_2 |w|_V^2 \geq a(u;w,w) \geq \lambda_1 |w|_V^2, \quad w \in V.$$

Let the element $f \in V^*$ be given. Consider an elliptic system (Lions, 1971, pp.52–53)

$$A[u]y = f, \quad y \in V. \tag{2.1}$$

The parameter u in (1.1) is unknown. It is known only that it belongs to a convex bounded closed set $P \subset U$. It is necessary to construct an approximation u_h to the parameter u based on the result $\xi \in V$ of the measurement the solution y of system (2.1) with error not larger than h

$$|\xi - y|_V \leq h. \tag{2.2}$$

The approximation u_h should satisfy the condition

$$|u_h - u|_U \longrightarrow 0 \quad \text{as} \quad h \to 0.$$

Let us stale the problem more precisely. It is well known that in specific problems the same solution $y = y_u$ of the system (2.1) can be generated by different values of the parameter u. Let P_* be the set of all elements of P which generate the same solution $y = y_u$ of system (2.1) as the parameter u does

$$P_* = \{v \in P : \ A[v]y_u = f\}.$$

For an element $v \in P$ we put

$$J(v) = \inf_{u \in P_*} |v - u|_U.$$

The initial problem can be formulated now like this.

PROBLEM 2.1. *Construct an operator* $D_2 : \ (0,\infty) \times V \to P$ *with the property*

$$\sup_{\xi \in \Xi} J(D_2(h,\xi)) \longrightarrow 0 \quad \text{as} \quad h \to 0, \tag{2.3}$$

where Ξ *is the set of all element* $\xi \in V$ *satisfying* (2.2).

An operator D_2 satisfying property (2.3) will be called, as above, a reconstruction algorithm. The element $u_h = D_2(h,\xi)$ will be taken as an approximation to the set P_*. The algorithm D_2 reconstructs the unknown parameter u if the set P_* consists of only one element u.
Below it is assumed that the operator $A[u]$ satisfies the following additional condition.

PROPERTY 2.1. *If a sequence $\{v_k\}$ weakly converges in $\mathbf{L}^2(T;P)$ to v_0, then for all $w \in V$ and $t \in T$ the sequence $\{\int_{t_0}^t A[v_k(\tau)]w\,d\tau\}$ converges in V^* to the element $\int_{t_0}^t A[v_0(\tau)]w\,d\tau$.*

This property implies, in particular, the weak compactness of P_* and the weak continuity of the function

$$(w_1, A[\cdot]w_2)_{V^*} : P \to (-\infty, +\infty)$$

for all fixed $w_1 \in V^*$, $w_2 \in V$.
Let us construct an algorithm D_2. Fix an $h > 0$ and a $\xi \in V$ which satisfy (2.2). Partition the interval T by points t_i, as was made in the first part. Consider a control system (model) on T

$$\dot{z}(t) = A[v_h(t)]\xi - f, \quad z(t_0) = \xi, \quad t_0 \le t \le \vartheta. \tag{2.4}$$

Here $v_h : T \to P$ is a piecewise constant control defined by

$$v_h(t) = v_i^*, \quad t_i \le t < t_{i+1}, \quad i = 0, \ldots, m-1,$$

where v_i^* is an element of P which corresponds to the minimal value of the functional $G(v)$ on the set P.

$$G(v) = 2(z(t_i) - \xi,\ A[v]\xi)_{V^*} + \alpha(h)|v|_U^2,$$

$\alpha(h)$ is a non-negative function on $(0, \infty)$ with the property:

$$\alpha(h) \longrightarrow 0, \quad h/\alpha(h) \longrightarrow 0 \quad \text{as} \quad h \to 0.$$

Note that there always exist the minimizing elements v_i^* and

$$z \in \mathbf{C}(T; V^*).$$

Operator D_2 now can be defined as the map

$$D_2(h, \xi) = u_h = \frac{1}{\vartheta - t_0} \int_T v_h(t)\,dt \in P.$$

THEOREM 2.1. *Let Property 2.1 hold. Then*
1)

$$\sup_{\xi \in \Xi} \inf_{u \in U_*} |v_h - u|_{\mathbf{L}^2} \longrightarrow 0 \quad \text{as} \quad h \to 0,$$

where $U_* = \{u \in \mathbf{L}^2 : u(t) \in P_*$ for almost all $t \in T\}$;
2) *if the set P_* is convex, then the operator D_2 solves the Problem 2.1.*

The proof of the theorem is similar to that of Theorem 1.1. It is based on the estimate

$$\max_{t \in T} \Lambda(t) \le \gamma h,$$

$$\Lambda(t) = |y - z(t)|_{V^*}^2 + \alpha(h) \int_{t_0}^{t} |v_h(\tau)|_U^2\,d\tau - \alpha(h) \int_{t_0}^{t} |u_*|_U^2\,d\tau,$$

where u_* is an element of P_* with minimal norm.

Remark 2.1. Let $P_*(u) = \{v \in P : A[v]y_u = f\}$,

$$Q(u) = \{v \in \mathbf{L}^2 : v(t) \in P_*(u), \quad \text{A.I.}, \ t \in T\},$$

$$\Xi(h,u) = \{\xi \in V : |\xi - y_u|_V \le h\}.$$

If M is a compact subset of P, then

$$\sup_{u\in M} \sup_{\xi\in\Xi(h,u)} \inf_{v\in Q(u)} |v_h - v|_{\mathbf{L}^2} \longrightarrow 0 \quad \text{as} \quad h \to 0.$$

If, in addition, the sets $P_*(u)$ are convex, then

$$\sup_{u\in M} \sup_{\xi\in\Xi(h,u)} \inf_{v\in P_*(u)} |u_h - v|_U \longrightarrow 0 \quad \text{as} \quad h \to 0.$$

Remark 2.2. The properties of the algorithm D_2 can be characterized by the following estimates:

$$\sup_{u\in P_*} \sup_{t\in T} \left(\int_{t_0}^{t} |v_h(\tau)|_U^2 \, d\tau - \int_{t_0}^{t} |u|_U^2 \, d\tau \right) \le \gamma_1 h/\alpha(h),$$

$$\sup_{u\in P_*} \sup_{t\in T} \left| \int_{t_0}^{t} A[v_h(\tau)]y \, d\tau - \int_{t_0}^{t} A[u]y \, d\tau \right|_{V^*} \le \gamma_2 \sqrt{h + \alpha(h)},$$

where $\gamma_1 > 0$ and $\gamma_2 > 0$ are constants depending only on the known data of the initial problem, but not on h. If

$$\left| \int_{t_0}^{\vartheta} A[v_h(\tau)]y \, d\tau - \int_{t_0}^{\vartheta} A[u]y \, d\tau \right|_{V^*} \ge \mu |v_h(\cdot) - u|_{\mathbf{L}^2},$$

where $\mu > 0$ is some constant (probably, dependent on y, but not on h), then

$$\mu |u_h - u|_U \le \sqrt{\vartheta - t_0} \gamma_2 \sqrt{h + \alpha(h)}.$$

Remark 2.3. When dealing with concrete problems of restoration of coefficients in elliptic systems, one can often take for U the space of functions defined on the corresponding domain, where the coefficients of the elliptic operator are given. If U can be continuously imbedded into the space of all continuous functions, then Theorem 3.1 implies the uniform convergence of the coefficients being restored.

Remark 2.4. In the case when the set P_* does not consist of one element the algorithm D_2 provides the convergence of the elements u_h to the elements of minimal norm of the set P_*. The set P_* is convex if the operator $A[u]$ is linear in u.

Example. Consider the elliptic system

$$\begin{cases} \frac{\partial}{\partial x_1}(u_1(x_1,x_2)\frac{\partial y}{\partial x_1}) + 2\frac{\partial^2 y}{\partial x_1 \partial x_2} + \frac{\partial}{\partial x_2}(u_2(x_1,x_2)\frac{\partial y}{\partial x_2}) = -1, \\ \Omega: \quad 0 < x_1 < 1, \quad 0 < x_2 < 1, \\ y(0,x_2) = 0 = y(1,x_2), \quad 0 \le x_2 \le 1, \\ y(x_1,0) = 0 = y(x_1,1), \quad 0 \le x_1 \le 1. \end{cases}$$

The unknown coefficients to be restored are

$$u_1 = 1 + x_2(1-x_2), \quad u_2 = 1 + x_1(1-x_1).$$

We take the approximation

$$\xi(x_1,x_2) = y(x_1,x_2) + \frac{h}{kn}\sin kx_1 \sin nx_2,$$

where k and n are some numbers, and we take the regularization parameter in the form $\alpha(h) = \sqrt{h}$. Let $U = \mathbf{W}_2^1(\Omega) \times \mathbf{W}_2^1(\Omega)$ ($\mathbf{W}_2^1(\Omega)$ is the Sobolev space),

$$P = \{(u_1,u_2) \in U : 1 \le u_1(x_1,x_2) \le 3, \quad 1 \le u_2(x_1,x_2) \le 3,$$

$$\left|\frac{\partial u_1(x_1,x_2)}{\partial x_1}, \frac{\partial u_1(x_1,x_2)}{\partial x_2}, \frac{\partial u_2(x_1,x_2)}{\partial x_1}, \frac{\partial u_2(x_1,x_2)}{\partial x_2}\right| \le 1, \quad \text{A.I.} \quad (x_1,x_2) \in \Omega\}.$$

The results of calculation are shown in the figures. The numbers 1,2,3 denote the results of restoration for $h = 0.5$, $h = 0.2$, $h = 0.05$ respectively. Fig. 1 shows the dynamics of discrepancy

$$r(t) = |u_1 - v_{1h}(t)|_{C(\Omega)} + |u_2 - v_{2h}(t)|_{C(\Omega)}, \quad t \in T = [0,1].$$

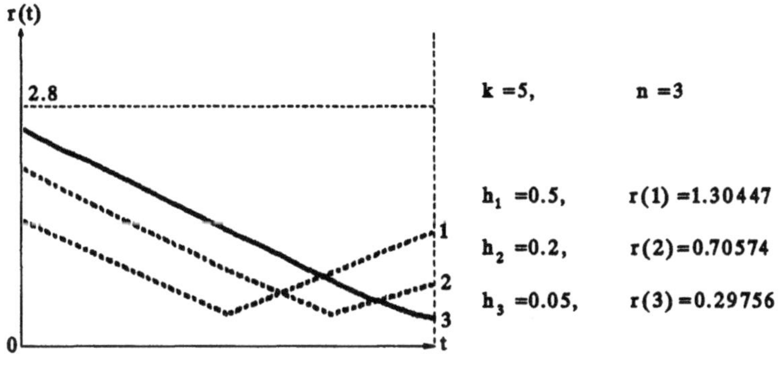

Fig. 1

The results of reconstruction of the coefficients u_1 and u_2 are depicted in Fig. 2 and Fig. 3 respectively, Fig. 2 shows the coefficient u_1 and its reconstructions

$u_{1h} = u_{1h}(0.476, x_2)$, Fig. 3 shows the coefficient u_2 and its reconstructions $u_{2h} = u_{2h}(x_1, 0.476)$, the numbers r_1, r_2, r_3 denote the corresponding discrepancies in the uniform norm.

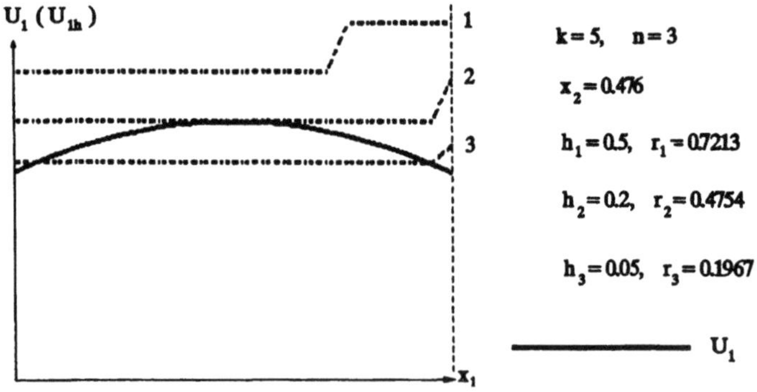

Fig. 2

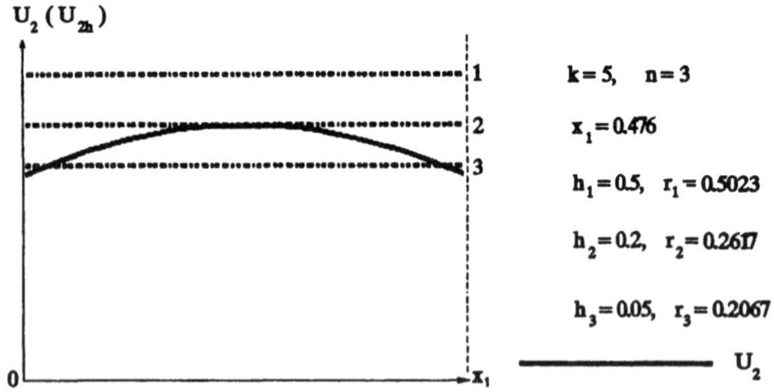

Fig. 3

REFERENCES

Aliphanov, O.M., Artyukhin, E.A., and Rumyantsev S.V. (1988). *Extremal Methods of Solution of Ill-Posed Problems.* Nauka, Moscow (in Russian).

Anikonov, Yu.E. (1978). *Some Methods of Investigation of Multi-Dimensional Inverse Problems.* Nauka, Novosibirsk (in Russian).

Bukhgeim, A.L. (1983). *Volterra Equations and Inverse Problems.* Nauka, Novosibirsk (in Russian).

Glasko, V.B. (1984). *Inverse Problems of the Mathematical Physics.* Moscow Univ. Press, Moscow (in Russian).

Ivanov, V.K., Vasin, V.V., and Tanana, V.P. (1978). *Theory of Linear Ill-Posed Problems and Applications*. Nauka, Moscow (in Russian).

Kim, A.V., Korotkii, A.I., and Osipov, Yu.S. (1990). Inverse problems of dynamics for parabolic system. *Appl. Math. Mech.* **54**, 754–759 (in Russian).

Krasovskii, N.N. and Subbotin, A.I. (1987). *Game-Theoretical Control Problems*. Springer-Verlag, New York.

Krutko, P.D. (1988). *Inverse Problems of Dynamics of Controlled Sustems*. Nauka, Moscow (in Russian).

Kryazhimskii, A.V. and Osipov, Yu.S. (1983). On modelling of control in a dynamical system. *Izv. Akad. Nauk SSSR, Ser. Tekhn. Kibern.* **2**, 51–60 (in Russian).

Kunisch, K. and White, L.N. (1987). Identifiability under approximation for an elliptic boundary value problem. *SIAM J. Control Optimization.* **25**, 279–297.

Kurzhanskii, A.B. (1977). *Control and Observation under Uncertainly*. Nauka, Moscow (in Russian).

Lattes, R. and Lions, J.-L. (1967). *Method of Quasi-Reversibility and Its Applications*. Dunod, Paris.

Lavrent'ev, M.M., Romanov, V.G., and Shishatskii, S.P. (1980). *Ill-Posed Problems of the Mathematical Physics and Analysis*. Nauka, Moscow (in Russian).

Lions, G.-L. (1971). *Optimal Control of Systems Described by Partial Differential Equations*. Springer-Verlag, Wiesbaden.

Maksimov, V.I. (1988). On dynamical modelling of the unknown disturbances in parabolic variational inequalities. *Appl. Math. Mech.* **52**, 743–750 (in Russian).

Morozov, V.A. (1974). *Regular Methods of Solution of Ill-Posed Problems*. Moscow Univ. Press, Moscow (in Russian).

Osipov, Yu.S. and Kryazhimskii, A.V. (1983). On dynamical solution of operator equations. *Soviet Math. Dokl.* **269**, 552–556 (in Russian).

Prilepko, A.I. (1973). Inverse problems of the potential theory. *Math. Notes.* **14**, 755–765 (in Russian).

Tikhonov, A.N. and Arsenin, V.Ya. (1977). *Solution of Ill-Posed Problems*. Wiley, New York.

Tikhonov, A.N., Goncharskii, A.V., Stepanov, V.V., and Yagola, A.G. (1990). *Numericale Methods of Solution of Ill-Posed Problems*. Nauka, Moscow (in Russian).

Ill-Posed Problems in Natural Sciences, pp. 118 – 133
A. Tikhonov (Ed.)
1992 VSP/TVP

EXTREME VALUE ESTIMATION, A METHOD FOR REGULARIZING ILL-POSED INVERSION PROBLEMS

P. PAATERO
University of Helsinki, Finland

ABSTRACT

The problem: A measurement is modeled by

$$Y_i = \sum K_{ij} C_j + E_i, \quad C_j \geq 0, \quad i = 1, \ldots, n, \quad j = 1, \ldots, m, \qquad (*)$$

where Y_i, K_{ij}, and $\sigma_i = \left(E(E_i^2)\right)^{0.5}$ are known, C_j and E_i are unknown. Usually $\text{Rank}(K) < \min(m, n)$, thus the mapping $C \Rightarrow Y$ is many-to-one if there are no errors E_i. The usual regularized solution (one-to-one $Y \Rightarrow C$) is satisfactory iff the true solution is *a priori* known to be the smoothest of all the least squares solutions of $(*)$. Define $\mathcal{D}$, the "set of acceptable solutions", to be the collection of all vectors C which solve $(*)$ "well enough" in the least squares sense. The set $\mathcal{D}$ is the primary solution according to the proposed method, one-to-many $Y \Rightarrow C$. Functionals of C are estimated by determining their supremum and infimum within $\mathcal{D}$, this is a Quadratic Programming problem. Estimation of systematically generated sequences of functionals results in "regularized" deconvolutes, having known statistical error limits with known confidence levels.
A set of artificial examples is generated by convoluting a line spectrum with a Gaussian kernel and by adding different amounts of noise. The numerical results by EVE demonstrate how the amount of information depends on noise level and how different functionals extract different information (cf. Heisenberg uncertainty principle). The method has been applied in different fields, especially in aerosol physics.

1. INTRODUCTION

Many problems lead to the constrained matrix equation

$$\mathbf{Y} = \mathbf{KC} + \mathbf{E}, \qquad (1)$$

$$C_j \geq 0, \quad j = 1, \ldots, m, \qquad (2)$$

where $\mathbf{C}$ is a vector of m unknowns c_j, Y_i $(i = 1, \ldots, n)$ are the observed or measured values, $\mathbf{K}$ is a known matrix of dimensions n, m, and E_i are unknown random "errors" with known standard deviations σ_i. The physicists would talk about "solving" equation (1), whereas the statisticians like to "estimate" the vector **C** in (1). (They would also call **C** a parameter and denote it by θ.) The distinction is not trivial: in the more difficult or ill-posed cases, we cannot simply "solve" equation (1) and have *the* solution **C**. We can only determine **C** to some accuracy, i.e. we get many possible solutions to (1). Thus the statistician way of thinking is highly recommended: we are only estimating the properties of the solutions, not "solving the equation".

Equation (1) arises when one is investigating an unknown function $C(t)$ by measurements. The argument t can be one-dimensional or many-dimensional. Usually it is a continuous variable (half-life, wave length, energy etc.) or a set of continuous variables (e.g. wave length and line width of spectral lines). In this work, only the one-dimesional case is considered. In some of the application-oriented references to EVE, two-dimensional problems are described. The unknown function $C(t)$ can be continuous or consist of "infinitely sharp" delta lines or be a combination of both. The experimental or measured function Y is an observed quantity (voltage, current, etc.) which depends on some experimental condition v (time, setting of the instrument, etc.), $Y = Y(v)$.

The dependence of $Y(v)$ on $C(t)$ is described by an integral equation. Equation (1) results when this integral equation is discretized so that only sets of discrete values of the physical parameter t and of the experimental variable v are considered. Throughout this work we assume that this discretization is dense enough. Note that the variable t does not appear explicitly in (1). However, each column of **K** corresponds to a value of t which is denoted by t_j. Thus we call t a *parameter* of (1).

The essential feature of the proposed new method is the following: a single *best fit* is not considered to be a useful *solution*. Instead, the collection of all the fits which produce a *good fit* is *the solution*. The realization of this idea leads to the following steps:

1. First, a best fit $\mathbf{C}_0$ is formed so that the feasibility constraint (2) is satisfied.

2. A criterion is selected for defining which fits to the measured data are considered good enough (not too much worse than the best fit). For each quantity to be estimated, the largest and smallest values of the quantity are found which correspond to good enough fits. These values form the confidence interval for the quantity. The "best fit" values for the quantities are ignored.

3. It is recognized that individual values of the function $C(t)$ usually cannot be determined because of the ill-posedness of the problem. Estimable quantities are represented by integrals,

$$f(C(t)) \doteq \int_a^b F(t)C(t)dt. \tag{3}$$

The program estimates such functionals f by determining their confidence intervals. There are ready-made methods for generating the coefficients of often-used functionals. Also, the user may specify functionals which are specific to any given problem. If the method would be integrated to some instrument control computer, then it would have built-in functionals specific to that instrument and the user might be quite unaware of the presence of any functionals.

4. Sometimes, the unknown function is represented by a set of user-specified known basis functions:

$$C(t) = \sum_k B_k Z_k(t), \tag{4}$$

$$B_k \geq 0, \quad k = 1, \ldots, h. \tag{5}$$

This arrangement enforces extra regularization, sometimes allowing individual values of $C(t)$ to be determined with meaningful accuracy.

2. PREVIOUS WORK

2.1. *Best Fit Methods*

In ill-posed curve fitting or inversion problems the best fit method is often augmented by a regularizing or penalty function which tends to enforce some desired property of the solution. Typical are smoothness and minimum norm requirements. Theoreticians do like regularization methods because there is a proven optimality property: if the magnitude of noise converges to zero, then the regularized result converges towards the correct value — except if some of the Singular Values of the Kernel matrix are zero! In real world situations the magnitude of errors is fixed. Then very little is known of the properties of the regularization methods. It is known that the regularized results are biassed. In many cases, the bias is so small that it does not matter. If the true unknown function is smooth enough, then regularization works. It is impossible to analyze a function which mostly consists of sharp peaks or to investigate the smoothness of a function by regularization techniques.
Another weakness of the best-fit methods is the lack of information about other possible solutions in addition to the best fit. Some curve fitting methods produce coarse estimates of covariance matrices for the results. However, the concept of a covariance matrix gets useless if some component of the solution can be either nonzero or constrained to zero, i.e., if there are alternative solutions with different (numbers of) free components.

2.2. *Bayes Methods*

Bayes methods (MB) are based on the assumption that the unknown $\mathbf{C}$ is a random vector with a known prior probability distribution. It has been suggested that the non-negativity criteria could be taken into account in the prior distribution of $\mathbf{C}$. There is a very deep problem here. The Bayes approach leads to rules which produce correct answers with a known frequency, in 95% of all cases, say — provided that $\mathbf{C}$ is random. In our case, however, $\mathbf{C}$ is fixed and usually contains a large, unknown number of zero components. Then it is possible that a Bayes rule could be wrong in 100% of repeated measurements if the true value of $\mathbf{C}$ happens to be a difficult one.

2.3. *The Work of Replogle, Holcomb, and Burrus*

An early paper by the mentioned authors contains most of the basic ideas of the present work, see (Replogle et al., 1967). The paper is based on the fundamental idea that any quantity should be estimated by determining the largest and smallest values of this quantity which are compatible with the measured result. The term "set of acceptable solution" was used. This is also the basis of the present work. The basic mathematical equations of the present work can be found in the mentioned paper. However, in the paper, there is no systematic approach for utilizing the idea in practice and useful functionals are not investigated. Also, the paper was decades ahead of its time: these computations require a lot of number crunching, they were too expensive in 1967.

2.4. *Characterization by the Likelihood Function L*

Different methods can be characterized by comparing them in terms of the likelihood function $L(\mathbf{Y}, \mathbf{C})$.

The shape of L is assumed known, this is the same as saying that the distribution of the error vector $\mathbf{E}$ is known. If $\mathbf{C}$ is regarded known, then L would express the distribution of $\mathbf{Y}$. Then one would not ordinarily talk of $L(\mathbf{Y}, \mathbf{C})$ but of the conditional distribution $P(\mathbf{Y}|\mathbf{C})$. In inversion problems, $\mathbf{Y}$ is given. Then $L(\mathbf{Y}, \mathbf{C})$ gives the plausibility or "likelihood" of any possible $\mathbf{C}$ as being the cause of the observed $\mathbf{Y}$.

The best-fit methods are usually maximum-likelihood (ML) methods: they determine

$$\mathbf{C}_0 = \arg\max_{\mathbf{C}} \left(L(\mathbf{Y}, \mathbf{C})\right), \tag{6}$$

where $\mathbf{Y}$ is given. Bayes methods typically form moment integrals of L over all the space where the prior distribution is nonzero. The proposed method "EVE" investigates the region where L is "large enough" and finds out how far this region extends in the directions of the functionals to be estimated.

3. THE PROPOSED METHOD

This work concentrates on such cases of equation (1) where the constraints (2) are essential information, i.e., where a significant fraction of the values C_j are exactly zero, as well in the unknown $\mathbf{C}_c$ as in the best solution $\mathbf{C}_0$. In ill-posed problems it often happens that the zero-valued components are not the same in $\mathbf{C}_c$ and $\mathbf{C}_0$. In such cases, both ML and MB estimation principles are unsatisfactory in producing error estimates for the result.

The principle for estimating $\mathbf{C}$ is the following: The primary estimate for $\mathbf{C}_c$ (for the correct value of $\mathbf{C}$) is the "solution set" $\mathcal{D}$ of all such points $\mathbf{C}$ which solve (1) "well enough". This set is also called "the set of acceptable solutions". Different definitions of $\mathcal{D}$ are possible. The simplest one is based on the decrease of likelihood function:

$$\mathbf{C} \in \mathcal{D} \iff \left\{ \begin{aligned} L(\mathbf{Y},\mathbf{C}) &\geq \max_{\mathbf{C}} \left(L(\mathbf{Y},\mathbf{C})\right) - \Delta L \\ &= L(\mathbf{Y},\mathbf{C}_0) - \Delta L = L_{\min} \\ C_j &\geq 0, \quad j = 1,\ldots,m. \end{aligned} \right\}. \tag{7}$$

For reasons of efficiency it is necessary to use the Least Squares formulation:

$$\mathbf{C} \in \mathcal{D} \iff \left\{ \begin{aligned} &Q(\mathbf{C}) < Q(\mathbf{C}_0) + \Delta Q \\ &C_j \geq 0, \quad j = 1,\ldots,m. \end{aligned} \right\}, \tag{8}$$

where

$$Q(\mathbf{C}) = \sum_{i=1}^{n} \frac{1}{\sigma_i^2} \left(Y_i - \sum_{j=1}^{m} \mathbf{K}_{ij} C_j \right)^2 \quad \text{and} \tag{9}$$

$$\mathbf{C}_0 = \arg\min_{\mathbf{C}} \left(Q(\mathbf{C})\right). \tag{10}$$

The quantity ΔQ (or ΔL) is a user-specified parameter, it determines the size of the set $\mathcal{D}$: with a larger ΔQ, more values $\mathbf{C}$ would be included in $\mathcal{D}$, whereby also the statement $\mathbf{C}_c \in \mathcal{D}$ would be true with higher probability.
If measurement errors E_i are normally distributed, then the two definitions (7) and (8) are equivalent. Otherwise, the LS definition is inferior to the definition based on the Likelihood function. It seems that definitions based on the Likelihood function would lead to very inefficient programs. They have not been attempted in the present work. The definition in (8) is mostly used in the present work. We shall briefly mention other variants which are also based on the least squares formulation.
The whole set $\mathcal{D}$, consisting of many-dimensional values $\mathbf{C}$, is of no practical value as an estimator. In practice, one must limit oneself to estimating various scalar functionals $f(\mathbf{C})$. Examples of $f(\mathbf{C})$:

$$f(\mathbf{C}) = C_k, \qquad \text{the } k'\text{th component of } \mathbf{C}, \tag{11}$$

$$f(\mathbf{C}) = \sum_{j=1}^{m} C_j, \qquad \text{the sum of all components of } \mathbf{C}. \tag{12}$$

According to best-fit and ML estimations, the estimate of $f(\mathbf{C})$ would be $f(\mathbf{C})$, the value of f at the best-fit point $\mathbf{C}_0$. In the unconstrained case, the uncertainty of this estimate is sometimes estimated

$$\operatorname{var}(f) = \left| \left(\frac{\partial f}{\partial C_1} \cdots \frac{\partial f}{\partial C_n} \right) \operatorname{cov}(\mathbf{C}) \left(\frac{\partial f}{\partial C_1} \cdots \frac{\partial f}{\partial C_n} \right)^T \right|_{\mathbf{C}=\mathbf{C}_0}. \tag{13}$$

This is a measure of the curvature of the Likelihood function at maximum, it is not meaningful in constrained cases.

According to the proposed method, we estimate any functional $f(\mathbf{C})$ by forming a confidence interval, consisting of an upper limit $f_{\max}$ and a lower limit $f_{\min}$ for $f(\mathbf{C})$:

$$f_{\max} = \sup_{C \in \mathcal{D}} f(\mathbf{C}), \tag{14}$$

$$f_{\min} = \inf_{C \in \mathcal{D}} f(\mathbf{C}). \tag{15}$$

The reasoning is as follows: we assume that the set $\mathcal{D}$ is large enough so that it is "almost certain" that the correct unknown solution $\mathbf{C}_c$ is one of the vectors $\mathbf{C}$ in $\mathcal{D}$. Thus the value of the functional at the correct value, $f(\mathbf{C}_c)$, is one of the values which $f(\mathbf{C})$ has in $\mathcal{D}$. It follows that the extreme values of $f(\mathbf{C})$ in $\mathcal{D}$ bracket the true value $f(\mathbf{C}_c)$ from above and below "almost certainly". We suggest that the method might be called Extreme Value Estimation or EVE.
We are not estimating any "best values" or expectation values for $f(\mathbf{C})$. Of course, the center point of the confidence interval,

$$f_{\text{best}} = 0.5\ (f_{\max} + f_{\min}), \tag{16}$$

might be taken as the "best" value for f. Quoting best values is, however, dangerous as it might lead to computing error estimates of secondary results in an incorrect way: assume that confidence intervals have been determined for some quantities a and b: $t < a < u$ and $v < b < w$. This could also be stated as: $a = \big((t+u)/2\big) \pm \big((u-t)/2\big)$ and similarly for b. How should one determine the confidence interval for the sum $a+b$? The correct method is to use the original inequalities: $t+v < a+b < u+w$. It would be incorrect to combine the errors by taking square root of sum of squares: $(a+b) = \big((t+u+v+w)/2\big) \pm 0.5\sqrt{(u-t)^2 + (w-v)^2}$ is not correct!

4. DETAILS OF THE EXTREME VALUE ESTIMATION METHOD

4.1. *Well-posedness of the approach*

We are interested in such practical problems where (1) is extremely ill-posed: There is no inverse matrix $\mathbf{K}^{-1}$, and the vector $\mathbf{C}$ is not uniquely determined by $\mathbf{Y}$ through a best fit of (1).
Solutions of (1) are unstable (ill-posed) if the partial derivatives of the result (C_j or $f(\mathbf{C})$) with respect to the measured values Y_i can get large values. More precisely, any norm $|J|$ of the matrix or vector J, defined by

$$J_{ji} = \frac{\partial C_j}{\partial Y_i} \qquad \text{or} \qquad J_i = \frac{\partial f}{\partial Y_i} \tag{17}$$

is a measure of the stability or well-posedness. In this sense, the constrained best fit solution is unstable: if the Kernel matrix approaches singularity and if worst-case values are selected for the measured vector $\mathbf{Y}$, then the norm of J, as defined by (17), can get arbitrarily large values. In practice this is observed so that if one tries to fit too many values, the results start to scatter arbitrarily.

It appears that there is an upper limit M for the norms of the corresponding partial derivative vectors J when evaluated for the extreme values:

$$J_i = \frac{\partial f_{\max}}{\partial Y_i} \qquad \text{or} \qquad J_i = \frac{\partial f_{\min}}{\partial Y_i}, \tag{18}$$

and the limit M is independent of how singular the Kernel is and what values there are in the measured vector $\mathbf{Y}$. This indicates that estimation of the extreme values is more stable than estimating the best-fit solution. In practice this is observed so that the results do not become unstable if one attempts an analysis with "too many" unknowns. Unfortunately there is no formal proof for this observation yet.

4.2. *Practical realization*

The computational algorithm is reduced to simple quadratic programming if the following two conditions are valid:

$$\text{– distribution of } \mathbf{E} \text{ is multinormal;} \tag{19}$$

$$\text{– } f \text{ is a linear function of } \mathbf{C}\text{, i.e., } f = \sum_{j=1}^{m} F_j C_j. \tag{20}$$

In practice, the Least Squares formulation is accepted even without any knowledge of the shape of distribution of $\mathbf{E}$. Until now, all practical work with the method has been based on (19) and (20).

The computational task is thus

$$\max_{\mathbf{C}} \sum_{j=1}^{m} F_j C_j$$

constrained by

$$C_j \geq 0, \qquad j = 1, \ldots, m, \tag{21}$$

$$Q(\mathbf{C}) < Q(\mathbf{C}_0) + \Delta Q, \qquad \text{where} \tag{22}$$

$$\mathbf{C}_0 = \arg\min_{\mathbf{C}} \big(Q(\mathbf{C})\big). \tag{23}$$

This could be solved by standard programs for Quadratic Programming. However, there are special properties in this task: the Kernel is often of banded or nearly banded shape and many similar problems are often solved consecutively, one after another. In order to take advantage of these properties, a special algorithm was written, see (Paatero, 1990). Depending on the problem, this advantage could exceed the factor of ten in computational speed.

How large problems may be solved? Till now, the largest problem had $m = 500$, $n = 900$ on a VAX or 386 computer. On a PC/XT computer, the largest problem had $m = 160$, $n = 200$. The complete analysis, estimation of some 50 functionals, took approximately ten minutes on the PC/XT.

4.3. *Confidence levels*

The selection of the confidence parameter ΔQ is a compromise between two aspects: a small ΔQ causes more narrow confidence intervals $(f_{\max} - f_{\min})$, but at the expense of the risk of error

$$r_f = P\big(f_c \notin (f_{\min}, f_{\max})\big) \tag{24}$$

becoming large. Ideally, we would like to select, once and for all, such a ΔQ which gives an acceptable risk, say $r_f = 0.05$. Another risk is defined by

$$r_D = P\big(\mathbf{C}_c \notin \mathcal{D}\big). \tag{25}$$

If $\mathbf{C}_c \in \mathcal{D}$, then estimations of *all* functionals would give correct results for the data in question. It might be tempting to define $\mathcal{D}$ so that it gives a suitable value for r_D, say $r_D = 0.05$. This is, however, unnecessary conservative. It would mean that the risk of generating at least one wrong result in estimation of any number of arbitrary functionals would be r_D. Only a very small part of the total family of all possible functionals would be interesting for anybody. Thus the risk for an error in this interesting part would be much smaller than r_D, and correspondingly the stated confidence intervals would be longer than necessary. It seems that r_D is not a useful quantity.

When $\mathbf{K}$ and $\mathbf{C}$ have been specified, it is possible to determine ΔQ by Monte-Carlo simulation so that $\max_f(r_f)$ gets a desired value for a given set of functionals. Monte-Carlo studies have revealed, however, that the risk r_f depends on the true value $\mathbf{C}_c$, which is unknown in practice. Thus ΔQ must be selected according to a worst-case principle: it must give an acceptably small risk even for the most sensitive alternatives of $\mathbf{C}_c$, for all of the functionals which are of interest. For other possible values of $\mathbf{C}_c$, the risk will then be much smaller.

In practice, the situation is not quite so complicated. Doubling or halving the value of ΔQ changes the results only a little. The value $\Delta Q = 3$ is a good first guess, giving $\max_f(r_f) \approx 0.1$ in many cases. By a few Monte-Carlo runs (with different $\mathbf{C}_c$!) one gets a "feel" of a possible need to increase ΔQ. One should not decrease ΔQ much below 3, even if the M-C runs seem to indicate so, because the likely reason is that the sensitive values of $\mathbf{C}_c$ have been missed. Only if exact probabilistic statements about the solution are needed, then there is need to undertake a full investigation of the correct value of ΔQ.

It has been observed that the definition of $\mathcal{D}$ in (8) is unsatisfactory in such cases where one column of Kernel matrix can be approximated as a linear combination of several (more than four, say) other columns. Then a good $\mathcal{D}$ should contain solutions with significantly different numbers of nonzero components and with Q values differing from the best-fit value by a larger amount than allowed by (8).

5. FUNCTIONALS TO BE ESTIMATED

In this chapter, we consider mainly problems of type "deconvolution" or "inversion of Laplace transform". As in all data analysis, here are two different aspects in solving ill-posed problems.

- Exploratory phase. The structure of the unknown distribution is investigated: what parameters (width, location, etc.) may be determined, are there peaks or a continuous distribution, how many peaks, etc.
- Evaluation phase. Values for well-defined quantities or parameters are computed: width of a distribution, distance between two peaks, ratio of their magnitudes, etc.

Quite different functionals are estimated in these two phases. In the exploratory phase, sequences of functionals (denoted by f^k, $k = 1, \ldots, p$) are computed. A sequence of values t^k is set up (automatically or by the user) for defining the sequence of functionals. These values are denoted by superscripts in order to better differentiate them from the parameter values t_j corresponding to the matrix rows. The set $(t^k,\ k = 1, \ldots, p)$ can be the same set as $(t_j,\ j = 1, \ldots, m)$, or a subset, or completely different values can be used.
The most often used definitions are:

$$f^k(t^k) = \sum_{j:\ |t_j - t^k| \leq \omega/2} C_j \qquad \text{(the window profile);} \tag{26}$$

$$f^k(t^k) = \sum_{j:\ t_j \leq t^k} C_j \qquad \text{(the cumulative profile).} \tag{27}$$

In the first definition, ω is a user-specified parameter which defines the width of the acceptance window. By looking at these profiles, it is possible to get a general idea of the unknown. It is also possible to find out such quantitative results as

- the minimum number of peaks which is compatible with this result;
- the uncertainty in the "mass" contained in some specified part of the distribution;
- whether it is possible to divide the distribution in separate parts having well-defined masses.

In the evaluation phase the functionals are specific to the problem. Example: Suppose that the EVE method would be used in a routine medical diagnosis which is based on some multicompartment model. During the development of the diagnostic method, one would determine indicators which discriminate between healthy and patological cases, and find linear or linearized expressions for these indicator quantities. During routine work, one would simply compute confidence intervals for these indicators and compare these against known limiting values.

Analysis of spectral lines or "peaks"
A sequence of window estimations (26) gives rough information about the locations of lines or "peaks": if the min estimate is positive, then there "almost certainly" is a spectral line inside the window (there may be several lines, too). Better information about the location of the peak(s) is obtained by estimating the location μ of the center-of-mass of all material inside the window:

$$\mu(t^k, \omega) = \frac{\int_{|t - t^k| \leq \omega/2} tC(t)dt}{\int_{|t - t^k| \leq \omega/2} C(t)dt}. \tag{28}$$

This expression is nonlinear in the unknown C and it cannot be estimated as such. There exists a suitable linearized approximation which is used instead. This approximation will be described in future publications of the EVE method. Similarly, the width of the peak mass distribution inside the window is estimated by means of another linearized expression.

The width ω of the acceptance window influences the results of window estimation: if ω is large (approximately equal to the peak width), then the confidence intervals for the expressions in (26) and (28) are small because the negative correlation of components inside the window cancels most of the variation. However, there is larger risk that there would be more than one true peak inside the window whence the computed results would not correspond to any single peak.

Making ω small (less than half of peak width, say) decreases the risk that there are several peaks inside the window. On the other hand, the confidence intervals get larger. With very small ω the min estimates approach zero, thus loosing all significance. We see here an analogy to the Heisenberg uncertainty principle: it is impossible to have optimal accuracy simultaneously in the abscissa variable (a narrow window) and in the ordinate (precise determination of mass inside the window).

6. PRESENTATION OF RESULTS

A systematic sequence of estimations produces a large amount of information which should be conveyed to the reader of a report in an understandable way.

In (Replogle et al., 1967) the individual components C_j were estimated with a single confidence parameter. The results were shown so that the sequences of max estimates were displayed as one curve, and the min estimates as another curve in the same diagram. The space between the curves indicated the allowed values for C_j. The authors correctly remark on p. 319 that the space indicated by this diagram is too large. For example, the vector corresponding to all the max estimates does not belong to $\mathcal{D}$. This is due to the fact that each max estimate corresponds to a different solution of equations (1) and (2). This problem is in fact a problem of convention: traditionally, points joined by a curve are thought to belong together. The technique of the next paragraph should make it clearer that the individual estimations are independent of each other. The problem here is very deep, it is related to the basic uncertainty which makes it impossible to single out one solution vector $\mathbf{C}$ and forces us to be satisfied with a set of solutions.

We prefer to display the results of sequences of estimations so that each pair of max and min estimates is joined by a vertical line which thus forms an "error bar" for the quantity. It is useful to perform the estimations with two different values of ΔQ in order to get some information about the sensitivity of the results with respect to a change in ΔQ. The two max and two min estimates are displayed by means of a single error bar as follows: the ends of the bar indicate the wider confidence interval (larger ΔQ, higher confidence). The narrower interval (smaller ΔQ, lower confidence) is indicated by small marks drawn across the bar.

As the abscissa scale, the quantity t is normally used. Each "error bar" is drawn onto the location $t = t^k$, where t^k defines the functional. Thus in window estimation the bars are drawn onto the center points of the estimated windows.

The method is often tested so that artificial mesurements are analyzed. Then the correct solution is known. It is useful to indicate the known correct values of the estimated functionals in the result diagram whose main contents are the confidence intervals for the functionals. We prefer to indicate the correct values by drawing a continuous curve through them. If all estimations are correct then this curve passes through all confidence intervals.

7. PRACTICAL EXPERIENCE

The method has been tested on several artificially generated examples and it has been in routine use in several different fields. After all the programming errors had been found, the program has not failed to solve any estimation! The application of the "allowed subspaces" definition of $\mathcal{D}$ very rarely produces unrealistically short confidence intervals.

Some of the successful real applications are described in the cited references. In a few other real cases useful physical results could not be obtained. In each of these cases it was finally found out that the assumed Kernel shape was not correct, e.g. the Kernel was not Gaussian although the experimentalist had been assuming so. Either there was no other Kernel shape available or it became likely that the peak shape was varying in an uncontrollable way within the spectrum. Thus the deconvolution attempts with EVE had to be abandoned in these few cases.

8. EXAMPLES

Extensive testing with Gaussian deconvolution and with separation of sums of exponentials has been performed. A few results are shown in Figures 1, 2, and 3 for a Gaussian example with different noise levels. This is an artificial example where the shape and the width of the Kernel are assumed known. The true solution is the same in all the examples, it consists of five Gaussians whose locations and heights are tabulated in the figures. Pseudorandom errors of rms amplitudes 0.2%, 1%, and 5% of the maximum data value were added to the correct data, thus producing the "measurements" consisting of 90 values, shown in Figures 1, 2, and 3, respectively. (The same pseudorandom sequence was used in all three examples.)

The deconvolution problem was set up with a Kernel matrix of dimensions 90×80. The parameter values (= locations of Gaussians) for the matrix columns were $t_j = 6, 7, \ldots, 85$. In all the estimations, the sequence $t^k = 10.5, 11.5, \ldots, 79.5$ was used for defining the functionals to be estimated. The estimations were based on the definition of $\mathcal{D}$ in (8). Two sets of estimates were computed, with values $\Delta Q = 3$ and $\Delta Q = 4.5$, roughly corresponding to worst-case confidence levels 0.9 and 0.96.

The lowest subfigures show the original "measured" curve and the best fit to it. Next to it is the cumulative estimation, cf. (27): each "error bar" shows the amount of peaks which is present to the left of the error bar. Thus the leftmost error bar (at $t^1 = 10.5$) estimates the combined peak intensity in peak positions 6 to 10.

The third subfigures show window estimation with window width $\omega = 4$, cf. (26).

Each "error bar" shows the amount of peaks which is present within distance = 2 of the error bar. Thus the leftmost error bar (at $t^1 = 10.5$) estimates the combined peak intensity in peak positions $t_j = 9, 10, 11, 12$. Finally, the topmost subfigures show window estimation with a wider window, $\omega = 8$. Comparison of these two results shows that the wider window gives shorter confidence intervals. On the other hand, in some locations (especially near $t = 60$) the wider window includes two peaks, thus giving rise to a misleading maximum value.

There is a continuous curve (drawn with a thin dotted line) in all those subfigures where results of estimations are shown. This line has been traced through the correct values of the functionals to be estimated. It is instructive to compare this "correct line" with the computed confidence intervals. If the confidence level would be 100%, then we would expect that the correct line stays everywhere within the confidence belt, i.e. it would cross all the vertical lines which represent confidence intervals. It was earlier indicated that the confidence is approximately 96% for the longer confidence intervals. Thus we might expect that a few of the estimations would be violated in each diagram: the correct line would go outside of the confidence belt in a few places in each diagram. In reality, there are no violations in any of the shown diagrams. This unexpected small number is caused because the confidence level 96% is a worst case estimate for the most critical estimations. If the demonstration run would be repeated with other pseudorandom numbers there would eventually also be cases with violations. This was a demonstration case with a known solution. In a real unknown case, one could thus have rather high confidence that the unknown "correct curve" really goes through the confidence belt.

In all of the present results, the two estimates with $\Delta Q = 3$ and $\Delta Q = 4.5$ agree very closely, indicating that the selection of ΔQ is not too critical. In more ill-posed situations (e.g., when analysing overlapping exponentials) these two estimates would differ somewhat more.

It is often important to determine the smallest number of peaks which is compatible with the measured result. The cumulative diagram indicates in all three cases that five peaks are required for fitting the data: it is impossible to traverse the confidence belt with less than five sharp "jumps". However, with 5% noise level the S-shaped band at $t = 60$ almost can be passed with one sharp jump. This indicates that fitting with four peaks is almost possible for this data. The shape of the bend of the confidence band in a cumulative diagram gives an approximate indication about the precision with which the location of the corresponding peak could be determined. A more accurate determination would require the estimation of special peak functionals. This was not done in this demonstration example.

9. DIRECTIONS FOR FURTHER RESEARCH

Better formulations for the Set of Acceptable Solutions $\mathcal{D}$ would make the confidence intervals shorter without sacrificing the associated confidence levels. Theoretical understanding of the statistical distribution of the constrained optimum point $\mathbf{C}_0$ and of the optimum value Q_0 is needed. The key issue here is how to utilize the number of free components (number of components with $C_j > 0$) in defining ΔQ, the allowed increase of Q within $\mathcal{D}$. Many promising definitions lead to non-convex shapes of $\mathcal{D}$. It is difficult or impossible to formulate a fast reliable algorithm for locating the extremal point of a functional within such a set.

Extensive Monte Carlo testing might be the decisive justification for different strategies of defining the set of acceptable solutions. Risks for producing false confidence intervals (= true value outside of the computed interval) would be evaluated for the strategies. The figures of merit for any strategy would be the worst (largest) risk level among all the different true solutions and all the interesting functionals on one hand, and the "volume" of the confidence band on the other hand. Work is in progress in this direction.

The EVE method has been applied successfully to two-dimensional problems. A special case is a spectrum of peaks on a smoothly varying background, see (Tapper and Paatero, 1990). There are some extra complications in these problems. This work will be described in another paper.

In some situations the number of nonzero components is well known *a priori*. This information cannot be incorporated in the present algorithm of EVE. Thus the results of present EVE could be less precise than the results of a conventional curve fitting program which fits the correct number of components. Measurements of the decay of a mixture of radioactive substances are an example, provided that the presence of impurities is absolutely impossible and the number of radioactive species is known. In such a situation the present EVE is a good first step in the analysis, but final values might be computed with a curve fitting program. It would be interesting to write an algorithm for performing the EVE task for the situation where the number of nonzero components is fixed *a priori*. The opposite situation is quite common: curve fitting programs with a fixed number of components are used in situations where the true number of components is not known or not even definable. This is typical of biological multicompartment models: the biological process runs at slightly different speeds in different parts of an organ, thus creating a continuous spectrum of half-lives for the depletion of some substance. In such a case the fitting with a fixed number of components is not justifiable. The results of present EVE would be more reliable.

10. DISCUSSION

Superficially, the results of the proposed Extreme Value Estimation method appear similar to the results of traditional regularization methods: both produce diagrams which appear smoother than the true unknown curve. There is, however, a deep distinction: regularization does not accept unsmooth features in its solution; instead, it produces a smoothed curve which is known to differ from the true solution but which often is a good approximation of it. Regularization pretends that this result is representative of the true solution. Comparison of experimental results with theory may be difficult because the influence of regularization (bias) is difficult to estimate.

On the other hand, EVE accepts all shapes of curves in the set of acceptable solutions, it will never reject the true curve because of sharp detail. Meaningful results are obtained when *smooth functionals* are estimated within this set. There is no danger of over-regularization here, the results are always correct. The results may not always be useful, however: e.g., if the window was made too wide in a sequence of window estimations, then the results do not correspond to single peaks. Comparison of EVE results of window estimation with theory is straightforward, it requires that the theoretical values are first convoluted by the window functional. Correspondingly, the results of cumulative estimation should be compared with the integral of the theoretical curve. In both cases, a necessary condition for agreement is that thetheoretically computed curve runs inside the "confidence belt" of EVE "almost everywhere".

A special merit of EVE is that it produces reliable error limits for all estimated quantities. This has lead to disappointments: the experimentalist did not like the EVE results because they showed that his data were not as significant as he had been believing. On the other hand, the EVE results do not possess anything resembling a covariance matrix, there is no information available about their correlation or lack of it. Sometimes one wishes to compute secondary results, e.g., the sum of areas of slightly overlapping peaks. The best way, losing no information, is to go back to the original data and to estimate the desired quantity directly with EVE by using a suitable special-purpose functional. Note that a comparable loss of information occurs

with all traditional computational methods if covariance matrices are not used when secondary results are computed, and

with EVE if several results of a sequence of estimations are combined in a secondary computation.

In the exploratory data analysis, the EVE method is demanding in use: one has to understand how to read the result diagrams and how to set up meaningful analyses. On the other hand, in a well-defined field it is possible to set up analysis packages which run fully automatically. This has been done for determining aerosol size distributions. Experience tells that the EVE approach is very robust: there do not seem to be any such special cases which would cause the program to fail, e.g., to get stuck in a local optimum.

Acknowledgements

The Monte-Carlo testing of the Gaussian and exponential examples was made by Tatjana Podossenova, from Moscow University.

REFERENCES

Replogle, J., Holcomb, B.D., and Burrus, W.R. (1967). The use of mathematical programming for solving singular and poorly conditioned systems of equations. *J. Math. Anal. Appl.* **20**, 310–324.

Paatero, P. and Raunemaa, T., (1989). Analysis of CO_2 thermograms by the new extreme-value estimation deconvolution principle. *Aerosol Sci. and Tech.* **10**, 365–369.

Paatero, P., Raunemaa, T., and Dod, R. L., (1988). Composition characteristics of carbonaceous particle samples, analyzed by the EVE deconvolution method. *J. Aerosol Sci.* **7**, 1223–1226.

Tapper, U. and Paatero, P. (1990). Application of the EVE deconvolution method to the analysis of PIXE spectra. *Nucl. Instr. Meth.* **B49**, 132–137.

Paatero, P. (1990). The extreme value estimation deconvolution method with applications in

aerosol research. *Univ. of Helsinki, Report Series in Physics.* **HU-P-250**, 1–41.
Honkimaki, V. and Suortti, P. (1991). Whole-pattern fitting in energy dispersive powder diffraction (to appear).

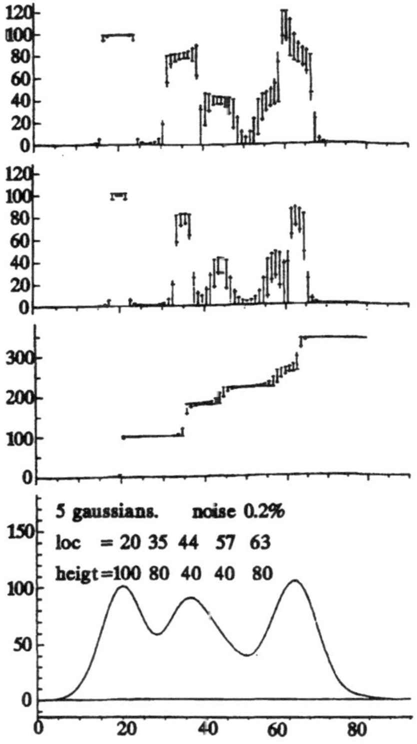

Fig. 1

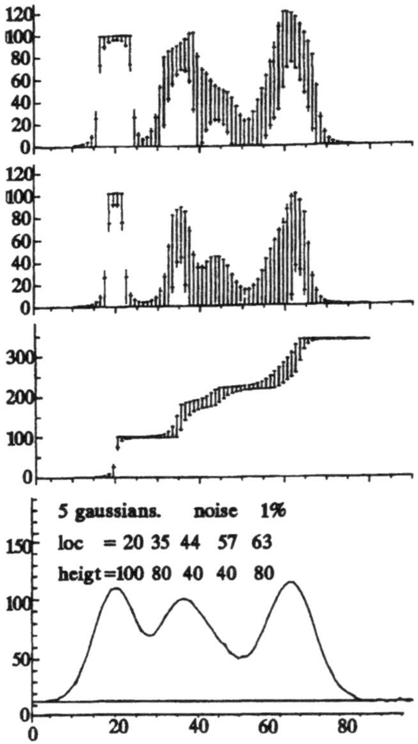

Fig. 2

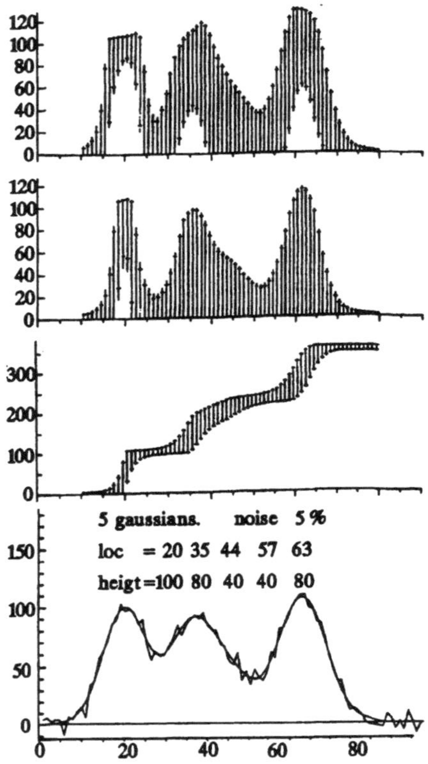

Fig. 3

Ill-Posed Problems in Natural Sciences, pp. 134 – 143
A. Tikhonov (Ed.)
1992 VSP/TVP

TWO ITERATIVE SCHEMES FOR SOLVING LINEAR NON-NECESSARILY WELL-POSED PROBLEMS IN BANACH SPACES

R. PLATO
Fachbereich Mathematik, Technische Universität Berlin, Strasse des 17. Juni, 135, D-1000 Berlin 12, Germany

ABSTRACT

For given Banach space X, bounded linear operator $A: X \to X$ and $y \in R(A)$, the range of A, we consider the implicit iterative method $(A + \mu I)x_{n+1} = \mu x_n + y$ and the explicit iterative method $x_{n+1} = (I - \mu A)x_n + y$ for solving the (non-necessarily well-posed) equation $Ax = y$; μ and I are a positive constant and the identity operator in X respectively. If in particular X is a Hilbert space and A is a selfadjoint positive semidefinite operator, then the convergence properties of these methods are well understood. In this paper these results are generalized to the case where the underlying space X is a Banach space.

1. INTRODUCTION

Let X be a Banach space, $A: X \to X$ be a bounded linear operator and $y \in X$. For solving the (non-necessarily well-posed) linear equation

$$Ax = y, \quad y \in R(A), \tag{1.1}$$

$R(A)$ denoting the range of A, we consider two stationary iterative methods. At first, we will investigate the implicit method

$$(A + \mu I)x_{n+1} = \mu x_n + y, \tag{1.2}$$

μ and I being a positive constant and the identity operator in X respectively.
Let us first consider the case where X is a Hilbert space and A is a selfadjoint positive semidefinite operater, i.e., $A = A^* \geq 0$. Then method (1.2) is an iterative variant of Lavrentiev's method. If in particular equation (1.1) results from a normalization of $Bx = a$, where $B: X \to Y$ is a bounded linear operator, Y is another Hilbert space, and $a \in R(B) + R(B)^\perp$, $R(B)^\perp$ denoting the orthogonal complement of the range of B, then method (1.2) is an iterative modification of Tikhonov's method which is considered, e.g., by Riley and by Fakeev (Riley, 1955; Fakeev, 1981) for the finite-dimensional case and more general by Kryanev

(Kryanev, 1973). The convergence properties are well-known, see, e.g., (Vainikko and Veretennikov, 1986) we always have $\|x_n - x_*\| \to 0 \ (n \to \infty)$, if $x_* \in \overline{R(A)} = N(A)^\perp$ solves (1.1), $x_0 = 0$ and $\mu > 0$ is fixed.

We will consider also the explicit iterative method (method of successive approximation)

$$x_{n+1} = (I - \mu A)x_n + y \tag{1.3}$$

for solving equation (1.1). If X is a Hilbert space and $A = A^* \geq 0$, then the convergence properties of this method are also well-known: $\|x_n - x_*\| \to 0 \ (n \to \infty)$, if $x_* \in \overline{R(A)} = N(A)^\perp$ solves (1.1), $x_0 = 0$ and $0 < \mu < \frac{2}{\|A\|}$, $\|A\| = \sup\{\|Ax\| : \|x\| = 1\}$, is fixed. For further details see, e.g., (Krasnoselskii, 1961) or (Krasnoselskii et al., 1972; Theorem 4.2.), for an early general analysis of this method or (Landweber, 1951; Fridman, 1958) who are considering Fredholm integral equations of the first kind.

In some papers Bakushinskii (see (Bakushinskii, 1973; 1976) or as general reference (Bakushinskii and Goncharskii, 1989)) considered the more general case, where the underlying space X is a Banach space, e.g., $X = \mathbf{L}^p(\Omega, \nu)$, where $1 \leq p \leq \infty$ and (Ω, ν) is some measure space, or $X = \mathbf{C}(D)$, where $\mathbf{C}(D)$ is the set of continuous functions on $D \subset \mathbf{R}^n$, supplied with the maximum norm. In these papers for the operator A the condition $(\lambda I - A)^{-1}$ is defined on X and fulfill $\|(\lambda I - A)^{-1}\| \leq \frac{C}{|\lambda|}$ for any λ outside a sector $S_\psi = \{re^{i\phi} : r \geq 0, \ |\phi| \leq \psi\}$ with some $0 < \psi < \frac{\pi}{2}$ is required. We also consider this more general case where the underlying space X is a Banach space while sometimes requiring weaker conditions on A.

Let us outline the structure of the paper. In §2 we provide a convergence theorem (Theorem 2.3) for the case of exact data and a general result on regularization methods (Theorem 2.5) for the case of noisy data. In the beginning of §3 we restrict ourselves to stationary iterative methods and obtain Corollaries 3.2 (exact data) and 3.3 (noisy data) while using a popular result of Katznelson and Tzafriri (Katznelson and Tzafriri, 1986). Finally, in items 3.1 and 3.2 we apply these results to algorithms (1.2) and (1.3). For example, assuming $I - \mu_0 A$ to be nonexpansive for some $\mu_0 > 0$ we present convergence results for these methods.

2. A GENERAL CLASS OF APPROXIMATION METHODS

Throughout this paper let X be an arbitrary Banach space. Our main aim of this section is to provide a convergence theorem (Theorem 2.3) for the case of exact data and a general result on regularization methods (Theorem 2.5) for the case of noisy data.We will do this in a more general approach which is motivated by the following observations.

Methods (1.2) and (1.3) are of the same structure: Multiplying equation (1.1) by some linear bounded one-to-one operator S: $X \to X$ and reformulating the resulting equation $SAx = Sy$ as fixed point equation $x = (I - SA)x + Sy$ yields the fixed point iteration $x_{n+1} = (I - SA)x_n + Sy$. Starting with $x_0 \in X$ and using the abbreviations

$$T = I - SA, \tag{2.1}$$

$$R_n = \sum_{j=0}^{n-1} T^j S, \tag{2.2}$$

the iterates x_n may be expressed as

$$x_n = x_0 - R_n(Ax_0 - y) = T^n x_0 + R_n y.$$

If $x_* \in X$ solves (1.1), then we have

$$x_n - x_* = H_n(x_0 - x_*), \tag{2.3}$$

where

$$H_n = I - R_n A = T^n. \tag{2.4}$$

In order that $\|x_n - x_*\| \to 0$ $(n \to \infty)$ for any $x_* \in X$ the operators H_n necessarily have to be uniformly bounded, or in other terminology, T has to be power bounded. As we are looking for sufficient conditions for convergence we therefore have to investigate the properties of H_n in more detail. We will do this in a more general context.

ASSUMPTION 2.1. *Let us assume the following for given operators* A, $H_n \in L_b(X) = \{\, B : X \to X : B$ *is linear and continuous* $\}$.
(1) *There is a constant* $M > 0$ *with* $\|H_n\| \le M$ *for any* $n \in \mathbf{N}$
(2) $H_n Az \to 0$ $(n \to \infty)$ *(strongly) for any* $z \in X$
(3) $H_n x \to x$ $(n \to \infty)$ *(strongly) for any* $x \in N(A)$.

Let us make a few comments on this assumption.
1. If $H_n = T^n$ as in (2.4), then condition (1) is equivalent to the power boundedness of the operator T, then condition (3) is fulfilled trivially and (2) means $(T^n - T^{n+1})z \to 0$ $(n \to \infty)$ (under the additional assumptions $SA = AS$ and $R(S) = X$) for any $z \in X$.
2. If the nullspace $N(A)$ of A is trivial, then condition (3) in the previous assumption and the following Theorem 2.2 hold trivially. This ergodic theorem relates $N(A)$ and $\overline{R(A)}$ in some sense and is not essential in providing convergence results for methods (1.2) and (1.3), therefore the reader may disregard it.

THEOREM 2.2. *Let Assumption* 2.1 *hold for arbitrary operators* A, $H_n \in L_b(X)$. *Then, for the operator* A, *the following necessarily hold:*
1) $\|x\| \le M\|x + y\|$ *for any* $x \in N(A)$, $y \in \overline{R(A)}$.
2) (*i*) $N(A) \cap \overline{R(A)} = \{0\}$,
(*ii*) $X_0 := N(A) \oplus \overline{R(A)}$ *is closed in* X.
3) *The projection* P: $X_0 \to X_0$ *onto* $\overline{R(A)}$ *with nullspace* $N(A)$ *is continuous*, $\|I - P\| \le M$.
4) $N(A) = N(A^2)$.
5) $X_0 = X$ *if and only if* $A(\overline{R(A)}) = \overline{R(A)}$.

Proof. We will give the proof for convenience of the reader despite the fact that this theorem is not really new. Let us first mention that Assumption 2.1 implies $H_n y \to 0$ $(n \to \infty)$ (strongly) for any $y \in \overline{R(A)}$.
1) For arbitrary $x \in N(A)$ and $y \in \overline{R(A)}$, we have

$$\|x\| = \|\lim_{n\to\infty} H_n(x+y)\| = \lim_{n\to\infty} \|H_n(x+y)\|$$
$$\leq \limsup_{n\to\infty} \|H_n\| \ \|x+y\| \leq M\|x+y\|.$$

2) (i) Let $x \in N(A) \cap \overline{R(A)}$, $-x \in \overline{R(A)}$ and 1) implies $\|x\| \leq M\|x+(-x)\| = 0$.
2) (ii) Suppose $x_n + y_n \to z \ (n \to \infty)$, $x_n \in N(A)$, $y_n \in \overline{R(A)}$. Then 1) implies

$$\|x_n - x_m\| \leq M\|x_n - x_m + y_n - y_m\| \to 0 \quad (n,\, m \to \infty),$$

hence $x_n \to x \ (n \to \infty)$ for some $x \in N(A)$ and then necessarily $y_n \to y \ (n \to \infty)$ for some $y \in \overline{R(A)}$. Hence X_0 is closed.
3) Let again $x \in N(A)$ and $y \in \overline{R(A)}$. Then $\|(I-P)(x+y)\| = \|x\| \leq M\|x+y\|$ which shows $\|I-P\| \leq M$.
4) We only have to show $N(A^2) \subset N(A)$. If $x \in N(A^2)$, then 2) (i) implies $Ax \in N(A) \cap \overline{R(A)} = \{0\}$.
5) Only "$\Leftarrow$" remains to show. For that let $x \in X$. Then $Ax \in R(A) = A(\overline{R(A)})$, hence $Ax = Az$ for some $z \in \overline{R(A)}$. From $x - z \in N(A)$ it follows $x \in X_0$.
Let us give two comments on Theorem 2.2. If $M = 1$ (as we will assume in the next section) in (1) of Assumption 2.1 and in 1) of Theorem 2.2, then $N(A)$ and $\overline{R(A)}$ are orthogonal subspaces in the sense of Birkhoff (Birkhoff, 1935) or James (James, 1947). Furthermore,

$$\|H_n x - (I-P)x\| \to 0 \quad (n \to \infty) \quad \text{for any } x \in X_0. \tag{2.5}$$

The following theorem supplies a convergence result for a general class of algorithms.

THEOREM 2.3. *Let A, $R_n \in L_b(X)$, $n \in \mathbf{N}$, such that the operators $H_n = I - R_n A$ fulfill Assumption 2.1. Let $X_0 := N(A) \oplus \overline{R(A)}$. Then for arbitrary right-hand side $y \in R(A)$ and starting vector $x_0 \in X$ such that $r_0 = Ax_0 - y \in A(X_0) = A(\overline{R(A)})$ we have $\|x_n - x_*\| \to 0$ as $n \to \infty$, where x_* is the unique solution of equation* (1.1) *with $x_0 - x_* \in \overline{R(A)}$ and*

$$x_n = x_0 - R_n(Ax_0 - y).$$

In order to substitute the condition $r_0 \in A(X_0)$ in Theorem 2.3 by the more convenient condition $r_0 \in R(A)$ (which is permitted if $X_0 = X$, see 5) in Theorem 2.2) we give the following lemma.

LEMMA 2.4. *Let H_n and A are commuting operators which fulfill Assumption 2.1 and the condition $R(I - H_n) \in \overline{R(A)}$ for each $n \in \mathbf{N}$. Then the following hold.*
1) Let $x \in X$. Then $(H_n x)_{n \in N'}$ converges weakly for some infinite $N' \subset \mathbf{N}$ if and only if $(H_n x)_{n \in \mathbf{N}}$ converges strongly if and only if $x \in X_0$.
2) If X is a reflexive Banach space, then $X_0 = X$.

Proof. 1) If $x \in X_0$, then $(H_n x)_{n\in\mathbf{N}}$ clearly converges as well strongly as weakly. Conversely, let us suppose that $H_n x \rightharpoonup z \ (n \to \infty,\ n \in N')$ for some $z \in X$ and some $N' \subset \mathbf{N}$. Then

$$0 \leftarrow H_n Ax = AH_n x \rightharpoonup Az \ \ (n \to \infty,\ n \in N'),$$

hence $z \in N(A)$. Since $(I-H_n)x \in \overline{R(A)}$ and $(I-H_n)x \rightharpoonup x - z \quad (n \to \infty)$ we have $x - z \in \overline{R(A)}$ and $x \in X_0$.
2) Let $x \in X$. Since $(H_n x)_{n\in\mathbf{N}}$ is bounded in X, we have $H_n x \rightharpoonup z \quad (n \to \infty,\ n \in N')$ for some $z \in X$ and some $N' \subset \mathbf{N}$ and the proof is finished.
Theorems 2.2 and 2.3 and Lemma 2.4 are similar to results of Shaw (Shaw, 1989). This author consider also unbounded operators A in Lebesgue spaces and characterizes the well-posedness of equation (1.1) by properties of H_n, see (Shaw, 1990, 1991). Next we investigate the case of noisy data. The following theorem, which supplies an a priori stopping rule, is a simple consequence of Theorem 2.2 and the triangle inequality.

THEOREM 2.5. *Let $A,\ R_n \in L_b(X),\quad n \in \mathbf{N}$, such that the operators $H_n = I - R_n A$ fulfill Assumption 2.1 und furthermore $\|R_n\| \le Cn$ for any $n \in \mathbf{N}$ and some constant $C > 0$. For arbitrary right-hand side $y \in R(A)$ and starting vector $x_0 \in X$ such that $r_0 = Ax_0 - y \in A(X_0) = A(\overline{R(A)})$ let*

$$x_{n,\delta} = x_0 - R_n(Ax_0 - y_\delta),$$

where $y_\delta \in X,\quad \|y - y_\delta\| \le \delta$. If $n_\delta \in \mathbf{N}$ is chosen such that

$$n_\delta \delta \to 0 \quad and \quad n_\delta \to \infty \quad as \quad \delta \to 0,$$

then we have $\|x_{n_\delta,\delta} - x_*\| \to 0$ as $\delta \to 0$, where x_* is the unique solution of equation (1.1) with $x_0 - x_* \in \overline{R(A)}$.

In the next section the results of §2 will be applied to some classes of operators.

3. APPLICATIONS TO TWO CLASSES OF OPERATORS

We now consider two classes of operators $A \in L_b(X)$ for which the implicit method (1.2) or the explicit method (1.3) are applicable. For that we need the following theorem on power bounded operators T which spectrum $\sigma(T)$ necessarily is a subset of the unit disc $B = \{\lambda \in \mathbf{C}\colon |\lambda| \le 1\}$. This theorem gives a sufficient condition for

$$\|T^n(I-T)x\| \to 0 \ \ (n \to \infty) \quad \text{for any} \quad x \in X. \tag{3.1}$$

For the relevance of (3.1) see (2.1)–(2.4), Assumption 2.1 and the first statement thereafter.

THEOREM 3.1 (Katznelson and Tzafriri). *Let X be a Banach space and $T \in L_b(X)$ be an arbitrary power bounded operator, i.e., there is a $M > 0$ with $\|T^n\| \le M$ for any $n \in \mathbf{N}$. Then $\|T^n - T^{n+1}\| \to 0\ (n \to \infty)$ if and only if $\sigma(T) \cap \{\lambda \in \mathbf{C} : |\lambda| = 1\} \subset \{1\}$.*

This theorem is due to (Katznelson and Tzafriri, 1986). Theorems of this type may also be found in papers (Allan et al., 1987; Esterle et al., 1990).
Before continuing we refer to two recent papers. In (Schock and Vu Quoc Phong) a theorem on power bounded operators T is proved which provides $\|T^n x\| \to 0\ (n \to \infty)$ for any $x \in X$ under other (and in some sense weaker) assumptions on the

spectrum of T and therefore is also very useful in the analysis of iterative methods for solving ill-posed problems. The second paper is (Nevanlinna, 1991) in which Theorem 3.1 is used in connexion with ill-posed problems. We also apply Theorem 3.1 in order to present a convergence result on stationary iterative method. This corollary is also a consequence of formulas (2.1)–(2.4) and Theorem 2.2.

COROLLARY 3.2. *Let $A, S \in L_b(X)$ be commuting operators, such that S is a bijection, $T = I - SA$ is power bounded and $\sigma(T) \cap \{\lambda \in \mathbf{C}\colon |\lambda| = 1\} \subset \{1\}$. Then for arbitrary right-hand side $y \in R(A)$ and starting vector $x_0 \in X$ such that $r_0 = Ax_0 - y \in A(\overline{R(A)})$ we have $\|x_n - x_*\| \to 0$ as $n \to \infty$, where x_* is the unique solution of equation (1.1) with $x_0 - x_* \in \overline{R(A)}$ and*

$$x_{n+1} = (I - SA)x_n + Sy.$$

If $\|T^n\| \leq M$ for any $n \in \mathbf{N}$, then (see (2.1)–(2.4)) $\|R_n\| \leq M\|S\|n$, hence under the assumptions of Corollary 3.2 we obtain the following corollary which is an application of Theorem 2.5 to iterative methods.

COROLLARY 3.3. *Let $A, S \in L_b(X)$ be commuting operators, such that S is a bijection, $T = I - SA$ is power bounded and $\sigma(T) \cap \{\lambda \in \mathbf{C}\colon |\lambda| = 1\} \subset \{1\}$. Then for arbitrary right-hand side $y \in R(A)$ and starting vector $x_0 \in X$ such that $r_0 = Ax_0 - y \in A(\overline{R(A)})$ let*

$$x_{n+1,\delta} = (I - SA)x_{n,\delta} + Sy_\delta,$$

where $y_\delta \in X$, $\|y - y_\delta\| \leq \delta$. If $n_\delta \in \mathbf{N}$ is chosen such that

$$n_\delta \delta \to 0 \quad \text{and} \quad n_\delta \to \infty \quad \text{as} \quad \delta \to 0,$$

then we have $\|x_{n_\delta,\delta} - x_*\| \to 0$ as $\delta \to 0$, where x_* is the unique solution of equation (1.1) with $x_0 - x_* \in \overline{R(A)}$.

If X is a reflexive Banach space then we may substitute the condition $r_0 \in A(X_0)$ in Corollaries 3.2 and 3.3 by the condition $r_0 \in R(A)$ since Lemma 2.4 is applicable. In the following two items these corollaries will be applied to two classes of operators A.

3.1. Accretive operators and the implicit method

We now apply Corollary 3.2 (convergence result in the case of exact data) and Corollary 3.3 (regularization in the case of noisy data) to accretive operators and the implicit method.

Definition 3.4. We call $A \in L_b(X)$ accretive if, for any $x \in X$ and $\mu \geq 0$, we have $\|x\| \leq \|(I + \mu A)x\|$.

If $A \in L_b(X)$ is accretive, then $\rho(-A) \supset (0, \infty)$, $\rho(-A)$ denoting the resolvent set of $-A$. This follows from the fact that $\lambda \in \mathbf{R}$, $\lambda > \|A\|$ implies $\lambda \in \rho(-A)$ and then for any $\mu \geq 0$ with $|\mu - \lambda| < \lambda$ we have $|\mu - \lambda| < \|(\lambda + A)^{-1}\|^{-1}$ and hence $\mu \in \rho(-A)$.

Let us mention that if X is a Hilbert space and $A \in L_b(X)$ is a selfadjoint positive semidefinite operator, then of course A is also accretive. Furthermore, we give some characterizations for arbitrary accretive operators in Banach spaces. For that let us introduce the following notations:
K is either **R** or **C**,
X^* is the dual space of X,
$[\cdot,\cdot] : X \times X^* \to \mathbf{K}$ is the corresponding bilinear form,
$(T(t))_{t\geq 0}$ is the C_0-semigroup with infinitesimal generator $-A$. Then we have A is accretive

$\Leftrightarrow$ $\rho(-A) \supset (0,\infty)$, $\quad \|(I+\mu A)^{-1}\| \leq 1 \quad$ for any $\quad \mu \geq 0$

$\Leftrightarrow$ $\rho(-A) \supset \{\lambda \in \mathbf{C}$: Re $\lambda > 0\}$ and $\|(\lambda I + A)^{-1}\| \leq \frac{1}{\text{Re}\,\lambda}$ for any $\lambda \in \mathbf{C}$ with Re $\lambda > 0$, i.e., A is m-accretive in the sense of Kato, (Kato, 1966)

$\Leftrightarrow$ (Pazy, 1983) for any $x \in X$ there is a $x^* \in F(x) = \{x^* \in X^*: [x, x^*] = \|x\|^2 = \|x^*\|^2\}$ such that $\text{Re}[Ax, x^*] \geq 0$

$\Leftrightarrow$ for any $x \in X$ and $x^* \in F(x)$ the estimate $\text{Re}[Ax, x^*] \geq 0$ holds

$\Leftrightarrow$ $-A$ is dissipative (Kato, 1966)

$\Leftrightarrow$ $I - A$ is pseudocontractive (co-accretive) (Browder, 1967)

$\Leftrightarrow$ $\|T(t)\| \leq 1$ for any $t \geq 0$.

For accretive operators A we consider the implicit method for which $S = (A+\mu I)^{-1}$ and hence $T = (I + \mu^{-1}A)^{-1}$ (for the notation see the beginning of §2). Here for any $t \geq 0$ we have $\sigma(I + tA) \subset \{ \lambda \in \mathbf{C} : \text{Re}\,\lambda \geq 1 \}$ and $\sigma(I + tA) \cap \{ \lambda \in \mathbf{C} : \text{Re}\,\lambda \geq 1 \} \subset \{ 1 \}$, hence $\sigma((I+tA)^{-1}) \subset \{ \lambda \in \mathbf{C}\text{: Re}\,\lambda \geq 0,\ |\lambda| < 1 \} \cup \{ 1 \}$. From Corollary 3.2 it follows

COROLLARY 3.5. *Let $A \in L_b(X)$ be accretive. Then, for $y \in R(A)$ and starting vector $x_0 \in X$ such that $Ax_0 - y \in A(\overline{R(A)})$, we have $\|x_n - x_*\| \to 0$ as $n \to \infty$, where x_* is the unique solution of equation* (1.1) *with $x_0 - x_* \in \overline{R(A)}$ and x_n is generated by the implicit method* (1.2).

At the end of this subsection we shortly consider regularization of equation (1.1) in the case of inexactly given right-hand side. Assuming A to be accretive we have $\|R_n\| \leq \mu^{-1}n$ (for the definition of R_n see (2.2) and the beginning of this subsection) and so we may apply the a priori stopping rule given in Corollary 3.3.

3.2. Nonexpansive operators and the explicit method

Once again we apply Corollary 3.2 (convergence result in the case of exact data) and Corollary 3.3 (regularization in the case of noisy data), this time to nonexpansive operators and the explicit method.

Definition 3.6. We call an operator $T \in L_b(X)$ nonexpansive if $\|T\| \leq 1$.
Nonexpansive operators are also called contractions. Let us again mention that if X is a Hilbert space and $A \in L_b(X)$ is a selfadjoint positive semidefinite operator, then $I - \mu A$ is nonexpansive for any $0 \leq \mu \leq 2/\|A\|$.

Example. Let l_∞ be the space of real bounded sequences $U = (U_k)_{k=-\infty}^{\infty}$, supplied with the maximum norm $\|U\|_\infty = \sup_k |U_k|$. Let S be the forward shift in l_∞ and $A = \frac{1}{h^2}(2I - S - S^{-1})$,

$$A = \frac{1}{h^2}\begin{pmatrix} \cdot & \cdot & \cdot & \cdot & \cdot & \cdot & \cdot \\ \cdot & \cdot & \cdot & \cdot & \cdot & \cdot & \cdot \\ \cdot & 2 & -1 & 0 & 0 & 0 & \cdot \\ \cdot & -1 & 2 & -1 & 0 & 0 & \cdot \\ \cdot & 0 & -1 & 2 & -1 & 0 & \cdot \\ \cdot & 0 & 0 & -1 & 2 & -1 & \cdot \\ \cdot & \cdot & \cdot & \cdot & \cdot & \cdot & \cdot \\ \cdot & \cdot & \cdot & \cdot & \cdot & \cdot & \cdot \end{pmatrix}$$

For any $0 < \mu \le \frac{1}{2}h^2$, the operator $I - \mu A$ is nonexpansive in l_∞ and also in other l_p-spaces. Equation $AU = F$ may be interpreted as a discretization of $-u''(x) = f(x)$, $x \in \mathbf{R}$, with central differences, and is itself an ill-posed equation, i.e., the solution U does not depend continuously on the right-hand side F even if it exists.

Example. Let us more generally consider an arbitrary bounded linear operator $A : l_\infty \to l_\infty$ which may be represented by an infinite matrix $(a_{i,j})_{i,j}$. Then it is not difficult to show that the operator $I - \mu A$ is nonexpansive for some $0 < \mu$ if and only if the logarithmic norm of $-A$

$$\mu_\infty(-A) = \sup_i\{-a_{i,i} + \sum_{j\neq i} |a_{i,j}|\}$$

is nonpositive. This indicates that the class of operators which are considered in this paper is rather small. Finally, after a more detailed examination of logarithmic norms one can show that $\mu_\infty(-A) \le 0$ again is equivalent to the case that A is accretive, i.e., for this example a conversion of the following Remark 3.7 holds.
Let us now consider again the case of general operators A. First we present a simple

Remark 3.7. If $A \in L_b(X)$ and for given $\mu_0 > 0$ the operator $I - \mu_0 A$ is nonexpansive, then for any $0 \le \mu \le \mu_0$ the operator $I - \mu A$ is nonexpansive and furthermore A is accretive.

Proof. If

$$T_\omega := I - \omega\mu_0 A, \tag{3.2}$$

then $T_\omega = (1-\omega)I + T_1$ and hence $\|T_\omega\| \le 1$ for $0 \le \omega \le 1$. Furthermore, for any $t > 0$ and $x \in X$ with $s = \frac{t}{\mu_0}$,the inequality

$$\|(I + tA)x\| = \|(1+s)x - s(I - \mu_0 A)x\| \ge (1+s)\|x\| - s\|(I - \mu_0 A)x\|$$

$$\ge (1+s)\|x\| - s\|x\| = \|x\|$$

holds. This argument can also be found in (Browder, 1967).

Therefore, if A has the property that $I - \mu_0 A$ is nonexpansive for some $\mu_0 > 0$, then we can apply the implicit method with an arbitrary $\mu > 0$ to solve (1.1). But now we are interested in the explicit method. Here $S = \mu I$ and $T = I - \mu A$ (for the notation see the beginning of §2). Choosing $0 < \mu < \mu_0$, defining $\omega := \mu/\mu_0$ (then $T = T_\omega$) and using (3.2) and $\sigma(T_1) = \sigma(I - \mu_0 A) \subset \{ \lambda \in \mathbf{C}: |\lambda| \leq 1 \}$, we see that $\sigma(T) = \sigma(T_\omega) = \sigma(I - \mu A) \subset 1 - \omega + \{ \lambda \in \mathbf{C}: |\lambda| \leq \omega \} \subset \{ \lambda \in \mathbf{C}: |\lambda| < 1\} \cup \{ 1 \}$, finally we have

COROLLARY 3.8. *Let $A \in L_b(X)$ and $\mu_0 > 0$ such that $I - \mu_0 A$ is nonexpansive. Then, for $y \in R(A)$ and starting vector $x_0 \in X$ such that $Ax_0 - y \in A(\overline{R(A)})$, we have $\|x_n - x_*\| \to 0$ as $n \to \infty$, where x_* is the unique solution of equation* (1.1) *with $x_0 - x_* \in \overline{R(A)}$ and x_n is generated by the explicit method* (1.3) *with $0 < \mu < \mu_0$.*

Finally, let us again consider the regularization of equation (1.1) in the case of inexactly given right-hand side. Under the given assumptions we have $\|R_n\| \leq \mu n$ (for the definition of R_n again see (2.2) and the beginning of this subsection) and so we may apply the a priori stopping rule given in Corollary 3.3.

REFERENCES

Allan, G.R., O'Farrell, A.G., and Ransford, J. (1987). A tauberian theorem arising in operator theory. *J. Funct. Anal.* **94**, 273–287.

Bakushinskii, A.B. (1973). The problem of constructing linear regularizing algorithms in Banach spaces. *USSR Comput. Math. and Math. Phys.* **13**, 261–270.

Bakushinskii, A.B. (1976). The stability and maximum domain of convergence of some regularizing algorithms. *USSR Comput. Math. and Math. Phys.* **16**, 219–224.

Bakushinskii, A.B. and Goncharskii, A.V. (1989). *Iterative Methods for Solving Incorrectly Posed Problems.* Nauka, Moskow (in Russian).

Birkhoff, G. (1935). Orthogonality in linear metric spaces. *Duke Math. J.* **1**, 169–172.

Browder, F.E. (1967). Nonlinear mappings of nonexpansive and accretive type in Banach spaces. *Bull. Amer. Math. Soc.* **73**, 875–882.

Esterle, J., Strouse, E., and Zouakia, F. (1990). Theorems of Katznelson–Tzafriri type for contractions. *J. Funct. Anal.* **94**, 273–287.

Fakeev, A.G. (1981). A class of iterative processes for solving degenerate systems of linear algebraic equations. *USSR Comput. Math. and Math. Phys.* **21**, 15–22.

Fridman, V.M. (1958). The method of successive approximation for the Fredholm integral equation of the first kind. *Russian Math. Surveys* **11** (in Russian).

James, R.C. (1947). Orthogonality and linear functionals in normed linear spaces. *Trans. Amer. Math. Soc.* **61**, 265–292.

Kato, T. (1966). *Perturbation Theory for Linear Operators.* Springer-Verlag, Berlin, Heidelberg, New York.

Katznelson, Y., and Tzafriri, L. (1986). On power bounded operators. *J. Funct. Anal.* **68**, 313–328.

Krasnoselskii, M.A. (1961). On solutions of equations with selfadjoint operators by the method of successive approximations. *Russian Math. Surveys* **15** (in Russian).

Krasnoselskii, M.A., Vainikko, G.M., Zabreiko, P.P., Rutitskii, Ya.B., and Stetsenko, V.Ya. (1972). *Approximate Solution of Operator Equations.* Wolters–Noordhoff Publishing, Groningen.

Kryanev, A.V. (1973). An iterative method for solving incorrectly posed problems. *USSR Comput. Math. and Math. Phys.* **14**, 24–35.

Landweber, L. (1951). An iteration formula for Fredholm integral equations of the first kind. *Amer. J. Math.* **73**, 615–624.

Nevanlinna, O. (1991). *Lectures on Iterative Methods: The Sublinear Case.* Preprint C8, Helsinki University of Technology, Helsinki.

Pazy, T. (1983). *Semigroups and Applications to Partial Differential Operators.* First Ed. Springer-Verlag, Berlin, Heidelberg, New York.

Riley, J.D. (1955). Solving systems of linear equations with a positive definite, symmetric, but possibly ill-conditioned matrix. *Math. Tables Aids Comp.* **9**, 96–101.

Schock, E. and Phong Vu Quoc. Regularization of ill-posed problems involving unbounded operators in Banach spaces. *Hokkaido Math. J.* (to appear).

Shaw, S. (1989). Mean ergodic theorems and linear functional equations. *J. Funct. Anal.* **87**, 428–441.

Shaw, S. (1990). Solvability of linear functional equations in Lebesgue spaces. *Publ. RIMS, Kyoto Univ.* **26**, 691–699.

Shaw, S. (1991). Uniform convergence of ergodic limits and approximate solutions. (to appear).

Vainikko, G.M. and Veretennikov, A.Y. (1986). *Iteration Procedures in Ill-Posed Problems.* Nauka, Moskow (in Russian).

Ill-Posed Problems in Natural Sciences, pp. 144 – 148
A. Tikhonov (Ed.)
1992 VSP/TVP

ON THE REGULARIZATION PARAMETER CHOICE FOR ILL-POSED PROBLEMS WITH A QUASISOLUTION

T. RAUS
Tartu University, Liivi 2–418, 202400 Tartu, Estonia

ABSTRACT

A class of regularization methods for solving ill-posed problems is considered in the case where the problem has only a quasisolution. For this class a priori and a posteriori regularization parameter choices, are proposed which provide convergence rates optimal with respect to data errors.

1. INTRODUCTION

Consider the operator equation

$$Au = f, \tag{1}$$

where $A \in \mathcal{L}(H, F)$ and H, F are Hilbert spaces. We assume that $Qf \in \underline{R(A)}$, where the range $R(A)$ is nonclosed and Q is the orthogonal projector onto $\overline{R(A)}$. If $q := \|f - Qf\| > 0$, then equation (1) has only a quasisolution (or quasisolutions if the kernel $N(A)$ is nontrivial). We also suppose that instead of f and A we know their approximations $f_\delta \in F$ and $A_\eta \in \mathcal{L}(H, F)$ such that $\|f_\delta - f\| \leq \delta$ and $\|A_\eta - A\| \leq \eta$.

2. A CLASS OF REGULARIZATION METHODS

To determine an approximation, we use the regularization methods which are generated by Borel measurable functions $g_r\colon [0, a] \to \mathbf{R}$, $r \geq 0$, $\|A_\eta\|^2 \leq a$ such that

$$\sup_{0 \leq \lambda \leq a} |g_r(\lambda)| \leq \gamma r, \; r \geq 0, \tag{2}$$

$$\sup_{0 \leq \lambda \leq a} \lambda^p |1 - \lambda g_r(\lambda)| \leq \gamma_p^{r-p}, \quad r > 0, \quad 0 \leq p \leq p_0 \tag{3}$$

where p_0, γ, and γ_p are positive constants (see (Vainikko and Veretennikov, 1986)). Let $u_0 \in H$ be an initial approximation. Then the approximation to the quasisolution u_* of (1), nearest to u_0, is given by the formula

$$u_r = \Big(I - A_\eta^* A_\eta g_r(A_\eta^* A_\eta)\Big)u_0 + g_r(A_\eta^* A_\eta)A_\eta^* f_\delta. \tag{4}$$

The well-known examples of this class are
1) the Tikhonov method: $u_r = (r^{-1}I + A_\eta^* A_\eta)^{-1} A_\eta^* f_\delta$;
2) the method of successive approximations (the explicit scheme):

$$u_r = u_{r-1} + \mu A_\eta^*(f_\delta - A_\eta u_{r-1}), \quad r = 1, 2, ...,$$

where $0 < \mu < 1/\|A_\eta^* A_\eta\|$;
3) the implicit scheme:

$$u_r = (\alpha I + A_\eta^* A_\eta)^{-1}(\alpha u_{r-1} + A_\eta^* f_\delta), \quad r = 1, 2, ...,$$

where $\alpha > 0$ is a constant.

3. REGULARIZATION PARAMETER CHOICE

First we formulate some results for the case where the parameter r is selected a priori.

THEOREM 1. *Let $Qf \in R(A), \|f_\delta - f\| \le \delta$ and $\|A_\eta - A\| \le \eta$.*
1. *If, for methods* (4), *we choose a parameter $r = r(\delta, \eta)$ such that*

$$r(\delta, \eta) \to \infty, \quad (\delta^2 + \eta) r(\delta, \eta) \to 0 \quad (\delta \to 0, \eta \to 0),$$

then

$$u_{r(\delta,\eta)} \to u_* \quad (\delta \to 0, \eta \to 0).$$

2. *If $u_* \in M_{p\rho u_0} = \{u \in H,\ u - u_0 = |A|^p v,\ \|v\| \le \rho\},\ \ |A| = (A^* A)^{1/2}$, $0 < p \le 2p_0$ and*

$$r(\delta, \eta) = d \min\Big[\delta^{-2/(p+1)},\ \eta^{-2/(p+2)}\Big], \quad d = \text{const},$$

then

$$\|u_{r(\delta,\eta)} - u_*\| \le c_{p\rho d}(\delta^{p/(p+1)} + \eta^{p/(p+2)}), \quad c_{p\rho d} = \text{const}.$$

Note that assertion **1** was proved in G.M. Vainikko (see (Vainikko, 1982)). In this paper an a posteriori parameter choice is also presented which provides the convergence rate $(\delta+\eta)^{p/(p+2)}$ $(0 < p \le 2p_0 - 2)$. For the case where the operator is known exactly, a parameter choice ensuring the optimal rate $\delta^{p/(p+1)}$ is considered in (Gfrerer, 1987; Engl and Gfrerer, 1988). Unfortunately, in this choice one must determine the element Qf_δ, which is generally a complicated problem.
Below we present an a posteriori parameter choice which leads to the same convergence rates as the a priori parameter choice. Let the operator $B_{r\eta} \in \mathcal{L}(F, F)$ be defined by

$$B_{r\eta} := \begin{cases} I, & \text{if } p_0 = \infty, \\ (I - A_\eta A_\eta^* g_r(A_\eta A_\eta^*))^1/p_0, & \text{if } p_0 < \infty. \end{cases}$$

Hence, we have $B_{r\eta} = (I + rA_\eta A_\eta^*)^{-1}$ for method **1)** ($\rho_0 = 1$) and $B_{r\eta} = I$ for methods **2)** and **3)** ($p_0 = \infty$). Denote

$$\varphi(r) := \sqrt{r}\|A_\eta^* B_{r\eta}(A_\eta u_r - f_\delta)\|, \quad \widetilde{\gamma}_p := (\gamma_{pp_0/(1+p_0)})^{1+1/p_0}.$$

Rule 1. *Let b_1, b_2, b_3 and b_4 be constants such that $b_2 \geq b_1 > \widetilde{\gamma}_{1/2}$ and $b_4 \geq b_3 > \widetilde{\gamma}_0$. If*

$$\varphi(r) < b_1\delta + b_3\sqrt{r}\eta\|f_\delta\| \tag{5}$$

for every $r \geq 0$, then choose $r(\delta, \eta) = 0$. In the opposite case choose $r = r(\delta, \eta)$ such that

$$\varphi\Big(r(\delta, \eta)\Big) \geq b_1\delta + b_3\sqrt{r(\delta, \eta)}\eta\|f_\delta\|$$

and, for every $r \geq r(\delta, \eta)$,

$$\varphi(r) \leq b_2\delta + b_4\sqrt{r}\eta\|f_\delta\|. \tag{6}$$

The following parameter selection rule may be applied to iteration procedures.

Rule 2. *Let d_1, d_2 and θ be the constants such that $d_1 > \widetilde{\gamma}_{1/2}, d_2 > \widetilde{\gamma}_0$ and $0 < \theta < 1$. If*

$$\varphi(r) < d_1\delta + d_2\sqrt{r}\eta\|f_\delta\| \tag{7}$$

for every $r \geq 0$, then choose $r(\delta, \eta) = 0$. In the opposite case choose $r = r(\delta, \eta)$ such that, for some $s \in [\theta r(\delta, \eta), r(\delta, \eta)]$ holds

$$\varphi(s) < d_1\delta + d_2\sqrt{s}\eta\|f_\delta\|$$

and for every $r \geq r(\delta, \eta)$, there holds

$$\varphi(r) < d_1\delta + d_2\sqrt{r}\eta\|f_\delta\|. \tag{8}$$

Since $\lim_{r\to\infty}\varphi(r) = 0$. Choosing the parameter $r(\delta, \eta)$ according to Rules 1 and 2 is possible. If we know some constants K, N such that $\|u_0 - u_*\| \leq K$, $\|u_*\| \leq N$ and

$$r_1 := \left(\widetilde{\gamma}_1 K / \max\left[0, (b_1 - \widetilde{\gamma}_{1/2})\delta - \widetilde{\gamma}_{1/2}\eta N\right]\right)^2,$$

$$r_2 := \left\{\widetilde{\gamma}_{1/2}N/\Big((b_2 - \widetilde{\gamma}_0)\|f_\delta\|\Big) + \sqrt{\widetilde{\gamma}_1 K/((b_2 - \widetilde{\gamma}_0)\eta\|f_\delta\|)}\right\}^2,$$

then for every r, $r \geq \min[r_1, r_2]$ we have $\varphi(r) \geq b_1\delta + b_2\sqrt{r}\eta\|f_\delta\|$ and hence we must check conditions (5)–(8) only on a finite interval.

Theorem 2. *Let $Qf \in R(A)$, $\|A_\eta - A\| \leq \eta$, $\|f_\delta - f\| \leq \delta$ and $f_\delta \neq 0$ or $u_0 = 0$ in the case $f_\delta = 0$. Suppose that, in addition to (2) and (3), the functions g_r satisfy the following conditions:*

1) *the function* $r \to |1 - \lambda g_r(\lambda)|$ *is decreasing for each* $\lambda \in [0, a]$;

2) *the function* $g_r(\lambda)$ *is continuously differentiable and for each* $\lambda \in [0, a]$

$$\left.\frac{\partial(g_s(\lambda))}{\partial s}\right|_{s=r} \le \bar{\gamma}\gamma\beta_r(\lambda)(1 - \lambda g_r(\lambda)), \quad \bar{\gamma} = \text{const}, \quad r > 0,$$

where

$$\beta_r(\lambda) := \begin{cases} 1, & \text{if } p_0 = \infty \\ (1 - \lambda g_r(\lambda))^{1/p_0}, & \text{if } p_0 < \infty. \end{cases}$$

1. *If, for methods* (4), *the parameter* $r = r(\delta, \eta)$ *is chosen according to Rule* 1 *or Rule* 2, *then*

$$u_{r(\delta,\eta)} \to u_* \quad (\delta \to 0, \eta \to 0).$$

2. *If* $u_* \in M_{p\rho u_0}$, $0 < p \le 2_{p_0}$, *then*

$$\|u_{r(\delta,\eta)} - u_*\| \le c(\delta^{p/(p+1)} + \eta^{p/(p+2)}), \quad c = \text{const}.$$

Note that for methods 1)–3) conditions 1) and 2) are satisfied with $\bar{\gamma} = 1$, $\bar{\gamma} = 1/(1 - \mu\|A_\eta^* A_\eta\|)$ and $\bar{\gamma} = 1$, respectively.

4. ON THE OPTIMALITY OF METHODS

Let $\mathfrak{A}$ be a subset of the set of all mappings of H into F. Now any mapping $P : F \times \mathfrak{A} \to H$ can be treated as a method to solve equation (1) and the approximate solution is given by $P(f_\delta, A_\eta) \in H$. Let $M \subset H$ be a given set and $q = \|f - Qf\| > 0$. Define the maximal error of the method P on the set M :

$$\Psi(\delta, \eta, M, q, P)$$

$$:= \sup_{\substack{u \in M, f_\delta \in F, f \in F, A_\eta \in \mathfrak{A}, \|A_\eta - A\| \le \eta, \\ \|f_\delta - f + Au - A_\eta u\| \le \delta + \eta b_M, \|A_\eta^*(f - Au)\| \le \eta q}} \|P(f_\delta, A_\eta) - u\|,$$

where $b_M := \sup_{u \in M} \|u\|$.

The method $P_{\delta\eta} : F \times \mathfrak{A} \to H$ is called (see (Vainikko and Veretennikov, 1986)) $\mathfrak{A}$-optimal on M if

$$\Psi(\delta, \eta, M, q, P_{\delta\eta}) = \inf_P \Psi(\delta, \eta, M, q, P)$$

and $\mathfrak{A}$-quasioptimal on M if

$$\Psi(\delta, \eta, M, q, P_{\delta\eta}) \le c \inf_P \Psi(\delta, \eta, M, q, P),$$

$$0 < \delta \le \delta_0, \ 0 < \eta \le \eta_0, \ c = \text{const}.$$

THEOREM 3. *For any method* P *and for any centrally-symmetrical set* $M \subset H$ *there holds*

$$\Psi(\delta, \eta, M, q, P)$$

$$\geq \max\left[\sup_{u\in M,\|Au\|\leq\delta+\eta b_M}\|u\|,\ \sup_{u\in M,\|A^*Au\|\leq\eta q}\|u\|\right].$$

THEOREM 4. *For any method P and for any set $M_{p\rho u_0}$ there holds*

$$\Psi(\delta,\eta,M_{p\rho u_0},q,P)\geq\max\left[\rho^{1/(p+1)}(\delta+\eta b_{p\rho})^{p/(p+1)},\rho^{2/(p+2)}(\eta q)^{p/(p+2)}\right].$$

COROLLARY. *The methods (4) with a priori or a posteriori (Rules 1, 2) parameter choice are $\mathfrak{A}$-quasioptimal on the sets $M_{p\rho u_0}$.*

REFERENCES

Engl, H.W. and Gfrerer, H. (1988). A posteriori parameter choice for general regularization methods for solving linear ill-posed problems. *Appl. Numer. Math.* **4**, 395–417.

Gfrerer, H. (1987). An a posteriori parameter choice for ordinary and iterated Tikhonov regularization of ill-posed problems leading to optimal convergence rates. *Math. Comput.* **49**, 507–522.

Vainikko, G.M. (1982). *Methods for Solving Linear Ill-Posed Problems in Hilbert Spaces.* Tartu Univ. Press, Tartu.

Vainikko, G.M. and Veretennikov, A.Y. (1986). *Iteration Procedures in Ill-Posed Problems.* Nauka, Moscow (in Russian).

Ill-Posed Problems in Natural Sciences, pp. 149 – 154
A. Tikhonov (Ed.)
1992 VSP/TVP

AN OPERATOR METHOD OF REGULARIZATION OF NONLINEAR MONOTONE ILL-POSED PROBLEMS

I. RJAZANCEVA
Nizhnii Novgorod Polytechnical Institute, Kasjanov str., 4/2, 603163, Niznii Novgorod, Russia

ABSTRACT

Sufficient conditions of convergence for an operator method of regularization of nonlinear ill-posed equations and variational inequalities with monotone operators have been obtained. Methods for choosing the regularization parameter are considered.

1. INTRODUCTION

We will consider nonlinear equations of the form

$$Ax = f \tag{1}$$

that belong to the class of ill-posed equations. For linear ill-posed problems, an operator method of regularization (Lavrent'ev, 1962 and Morozov, 1988) and variational methods such as A.N.Tikhonov's method (Tikhonov and Arsenin, 1979), V.K. Ivanov's method of quasi-solutions and a residual method (Ivanov et al., 1978) have been developed and proved to be efficient. However one fails to obtain similar results for the vast multitude of nonlinear problems. Therefore a class of monotone problems was chosen.
Let X be a Banach E-space, X^* its conjugate strictly convex space, and H be a Hilbert space. Without losing the generality we assume X^* to be a strictly convex space. A: $X \to 2^{X^*}$ means that the multivalued operator A acts from X to X^*.

Definition 1. An operator A: $X \to 2^{X^*}$ is called monotone if for any x_1 and x_2 from $D(A)$

$$\langle y_1 - y_2,\ x_1 - x_2 \rangle \geq 0, \quad \forall y_1 \in Ax_1, \quad \forall y_2 \in Ax_2,$$

is met.

Definition 2. A monotone operator A: $X \to 2^{X^*}$ is called maximal monotone if its graph is not a regular part of the graph of any other monotone operator from X to X^*.

As a mathematical apparatus we will employ the modern theory of monotone operators (see, for example, (Kluge, 1974)).
For a nonlinear operator A, variational methods lose their important properties and benefits (Liskovec, 1981), therefore we will be concerned with the operator method of regularization. The idea of this method goes back to the papers of (Lavrent'ev, 1962) and consists in replacing equation (1) by a close one-parametric equation that belongs to the class of correct equations.

2. OPERATOR METHOD FOR EQUATIONS

Let $A\colon X \to 2^{X^*}$ be a maximal monotone operator and $N \neq \emptyset$ be a set of solutions of (1). Assume that instead of the element f a sequence of approximations $\{f^\delta\}, \|f - f^\delta\| \leq \delta, \delta > 0$ is specified and instead of A maximal monotone approximations $A^h\colon X \to 2^{X^*}$ are known, with $D(A) = D(A^h)$ and

$$r^*(Ax, A^h x) \leq g(\|x\|)h, \tag{2}$$

where Ax and $A^h x$ denote the set of values of for the operators A and A^h at the point x, $r^*(P,Q)$ is the Hausdorff distance between the sets P and Q in X^*, and $g(t)$ is a nonnegative continuous non-decreasing function of the argument $t \geq 0$.
Consider a regularized equation (Browder, 1966; Cruceanu, 1971; Bakushinsky and Polyak, 1974):

$$A^h x + \alpha U^\mu x = f^\delta, \quad \alpha > 0, \tag{3}$$

where U^μ is the dual mapping in X with the scale function $\mu(t)$ (Kluge, 1974).

THEOREM 1. *If* $(\delta + h)/\alpha \to 0$ *as* $\alpha \to 0$, *then under assumptions the sequence* $\{x_\alpha^\gamma\}, \gamma = \{\delta, h\}$, *of solutions of* (3) *converges in the norm of* X, *as* $\alpha \to 0$, *to a solution* x^* *(minimal in the norm) of equation* (1).

Suppose that $A^h\colon X \to 2^{X^*} (h \geq 0)$ are not maximal monotone operators or their maximal monotonicity is difficult to establish. Then equation (3) will also be considered as a regularized one, and solutions of (1) and (3) will be defined as follows (Abramov and Gaipova, 1972).

Definition 3. An element $x_0 \in X$ is called a solution of equation (1) if for all $x \in D(A), y \in Ax$ the inequality $\langle y - f, x - x_0 \rangle \geq 0$ is satisfied.
It is easily proved that if $D(A)$ is a convex and closed set in X, $\operatorname{int} D(A) \neq \emptyset$, then equation (3) has a unique solution in $D(A)$. Theorem 1 remains valid if (2) is replaced by the inequality $r^*(\overline{A}x, \overline{A}^h x) \leq g(\|x\|)h$, where $\overline{A}^h$ is a maximal monotone extension of $A^h, h \geq 0$, and $D(\overline{A}^h) = D(\overline{A}) = D(A)$.

3. REGULARIZATION PARAMETER

Sufficient conditions for the convergence of the operator method of regularization (Theorem 1) allow for a priori determination of α in (3). Other methods of finding α are also widely used in linear problems. Consider the residual principle.
Let $A\colon X \to 2^{X^*}$ be a maximal monotone operator and the regularized equation has the form

$$A^h x + \alpha U x = f^\delta, \ \alpha > 0, \tag{4}$$

x_α^γ being its unique solution. Here U is the normalized dual mapping of $\mu(t) = t$. Specifying γ we construct the residual function $\rho(\alpha) = \alpha\|x_\alpha^\gamma\|$.

LEMMA 1. *$\rho(\alpha)$ is a single-valued continuous function for $\alpha \geq \alpha_0 > 0$, and if $0 \in D(A)$, $\rho(\alpha) \to \|y_*^h - f^\delta\|$, as $\alpha \to \infty$, where $y_*^h \in A^h(0)$ and*

$$\|y_*^h - f^\delta\| = \min\{\|y^h - f^\delta\|: \; y^h \in A^h(0)\}, \tag{5}$$

if $0 \notin D(A)$, then $\rho(\alpha) \to +\infty$ as $\alpha \to \infty$.

THEOREM 2 (residual principle). *Let X be a E-space, A^h: $X \to 2^{X^*}$ be maximal monotone operators, $D(A) = D(A^h)$, and assume that condition (2) is satisfied, let $g(t)$ be an increasing function, $0 < \delta + h \leq 1$. If $0 \in D(A)$, the inequality*

$$\|y_*^h - f^\delta\| > [k + g(0)](\delta + h)^p, \quad k > 1, \quad p \in (0,1), \tag{6}$$

is also assumed to be valid. Here the element $y_h^* \in A^h(0)$ and satisfies (5). Then there exists a unique solution $\bar{\alpha}$ of the equation

$$\rho(\bar{\alpha}) = [k + g(\|x_{\bar{\alpha}}^\gamma\|)](\delta + h)^p, \tag{7}$$

where $x_{\bar{\alpha}}^\gamma$ is the solution of (4), when $\alpha = \bar{\alpha}$. Besides, $\bar{\alpha} \to 0$, $(\delta + h)/\bar{\alpha} \to 0$ and $x_{\bar{\alpha}}^\gamma \to x^*$ as $\gamma \to 0$.

For $\mu(t) = t^{S-1}$, $S > 1$, we can also construct a procedure for choosing α of the form (7), if $\rho_S(\alpha) = \alpha\|x_\alpha^\gamma\|^{S-1}$ is taken as a residual function.
In order to apply the residual principle (7) we need to know the error level for δ and h. This requirement is not always fulfilled in the solution of practical problems. Variables h and δ are chosen from the quasioptimal principle and from the principle of minimal residual. Bakushinsky and Goncharsky (Bakushinsky and Goncharsky, 1989) found out that with unknown γ strong convergence of regularized solutions may be obtained only under some limitations on approximate data of the problem. The algorithms for finding α with unknown γ were studied for nonlinear monotone equations by Rjazanceva (Rjazanceva, 1986, 1989, 1991).

4. VARIATIONAL INEQUALITIES

Consider the variational inequality

$$\langle Ax - f, z - x\rangle \geq 0, \quad \forall z \in \Omega, \tag{8}$$

where A: $X \to 2^{X^*}$ is a maximal monotone operator, Ω is a set that is convex and closed in $X, \Omega \subset D(A)$, and $\text{int}\Omega \neq \emptyset$ or $\Omega \bigcap \text{int}D(A) \neq \emptyset$; if $\text{int}\Omega = \emptyset$, then the set $\Omega \bigcap \text{int}D(A^h)$ is assumed to be nonempty, and $N \neq \emptyset$ is the set of solutions of (8).
An approximation x_α^γ of the element $x^* \in N$ that is minimal in norm can be found from the regularized variational inequality

$$\langle A^h x + \alpha U^\mu x - f^\delta, z - x\rangle \geq 0, \quad \forall z \in \Omega, \quad x \in \Omega. \tag{9}$$

When $\alpha \to 0$, Theorem 1 is valid.
Let $\mu(t) \equiv t$ in (9) and $\rho(\alpha) = \alpha\|x_\alpha^\gamma\|$ be again the residual function.

LEMMA 2. *$\rho(\alpha)$ is a single-valued and continuous function, when $\alpha \geq \alpha_0 > 0$; if $0 \notin \Omega$, then $\lim_{\alpha\to\infty} \rho(\alpha) = \infty$, and if $0 \in \Omega$, then $\lim_{\alpha\to\infty} \rho(\alpha) = \|y_*^h - f^\delta\|$. In addition, if $0 \in \text{int}\Omega$, then y_*^h is determined by (5) while for $0 \in \partial\Omega$ we have*

$$\|y_*^h - f^\delta\| = \min\{\|y^h - f^\delta\|, y^h \in B^h(0), B^h = A^h + \partial I_\Omega\},$$

where ∂I_Ω it the subdifferential of the indicator function I_Ω of the set Ω.

If for $0 \in \Omega$ the element y_*^h in (9) is determined from Lemma 2, then under our assumptions Theorem 2 is also valid for the variational inequalities (8) and (9).
Let Ω also be an approximate set, i.e., instead of Ω a sequence $\{\Omega_\sigma\}, \sigma > 0$ of convex closed sets is known, with either $\text{int}\Omega_\sigma \neq \emptyset$ or $\Omega_\sigma \bigcap \text{int}D(A^h) \neq \emptyset, \Omega_\sigma \subset D(A^h)$, and the condition (2) is met for $x \in \Omega \bigcap \Omega_\sigma$ for each chosen pair Ω_σ and $A^h, \gamma = \{\delta, h, \sigma\}$.
Assume that $X = H$ is a Hilbert space and each single-valued cross-section of the operator $A^h (h \geq 0)$ possesses the property

$$\|y^h - f^\delta\| \leq c(\|x\| + 1), \quad c > 0, \quad y^h \in A^h x, \quad x \in \Omega\sigma, \tag{10}$$

and the sets Ω_σ "uniformly" approximate Ω in the Hausdorf metric r.

$$r(\Omega, \Omega_\sigma) \leq \sigma \tag{11}$$

The regularized solutions x_α^γ are found from the variational inequality

$$(A^h x + \alpha x - f^\delta, z - x) \geq 0, \quad \forall z \in \Omega_\sigma, \quad x \in \Omega_\sigma. \tag{12}$$

THEOREM 3. *If under our assumptions $\Omega \subset \Omega_\sigma, (\delta + h + \sigma)/\alpha \to 0$ as $\alpha \to 0$, then $x_\alpha^\gamma \to x^*$, when $\alpha \to 0$. If $\Omega_\sigma \subset \Omega$, then the solutions x_α^γ should be found from the variational inequality*

$$(A^h x + \alpha E^\mu x - f^\delta, z - x) \geq 0, \quad \forall z \in \Omega_\sigma, \quad x \in \Omega_\sigma, \tag{13}$$

where $E^\mu x = x\mu(\|x\|)/\|x\|, x \neq 0$, $E^\mu(0) = 0$, $\mu(t) > g(t) + 1$, when $t \geq t_0 > 0$ and $\mu(t)$ is an increasing function, $\mu(t) \to +\infty$ as $t \to \infty$. If (2) is valid for $x \in \Omega \cup \Omega_\sigma$, then the convergence of the solutions of the variational inequalities (12) and (13) to the solution of (8) holds without additional assumptions of the embeddobility of the sets Ω and Ω_σ.

The requirement (11) of the closeness of Ω and Ω_σ is natural for bounded sets. Examples of unbounded sets Ω and Ω_σ such that $r(\Omega, \Omega_\sigma) = \infty$ for all $\sigma > 0$ can be easily constructed. In this case we can specify conditious under which one can pass over from problems (8), (12) and (13) to variational inequalities for bounded sets $\Omega^R = \Omega \cap B(0, R), \Omega_\sigma^R = \Omega_\sigma \cap B(0, R), B(0, R) = \{x : \|x\| \leq R\}$, where R is a sufficiently large number. If we fail to find a value of R that meets these reguirements, then we define the value $l(R, \Omega, \Omega_\sigma) = \sup_{v\in\Omega^R} \inf_{u\in\Omega_\sigma} \|u - v\|$, $R \geq 0$, assuming if $\Omega^R = \emptyset$ that $l(R, \Omega, \Omega_\sigma) = 0$, and if at least one of the sets Ω^R or Ω^R_σ is nonempty we find $\tau(R, \Omega, \Omega_\sigma) = \max\{l(R, \Omega, \Omega_\sigma), l(R, \Omega_\sigma, \Omega)\}$. Let

$$\tau(R, \Omega, \Omega_\sigma) \leq a(R)\sigma, \tag{14}$$

where $a(R)$ is a nonnegative continuous and nondecreasing function such that $a(R) \to \infty$ as $R \to \infty$.

THEOREM 4. *If in* (13) *the function* $\mu(t)$ *is defined by the inequality* $\mu(t) > \max\{a(t), g(t)\} + 1$ *and under the conditions of Theorem 3 the reguirement* (11) *is replaced by* (14), *then the sequence of solutions of variational inequality* (13) *will converge as* $\alpha \to 0$ *to* $x^* \in N$.

Let now A: $X \to 2^{X^*}$ (the case of a Banach space) be a maximal monotone operator and instead of (10) the condition

$$\|y^h - f^\delta\| \leq \varkappa(\|x\|), \quad y^h \in A^h x, \quad \forall x \in \Omega_\sigma,$$

is satisfied, where $\varkappa(t)$ belongs to the some class of functions as $g(t)$. Assume that $\mu(t) > \max\{a(t), g(t), \varkappa(t)\} + 1$ for $t \geq t_0 > 0$ and the regularized solutions x^γ_α are found from the variational inequality

$$\langle A^h x + \alpha U^\mu x - f^\delta, z - x\rangle \geq 0, \quad \forall z \in \Omega_\sigma, \quad x \in \Omega_\sigma. \tag{15}$$

Then the convergence $x^\gamma_\alpha \to x^*$ holds as $\alpha \to 0$, if $(\delta + h + \sigma)/\alpha \to 0$ as $\alpha \to 0$.
In a Hilbert space when condition (2) is satisfied in set $\Omega \cup \Omega_\sigma$, under certain conditions (see Theorem 2) the regularization parameter α in (15) may be chosen from the residual principle in the following form

$$\rho(\bar{\alpha}) = \left[k + c + g(\|x^\gamma_{\bar{\alpha}}\|)\right](\delta + h + \sigma)^p, \quad k > 1, \quad \rho \in (0, 1),$$

with $\bar{\alpha} \to 0$, $(\delta + h + \sigma)/\bar{\alpha} \to 0$, $x^\gamma_{\bar{\alpha}} \to x^*$ as $\gamma \to 0$.

Definition 4. A solution of the variational inequality (8) with an arbitrary monotone operator A is an element $x \in \Omega$ for which there exists $y \in \bar{A}x$ such that for all $z \in \Omega$ the inequality $\langle y - f, z - x\rangle \geq 0$ holds, where $\overline{A}$ is a maximal monotone extension of the map A.
Results of §4 also hold for variational inequalities with monotone operators that are neither continuous nor maximal monotone provided that their solutions are understood in the sense of Definition 4.

REFERENCES

Abramov, A. and Gaipova, A. (1972). On the solution of some equations containing discontinuous transforms. *USSR Compit. and Math. Phys.* **12**, 204–207 (in Russian).

Bakushinsky, A. and Polyak, B. (1974). On the solution of variational inequlities. *Dokl. Akad. Nauk USSR* **219**, 1038–1041 (in Russian).

Bakushinsky, A. and Goncharsky, A. (1989). Basic concepts of regularization theory. In: *Iteration Methods for the Solution of Incorrect Problems.* Nauka, Moscow, pp. 9–25 (in Russian).

Browder, F. (1966). Existence and approximation of solutions of nonlinear variational enequalities. *Proc. Nat. Acad. Sci. USA* **56**, 1080–1086.

Cruceanu, S. (1971). Regularization pour les methode de Galerkine. *Comment. Math. Univ.* Carolinae **12**, 1–12.

Ivanov, V., Vasin, V. and Tanana, V. (1978). *Theory of Linear Incorrect Problems.* Nauka, Moscow.

Kluge, R. (1978). *Nichtlinear Variationsungleihungen und Extremalaufgaben. Theorie und Naherugsverfahren.*. Dtsch. Verl. Wiss., Berlin.

Lavrent'ev, M. (1962). *On Some Incorrect Problems of Mathematical Physics.* Siberian Branch Akad. Sci. USSR., Novosibirsk (in Russian).

Liskovec, O. (1981). *Variational Methods for the Solution of Unstable Problems.* Nauka i Tekhnika, Minsk (in Russian).

Morozov, V. (1988). *Methods for the Regularization of Unstable Problems.* Moscow University Press., Moscow (in Russian).

Rjazanceva, I. (1986). On quasioptimal choice of a regularization parameter for solving the nonlinear equations with monotone operators. *USSR Comput. Math. and Math. Phys.* **26**, 1731–1735 (in Russian).

Rjazanceva, I. (1989). On one algorithm for solving nonlinear monotone equations with an unknown estimate of input errors. *Zh. Vychisl. Mat. i Mat. Fiz.* **29**, 1572–1576 (in Russian).

Rjazanceva, I. (1991). On the principle of minimal residues in nonlinear mathematical monotone problems. *Zh. Vychisl. Mat. i Mat. Fiz.* **31**, 771–781 (in Russian).

Tikhonov, A. and Arsenin, V. (1979). *Methods for the Solution of Incorrect Problems.* Nauka, Moscow (in Russian).

Ill-Posed Problems in Natural Sciences, pp. 155 – 165
A. Tikhonov (Ed.)
1992 VSP/TVP

ON THE NUMERICAL INVERSION OF THE LAPLACE TRANSFORM WITH BOUNDARY CONSTRAINTS

G. RODRIGUEZ and S. SEATZU
Dipartimento di Matematica, Universitá di Cagliati, viale Merello 92, 09123 Cagliari, Italy

ABSTRACT

In this paper we propose a method for the numerical inversion of the Laplace transform into three different Hilbert spaces, based on the hypothesis that the recovering function is continuous and that the values at the ends of its range are known. The solution is given by a weighted linear combination of Jacobi polynomials whose coefficient are expressed in terms of the Laplace transform evaluated on equally spaced points. The effectiveness of the method is illustrated by the recovery of a number of functions for the most part already proposed in the literature.

1. INTRODUCTION

In the literature there is a very wide class of methods of the numerical inversion of the Laplace transform (see, e.g., (Davies and Martin, 1979; Piessens, 1975)). Recently some new methods have been proposed. However, in what follows we shall only refer to those papers which are of direct concern.

In particular, in (Rodriguez and Seatzu, 1991a) a method has been developed for solving this problem using a moderately high number of values of the Laplace transform on equally spaced data. It has been numerically verified that a wide class of functions can be recovered "very well" provided that the two parameters which characterize the mentioned sequence of values are suitably chosen for each test function. However, no criterion is furnished to obtain such values.

In this paper we consider the following problem of the numerical inversion of the Laplace transform with boundary constraints:

$$Lf - \mathbf{F}, \qquad (Lf)_i = \int_0^\infty c^{-s_i t} f(t)dt, \tag{1.1}$$

where $\mathbf{F} = (F_1, \cdots, F_m)^T$ is a known vector which gives the Laplace transform on a discrete set of equally spaced points ($F_i = F(s_i)$, $s_i = \hat{s} + (i-1)h$, $h \in \mathbf{R}^+, i = 1, \cdots, m$), and $f \in C[0, \infty]$ is an unknown function which satisfies the boundary constraint

$$Mf = \begin{bmatrix} f(0) \\ f(\infty) \end{bmatrix} = \mathbf{0}. \tag{1.2}$$

More precisely, supposing to know the analytical form of the Laplace trasform, in §2 we develop a method to solve this problem in three different Hilbert spaces. The method includes an effective criterion to fix the values of the two mentioned parameters for any model function.
In §3 we give an evaluation of the recovering errors for a number of test functions including the test functions introduced in (Davies and Martin, 1979; Essah and Delves, 1988) which satisfy the boundary condition $Mf = \mathbf{0}$.

2. THE METHOD

The mapping

$$x = 2e^{-ht} - 1$$

immediately transforms problem (1.1)–(1.2) into the following equivalent finite moment problem:

$$A\varphi = \mu, \qquad (A\varphi)_i = \int_{-1}^{1} (x+1)^{i-1}\varphi(x)\,dx, \;\; \mu_i = 2^{i-1}F_i, \;\; (i = 1, \ldots, m), \tag{2.1}$$

where the unknown function

$$\varphi(x) = \frac{2^{-\hat{s}/h}}{h}(x+1)^{(\hat{s}/h-1)} f\left(-\frac{1}{h}\log\frac{x+1}{2}\right) \tag{2.2}$$

satisfies the boundary constraint

$$E\varphi = \begin{bmatrix} \varphi(-1) \\ \varphi(1) \end{bmatrix} = \mathbf{0}. \tag{2.3}$$

Let $H^0 = L^2$ be the Hilbert space $L^2[-1,1]$ where, as usual, the inner product and the induced norm are $\langle u,v\rangle = \int_{-1}^{1} u(x)\; v(x)\; dx$, $\|u\|^2 = \langle u,u\rangle$ and let H^2 be the following function spaces $H^2 = \{f\colon\; f, f' \text{ absolutely continuous}, \; f'' \in L^2\}$. Besides, denoting by w the weight function $w(x) = 1 - x^2$, let $H_0^n (n = 0, 1, 2)$ be the following Hilbert spaces:

$$H_0^0 = \{\varphi = w\hat{\varphi}\colon\; \hat{\varphi} \in H^0\}, \quad \text{where} \quad \langle u,v\rangle_{H_0^0} = \langle \hat{u}, v\rangle = \langle \hat{w}^{-1}u, v\rangle;$$

$$H_0^1 = \{\varphi \in H^2\colon\; E\varphi = \mathbf{0}\}, \quad \text{where} \quad \langle u,v\rangle_{H_0^1} = \langle u', v'\rangle;$$

$$H_0^2 = \{\varphi \in H^2\colon\; E\varphi = \mathbf{0}\}, \quad \text{where} \quad \langle u,v\rangle_{H_0^2} = \langle u'', v''\rangle.$$

Let φ_n^+ be the normal solution in H_0^n of the finite moment problem (2.1)–(2.3), then φ_n^+ is uniquely characterized by the following property:

$$\|\varphi_n^+\|_{H_0^n} = \min\{\|\varphi\|_{H_0^n}\colon\; \varphi \in H_0^n,\; A\varphi = \mu\}. \tag{2.4}$$

Let A^*: $\mathbf{R}^m \to H_0^n$ be the adjoint operator of A: $H_0^n \to \mathbf{R}^m$, where $\mathbf{R}^m$ is equipped with the euclidean inner product $\langle \cdot, \cdot \rangle_2$.
In what follows, in order to emphasize that A and A^* change as n varies, we will write A_n and A_n^* instead of A and A^*, respectively.
Moreover, denoting by $N(A_n)$ the null space of A_n, by $N(A_n)^\perp$ its orthogonal complement, and by $R(A_n^*)$ the range of A_n^*, we will use the well-known property

$$\varphi_n^+ \in N(A_n)^\perp = R(A_n^*) \tag{2.5}$$

of the normal solution associated to a linear bounded operator from a Hilbert space onto a finite dimensional Hilbert space (Yosida, 1978).
The Riesz theorem, applied to the linear continuous operator A_n, implies the existence in H_0^n of m vectors $\{\eta_{ni}\}_{i=1}^m$ such that

$$(A_n\varphi)_i = \langle \eta_{ni}, \varphi \rangle_{H_0^n}, \quad i = 1, \ldots, m, \ \forall \varphi \in H_0^n. \tag{2.6}$$

Then, as an immediate consequence of the definition of A_n^*, one has

$$R(A_n^*) = \text{span } \{\eta_{ni}\}_{i=1}^m. \tag{2.7}$$

Moreover, following (Rodriguez and Seatzu, 1991b), we prove

THEOREM 2.1. *Denoting by $G_x(t)$ the Green function of the differential operator $\frac{d^2}{dx^2}$ with respect to the boundary operator E, the representers $\{\eta_{ni}\}$ may be expressed as follows:*

$$\eta_{0i}(x) = w(x)(x+1)^{i-1}, \quad i = 1, \ldots, m; \tag{2.8}$$

$$\eta_{ni}(x) = \int_{-1}^{1} G_x(t)\eta_{ni}''(t)\, dt, \quad n = 1, 2,\ i = 1, \ldots, m; \tag{2.9}$$

where

$$\eta_{1i}''(t) = -(t+1)^{i-1}; \tag{2.10}$$

$$\eta_{2i}''(t) = \frac{w(t)}{i(i+1)}\theta_{i-1}(t), \qquad \theta_{i-1}(t) = \sum_{j=1}^{i} 2^{i-j}(t+1)^{j-1}. \tag{2.11}$$

Relations (2.5)–(2.7) imply that $\varphi_n^+ = \sum_{i=1}^m \alpha_{ni}^+ \eta_{ni}$ where, by virtue of (2.1)–(2.6), α_n^+ is the solution of the Gram system $G_n\alpha = \mu$, $(G_n)_{ij} = \langle \eta_{ni}, \eta_{nj} \rangle_{H_0^n}$. This system cannot be solved for moderately high values of m, because, as m increases, the matrix G_n becomes severely ill-conditioned. Our numerical observations suggest that $\text{cond}(G_0) \cong e^{3.6m-4.6}$, $\text{cond}(G_1) \cong e^{3.8m-3.6}$, $\text{cond}(G_2) \cong e^{4m-1.8}$. In the sequel, in order to have a better conditioned representation of φ_n^+, we develop a numerically stable method to obtain a set $\{\hat{\eta}_{ni}\}$ of polynomials orthogonal in H_0^n such that $\text{span}\{\hat{\eta}_{ni}\} = \text{span}\{\hat{\eta}_{ni}\}$.
In this case the Gram–Schmidt method is not a good choice as an orthogonalization process, since, as m increases, it quickly becomes numerically unstable.

On the contrary, the following remark is very useful.
Theorem 2.1 essentially says that, the vectors $\{\eta_{ni}\}$ are polynomials characterized by the following properties:

$$\text{degree}(\eta_{ni}) = i+1, \qquad E\eta_{ni} = 0, \quad n = 0,1;$$

$$\text{degree}(\eta_{2i}) = i+3, \qquad E\eta_{2i} = 0 \quad \text{and} \quad E\eta''_{2i} = 0.$$

Therefore, if by Π_k the space of polynomials of degree $\le k$, Theorem 2.1 states in particular that:

(1) $\varphi_n^+ \in \Pi_{m+1}, n = 0,1;\ \varphi_2^+ \in \Pi_{m+3}$;

(2) while φ_0^+ and φ_1^+ only satisfy the boundary constraints $E\varphi_0^+ = E\varphi_1^+ = 0, \varphi_2^+$ also satisfies the additional constraint $E(\varphi_2^+)'' = 0$.

Let us denote, as usual, by $\{P_{j-1}^{(\alpha,\beta)}\}$ the Jacobi polynomials orthogonal in $[-1,1]$ with respect to the weight function $w^{(\alpha,\beta)}(x) = (1-x)^\alpha(1+x)^\beta$ $(\alpha,\beta > -1)$.

THEOREM 2.2. *The orthogonal polynomials $\{\hat\eta_{ni}\}$ are expressible in terms of the Jacobi polynomials in the following way:*

$$\hat\eta_{0i}(x) = w(x)P_{i-1}^{(1,1)}(x); \tag{2.12}$$

$$\hat\eta_{1i}(x) = \frac{i+1}{2}\int_{-1}^{1} G_x(t)P_{i-1}^{(1,1)}(t)\,dt; \tag{2.13}$$

$$\hat\eta_{2i}(x) = \int_{-1}^{1} G_x(t)w(t)P_{i-1}^{(2,2)}(t)\,dt. \tag{2.14}$$

Proof. Relation (2.12) is an immediate consequence of (2.8) combined with the definition of the Jacobi polynomials and that of the inner product in H_0^0.
Relation (2.10) and the definition of the inner product in H_0^1, imply in particular that $\hat\eta'_{1i} = P_i^{(0,0)}$, so that, by virtue of the classical result on the derivatives of the Jacobi polynomials (Szego, 1967), namely

$$\frac{d}{dx}P_i^{(\alpha,\beta)} = \frac{1}{2}(\alpha+\beta+i+1)P_{i-1}^{(\alpha+1,\beta+1)},$$

one has $\hat\eta''_{1i} = \frac{i+1}{2}P_{i-1}^{(1,1)}$. Then (2.13) immediately follows from the definition of the Green function G_x.
Relation (2.11) and the definition of inner product in H_0^2 imply the equality $\hat\eta''_{2i} = wP_{i-1}^{(2,2)}$, from which (2.14) follows.
The orthogonal polynomials $\{\hat\eta_{ni}\}$ can be obtained with a numerically stable algorithm. Indeed, the Jacobi polynomials $\{P_{i-1}^{(1,1)}\}$ and $\{P_{i-1}^{(2,2)}\}$ can be generated by the classical three-term recurrence relation and the integrals (2.13) and (2.14) can be evaluated analytically. Since $\varphi_n^+ \in R(A_n^*) = \text{span}\{\hat\eta_{ni}\}$, the normal solution φ_n^+ can be expanded as follows:

$$\varphi_n^+ = \sum_{i=1}^{m} \beta_{ni}^+\hat\eta_{ni}, \tag{2.15}$$

where, by virtue of (2.1)–(2.6), β_n^+ is the solution of the system

$$\widehat{G}_n\beta = \mu, \qquad (\widehat{G}_n)_{ij} = \hat{g}_{ij}^n = \langle \eta_{ni}, \hat{\eta}_{nj}\rangle_{H_0^n}. \tag{2.16}$$

The orthogonality of $\{\hat{\eta}_{ni}\}$ implies that $\widehat{G}_n$ is a lower triangular matrix.
Proceeding as in (Rodriguez and Seatzu, 1991c) we state the following theorem concerning the condition number of $\widehat{G}_n$.

THEOREM 2.3. *Let $D_n = \text{diag}\{d_{n1}, \dots, d_{nm}\}$ for $n = 0, 1, 2$, be the diagonal matrix defined by $d_{ni} = \|\hat{\eta}_{ni}\|_{H_0^n}^{-1}$, $\|\cdot\|_{H_0^n}^2 = \langle\cdot,\cdot\rangle_{H_0^n}$, $i = 1, \dots, m$. Then one has* $\text{cond}(\widehat{G}_n D_n) = \sqrt{\text{cond}(G_n)}$.

Each matrix $\widehat{G}_n$ ($n = 0, 1, 2$) is a lower triangular matrix, so system (2.16) may be solved directly by straightforward substitution. However, since $\text{cond}(\widehat{G}_0) \cong e^{2m-3}$, $\text{cond}(\widehat{G}_1) \cong e^{2m-2}$, $\text{cond}(\widehat{G}_2) \cong e^{2m-1}$, it is crucial to compute the elements of $\widehat{G}_n$ very carefully.
For moderately high values of m, we achieve a satisfactory precision in β by making use the threeterm recurrence relationship between the Jacobi polynomials. Indeed, setting $\widehat{G}_n = D\widetilde{G}_n$ ($D = \text{diag}(1, 2, \dots, 2^{m-1})$) and proceeding as in (Rodriguez and Seatzu, 1990), we obtain the following recurrence relations:

$$\begin{cases} 2b_j^{(1,1)}\tilde{g}_{i+1,j+1}^0 = a_j^{(1,1)}\tilde{g}_{i,j+2}^0 + b_j^{(1,1)}\tilde{g}_{i,j+1}^0 + d_j^{(1,1)}\tilde{g}_{ij}^0 \\ \tilde{g}_{i1}^0 = \frac{8}{(i+1)(i+2)}, \end{cases}$$

$$\begin{cases} 2b_j^{(0,0)}\tilde{g}_{i+1,j+1}^1 = a_j^{(0,0)}\tilde{g}_{i,j+2}^1 + b_j^{(0,0)}\tilde{g}_{i,j+1}^1 + d_j^{(0,0)}\tilde{g}_{ij}^1 \\ \tilde{g}_{i1}^1 = -\frac{4}{(i+1)(i+2)}, \end{cases}$$

$$\begin{cases} \widetilde{G}_2 = L\Sigma, \quad L,\ \Sigma \text{ are lower triangular matrices} \\ l_{ij} = \frac{2}{i(i+1)}, \quad \sigma_{ij} = 2^{-i}\langle w^{(2,2)}(x)\,(x+1)^{i-1}, P_{j-1}^{(2,2)}\rangle \\ 2b_j^{(2,2)}\sigma_{i+1,j+1} = a_j^{(2,2)}\sigma_{i,j+2} + b_j^{(2,2)}\sigma_{i,j+1} + d_j^{(2,2)}\sigma_{ij} \\ \sigma_{i1} = \frac{32}{(i+2)(i+3)(i+4)}, \end{cases}$$

where $\{a_j^{(\alpha,\beta)}, b_j^{(\alpha,\beta)}, d_j^{(\alpha,\beta)}\}$ are the coefficients of the three-term recurrence relation between the Jacobi polynomials normalized by the condition $P_j^{(\alpha,\beta)}(1) = \binom{\alpha}{j}$.
Since, as stated in (2.1), $\mu = D\mathbf{F}$, the solution of system (2.16) is now obtained by solving the system $\widetilde{G}_n\beta = \mathbf{F}$.

A normal solution of the Laplace transform inversion problem.
The function φ_n^+, that is the normal solution of the finite moment problem (2.1)–(2.3) in the Hilbert space H_0^n gives, by virtue of (2.2), the solution $f_n(t) = 2he^{(\hat{s}/h-1)ht}\varphi_n^+(x(t))$, $x(t) = 2e^{-ht} - 1$ of problem (1.1)–(1.2).
However, while the mapping $x(t) = 2e^{-ht} - 1$ states in a very natural way the equivalence between H_0^n and the Hilbert space $W_0^n = \{f \in C[0,\infty]\colon f(t(x)) \in$

H_0^n, $\langle f,g\rangle_{W_0^n} = \langle f(t(x)), g(t(x))\rangle_{H_0^n}\}$, f_n is not the normal solution in W_0^n of problem (1.1)–(1.2). Indeed, f_n is the normal solution in W_0^n only if the inner product of W_0^n is weighted by the function $p(t) = \frac{1}{2h}e^{(1-\hat{s}/h)ht}$.
To obtain the normal solution of the initial problem in W_0^n, it is necessary to solve in H_0^n the following weighted finite moment problem:

$$B\psi = \widehat{\mu}, \qquad (B\psi)_i = \int_{-1}^{1} \hat{p}(x)(x+1)^{i-1}\psi(x)dx, \qquad \widehat{\mu}_i = h2^{(\hat{s}/h+i-1)}\mu_i, \quad (2.17)$$

where

$$\hat{p}(x) = (1+x)^{(\hat{s}/h-1)}, \qquad \psi(x) = f\Big(-\frac{1}{h}\log\frac{x+1}{2}\Big) \quad \text{and} \quad E\psi = 0.$$

Although it is in principle possible to adapt our present method to solve the weighted problem, the method becomes more expensive because of the need to change the set of Jacobi polynomials for each choice of $(\hat{s}, h)$. We have therefore preferred to solve the finite moment problem in the absence of the weight function $\hat{p}(x)$, but we intend to ascertain in the future if these two alternative finite moment problems give the same precision in the numerical inversion of the Laplace transform.

The choice of the parameters $m, \hat{s}$ and h.
Since cond $(\widetilde{G}_n)$ increases exponentially with m, the number m cannot be arbitrarily large. The maximum value of m essentially depends on the precision obtained in the calculation of the elements of $\widetilde{G}_n$.
Since all the calculations were made on a 386-based computer equipped with MATLAB (Mathworks, 1989), we considered only values $m \leq 20$.
In order to obtain numerically an "optimum" set of values for the parameters $m, \hat{s}$ and h, we adopted an index of "a posteriori validation", depending from a sample vector ξ chosen in the domain of the Laplace transform.
More precisely, denoting, for each set $(m, \hat{s}, h)$, by $F(\xi)$ and $F_n(\xi)$ the vectors generated by the Laplace transform of the test function f and its recovery f_n, respectively, i.e.,

$$F(\xi) = \big(F(\xi_0), \ldots, F(\xi_r)\big)^T, \qquad F(\xi_i) = \int_0^\infty e^{-\xi_i t} f(t)dt, \quad i = 0, 1, \ldots, r,$$

$$F_n(\xi) = \big(F_n(\xi_0), \ldots, F_n(\xi_r)\big)^T, \qquad F_n(\xi_i) = \int_0^\infty e^{-\xi_i t} f_n(t)dt, \quad i = 0, 1, \ldots, r,$$

we adopted the following index of "a posteriori validation":

$$R_n = R_n(f; m, \hat{s}, h; \xi) = \sum_{i=0}^{r}{}' \Big(1 - \frac{F_n(\xi_i)}{F(\xi_i)}\Big)^2,$$

where the dash specifies that in the case $F(\xi_i) = 0$, the addendum $1 - F_n(\xi_i)/F(\xi_i)$ is overlooked. We notice that, whatever is the sample vector $\xi, F_n(\xi)$ can be evaluated analytically.
Therefore, for a fixed ξ, we minimize R_n with respect to $\hat{s}$ and h for $m \leq 20$, and we consider as "optimum" the set $(m^*, \hat{s}^*, h^*)$ which produces the minimum value of R_n.
To minimize R_n, we used the routine FMINS of MATLAB, which is based on the simplex method of Nelder–Mead. This method, same as any other algorithm used in the global minimization of functions analytically unknown, is sensitive to the initial guess of $\hat{s}$ and h. For this reason we always used three different initial guesses for $\hat{s}$ and h in the minimization of R_n.

The choice of the sample vector ξ.
The optimum set $(m^*, \hat{s}^*, h^*)$ obviously depends on the sample vector ξ, which is a good sample vector whenever the points $\{(\xi_j, F(\xi_j))\}_{j=0}^{r}$ are representative of the behaviour of the Laplace transform.
Since the nodes $\{s_i\}$ used to obtain the moments are equally spaced, the sample values $\{\xi_i\}$ must be unequally spaced in order to use correctly the validation index R_n.
As a consequence of our numerical results, we believe that a good sample vector ξ is obtained by considering $\xi \in R^{15}$ and selecting its components by the following procedure:

(1) first select the support interval $[\xi_0, \xi_{14}]$ depending on the behaviour of the Laplace transform:

(2) then evaluate recursively the other components of ξ imposing that they satisfy the equation

$$\int_{\xi_{i-1}}^{\xi_i} |F(s)|ds = \frac{1}{14} \int_{\xi_0}^{\xi_{14}} |F(s)|\, ds, \quad i = 1, \ldots, 13.$$

3. NUMERICAL RESULTS

To assess the numerical accuracy of the method, we give some results on the recovery in H_0^n of each of the functions listed in Table 1. These functions include all the test functions studied in (Davies and Martin, 1979) and (Essah and Delves, 1988) that satisfy the boundary constraint $Mf = 0$.
Denoting by f_n the recovery in H_0^n of the model function f, we adopt the following "quality index" to assess the accuracy of the numerical results:

$$E_n(f) = E_n(f; m, \hat{s}, h; \xi) = \frac{\|f - f_n\|_\infty}{\|f\|_\infty},$$

where $\|g\|_\infty = \max\{|g(ih)|: \ i = 1, \ldots, 90, h = \frac{1}{6}\}$. The index E_n measures the accuracy of the recovery on the interval [0, 15], usually chosen (see (Davies and

Martin, 1979; Essah and Delves, 1988)) as a test interval in the numerical inversion of the Laplace transform.

Table 1. Test functions

$f_1(t) = te^{-t}$	$F_1(p) = \frac{1}{(p+1)^2}$
$f_2(t) = t^2 e^{-t/2}$	$F_2(p) = \frac{2}{(p+1/2)^3}$
$t_3(t) = t^3 e^{-t}$	$F_3(p) = \frac{6}{(p+1)^4}$
$f_4(t) = e^{-t} \sin t$	$F_4(p) = \frac{1}{(p+1)^2+1}$
$f_5(t) = e^{-t/3} \sin t$	$F_5(p) = \frac{1}{(p+1/3)^2+1}$
$f_6(t) = e^{-t/5} \sin t$	$F_6(p) = \frac{1}{(p+1/5)^2+1}$
$f_7(t) = \frac{2e^{-4/t}}{\sqrt{\pi t^3}}$	$F_7(p) = e^{-4\sqrt{p}}$
$f_8(t) = \frac{e^{-1/16t}}{\sqrt{16\pi t^3}}$	$F_8(p) = e^{-\sqrt{p}/2}$
$f_9(t) = \frac{2}{t}(1 - \cos t)$	$F_9(p) = \log \frac{p^2+1}{p^2}$

Davies–Martin and Essah–Delves, in their quoted papers, adopted the following two measures of accuracy:

$$L_n(f) = \left\{ \frac{1}{30} \sum_{i=1}^{30} \left[f\left(\frac{i}{2}\right) - f_n\left(\frac{i}{2}\right) \right]^2 \right\}^{\frac{1}{2}};$$

$$L_{n,e}(f) = \left\{ \sum_{i=1}^{30} \left[f\left(\frac{i}{2}\right) - f_n\left(\frac{i}{2}\right) \right]^2 e^{-i/2} \Big/ \sum_{i=1}^{30} e^{-i/2} \right\}^{\frac{1}{2}}.$$

Therefore, in addition to the index E_n, which we consider more meaningful, we adopted also the above measures, in order to make a comparison of our results with those reported in (Davies and Martin, 1979) and (Essah and Delves, 1988). Denoting by E_n^*, L_n^* and $L_{e,n}^*$ the values of E_n, L and L_e that correspond to the parameter values $(m^*, \hat{s}^*, h^*)$ obtained by the minimization in H_0^n of R_n, we list the most significant numerical results in Table 2, where, for each test function, those values of E_n^*, L_n^* and $L_{e,n}^*$ are listed which correspond to the space H_0^n for which R_n is minimum. Note that the numbers in parentheses indicate decimal exponents.

All the numerical results must be viewed in the light of the following limitation: the floating point relative accuracy of 386-Matlab is roughly 10^{-16}, so that, as a consequence of the growth of cond $(\widetilde{G}_n)$ with respect to m, the numerical accuracy of the method has been investigated only for $m \leq 20$.

Table 2 indicates that: (1) The criterion adopted to select the values of the parameters $\hat{s}$ and h is effective. (2) Since E_n generally decreases as R_n decreases, it is very useful to carry out the numerical inversion of the Laplace transform in more than one Hilbert space. (3) The constraint which originates from the bound $m \leq 20$ is not restrictive in practice.

Table 2. Best results obtained in H_0^n $(n = 0, 1, 2)$

f	f_1	f_2	f_3	f_4	f_5
n	1	0	1	0	0
E_n^*	1.4(−10)	2.8(−09)	7.5(−09)	5.0(−06)	4.0(−03)
L_n^*	3.0(−11)	3.7(−09)	5.8(−09)	8.1(−07)	1.4(−03)
$L_{e,n}^*$	1.8(−11)	1.8(−09)	3.9(−09)	7.3(−07)	4.4(−04)

f	f_6	f_7	f_8	f_9
n	0	1	0	2
E_n^*	3.4(−02)	1.3(−02)	3.5(−02)	4.1(−02)
L_n^*	1.4(−02)	1.7(−04)	2.3(−02)	2.0(−02)
$L_{e,n}^*$	6.2(−03)	2.3(−04)	3.3(−02)	1.5(−03)

We note that numerical results have been presented in (Davies and Martin, 1979) for examples 1, 4, 6, 7 and in (Essah and Delves, 1988), for examples, 1, 2, 4, 8. For examples 1, 4, 6 and 7, the accuracy we achieve is lower than the ones quoted in Table IV of the paper written by Davies–Martin and obtained by the algorithms of Weeks, Piessens and Piessens–Branders which are the most successful.
However in the implementation of these algorithms Davies–Martin used requirements on the Laplace transform which are not always available. Moreover they reported in the aforementioned Table the maximum accuracy achievable by each devised method in the recovery of each function, whereas this is only really attainable if the function to be recovered is known in advance.
In contrast, the results quoted in Table V of the same paper, which are obtained by the same methods but by ignoring the functional form of the test functions, are no better than ours.

We note in particular that, while the method of Piessens–Branders which is one of the most efficient, fails in the recovery of the function f_7, our method still gives a good result, and indeed the function and the fit are visually almost undistinguishable (Figure 1).
Our overall results for examples 1, 2, 4, 8, appear to be better than those presented by Essah–Delves. In particular while our method gives an acceptable recovery of the peak which characterize example 8, as Figure 2 shows, the method of Essah–Delves fails in the recovery of the peak of this test function.

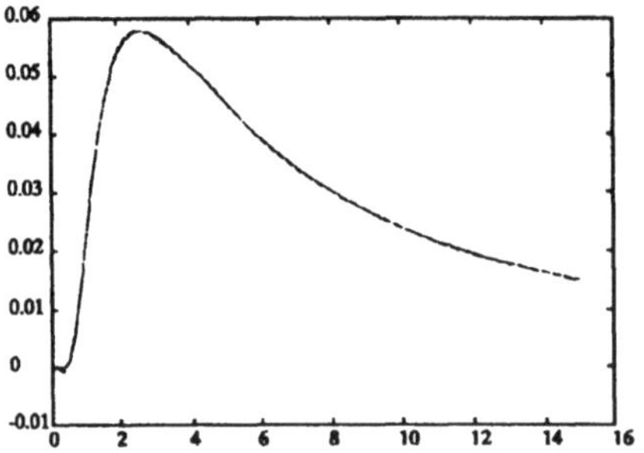

Fig. 1. Results for example 7: test function (solid line), recovery (dashed line)

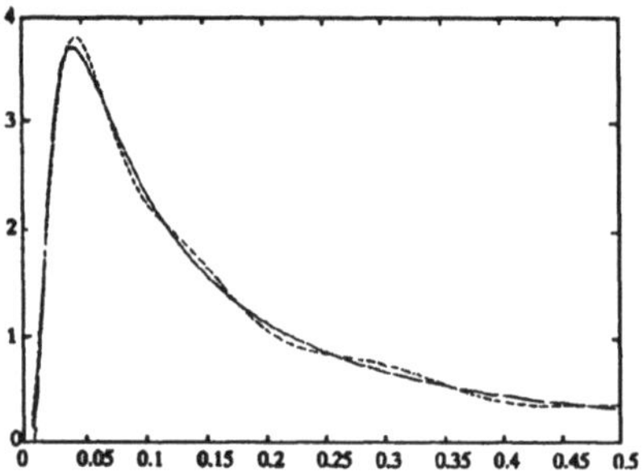

Fig. 2. Results for example 8: test function (solid line), recovery (dashed line)

In agreement with Davies–Martin (see (Davies and Martin, 1979)) we believe that an all-purpose method for the numerical inversion of the Laplace transform does not exist, but, in agreement with Essah–Delves (see (Essah and Delves, 1988)), we consider too pessimistic the idea expressed by Varah in (Varah, 1983) implying the existence of pitfalls in any numerical method of solution of integral equations of the first kind.

Acknowledgments
This work was supported by the Italian Ministry of University and Scientific and Technological Research and by the Italian Research Council.

REFERENCES

Davies, B. and Martin, B. (1979). Numerical inversion of the Laplace transform: a survey and comparison of methods. *J. Comp. Physics.* **33**, 1–32.

Essah, W.A. and Delves, L.M. (1988). On the numerical inversion of the Laplace transform. *Inverse Problems.* **4**, 705–724.

The MathWorks Inc. (1989). In: *MATLAB-386 Version 3.5e.* The MathWorks Inc., South Natick, MA.

Piessens, R. (1975). A bibliography on numerical inversion of the Laplace transform and applications. *J. Comp. Appl. Math.* **1**, 115–128.

Rodriguez, G. and Seatzu, S. (1990). Numerical solution of the finite moment problem in a reproducing kernel Hilbert space. *J. Comp. Appl. Math.* **33**, 233–244.

Rodriguez, G. and Seatzu, S. (1991a). On the numerical inversion of the Laplace transform with equally spaced data. *Atti Sem. Mat. Fis. Univ. Modena* XXIX, 401–410.

Rodriguez, G. and Seatzu, S. (1991b). On the solution of the finite moment problem. *J. Math. Anal. Appl. (to appear).*

Rodriguez, G. and Seatzu, S. (1991c). On the numerical inversion of the Laplace transform in reproducing kernel Hilbert spaces (to appear).

Szegő, G. (1967). *Orthogonal Polynomials.* 3rd. Ed. AMS Colloq. Publ. **23**, New York.

Varah, J.M. (1983). Pitfalls in the numerical solution of linear ill-posed problems. *SIAM J. Sci. Stat. Comp.* **4**, 164–176.

Yosida, K. (1978). *Functional Analysis.* Springer-Verlag., Berlin.

Ill-Posed Problems in Natural Sciences, pp. 166 – 173
A. Tikhonov (Ed.)
1992 VSP/TVP

REGULARIZATION OF DIFFERENCE SCHEMES

A.A. SAMARSKII and P.N. VABISHCHEVICH
All-Union Center for Mathematical Modelling, Russia Academy of Sciences, Miusskaya sq. 4, 125047 Moscow, Russia

ABSTRACT

A great attention in the theory of difference schemes has been paid to their stability with respect to both initial data and right-hand side. To achieve the stability, a regularization principle has been suggested. The principle is based upon the construction of some simplest difference scheme, the representation of it in a canonic form. Then the difference operators involved are perturbed in such a way as to ensure that the resulting difference scheme be stable.

The regularization principle is applied to the construction of stable difference schemes for classical nonstationary problems occurring in mathematical physics. Two and three level difference schemes are considered for evolution equations of both first and second order.

A similar approach is suggested for deriving difference schemes for evolution problems which are ill-posed in the usual sense. The investigation of the stability of the difference schemes for such unstable problems is based on the general theory of ρ-stability of difference schemes with $\rho > 1$ (i.e., when a limited growth of the norm of the solution is allowed). It is shown that the use of the regularization principle allows us to derive the difference schemes corresponding matching either to the already known or to some new quasi-inversion methods.

1. INTRODUCTION

A great attention in the theory of difference schemes in (Samarskii, 1983) has been paid to studying the stability of difference schemes for non-stationary problems occurring in mathematical physics. Necessary and sufficient conditions were obtained in (Samarskii and Gulin, 1975) for the stability of a wide class of two and multilevel schemes. The technique developed there is based upon a canonic form of a difference scheme, which is invariant with respect to the underlying differential equation.

The difference scheme regularization principle, formulated in (Samarskii, 1976), provides a basis for the construction of stable difference schemes. At first, some simple difference scheme is proposed, not necessarily possessing desirable properties. Then this scheme is written in canonical form and modified by perturbing the difference scheme operators in order to improve its quality. The difference scheme

regularization theory serves here as an instrument to do this. The technique is essentially the introduction of additional summands to the difference scheme operators.
The difference scheme regularization principle is a tool used to construct stable difference schemes for both well-posed and ill-posed problems for evolution equations. Both direct and inverse problems for a second-order parabolic equation are considered. Different kinds of regularizators are studied, which are shown to lead to some known or new difference schemes. For unstable evolution problems the regularization principle was originally proposed in (Samarskii and Vabishchevich, 1990).

2. DIRECT PROBLEM

Let the simplest boundary value problem for a second order parabolic equation serve as an example. Let Ω denote a bounded region in the m-dimensional space $\mathbf{R}^m$ with smooth enough boundary $\partial\Omega$. Let us consider in $\mathbf{R}^m \times \{-\infty < t < \infty\}$ the bounded cylinder

$$Q = \{(x,t) \mid x \in \Omega,\ 0 < t < T\}, \quad T > 0,$$

where $x = (x_1, x_2, \ldots, x_m)$, and

$$\Gamma = \{(x,t) \mid x \in \partial\Omega,\ 0 < t < T\}.$$

denotes its lateral surface.
Let us define for every $x \in \Omega$ the uniformly elliptic self-adjoint operator

$$Lu \equiv -\sum_{i,j=1}^{m} \frac{\partial}{\partial x_i}\Big(a_{ij}(x)\frac{\partial u}{\partial x_j}\Big) \tag{1}$$

with sufficiently smooth coefficients $a_{ij}(x) = a_{ji}(x),\ x \in \Omega$.
We are looking for the solution $u(x,t)$, of

$$\frac{\partial u}{\partial t} + Lu = 0, \quad (x,t) \in Q \tag{2}$$

satisfying the boundary conditions

$$u(x,t) - 0, \quad (x,t) \in \Gamma \tag{3}$$

and the initial data

$$u(x,0) = \Phi(x), \quad x \in \Omega. \tag{4}$$

Let $\|\cdot\|$ denote the norm in $L_2(\Omega)$, i.e.,

$$\|v(x)\|^2 = \int_\Omega v^2(x)\,dx.$$

The following estimate states the stability of the solution of (1)–(4) with respect to the initial data:

$$\|u(x,t)\| \le \|\Phi(x)\|. \tag{5}$$

When solving (1)–(4) approximately, it is natural to require that an estimate of type (5) be valid for the approximate solution too. Let us introduce a grid ω_h covering Ω. Without loss of generality we shall assume that the grid ω_h is regular in each direction. Let h_i stand for the grid-step in the direction x_i, $i = 1, 2, \ldots, m$. Suppose the operator L, defined by(1) together with (3) is approximated by the corresponding difference operator L_h. The particular choice of L_h is based upon the theory of difference schemes (Samarskii, 1983) and finite element methods (Norrie and Vries, 1978). We shall outline only some most essential properties the operator L_h should possess. Let us consider the space $H = L_2(\omega_h)$ of grid functions defined on the mesh ω_h, with the inner product

$$(y,z) = \sum_{x \in \omega_h} y(x)\, z(x)\, h_1 h_2 \ldots h_m.$$

The operator L_h defined in this way is self-adjoint and positive definite in H, and the following inequality holds:

$$\delta E \le L_h \le \Delta E, \tag{6}$$

with $\delta = O(1) > 0$, $\Delta = O(h^{-2})$, where h is the mesh parameter.
The original problem (1)–(4) gives rise to the following differential discrete problem. We are to solve the equation

$$\frac{dy}{dt} + L_h y = 0, \tag{7}$$

with the initial data

$$y(x,0) = \Phi(x), \quad x \in \omega_h. \tag{8}$$

In t-variable we introduce the regular grid $\omega_\tau = \{t \mid t = n\tau,\ n = 0, 1, \ldots, N,\ N\tau = T\}$ with the time-step denoted by τ, $\tau > 0$.
The stability theory of difference schemes is based on the canonical form of the difference scheme. In the case of two-level homogeneous difference schemes the canonical form can be written as (Samarskii, 1983; Samarskii and Gulin, 1975)

$$B\frac{y_{n+1} - y_n}{\tau} + Ay_n = 0, \quad n = 0, 1, \ldots\ , \tag{9}$$

where $y_0(x) = \Phi(x)$ is a given function, $x \in \omega_h$.
For well-posed evolution problems the usual stability condition with respect to the initial data in H_D is

$$\|y_{n+1}\|_D \le \|y_n\|_D, \tag{10}$$

where $\|y\|_D^2 = (Dy, y)$, and $D = D^* > 0$. The above inequality is a difference analog of the estimate (5).
Now we state the main result in the stability theory of the two-level difference schemes (Samarskii, 1983; Samarskii and Gulin, 1975).

THEOREM 1. *Let the operators A and B in (9) be independent of n, self-adjoint and positive definite ($A = A^* > 0$, $B = B^* > 0$). Then the operator inequality*

$$B \geq (\tau/2)A \tag{11}$$

is a necessary and sufficient condition for the stability of difference scheme (9) in H_A and H_B.

In (Samarskii, 1967), a difference scheme regularization principle is proposed and illustrated by numerous examples. Remedies for improving the quality (stability) of two and three level schemes of the general form are formulated. In this report, we shall apply the ideas of regularization first to the above formulated well-posed problem (1)–(4).

Consider the following simplest explicit difference scheme for the differential-discrete problem (7), (8):

$$\frac{y_{n+1} - y_n}{\tau} + L_h y_n = 0, \quad n = 0, 1, \dots . \tag{12}$$

It is well known (Samarskii, 1983; Samarskii and Gulin, 1975) that this scheme is stable only when the time-step τ is small enough ($\tau \leq O(h^2)$). We shall show how the regularization principle can be used to obtain absolutely stable difference schemes, based on the explicit scheme (12).

The scheme (12) becomes (9) if $B = E$, where E is the unity operator, and $A = L_h$. The regularization procedure is essentially the transition from the original scheme to an another (perturbed) one. Let $R = R^* > 0$ be a difference regularization operator and let α denote the regularization (perturbation) parameter. In accordance with the necessary and sufficient condition (11), it is natural to relate the regularization to some additive perturbation of the difference operator B. Therefore, the regularized scheme can be written as

$$(E + \alpha R)\frac{y_{n+1} - y_n}{\tau} + L_h y_n = 0, \quad n = 0, 1, \dots . \tag{13}$$

For regularized scheme (13) the following theorem holds.

THEOREM 2. *Difference scheme (13) with $L_h = L_h^* > 0$ is unconditionally stable in H_A and H_B provided that either the regularizator is chosen as $R = L_h$, and then $\alpha \geq \tau/2 - \|L_h\|^{-1}$, or the regularizator is $R = L_h^2$, and then $\alpha \geq \tau^2/16$.*

The choice $R = L_h^2$ for the regularizator has rarely been considered, though, in our opinion, it is of great interest. The proof of the theorem is based on the verification of necessary and sufficient condition of stability (11). For scheme (13) we have

$$B = E + \alpha R, \qquad A = L_h. \tag{14}$$

Taking into account $L_h \leq \|L_h\| E$ with $R = L_h$ and substituting (14) into (11) we have $(\|L_h\|^{-1} + \alpha - \tau/2)L_h \geq 0$, whence the assertion of the theorem follows. Note that by inequalities (6), $\|L_h\| \leq \Delta$.

When the regularizator $R = L_h^2$ is used, we see that, by (14), inequality (11) takes the form $B - \tau/2A = E + \alpha L_h^2 - \tau/2L_h = (\alpha^{1/2}L_h - \tau/(4\alpha^{1/2})E)^2 + (1$

$-\tau^2/(16\alpha))E \geq 0$. Hence we conclude that for $R = L_h^2$ and $\alpha \geq \tau^2/16$ regularized difference scheme (13) is absolutely stable.
Let us note that for $\alpha = \sigma\tau$ the choice $R = L_h$ corresponds to the usual weighted scheme. Such regularized schemes for equations of parabolic type were studied as early as in (Samarskii, 1967).
We have improved the quality of explicit scheme (12) by perturbing the difference operator B. At the same time the perturbance of the operator A may also be exploited in the context of regularization when two-level difference schemes are treated. Let us consider, instead of (12), the following scheme:

$$\frac{y_{n+1} - y_n}{\tau} + (L_h - \alpha R)y_n = 0, \quad n = 0, 1, \dots . \tag{15}$$

THEOREM 3. *Difference scheme* (15) *with* $L_h = L_h^* > 0$ *is stable in* H_A *and* H_B, *if the regularizator is chosen as* $R = L_h^2$, *provided that the regularization parameter satisfies* $1/\Delta > \alpha \geq \tau/8$.

The requirement that the operator A is positive definite leads to the restriction $1/\Delta > \alpha$, and (11) is transformed into $B - \tau/2A = E + \tau/2(\alpha L_h^2 - L_h) = \tau/2(\alpha^{1/2}L_h - 1/(2\alpha^{1/2})E)^2 + (1 - \tau/(8\alpha))E \geq 0$. Hence, the restriction $\alpha \geq \tau/8$ follows.
Explicit scheme (15) is conditionally stable. It follows from the theorem stated above, that the time-step should be restricted as $\tau < 8/\Delta$. Comparing this with the stability condition for the original explicit scheme ($\tau \leq 2/\Delta$), one can conclude that regularized scheme (15) allows the time-step as large as four times the time-step allowed in the original scheme. We note that some other schemes of type (15) are considered in (Sauljev, 1960).

3. INVERSE PROBLEM

Consider now some aspects of constructing stable difference schemes for ill-posed evolution problems. One of the simplest problems of this kind is the problem with the backward time for a parabolic equation. We wish to solve for u the equation

$$\frac{\partial u}{\partial t} - Lu = 0, \quad (x,t) \in Q, \tag{16}$$

together with boundary and initial data (3), (4). The problem (1), (3), (4), (16) is ill-posed due to the unstability of the solution with regard to small perturbations in the initial data (the function $\Phi(x)$). This problem gives rise to the differential-discrete problem for the equation

$$\frac{dy}{dt} - L_h y = 0. \tag{17}$$

To solve the ill-posed problems like the above-mentioned ones, a more accurate definition of the solution to the problem is needed. If the problem is conditionally ill-posed, i. e., well-posed according to Tikhonov (Tikhonov and Arsenin, 1986), then the solution has to belong to some prescribed set. The restriction to such a set allows us in some cases to deal with a well-posed problem instead of an ill-posed one.

A priori restrictions which are to be imposed to define a particular set may be of different classes. For the retrospective inverse problem under consideration (1),(3), (4) and (16), the correctness holds for bounded solutions.

When treating the difference schemes for ill-posed problems like (1), (3),(4) and (16), the stability condition like (10) is not suitable. The cause of it is that the solution of the inverse problem (its norm) for the heat transfer equation is growing, in contrast with the solution of the direct problem. The positive definitness of the operator L_h reflects this feature in the corresponding differential-discrete problem (8), (17).

The reason the problem under consideration is ill-posed is that the growth of spurious high-frequency harmonics is overwelming. It is required that the low-frequency harmonics be allowed to grow while the growth of the high-frequency ones be limited. If the growth of the solution is allowed, what is usual, let us remind, for the inverse retrospective problem (1), (3), (4), (16), then the stability condition of a difference scheme (10) should be made more precise. Thus, the notion of ρ-stability of difference schemes naturally arises.

The difference scheme (9) is said to be ρ-stable (Samarskii, 1983; Samarskii and Gulin, 1975), if the condition

$$\|y_{n+1}\|_D \leq \rho \|y_n\|_D, \tag{18}$$

is fulfilled with $\rho > 0$. The characteristic feature of ill-posed problems (the norm of the solution grows in time) is reflected in the requirement that the values $\rho > 1$ are to be chosen.

Two particular ways of determining ρ can be suggested:

$$\rho = \exp(c, \tau), \qquad \rho = 1 + c\tau,$$

where c is a constant independent of the mesh (i.e., of neither τ nor h). For the values of ρ indicated above, from (18) we derive the following stability estimate for the discrete solution with respect to the initial data:

$$\|y_{n+1}\|_D \leq \exp(ct_{n+1})\|y_0\|_D.$$

One of the special feature of the difference schemes for ill-posed problems is that the operator A is no more positive definite. This is why the difference schemes are studied in H_B.

THEOREM 4. *Let the operators A and B in (9) be independent of n and self-adjoint; let B be positive definite. Then the inequalities*

$$\frac{1-\rho}{\tau} B \leq A \leq \frac{1+\rho}{\tau} B \tag{19}$$

constitute necessary and sufficient conditions for the stability of difference scheme (9) in H_B.

Now we shall apply the regularization principle to the construction of difference schemes for ill-posed problem (8),(17). Consider first the explicit difference scheme for (8), (17):

$$\frac{y_{n+1} - y_n}{\tau} - L_h y_n = 0, \quad n = 0, 1, \dots \tag{20}$$

Its canonical form as in (9) shows that $B = E$, $A = -L_h$, i.e., $A = A^* < 0$.

THEOREM 5. *Explicit scheme (20) is ρ-stable in H with*

$$\rho = 1 + \Delta\tau, \tag{21}$$

where Δ is defined in (6).

Theorem 5 reveals that the usage of common difference schemes for solving approximately ill-posed problems produces the effect of regularization. This feature is closely related to the boundness of the discrete operator L_h. Noticing that $\Delta = O(h^{-2})$, one can conclude that the spatial mesh parameter bounds the growth of the solution, i.e., acts as a regularization parameter.
The proof is based upon the verification of ρ-stability condition (19) for scheme (20). For $B > 0$, $A < 0$ and $\rho > 1$ the right inequality in (19) is valid for all $\tau > 0$. The left one implies $(\rho - 1)/\tau E \geq L_h$, which is true when ρ is chosen as in (21).
Proceeding in the same way as in the case of the direct problem (see (14)), we represent the regularized scheme in the canonical form (9) with

$$B = E + \alpha R, \qquad A = -L_h. \tag{22}$$

THEOREM 6. *Regularized scheme (9), (22) is ρ-stable in H_B with*

$$\rho = 1 + \tau/\alpha, \tag{23}$$

if $R = L_h$, and with

$$\rho = 1 + 2^{-1}\alpha^{-1/2}\tau \tag{24}$$

if $R = L_h^2$.

The proof is essentially the verification of the left inequality in (19), which in view of (22) becomes

$$(\rho - 1)/\tau(E + \alpha R) \geq L_h. \tag{25}$$

The last inequality holds for $R = L_h$ when ρ is chosen as in (23).
For $R = L_h^2$, (25) becomes

$$E + \alpha L_h^2 - \frac{\tau}{\rho - 1}L_h = (\alpha^{1/2}L_h - \frac{\tau}{2(\rho - 1)}\alpha^{-1/2}E)^2 +$$
$$+\left(1 - \frac{\tau^2}{4(\rho - 1)^2}\alpha^{-1}\right)E \geq 0.$$

The above holds with ρ form (24). The theorem is proved.

4. GENERALIZATIONS

In the previous sections we confined ourselves to some simplest illustrations of the application of the regularization principle to the derivation of stable difference schemes for well- and ill-posed problems. Now let us mention some other investigations in this area.

In the stability theory of difference schemes (Samarskii, 1983; Samarskii and Gulin, 1975) basic results have been obtained for both two and three level difference schemes. Based on the regularization principle, stable three level difference schemes have been developed for boundary value problems for the second order parabolic equations.
Moreover, well-posed and ill-posed problems for the second order evolution equations have been studied. For example, regularized difference schemes have been considered for the Cauchy problem for the elliptic equation

$$\frac{\partial^2 u}{\partial t^2} - Lu = 0, \quad (x,t) \in Q,$$

subject to conditions (3),(4), complemented by the second initial data

$$\frac{\partial u}{\partial t}(x,0) = \psi(x), \quad x \in \Omega.$$

When solving approximately multidimensional problems in mathematical physics, a great attention is paid to the construction of economic schemes. Based upon the regularization principle, factorized difference schemes and schemes of integral approximation have been developed.
We have restricted ourselves to the consideration of problems with a self-adjoint operator L_h. There have been some interesting generalizations covering the case of an arbitrary (inclusive indefinite) operator L_h.

REFERENCES

Norrie, D.H. and de Vries, G. (1978). *An Introduction to Finite Element Analysis*. Academic Press, New York.

Samarskii, A.A. (1983). *The Theory of Difference Schemes*. Nauka, Moscow (in Russian).

Samarskii, A.A. and Gulin, A.V. (1975). *Stability of Difference Schemes*. Nauka, Moscow (in Russian).

Samarskii, A.A. (1967). On the regularization of difference schemes. *J. Comp. Math. and Math. Phys.* **7**, 62–93 (in Russian).

Samarskii, A.A. and Vabishchevich, P.N. (1990). Difference schemes for unstable problems. *Math. Mod.* **2**, (11), 89–98 (in Russian).

Sauljev, V.K. (1960). *Integrating Equations of Parabolic Type by the Net-Based Methods*. Fizmatgiz, Moscow (in Russian).

Tikhonov, A.N. and Arsenin, V.Ya. (1986). *Methods for Solving Ill-Posed Problems*. Nauka, Moscow (in Russian).

Ill-Posed Problems in Natural Sciences, pp. 174 – 183
A. Tikhonov (Ed.)
1992 VSP/TVP

ILL-POSED PROBLEMS OF ON-LINE CONDITIONALLY OPTIMAL FILTERING

I.N. SINITSYN
Institute of Informatic Problems, Russia Academy of Science, Vavilova str. 30/6, 117900 Moscow, Russia

ABSTRACT

Modern filtering problems or the problems of on-line estimation of variables and parameters in nonlinear stochastic hereditary systems using current continuous and discrete observation often arise in scientific experiment. The main idea of conditionally optimal filtering (in Pugachev sense) consists in restricting the class of admissible (computationally simple) filters in such a way that any admissible filter could realize on-line estimation directly in experiment. The main factors that lead to the ill-posedness are the unsufficient smoothness of observations, hereditary kernels, nonlinear functions in Volterra integral equations of the first kind and the non-Markov behaviour of random processes. Well-posed conditionally optimal continuous and discrete solutions based on the mean square error and complex statistical criteria of ill-posed on-line nonlinear filtering problems are considered. Specialized software (intellectualized package for the analysis and modelling of stochastic processes computer-aided filter and identificator design on IBM PC XT/AT "StS-filter") and typical applications in automization and control theory are discussed. It is noted that continuous and discrete conditionally optimal filtering may be directly used for numerical solutions of nonlinear ill-posed stochastic problems. In conclusion some generalizations are considered.

1. INTRODUCTION

Problems of estimation of the variables and parameters in various hereditary systems using some observations often arise in practice. All estimation problems in heriditary systems may be divided into two classes. One class covers the problems in which data processing may be performed only after the experiments when all observations are finished. The other covers the problems of on-line estimation in which data processing must be performed directly in the process of experiment while receiving the results of observations. The characteristic feature of estimation problems of the first class is the usage of all the results of observations. The characteristic feature of estimation problems of the second class is the possibility of using at any instant only those results of observations which occured before this instant. It should be emphasized that the use of the methods designed for solving the problems of on-line estimation for posterior data processing is senseless. By the same reason only such methods of solving the ptoblems of the second class are of practical use which may be realized in real time.

We shall consider here the problem of on-line estimation of the hereditary system state and parameter identification based on the past observations. The requirement that the estimation should be performed in on-line regime imposes the following rigid constraint: the calculation of estimates should be sufficiently easy for on-line realization. This requirement gave rise to the theory of on-line conditionally optimal estimation. The problem of estimation of the current state and the parameters of a system is generally called the problem of filtering. The theory of on-line conditionally optimal filtering (COF) was first developed by Pugachev (Pugachev, 1978a,b). The main idea of COF consists in restricting the class of admissible filters in such a way that any admissible filter could realize the on-line estimation. Thus, we give up the absolute optimality of estimators for the sake of their computational simplicity. Filters determining estimates as the solutions of rather simple differential or difference equations may serve as examples of filters satisfying the requirement of computational simplicity. For the current state of the art in the COF theory and applications see (Pugachev and Sinitsyn, 1985, 1987, 1990; Pugachev et al., 1986, 1987, 1991a,b; Raol and Sinha, 1986; Sinitsyn, 1990). The COF theory and software for nonlinear stochastic sampled data systems (SSDS), stochastic differential systems (SDS), mixed sampled data and differential systems are well developed. The COF theory for stochastic hereditary systems (SHS) and software are developed only for some classes of systems reducible to SDS (Sinitsyn 1986a,b; Pugachev and Sinitsyn, 1990).

2. PROBLEM STATEMENT

A wide range of scientific research and automated design and control problems for hereditary systems (e.g., systems which contain elements described by hereditary mechanics laws, ecological, medico-biological and man-machine systems) under the conditions of random disturbances and stochastic hereditary leads to mathematical models, described by the Ito stochastic integro-differential equations of two forms:

$$Y = \int_{t_0}^{t} [w_1(t,s)\varphi_1(Z,s)ds + w_2(t,s)\varphi_2(Z,s)]dW(s)$$

$$+ \int_{\mathbf{R}^q} w_3(t,s)\varphi_3(Z,s,v,)P^0(ds,dv), \tag{1}$$

$$dZ = a(Y,Z,t)dt + b(Y,Z,t)dW + \int_{\mathbf{R}^q} c(Y,Z,t,v)P^0(dt,dv), \tag{2}$$

$$dY = a_1(Y,Z,t)\,dt + b_1(Y,Z,t)dW$$

$$+ \int_{\mathbf{R}^q} c_1(Y,Z,t,v)P^0(dt,dv), \tag{3}$$

$$dZ = a(U,Y,Z,t)dt + b(U,Y,Z,t)dW$$

$$+\int_{\mathbf{R}^q} c(U,Y,Z,t,v)P^0(dt,dv), \tag{4}$$

$$U=\int_{t_0}^{t}[w_1(t,s)\varphi_1(Y,Z,s)ds+w_2(t,s)\varphi_2(Y,Z,s)dW(s)$$

$$+\int_{\mathbf{R}^q} w_3(t,s)\varphi_3(Y,Z,s,v)P^0(ds,dv)] \tag{5}$$

for continuous observations and mixed stochastic integro-differential and difference equations of the form (4) and (5) and

$$\begin{aligned} Y_n &= \omega_n(Z_n,W_n)\equiv a_{1n}(Z_n)+b_{1n}(Z_n)W_n,\\ Y_n &= Y(t_n),\quad Z_n=Z(t_n),\quad t_n=hn,\quad n=0,1,2,\cdots \end{aligned} \tag{6}$$

for discrete observations. Here $Z=Z(t)$, $Z_n\in\mathbf{R}^p$ is the unobservable state vector, $Y=Y(t)$, $Y_n\in\mathbf{R}^m$ is the observation vector; $U=U(t)\in\mathbf{R}^h$; $W=W(t)\in\mathbf{R}^k$ is the Wiener process with intensity $\nu=\nu(t)$; $P^0(\Delta,\Delta_q)$ is the centred Poisson measure, $\Delta=(t_1,t_2)$, $\Delta_q\in\mathbf{R}^q$, $v\in\mathbf{R}^q$. Thus,

$$P^0(\Delta,\Delta_q)=P(\Delta,\Delta_q)-\mathbf{E}P(\Delta,\Delta_q)\equiv P(\Delta,\Delta_q)-\mu(\Delta,\Delta_q)$$

and the integral $\int_\Delta(dt,\Delta_q)$ is the number of jumps in the interval Δ of the Poisson process $P(t,\Delta_q)$ and $\nu_P(t,\Delta_q)$ is its intensity. The processes $W(t)$ and $P(t,\Delta_q)$ on any Borel set Δ_q are supposed to be mutually independent and independent of the initial values Y_0 and Z_0. The integrals in (1) and (5) with respect to v are defined over the whole space $\mathbf{R}^q$ with removed origin; $\{W_n\}$ is a sequance of independent random variables with a known distribution; $\omega_n=\omega_n(z_n,v_n)$ are known functions which map $\mathbf{R}^{p+r}$ into $\mathbf{R}^m$; h is the interval discrete observations of t_0 is an instant (not necessarily initial).

We assume that in (1) and (2) $w_i=w_i(t,s)$, $i=1,2,3$, are $(m\times m)$ are matrix hereditary kernels; $\varphi_1=\varphi_1(z(s),s)$, $\varphi_2=\varphi_2(z(s),s)$, $\varphi_3=\varphi_3(z(s),s,v)$ are generally nonlinear $(m\times 1)$, $(m\times k)$, $(m\times 1)$ matrix functions; $a=a(y(t),z(t),t)$, $b=b(y(t),z(t),t)$ and $c=c(y(t),z(t),t,v)$ are functions whicn map $\mathbf{R}^{m+p+1}$ and $\mathbf{R}^{m+p+q+1}$ into $\mathbf{R}^p$, $\mathbf{R}^{pk}$, $\mathbf{R}^p$ respectively and in (3)–(5) $w_i=w_i(t,s)$, $i=1,2,3$, are $(h\times h)$ matrix heriditary kernels; $\varphi_1=\varphi_1(y(s),z(s),s)$, $\varphi_2=\varphi_2(y(s),z(s),s)$, $\varphi_3=\varphi_3(y(s),z(s),s,v)$ are generally nonlinear $(h\times 1)$, $(h\times k)$, $(h\times 1)$ matrix functions,

$$a=a(u(t),y(t),z(t),t),\quad a_1=a_1(y(t),z(t),t),$$

$$b=b(u(t),y(t),z(t),t),\quad b_1=b_1(y(t),z(t),t),$$

$$c=c(u(t),y(t),z(t),t,v),\quad c_1=c_1(y(t),z(t),t,v),$$

are functions which map $\mathbf{R}^{h+m+p+1}$, $\mathbf{R}^{h+m+p+q+1}$ into $\mathbf{R}^p$, $\mathbf{R}^m$, $\mathbf{R}^{pk}$, $\mathbf{R}^{mk}$, $\mathbf{R}^p$, $\mathbf{R}^m$.

Let random processes U, Y and Z represent a strong solutions of (1), (2), (3)–(5), (4)–(6) (for sufficient conditions see (Ponosov, 1990; Azbelev et al., 1991)).

According to the main idea of the COF theory in the case of continuous observations, the class of admissible conditionally optimal filters satisfies the following filter equation

$$d\hat{Z} = \alpha\xi(Y, \hat{Z}, t)dt + \beta\eta(Y, \hat{Z}, t)dY + \gamma dt. \tag{7}$$

where $\xi = \xi(y, \hat{z}, t)$ and $\eta = \eta(y, \hat{z}, t)$ are known functions which map $\mathbf{R}^{m+p+1}$ into $\mathbf{R}^r$ and $\mathbf{R}^{sm}$; $\alpha = \alpha(t)$, $\beta = \beta(t)$ and $\gamma = \gamma(t)$ are unknown $(p\times r)$, $(p\times s)$, $(p\times 1)$ coefficients determined from the mean square error criterion

$$R_e(t) = \mathbf{E}|e(t)|^2 = \min, \quad e(t) = \hat{Z}(t) - Z(t) \tag{8}$$

at any t and at infinitesimally close next t', $t' \to t$. In the case of discrete observations the filter is taken in the form

$$X_n = \alpha_n\xi_n(\hat{Z}_n) + \beta_n\eta_n(\hat{Z}_n)Y_n + \gamma_n, \tag{9}$$

where $\xi_n = \xi_n(\hat{Z}_n)$ and $\eta_n = \eta_n(\hat{Z}_n)$ are known functions ; the unknown coefficients α_n, β_n, γ_n are found from the criterion

$$R_{e_{n+1}} = \mathbf{E}|e_{n+1}|^2 = \min, \quad e_{n+1} = \hat{Z}_{n+1} - Z_{n+1} \tag{10}$$

for each n and the next instant t^{n+1}.
The COF problem for SHS described by equations (1), (2), ((3)–(5) or (4)–(6)) for arbitrary hereditary kernels w_i and nonlinear functions φ_i is ill-posed. The main factors that lead to the ill-posedness are the unsufficient smoothness of the kernels w_i, the nonlinearity of the functions φ_i, and the observations Y and also the fact that for arbitrary kernels w_i the processes Y, Z (U, Y, Z) are nonmarkovian (even for linear systems). Thus we state the problem to find well-posed on-line COF methods for SHS with jump component reducible to SDS and SSDS of autoregression type and corresponding software.

3. CONTINUOUS OBSERVATIONS

There are exist practically important classes of hereditary kernels for which (1), (2) (or (3)–(5)) may be reduced to Ito stochastic differential equations for the extended vectors $\bar{Y}$ and $\bar{Z}$

$$d\bar{Y} = \bar{a}_1(\bar{Y}, \bar{Z}, t)dt + \bar{b}_1(\bar{Y}, \bar{Z}, t)dW + \int_{\mathbf{R}^q} \bar{c}_1(\bar{Y}, \bar{Z}, t, v)P^0(dt, dv),$$

$$d\bar{Z} = \bar{a}(\bar{Y}, \bar{Z}, t)dt + \bar{b}(\bar{Y}, \bar{Z}, t)dW + \int_{\mathbf{R}^q} \bar{c}(\bar{Y}, \bar{Z}, t, v)P^0(dt, dv). \tag{11}$$

Equation (11) determines a Markov process. The condition that (1), (2) ((3)–(5)) reduce to (11) is equivalent to the condition that the Volterra integral equations (1) (or (5)) of the 1st kind can be transformed to the Volterra equations of the 2nd kind (i.e., to a well-posed problem).

Following (Pugachev and Sinitsyn, 1987, 1990) for (11) (with bars omitted) from the equations for optimal COF coefficients α, β and γ in (7)

$$\alpha m_1 + \beta m_2 + \gamma = m_0, \quad \beta \varkappa_{22} = \varkappa_{02},$$

$$\alpha \varkappa_{11} + \mathbf{E}(\hat{Z} - Z)(\xi^T \alpha^T + \gamma^T)\frac{\partial \xi^T}{\partial \hat{Z}} = \varkappa'_{01} - \beta \varkappa'_{21}$$

with coefficients defined by

$$m_0 = \mathbf{E}a, \quad m_1 = \mathbf{E}\xi, \quad m_2 = \mathbf{E}\eta a_1, \quad \varkappa_{01} = \mathbf{E}(a - m_0)\xi^T,$$

$$\varkappa_{11} = \mathbf{E}(\xi - m_1)\xi^T, \quad \varkappa_{22} = \mathbf{E}\eta b_1 \nu b_1^T \eta^T,$$

$$\varkappa_{21} = \mathbf{E}(a_1 - m_2)\xi^T, \quad \varkappa'_{21} = \mathbf{E}(\eta b_1 - m_2)\xi^T,$$

$$\varkappa'_{01} = \varkappa_{01} + \mathbf{E}(Z - \hat{Z})\frac{\partial \xi^T}{\partial t} + \mathbf{E}\{(Z - \hat{Z})[a_1^T - \int_{\mathbf{R}^q} c_1^T \nu_P(t,v)\, dv]$$

$$+ b\nu b_1^T - \beta \eta b_1 \nu b_1^T\} \left(\frac{\partial}{\partial y} + \eta^T \beta^T \frac{\partial}{\partial \hat{Z}}\right) \xi^T$$

$$+ \frac{1}{2}\mathbf{E}(Z - \hat{Z})\left\{\mathrm{tr}\left[b_1 \nu b_1^T \left(\frac{\partial}{\partial y} + 2\eta^T \beta^T \frac{\partial}{\partial \hat{Z}}\right) \frac{\partial^T}{\partial y}\right]\right.$$

$$\left. + \mathrm{tr}\left[\beta \eta b_1 \nu b_1^T \eta^T \beta^T \frac{\partial}{\partial \hat{Z}} \frac{\partial^T}{\partial \hat{Z}}\right]\right\} \xi^T$$

$$+ \int_{\mathbf{R}^q} \mathbf{E}[Z - \hat{Z} + c - \beta \eta c_1][\xi(Y + c, \hat{Z} + \beta \eta c_1, t)^T - \xi^T]\nu_P(t,v)\, dv,$$

$$\varkappa_{02} = \mathbf{E}(Z - \hat{Z})a_1^T \eta^T + \mathbf{E}b\nu b_1^T \eta^T,$$

we obtain the following theorem.

THEOREM 1. *Let the hereditary kernels admit the reduction of the nonlinear equations (1), (2) (or (3)–(5)) to equation (11), random processes Y and Z be strong solutions of (11), matrices $\varkappa_{11}$ and $\varkappa_{22}$ be invertible. Then the COF problem for SHS reducible to SDS is well-posed.*

Following (Sinitsyn and Shin, 1991a), we have an analogous result for the complex statistical criterion

$$I = I(m_e, R_e) = \min, \quad m_e = \mathbf{E}e, \quad R_e = \mathbf{E}ee^T.$$

4. DISCRETE OBSERVATIONS

Using the Euler discretization procedure at the observation interval h (see (6)) we obtain the following stochastic difference equations for the so-called strong approximation of equations (4), (5):

$$Z_{n+1} = a_n(U_n, Y_n, Z_n) + b_n(U_n, Y_n, Z_n)V_n, \tag{12}$$

$$U_n = a'_n(Y_n, Z_1, \cdots, Z_n) + b'_n(Y_n, Z_1, \cdots, Z_n)V_n. \tag{13}$$

Here it is assumed that

$$a_n = z_n + ha(U_n, Z_n, hn), \quad b_n = [b_{1n} \cdots b_{N+1,n}],$$

$$b_{1n} = b_1(U_n, Y_n, Z_n, hn), \quad b_{in} = c_{i-1}(Z_n, hn), \quad i = 2, \cdots, N+1,$$

$$a'_n = a'_n(Y_n, Z_1, \cdots, Z_n) = h \sum_{j=1}^{n} w_{1j}(hn, hj)\varphi_j(Z_j, hj),$$

$$b'_n = b'_n(Y_n, Z_1, \cdots, Z_n) = [b'_{1n} \cdots b'_{N+1,n}],$$

$$b'_{1n} = \sum_{j=1}^{n} w_2(hn, hj)\varphi_2(Z_j, hj),$$

$$b'_{in} = \sum_{j=1}^{n} w_3(hn, hj)\varphi_3(Z_j, hj, v_i), \quad v_i \in \Delta_q^i \in \Delta_q,$$

$$V_n = [V_{1n}^T V_{2n} \cdots V_{N+1,n}]^T,$$

$$V_{1n} = W(h(n+1)) - W(hn), \quad V_{in} = P_{i-1}^0(h(n+1)) - P_{i-1}^0(hn).$$

The vector blocks V_{1n} have the normal distribution

$$\mathcal{N}(0, \bar{G}_n), \quad \bar{G}_n = \int\limits_{hn}^{h(n+1)} \nu(s)\, ds \approx h\nu(hn),$$

and the scalar blocks V_{in} $(i = 2, \cdots, N+1)$ have the Poisson distribution with the parameters

$$\mu_{in} = \int_{hn}^{h(n+1)} \nu_i(s)\, ds \approx h\nu_i(hn),$$

$\nu_i(t) = (\nu(t), \Delta_q^i) = d\mu(\Delta, \Delta_q^i)/dt$ being the intensities of the centred simple Poisson processes $P_i^0(t) = P^0(\Delta, \Delta_q^i) - \mu(\Delta, \Delta_q^i)$. The covariance matrix G_n of the mutually independent vectors V_n has the block-diagonal form

$$G_n = [\bar{G}_n \operatorname{diag}(\mu_{2n} \cdots \mu_{N+1,n})].$$

Equations (6), (12), and (13) determine Y_{n+1}, Z_{n+1} for $Y_n = y_n$, $Z_n = z_n$ up to h at a_n, a'_n and $h^{1/2}$ at b_n, b'_n. For a more accurate approximation of equations (12), (13) (up to the order of h^3 and $h^{5/2}$ and higher), see (Kloeden and Plathen, 1989).

Following (Pugachev, 1984) for SHS described by (6), (12), (13) and reducible to SSDS of the autoregression type

$$\bar{Y}_n = \bar{a}_{1n}(\bar{Z}_n) + \bar{b}_{1n}(\bar{Z}_n)V_n, \quad \bar{Z}_{n+1} = \bar{a}_n(\bar{Y}_n, \bar{Z}_n) + \bar{b}_n(\bar{Y}_n, \bar{Z}_n)V_n \tag{14}$$

from the equations for optimal coefficients α_n, β_n and γ_n in (9)

$$\alpha_n \varkappa_{11}^{(n)} + \beta_n \varkappa_{12}^{(n)} = \varkappa_{01}^{(n)}, \quad \alpha_n \varkappa_{21}^{(n)} + \beta_n \varkappa_{22}^{(n)} = \varkappa_{02}^{(n)},$$

$$\gamma_n = m_n - \alpha_n l_1^{(n)} - \beta_n l_2^{(n)}$$

with coefficients defined by

$$\varkappa_{01}^{(n)} = \mathbf{E}(a_n - m_n)\xi_n^T, \quad m_n = \mathbf{E}a_n, \quad l_1^{(n)} = \mathbf{E}\xi_n,$$

$$l_2^{(n)} = \mathbf{E}\eta_n a_{1n}, \quad \varkappa_{11}^{(n)} = \mathbf{E}(\xi_n - l_1^{(n)})\xi_n^T,$$

$$\varkappa_{02}^{(n)} = \mathbf{E}[(a_n - m_n)a_{1n}^T\eta_n + b_{1n}G_n b_{1n}^T\eta_n^T],$$

$$\varkappa_{12}^{(n)} = \varkappa_{21}^{(n)T} = \mathbf{E}(\xi_n - l_1^{(n)})a_{1n}^T\eta_n^T,$$

$$\varkappa_{22}^{(n)} = \mathbf{E}[(\eta_n a_{1n} - l_2^{(n)})a_{1n}^T\eta_n^T + \eta_n b_{1n}G_n b_{1n}^T\eta_n^T]$$

and bars omitted we obtain the following theorem.

THEOREM 2. *Let hereditary kernels admit the reduction of the nonlinear equations (6), (12), (13) to equation (14), the random processes Y_n and Z_n be strong solutions of (14) the matrices $\varkappa_{11}^{(n)}$ and $\varkappa_{22}^{(n)}$ be invertible. Then the COF problem for SHS reducible to the autoregression SSDS is well-posed.*

Following (Sinitsyn and Shin, 1991b), we get an analogous result for the complex statistical criterion

$$I_{n+1} = I(m_{e_{n+1}}, R_{e_{n+1}}) = \min, \quad m_{e_{n+1}} = \mathbf{E}e_{n+1}, \quad R_{e_{n+1}} = \mathbf{E}e_{n+1}e_{n+1}^T.$$

5. SOFTWARE

On practice, the following two main types of on-line COF algorithms are recommended for continuous and discrete observations, respectively:

— approximation of the hereditary kernels in (1) (or (5)) by finite sums of degenerate kernels or kernels which are weigth functions of some linear differential systems and the reduction of SHS to SDS by a finite extension of the state and observation vectors;

— transformation of equations (1) (or (5)) to the integral Volterra equations of the 2nd kind using integration by parts, differentiation, deterministic or statistical regularization and discrete reduction of SHS to SSDS of autoregression type.

For some practically useful details of continuous and discrete reduction see, e.g., (Tikhonov and Arsenin, 1979; Sinitsyn, 1986a,b; Pugachev and Sinitsyn, 1990; Fedotov, 1991; Azbelev et al., 1991; etc).

The calculations necessary to construct conditionally optimal filters are rather cumbersome and time consuming. But these calculations are based only on the prior data and do not involve observations. Consequently, all these calculations may be fulfilled beforehand while designing the filter. Nevertheless, these calculations cannot be carried out without corresponding software especially for problems of high dimension. It is easily seen from § 3 and 4 that the software for analysis problems should be included into the software for filter design. This software should provide an automatic construction of the equations for appropriate parameters of distributions and their solution together with the equations for the optimal coefficients of filter equations.

For the specialized software packages "KERNEL", "StS-ANALYSIS", "MOMENT", "StS-FILTER", see (Pugachev and Sinitsyn, 1990; Sinitsyn, 1990; Pugachev et al., 1987, 1991a,b).

6. "StS-FILTER: AN INTELLECTUALIZED PACKAGE FOR ANALYSIS AND MODELLING OF STOCHASTIC PROCESSES, COMPUTER-AIDED FILTER AND IDENTIFICATOR DESIGN ON IBM PC/XT/AT

The intellectualized package "StS-filter" (version 1.1) for IBM PC/XT/AT compatible computers realizes the methods of statistical analysis, modelling, conditionally on-line optimal filtering and identification of processes and parameters in stochastic systems (StS). It facilitates the statistical research and design higher-order linear and nonlinear StS on personal computers. The "StS-filter" may be used for modelling, accuracy and sensivity analysis, parameter identification and optimization, computer-aided optimal filters and identificators design on PC. The software may be used in subject-oriented courses like "Probability Theory and Mathematical Statistics", "Random Processes", "Signal Processing", "Modelling and Simulation", "CAD/CAM/CAE" and in studying statistical aspects of informatics, control theory, mechanics, physics, economy, biology, ecology, medicine etc. (See (Pugachev and Sinitsyn, 1991b)).

The package represents the integrated environment including the following modules

1. An original mathematical word processor allowing
   — input, editing mathematical models in the conventional mathematical form;
   — performance of the syntax and semantic analysis;
   — input of text information for models.
2. A compiler.
3. A mathematical module which implements
   — installation of initial data;
   — calculation of probabilistic characteristics of the model;
   — calculation of filters and identificators coefficients;
   — modelling;
   — calculation of mean square errors;
   — saving the results of calculation on disk.
4. A graphic visualization module providing

— simultaneous overview of several characteristics during the computer experiment;
— output of several graphs for required characteristics;
— scaling and microscope regime;
— hard copy reception of any graphs;
— output of results table on printer.

Modules control, selection, copying and deleting of models is performed by using the hierarchical menu systems.
"StS-filter" functions under control of the operational systems MS-DOS of the versions 3.00 and higher. It is realized on C programming language. A color graphical display with an E←A or VGA adapter is required for package functioning.

7. CONCLUSIONS

The well-posed on-line COF methods considered above are easily generalized for (1), (5) with arbitrary Hammerstein integrands (both deterministic and statistical) using canonical expansions of random functions (see (Pugachev and Sinitsyn, 1990)). The further development of the COF theory must include investigation of more complex stochastic hereditary operators with autocorrelated noises and parallel numerical algorithms oriented on classical and new computer architectures. It should be noted that the COF theory may be used directly for solution of nonlinear ill-posed stochastic problems.

REFERENCES

Azbelev, N.V., Maksimov, V.P., and Rakhmatullina, L.F. (1991). *Introduction to the Theory of Functional Differential Equations*. Nauka, Moscow (in Russian).

Fedotov, A.M. (1990). *Ill-Posed Problems with Random Errors Data*. Nauka, Novosibirsk (in Russian).

Kloeden, P.E. and Platen, E.A. (1989). A survey of numerical methods for stochastic differential equations. *Stochastic Hydrology and Hydraulics*. **3**, 155–178.

Ponosov, A.V. (1990). To the theory of reduction of stochastic functional-differential equations on semimartingale. In: *Functional-Differential Equations*. Perm' Univ. Press, Perm'. pp. 111–120 (in Russian).

Pugachev, V.S. (1965). *Theory of Random Functions and Its Applications to Control Problems*. Pergamon Press, London.

Pugachev, V.S. (1978a). Estimation of variables and parametes in stochastic systems described by difference equations. *Soviet Math. Docl.* **19**, 967–971.

Pugachev, V.S. (1978b). Recursive estimation of variables and parameters in stochastic systems described by difference equations *Soviet Math. Docl.* **19**, 1495–1497.

Pugachev, V.S. (1984). *Probability Theory and Mathematical Statistics for Engineers*. Pergamon Press, New York.

Pugachev, V.S. and Sinitsyn, I.N. (1987). *Stochastic Differential Systems. Analysis and Filtering*. John Wiley, Chichester.

Pugachev, V.S. and Sinitsyn, I.N (1990). *Stochastic Differential Systems. Analysis and Filtering.* Second Ed. Nauka, Moscow (in Russian)

Pugachev, V.S., Sinitsyn, I.N., and Shin, V.I. (1986). Problems of analysis and on-line conditionally optimal filtering of processes in nonlinear stochastic systems. In: *Preprints of the Second IFAC Symposium on Stochastic Control (Vilnius, USSR, 19–23 May, 1986),* pp. 4–18.

Pugachev, V.S., Sinitsyn, I.N., and Shin, V.I (1987). Problems of analysis and on-line conditionally optimal filtering of processes in nonlinear stochastic systems (Survey). *Automation and Remote Control.* **2**, 62–68.

Pugachev, V.S., Sinitsyn, I.N., Cherednichenko A.A., and Shin, V. I. (1991a). Software for discrete conditionally optimal filtering design. In: *Informatics and Computer Systems.* Nauka, Moscow (to appear).

Pugachev, V.S., and Sinitsyn, I.N.(eds) (1991b). *Intellectualized Software Design Concepts for Conditionally Optimal Filters Package of Applied Programs "StS-filter" (Version 1).* Preprint. Institute of Informatics Problems, USSR Academy of Sciences (in Russian).

Raol, J.I., and Sinha, N.K. (1986). A nonlinear filter for estimation of state of a continuous-time system with discrete measurement. In: *Preprints of the Second IFAC Symposium on Stochastic Control* (Vilnius, USSR, 19–23 May, 1986), Part 1, pp. 211–216.

Sinitsyn, I.N. (1986a). Stochastic hereditary control systems. *Probl. Control Inform. Theory.* **15**, 287–298.

Sinitsyn, I.N. (1986b). Finite-dimensional distributions of the processes in stochastic integral and integro-differential systems. In: *Preprints of the Second IFAC Symposium on Stochastic Control* (Vilnius, USSR, 19–23 May, 1986), N 1, pp. 149–153.

Sinitsyn, I.N. (1990). Problems of signal analysis and conditionally optimal processing in stochastic differential systems. In: *Proceedings Latvian Signal Processing International Conference,* Riga. 24–26 April. Vol. **2**, pp. 60–64.

Sinitsyn, I.N. and Shin, V.I. (1991a). Conditionally optimal filtering for processes in stochastic differential systems on complex statistical criteria. *Soviet Math. Dokl.* **320** (to appear).

Sinitsyn, I.N. and Shin, V.I. (1991b). State and parameters estimation in sampled data stochastic system on complex statistical criteria. *Soviet Math. Dokl.* **320** (to appear).

Tikhonov, A.N. and Arsenin, V.Ya. (1979). *The Methods for Solving the Ill-Posed Problems.* Second ed. Nauka, Moscow (in Russian).

Verlan, A.F. and Sizikov, V.S. (1986). *Integral Equations. Methods, Algorithms, Programs.* Kiev, Naukova Dumka (in Russian).

Ill-Posed Problems in Natural Sciences, pp. 184 – 190
A. Tikhonov (Ed.)
1992 VSP/TVP

THE ESTIMATION OF THE ERROR OF THE REGULARIZATION METHOD FOR ILL-POSED PROBLEMS WITH NONINJECTIVE OPERATOR

V.P. TANANA and T.N. RUDAKOVA
Chelyabinsk State University, 454130 Chelyabinsk, Russia

ABSTRACT

Let H be a separable Hilbert space, $A\colon H \to H$ be a linear self-conjugate compact noninjective operator. The range of the operator A is nonclosed.
Let us consider the operator equation:

$$Au = f, \quad u, f \in H.$$

Assume that the set of exact solutions of equation (1) has an nonempty intersection with the set $M_r = BS_r$, where $S_r = \{v \in H\colon \|v\| \le r\}, B$ is a linear self-conjugate compact injective operator, $B = A^p (0 < p \le 1)$ on the orthogonal subspace to the kernel of A.
Let f_δ such that $\|f - f_\delta\| \le \delta$ and a linear bounded self-conjugate one-to-one operator $A_h\colon H \to H$ such that $\|A - A_h\| \le h$ be known.
An approximation to the normal solution of equation (1) is $u_{\delta h}^\alpha = R_\alpha f_\delta$, where

$$R_\alpha = R_\alpha[A_h] = B(C_h^* C_h + \alpha E)^{-1} C_h^*, \qquad C_h = A_h B, \qquad \alpha = \alpha(\delta, h).$$

The error of the solution is defined as:

$$\Delta(R_\alpha) = \sup_{u, A_h, f_\delta} \{\|u_{\delta h}^\alpha - u\|\colon u \in M_r \cap L,\ \|A_h - A\| \le h,$$

$$\|f_\delta - Au\| \le \delta\}.$$

The exact in order estimate $\Delta(R_\alpha) \le C(r\|B\|h + \delta)^{2p/(3p+2)}$ for the method $R_\alpha[A_h]$ with the best regularization parameter $\alpha = (r\|B\|h + \delta)^{4(p+1)/(3p+2)}$ is valid.

1. THE PROBLEM

Let H be a separable Hilbert space. Suppose A is a linear self-conjugate compact operator of H into H . The operator A is noninjective and the range $R(A)$ of the operator A is nonclosed.

Let us consider the operator equation of the first kind:

$$Au = f, \quad u, f \in H, \tag{1}$$

Let N be the kernel of the operator A and L be the orthogonal subspace of H to N.
Assume that the set of exact solutions of equation (1) has an nonempty intersection with the set M_r, where $M_r = BS_r$, $S_r = \{v \in H\colon \|v\| \leq r\}$, and B is a linear self-conjugate compact injective operator. The restriction $B|_L$ of B to the subspace L is a function of A, that is $B|_L = g(A)$. In this work we suppose that $g(\lambda) = \lambda^p$, $0 < p \leq 1$.
Assume that f and the operator A are unknown and we know only f_δ such that $\|f - f_\delta\| \leq \delta$ and a linear bounded self-conjugate one-to-one operator A_h of H into H such that $\|A - A_h\| \leq h$.
The main problem is to find the approximation solution to the normal solution of equation (1) using the approximation data $\{A_h, f_\delta, h, \delta\}$ (the normal solution is a solution with the minimum norm).
The approximation to the normal solution $u_0 \in M_r \cap L$ of equation (1) will be found in the form (Tikhonov A.N. (1963)):

$$u_{\delta h}^{\alpha} = R_\alpha f_\delta \tag{2}$$

$$R_\alpha = R_\alpha[A_h] = B(C_h^* C_h + \alpha E)^{-1} C_h^*, \qquad C_h = A_h B, \ \alpha = \alpha(\delta, h). \tag{3}$$

The error of the regularization solution (2) is defined as (Tanana et al., 1987):

$$\Delta(R_\alpha) = \sup_{u, A_h, f_\delta} \{\|u_{\delta h}^{\alpha} - u\|\colon u \in M_r \cap L,$$

$$\|A_h - A\| \leq h, \ \|f_\delta - Au\| \leq \delta\}. \tag{4}$$

2. THE ESTIMATION OF THE ERROR OF REGULARIZATION FOR THE PROBLEM WITH INJECTIVE OPERATOR

If the operator A in equation (1) is injective the method of regularization with the parameter $\alpha = (\delta + h)^2$ is optimal in order. If $\delta \leq h, \|B\| \leq 1, \ r \geq 1$ one has an exact in order estimate of error (4) of regularization solution (2):

$$\Delta(R_\alpha) \leq 8w(r\|B\|h + \delta, r), \tag{5}$$

where

$$\bar{w}(\tau, r) = \sup\{\|u\|\colon u \in M_r, \quad \|A_h u\| \leq \tau\}. \tag{6}$$

It was obtained in (Ivanov et al., 1978). The form of the modula of continuity (6) for $g(\lambda) = \lambda^p$ gives (Ivanov, 1974):

$$\Delta(R_\alpha) \leq C(r\|B\|h + \delta)^{p/(p+1)}, \tag{7}$$

where C is some constants. In this paper we will use letter C to denote some constants when their values are not impotant to us. These constants do not depend on δ and h and have different values in different formulas.
If the operator A is noninjective the estimate (7) is not valid.
Let $e_0(\|e_0\| = 1)$ be the ligen vector from N corresponding to an ligen value $\lambda_0 \neq 0$ of the operator B.
Put $u = 0,\ f_\delta = \delta e_0,\ A_h|_L = A,\ A_h|_N = hE$, then

$$\Delta(R_\alpha) \geq \|B(h^2B^2 + \alpha E)^{-1} h\delta B e_0\| = \frac{h\delta\lambda_0^2}{h^2\lambda_0^2 + \alpha}. \tag{8}$$

Substituting $\alpha = (\delta + h)^2$ and $\delta = h$ in the right part of inequality (8) we obtain that

$$\Delta(R_\alpha) \geq \lambda_0^2/(\lambda_0^2 + 4). \tag{9}$$

Moreover, since the right part of inequality (9) is a constant. It not depending on δ and h method (3) with the parameter $\alpha = (\delta + h)^2$ is not regularization for problem (1) with an noninjective operator. We shall find a parameter α such that the error $\Delta(R_\alpha)$ defined by (4) will be minimum.

3. AUXILIARY RESULTS

Let us formulate two well known lemmas.

LEMMA 1. (Vainikko, 1982). *Let A_h, A be a linear bounded self-conjugate non-negative operators of H into H and $\|A - A_h\| \leq h$.
Then for any positive number p the following upper bound is valid*

$$\|A^p - A_h^p\| \leq C_p h^{\min(1,p)}$$

where C_p is a some constant not depending on h.

We introduce the notation

$$C(s,q) = \begin{cases} (\frac{s}{q})^{\frac{s}{q}}(1 - \frac{s}{q})^{(1-\frac{s}{q})}, & s < q \\ 1, & s = q \end{cases}$$

where $q \geq s \geq 0,\ q > 0$.

LEMMA 2. (Tanana et al., 1978).
*Let $\Psi(\lambda, \alpha) = \lambda^s/(\lambda^q + \alpha)$ where q and s are real numbers such that $q, \alpha > 0, s \geq 0$, $0 < a < \infty$
Then*

$$\sup_{\lambda \in [0,a]} \Psi(\lambda, \alpha) \leq \begin{cases} C(s,q)\alpha^{\frac{s}{q}-1}, & s \leq q \\ a^{s-q}, & q < s \end{cases}$$

for sufficiently small α.

4. THE CALCULATION OF THE OPTIMAL PARAMETER OF THE REGULARIZATION

Let $\delta \leq h,\ \|B\| \leq 1,\ r \geq 1$. The regularization parameter α will be found in the form

$$\alpha = \tau^x, \quad \tau = r\|B\|h + \delta. \tag{10}$$

Denote

$$\Delta_L = \Delta_L(\alpha, \delta, h)$$

$$= \sup_{u, A_h, f_\delta} \{\|u_{\delta h}^\alpha - u\|: \ u \in M_r \cap L, \|A_h - A\| \leq h, \ f_\delta \in L, \ \|f_\delta - Au\| \leq \delta\},$$

$$\Delta_N = \Delta_N(\alpha, \delta, h)$$

$$= \sup_{u, A_h, f_\delta} \{\|u_{\delta h}^\alpha - u\|: \ u \in M_r \cap L, \|A_h - A\| \leq h, \ f_\delta \in N, \ \|f_\delta - Au\| \leq \delta\}.$$

Then the following obvious estimates are valid for error (4):

$$\Delta(R_\alpha) \leq \Delta_L + \Delta_N,$$

$$\Delta(R_\alpha) \geq \max\{\Delta_L, \Delta_N\} \geq \frac{\Delta_L + \Delta_N}{2}.$$

Hence

$$\Delta(R_\alpha) \sim \Delta_L + \Delta_N, \qquad \delta, h \to 0. \tag{11}$$

We find an upper bound for $\Delta_L = \Delta_L(\alpha, \delta, h)$:

$$\Delta_L \leq \Delta_L^1 + \Delta_L^2 \tag{12}$$

where

$$\Delta_L^1 = \Delta_L^1(\alpha) = \sup_u\{\|u - R_\alpha A_h u\|: \ u \in M_r \cap L\},$$

$$\Delta_L^2 = \Delta_L^2(\alpha, \delta, h) = \sup_{u, A_h, f_\delta} \{\|R_\alpha A_h u - R_\alpha f_\delta\|:$$

$$u \in M_r \cap L, \ \|A_h - A\| \leq h, \ f_\delta \in L, \ \|f_\delta - Au\| \leq \delta\}.$$

It is clear that

$$\|u - R_\alpha A_h u\| = \|Bv - B(BA_h^2 B + \alpha E)^{-1} BA_h^2 Bv\|$$

$$= \|B(BA_h^2 B + \alpha E)^{-1} \alpha v\| \leq \alpha r \|B(BA_h^2 B + \alpha E)^{-1}\|,$$

for any $u \in M_r \cap L, \ u = Bv, \ v \in S_r \cap L$.
It is evident that $B = (BA^2B)^{p/(2p+2)}$ on L. Hence the inequality

$$\|u - R_\alpha A_h u\| \leq \alpha r \Big(\|(BA_h^2 B)^{p/(2p+2)} (BA_h^2 B + \alpha E)^{-1}\|$$

$$+ \|((BA_h^2 B)^{p/(2p+2)} - (BA^2 B)^{p/(2p+2)})(BA_h^2 B + \alpha E)^{-1}\| \Big)$$

and Lemmas 1 and 2 from §3 in combination with condition (10) imply the inequality

$$\|u - R_\alpha A_h u\| \leq C\tau^{xp/(2p+2)}$$

and the upper bound for Δ_L^1

$$\Delta_L^1 \leq C\tau^{xp/(2p+2)}. \tag{13}$$

The evident inequality

$$\|R_\alpha A_h u - R_\alpha f_\delta\| \le \|R_\alpha A_h u - R_\alpha A u\| + \|R_\alpha A u - R_\alpha f_\delta\|$$

implies

$$\Delta_L^2 \le \|R_\alpha|_L\|\tau, \tag{14}$$

$$\|R_\alpha|_L\| = \|B(BA_h^2B + \alpha E)^{-1}BA_h|_L\|$$

$$\le \|B(BA_h^2B + \alpha E)^{-1}BA|_L\| + \|B(BA_h^2B + \alpha E)^{-1}B(A_h - A)|_L\|. \tag{15}$$

The second summand in the right part of (15) is estimated as

$$\|B(BA_h^2B + \alpha E)^{-1}B(A_h - A)|_L\| \le \|B\|^2\frac{h}{\alpha}.$$

To find an upper bound for the first summand in the right part of inequality (15) one can use the obvious equation $BA = (BA^2B)^{1/2}$

$$\|B(BA_h^2B + \alpha E)^{-1}BA|_L\| \le d_1 + d_2 + d_3 + d_4$$

where

$$d_1 = \|((BA^2B)^{p/(2p+2)} - (BA_h^2B)^{p/(2p+2)})(BA_h^2B + \alpha E)^{-1}$$

$$\times((BA^2B)^{1/2} - (BA_h^2B)^{1/2})\|,$$

$$d_2 = \|(BA_h^2B)^{p/(2p+2)}(BA_h^2B + \alpha E)^{-1}((BA^2B)^{1/2} - (BA_h^2B)^{1/2})\|,$$

$$d_3 = \|(BA_h^2B)^{p/(2p+2)}(BA_h^2B + \alpha E)^{-1}(BA_h^2B)^{1/2}\|,$$

$$d_4 = \|((BA_h^2B)^{p/(2p+2)} - (BA^2B)^{p/(2p+2)})(BA_h^2B + \alpha E)^{-1}(BA_h^2B)^{1/2}\|.$$

Applying Lemmas 1 and 2 we obtain

$$d_1 \le C\tau^{-x+(2p+1)/(p+1)},$$

$$d_2 \le C\tau^{1-x(p+2)/(2p+2)},$$

$$d_3 \le C\tau^{-x/(2p+2)},$$

$$d_4 \le C\tau^{-(x/2)+p/(p+1)}.$$

Hence

$$\|B(BA_h^2B + \alpha E)^{-1}BA\| \le C\tau^{-x/(2p+2)}$$

Taking into account (10) and using we arrive to inequalities (13) and (14) the following inequality

$$\Delta_L^2 \le C(\tau^{2-x} + \tau^{-x/(2p+2)}) \tag{16}$$

From (12), (13) and (16) it follows

$$\Delta_L \le C(\tau^{xp/(2p+2)} + \tau^{2-x}). \tag{17}$$

In a similar way one can find the following upper bound for Δ_N

$$\Delta_N \leq C(\tau^{xp/(2p+2)} + \tau^{2-x}). \tag{18}$$

The upper bound for $\Delta(R_\alpha)$ is minimum if $x = 4(p+1)/(3p+2)$ (see (11), (17), (18)), that is if

$$\alpha = (r\|B\|h + \delta)^{4(p+1)/(3p+2)}.$$

5. THE ESTIMATION OF THE ERROR OF REGULARIZATION FOR THE PROBLEM WITH NONINJECTIVE OPERATOR

THEOREM 1. *Let the operators A, A_h, B satisfy all the conditions formulated in §1. Then the following upper bound of the regularization solution error $\Delta(R_\alpha)$ for the regularization method with parameter $\alpha = (r\|B\|h+\delta)^{4(p+1)/(3p+2)}$ on the set M_r is valid:*

$$\Delta(R_\alpha) \leq C(r\|B\|h + \delta)^{2p/(3p+2)} \tag{19}$$

Proof follows from (11), (17), (18).

THEOREM 2. *Let the operators A, A_h, B satisfy all the conditions formulated in §1. Then for any function $\alpha = \alpha(\delta, h)$ such that $\alpha \geq 0$ and $\lim_{h,\delta\to 0} \alpha(\delta, h) = 0$ there eseist infinitesimal seguences $\{h_n\}_{n=1}^\infty$ and $\{\delta_n\}_{n=1}^\infty$ such that the following lower bound of the error $\Delta(R_\alpha)$ of the regularization solution for the regulation method with parameter α is valid*

$$\Delta(R_\alpha) \geq C(r\|B\|h_n + \delta_n)^{2p/(3p+2)}.$$

Proof. Let $\{\lambda_n\}_{n=1}^\infty$ be a non-negative sequence of ligen values of the operator $A, \lambda_n \to 0$ as $n \to \infty$, and let $\{e_n\}_{n=1}^\infty \subset L(\|e_n\| = 1)$ be the corresponding sequence of eigen vectors, that is $Ae_n = \lambda_n e_n$ for any positive integer n. The function $\alpha = \alpha(\delta, h)$ is non-negative and $\lim_{h,\delta\to 0} \alpha(\delta, h) = 0$. Define $\{h_n\}_{n=1}^\infty$ and $\{\delta_n\}_{n=1}^\infty$ as follows

$$h_n = \delta_n = \frac{\lambda_n^{2(p+1)/x}}{r\|B\| + 1}, \quad x = \frac{4(p+1)}{3p+2}. \tag{20}$$

Denote

$$\tau_n = r\|B\|h_n + \delta_n, \quad \alpha_n = \alpha(\delta_n, h_n). \tag{21}$$

Let us consider two cases.

Case 1. The sequence $\{\alpha_n/\tau_n^x\}_{n=1}^\infty$ is separated from the origin, that is the inequalities

$$\alpha_n/\tau_n^x \geq C \tag{22}$$

are valid for all positive integers n.
If $A_{h_n} = A,\ u_0^n = Bre_n,\ f_{\delta_n} = A_{h_n} u_0^n$, then

$$\Delta(R_\alpha) \geq \|R_{\alpha_n} Au_0^n - u_0^n\|$$

$$= r\|(A^{2(p+1)} + \alpha_n E)^{-1}\alpha_n A^p e_n\| = r\alpha_n\lambda_n^p/(\lambda_n^{2(p+1)} + \alpha_n). \tag{23}$$

Using (29) and (23) we obtain

$$\Delta(R_\alpha) \geq r\frac{\alpha_n}{\tau_n^x + \alpha_n}\tau_n^{2p/(3p+2)} = r(1 - \frac{1}{1+\alpha_n/\tau_n^x})\tau^{2p/(3p+2)}.$$

By (21), (22) we find that

$$\Delta(R_\alpha) \geq C(r\|B\|h_n + \delta_n)^{2p/(3p+2)}.$$

Case 2. There exists a subsequence $\{\alpha_{n_k}/\tau_{n_k}^x\}_{k=1}^{\infty}$ of the sequence $\{\alpha_n/\tau_n^x\}_{n=1}^{\infty}$ tending to zero as $n \to \infty$. We can assume without lass of generality that

$$\frac{\alpha_n}{\tau_n^x} \to 0, \quad n \to \infty. \tag{24}$$

Since $B|_N \neq 0$ and $B(N) \subset N$. Then one can find in N an eigenvector e_0 corresponding to eigenvalue $\lambda_0 \neq 0$ of the operator B.
Set $u_n = 0$, $f_{\delta_n} = \delta_n e_0$, $A_{h_n}|_L = A$, $A_{h_n}|_N = h_n E$, then

$$\Delta(R_\alpha) \geq \|B(h_n^2 B^2 + \alpha_n E)^{-1} B h_n^2 e_0\| = \frac{h_n^x/\alpha_n + \lambda_0^2}{h_n^x\lambda_0^2/\alpha_n + 1} h_n^{1-x}.$$

Using (24) we obtain

$$\Delta(R_\alpha) \geq C(r\|B\|h_n + \delta_n)^{2p/(3p+2)}.$$

The theorem is proved.

COROLLARY 1. *Estimate (19) for the regularization method $R_\alpha[A_h]$ with the best parameter $\alpha \doteq (r\|B\|h + \delta)^{4(p+1)/(3p+2)}$ is an exact in order estimate.*

An exact in order error estimate for the optimal method P_{opt} of solving problem (1) with noninjective operator was obtained in (Tanana, et al., 1987)

$$\Delta(P_{\text{opt}}) \leq C(r\|B\|h + \delta)^{p/(p+1)}$$

This result proves the following statement.

COROLLARY 2. *For any parameter α the regularization method $R_\alpha[A_h]$ for problem (1) with noninjective operator is not optimal in order method.*

REFERENCES

Vainikko, G.M. (1982). *The Methods of the Ill-Posed Problems Solving in Hilbert Spaces.* Tartu Univ. Press, Tartu (in Russian).

Ivanov, V.K., Vasin, V.V., and Tanana, V.P. (1978). *The Theory of Linear Ill-Posed Problems and its Applications.* Nauka, Moscow (in Russian).

Ivanov, V.K. (1974). On the estimation of ill-posed problems solving stability. In: *The Ill-Posed Problems Solving Methods and They Application.* Moscow State Univ. Press, Moscow, pp. 17–19 (in Russian).

Tanana, V.P., Rekant, M.Ya., and Yanchenko, S.I. (1987). *The Optimization of the Methods of Operator Equations Solving.* Ural Univ. Press, Sverdlovsk (in Russian).

Tikhonov, A.N. (1963). On solving the ill-posed problems and the regularization method. *Dokl. Acad. Nauk SSSR* **151**, 501–504 (in Russian).

Ill-Posed Problems in Natural Sciences, pp. 191 – 201
A. Tikhonov (Ed.)
1992 VSP/TVP

SOME NUMERICAL METHODS OF PARAMETER IDENTIFICATION IN NONLINEAR MODELS

O. VAARMANN
Institute of Cybernetics, Estonian Academy of Sciences,
Akademia tee 21, 200108 Tallinn, Estonia

ABSTRACT

It is a principal question for methods of reqularization in what sense one should understand an approximate solution to ill-posed problems since the solution of problem depends on the spaces considered. In this paper for solving a nonlinear equation in the least squares sense in terms of IU-weighted pseudoinverse (where I denotes the identity mapping and U is an arbitrary nonsingular linear operator in Hilbert space considered) a general family of methods is studied which includes as particular cases the Gauss–Newton, Ben-Israel (generalized Gauss–Newton), Levenberg–Marquardt (regularized Gauss–Newton) method and methods with successive approximation of pseudoinverse operator via recurrence formula. A local convergence theorem is proved for this family of methods. Computational aspects are discussed briefly as well.

INTRODUCTION

In recent years rapid developments have taken place in computing methods for inverse problems of mathematical physics. The numerical treatment of inverse problems often lead to the solution of ill-posed or singular operator equations whose solution is seeked in the least squares sense. In many scientific and engineering applications, where differential or integral equations models are involved, one may distinguish several different types of inverse problems. All these types of inverse problems have not been equally investigated. More attention has been paid to parameter indentification (PI): given measured data and associated theoretical model, unknown parameters are to be determined or unknown functions are to be parametrized in such a way that some "measure of fit" between data and model is minimal. In the actual computation of a solution for a inverse problem of this kind, the following typical difficulties arise: 1) The solution of model equation repeatedly may be rather time and storage-consuming. 2) Often the inverse problem is ill-conditioned (the solution does not continuosly depend on the data). In order to solve the inverse problem in the presence of noisy or incomplete data or inevitable discretization errors, special purpose (stable) techniques are necessary.

Further we shall consider the least squares problem in the form more suitable for mathematies, i.e., x stands for a desired quantity (parameter vector).

BASIC THEOREM

Let $F(x)$ be a differentiable (how many times as necessary) operator from a Hilbert space H_1 in to another H_2 and such that the range of its derivative $R(F'(x)) \subset H_2$ is closed. In the general case the equation

$$F(x) = 0 \tag{1}$$

may not have any solution in the classical sense, or may have several solutions. This paper is devoted to a study of computing an approximate solution of the equation (1), if it exists, or otherwise, a solution of (1) in the least squares sense, i.e., of minimizing the functional $\varphi(x) = \frac{1}{2}\|F(x)\|^2$. A necessary condition for x^* to be a local minimum point of $\varphi(x)$ is

$$\left[F'(x^*)\right]^* F(x^*) = 0, \tag{2}$$

where $[\cdot]^*$ denotes the dual mapping. In fact, calculating on a computer with finite wordlength, F' cannot be obtained exactly, so that in practice an approximation to F' is used. We will therefore consider a family of methods which are known as approximate or inexact variants of standard methods. In general, by an approximate variant of a standard method the author means one in which certain approximations for the derivatives and pseudoinverse operator are exploited or the corresponding linear equations are solved approximately by taking finitely many steps of an iterative method. More precisely for solving the operator equation (2) we consider iterative methods of the type

$$x_{k+1} = x_k - Q(x_k, A_k), \quad k = 0, 1, \ldots, \tag{3}$$

where $Q(\cdot, A_k)$ is an operator from H_1 into itself and A_k is an approximation to the pseudoinverse occuring in the corresponding exact method $x_{k+1} = x_k - Q(x_k, \cdot)$ having the convergence rate $p \geq 2$.

The family methods (3) includes, as special cases, Newton type methods

$$x_{k+1} = x_k - A_k F(x_k), \tag{4}$$

where A_k approximates the operator $[\nabla^2\varphi(x_k)]^{-1}[F'(x_k)]^*$, and $\nabla^2\varphi(x) = [F'(x)]^* F'(x) + [F''(x)]^* F(x)$. The evaluation of the latter operator $\nabla^2\varphi$ is as a rule a labour-consuming procedure. That is why one often uses inexact Newton methods that instead of the exact $\nabla^2\varphi$ implement its readily evaluable approximation. In particular, if $A_k = [B_k + S_k]^{-1}[F'(x_k)]^*$ with $B_k = B(x_k) = [F'(x_k)]^* F'(x_k)$ and $S_k = S(x_k) \approx [F''(x_k)]^* F(x_k)$ then from (4) so called augmented Gauss–Newton method follows. If the residual is expected to be small then rather simple rules of updating the term S_k in A_k can be exploited.

Putting $H = U^*U$ and $S_k = \alpha_k H$ with $\alpha_k > 0$ (4) turns out to be a regularized Gauss–Newton (Levenberg–Marquardt) method while $\lim_{\alpha_k \to \infty} A_k = U^{-1}[F' \times$

$(x_k)U^{-1}]^+$. If $A_k = U^{-1}[F'(x_k)U^{-1}]^+$, then one gets an extension of Ben–Israel method which coincides with the usual Ben–Israel one provided $U = I$.
Since F' has the closed range $R(x) = R(F'(x))$ and U is a nonsingular linear operator from H_1 into itself then there exists the bounded IU-weighted pseudoinverse $U^{-1}(F'(x)U^{-1})^+$.
Let

$$\Phi(x) = \left[F'(x)U^{-1}\right]^+, \qquad \Psi(x) = \Phi^+(x) = F'(x)U^{-1} \tag{5}$$

and let $P_{R(x)} = P_{R(\Psi(x_k))}$, $P_k = P_{R(\Psi(x_k))}$ denotes the orthogonal projector of H_2 onto $R(\Psi(x))$ and let c_i, h_i, ω_i be certain nonnegative constants except for $c_p > 0$. Assume that $\overline{A}_k$ is an approximation to $[F'(x_k)U^{-1}]^+$, i.e., $\overline{A}_k = UA_k$ and the following relations hold

$$\overline{A}_k = \overline{A}_k P_k, \qquad \|\overline{A}_k\| \le \bar{\lambda}_k \le \bar{\lambda} < \infty, \qquad \|P_k - \Psi_k \overline{A}_k\| \le \bar{\gamma}_k,$$

$$\|A_k\| \le \lambda_k, \tag{6}$$

$$\|P_{k+1}F_{(x_{k+1})}\| \le \sum_{\nu=1}^{h_1} \omega_\nu \bar{\gamma}_k^{a_\nu} \|P_k F(x_k)\|^{u_\nu}$$

$$+\sum_{i=p}^{h_2} c_i \|P_k F(x_k)\ \|^i \ (a_\nu, u_\nu, \ge 0) \tag{A}$$

and the quantities $\bar{\gamma}_k$ can be expressed in the form

$$\bar{\gamma}_{k+1} = \varphi(\bar{\gamma}_k, \|P_k F(x_k)\|), \tag{B}$$

where φ is a positive monotonically decreasing scalar function of its arguments or $\bar{\gamma}_k$ vanish in the limit as $k \to \infty$ according to another rule, e.g.,

$$\bar{\gamma}_{k+1} = b_{k+1} = \bar{d} b_k^q, \qquad \bar{d} > 0, \quad b_0 < 1, \quad q \ge 2, \tag{C}$$

or $\bar{\gamma}_k \le \bar{\gamma} < 1$ while $\bar{\gamma}$ can as well amount to zero. If the quantities $\bar{\gamma}_k$ are defined by

$$\bar{\gamma}_{k+1} = \varphi\big(\bar{\gamma}_k, \|P_k F(x_k)\|\big) = \left(\bar{\gamma}_k + h\|P_k F(x_k)\|\right)^q, \qquad \bar{\gamma}_0 + h\|P_0 F(x_0)\| < 1 \tag{B'}$$

we shall introduce the sequences $\{b_k\}$ and $\{d_k\}$. Let be $s = \min_\nu u_\nu$ and $l = \min_{\nu \in V} a_\nu$, $V = \{\nu\colon\ u_\nu = s\}$ for all terms in (**A**) with $\omega_\nu \ne 0$. We define for $a_\nu + u_\nu \ge p$ with $\omega_\nu \ne 0$ and $q \ge p$

$$b_k \ge \max\left\{\bar{\gamma}_k, \|P_k F(x_k)\|\right\} \tag{7}$$

$$d_k = \max\left\{\sum_{\nu=1}^{h_1} \omega_\nu b_k^{a_\nu + u_\nu - p} + \sum_{i=p}^{h_2} c_i b_k^{i-p}, (1+h)^q b_k^{q-p}\right\}, \tag{8}$$

and for $l < p - s$ and for q sufficiently large that $q \ge (p(p-s))/l$

$$b_k \ge \max\left\{\bar{\gamma}_k^{\frac{l}{p-s}}, \|P_k F(x_k)\|\right\}, \tag{9}$$

$$d_k = \max\left\{ \sum_{\nu=1}^{h_1} \omega_\nu b_k^{((p-s)a_\nu + l(u_\nu - p))/l} \right.$$

$$\left. + \sum_{i=p}^{h_2} c_i b_k^{i-p}, (1+h)^{ql/(p-s)} b_k^{ql/(p-s)-p} \right\} \tag{10}$$

Generally speaking the quantities ω_ν and c_i may depend on index k but for the sake of simplicity it is dropped here.
In the case or $\bar{\gamma}_k = 0$ we denote by

$$d_k = c_p + \sum_{i=p+1}^{h_2} c_i b_k^{i-p}. \tag{11}$$

If $p \geq 2$, $\min\limits_{\nu} u_\nu = u \geq 1$ and $0 < \bar{\gamma}_k < 1$ we take

$$\delta_k^{(1)} = \sum_{\nu=1}^{h_1} \omega_\nu \bar{\gamma}_k^{a_\nu} \|P_k F(x_k)\|^{u_\nu - 1} + \sum_{i=p}^{h_2} c_i \|P_k F(x_k)\|^{i-1}. \tag{12}$$

THEOREM. *Let $x_0 \in H_1$, $S = \{x \in H_1 \colon \|x - x_0\| \leq \rho\}$ and $F(x)$ be a differentiable operator fulfilling the condition* (**A**) *and the inequality $\|Q(x_k, A_k)\| \leq \lambda \|P_k F(x_k)\|$, $0 < \lambda < \infty$, be valid in S.*
Then 1) If $p \geq 2$, $u \geq 1$, $\bar{\gamma}_k \leq \bar{\gamma} \neq 0$, $r_1 = \lambda \|P_0 F(x_0)\|/(1-\delta) \leq \rho$, $\delta = \delta_0^{(1)} < 1$, $\bar{\gamma}_0 = \bar{\gamma}$, then the equation $[F'(x)]^* F(x) = 0$ has a solution x^* in S, $\|x^* - x_0\| \leq r_1$, to which the sequence (3) converges with

$$\|x_k - x^*\| \leq r_1 \delta^k;$$

if operator $P_{R(x)}$ is independent of x and $p \geq 2$ provided $\bar{\gamma} \geq \bar{\gamma}_0 \geq \ldots \geq \bar{\gamma}_k \geq 0$ and $\bar{\gamma}_k \to 0$ as $k \to \infty$, then the sequence (3) converges at least ($u = 1$) superlinearly with

$$\|x_k - x^*\| \leq r_1 \prod_{i=0}^{k-1} \delta_i, \qquad \delta_i \to 0 \quad \text{as } i \to \infty.$$

2) If operator $P_{R(x)}$ is independent of x and $p \geq 2$ and the quantities $\bar{\gamma}_k$ are expressed in the form (**B′**) *or* (**C**) *or $\bar{\gamma}_k = 0$ $(k = 0, 1, \ldots)$, $\delta = \delta_0^{(p)} = \sqrt[p-1]{d_0}\, b_0 < 1$, $r_p = \lambda H_0(\delta)/\sqrt[p-1]{d} \leq \rho$, where $\lim\limits_{k\to\infty} d_k = d > 0$ and $H_k(\delta) = \sum_{i=k}^{\infty} \delta^{p^i}$, then the equation $[F'(x)]^* F(x) = 0$ has a solution x^* in S, $\|x^* - x_0\| \leq r_p$, to which the sequence (3) converges with*

$$\|x_k - x^*\| \leq (\lambda/\sqrt[p-1]{d}) H_k(\delta).$$

Proof. 1) Taking into account relation $\|x_{k+1} - x_k\| \leq \lambda \|P_k F(x_k)\|$, $\delta = \delta_0^{(1)} < 1$ and (**A**) we obtain.

$$\|P_{k+1} F(x_{k+1})\| \leq \delta_k^{(1)} \|P_k F(x_k)\| \leq \|P_0 F(x_0)\| \prod_{i=0}^{k} \delta_i^{(1)} \leq \|P_0 F(x_0)\| \delta^{k+1},$$

$$\|x_{k+1} - x_k\| \leq \lambda \|P_0 F(x_0)\| \delta^k,$$

$$\|x_m - x_k\| \leq \sum_{i=k}^{m-1} \|x_{i+1} - x_i\| \leq r_1(\delta^k - \delta^m) \ (m \geq k).$$

This means that the sequence (3) is fundamental and consequently, $x^* = \lim_{k\to\infty} \times x_k, \|x^* - x_k\| \leq r_1\delta^k$, $\lim_{k\to\infty} \|P_k F(x_k)\| = 0$, $\|x_k - x_0\| \leq r_1(1-\delta^k) \leq r_1 \leq \rho$, i.e., all $x_k \in S$.
Observing that

$$\|[F'(x)]^* F(x)\| \leq \|H\| \|H^{-1}[F'(x)]^* F(x)\| \leq \|H\| \|U^{-1}(F'(x)U^{-1})^* F(x)\|$$

$$\leq \|H\| \|U^{-1}(F'(x)U^{-1})^* \Psi(x)\Phi(x)F(x)\|$$

$$\leq \|H\| \|U^{-1}(F'(x)U^{-1})^*\| \|P_{R(x)} F(x)\|$$

and under the assumptions of the theorem the operator $R_{R(x)}F(x)$ is continuous, it is evident that $[F'(x^*)]^* F(x^*) = 0$. In the case $\bar{\gamma}_k \to 0$ as $k \to \infty$ and the operator $P_{R(x)}$ being independent of x, obviously, $\delta_k^{(1)} \to 0$ as well and therefore the sequence (3) converges at least ($u = 1$) superlinearly to a solution of (2).
2) If the operator $P_{R(x)}$ is independent of x and relation (**B′**) or (**C**) is valid or $\bar{\gamma}_k = 0$ then on basis of (7)—(12) one concludes that in the capacity of b_{k+1} one can take $d_k b_k^p$, where

$$b_{k+1} = d_k b_k^p \leq d^{-1/(p-1)}(d_k^{1/(p-1)} b_k)^p \leq \ldots \leq d^{-(1/(p-1))}\delta^{k+1}. \tag{13}$$

We shall show the correctness of this assertion for $l < p - s$ while $\bar{\gamma}_k$ is given by (**B′**).
Owing to $\bar{\gamma}_k \leq b_k^{(p-s)/l}$ it follows from (**A**) that

$$\|P_{k+1}F(x_{k+1})\| \leq \left(\sum_{\nu=1}^{h_1} \omega_\nu b_k^{(p-s)a_\nu + l(u_\nu - p)/l} + \sum_{i=p}^{h_2} c_i b_k^{i-p} \right) b_k^p, \tag{14}$$

while $\frac{a_\nu(p-s)}{l} + (u_\nu - p) \geq (p - s) + (s - p) = 0$ by reason of $a_\nu \geq l$ and $u_\nu \geq s$.
Analogously we have

$$\bar{\gamma}_{k+1} l/(p-s) = (\bar{\gamma}_k + h\|P_k F(x_k)\|)^{ql/(p-s)} \leq (b_k^{(p-s)/l} + h b_k^{ql/(p-s)}) \tag{15}$$

$$\leq (1+h)^{ql/(p-s)-p} b_k^p \leq d_k b_k^p.$$

From $\delta = \delta_0^{(p)} < 1$ it follows that $b_1 = d_0 b_0^p = d_0 b_0^{p-1} b_0 = \delta^{p-1} b_0 \leq \delta b_0 \leq b_0$ for $p \geq 2$ and then also $d_1 \leq d_0$. Further we have by induction that $b_0 \geq b_1 \geq \ldots \geq b_k \geq \ldots \geq 0$ and $d_0 \geq d_1 \geq \ldots \geq d_k$. On the other hand the sequence $\{d_k\}$ is bounded below. Indeed because of the assumption $c_p > 0$ we have $\sum_{i=p}^{h_2} c_i b_k^{i-p} \geq c_p > 0$. Therefore there exists a limit $d = \lim_{k\to\infty} d_k$. Combining (14) with (15) yields (13). In a similar way we may be convinced that (13) hold for $l \geq p - s$ when the relation (**B′**) holds for $\bar{\gamma}_k$.

In the case of $\bar{\gamma}_k = 0$ the expression for d_k is given by (12). If (C) holds for $\bar{\gamma}_k$, then the expression for d_k can also be easily deduced from (**A**).
Thus

$$\|x_{k+1}-x_k\| \leq \lambda d^{-1/(p-1)}\delta^{p^k}, \qquad \|x_n-x_k\| \leq \lambda d^{-1/(p-1)}[H_k(\delta)-H_n(\delta)] \ (n \geq k),$$

i.e., the sequence $\{x_k\}$ is fundamental and therefore it has a limit point $x^* = \lim_{k\to\infty} x_k$ with

$$\|x_k - x^*\| \leq \lambda d^{-\frac{1}{p-1}} H_k(\delta) \leq \rho, \qquad \|x_0 - x^*\| \leq \lambda d^{-\frac{1}{p-1}} H_0(\delta).$$

Further we shall demonstrate some applications of this Theorem.

VARIANTS OF QUADRATICALLY CONVERGENT METHODS

First of all let us present without proof some auxiliary statements. Using the results (Krasnosel'skii et al., 1969) it is not difficult to prove the following assertion (see also (Vaarmann and Lomp, 1982)).

LEMMA. *Let $F'(x)$ be Lipschitz continuous and $\Phi(x)$ be uniformly bounded on some set S, i.e.,*

$$\|F'(x) - F'(y)\| \leq L_1\|x - y\|, \quad \forall x, y \in S, \tag{16}$$

$$\|\Phi(x)\| \leq \overline{C}, \quad \forall x \in S, \tag{17}$$

then

$$\|P_{R(x)} - P_{R(y)}\| \leq \overline{L}_0\|x - y\|, \quad \forall x, y \in S, \tag{18}$$

where $\overline{L}_0 = \overline{C}L_1\|U^{-1}\|$.

Remark 1. From Lemma it follows that there exist scalars N, N' and N'' such that

$$\|(P_{R(y)} - P_{R(y)}P_{R(x)})F(x)\| = \|(P_{R(y)} - P_{R(x)})(I - P_{R(x)})F(x)\|$$

$$\leq N'\|x - y\|, \qquad x, y \in S,$$

and

$$\|(P_{k+1} - P_{k+1}P_k)F(x_k)\| \leq N\|P_kF(x_k)\|, \quad x_k, x_{k+1} \in S, \tag{19}$$

where $\sup_{x\in S}\|(I - P_{R(x)})F(x)\| \leq N''$, $N' = N''L_0$ and $N = N'\lambda$. It can be readily shown that if F' satisfies Lipschitz condition with some constant L_1 then Ψ satisfies Lipschitz condition with the constant $\overline{L}_1 = L_1\|U^{-1}\|$.
If instead of $F'(x)$ to put its first order divided difference $F(y_k; x_k)$ where the basic elements x_k and y_k satisfy the condition $\|y_k - x_k\| \leq C_2\|P_kF(x_k)\|$ $(C_2 > 0)$, symmetric second order divided difference $F(y_k; x_k; x_k)$ is bounded, then one can take

$$\bar{\gamma}_k = C_3\|P_kF(x_k)\| \ (C_3 > 0) \tag{20}$$

provided $R(F'(x_k)) = R(F(y_k; x_k))$, $R([F'(x_k)]^*) = R([F(y_k; x_k)]^*$ (Ortega and Rheiboldt, 1970; Boggs and Dennis, 1976; Vaarmann, 1989).
If $P_{R(x)}$ is independent of x, e.g., $R(F'(x)) = H_2$, then the approximations to $\Phi(x)$ can be defined by

$$\overline{A}_{k+1} = \overline{A}_k \sum_{i=0}^{q-1} \left(I - \Psi(x_{k+1})\overline{A}_k\right)^i, \qquad q \geq 2, \qquad k = 0, 1, \ldots, \tag{21}$$

while

$$\|P_k - \Psi(x_{k+1})\overline{A}_k\| \leq \|P_k - \Psi(x_k)\overline{A}_k\| + \|[\Psi(x_{k+1}) - \Psi(x_k)]\overline{A}_k\|$$
$$\leq \bar{\gamma}_k + h\|P_k F(x_k)\|,$$

where $h = \overline{\lambda} L_1 \|U^{-1}\|$ and

$$\|P_{k+1} - \Psi(x_{k+1})A_{k+1}\| \leq \|P_k - \Psi(x_{k+1})\overline{A}_k\|^q$$
$$\leq \left(\bar{\gamma}_k + h\|P_k F(x_k)\|\right)^q = \bar{\gamma}_{k+1}.$$

Let us consider the methods of the type

$$x_{k+1} = x_k - A_k F(x_k), \tag{22}$$

i.e., $Q(x_k, A_k) = A_k F(x_k)$. Then on the basis of Taylor expansion one obtains

$$P_{k+1}F(x_{k+1}) = (P_{k+1} - P_{k+1}P_k)F(x_k) + P_{k+1}\Big\{\left(P_k - \Psi(x_k)\overline{A}_k\right)F(x_k)$$
$$+ \int_0^1 \left[F'(x_k) - F'\left(x_k + t(x_{k+1} - x_k)\right)\right] A_k F(x_k) dt\Big\}. \tag{23}$$

If F satisfies (19) and $\bar{\gamma}_k$ is given by (20), then

$$\|P_{k+1}F(x_{k+1})\| \leq N + \left(C_2 + \frac{1}{2}\lambda_k^2 L_1\right)\|P_k F(x_k)\|^2. \tag{24}$$

According to the Part 1 of the theorem (see also (12)) the sequence (3) has only the linear rate of convergence. In the case of (20) one can take $\bar{d} = C_3$ and $b_k = \|P_k F(x_k)\|$. If moreover $P_{R(x)}$ is independent of x, then $N = 0$, and the sequence (3) converges quadratically to a solution of (2). If the approximations $\{A_k\}$ are defined by (21) we have $\|P_{k+1}F(x_{k+1})\| \leq \bar{\gamma}_k\|P_k F(x_k)\| + \frac{1}{2}\lambda_k^2 L_1 \|P_k F(x_k)\|^2$, i.e., $a_\nu + u_\nu = 2$. Thus $q \geq 2$ is sufficient to ensure the quadratic rate of convergence for (3) in this case. In more general case if $P_{R(x)}$ depends on x the approximations generated by (21) need not to satisfy the condition $U\overline{A}_k = A_k = A_k P_k$. Hence it is reasonable to seek for other possibilities for building up A_k.
In (Vaarmann, 1978) one considers an approximate variant of Gauss–Newton method

$$x_{k+1} = x_k - A_k F(x_k), \qquad A_k = D_k\left[F'(x_k)\right]^*, \tag{25}$$

$$D_{k+1} = 2D_k - D_k B(x_{k+1}) D_k, \tag{26}$$

where $B(x) = [F'(x)]^* F'(x)$, $\overline{A}_k = A_k$ and the starting operator D_0 is close enough to $[B'(x_0)]^*$. Obviously, $A_k = A_k P_k$. If $F'(x)$ and $F'(y)$ are in the acute case, then

$$\|P_{R(x)} - P_{R(y)}\| \le \widehat{N} < 1, \qquad \|A_k\| \le \lambda_k,$$

$$\lambda_k = C(1+\gamma_k)/(1-\widehat{N}), \tag{27}$$

$$\gamma_k = \Big(\gamma_{k-1} + \widehat{N} + \big[\lambda^2 L_1 + (C'/(1-\widehat{N}))\big] \|P_{k-1} F(x_{k-1})\|\Big)^2$$

$$(C, C' > 0). \tag{28}$$

Here $N = \widehat{N}$ and from (24), (27) and (28) it follows that the sequence $\{x_k\}$ defined by (25), (26) converges linearly. If $P_{R(x)}$ does not depend on x, then $\widehat{N} = 0$ and on the basis of (24) and (28) we conclude that the sequence $\{x_k\}$ defined by (25), (26) converges quadratically.
For solving (2) often the Levenberg–Marquardt type methods

$$x_{k+1} = x_k - D_k \big[F'(x_k)\big]^* F(x_k), \qquad A_k = D_k \big[F'(x_k)\big]^* \tag{29}$$

are used. Here D_k is an approximation to M_k^{-1}, $M_k = B(x_k) + \alpha_k H$. In particular, D_k might be generated by

$$D_{k+1} = D_k \sum_{i=0}^{q-1} (I - M_{k+1} D_k)^i. \tag{30}$$

If F satisfies the condition (19), then, by the theorem, the sequence $\{x_k\}$ defined by (29) has in general the linear rate of convergence. If $P_{R(x)}$ is independent of x, then depending on the choice of α_k and the formula for A_k the sequence (29) has the linear, superlinear or quadratic rate of convergence (Vaarmann, 1980, 1989).

VARIANTS OF CUBICALLY CONVERGENT METHODS

For the following modification of tangent parabolas

$$x_{k+1} = x_k - A_k F(x_k) - \frac{1}{2} A_k F''(x_k) \big(A_k F(x_k)\big)^2 \tag{31}$$

we have

$$P_{k+1} F(x_{k+1}) = (P_{k+1} - P_{k+1} P_k) F(x_k) + P_{k+1} \big(P_k - \Psi(x_k) \overline{A}_k\big) F(x_k)$$

$$+ \frac{1}{2} P_{k+1} \big(P_k - \Psi(x_k) \overline{A}_k\big) F''(x_k) \big(A_k F(x_k)\big)^2$$

$$+ \frac{1}{2} (P_{k+1} - P_{k+1} P_k) F''(x_k) \big(A_k F(x_k)\big)^2$$

$$+P_{k+1}\int_0^1 \Big[F''(x_k+t(x_{k+1}-x_k))-F''(x_k)\Big](x_{k+1}-x_k)^2(1-t)dt,$$

$$\|P_{k+1}F(x_{k+1})\| \le N\|P_kF(x_k)\| + \overline{\gamma}_k\|P_kF(x_k)\|$$

$$+\frac{1}{2}\overline{\gamma}_k L_2\lambda_k^2\|P_kF(x_k)\|^2 + \frac{\lambda^3}{2}(\overline{L}_0L_2+\frac{1}{3}L_2)\|P_kF(x_k)\|^3,$$

$$\text{where} \quad \|F''(x)\| \le L_2.$$

If F satisfies the condition (19), then the last relation ensures only linear rate of convergence for (31). In order to guarantee the cubic convergence rate for it with A_k defined by (21) and the operator $P_{R(x)}$ being independent of x the inequality $q \ge p(p-s)/l = 6$ must be valid since here $p = 3$ and $s = l = 1$ or $\overline{\gamma}_k = O(\|P_kF(x_k)\|^2)$ if A_k is determined by another rule.
Using in (31) the following approximation

$$F''(x_k)\big(A_kF(x_k)\big)^2 = F''(x_k)\Big[(x_k - A_kF(x_k)) - x_k\Big]$$

$$\approx 2\Big[F(x_k - A_kF(x_k)) - F(x_k) - A_k^{-1}\Big((x_k - A_kF(x_k)) - x_k\Big)\Big]$$

one gets

$$x_{k+1} = x_k - A_kF(x_k) - A_kF(y_k), \qquad y_k = x_k - A_kF(x_k), \tag{32}$$

If $P_{R(x)}$ is independent of x, then in a similar way as before one obtains

$$\|P_{k+1}F(x_{k+1})\| \le \overline{\gamma}_k^2\|P_kF(x_k)\| + \omega_2\overline{\gamma}_k\|P_kF(x_k)\|^2 + C_3\|P_kF(x_k)\|^3.$$

In this case $a_\nu + u_\nu = 3$ and therefore $q \ge 3$ provides the cubic convergence for the sequence $\{x_k\}$ defined by (21), (32).

CONCLUDING REMARKS

Making use of approximations instead of exact pseudoinverses is theoretically equivalent to solving the corresponding equations at each step within certain tolerance say $\eta_k\|P_kF(x_k)\|$ ($\eta_k > 0$). Therefore the theorem under discussion is applicable also in the case when for solving corresponding linear equations iterative methods are employed.
To obtain a method that is both reliable and computationally convenient this is far from a trivial task in practice. That is why in last years one prefers to use a flexible polyalgorithmic approach for solving nonlinear equation which combines various methods. In this case it is required that a single numerical algorithms involving in the combination is efficient in a stage of computational process or for a class of problems.
The results of computer comparison show that methods with successive approximation of inverse or pseudoinverse operator may be methods of choice. This paper

is not intended to give a numerical comparative study of methods under consideration with the other related methods. Some results about numerical behavior of these methods can be found in (Vaarmann et. al., 1988). Further let us discuss some computational aspects of these methods on the theoretical basis.
The main disadvantage of methods with successive approximation the (pseudo-) inverse operator based on the formulas (21) or (30) as applied to solving a system of nonlinear equations is the great amount of arithmetic necessary for computing the next approximation to (pseudo-)inverse matrix.
The situation may be different for abstract spaces. For instance, the fast Fourier transformations permit to calculate the needed multiplications of linear operators with the smaller cost of computation than the solution of corresponding linear operator equations.
An important advantage of formula (30) is its unsusceptibility to small perturbations in the operator to be inverted since it is self-correcting for surjective operators.
Formulas (21) and (30) avoid the need to solve linear equations at every iteration and permit to obtain an estimate for (pseudo-)inverse operator. Knowing the inverse (or its approximation) is benefical in many cases. It facilitates to compute a condition number, to obtain error bounds, to pass easily on to interval-arithmetic versions etc.
The methods with successive approximation inverse (pseudoinverse) operator have some attractive features in finite dimensional case as well:
Matrix multiplications are attainable by means of scalar products that are independent and therefore can be performed simultaneously on a parallel computer.
The scalar product turns to be a key operation for obtaining highly accurate results in scientific computation. The higher computer arithmetic ("advanced computer arithmetic") combines the interval analysis with maximally accurate scalar products.

REFERENCES

Boggs, P.T., and Dennis, J.E. (1976). A stability analysis for perturded nonlinear iterative method. *Math. Comput.* **30**, 199–215.

Krasnosel'skii, M.A., Vainikko, G.M., Zabreiko, P.P., Rutit'skii, J.B., and Stetsenko, B.J. (1969). *The Approximate Solution of Operator Equations.* Nauka, Moscow (in Russian).,

Ortega, J.M. and Rheinboldt, W.C. (1970). *Iterative Solution of Nonlinear Equations in Several Variables.* Academic Press, New York, London.

Vaarmann, O. (1978). Certain iterative methods for the nonlinear least squares problem. *Eesti NSV Tead. Akad. Toimetised. Füüs.-Mat.* **27**, 251–258.

Vaarmann, O. (1980). On a class of Levenberg–Marquardt-type methods. *Eesti NSV Tead. Akad. Toimetised. Füüs.-Mat.* **29**, 233–240.

Vaarmann, O. and Lomp, M. (1982). On enlarging the domain of convergence of iterative methods with approximation pseudoinverse operator. *Eesti NSV Tead. Akad. Toimetised. Füüs.-Mat.* **31**, 410–417 (in Russian).

Vaarmann, O., Kulla, V., Lomp, M., and Meressoo, T. (1988). Numerical comparison of some methods for solving nonlinear equations. In: *Numerical Methods and Optimization*, Valgus,

Tallinn, pp. 46–55 (in Russian).
Vaarmann, O. (1989). Solution of nonlinear least squares problems by Levenberg–Marquardt type methods. *Eesti NSV Tead. Akad. Toimetised. Füüs.-Mat.* **38**, 146–153.

Ill-Posed Problems in Natural Sciences, pp. 202 – 213
A. Tikhonov (Ed.)
1992 VSP/TVP

IDENTIFICATION OF FILTRATION COEFFICIENT

G. VAINIKKO
Department of Mathematics, Tartu University, Ülikooli str.,18, 202400 Tartu, Estonia

ABSTRACT

This lecture is devoted to the coefficient inverse problem for an elliptic differential equation describing the filtration of ground water. As a starting point for discretization and regularization, we use a weak formulation of the problem. It happens to be equivalent to the operator equation $Ta = \nabla\psi$ with a noncompact operator T having a nonclosed range. A specific feature of the problem is than the evaluation of Ta at an $a \in \mathbf{L}^2(\Omega)$ is a more expensive operation than the evaluation of $T^*\nabla w$ at a $w \in \mathbf{H}^1(\Omega)$, therefore approximate methods should use evaluations of the abjoint operator only. A more serious trouble is caused by noncompactness of T. A suitable discretization is obtained by a special projection method (the method of minimal error). In the case of exact data this method converges, and in the case of polluted data it has a regularizing feature; combined with the Tikhonov and/or iterative regularization it provides convenient numerical schemes adwitting independent limiting processes in discretization and regularization parameters and the choice of the regularization parameter by the residual principle.
Our consideration allows discontinuities of the filtration coefficient; the only a priori assumption is that $a \in \mathbf{L}^2(\Omega)$. All the methods converge to the normal solution (to the coefficient a with the minimal $\mathbf{L}^2(\Omega)$ norm). Here the problem of identifiability of functions of the class $\mathbf{L}^2(\Omega)$ arises. This problem is tackled in a slightly more general setting, namely the identifiability of function of the class $\mathbf{L}(\Omega)$ is discussed.
This is a preliminary publication on the subject. The exposition of ideas and results is comprehensive and self-contained but owing to space limitation, the proofs of theorems are omitted.

1. INVERSE PROBLEM

1.1. *The boundary value problem formulation*
Let $\Omega \subset \mathbf{R}^n (n \geq 2)$ be an open bounded region with a piecewise smooth boundary $\partial\Omega$; we denote by ν the outer unit normal to $\partial\Omega$. Let $\Gamma \subset \partial\Omega$ be a relatively open set having a piecewise smooth boundary in $\partial\Omega$. We shall deal with the following inverse problem: find a coefficient $a \in \mathbf{L}^2(\Omega)$ such that

$$\begin{aligned} -\text{div}[(a(x)\nabla u(x)] = f(x), \quad x \in \Omega, \\ [a(x)\nabla u(x)] \cdot \nu(x) = g(x), \quad x \in \Gamma, \end{aligned} \tag{1.1}$$

where $f \in \mathbf{L}^2(\Omega)$, $g \in \mathbf{L}^2(\Gamma)$, $u \in \mathbf{W}^{1,\infty}(\Omega)$ are given functions. Physically, u can be interpreted as the piezometrical head of the ground water in Ω; the function f characterizes the sources and sinks in Ω and the function g characterizes the inflow and outflow through $\Gamma \subset \partial\Omega$. The filtration (transmissivity) coefficient a is, physically, positive and piecewise smooth with possible discontinuities of the first kind on some surfaces in Ω.

We do not exclude the case $\Gamma = \varnothing$. The boundary condition is omitted in (1.1) in this case.

Conditions (1.1) can be understood in the sense of distributions. We prefer to deal with the weak formulation of the problem.

1.2. *The weak formulation*

Let us temporarily assume that functions a and u are smooth (e.g., $a \in \mathbf{H}^1(\Omega)$, $u \in \mathbf{W}^{2,\infty}(\Omega)$). Multiplying the first equation in (1.1) by $w \in \mathbf{H}^1(\Omega)$, integrating by parts and using the second equation we obtain

$$\int_\Omega a\nabla u \cdot \nabla w \, dx = \int_\Omega fw \, dx + \int_\Gamma gw \, dS + \int_{\partial\Omega\setminus\Gamma} (a\nabla u)\cdot \nu w \, dS.$$

Introduce the subspace

$$\mathbf{H}^1(\Omega, \Gamma) = \{w \in \mathbf{H}^1(\Omega) \colon w(x) = 0 \quad \text{for} \quad x \in \partial\Omega\setminus\Gamma\} \subset \mathbf{H}^1(\Omega).$$

We obtain the following weak formulation of inverse problem (1.1): find $a \in \mathbf{L}^2(\Omega)$ such that

$$\int_\Omega a\nabla u \cdot \nabla w \, dx = \int_\Omega fw \, dx + \int_\Gamma gw \, dS, \qquad \forall w \in \mathbf{H}^1(\Omega, \Gamma). \tag{1.2}$$

The same formulation can be obtained in the case of piecewise smooth a and u. Problem (1.2) has sense for $u \in \mathbf{W}^{1,\infty}(\Omega)$, $a \in \mathbf{L}^2(\Omega)$.

1.3. *The operator equation formulation*

Let us denote by G the space of gradients of functions $w \in \mathbf{H}^1(\Omega, \Gamma)$:

$$G = G(\Omega, \Gamma) = \{\nabla w \colon w \in \mathbf{H}^1(\Omega, \Gamma)\} \subset (\mathbf{L}^2(\Omega))^n.$$

Let Q_G denote the orthoprojector in $(\mathbf{L}^2(\Omega))^n$ corresponding to G. We observe that Problem (1.2) is equivalent to the equation

$$Ta = \nabla\psi, \tag{1.3}$$

where the operator $T = T_u \in \mathcal{L}(\mathbf{L}^2(\Omega), G)$ is defined by the formula

$$Ta = Q_G(a\nabla u), \qquad a \in \mathbf{L}^2(\Omega), \tag{1.4}$$

and $\psi = \psi_{f,\,g}$ is a solution of the direct problem

$$-\Delta\psi(x) = f(x), \qquad x \in \Omega,$$

$$\nabla\psi(x)\cdot\nu(x) = g(x),\ x \in \Gamma, \qquad \psi(x) = 0, \qquad x \in \partial\Omega\backslash\Gamma. \tag{1.5}$$

Indeed, ψ satisfies the variational equality

$$\int_\Omega \nabla\psi\cdot\nabla w\,dx = \int_\Omega fw\,dx + \int_\Gamma gw\,dS, \qquad \forall w \in \mathbf{H}^1(\Omega,\,\Gamma), \tag{1.6}$$

thus (1.2) takes the form

$$\int_\Omega a\nabla u\cdot\nabla w\,dx = \int_\Omega \nabla\psi\cdot\nabla w\,dx, \qquad \forall w \in \mathbf{H}^1(\Omega,\,\Gamma),$$

and this is equivalent to (1.3) since $Q_G\nabla\psi = \nabla\psi$.
In case $\Gamma \neq \partial\Omega$, problem (1.5) is uniquely solvable. In the case $\Gamma = \partial\Omega$, (1.5) is solvable if $\int_\Omega f dx + \int_{\partial\Omega} g dS = 0$; this condition is necessary for the solvability of inverse problem (1.2), too, if $\Gamma = \partial\Omega$.

1.4. *Ill-posedness of the inverse problem*
The operator $T \in \mathcal{L}(\mathbf{L}^2(\Omega),\ G)$ has a very simple adjoint operator $T^* \in \mathcal{L}(\mathbf{L}^2(\Omega), G)$:

$$T^*\nabla w = \nabla u\cdot\nabla w, \qquad \nabla w \in G. \tag{1.7}$$

It is easy to see that the range $\mathcal{R}(T^*) \subset \mathbf{L}^2(\Omega)$ is nonclosed in $\mathbf{L}^2(\Omega)$ even if $|\nabla u| \geq c_0 > 0$ in Ω (in this our case $n \geq 2$ essentially differs from the case $n = 1$). It is also clear that T^* is noncompact. Consequently, $T \in \mathcal{L}(\mathbf{L}^2(\Omega), G)$ has a nonclosed range $\mathcal{R}(T) \subset G$ and T is noncompact too. Thus, (1.3) is an ill-posed problem with a noncompact operator. For ill-posed problems with compact operators, discretization and regularization algorithms are rather fully developed and investigated, see, e.g., (Vainikko and Hämarik, 1985; Vainikko and Veretennikov, 1986; Neubauer, 1988; Engl and Gfrerer, 1988; Plato and Vainikko, 1990) but we must revise those methods since T is noncompact in our case.
For a $z \in (\mathbf{L}^2(\Omega))^n$ the projection $Q_G z = \nabla\psi$ can be found by solving direct problem (1.5) with $f = -\operatorname{div} z, g = z\cdot\nu$. Thus, an evaluation of Ta for a $a \in \mathbf{L}^2(\Omega)$ is a more expensive operation than the evaluation of $T^*\nabla w$ for a $\nabla w \in G$, cf. (1.4) and (1.7). Therefore, we are interested in discretization and regularization schemes which can be out by applying T^* and not using T itself.

2. DISCRETIZATION

2.1. *Projection methods*
A natural way to discretize inverse problem (1.1) is to apply finite element approximations to weak formulation (1.2) of the problem. Let us introduce finite

dimensional subspaces $S_h^0 \subset \mathbf{L}^2(\Omega)$ and $S_h^1 \subset \mathbf{H}^1(\Omega, \Gamma)$, where $h > 0$ is discretization parameter. We can construct the following discrete version of problem (1.2):

$$\text{find} \quad a_h \in S_h^0 \quad \text{such that}$$

$$\int_\Omega a_h \nabla u \cdot \nabla w_h \, dx = \int_\Omega f w_h \, dx + \int_\Gamma g w_n \, dS, \qquad \forall w_n \in S_h^1. \tag{2.1}$$

Using operator equation formulation (1.3)–(1.6) we see that (2.1) is equivalent to the following problem

$$\text{find} \quad a_h \in S_h^0 \quad \text{such that} \quad Q_h T a_h = Q_h \nabla \psi, \tag{2.2}$$

where Q_h is the orthoprojector in G corresponding to the subspace $G_h = \{\nabla w_h\colon w_h \in S_h^1\} \subset G$.

2.2. *The method of minimal error*

Here only the subspaces $S_h = S_h^1 \subset \mathbf{H}^1(\Omega, \Gamma)$ are needed and the following discrete version of problem (1.2) is introduced

$$\text{find} \quad a_h \in \mathbf{L}^2(\Omega) \quad \text{of the minimal} \quad \mathbf{L}^2(\Omega) \quad \text{norm such that}$$

$$\int_\Omega a_h \nabla u \cdot \nabla w_h \, dx = \int_\Omega f w_h \, dx + \int_\Gamma g w_h \, dS, \qquad \forall w_h \in S_h. \tag{2.3}$$

This problem is uniquely solvable if its restrictions are admissible, e.g., in the following two cases: (i) if (1.2) is solvable; (ii) if $w_h \in S_h$, $\nabla u \cdot \nabla w_h = 0$ imply $\nabla w_h = 0$. The solution of (2.3) has the form $a_h = \nabla u \cdot \nabla v_h$, $v_h \in S_h$. Thus, (2.3) is a special case of projection method (2.1) with $S_h^0 = \{a_h \in \mathbf{L}^2(\Omega)\colon a_h = \nabla u \cdot \nabla w_h,\ w_h \in S_h\}$; let us denote by $P_h = P_{h,\,u}$ the corresponding orthoprojector in $\mathbf{L}^2(\Omega)$.

The operator equation formulation of method (2.3) takes the form

$$\text{find} \quad v_h \in S_h \quad \text{such that} \quad Q_h T T^* \nabla v_h = Q_h \nabla \psi$$

$$\text{and put} \quad a_h = T^* \nabla v_h = \nabla u \cdot \nabla v_h, \tag{2.4}$$

where Q_h is the orthoprojector in G corresponding to the subspace $G_h = \{\nabla w_h\colon \nabla w_h \in S_h\} \subset G$. Choosing a basis $w_j = w_{j,\,h}$ $(j = 1, \cdots, l = l_h)$ in S_h, a practical application of method (2.3) leads reduces to the following:

$$\text{find} \quad a_h = \sum_{j=1}^{l} c_j \nabla u \cdot \nabla w_j \quad \text{by solving the system}$$

$$\sum_{j=1}^{l} \int_\Omega (\nabla u \cdot \nabla w_j)(\nabla u \cdot \nabla w_i) \, dx c_j = \int_\Omega f w_i \, dx + \int_\Gamma g w_i \, dS, \qquad i = 1, \cdots, l. \tag{2.5}$$

2.3. *Convergence*
From the equalities

$$(P_h a,\ \nabla u \cdot \nabla w_h) = (a,\ \nabla u \cdot \nabla w_h), \qquad w_h \in S_h,$$

it follows that if a is a solution of (1.2), then $P_h a$ is the (unique) solution of (2.3). As an easy consequence from this observation we obtain the following result.

THEOREM 2.1. *Assume that S_h is full in $\mathbf{H}^1(\Omega,\ \Gamma)$ as $h \to 0$, i.e., for any $w \in \mathbf{H}^1(\Omega,\ \Gamma)$, there exist $w_h \in S_h$ such that $w_h \to w$ in $\mathbf{H}^1(\Omega)$ as $h \to 0$. If for given $u \in \mathbf{W}^{1,\infty}(\Omega)$, $f \in \mathbf{L}^2(\Omega)$, $g \in \mathbf{L}^2(\Gamma)$, problem (1.2) is solvable and $a \in \mathbf{L}^2(\Omega)$ denotes the normal solution, then problem (2.3) is solvable too, and $a_h \to a$ in $\mathbf{L}^2(\Omega)$ as $h \to 0$. Conversely, if problem (1.2) is nonsolvable in $\mathbf{L}^2(\Omega)$ but (2.3) is solvable, then $\|a_h\|_{\mathbf{L}^2(\Omega)} \to \infty$ as $h \to 0$.*

2.4. *Regularizing property of the discretization by the method of minimal error*
Now consider the case where, instead of exact data denoted in the sequal by u_0, f_0 and g_0 and corresponding to a filtration coefficient $a_0 \in \mathbf{L}^2(\Omega)$, we have at our disposal polluted data $u \in \mathbf{W}^{1,\infty}(\Omega)$, $f \in \mathbf{L}^2(\Omega)$ and $g \in \mathbf{L}^2(\Gamma)$. We assume that

$$\|\nabla\psi - \nabla\psi_0\|_{(L^2(\Omega))^n} \le \delta, \tag{2.6}$$

$$\sup_{x\in\Omega} |\ \nabla u(x) - \nabla u_0(x)\ | \le \eta, \tag{2.7}$$

where ψ_0 and ψ are the solution to problem (1.5) with right hand terms f_0, g_0 and f, g, respectively, and δ and η are small positive numbers. It turns out that, under some conditions, discretization (2.3) acts as a reqularizer. Let as introduce the instability numbers

$$\varkappa_h^0 = \sup_{0\neq\nabla w_h\in G_h} \|\nabla w_h\|/\|\nabla u_0 \cdot \nabla w_h\| \le \infty,$$

$$\varkappa^h = \sup_{0\neq\nabla w_h\in G_h} \|\nabla w_n\|/\|\nabla u \cdot \nabla w_h\| \le \infty.$$

The determinant of the system in (2.5) is nonvanishing if and only if $\varkappa_h < \infty$. It is easy to verify that

$$\varkappa_h^0 \le \varkappa_h(1-\varkappa_h\eta)^{-1} \quad \text{if} \quad \varkappa_h < \infty \quad \text{and} \quad \varkappa_h\eta < 1,$$

$$\varkappa_h \le \varkappa_h^0(1-\varkappa_h^0\eta)^{-1} \quad \text{if} \quad \varkappa_h^0 < \infty \quad \text{and} \quad \varkappa_h^0\eta < 1;$$

note that $\varkappa_h^0 \to \infty$, $\varkappa_h \to \infty$ as $h \to 0$. For practical purposes note also that $\varkappa_h = \lambda_h^{-1/2}$, where λ_h is the smallest eigenvalue of the algebraic eigenvalue problem

$$Ac = \lambda Bc \tag{2.8}$$

with Gram matrices $A = (a_{ij})$ and $B = (b_{ij})$,

$$a_{ij} = \int_\Omega (\nabla u \cdot \nabla w_j)(\nabla u \cdot \nabla w_i)\, dx, \qquad b_{ij} = \int_\Omega \nabla w_j \cdot \nabla w_i\, dx; \tag{2.9}$$

here $w_j = w_{j,h}$ $(j = 1, \cdots, l = l_h)$ is a basis in S_h as in (2.5).

THEOREM 2.2. *Assume that S_h is full in $\mathbf{H}^1(\Omega, \Gamma)$ as $h \to 0$ and $\varkappa_h^0 < \infty$ for all small $h > 0$. Assume that problem (1.2) corresponding to exact data $u_0 \in \mathbf{W}^{1,\infty}(\Omega)$, $f_0 \in \mathbf{L}^2(\Omega)$, $g_0 \in \mathbf{L}^2(\Gamma)$ is solvable; let $a_0 \in \mathbf{L}^2(\Omega)$ denote the normal solution. Let $a_{h,\delta,\eta}$ denote the (unique) solution of problem (2.3) corresponding to polluted data $u \in \mathbf{W}^{1,\infty}(\Omega)$, $f \in \mathbf{L}^2(\Omega)$, $g \in \mathbf{L}^2(\Gamma)$ that satisfy (2.6) and (2.7). Choose $h = h(\delta, \eta)$ so that $h \to 0$, $(\delta + \eta)\varkappa_h \to 0$ as $\delta \to 0$, $\eta \to 0$. Then $a_{h,\delta,\eta} \to a_0$ in $\mathbf{L}^2(\Omega)$ as $\delta \to 0$, $\eta \to 0$ and*

$$\|a_{h,\delta,\eta} - P_{h,u_0} a_0\|_{\mathbf{L}^2(\Omega)} \le \varkappa_h \big(\delta + 2\|a_0\|_{\mathbf{L}^2(\Omega)} \eta\big).$$

Unfortunately, it may happen that $\varkappa_h^0 = \infty$ for small h even if $|\nabla u_0(x)| \ge c_0 > 0$ $(x \in \Omega)$, and Theorem 2.2 fails in this case. It is rather risky to hope to have the regularizing properties of a bare discretization only. On the other hand, the regularizing property of method (2.3) influences favourably even if a supplementary reqularization is used.

3. REGULARIZATION

3.1. *A class of regularization methods*

Let E and G be Hilbert spaces (here arbitrary), $\psi_0, \psi_\delta \in G$, $T, T_\varepsilon \in \mathcal{L}(E, G), T_\varepsilon$ is finite dimensional,

$$\|\psi_\delta - \psi_0\| \le \delta,$$

$$T_\varepsilon^* \varphi \to T_0^* \varphi \quad \text{as} \quad \varepsilon \to 0 \ (\forall \varphi \in G), \qquad \|T_\varepsilon\|^2 \le \lambda_* = \text{const.} \tag{3.1}$$

We try to solve the problem $T_0 a = \psi_0$ using the data $T_\varepsilon, \psi_\delta$. Introduce a family of (generating) functions g_r: $[0, \lambda_*] \to \mathbf{R}$ depending on a (regularization) parameter $r > 0$ and satisfying the conditions

$$\sup_{0 \le \lambda \le \lambda_*} |1 - \lambda g_r(\lambda)| = 1, \qquad \sup_{0 \le \lambda \le \lambda_*} \lambda |1 - \lambda g_r(\lambda)| \le \gamma_1 r^{-1},$$

$$\sup_{0 \le \lambda \le \lambda_*} \lambda^{1/2} |g_r(\lambda)| \le \gamma_* r^{1/2}, \qquad \forall r > 0, \tag{3.2}$$

where the constants γ_1, γ_* do not depend on r. A regularized approximate solution of the problem $T_0 a = \psi_0$ will be defined by the formula

$$a_r = a_{r,\varepsilon,\delta} = \big(I_E - T_\varepsilon^* T_\varepsilon g_r(T_\varepsilon^* T_\varepsilon)\big) T_\varepsilon^* \varphi_0 + g_r(T_\varepsilon^* T_\varepsilon) T_\varepsilon^* \psi_\delta, \tag{3.3}$$

where I_E is the identity operator in E and $\varphi_0 \in G$ is an initial guess, e.g., $\varphi_0 = 0$ (in this case, (3.3) simplifies to $a_r = g_r(T_\varepsilon^* T_\varepsilon) T_\varepsilon^* \psi_\delta$). For practical purposes, (3.3) can be rewritten in an equivalent form

$$a_r = T_\varepsilon^* \varphi_r, \qquad \varphi_r = \big(I_G - T_\varepsilon T_\varepsilon^* g_r(T_\varepsilon T_\varepsilon^*)\big) Q_\varepsilon \varphi_0 + g_r(T_\varepsilon T_\varepsilon^*) Q_\varepsilon \psi_\delta, \tag{3.4}$$

where Q_ε is an orthoprojector in G such that $Q_\varepsilon T_\varepsilon = T_\varepsilon$.

THEOREM 3.1. *Let conditions (3.1) and (3.2) be fulfilled. Assume that the problem $T_0 a = \psi_0$ is solvable and $a_0 \in E$ is the normal solution. Assume that*

$$\|Q_\varepsilon(T_\varepsilon - T_0)a_0\| \le \tau_\varepsilon \|a_0\| \to 0 \quad as \quad \varepsilon \to 0. \tag{3.5}$$

In (3.4), let us choose $r = r(\varepsilon, \delta)$ so that

$$r(\varepsilon,\delta) \to \infty, \qquad r(\varepsilon,\delta)(\delta^2 + \tau_\varepsilon^2) \to 0 \quad as \quad \varepsilon \to 0, \delta \to 0. \tag{3.6}$$

Then $\|a_{r(\varepsilon,\delta)} - a_0\| \to 0$ as $\varepsilon \to 0, \delta \to 0$.

THEOREM 3.2. *Let conditions of Theorem 3.1 be fulfilled. In (3.4) choose $r = r(\varepsilon, \delta)$ by residual so that*

$$b_1(\delta + \|a_r\|\tau_\varepsilon) \le \|Q_\varepsilon(T_\varepsilon a_r - \psi_\delta)\| \le b_2(\delta + \|a_r\|\tau_\varepsilon), \tag{3.7}$$

where b_1 and b_2 are some constants, $b_2 \ge b_1 > 1$. Then

$$\|a_{r(\varepsilon,\delta)} - a_0\| \to 0 \quad as \quad \varepsilon \to 0, \ \delta \to 0.$$

In the case $\|T_\varepsilon - T_0\| \to 0$ (implying the compactness of T_0) such results are well known, see (Vainikko and Veretennikov, 1986). In the case of pointwise convergence $T_\varepsilon^* \to T_0^*$ assumed in (3.1), some trouble is caused by condition (3.5), in general we do not know the approximation properties of $T_\varepsilon - T_0$ on an unknown exact solution a_0. This is the case, e.g., if a general projection method of type (2.1) is applied. On the other hand, if the discretization is performed by applying the method of minimal error, i.e., $T_\varepsilon = Q_\varepsilon T_0$ or, more generally, $T_\varepsilon = Q_\varepsilon T$, where T is an available approximation to T_0, $\|T - T_0\| \le \tau_\varepsilon$, then (3.5) is fulfilled. This is the reason of a special interest to the method of minimal error.

Let us present some examples of the reqularization methods of type {(3.4), (3.2)}. The Tikhonov method

$$a_\alpha = T_\varepsilon^* \varphi_\alpha, \qquad \varphi_\alpha = (\alpha I + T_\varepsilon T_\varepsilon^*)^{-1} Q_\varepsilon \psi_\delta$$

corresponds to $g_r(\lambda) = (r^{-1} + \lambda)^{-1}$, $r = 1/\alpha$, $\varphi_0 = 0$. The iterated Tikhonov method

$$a_\alpha^m = T_\varepsilon^* \varphi_\alpha^m, \qquad \varphi_\alpha^0 = Q_\varepsilon \varphi_0, \qquad \varphi_\alpha^k = (\alpha I + T_\varepsilon T_\varepsilon^*)^{-1}(\alpha \varphi_\alpha^{k-1} + Q_\varepsilon \psi_\delta),$$

$$k = 1, \cdots, m,$$

corresponds to the generating function $g_\alpha^m(\lambda) = \sum_{k=0}^{m-1} \alpha^k (\alpha + \lambda)^{-1-k}$; here as a regularization parameter can be taken $r = 1/\alpha$ ($m \ge 1$ fixed), $r = m$ ($\alpha > 0$ fixed) or $r = m/\alpha$ (both α and m may vary). The iteration method

$$a_r = T_\varepsilon^* \varphi^r, \qquad \varphi^0 = Q_\varepsilon \varphi_0, \qquad \varphi^k = \varphi^{k-1} - \mu(T_\varepsilon T_\varepsilon^* \varphi^{k-1} - Q_\varepsilon \psi_\delta),$$

$$k = 1, \cdots, r,$$

with $0 < \mu < 2/\lambda_*$ corresponds to $g_r(\lambda) = \sum_{k=0}^{m-1}(1-\mu\lambda)^k\mu$. For these methods, Theorem 3.2 can be extended to the case $b_1 \geq 1$ in (3.7). We refer to (Vainikko and Veretennikov, 1986) for detailed discussions and further examples; one can find there information on properties of norms $\|Q_\varepsilon(T_\varepsilon a_r - \psi_\delta)\|$ and $\|a_r\|$ (limits as $r \to 0$ and $r \to \infty$, continuity and monotony in r), too.

3.2. *Application to the inverse problem*
Now we return to the case where $T_0 = Q_G(.\nabla u_0), E = \mathbf{L}^2(\Omega)$ and $G = G(\Omega, \Gamma)$ is the space of gradient vector functions; respectively, instead of φ, ψ and other elements of G we shall write $\nabla\varphi, \nabla\psi$, etc. The problem $T_0 a = \nabla\psi_0$ now means the operator equation formulation (1.3) of inverse problem (1.1) with exact data $u_0 \in \mathbf{W}^{1,\infty}(\Omega), f_0 \in \mathbf{L}^2(\Omega), g_0 \in \mathbf{L}^2(\Gamma)$. Let again $u = u_\eta \in \mathbf{W}^{1,\infty}(\Omega)$, $f = f_\delta \in \mathbf{L}^2(\Omega)$, $g = g_\delta \in \mathbf{L}^2(\Gamma)$ be available approximate data such that (2.6) and (2.7) hold. Further, let $S_h \subset \mathbf{H}^1(\Omega, \Gamma)$ be a family of finite dimensional subspaces which is full in $\mathbf{H}^1(\Omega, \Gamma)$ as $h \to 0$; we denote by $w_j = w_{j,h}$ $(j = 1, \cdots, l = l_h)$ a basis in S_h. We put $T_\varepsilon = Q_h Q_G(.\nabla u_\eta), \varepsilon = (\eta, h), Q_\varepsilon = Q_h$, where as before Q_h is the orthoprojector in $G \subset (\mathbf{L}^2(\Omega))^n$ corresponding to subspace $G_h = \{\nabla w_h\colon w_h \in S_h\}$.
It follows from (2.7) that

$$\|Q_G(.\nabla u_\eta) - Q_G(.\nabla u_0)\| \leq \sup_{x\in\Omega} |u_\eta(x) - u_0(x)| \leq \eta,$$

thus (3.5) is fulfilled with $\tau_\varepsilon = \eta$. It is clear that (3.1) is fulfilled, too, and we can apply Theorems 3.1 and 3.2. We shall specify the computation schemes and parameter choices (3.6) and (3.7) for concrete examples of regularization methods.
3.2.1. The Tikhonov method takes the form

$$a_\alpha = \sum_{j=1}^{l} c_{j,\alpha} \nabla u \cdot \nabla w_j, \qquad (\alpha B + A)\mathbf{c}_\alpha = \mathbf{d},$$

where A and B are $l \times l$-matrices with elements (2.9) and $\mathbf{c}_\alpha$ and $\mathbf{d}$ are l-vectors with components $c_{j,\alpha}$ and

$$d_i = \int_\Omega \nabla\psi \cdot \nabla w_i dx = \int_\Omega f w_i dx + \int_\Gamma g w_i dS, \qquad i = 1, \cdots, l.$$

A priori parameter choice:

$$\alpha = \alpha(h, \delta, \eta) \to 0, \qquad (\delta^2 + \eta^2)/\alpha(h, \delta, \eta) \to 0 \quad \text{as} \quad h \to 0, \delta \to 0, \eta \to 0.$$

Parameter choice by residual:

$$b_1\big(\delta + \langle A\mathbf{c}_\alpha, \mathbf{c}_\alpha\rangle^{1/2}\eta\big) \leq \big\langle A\mathbf{c}_\alpha - d,\ B^{-1}(A\mathbf{c}_\alpha - d)\big\rangle^{1/2} \leq b_2\big(\delta + \langle A\mathbf{c}_\alpha, \mathbf{c}_\alpha\rangle^{1/2}\eta\big),$$

$$1 \leq b_1 \leq b_2,$$

where $\langle A\mathbf{c}_\alpha, \mathbf{c}_\alpha\rangle$ and $\langle A\mathbf{c}_\alpha - d, B^{-1}(A\mathbf{c}_\alpha - d)\rangle$ are scalar products in $\mathbf{R}^l$.

3.2.2. The iterated Tikhonov method:

$$a_\alpha^m = \sum_{j=1}^{l} c_{j,\alpha}^m \nabla u \cdot \nabla w_j,$$

$$B\mathbf{c}_\alpha^0 = \mathbf{d}^0, \qquad (\alpha B + A)\mathbf{c}_\alpha^k = \alpha B\mathbf{c}_\alpha^{k-1} + \mathbf{d}, \qquad k = 1, \cdots, m,$$

where $A, B, \mathbf{d}$ are the same as in 3.2.1 and the l-vector $\mathbf{d}^0$ corresponds to the initial guess $\nabla\varphi_0 \in G$:

$$d_i^0 = \int_\Omega \nabla\varphi_0 \cdot \nabla w_i dx, \qquad i = 1, \cdots, l.$$

A priori choice of parameters $\alpha = \alpha(h, \delta, \eta), m = m(h, \delta, \eta)$ is

$$m(h,\delta,\eta)/\alpha(h,\delta,\eta) \to \infty, \qquad (\delta^2 + \eta^2)m(h,\delta,\eta)/\alpha(h,\delta,\eta) \to 0$$

$$\text{as} \quad h \to 0, \delta \to 0, \eta \to 0.$$

Parameters choice by residual principle is

$$b_1(\delta + \langle A\mathbf{c}_\alpha^m, \mathbf{c}_\alpha^m\rangle^{1/2}\eta) \le \langle A\mathbf{c}_\alpha^m - d, B^{-1}(A\mathbf{c}_\alpha^m - d)\rangle^{1/2} \le b_2(\delta + \langle A\mathbf{c}_\alpha^m, \mathbf{c}_\alpha^m\rangle^{1/2}\eta),$$

$$1 \le b_1 \le b_2.$$

3.2.3. The iteration method:

$$a_r = \sum_{j=1}^{l} c_j^r \nabla u \cdot \nabla w_j, \qquad B\mathbf{c}^0 = \mathbf{d}^0, \qquad B\mathbf{c}^k = B\mathbf{c}^{k-1} - \mu(A\mathbf{c}^{k-1} - \mathbf{d}),$$

$$k = 1, \cdots, r, \qquad 0 < \mu < 2/\sup_{x\in\Omega} |\nabla u(x)|^2.$$

The a priori stopping rule is

$$r = r(h,\delta,\eta) \to \infty, \quad (\delta^2 + \eta^2)r(h,\delta,\eta) \to 0 \quad \text{as} \quad h \to 0, \quad \delta \to 0, \quad \eta \to 0.$$

The residual stopping on the first $r = r(h, \delta, \eta)$ such that

$$\langle A\mathbf{c}^r - \mathbf{d}, B^{-1}(A\mathbf{c}^r - \mathbf{d})\rangle^{1/2} \le b(\delta + \langle A\mathbf{c}^r, \mathbf{c}^r\rangle^{1/2}\eta), \qquad b = \text{const} \ge 1.$$

3.3. *Remarks*

3.3.1. Assumption (2.7) turns into a somewhat unrealistic one if ∇u_0 has discontinuities on some unknown surfaces $\mathcal{M}_i$ in Ω (the case of composite soil). The results concerning a priori parameter choice remain valid if, instead of (2.7), we have

$$a_0 \in \mathbf{L}^\infty(\Omega), \qquad \|\nabla u_\eta - \nabla u_0\|_{(\mathbf{L}^2(\Omega))^n} \le \eta, \qquad \sup_{x\in\Omega} |\nabla u(x)|^2 \le \lambda_*.$$

These conditions preserve the pointwise convergence $T_\varepsilon^* \to T_0^*$ as $\varepsilon \to 0$ and the inequality $\|Q_\varepsilon(T_\varepsilon - T_0)a_0\| \leq \eta \sup_{x\in\Omega} |a_0(x)|$ (recall that $\varepsilon = (\eta, h)$).

3.3.2. The results can be somewhat improved if $\varkappa_h^0 < \infty$ (see § 2.4). Define

$$\rho(r,h) = \sup_{\lambda_h \leq \lambda \leq \lambda_*} \lambda^{1/2}|g_r(\lambda)|,$$

where λ_h is the smallest eigenvalue of problem (2.8). It is clear that $\rho(r,h) \leq \gamma_* r^{1/2}$. The assertion of Theorem 3.1 remains true for the parameter choice $r = r(\varepsilon, \delta) \to \infty$, $(\delta + \tau_\varepsilon)\rho(r,h) \to 0$ as $\varepsilon \to 0, \delta \to 0$. For instance, a sufficient a priori parameter choice in the Tikhonov method (§ 3.2.1) is

$$\alpha = \alpha(h,\delta,\eta) \to 0, \quad (\delta^2 + \eta^2)\min\{\alpha^{-1}, \lambda_h^{-1}\} \to 0 \quad \text{as} \quad h \to 0, \delta \to 0, \eta \to 0.$$

3.3.3. The methods considered in this work need an evaluation of the first derivatives of u. Let us compare them with other methods. In the method of output least squares no derivatives of u are needed but one has to solve a nonlinear operator equation. In the equation error method and in the method of characteristics, the second derivatives of u must be evaluated. For these methods, we refer to (Kunisch, 1988; Yeh, 1986 and Richter, 1981). It is not clear how to regularize these methods, they are based on the hope that they are self-regularized by the discretization.

4. IDENTIFIABILITY

4.1. *Introduction*

All the methods considered in §§ 2 and 3 converge to the normal solution of problem (1.2). The question about the uniqueness (identifiability) in $\mathbf{L}^2(\Omega)$ of the filtration coefficient $a \in \mathbf{L}^2(\Omega)$ arises. For smooth data and a, the identifiability is sufficiently fully analyzed, see (Richter, 1981; Chicone and Gerlach, 1987; Kunisch, 1988). It turns out that, with natural changes, these results can be extended to the case of $a \in \mathbf{L}^2(\Omega)$ and even $a \in \mathbf{L}(\Omega)$ under some rather realistic assumptions on data u, f and g.

We slightly generalize our inverse problem: having some functions $u \in \mathbf{W}^{1,\infty}(\Omega)$, $f \in \mathbf{L}(\Omega)$ and $g \in \mathbf{L}(\Gamma)$, we look for a $a \in \mathbf{L}(\Omega)$ such that

$$\int_\Omega a\nabla u \cdot \nabla w \, dx = \int_\Omega fw \, dx + \int_\Gamma gw \, dS, \qquad \forall w \in \mathbf{W}^{1,\infty}(\Omega,\Gamma), \tag{4.1}$$

where

$$\mathbf{W}^{1,\infty}(\Omega,\Gamma) = \{w \in \mathbf{W}^{1,\infty}(\Omega)\colon\ w(x) = 0 \quad \text{for} \quad x \in \partial\Omega\backslash\Gamma\}.$$

We say that a is L-identifiable by (4.1) on a subregion $\Omega' \subseteq \Omega$ if, for any solutions $a_1 \in \mathbf{L}(\Omega)$ and $a_2 \in \mathbf{L}(\Omega)$ of problem (4.1), we have $a_1(x) = a_2(x)$ for a.e. $x \in \Omega'$. Our purpose is to describe subregions $\Omega' \subseteq \Omega$ where a is L-identifiable.

4.2. *The case of smooth u*

Here we assume that, for any $\varepsilon > 0$, $u \in \mathbf{W}^{2,\infty}(\Omega_\varepsilon)$, where Ω_ε consists of all points $x \in \Omega$ such that the distance from x to the nearest nonsmoothness point of

$\partial\Omega$ is greater than ε; if $\partial\Omega$ is smooth, then $u \in \mathbf{W}^{2,\infty}(\Omega)$. Let us introduce flow curves $x = \varphi(t, y)$ as the maximal solution (solution on the maximal time interval) of the Cauchy problem

$$dx/dt = -\nabla u(x), \qquad x(0) = y \in \Omega. \tag{4.2}$$

Physically, the ground water flows along these curves but with a speed depending on a (Darcy's Law).
Due to the condition $u \in \mathbf{W}^{2,\infty}(\Omega_\varepsilon)$, ∇u is bounded and locally Lipshitz continuous on Ω, with possible singularities of the Lipshitz coefficient only at nonsmoothness points of $\partial\Omega$. Thus, problem (4.2) is uniquely solvable and $\varphi(t, y)$ is defined on a finite or infinite time interval (t_y^-, t_y^+). Introduce the following sets:

$$\Omega^+ = \{y \in \Omega\colon \nabla u(y) \neq 0,\ t_y^+ = \infty\},$$

$$\Omega^- = \{y \in \Omega\colon \nabla u(y) \neq 0,\ t_y^- = -\infty\},$$

$\Omega_\Gamma^+ = \{y \in \Omega\colon \nabla u(y) \neq 0, t_y^+ < \infty, \varphi(t, y)$ transversely reaches a smoothness point of $\Gamma \subseteq \partial\Omega$ as $t \uparrow t_y^+\}$;

$\Omega_\Gamma^- = \{y \in \Omega\colon \nabla u(y) \neq 0,\ t_y^- > -\infty, \varphi(t, y)$ transversely reaches a smoothness point of $\Gamma \subseteq \partial\Omega$ as $t \downarrow t_y^-\}$.

Recall that $\Gamma \subset \partial\Omega$ is a given relatively open set, thus Ω_Γ^+ and Ω_Γ^- are open subsets of Ω.

THEOREM 4.1. *Assume that $f \in \mathbf{L}(\Omega), g \in \mathbf{L}(\Gamma), u \in \mathbf{W}^{1,\infty}(\Omega)$ and $u \in \mathbf{W}^{2,\infty}(\Omega_\varepsilon)$ for any $\varepsilon > 0$. Then a is L-identifiable by problem (4.1) on sets* $\operatorname{int} \Omega^+, \operatorname{int} \Omega^-, \Omega_\Gamma^+$ *and* Ω_Γ^-.

It happens that in general L-identifiability cannot be extended to the closures of $\operatorname{int} \Omega^+, \operatorname{int} \Omega^-, \Omega_\Gamma^+, \Omega_\Gamma^-$. Theorem 4.1 fails also if we replace the assumption $u \in \mathbf{W}^{2,\infty}(\Omega_\varepsilon)$ by the assumption $u \in \mathbf{W}^{2,p}(\Omega), 1 \leq p < \infty$.

4.3. *The case of piecewise smooth u*

Now consider the case where u remains continuous on Ω but ∇u may jump on some piecewise smooth surfaces $\mathcal{M}_i (i = 1, \cdots, m)$ in Ω. Physically, $\mathcal{M}_i$ are the surfaces between different kinds of soil. Denote $\mathcal{M} = \cup_{i=1}^m \mathcal{M}_i$. We assume that, for any $\varepsilon > 0, u \in \mathbf{W}^{2,\infty}(\Omega_\varepsilon \backslash \mathcal{M})$, where Ω_ε consist of points $x \in \Omega$ such that the distance from x to the nearest nonsmoothness point of $\partial\Omega$ and $\mathcal{M}_i (i = 1, \cdots, m)$ as well to intersection of these surfaces is greater than ε. Further, we assume that there is a strictly positive piecewise smooth function a_{test}, with possible breaks on $\mathcal{M}$, such that $\operatorname{div}(a_{\text{test}} \nabla u) \in \mathbf{L}(\Omega)$, where the derivatives are understood in the sense of distributions. This implies the standard break law for ∇u and for flow curves on $\mathcal{M}$. Omitting the detailed formulation we present only the following corollary: if a flow curve transversely reaches at a finite time moment a smoothness point of $\mathcal{M}_i$, then this flow curve passes $\mathcal{M}_i$ transversely to the other side of $\mathcal{M}_i$ and continues there. We can define sets $\Omega^+, \Omega^-, \Omega_\Gamma^+, \Omega_\Gamma^-$ as in § 4.2 adding a requirement about the transversal cutting of $\mathcal{M}$, e.g.,

$\Omega^+ = \{y \in \Omega$: $\nabla u(y) \neq 0, t_y^+ = \infty$ and, for $0 \leq t < \infty, \varphi(t,y)$ meets $\mathcal{M}$ not more than finite times and every meating is transversal and takes place at a smoothness point of a $\mathcal{M}_i, 1 \leq i \leq m\}$;

$\Omega_\Gamma^- = \{y \in \Omega$: $\nabla u(y) \neq 0, t_y^- > -\infty, \varphi(t,y)$ transversely reaches a smoothness point of $\Gamma \subset \partial\Omega$ as $t \downarrow t_y^-$ and, for $t_y^- < t \leq 0, \varphi(t,y)$ meets $\mathcal{M}$ not more than finite times whereby every meating is transversal and takes place at a smoothness point of a $\mathcal{M}_i, 1 \leq i \leq m\}$.

THEOREM 4.2. *Assume that* $f \in \mathbf{L}(\Omega), g \in \mathbf{L}(\Gamma), u \in \mathbf{W}^{1,\infty}(\Omega)$ *and that the piecewise smoothness conditions formulated above are fulfilled. Then, with the modification of definitions introduced above, the filtration coefficient* a *is L-identifiable by (4.1) on sets* int Ω^+, int $\Omega^-, \Omega_\Gamma^+$ *and* Ω_Γ^-.

REFERENCES

Chicone, C. and Gerlach, J. (1987). Identifiability of distributed parameters. In: *Inverse and Ill-Posed Problems*, Ed. by Engl H.W. and Groetsch C.W., Academic Press, Boston etc, pp. 513–521.

Engl, H.W. and Gfrerer, H. (1988). A posteriori parameter choice for generalized regularization methods for solving linear ill-posed problems. *Appl. Numer. Math.* **4**, 395–417.

Kunisch, K. (1988). A review of some recent results on the output least squares formulation of parameter estimation problem. *Automatica* **24**, 531–539.

Neubauer, A. (1988). A posteriori parameter choice for Tikhonov regularization in the presence of modeling error. *Appl. Numer. Math.* **4**, 507–519.

Plato, R. and Vainikko, G. (1990). On the regularization of projection methods for solving ill-posed problems. *Numer. Math.* **57**, 63–79.

Richter, G.R. (1981). An inverse problem for the steady state diffusion equation. *SIAM J. Appl. Math.* **41**, 210–221.

Vainikko, G.M. and Hämarik, U.A. (1985). Projection methods and self-regularization in ill-posed problems. *Sov. Math.* **29**, 1–20. Vainikko, G.M. and Veretennikov, A.J. (1986). *Iteration Procedures in Ill-Posed Problems.* Nauka, Moscow (in Russian).

Yeh, W.W.-G. (1986). Review of parameter identification procedures in groundwater hydrology: the inverse problem. *Water Resources Res.* **22**, 95–108.

Ill-Posed Problems in Natural Sciences, pp. 214 – 223
A. Tikhonov (Ed.)
1992 VSP/TVP

ILL-POSED PROBLEMS AND ITERATIVE APPROXIMATION OF FIXED POINTS OF PSEUDO-CONTRACTIVE MAPPINGS

V.V. VASIN
Institute of Mathematics and Mechanics,
Ural Branch of Russia Academy Sciences, Kovalevskoi str., 16, 620219,
Sverdlovsk, Russia

ABSTRACT

The object of this paper is the iterative methods of the following type

$$x^{n+1} = k_{n+1}T(x^n) + (1 - k_{n+1})v_0, \quad n = 0, 1, \dots,$$

where T is a nonexpansive or pseudo-contractive mapping of an infinite dimensional Hilbert space X into itself, the sequence of real number $k_n \to 1$ as $n \to \infty, v_0$ is an arbitrary point in X. The main purpose of the paper is to survey, systematize and extend some results concerning the strong convergence to a fixed point of T and the stability of the iterative sequence and also to investigate applications to ill-posed operator equations with a priori information and problems of convex programming.

1. INTRODUCTION AND PRELIMINARY RESULTS

In the construction of regular algorithms for solving ill-posed problems on the basis of iterative methods the initial problem is usually reduced to an equivalent problem of finding the fixed point of a (non)linear mapping T in an infinite dimensional Hilbert space X

$$x = T(x), \tag{1.1}$$

where T: $X \to X$ is the step operator of the successive iteration method

$$x^{n+1} = T(x^n), \quad x^0 \in X, \quad n = 0, 1, \dots, \tag{1.2}$$

or of the simple iteration method

$$x^{n+1} = \lambda T(x^n) + (1 - \lambda)x^n, \quad \lambda \in (0, 1), \quad x^0 \in X, \; n = 0, 1, \dots, \tag{1.3}$$

The subject of the present paper is to survey, systematize and extend some results concerning the strong convergence and stability of iteration methods (1.2), (1.3) and their modification by means of correcting multiplies and also to investigate applications to ill-posed problems.

Let us introduce the notation Fix(T) for the set of fixed points of T, i.e., for the set of solutions of equation (1.1). In what follows we denote the weak and strong convergence in the Hilbert space. X by "$\rightharpoonup$" and "$\to$", respectively.
For mapping T: $X \to X$ we define five classes which admit iterative methods for the approximation of their fixed points (see Petryshyn and Willamson, 1973; Maruster, 1977; Eremin, 1969; Vasin, 1987):

$\mathcal{K}_Q = \{T$: T is Q-quasi-nonexpansive, i.e., $Q = \mathrm{Fix}(T) \neq \emptyset$ and $\|T(x) - z\| \leq \|x - z\|$ for any $x \in X$ and $z \in Q\}$,

$\mathcal{F}_Q = \{T$: T is strictly Q-quasi-nonexpansive, i.e., $Q = \mathrm{Fix}(T) \neq \emptyset$ and $\|T(x) - z\| < \|x - z\|$ for any $z \in Q,\ x \notin Q,\ x \in X\}$,

$\mathcal{P}^{\nu}_Q = \{T$: T is Q-pseudo-contractive, i.e., $Q = \mathrm{Fix}(T) \neq \emptyset$ and a constant exists $0 < \nu \leq 1$ such that $\|T(x) - z\|^2 \leq \|x - z\|^2 - \|T(x) - x\|$ for any $z \in Q,\ x \in X\}$,

$\mathcal{P}^{\nu} = \{T$: T is pseudo-contractive, i.e., a constant ν exists such that $\|T(x) - T(z)\|^2 \leq \|x - z\|^2 - \nu\|(x - T(x)) - (z - T(x))\|^2$ for any $x, z \in X\}$,

$\mathcal{K} = \{T$: T is nonexpansive, i.e., $\|T(x) - T(z)\| \leq \|x - z\|\ \forall x, z \in X\}$.
It follows from the definition that on the set of all mappings of T with Fix(T) $\neq \emptyset$ the strict imbeddings $\mathcal{P}^{\nu} \subset \mathcal{P}^{\nu}_Q \subset \mathcal{F}_Q \subset \mathcal{K}_Q$, $\mathcal{K} \subset \mathcal{K}_Q, \mathcal{P}^{\nu} \subset \mathcal{K}$ are valid. In addition the following relations

$$\mathcal{P}^{\nu}_Q = \frac{1}{1+\nu}\mathcal{K}_Q + \frac{\nu}{1+\nu}I, \qquad \mathcal{P}^{\nu} = \frac{1}{1+\nu}\mathcal{K} + \frac{\nu}{1+\nu}I \tag{1.4}$$

hold. Hence the scheme (1.2) for $T \in \mathcal{P}^{\nu}_Q$ (or T $\in \mathcal{P}^{\nu}$) is equivalent to the scheme (1.3) for $T \in \mathcal{K}_Q$ (or $T \in \mathcal{K}$) and moreover

$$\mathrm{Fix}(T) = \mathrm{Fix}\big(\lambda T + (1 - \lambda)I\big).$$

The belonging of T to one of the above clases does not imply strict contractibility ($\|T(x) - T(y)\| \leq q\|x - y\|$, $q < 1$, for any $x, y \in X$) and does not guarantee the strong convergence of iterations (1.2), (1.3).
As is known (Browder and Petryshyn, 1967; Martinet, 1972; Vasin, 1987) if T lies in $\mathcal{P}^{\nu}, \mathcal{P}^{\nu}_Q$ (or $T \in \mathcal{K}, \mathcal{K}_Q$), then in general it is possible only to obtain the weak convergence of iterations (1.2) (or (1.3)) (see an example in (General and Lindenstrauss, 1975)). However, in some particular cases under additional assumptions (on the space X or the operator T) the strong convergence holds. For example, if $T \in \mathcal{F}_Q$ and X is finite-dimensional space (Maruster, 1977) or if $T \in \mathcal{F}_Q$ and T is a compact operator (Eremin, 1969), then process (1.2) converges strongly to a fixed point of T. Also the condition $T \in \mathcal{K}$ and the semicompactness of the operator T (Browder and Petryshyn, 1967) guarantee the strong convergence of process (1.3). Here T is a nonlinear operator from X into X. If T is a linear operator from the class $\mathcal{K}$ then the simple iteration method (1.3) converges strongly to $x \in \mathrm{Fix}(T)$ for any $x^0 \in X$.
The main problem consists in modifying (iterativly regularizing) processes (1.2), (1.3) in such a way as to obtain the strong convergence of iterations.

To attain this purpose we use the Browder–Halperin technique diveloped for an operator $T \in \mathcal{K}$ and based on the introduction of correcting (damping) multipliers of the following form

$$x^{n+1} = k_{n+1}T(x^n) + (1 - k_{n+1})v_0, \quad n = 0, 1, \ldots, \tag{1.5}$$

where v_0 is a given element (trial solution) from X and $k_n \to 1$ as $n \to \infty$.
The first fundamental result on the strong convergence of (1.5) was the following theorem (Browder, 1967).

THEOREM 1.1 *Let T: $X \to X$ be a nonexpansive mapping in Hilbert space. Suppose that there exists a bounded closed convex subset D of X mapped by T into itself. Let v_0 be arbitrary point of D, and, for each k with $0 < k < 1$, let $T_k(x) = kT(x) + (1 - k)v_0$. Then T_k has an unique fixed point z_k in D, and z_k converges as $k \to 1$ strongly in X to the fixed point $\hat{x}_0$ of T in D closest to v_0.*

After that Halperin (Halperin, 1967) introduced the notion of acceptable sequence of real numbers $\{k_n\}$.

Definition 1.1. A sequence $\{k_i\}$ is called acceptable if the following conditions are fulfilled: 1) $0 < k_i < 1, i = 1, 2, \ldots,$ 2) $k_i < k_{i+1}$, $i = 1, 2, \ldots,$ 3) $\lim_{i\to\infty} k_i = 1$.
A sequence of numbers $n(i)$ exists such that: 4) $n(i + 1) > n(i)$, $i = 1, 2, \ldots,$ 5) $\lim_{i\to\infty} \varepsilon_{i+n(i)}\varepsilon_i^{-1} = 1$, 6) $\lim_{i\to\infty} n(i)\varepsilon_i = \infty$, where $\varepsilon_i = 1 - k_i$.
Using Theorem 1.1 Halperin (Halperin, 1967) proved the following.

THEOREM 1.2. *Let the conditions of Theorem 1.1 be fulfilled. Then, for any starting point $x^0 \in D$ and for any acceptable sequence k_n, the iterations x^n from (1.5) converge strongly to the fixed point $\hat{x}_0$ of T which is closest to v_0.*

Remark 1.1. The sequence of numbers $k_i = 1 - i^{-p}$ is acceptable.
Below in §2, Theorem 1.2 is extended to the case of pseudo-contractive operator T. moreover the regularizing properties of process (1.5) are established. In § 3 some classes of ill-posed problems are considered and concrete iteration schemes with pseudo-contractive step operator T are discussed. In concluding §4 the results of some numerical experiments concerning the numerical solution of the finite moment problem and the problem of gravimetry are analysed.

2. CONVERGENCE AND STABILITY OF ITERATIONS FOR PSEUDO-CONTRACTIVE OPERATOR

The assumption $T \in \mathcal{K}$ (see Theorems 1.1 and 1.2) is not satisfied for certain classes of iterative methods (Vasin, 1987, 1988), even though the condition $T \in \mathcal{P}_Q^\nu$ holds. Therefore it is important to extend Theorem 1.2 to this case (Vasin, 1991).

Remark 2.1. If $T \in \mathcal{F}_Q$ or $T \in \mathcal{P}_Q^\nu$, then Fix(T) is a convex closed set (Eremin, 1969). Hence, for any given $v_0 \in X$, there exists the unique solution $\hat{x}_0$ of the problem

$$\min\{\|v_0 - z\|: z \in \mathrm{Fix}(T)\},$$

which $\hat{x}_0$ is called v_0-normal solution of problem (1.1).

THEOREM 2.1. *Let $T \in \mathcal{P}_Q^{\nu}$ and the following relation be fulfilled*

$$x_n \rightharpoonup x \ \&\ T(x_n) - x_n \to 0 \quad \Rightarrow \quad x \in Q = \mathrm{Fix}(T). \tag{2.1}$$

Then, for any $x^0 \in X$ and any acceptable sequence $\{k_n\}$ (Def. 1.1), the iterations $\{x^n\}$ defined by process (1.5) converge strongly to $\hat{x}_0 \in \mathrm{Fix}(T)$.

COROLLARY 2.1. *Let $S \in \mathcal{K}_Q$ and for S relation (2.1) be satisfied. Then, for the operator $T = \lambda S + (1-\lambda)I$ ($\lambda \in (0,1)$), the assertion of Theorem 2.1 is valid.*

Remark 2.2. If $T \in \mathcal{K}$ or $T \in \mathcal{P}^{\nu}$, relation (2.2) is automatically fulfilled (Browder, 1967).

Remark 2.3. The approach of Browder–Halperin is one of the variants of the iterative regularization principle (Bakushinskii and Polyak, 1974). This principle has wider field of application (variational inequalities) and apparently greater possibility for the regularization of the operator T (not necessary in the form (1.5)). However, the first approach is more simple and universal. It should be noted that the principle of iterative regularization does not cover the case $T \in \mathcal{P}_Q^{\nu}$.
Now we investigate the question of stability of iterations (1.5) and prove that process (1.5) generates a regularizing algorithm.
Together with the exact scheme (1.5) we consider the perturbed one

$$\tilde{x}^{n+1} = k_{n+1}\tilde{T}(\tilde{x}^n) + (1-k_{n+1})v_0, \quad n = 0, 1, \dots, \tag{2.2}$$

with the same starting point $x^0 \in X$.
Let δ characterize the errors of the input data. We assume that the following properties of approximation hold

$$\|\tilde{T}(\tilde{x}^n) - T(\tilde{x}^n)\| \le \varphi(\delta), \quad n = 0, 1, \dots, \tag{2.3}$$

or

$$\|\tilde{T}(\tilde{x}^n) - T(x^n)\| \le \varphi(\delta), \quad n = 0, 1, \dots, \tag{2.4}$$

where $\varphi(\delta) \to 0$ as $\delta \to 0$ (usually $\varphi(\delta) = c\delta$ see § 3).

THEOREM 2.2. *Let the assumption of Theorem 1.2 and condition (2.3) or (2.4) be fulfilled. If $n(\delta)\varphi(\delta) \to 0$ as $\delta \to 0$, then the sequence $\tilde{x}^{n(\delta)}$ converges to the v_0-normal solution $\hat{x}_0$.*

Let us denote the class of operators T: $X \to X$ with Lipshitz condition by $L^{c,1}$ ($\|T(x) - T(y)\| \le c\|x - y\| \quad \forall x, y \in X$).

THEOREM 2.3. *Let $T \in \mathcal{P}_Q^{\nu} \cap L^{c,1}$, $\tilde{T} \in L^{c,1}$ and relations (2.1), (2.3) or (2.4) be satisfied. If $n(\delta)$ is such that $\varphi(\delta)\sum_{i=0}^{n(\delta)}(\prod_{\nu=0}^{i} k_{n(\delta)+1-\nu})c^i \to 0$ as $\delta \to 0$, then the sequence $\{\tilde{x}^{n(\delta)}\}$ defined by process (2.2) converges strongly to the v_0-normal solution $\hat{x}_0$ of (1.1).*

Thus under the assumptions of Theorems 2.2, 2.3 the process (1.5) generates a regularizing algorithm and $\{\tilde{x}^{n(\delta)}\}$ is a regularized family of approximate solutions of problem (1.1).

Remark 2.4. In our consideration X is a Hilbert space. The assertions of Theorems 1.1, 1.2, 2.2 remain true in the case of a smooth space (Opial, 1967), admitting a dual mapping which is weakly continuous, see (Reich, 1973; Schu, 1990).

Remark 2.5. In Theorem 2.3 the condition $T \in L^{c,1}$ is sufficient but not necessary. For example, for α-processes the step operator T does not satisfy Lipshitz condition (Eicke, 1991) but $T \in \mathcal{P}_M^1$, $M = \{x\colon Ax = y\}$ (Vasin, 1987) and the regularizing properties of iterations hold (Sarv, 1982, Gilyazov, 1987).

3. APPLICATION TO THE PROBLEMS OF CONVEX PROGRAMMING AND THE OPERATOR EQUATIONS OF THE FIRST KIND

3.1. *Finding the solution of a system of convex inequalities*
The problem consists in finding a point $x \in Q$, where

$$Q = \{x \in X\colon g_j(x) \le 0, \quad j = 1, 2, \ldots, m\}, \tag{3.1}$$

and g_j are convex differentiable lower semi-continuous functionals. Let us define the mappings (Eremin, 1969; Vasin, 1988, 1991)

$$\begin{cases} T(x) = x - \lambda d(x)e(x)/\|e(x)\|^2, \\ d(x) = \sum\limits_{j=1}^{m} [g_j^+(x)]^\mu, \quad e(x) = \sum\limits_{j=1}^{m} [g_j^+(x)]^{\mu-1}\nabla g_j(x), \end{cases} \tag{3.2}$$

where $\lambda \in (0,1)$, $\mu \ge 1$, $g_j^+(x) = \max\{0, g_j(x)\}$, ∇g_j is gradient, which is a bounded mapping.

LEMMA 3.1. *Under the above assumptions the mappings T defined by (3.2) belong to the class $\mathcal{P}_Q^1$ for the set Q from (3.1). Moreover relation (2.1) for the mapping T is fulfilled.*

From Theorem 2.1 and Lemma 3.1 we obtain the following.

COROLLARY 3.1. *For any acceptable sequence $\{k_n\}$ and for any $x^0 \in X$, iteration process (1.5) with the operator T from (3.2) converges strongly to the v_0-normal solution $\hat{x}_0$ of system (3.1).*

To investigate the stability of the iteration method we restrict ourself to the case of affine functionals in (3.1), i.e., $g_j(x) = \langle h_j, x\rangle - b_j$. Let the perturbations of the functionals g_j and the operator T arise from the errors δ_j in the elements h_j and numbers b_j, i.e.,

$$\tilde{g}_j(x) = \langle \tilde{h}_j, x\rangle - \tilde{b}_j, \quad \|\tilde{h}_j - h_j\| \le \delta_j, \quad |\tilde{b}_j - b_j| \le \delta_j, \qquad j = 1, 2, \ldots, m. \tag{3.3}$$

LEMMA 3.2. *Let relations (3.3) for g_j and $\tilde{g}_j$ be fulfilled. Then for the operators $T, \tilde{T}$ defined by (3.2) (with $\mu = 1$) estimate (2.4) is valid. Here* $\varphi(\delta) = [a\tilde{a}(c+1) + (c\sqrt{m}h + b)(a + \sqrt{m}(a+\tilde{a}))]\sqrt{m}\delta/\tilde{a}^2 a$, $a = \|\sum_{j=1}^m h_j\|$, $\tilde{a} = \|\sum_{j=1}^m \tilde{h}_j\|$, $b = \sum_{j=1}^m |b_j|$, $h = [\sum_{j=1}^m \|h_j\|^2]^{1/2}$, $\|x^n\| \le c$, $\delta = [\sum_{j=1}^m \delta_j^2]^{1/2}$.

It is easy to verify that the operators $T, \tilde{T} \in L^{c,1}$. It means that iterative method (1.5) with the step operator from (3.2) generates a regularizing algorithm (see Theorem 2.3).

Remark 3.1. For $k_n \equiv 1$, $v_0 = 0$, the strong convergence of iterations (1.5) with the operator T defined by (3.2) was proved by Raik (Raik, 1967) for correct system (3.1). The introduction of the damping multipliers k_n permits to prove the strong convergence in the general case.

3.2. *Prox-method*

Let $f\colon X \to R^1$ be a lower semi-continuous convex functional and Q be a convex closed subset of a Hilbert space X. Consider the extremal problem

$$\min \{ f(x)\colon x \in Q \}. \tag{3.4}$$

Definition 3.1. The operator S_Q^f defined by the relation $S_Q^f\colon v \to \operatorname{argmin} \{f(x) + (1/2)\|x - v\|^2\colon x \in Q\}$ is called prox-mapping (Moro, 1965).
As is well known (Moro, 1965; Martinet, 1972; Vasin, 1982) $S_Q^f \in \mathcal{P}^1$ and $\mathrm{Fix}(S_Q^f) = M$, where M is the optimal set of problem (3.4). Thus from Theorem 1.2 we have the following.

COROLLARY 3.2. *For any acceptable sequence* $\{k_n\}$ *and for any* $x^0 \in X$, *the modified prox-method*

$$x^{n+1} = k_{n+1} S_Q^f(x^n) + (1 - k_n)v_0, \qquad n = 0, 1, \ldots, \tag{3.5}$$

converges strongly to the v_0*-normal solution of problem* (3.4) *(for the* $Q = X$, $v_0 = 0$ *see (Bakushinskii, 1977)).*

It should be noted that for $k_n \equiv 1$ and an infinite dimensional space X it is possible to guarantee only weak convergence of method (3.5) (Martinet, 1972; Vasin, 1982).

3.3. *The gradient projection method*

Let us investigate problem (3.4) for the quadratic functional $f(x) = \|Ax - y\|^2$, where $A \in L(X, Y)$. The step operator T in the gradient projection method (GPM)

$$x^{n+1} = P_Q\big[(I - \beta A^*A)x^n + \beta A^*y\big] \equiv T(x^n), \qquad n = 0, 1, \ldots, \tag{3.6}$$

belongs to the class $\mathcal{P}^\nu$ with $\nu = 1/\beta\|A\|^2 - 1/2$, $0 < \beta \le 2\|A\|^{-2}$ and $\mathrm{Fix}(T)$, where M is the optimal set in (3.4). Moreover, if $\|A - \tilde{A}\| \le \delta$, $\|y - \tilde{y}\| \le \delta$, then for the operator T condition (2.4) with $\varphi(\delta) = \beta[c(\|\tilde{A}\| + \|A\|) + \|\tilde{y}\| + \|A\|]\delta$ is fulfilled. Hence for the modified GPM

$$\tilde{x}^{n+1} = k_{n+1} P_Q\big[(1 - \beta A^*A)\tilde{x}^n + \beta A^*y\big] + (1 - k_{n+1})v_0 \tag{3.7}$$

Theorem 2.2 is true.
Eicke (Eicke, 1991) considered some what other regularization of GPM

$$x^{n+1} = P_Q\Big\{k_n\big[(I - \beta A^*A)x^n + \beta A^*y\big]\Big\} + (1 - k_n)v_0 \tag{3.8}$$

and proved the strong convergence of $\{x^n\}$ to the normal solution for $k_n = 1/(1+n^{-p})$, $v_0 = 0$. Note that for method (3.8), as for process (3.7), Theorems 1.2 and 2.2 hold.

3.4. *Operator equations with a priori information*
One of the possible approaches to the construction of regularizing algorithms for solution of an operator equation with a priori convex constraints

$$Ax = y, \quad x \in Q,$$

is the use of iterative schemes of the form

$$x^{n+1} = k_{n+1}P(T(x^n)) + (1 - k_{n+1})v_0, \qquad n = 0, 1, \ldots, \tag{3.9}$$

$$x^{n+1} = k_{n+1}\left[\lambda P(x^n) + (1-\lambda)T(x^n)\right] + (1 - k_{n+1})v_0, \tag{3.10}$$

where T is the step operator of a certain base iteration procedure from the class $\mathcal{P}_M^{\nu_1}$, $M = \{x\colon\ Ax = y\}$ and $P \in \mathcal{P}_Q^{\nu_2}$. If $M \cap Q \neq \varnothing$ and, for P, T, relation (2.1) is holds true, then $PT, \lambda P + (1-\lambda)T \in \mathcal{P}_{M\cap Q}^{\min\{\nu_1,\nu_2\}}$ and therefore, accoding to Theorem 2.1, the sequence $\{x^n\}$ in (3.9), (3.10) converges strongly to the point $\hat{x}_0 \in M \cap Q$ closest to v_0. For example, it is possible to take explicit and nonexplicit iterative schemes, α-processes, the conjugate gradient method, methods of Chebyshev type as the base iteration procedure and as P one can take mappings defined by formulas (3.2) (Vasin 1987, 1988).

4. ITERATIVE APPROXIMATION OF THE SOLUTION OF "THE FINITE MOMENT PROBLEM AND THE PROBLEM " OF GRAVIMETRY

4.1. The finite moment problem (FMP) is formulated as the problem of reconstruction of a function $x \in X$ from the following system of linear equations

$$\langle u_i, x\rangle = \mu_i, \qquad i = 1, 2, \ldots, m, \tag{4.1}$$

where the functions $\{u_i\}$ are given and $\mu = (\mu_1, \ldots, \mu_m)$ is the known vector of moments. As is known if the FMP is solvable for a given μ then the solution is nonunique and the system is very ill-conditioned, i.e., the FMP is a typical ill-posed problem.
For the case $u_i = t^{i-1}$, $X = L_2[-1,1]$ we consider the iterative process

$$x^{n+1} = P_m\Big(P_{m-1}\ldots(P_1(x^n))\Big) \equiv T(x^n), \qquad n = 0, 1, \ldots, \tag{4.2}$$

where $P_i(x) = x - (\langle u_i, x\rangle - \mu_i)u_i/\|u_i\|^2$ is the projection on the hyperplane $M_i = \{x\colon\ \langle u_i, x\rangle = \mu_i\}$. For any $x^0 \in \operatorname{span}\langle u_1, \ldots, u_m\rangle$, method (4.2) converges to the normal solution $\hat{x}_0$ of system (4.1) (Vasin, 1991).
For three model functions (see Rodriguez and Seatzu, 1990)

$$x_1(t) = (1+t)^3(1-t)^u, \quad x_2(t) = (1+t)^2(1-t)^2t^{10}, \quad x_3(t) = (1+125t^2)^{-1}$$

the moment vector μ was calculated and then iterative scheme (4.2) was applied to system (4.1) after its discretization by the quadrature or projection method. For the relative error $L_i(m) = (\|x_i - \tilde{x}_i\|)/\|x_i\|$, where $\|x\| = \{(1/30)\sum_{j=0}^{30}[x(t_j)]^2\}^{1/2}$, $t_j = -1 + 2j/30$, and $\tilde{x}$ is the computer solution, the following numerical values were obtained after 10 000 iterations (as $x^0 = 1$)

$$L_1(15) = 3.70(-3), \quad L_2(15) = 7.16(-4), \quad L_3(15) = 1.28(-1).$$

Since $\hat{x}_0 \in \mathrm{span}\langle u_1, \ldots u_m\rangle \not\ni x_3$ it is possible that for the third case the scheme converges to an another solution different from x_3. To improve the quality of the solution and to obtain convergece to x_3 in this case the iteration method

$$x^{n+1} = k_{n+1}T(x^n) + (1 - k_{n+1})v_0, \qquad n = 0, 1, \cdots$$

was used. For $k_n = (n^{9/10} - 1)/n^{9/10}$ and $v_0^1 = 0.5x_3$, $v_0^2 = 0.75x_3$, $v_0^3 = x_3$ we obtained respectively

$$L_3^1(15) = 6.34(-2), \quad L_3^2(15) = 3.16(-2), \quad L_3^3(15) = 7.14(-4).$$

These results are much better than obtained by method (4.2) (see $L_3(15)$).
4.2. *The problem of the gravimetry.*
The determination of the surface of interface between two media (with different densities ρ_1 and ρ_2) by the anomaly of the gravity is reduced to the solution of the nonlinear integral equation

$$A[x](t) \equiv \frac{\rho_1 - \rho_2}{4\pi} \int_{-1}^{1} \log \frac{H^2 + (t-s)^2}{(t-s)^2 + (H - x(s))^2} ds = y(t). \tag{4.3}$$

First, for $\rho_1 - \rho_2 = 1$, $H = 2$, $-1 \le t \le 1$, and the model solution $\bar{x}(t) = (1 - t^2)^2$ we calculate the right-hand side $\bar{y}(t) = A[\bar{x}](t)$ and regularize (4.3) as

$$A[x] + \alpha x = \bar{y}. \tag{4.4}$$

Next, we discretize regularized equation (4.4) by quadrature method and use the iterative methods ($F(x) = A[x] + \alpha x - y$)

$$x^{n+1} = x^n - [F'(x^n)]^{-1} f(x^n) \equiv T(x^n), \tag{4.5}$$

$$x^{n+1} = [\lambda T(x^n) + (1 - \lambda)x^n] \equiv U(x^n), \tag{4.6}$$

$$x^{n+1} = k_{n+1}U(x^n) + (1 - k_{n+1})v_0 \tag{4.7}$$

for the starting points $x_1^0(t) \equiv 1$, $x_2^0(t) \equiv 0.5$, $x_3^0(t) = 0.5(1 - |t|)$, $x_4^0(t) \equiv 0.1$ and $k_n = 1 - 1/n^{\frac{9}{10}}$, $v_0 = 0.75\bar{x}$, $\alpha = 10^{-2}$.
Newton–Kantorovich method (4.5) diverges for all starting points $x_i^0, i = 1, 2, 3, 4$. Below for processes (4.6), (4.7) the maximal (componentwise) relative errors Δ_i of the computer solution after 5 iterations are presented.
Scheme (4.6): $\Delta_1 = 5.76(-2)$, $\Delta_2 = 3.85(-2)$, $\Delta_3 = 3.7(-2)$, $\Delta_4 = 3.69(-2)$.

Scheme (4.7): $\Delta_1 = 8.54(-2)$, $\Delta_2 = 8.54(-2)$, $\Delta_3 = 8.53(-2), \Delta_4 = 8.51(-2)$.

REFERENCES

Bakushinskii, A.B. (1977). Methods of the solution of monotone variational inequalities based on the principle of iterative regularization. *Zh. Vychisl. Mat. i Mat.Fiz.* **17**, 1350–1362 (in Russian).

Bakushinskii, A.B. and Polyak, B.T. (1974). On the solution of variational inequalities. *Dokl. Akad. Nauk SSSR* **219**, 1038–1041 (in Russian).

Browder, F.E. (1967). Convergence of approximants to fixed points of nonexpansive nonlinear maps in Banach spaces. *Arch. Rational Mech. Anal.* **24**, 82–90.

Browder, F.E. and Petryshyn, W.V. (1967). Construction of fixed points of nonlinear mappings in Hilbert space. *J.Math. Appl.* **20**, pp. 197–228.

Dotson, W.G. (1970). On the Mann iterative process. *Trans. Amer. Math. Soc.* **149**, 65–73.

Eicke, B. (1991). Iterationsverfahren für restringierte Probleme In: *Doktor Thesis.*, Techn. Univ. Berlin., Berlin, 29–56.

Eremin, I.I. (1969). A general theory of Feier mappings. *Mat. Zapiski Ural. Gos. Univ.* **7**, 50–58 (in Russian).

Genel, A. and Lindenstrauss, L. (1975). An example concerning fixed points. *Isreal J. Math.* **22**, 81–86.

Gilyazov, S.F. (1987). Regularizing algorithm based on the method of steepest descent In: *Methods of the solution of linear ill-posed problems.*, Moskow Univ. Press, Moscow, 45–74 (in Russian).

Halperin, B. (1967). Fixed points of nonexpanding maps. *Bull. Amer. Math. Soc.* **73**, 957–961.

Martinet, B. (1972). Determination approchèe d'un point fixe d'une applications pseude-contractante. *C.R. Acad. Sci. Paris* **274**, 163–165.

Maruster, S. (1977). Quasi-nonexpansivity and two classical methods for solving nonlinear equations. *Proc. Amer. Math. Soc.* **62**, 119–123.

Moreau, J.J. (1965). Proximitè et dualitè dans un espace Hilbertien. *Bull. Soc. Math. France.* **93**, 273–299.

Opial, Z. (1967). Weak convergence of the successive approximations for nonexpansive mappings in Banach spaces. *Bull. Amer. Math. Soc.* **73**, 591–597.

Petryshyn, W.V. and Willamson, T.E. (1973). Strong and weak convergence of the sequence of successive approximations for quasi-nonexpansive mappings. *J. Math. Anal. Appl.* **43**, 459–497.

Raik, E. (1967). The methods of Feier's type in Hilbert space. *Izv. Akad. Nauk Eston.SSR, ser. Mat.* **16**, 286–293 (in Russian).

Reich, S. (1973). Asymptotic behavoir of contractions in Banach spaces. *J. Math. Anal. Appl.* **44**, 57–70.

Sarv, L. (1982). A family of nonlinear iterative methods for the solution of ill-posed problems. *Izv. Akad. Nauk Eston.SSR, ser. Mat.* **31**, 261–269 (in Russian).

Schu, J. (1990). Iterative approximation of fixed points of nonexpansive mappings with star-shaped domain. *Comment Math. Univ. Caroline* **31**, 277–282.

Rodriguez, G. and Seatzu, S. (1990). Numerical solution of the finite moment problem in a reproducing kernel Hilbert space. *J. Comp. Appl. Math.* **33**, 233-244.

Vasin, V.V. (1982). *Prox-method with projection for problems of convex programming.* Preprint. Uralskii Center Akad. Sci. USSR, Sverdlovsk (in Russian).

Vasin, V.V. (1987). Iterative methods for the approximate solution of ill-posed problems with a priori information and their applications. *Inverse and Ill-Posed Problems. Ed by H.W. Engl. and C.W. Groetsch.* Academic Press, Boston etc., pp. 211–229.

Vasin, V.V. (1988). Iterative methods for solving of ill-posed problems with a priori information in Hilbert spaces. *Zh. Vychisl. Mat. i Mat. Fiz.* **28**, 971–980 (in Russian).

Vasin, V.V. (1991). Iterative approximation of the solution for the finite moment problem. *Dokl. Akad. Nauk SSSR* **318**, 1042–1045 (in Russian).

Part Two

INVERSE PROBLEMS IN MATHEMATICAL PHYSICS

Ill-Posed Problems in Natural Sciences, pp. 227 – 232
A. Tikhonov (Ed.)
1992 VSP/TVP

NONLINEAR INVERSE PROBLEMS OF ACOUSTIC POTENTIAL

G.V. ALEKSEYEV and A.Yu. CHEBOTAREV
Institute of Applied Mathematics, For Eastern Division of USSR Academy of Sciences, Vladivostok, Russia

ABSTRACT

Two nonlinear inverse problems of acoustic potential in the three-dimensional space are considered. The first one (Problem 1) consists in finding an unknown radiating system (a simply connected domain in the space) with a given distribution of the source density from given values of the outher potential of the acoustic field generated by this radiating system. Problem 2 consists in finding a radiating system generating an additional acoustic field which minimizes (or suppresses) over a certain domain of the space a parasitical sound field created by a noising object. To study the both problems, we first find a nonlinear integral equation of the first kind for the function which describes the shape of the unknown boundary of a radiating system and show that Problem 2 is equivalent within a certain class of domains to the minimization of the discrepancy functional for this integral equation. Then we prove the existence of a solution of Problem 2 (i.e., the existence of a minimizer of the discrepancy functional) in the Sobolev space, study some properties of the solution and establish sufficient conditions for a solution of Problem 2 to be a solution of Problem1.

1. INTRODUCTION

The investigation of the ocean by acoustic techniques calls for solving inverse source problems of acoustic potential. Among these problems, an important role is played by inverse source problems in which it is required to determine an unknown radiating system from the outer potential of the radiated field. In this paper, we consider two nonlinear inverse source problems of acoustic potential and develop a method for analyzing their solvability, which is based on the theory of extreme value problems and on analytical properties of the acoustic potentials.

2. STATEMENT OF THE DIRECT PROBLEM OF SOUND RADIATION

Let f be a given function on the free space $\mathbf{R}^3$ which can be treated as the volume density of the sound source. The direct sound radiation problem consists in finding the acoustic field potential Φ from the conditions

$$L\Phi \equiv \Delta\Phi + k^2\Phi = -f \text{ in } \mathbf{R}^3, \quad \Phi \in \mathfrak{m}(\mathbf{R}^3). \tag{1}$$

Here, $k \neq 0$ is a constant wave number, $\mathfrak{m}(\mathbf{R}^3)$ denotes the class of functions in $\mathbf{R}^3$ satisfying the Sommerfeld radiation condition as $|x| \to \infty$. Let supp$f \subset \Omega$, $f \in L^p(\Omega), p > 1$, where Ω is a bounded domian in $\mathbf{R}^3$. Then the solution $\Phi \in W^{2,p}_{\text{loc}}(\mathbf{R}^3)$ of problem (1) exists, is unique (Alekseyev, 1991) and has the form

$$\Phi(x) = \int_\Omega G(x,y)f(y)dy, \quad x \in \mathbf{R}^3. \tag{2}$$

Here, $G(x,y) = e^{ik|x-y|}/4\pi|x-y|$. In what follows, the pair (Ω, f) is called a (volume) radiating system and the function Φ of form (2) is called the acoustic potential of the pair (Ω, f).

3. STATEMENT OF NONLINEAR INVERSE SOURCE PROBLEMS

There are linear and nonlinear inverse source problems of acoustic potential. A linear problem consists in finding the density f involved in equation (2) in a given domain Ω outher the from potential values Φ_l of the radiated wave field. For results on existence, uniqueness and nonuniqueness of a solution of the linear inverse problem, see (Alekseyev, 1991) and references therein.
Below we consider two nonlinear inverse source problems. Let us formulate these problems. Let $B = \{x \in \mathbf{R}^3\colon |x| < R\}$ be a ball in $\mathbf{R}^3$ of radius R, f be a given function in B, and Φ_l be a given outer potential in $B_l = \mathbf{R}^3 \backslash B$.

PROBLEM 1. *Given a ball B and two functions, f in B and Φ_l in B_l, find a simply connected domain $\Omega \in B$ such that a solution of the direct problem*

$$L\Phi = -f_\Omega \text{ in } \mathbf{R}^3, \quad \Phi \in \mathfrak{m}(\mathbf{R}^3) \tag{3}$$

satisfies the condition

$$\Phi = \Phi_l \text{ in } B_l.$$

Here, the function f_Ω is equal to f on Ω and Q on $\Omega_l = \mathbf{R}^3 \backslash \Omega$.

We denote by $P = P(R, f)$ the set of all functions (potentials) Φ, satisfying (3) for some domain $\Omega \subset B$. Let $Q \subset B_l$ be a bounded domain and $-\Phi_0$ be a given function in Q which has the physical meaning of a parasitical acoustic field generated in Q by a noising object.

PROBLEM 2. (Inverse problem of active minimization of sound). *Given domains B, Q and two functions, f in B and Φ_0 in Q, find a domain $\Omega \subset B$ such that a solution Φ of (3) satisfies the condition*

$$\|\Phi - \Phi_0\|_{L^2(Q)} \leq \|\Psi - \Phi_0\|_{L^2(Q)} \quad \forall \Psi \in P. \tag{4}$$

Remark 1. In the particular case of $k = 0$, Problem 1 turns into a nonlinear gravitational potential inverse problem which was studied by many authors (see, e.g., (Novikov, 1938; Prilepko, 1966a; Cherednichenko, 1976)). When $k = ic$, $c > 0$, Problem 1 is a nonlinear methagarmonical potential inverse problem which was first studied in (Prilepko, 1966b).

4. REDUCTION TO NONLINEAR FIRST KIND INTEGRAL EQUATION

Now we introduce an admissible class of domains in which we shall seek for solutions of Problems 1 and 2. Let Σ be the unit sphere in $\mathbf{R}^3$:

$$\Sigma = \{x = (r,\sigma)\colon\ r = 1, \quad \sigma = (\theta,\varphi), \quad \theta \in [0,\pi], \quad \varphi \in [0,2\pi)\},$$

$H^S = H^S(\Sigma)$ be the Sobolev function space on Σ with norm $\|\cdot\|_s$. Define a closed convex set K in H^S:

$$K = K_M^S = \{v \in H^S\colon\ v \geq 0,\ \|v\|_s \leq M\}, \quad S > 1.$$

Here, M is a constant such that by the Sobolev embedding theorem (Lions and Megenes, 1968) the following inequality holds

$$\|v\|_{C(\Sigma)} \leq C_S M < R \quad \forall v \in K, \quad C_S = \text{const.} \tag{5}$$

Let F be a map which puts in correspondence with any function $v \in K$ the domain

$$\Omega_v = \{x = (r,\sigma)\colon\ 0 < r < v(\sigma), \quad \sigma \in \Sigma\} \subset B$$

with a boundary $\Gamma_v = \partial\Omega_v$. Define the admissible class T of domains for Problems 1 and 2 as

$$T = T_M^S = R(F) = \{Fv\colon\ v \in K_M^S\}.$$

Then Problem 1 is equivalent to finding $u \in K$ such that

$$[Au](x) = \Phi_0(x), \quad x \in B_l. \tag{6}$$

Here, $A\colon H^s \to L^2(Q)$ is a nonlinear integral operator defined by

$$[Av](x) = \int_\Sigma \int_0^{v_\sigma} G(x,y) f(y) I(y)\, dr\, d\sigma, \tag{7}$$

where $v_\sigma = v(\sigma)$, $y = (r,\sigma)$, $0 \leq r < v_\sigma$ and I is the Jacobian of the transformation to the spherical coordinates. Similarly, Problem 2 in the class T is equivalent to finding a function $u \in K$ such that

$$J(u) \leq J(v) \equiv \|Av - \Phi_0\|_{L^2(Q)} \quad \forall v \in K. \tag{8}$$

Remark 2. Denote by $K^2(Q)$ the (closed) subspace in $L^2(Q)$ which consists of the functions Ψ satisfying the condition $L\Psi = 0$ in Q (in the distribution sense). We note that $Av|_Q \in K^2(Q)$ for every $v \in K$ since by (Colton and Kpess, 1983) the function Av is analytic in B_l and satisfies the condition $L(Av) = 0$ in B_l.

Remark 3. Let u be a solution of Problem 2 and let

$$J(u) = 0 \quad (Au = \Phi_0 \text{ in } Q). \tag{9}$$

Then, by Remark 2, the function Φ_0 is analytic in Q, satisfies the condition $L\Phi_0 = 0$ in Q and admits a (unique) analytic continuation Φ_l into B_l which equals $Au|B_l$. So the function u is also a solution of Problem 1 for the function Φ_l.

5. MAIN RESULTS

Let the following condition be satisfied

$$\text{(i)}\ \ \overline{B} \cap \overline{Q} = \varnothing, \quad f \in C(\overline{B}), \quad \Phi_0 \in L^2(Q).$$

LEMMA 1. *Under condition* (i) *the operator* A *in* (7) *and the functional* J *in* (8) *are continuous and Frechet differentiable on the set* K *with respect to the relative topology of the space* H^s. *In particular,*

$$[J'(u)](\sigma) = 2\text{Re} \int_Q G(x, y_{u_\sigma})[\overline{Au - \Phi}](x)\, dx\, f(y_{u_\sigma}) u_\sigma^2. \tag{10}$$

Here, $y_{u_\sigma} \equiv (\sigma, u_\sigma) \in \Gamma_u\ \forall\sigma \in \Sigma,\ u \in K$.

Proof. The statement of Lemma 1 is a direct consequence of (7) and (8) and condition (i).

THEOREM 1. *There exists at least one solution of Problem* 2.

Proof. The proof is based on standard techniques, see, e.g., (Cea, 1971).

THEOREM 2. *Let conditions* (i) *hold and let*

$$\text{(ii)}\quad f > 0 \text{ in } B,$$

$$\text{(iii)}\quad K^2 > 0, \quad \partial Q \in C^{0,1}, \quad \Phi_0 \in K^2(Q).$$

If a solution u *of Problem* 2 *satisfies the conditions*

$$\text{(iv)}\quad u > 0 \ \text{ on } \ \Sigma, \quad \|u\|_s < M,$$

(v) K^2 *is not an eigenvalue of the operator* $-\Delta$ *in* Ω_u. *Then* u *satisfies condition* (9).

Proof. Let $g = g_1 + ig_2 = \overline{Au - \Phi_0}$,

$$\Psi(y) = \Psi_1(y) + i\Psi_2(y) = \int_Q G(x,y)g(x)\,dx, \quad y \in \mathbf{R}^3. \tag{11}$$

The function Ψ in (11) is a potential of the pair (Q, g). So Ψ is analytic in $Q_l = \mathbf{R}^3|Q$ and satisfies the conditions (Alekseyev, 1991)

$$L\Psi = -g \text{ in } Q, \quad L\Psi = 0 \text{ in } Q_l, \tag{12}$$

$$(\partial\Psi(x)/\partial|x|) - ik\Psi(x) = 0(1/|x|) \quad \text{as} \quad |x| \to \infty. \tag{13}$$

Since by (iv) the solution u of Problem 2 is an interior point of the set K, we have $J'(u) = 0$ according to (Cea, 1971). This means, by (10) and (11), that $\Psi_1 = 0$ on $\Gamma u = \partial\Omega_u$. Since $\operatorname{Im} k^2 = 0$ by (iii), it follows from (12) that $L\Psi_1 = 0$ in Q_l. Therefore, the restriction $\Psi_1|_{\Omega_u}$ of the function Ψ_1 to the domain Ω_u satisfies the conditions

$$L\Psi_1 = 0 \text{ on } \Omega_u \text{ and } \Psi_1 = 0 \text{ on } \Gamma_u. \tag{14}$$

Using (v), we deduce from (14) that $\Psi_1 = 0$ in Ω_u and, since Ψ_1 is analytic in Q_l we conclude that

$$\Psi_1 \equiv \operatorname{Re}\Psi = 0 \text{ on } Q_l. \tag{15}$$

Combining (13) and (15), we get

$$\Psi(x) \equiv \Psi_2(x) = o\left(\frac{1}{|x|}\right) \text{ as } |x| \to \infty. \tag{16}$$

Now it follows from (Colton et al., 1983) that $\Psi = 0$ in Q_l. This means that the function Ψ is a zero outer potential of the pair (Q, g). Finally since $g \in K^2(Q)$ by (iii) and Remark 2, we deduce, taking into account (Alekseyev et al., 1985), that $g = 0$ in Q, $\Psi = 0$ in $\mathbf{R}^3$. The theorem is proved.

Remark 4. It follows from Theorem 2 and Remark 3 that, under conditions (i)–(iii) for the data and conditions (iv) and (v) for the solution u of Problem 2, the given function Φ_0 admits an analytical continuation $\Phi_l = Au|_{B_l}$ into B_l and, moreover the function u is also a solution of Problem 1 with the outer potential Φ_l. One can use this fact to the prove the local (near a given "body") solvability of inverse Problem 1. The corresponding theory will be developed in a forthcoming paper by the authors.

REFERENCES

Alekseyev, G.V. (1991). *Inverse Problems of Sound Radiation and Sygnal Theory.* Far East. State University, Vladivostok (in Russian).

Alekseyev, G.V. and Chebotarev, A.Yu. (1985). Inverse problems of acoustic potential *J. Comput. Math. and Math. Phys.* **25**, 1189–1199 (in Russian).

Cea, J. (1971). *Optimisation. Theorie et Algorithms.* Dunod, Paris.

Cherednichenko, V.G. (1976). On the solvability "in small" of inverse potential problem with a variable density in two-dimensional case. *Siberian Math. J.* **17**, 1168–1176 (in Russian).

Colton, D. and Kpess, R. (1983). *Integral Equation Methods in Scattering Theory.* John Wiley, New York.

Lions, J.L. and Magenes, E. (1968). *Problems aux non Homogenes et Applications.* Dunod, Paris.

Novikov, P.S. (1938). On the uniqueness for the inverse problem of potential theory. *Dokl. Akad. Nauk SSSR* **18**, 165–168 (in Russian).

Prilepko, A.I. (1966a). On the uniqueness of solution to outer inverse problem of Newtonian potential. *Different. Equat.* **2**, 107–124 (in Russian).

Prilepko, A.I. (1966b). On the uniqueness of solution to inverse problems of methagarmonic potentials. *Different. Equat.* **2**, 194–204 (in Russian).

Ill-Posed Problems in Natural Sciences, pp. 233 – 243
A. Tikhonov (Ed.)
1992 VSP/TVP

A MEASURE THEORETIC APPROACH FOR STUDYING INVERSE PROBLEMS OF THE HEAT EQUATION AND THE WAVE EQUATION

G. ANGER
Martin Luther University Halle-Wittenberg, Department of Mathematics,
D-O-4010 Halle, Germany.

ABSTRACT

If differential equations are written in integral form (weak solutions) it is possible to replace the mass density and other physical parameters by measures. These measures can be concentrated on surfaces or on other subsets. The aim of this paper is to study such more general differential equations and to present first results. By using the classical separation approach it is possible to construct solutions. In the case of evolution equations one equation obtained is the Helmholtz equation containing measures as coefficients. In 1989 W. Hansen and H. Hueber developed a complete theory for the more general Helmholtz equation using the theory of harmonic spaces. Especially they developed a complete theory for its eigenfunctions. There are no restrictions on the boundaries of the domains considered. The results of Hansen and Hueber are applied to construct solutions of evolution equations by using eigenfunctions of the Helmholtz equation. Further, we give some applications to inverse problems. Most problems of this new approach remain unsolved in this paper. The results were obtained in co-operation with J. Gottlieb. In his paper one can find further results.

1. WEAK SOLUTIONS RELATIVE TO A MEASURE

1.1. *Some basic evolution equations*
In applications the following evolution equations play a fundamental role

$$\rho(x)u_t = \operatorname{div}\big(a(x)\operatorname{grad} u\big) + f(x,t), \tag{1}$$

$$\rho(x)u_{tt} = \operatorname{div}\big(a(x)\operatorname{grad} u\big) + f(x,t), \tag{2}$$

where $\rho(x)$ is the mass density, a is a scalar function in the isotropic case and $a = (a_{jk})$ is a matrix function in the anisotropic case. These functions are defined on an open set $\Omega \subset \mathbf{R}^n$, f is a function depending on $x \in \Omega$ and the time coordinate t. In many problems a is a function of the solution u or of $|\operatorname{grad} u|$ (see (Anger, 1990)).

The density ρ defines a measure (mass distribution) μ_ρ,

$$\mu_\rho(f) = \int f(y)\rho(y)\,dy. \tag{3}$$

Further, we shall consider measures μ_σ whose support is on a (smooth) surface $S \subset \mathbf{R}^n$,

$$\mu_\sigma(f) = \int f(y)\sigma(y)\,dS(y), \quad dS \text{ surface element.} \tag{4}$$

The consideration of these measures is necessary, for instance, in electricity. A further example is the solution of the Dirichlet problem relative to the Laplace equation $\Delta u = 0$,

$$u(x) = \int h(y)\,d\mu_X(y) = \int_{\partial\Omega} h(y)\frac{\partial G_L^\Omega(x,y)}{\partial n}\,dS(y).$$

Here G_L^Ω is the Green function of Ω with smooth boundary $\partial\Omega$. The so-called harmonic measure μ_X with support on $\partial\Omega$ exists for each bounded domain $\Omega \subset \mathbf{R}^n$ (see, for instance, (Landkof, 1972; Hansen and Hueber, 1989)). In the following it is necessary to use the Newtonian potential $G_L\nu$ relative to a measure ν, where

$$G_L\nu(x) = \int G_L(x,y)\,d\nu(y). \tag{5}$$

Here $G_L(x,y)$ is the Newtonian kernel if $n \geq 3$, the logarithmic kernel if $n = 2$. If ν generates a continuous potential $G_L\nu$, then the Green function G_L^Ω generates also a continuous potential $G_L^\Omega\nu$, where

$$G_L^\Omega\nu(x) = \int G_L^\Omega(x,y)\,d\nu(y).$$

The measures (3) and (4) with bounded density generate continuous Newtonian potentials.

1.2. *Separation of weak solutions*

Let $f(x,t) = \rho(x)f_0(x,t)$. In order to study equation (2) or (1) relative to a measure ν, we write these equations in integral form (weak solutions), where $\psi \in \mathbf{C}_0(\Omega \times (0,t))$,

$$\int_0^t \left(\int_\Omega u_{\tau\tau}(x,\tau)\psi(x,\tau)\rho(x)\,dx \right) d\tau$$

$$= \int_0^t \left(\int_\Omega \operatorname{div}\left(a(x)\operatorname{grad} u(x,\tau)\right)\psi(x,\tau)\,dx \right) d\tau$$

$$+\int_0^t \left(\int_\Omega f_0(x,\tau)\psi(x,\tau)\rho(x)\,dx \right) d\tau. \tag{6}$$

Integration by parts and replacing $\rho(x)dx$ by a measure $d\nu(x)$ yields

$$\int_0^t \left(\int u(x,\tau)\psi_{\tau\tau}(x,\tau)\,d\nu(x) \right) d\tau$$

$$= \int_0^t \left(\int_\Omega u(x,\tau)\mathrm{div}\,(a(x)\mathrm{grad}\,\psi(x,\tau))\,dx \right) d\tau$$

$$+\int_0^t \left(\int f_0(x,\tau)\psi(x,\tau)\,d\nu(x) \right) d\tau. \tag{7}$$

We determine solutions u of (7) having the form

$$u(x,t) = v(x)T(t). \tag{8}$$

This approach is the separation, in classical analysis well known (Tikhonov and Samarskii, 1963). In this case we consider special $f_0(x,t)$ having the form $f_0(x,t) = v(x)f(t)$ and special $\psi(x,t) = \varphi(x)\chi(t)$, where $\varphi \in \mathbf{C}_0^\infty(\Omega)$ and $\chi \in \mathbf{C}_0^\infty((0,t))$. Then we get

$$\left(\int_0^t T(\tau)\chi_{\tau\tau}(\tau)\,d\tau \right) \left(\int v(x)\varphi(x)\,d\nu(x) \right)$$

$$= \left(\int_0^t T(\tau)\chi(\tau)\,d\tau \right) \left(\int_\Omega v(x)\mathrm{div}\,(a(x)\mathrm{grad}\,\varphi(x))\,dx \right)$$

$$+\left(\int_0^t f(\tau)\chi(\tau)\,d\tau \right) \left(\int v(x)\varphi(x)\,d\nu(x) \right). \tag{9}$$

By separation we obtain

$$\int_\Omega v(x)\mathrm{div}\,(a(x)\mathrm{grad}\,\varphi(x))\,dx = \lambda \int v(x)\varphi(x)\,d\nu(x) \tag{10}$$

and

$$\int_0^t (T(\tau)\chi_{\tau\tau}(\tau) - f(\tau)\chi(\tau))\,d\tau = \lambda \int_0^t T(\tau)\chi(\tau)\,d\tau. \tag{11}$$

For simplicity let $a(x) \equiv 1$ for every $x \in \Omega$. Then

$$\int v \Delta\varphi \, dx = \lambda \int \varphi \, d\nu, \tag{12}$$

which can be formally written as

$$\Delta v = \lambda \nu v. \tag{12'}$$

If $v|_{\partial\Omega} = 0$ on $\partial\Omega$, then (12) is an eigenvalue problem. Equation (11) can be written as

$$T'' - \lambda T = f. \tag{13}$$

In the case of (1) equation (13) is to be replaced by

$$T' - \lambda T = f. \tag{14}$$

In (Hansen, 1989; Hansen and Hueber, 1989) the following theorem was proved by using the theory of harmonic spaces. We note that the differential equation in Theorem 1 is written in the weak form (12) and that μv is defined by

$$\mu v(\varphi) = \int v\varphi \, d\mu.$$

Let $\Omega \subset \mathbf{R}^n$, $n \geq 1$, be a nonempty bounded domain, $\mu = \mu^+ - \mu^-$ a signed measure, such that $|\mu| = \mu^+ + \mu^-$ generates a continuous potential $G_L|\mu|$. In the classical case (Schrödinger operator) we have $d\mu(y) = q(x)dx$. Further, let ν be a positive measure on Ω such that $G_L\nu$ is continuous, $\mathbf{L}^2(\Omega, \nu)$ the space of all real functions, for which $|f|^2$ is ν-integrable,

$$(f, g) = \int fg \, d\nu$$

the scalar product on $\mathbf{L}^2(\Omega, \nu)$. Let G_L^Ω be the Green function of the Laplace operator. Then

$$K_\Omega v(x) = \int G_L^\Omega(x, y) v(y) \, d\nu(y), \tag{15}$$

where $\Delta K_\Omega v = -\nu v$ in the weak sense (12), defines a compact operator on the space $\mathbf{C}_b(\Omega)$ of all functions which are bounded and continuous on Ω. Further, the operator (15) defines a compact operator on $\mathbf{L}^2(\Omega, \nu)$ (see (Hansen, 1989)). The eigenfunctions v_k, satisfying

$$v_k(x) = -\lambda_k \int G_L^\Omega(x, y) v_k(y) \, d\nu(y), \tag{16}$$

are orthogonal relative to the scalar product (f, g), since for $\alpha_k = -1/\lambda_k$ we obtain

$$\alpha_k(v_k, v_l) = (\alpha_k v_k, v_l) = (K_\Omega v_k, v_l) = (v_k, K_\Omega v_l) = (v_k, \alpha_l v_l) = \alpha_l(v_k, v_l),$$

or

$$(\alpha_k - \alpha_l)(v_k, v_l) = 0.$$

THEOREM 1. *The set of all eigenvalues of*

$$(\Delta - \mu)v = \lambda \nu v, \quad v\big|_{\partial\Omega} = 0$$

is a nonempty upper bounded subset of $\mathbf{R}$, *which has no accumulation point. All eigenspaces have finite dimension. The eigenspace of the greatest eigenvalue is one-dimensional. None of the other eigenspaces contain any positive eigenfunction. If* $\nu(U) > 0$ *for each open set* $U \subset \Omega$, *then the* ν*-eigenfunctions form an orthonormal basis in* $\mathbf{L}^2(\Omega, \nu)$.

Remark 1. The results hold if Δv is replaced by a linear second order differential operator $P(x, \partial)v$ such that the continuous solutions of $P(x, \partial)v = 0$ generate an elliptic harmonic space having a Green function. This includes uniformely elliptic operators with Hölder continuous coefficients, uniformely elliptic operators in divergence form with bounded measureable coefficients and even a large class of degenerate elliptic differential operators. Moreover, the results have immediate applications to linear second order differential operators corresponding to harmonic spaces which are not necessarily elliptic.
Let T'' and f be continuous. Then from (11) it follows that T satisfies

$$T'' - \lambda T = f.$$

Its solution T is for $\lambda < 0$ given by

$$T(t) = \int_0^t f(\tau)\{\frac{1}{\sqrt{-\lambda}} \sin \sqrt{-\lambda}(t-\tau)\}\, d\tau + a \cos \sqrt{-\lambda} t + b \sin \sqrt{-\lambda} t.$$

Remark 2. If $\varphi \in \mathbf{C}_0^\infty(\Omega)$, $v \in \mathbf{C}^1(\Omega)$ and $a(x) = \big(a_{jk}(x)\big)$, then from (10) it follows that

$$-\int_\Omega v(x) \mathrm{div}\, \big(a(x) \mathrm{grad}\, \varphi(x)\big)\, dx = \int_\Omega \Big(\sum_{j,k=1}^n a_{jk}(x) \frac{\partial v}{\partial x_k} \frac{\partial \varphi}{\partial x_j} \Big)\, dx.$$

We can replace $a_{jk}(x)dx$ by a measure $d\nu_{jk}$ and define the operator A by

$$(Av, \varphi) = \int_\Omega \Big(\sum_{j,k=1}^n \frac{\partial v}{\partial x_k} \frac{\partial \varphi}{\partial x_j}\, d\nu_{jk} \Big).$$

The measures ν_{jk} can be concentrated on surfaces S, for instance, $d\nu_{jk}(x) = \sigma_{jk}(x)dS(x)$. In physics and biology properties of surfaces play an important role, see also the author's results on extremal elements (Anger, 1990).

2. THE SOLUTION OF THE DIRECT PROBLEM

2.1. Representation of solutions by means of particular solutions
Let λ_k be an eigenvalue of equation (12). In this case $v_k(y) = 0$ on $\partial\Omega$. In (Hansen and Hueber, 1989) it was proved, that $\lambda_k < 0$ and $\lambda_k \to -\infty$ for $k \to \infty$. A particular solution of (12) is given by $(f_0(x,t) = 0)$

$$u(x,t) = \sum_{k=1}^{\infty} v_k(x)T_k(t) = \sum_{k=1}^{\infty} v_k(x)(a_k \cos\sqrt{-\lambda_k}t + b_k \sin\sqrt{-\lambda_k}t).$$

Let

$$f_0(x,t) = \sum_{1}^{\infty} f_k(t)v_k(x), \quad f_k(t) = \int f_0(x,t)v_k(y)\,d\nu(y), \tag{17}$$

where the v_k are the eigenfunctions of (12), $v_k \in \mathbf{L}^2(\Omega,\nu)$. From (15) it follows that $f_0(y,t) = 0$ for $y \in \partial\Omega$. Then

$$u(x,t) = u_1(x,t) + u_2(x,t),$$

where

$$u_1(x,t) = \int_0^t \Big(\int G(x,y;t-\tau,0)f_0(y,\tau)\,d\nu(y)\Big)\,d\tau$$

$$= \int_0^t \Big(\int \Big\{\sum \frac{1}{\sqrt{-\lambda_k}} \sin\sqrt{-\lambda_k}(t-\tau)v_k(x)v_k(y)\Big\} f_0(y,\tau)\,d\nu(y)\Big)\,d\tau \tag{18}$$

is a solution of the inhomogeneous equation (7) with $a(x) \equiv 1$, and

$$u_2(x,t) = \sum_{k=1}^{n} \{a_k \cos\sqrt{-\lambda_k}t + b_k \sin\sqrt{-\lambda_k}t\}v_k(x)$$

is a solution of (7) with $f_0(x,t) = 0$. The coefficients a_k and b_k can be determined by the initial values

$$u(x,0) = h_1(x) \quad \text{and} \quad u_t(x,0) = h_2(x), \quad x \in \Omega.$$

In the case of the heat equation $\big(a(x) \equiv 1\big)$ we obtain

$$u(x,t) = u_1(x,t) + u_2(x,t),$$

where

$$u_1(x,t) = \int_0^t \Big(\int G_L(x,y;t-\tau,0)f_0(y,\tau)\,d\nu(y)\Big)\,d\tau$$

$$= \int_0^t \Big(\int \Big\{\sum_{k=1}^{\infty} e^{\lambda_k(t-\tau)}v_k(x)v_k(y)\Big\} f_0(y,\tau)\,d\nu(y)\Big)\,d\tau,$$

and

$$u_2(x,t) = \sum c_k e^{\lambda_k t} v_k(x).$$

The coefficients can be determined by the initial values

$$u(x,0) = h(x), \quad x \in \Omega.$$

2.2. *Some further properties of the eigenfunctions*

Let ν be a positive measure on Ω such that the Newtonian potential $G_L\nu$ is continuous. Following (Hansen, 1989) the set of all eigenfunctions of (12) forms an orthonormal basis in $\mathbf{L}^2(\Omega, \nu)$ if $\nu(U) > 0$ for all open sets $U \subset \Omega$. This condition is fulfilled if ν is the Lebesgue measure. The linear operator K_Ω: $\mathbf{L}^2(\Omega,\nu) \to \mathbf{L}^2(\Omega,\nu)$ defined in (15) is compact. From this fact it follows that K_Ω is also continuous.

If all eigenfunctions are known then each $v \in \mathbf{L}^2(\Omega,\nu)$ can be represented in the form

$$v = \sum_{k=1}^{\infty} c_k v_k.$$

Especially for the characteristic function $\mathbf{1}_B$ of a Borel set B, i.e., $\mathbf{1}_B(x) = 0$ for $x \in B$, $\mathbf{1}_B(x) = 0$ otherwise, we obtain for

$$v(y) = \mathbf{1}_B(y)/G_L^\Omega(x,y) = \sum_{k=1}^{\infty} c_k v_k(y)$$

from (16)

$$\sum_{k=1}^{\infty} c_k \frac{v_k(x)}{\lambda_k} = \int G_L^\Omega(x,y) v(y)\, d\nu(y)$$

$$= \int \left(G_L^\Omega(x,y) \sum_{k=1}^{\infty} c_k v_k(y) \right) d\nu(y) = \int \mathbf{1}_B(y)\, d\nu(y) = \nu(B). \tag{19}$$

From (19) it follows that the measure ν is uniquely determined. If only finitely many eigenvalues are known, then different ν can satisfy the same equivalence relation

$$\int G_L(x,y) v(y)\, d\nu(y) = \int G_L(x,y) v(y)\, d\nu^*(y), \tag{20}$$

where $v = \sum_{k=1}^{n} c_k v_k$.

In applications such equivalence relations are well known (see (Anger, 1990)), which can be solved only by additional information coming from experiments. Section 3 deals with some inverse problems and its equivalence relations.

In the following we should like to calculate some special eigenfunctions for a special domain and a special positive measure ν. Let $B(0,1) = \{x \in \mathbf{R}^n\colon |x| < 1\}$ be a ball. If $n = 2$ the Green function relative to $\Omega = B(0,1)$ has the form (Stakgold,

1968), where $x, y \in B(0,1)$, $x = (r, \beta)$, $y = (r_0, \beta^*)$, $r_> = \max(r, r_0)$, $r_< = \min(r, r_0)$,

$$G_L^{\Omega}(x,y) = -\frac{1}{2\pi}\log r_> - \frac{1}{2\pi}\sum_1^{\infty}\frac{\cos m(\beta-\beta^*)}{m} r_<^m (r_>^m - r_<^{-m}).$$

Let $r < r_0$. Then

$$G_L^{\Omega}(x,y) = -\frac{1}{2\pi}\log r_0 - \frac{1}{2\pi}\sum_1^{\infty}\frac{\cos m(\beta-\beta^*)}{m} r^m (r^m - r^{-m}). \tag{21}$$

If $d\nu(y) = (2\pi)^{-1} d\beta^*$ the solution of the Dirichlet problem of the Laplace equation $\Delta u = 0$ with (sufficiently smooth) boundary values h on $\partial B(0, r_0)$, $0 < r_0 < 1$, where

$$h(\beta) = \sum_0^{\infty}(a_j \cos j\beta + b_j \sin j\beta) \tag{22}$$

is for $r < r_0$ given by, $|x| < r_0$,

$$u(x) = \sum_0^{\infty}(a_j \cos j\beta + b_j \sin j\beta)(r/r_0)^j. \tag{23}$$

The potential

$$x \mapsto -\lambda_k \int G_L(x,y) v_k(y)\, d\nu(y) = -\frac{\lambda_k}{2\pi}\int_0^{2\pi} G_L(x,y) v_k(y)\, d\beta^* \tag{24}$$

is harmonic for all $x \notin \partial B(0,r)$ and continuous on $\partial B(0,r_0)$. Such a function must be represented by (23). Substituting (21) and (22) in (24) we obtain for $|x| < r_0$

$$\sum_0^{\infty}(a_j \cos j\beta + b_j \sin j\beta)(r/r_0)^j$$

$$= \frac{\lambda_k}{2\pi}\int_0^{2\pi}\left(\frac{1}{2\pi}\log r_0 + \frac{1}{2\pi}\sum_1^{\infty}\frac{\cos m(\beta-\beta^*)}{m} r^m (r^m - r^{-m})\right)$$

$$\times\left(\sum_0^{\infty}(a_j \cos j\beta^* + b_j \sin j\beta^*)\right) d\beta^*.$$

The orthogonality yields on the right-hand side

$$\frac{\lambda_k}{2\pi}(a_0 \log r_0 + \frac{1}{2}\sum_0^{\infty}\frac{1}{j}(a_j \cos j\beta + b_j \sin j\beta) r^j (r_0^j - r^{-j})).$$

If r tends to r_0, we get

$$\sum_0^\infty (a_j \cos j\beta + b_j \sin j\beta)$$

$$= \frac{\lambda_k}{2\pi}\Big(a_0 \log r_0 + \sum_1^\infty (a_j \cos j\beta + b_j \sin j\beta)(r_0^{2j} - 1)\Big).$$

From this equation it follows that

$$\lambda_0 = 2\pi/\log r_0, \quad \lambda_k = 2\pi k/(r_0^{2k} - 1), \quad r_0 < 1.$$

Similar calculations are possible in the case $n \geq 3$, for toroidal functions (Lense, 1950) and for other special coordinates.

3. SOME RESULTS ON INVERSE PROBLEMS

3.1. *Some basic inverse problems*

The main system in geophysics, which is to be studied, reads (Alekseev and Bubnov, 1984)

$$\rho(x)u_{tt} = -\operatorname{div}\big(\lambda(x)\operatorname{grad} u\big) = f(x,y), \tag{25}$$

$$\Delta w = -4\pi\rho(x), \tag{26}$$

where $\rho(x)$ is the mass density and $\lambda(x)$ the Lamé coefficient. Further, we have the following initial conditions and boundary conditions

$$u(x,0) = 0, \quad u_t(x,0) = 0, \quad x \in \Omega, \tag{27}$$

$$\frac{\partial u(x,t)}{\partial n} = h_1(x,t), \quad x \in \partial\Omega, \quad 0 < t, \tag{28}$$

or

$$u(x,t) = h_0(x,t), \quad x \in \partial\Omega, \quad 0 < t. \tag{29}$$

In applications often $h_1(x,t) = 0$, and $h_0(x,t)$ can be measured. For (26) the values

$$w(x) = w_0(x), \quad x \in \Omega, \tag{30}$$

are of interest in applications. The functions w_0 and h_0 are additional information. The inverse problems relative to this system consists in determining, if possible, the mass density $\rho(x)$ and the Lamé parameter $\lambda(x)$.

In soil mechanics the following differential equation plays a fundamental role

$$u_t - \operatorname{div}\big(k(x)\operatorname{grad} u\big) = f(x,t). \tag{31}$$

Here $k(x)$ is the diffusion coefficient. For simplicity, let $\Omega = I$ be an open interval in $\mathbf{R}^3$. On the upper boundary we have $u(x,t) = 0$, on the other parts of ∂I

$$k(x)\frac{\partial u(x,t)}{\partial n} = 0.$$

In finitely many bore holes $f(x,t)$ is given, in some other $u(x,t)$ and $k(x)\frac{\partial u(x,t)}{\partial n}$ can be measured. From these measurements $k(x)$ or some averages of $k(x)$ are to be determined. Similar problems were studied by A. Calderon and other scientists (see (Sylvester and Uhlmann, 1990)).
One can find further inverse problems in geophysics in (Alekseev et al., 1981; Tikhonov and Goncharskii, 1987).

3.2. *Some results on inverse problems*
Most inverse problems cannot be uniquely solved. For the inverse source problem of the Laplace equation the author introduced relative to a positive measure μ concentrated on $\bar{\Omega}$ the following set

$$B(\mu) = \{\nu \geq 0\colon \operatorname{supp}\nu \subset \bar{\Omega},\quad G_L\nu(x) = G_L\mu(x),\quad x \notin \Omega\}. \tag{32}$$

In 1911 G. Lauricella proved for densities μ_ρ that (32) can be defined as (see (Moritz, 1990))

$$B(\mu) = \{\nu = 0\colon \operatorname{supp}\nu \subset \bar{\Omega},\ \int u\,d\nu = \int u\,d\mu,\ u \in \mathbf{C}(\bar{\Omega}),\ \Delta u = 0 \text{ in } \Omega\}.$$

In this way we get $B(\mu) \subset \mathbf{C}^*(\bar{\Omega})$, where $\mathbf{C}^*(\bar{\Omega})$ is the set of all measures γ satisfying $\operatorname{supp}\gamma \subset \bar{\Omega}$. Further, following (Riesz, 1909) $\mathbf{C}^*(\bar{\Omega})$ is the dual space of $\mathbf{C}(\bar{\Omega})$. Therefore the results of functional analysis, especially the results on convex sets, can be applied to inverse problems.
Let u be a solution of (2) with $f(x,t) = 0$ satisfying $u(z,t) = 0$ for $z \in \partial\Omega$, $\partial\Omega$ sufficiently smooth. Then $v(z) = 0$. Special solutions v_k satisfy equation (16). If the additional information on Ω is

$$\frac{\partial v_k(z)}{\partial n} = g_k(z),$$

then from (16) it follows that

$$g_k(z) = -\lambda_k \int P(z,y)v_k(y)\,d\nu(y), \tag{33}$$

where

$$P(z,y) = \frac{\partial G_L^{\Omega}(y,z)}{\partial n}$$

is the Poisson kernel. Integrating (33) with respect to $h(z)dS(z)$, we get

$$\int hg_k\,dS = -\lambda_k \int w(z)v_k(z)\,d\nu(z).$$

Here $w \in \mathbf{C}(\bar{\Omega})$ is an arbitrary harmonic function with boundary values h. If the support of ν is concentrated on the boundary of a domain $\Omega_1 \subset \Omega$, then ν is uniquely determined (see (Anger, 1990)). In general we have the equivalence relation

$$\int w(z)v(z)\,d\nu(z) = \int w(z)v(z)\,d\nu^*(z),$$

where

$$v(z) = \sum_{1}^{m} c_k \lambda_k v_k(z).$$

There are relations to the results of G. Anger. One can find more results on such inverse problems with applications to soil mechanics in (Gottlieb, 1991).

REFERENCES

Alekseev, A.S. and Bubnov, B.A. (1981). On a combined statement of inverse seismics and gravimetry problems. *Dokl. Akad. Nauk SSSR* **261**, 1086–1090 (in Russian).

Alekseev, A.S., Lavrentiev, M.M., Romanov, V.G., and Romanov, M.E. (1990). Theoretical and computational aspects of seismic tomography. *Surveys in Geophysics* **11**, 395–409.

Anger, G. (1990). *Inverse Problems in Differential Equations.* Akademie-Verlag, Plenum Press, London.

Gottlieb, J. (1991). Identification of measures in differential equations (to appear).

Hansen, W. (1989). Valeurs propres pour l'operateur de Schrödinger. *Lect. Notes Math.* **1393**, 117–134.

Hansen, W. and Hueber, H. (1989). Eigenvalues in potential theory. *J. Different. Equat.* **73**, 133–152.

Landkof, N.S. (1972). *Foundations of Modern Potential Theory.* Springer-Verlag, New York, Berlin.

Lense, J. (1950). *Kugelfunktionen.* Akademische Verlagsgesellschaft Geest & Portig, Leipzig.

Moritz, H. (1990). *The Figure of the Earth.* Wichmann, Karlsruhe.

Stakgold, I. (1968). *Boundary Value Problems of Mathematical Physics II.* Macmillan, New York, London.

Sylvester, J. and Uhlmann, G. (1990). The Dirichlet to Neumann map and applications. In: *Inverse Problems in Partial Differential Equations.* Ed. by D.Colton, R.Ewing and W.Rundell. SIAM Proc. Ser., Philadelphia, pp. 99–139.

Tikhonov, A.N. and Samarskii, A.A. (1963). *Equations of Mathematical Physics.* Pergamon Press, London, New York.

Tikhonov, A. and Goncharskii, A. (1987). *Improperly Posed Problems in Natural Sciences.* Mir, Moscow.

Ill-Posed Problems in Natural Sciences, pp. 244 – 253
A. Tikhonov (Ed.)
1992 VSP/TVP

NEW METHODS AND RESULTS IN MULTIDIMENSIONAL INVERSE PROBLEMS FOR KINETIC EQUATIONS

Ju.E. ANIKONOV
Institute of Mathematics, Siberian Branch of Russia Academy of Sciences, University Aveny str. 4, 630090 Novosibirsk, Russia

INTRODUCTION

Problems of finding the coefficients of differential equations from information about their solution are usually called inverse problems for differential equation. As a rule, inverse problems are nonlinear and their study, especially in the multidimensional case, is often connected with significant mathematical difficulties. In this report I want to present some methods for studing the existence and uniqueness of solutions of inverse problem for kinetic equations, particularly for transport equations.

Kinetic equations are important equations. Often they are called "master equations", because they characterize the continuous movement of subjects.

Examples of kinetic equations are the equations of Vlasov, Maxwell, Boltzmann, Liouvill, Bogolubov and others.

Below $F(x,p,t)$ is a distribution function.

1) Liouville equation:

$$\frac{\partial F}{\partial t} + \{F,H\} = 0, \ \{F,H\} = \sum_{j=1}^{n} \Big(\frac{\partial F}{\partial x_j}\frac{\partial H}{\partial p_j} - \frac{\partial F}{\partial p_j}\frac{\partial H}{\partial x_j}\Big).$$

2) Boltzmann equation:

$$\frac{\partial F}{\partial t} + \sum_{j=1}^{n} \frac{\partial F}{\partial x_j} p_j = \mathrm{St}F.$$

3) Transport equation:

$$\frac{\partial F}{\partial t} + \Omega \mathrm{grad}_X F + \sigma F = \int_{\omega} \mathcal{J}(x,t,\Omega,\Omega')F\, d\Omega' + q.$$

4) Transport equation in tomography:

$$\frac{\partial F}{\partial t} + \Omega \mathrm{grad}_X F + \sigma F = \int_{E_1}^{E_2}\int_{\omega} \mathcal{J}(x,t,E',\Omega,\Omega')F\,dE'\,d\Omega' + q.$$

5) Kinetic equations in problems of integral geometry:

$$\mathrm{grad}_X F\Omega + \sigma F = 0, \qquad \mathrm{ln}F = \Psi \rightarrow \mathrm{grad}_X \Psi\Omega = -\sigma,$$

$$\frac{\partial F}{\partial x}\cos\varphi + \frac{\partial F}{\partial y}\sin\varphi + K(x,y,\varphi)\frac{\partial F}{\partial \varphi} = \lambda(x,y,\varphi),$$

K is first curvature.

6) Fokker–Plank equation:

$$\frac{\partial F}{\partial t} + \{F,H\} = \int_{R^n} [W(p+q,q)F(x,p+q,t) - W(p,q)F(x,q,t)]\,dq.$$

7) Quantum equation:

$$\frac{\partial F}{\partial t} + \{F,H\} + \sigma F = \frac{i}{(2\pi)^n h}\int_{R^{2n}} [\Phi(x - \frac{1}{2}hy,t) - \Phi(x + \frac{1}{2}hy,t)]e^{iy(p-p')}$$

$$\times F(x,p',t)\,dp'\,dy + \mathrm{St}F + q.$$

Analysis of physical problems connected with kinetic equations (for example, the problem of tomography) shows that, in addition to the distribution function $F(x,p,t)$ some coefficients involved in this equations can be considered to be unknown and to be determined.

The potential Φ or a function connected with the collision integral StF which describes the scattering of particles may be such coefficients. The sources may be unknown also.

Our basic information in inverse problems for kinetic equations is the function $F_0(s,p,t) = F(x,p,t)\,|_\Gamma$, $s \in \Gamma$, $p \in D \subset \mathbf{R}^n$, $t_1 \le t \le t_2$, where Γ is a manifold in $\mathbf{R}^n$, for example, $\Gamma = \{x\colon |x| = 1 \text{ or } x\colon x_n = 0\}$.

In other words we must know the trace function of the distribution $F(x,p,t)$ on Γ.

Information about such inverse problems for kinetic equations is contained in (Anikonov, 1978; Anikonov and Amirov, 1983; Anikonov and Bubnov, 1988; Aniconov, 1986; Anikonov and Pestov, 1990; Anikonov, 1987, 1988, 1990).

Example. We consider the inverse problems of determining two functions $(F, q(x,t))$ or $(F, \lambda(x,t)), \partial q/\partial p = 0$, $\partial\lambda/\partial p = 0$ in the domain $|p| < \infty$, $|x| < a$, $|t - t_0| < b$, $a > 0$ and $b > 0$,

$$\frac{\partial F}{\partial t} + p\frac{\partial F}{\partial x} + \lambda(x,t)\frac{\partial F}{\partial p} = q(x,t), \tag{1}$$

$$F\mid_{x=0}=F_0(p,t), \qquad |p|<\infty, \qquad |t-t_0|<b. \tag{2}$$

Equation (1) is the one-dimensional kinetic equation (Vlasov equation for $q=0$), $\lambda(x,t)=-\partial\Phi/\partial x$ and Φ is the potential.

THEOREM 1. *Suppose that $F_0(p,t)$ is an analytic function in the domain $|t-$ $-t_0|<\delta$, $|p|<\infty$, and for any t, F_0 is an entire function of exponential type in p which belongs to the space $\mathbf{L}_2$. Then there exist numbers $a_1>0$ and $b_1>0$, such that the inverse problem (1), (2) of determing (F,q) has an analytic solution in the domain $|x|<a_1$, $|t-t_0|<b_1$, $|p|<\infty$.*

THEOREM 2. *Let the conditions of Theorem 1 are satisfied and $\partial F_0/\partial p\mid_{p=0}\neq$ $\neq 0$. Then there exist numbers $a_1>0$, $b_1>0$, such that the inverse problems (1), (2) of determing (F,λ) has an analytic solution in the domain $|x|<a_1$, $|t-t_0|<$ b_1, $|p|<\infty$.*

ONE-DIMENSIONAL INVERSE PROBLEMS IN THE STATIONARY CASE

Here we consider the equation

$$\frac{\partial F}{\partial x}p-\Phi'(x)\frac{\partial F}{\partial p}=\lambda(x)R(F), \qquad R\neq 0. \tag{3}$$

In this case the following relations are important for inverse problems.

THEOREM 3. *Let $\partial B/\partial z=R(B)$, $z\in\mathbf{R}^1$ and $f(z),g(z)$ belong to $\mathbf{C}^1(\mathbf{R}^1)$. If $\lambda(x)=-2\Phi'(x)g(\Phi(x))$, then*

$$F(x,p)=B\Big(f\Big(\frac{1}{2}p^2+\Phi(x)\Big)\Big)+\int_{-p}^{p}g\Big(\frac{1}{2}p^2-\frac{1}{2}\eta^2+\Phi(x)\Big)\,d\eta.$$

is a solution of (3).

This theorem can be used for determining F and R or λ, or Φ if we have the information

$$F_0(p)=F\mid_{x=0}. \tag{4}$$

Example $(R=z)$.

THEOREM 4. *Consider functions*

$$\hat f(p)=\frac{\log F_0(p)+\log F_0(-p)}{2}, \qquad \tilde g(p)=\frac{\log F_0(p)-\log F_0(-p)}{2},$$

$$f(p)=\hat f(\sqrt{2p}), \qquad g(p)=\tilde g(\sqrt{2p}),$$

$$b(p)=\frac{1}{2\pi}\frac{d}{dp}\int_0^p\frac{g(z)\,dz}{\sqrt{p-z}}\ \ (F_0>0, F_0\in\mathbf{C}^1(\mathbf{R}^1)),$$

$$\Phi(x) \geq 0, \qquad \Phi(0) = 0.$$

Then inverse problem (3)–(4) of determining $F(x,p)$ and $\lambda(x)$ has the following solution

$$\lambda(x) = -2b(\Phi(x))\Phi'(x),$$

$$F(x,p) = \exp\left[f\left(\frac{1}{2}p^2 + \Phi(x)\right) + \int_{-p}^{p} b\left(\frac{1}{2}p^2 - \frac{1}{2}\eta^2 + \Phi(x)\right) d\eta\right].$$

Some results in the analytic case

In the Euclidean space $\mathbf{R}^{2n+1}, n \geq 2$, we consider the kinetic equation

$$\frac{\partial F}{\partial t} + \sum_{j=1}^{n}\left(\frac{\partial F}{\partial x_j}p_j + \frac{\partial F}{\partial p_j}\frac{\partial \Phi}{\partial x_j}\right) = \mathrm{St}F. \tag{5}$$

Below it will be convenient for us to represent the collision integral $\mathrm{St}F$ as: $\mathrm{St}F =$ $= aF + b$, where a and b are function of the variables $(\bar{x}, \bar{p}, t)$, which are possibly generated by the distribution $F(\bar{x}, \bar{p}, t)$.

Let $\varphi = \arctan\left(p_n/p_{n-1}\right)$, $q = \sqrt{p_n^2 + p_{n-1}^2}$, $x = x_{n-1}$, $y = x_n$,

$LF = \sum_{j=1}^{n-2}\left(\partial p/\partial x_j p_j + \partial F/\partial p_j \cdot \partial\Phi/\partial x_j\right)$.

In this variables (5) can be written as:

$$\frac{\partial F}{\partial t} + q\left(\frac{\partial F}{\partial x}\cos\varphi + \frac{\partial F}{\partial y}\sin\varphi\right) + \frac{\partial F}{\partial q}\left(\frac{\partial \Phi}{\partial x}\cos\varphi + \frac{\partial \Phi}{\partial y}\sin\varphi\right) \tag{6}$$

$$-\frac{1}{q}\frac{\partial F}{\partial \varphi}\left(\frac{\partial \Phi}{\partial y}\cos\varphi - \frac{\partial \Phi}{\partial x}\sin\varphi\right) + LF = aF + b.$$

Below it is assumed that the function Φ, a and b in (6) do not depend on the variable φ, i.e.,

$$\frac{\partial \Phi}{\partial \varphi} = 0, \quad \frac{\partial a}{\partial \varphi} = 0, \quad \frac{\partial b}{\partial \varphi} = 0.$$

We formulate the following inverse problems.

PROBLEM I. *Find the function F, a and b if the function $F_0 =$ $= F\,|_{y=0}$ and Φ are known.*

PROBLEM II. *Find the functions F, Φ and b, if the functions $\Phi_0 =$ $= \Phi\,|_{y=0}, F_0 = F\,|_{y=0}, a$ are known.*

PROBLEM III. *Find the functions F, Φ and a if functions $F_0 = F\,|_{y=0}, \Phi_0 =$ $= \Phi\,|_{y=0}$ and b are known.*

All functions considered below are assumed to be analytic in some neighbourhood of a point $(\bar{0}, q_0), q_0 > 0$, of the space $\mathbf{R}^{2n+1}$ and periodic in φ with period 2π.

THEOREM 5. *Suppose that, for each inverse problem I–III, the condition with the same number below is satisfied*

1) $F_0 \mid_{\varphi=0} - F_0 \mid_{\varphi=\pi} \neq 0$,

2) $\frac{\partial F_0}{\partial \varphi} \mid_{\varphi=0} + \frac{\partial F_0}{\partial \varphi} \mid_{\varphi=\pi} \neq 0$,

3) $\frac{\partial F_0}{\partial \varphi} F_0 \mid_{\varphi=0} + \frac{\partial F_0}{\partial \varphi} F_0 \mid_{\varphi=\pi} \neq 0$.

Then the uniqueness of solutions of these problems is guaranteed in the class of analytic functions.

THEOREM 6. *Suppose that*

$$F_0 = \sum_{k=0}^{\infty} u_k^0 \cos k\varphi + v_k^0 \sin k\varphi,$$

where $|u_k^0| < Ae^{-pk}$, $|v_k^0| < Ae^{-pk}$, $k = 0, 1, 2, \ldots$, $A > 0$, $p > 0$ *are constants and conditions* 1)–3) *of Theorem* 5 *are satisfied. Then there exist analytic solutions of inverse problems* I–III.

Consider the inverse problem for kinetic equation in the multidimensional case

$$\frac{\partial F}{\partial t} + \frac{\partial F}{\partial y} q - \frac{\partial F}{\partial q} a + \{F, H\} = \sigma F + \mathrm{St} F + f.$$

The problem is to find $F(x, y, q, p, t)$, $\sigma(x, p, t)$, if the function $F \mid_{y=0} = F_0(x, q, p, t)$ is known (a, f and St are known also).

LEMMA. *The function* $F(x, y, q, p, t)$ *is a solution of the problem*

$$\begin{cases} \dfrac{\partial F}{\partial y} = F \displaystyle\int_0^1 \frac{\partial}{\partial q}\Big\{\frac{1}{F}\Big[F_t + \frac{\partial F}{\partial q} a - \{F, H\} + \mathrm{St} F + F\Big]\Big\}(q \cdot \eta)\, d\eta, \\ F \mid_{y=0} = F_0. \end{cases}$$

THEOREM 7. *In the analytic case the inverse problem of finding* $F > 0$ *and* σ *has only one solution.*

IDENTITIES

Differential identities play an important role in the multidimensional inverse problems for kinetic equations. We write some of them.

THEOREM 8. *Let* $F(x, p, t) \in \mathbf{C}^2(Q)$, $H(x, p) \in \mathbf{C}^2(D \times D_1)$, $x \in D \subset \mathbf{R}^n$, $p \in D_1 \subset \mathbf{R}^n$, $t_1 < t < t_2$, $Q = D \times D_1 \times (t_1, t_2)$ *and* $LF = \frac{\partial F}{\partial T} + \{F, H\}$, $BF = \sum_{j=1}^{n} \frac{\partial F}{\partial x_j} \frac{\partial F}{\partial p_j}$, *where* $\{F, H\}$ *is Poisson's bracket. Then the following identity holds*

$$LBF = \Big[\sum_{j=1}^{n} \frac{\partial F}{\partial p_j} \frac{\partial LF}{\partial x_j} + \frac{\partial F}{\partial x_j} \frac{\partial LF}{\partial p_j}\Big]$$

$$+\sum_{j=1}^{n}\sum_{i=1}^{n}\Big[\frac{\partial^2 H}{\partial x_i \partial x_j}\frac{\partial F}{\partial p_i}\frac{\partial F}{\partial p_j}-\frac{\partial^2 H}{\partial p_i \partial p_j}\frac{\partial F}{\partial x_i}\frac{\partial F}{\partial x_j}\Big].$$

THEOREM 9. *If functions $F(x,y,z)$, $Q(x,y,z)$, $d(x,y)$, $c(x,y)$ belong to $\mathbf{C}^2(D)$, $D \in \mathbf{R}^3$ and*

$$MF=\Big(\frac{\partial F}{\partial x}-c\Big)\cos\theta+\Big(\frac{\partial F}{\partial y}-d\Big)\sin\theta,$$

$$M'F=\Big(\frac{\partial F}{\partial x}-c\Big)\sin\theta-\Big(\frac{\partial F}{\partial y}-d\Big)\cos\theta,$$

then the following identity holds

$$M'F\frac{\partial}{\partial z}MF-MF\frac{\partial}{\partial z}M'F=-\frac{\partial\theta}{\partial z}\Big[\Big(\frac{\partial F}{\partial x}-c\Big)^2+\Big(\frac{\partial F}{\partial y}-d\Big)^2\Big]$$

$$+\frac{\partial}{\partial y}\Big(\frac{\partial F}{\partial x}\frac{\partial F}{\partial z}\Big)-\frac{\partial}{\partial x}\Big(\frac{\partial F}{\partial y}\frac{\partial F}{\partial z}\Big)-\frac{\partial}{\partial z}\Big(c\frac{\partial F}{\partial y}-d\frac{\partial F}{\partial x}\Big).$$

Here is yet another result

THEOREM 10. *Let $LF=\lambda(x,p,t)R(F)$, $R(z)>0$, $R(z)\in \mathbf{C}^2(-\infty,\infty)$, $\sum_{j=1}^{n} \partial^2\lambda/\partial x_j\partial p_j=0$ and matrices $\partial^2 H/\partial p_i\partial p_j, -\partial^2 H/\partial x_i\partial x_j$ be positive definite, $i,j=1,2,\cdots,n$. Then inverse problem to find $F\in \mathbf{C}^2(\overline{Q})$ and $\lambda\in \mathbf{C}^2(\overline{Q})$ in the domain $Q=D\times D_1\times(t_1,t_2)$ for $F\mid_{\partial Q}=F_0$ has no more then one solution.*

For other results see (Anikonov, 1978; Anikonov and Amirov, 1983; Anikonov, 1986; Anikonov and Pestov, 1990; Anikonov, 1987, 1990).

INVERSE PROBLEMS IN TRANSPORT THEORY

Here we pose inverse problems for transport equation and present uniquenees and existence results for solutions of such problems. Consider the boundary value problem for the transport equation

$$\frac{\partial F}{\partial t}+(\omega,\nabla_X F)+\sigma(x,\omega)F=\sigma_s(x,\omega)\int_\Omega g(x,\omega,\omega')$$

$$\times F(x,t,\omega')\,d\omega'+f(x,\omega)h(x,t,\omega), \tag{7}$$

$$F\mid_{\Gamma_-}=\varphi, \tag{8}$$

where $t\in\mathbf{R}$, $x\in D\subset\mathbf{R}^n$, D is a strictly convex bounded domain with boundary $\gamma\in\mathbf{C}^2$, $\omega\in\Omega$, $|\Omega|=1$ and $\Gamma_-=\{(x,t,\omega)\colon t\in\mathbf{R},\ x\in\gamma,\ (n,\omega)<0\}$ is the unit vector of the outer normal to γ.

We pose three inverse problems:

1) Find the functions F,σ $(\partial\sigma/\partial t=0)$

2) Find the functions F,σ_s $(\partial\sigma_s/\partial t=0)$

3) Find the functions F, f $(\partial f/\partial t = 0)$

If conditions (7), (8) are supplemented by data for $t = 0$ and if we assume that the solution F behave regulary with respect to the variable t on $(-\infty, \infty)$, i.e., following conditions

$$F\mid_{t=0}= F^0, \quad F^0\mid_{\gamma^-} = \varphi\mid_{t=0},$$

$$\gamma^- = \{(x,\omega)\colon\ x \in \gamma, (n,\omega) < 0\},$$

$$\max_{x,\omega} \int_{-\infty}^{\infty} (1+|z|)^{2p}(|\hat{F}|^2 + |\nabla_x \hat{F}|^2)\,dz < \infty, \qquad p \geq 2,$$

where $\hat{F}$ is the Fourier transform of $F(x,t,\omega)$ with respect to the variable t, then inverse problems 1)–3) posed above turn out to reduce to the investigation of boundary value problems for integrodifferential equations. This reduction is formulated in the following lemmas.

LEMMA 1. *If $v(\lambda, x, \omega)$ is a solution of the boundary value problem*

$$-i\lambda v + (\omega, \nabla_x v) + v[Q_0 + \frac{i}{F^0}\int_{-\infty}^{\infty} \lambda v d\lambda] = \sigma_s \int_{\Omega} gv\,d\omega' + f\hat{h},$$

$$v\mid_{\gamma^-} = \varphi, \quad |F^0| \geq \delta > 0,$$

$$Q_0 = \sigma_s \int_{\Omega} gF^0\,d\omega' - h(0,x,\omega)f - (\omega, \nabla_x, F^0)$$

from the class

$$\max_{x,\omega} \int_{-\infty}^{\infty} (1+|z|)^{2p} (|v|^2 + |\nabla_x v|^2)\,dz, \quad p \geq 2,$$

then functions $F(x,t,\omega)$ and $\sigma(x,\omega)$ which are to be found in inverse problem 1) are

$$F(x,t,\omega) = \int_{-\infty}^{\infty} ve^{-izt}\,dz, \quad \sigma(x,\omega) = \frac{Q_0}{F^0} + \frac{i}{F^0}\int_{-\infty}^{\infty} zv\,dz.$$

LEMMA 2. *If $v(\lambda, x, \omega)$ is solution of the boundary value problem*

$$-i\lambda v + (\omega, \nabla_x v) + \sigma v = \Big[F_1 - \frac{i}{\int_{\Omega} gF^0\,d\omega'}\int_{-\infty}^{\infty} \lambda v\,d\lambda\Big] \int_{\Omega} gv\,d\omega' + f\hat{h},$$

$$v\mid_{\gamma^-} = \varphi, \quad \Big|\int_{\Omega} gF^0\,d\omega\Big| \geq \delta > 0,$$

$$F_1 = \frac{(\omega, \nabla_x F^0) + \sigma F^0 - f h(0,x,\omega)}{\int\limits_\Omega g F^0 \, d\omega'}$$

from the class

$$\max_{x,\omega} \int\limits_{-\infty}^{\infty} (1+|z|)^{2p} \big(|v|^2 + |\nabla_x v|^2\big) \, dz < \infty, \quad p \geq 2,$$

then the function $F(x,t,\omega)$ *and* $\sigma_s(x,\omega)$ *which are to be found in sought inverse problem 2) are*

$$F = \int\limits_{-\infty}^{\infty} v e^{-izt} \, dz, \quad \sigma_s = F_1 - \frac{i}{\int\limits_\Omega g F^0 \, d\omega'} \int\limits_{-\infty}^{\infty} zv \, dz.$$

LEMMA 3. *If* $v(\lambda, x, \omega)$ *is a solution of the boundary value problem*

$$-i\lambda v + (\omega, \nabla_x, v) + \sigma v$$

$$= \sigma_s \int\limits_\Omega g v \, d\omega' + \frac{\hat{h}}{h(0,x,\omega)} \Big[-i \int\limits_{-\infty}^{\infty} \lambda v \, d\lambda + (\omega, \nabla_x F^0) + \sigma F^0 - \sigma_s \int\limits_\Omega g F^0 \, d\omega \Big],$$

$$v \,|_{\gamma^-} = \varphi, \quad |\, h(0,x,\omega) \,| \geq \delta > 0$$

from the class

$$\max_{x,\omega} \int\limits_{-\infty}^{\infty} (1+|z|)^{2p} \big(|v|^2 + |\nabla_x v|^2\big) \, dz < \infty, \quad p \geq 2,$$

then the functions $F(x,t,\omega)$ *and* $f(x,\omega)$ *which are to be found in inverse problem 3) are*

$$F(x,t,\omega) = \int\limits_{-\infty}^{\infty} v e^{-izt} \, dz,$$

$$f(x,\omega) = \frac{1}{h(0,x,\omega)} \Big[(\omega, \nabla_x F^0) + \sigma F^0 - \sigma_s \int\limits_\Omega g F^0 \, d\omega' - \int\limits_{-\infty}^{\infty} zv \, dz \Big].$$

Uniqueness and existence results can be obtained on the basis of Lemma 1–3 for a small domain D and with input data sufficiently small in norm, see (Anikonov and Bubnov, 1988).

Remark. The method is general. From the method it is clear that the above facts can be generalized for other evolution eguations and other boundary value problems.

THE METHOD OF SPHERICAL HARMONICS AND INVERSE PROBLEMS KINETIC EQUATIONS

Let

$$\frac{1}{v}\frac{\partial F}{\partial t} + \Omega \operatorname{grad}_{\bar{x}} F + \sigma F = c\sigma \int\limits_{\Omega} f(\bar{x}, t, \Omega\Omega')F\,d\Omega' + \frac{1}{2}Q$$

be a transport equation and let

$$y_{lm}(\Omega) = \left[\frac{2l+1}{4\pi}\frac{(l-|m|)!}{(l+|m|)!}\right]^{1/2}(-1)^{(m+|m|)/2}P_l(\cos\theta)e^{im\varphi}$$

be spherical harmonics, $\bar{x} = (x, y, z)$. If

$$F = \sum_{l=0}^{\infty}\sum_{m=-l}^{l}\left[\frac{2l+1}{4\pi}\right]^{1/2} F'_{lm}(\bar{x}, t)y_{lm}(\Omega),$$

$$Q = \sum_{l=0}^{\infty}\sum_{m=-l}^{l}\left(\frac{2l+1}{4\pi}\right)^{1/2} Q_{lm}(\bar{x}, t)y_{lm}(\Omega),$$

$$f(\bar{x}, t, \Omega\Omega') = \sum_{l=0}^{\infty}\left[\frac{2l+1}{4\pi}\right]^{1/2} f_l(\bar{x}, t)\mathcal{P}_l(\Omega\Omega'),$$

then the following well known relations hold:

$$\frac{1}{v}\frac{\partial F_{lm}}{\partial t} + \frac{1}{2l+1}\Big[(l+1+m)^{1/2}(l+1-m)^{1/2}\frac{\partial F_{l+1,m}}{\partial z} + (l-m)^{\frac{1}{2}}(l+m)^{\frac{1}{2}}\frac{\partial F_{l-1,m}}{\partial z}$$

$$-\frac{1}{2}\Big(\frac{\partial}{\partial x}+i\frac{\partial}{\partial y}\Big)\Big[(l+m)^{1/2}(l+m-1)^{\frac{1}{2}}F_{l-1,m-1}-(l-m+2)^{\frac{1}{2}}(l-m+1)^{\frac{1}{2}}F_{l+1,m-1}\Big]$$

$$-\frac{1}{2}\Big(\frac{\partial}{\partial x}+i\frac{\partial}{\partial y}\Big)\Big\{-(l-m)^{1/2}(l-m-1)^{\frac{1}{2}}F_{l-1,m-1}+(l+m+1)^{\frac{1}{2}}(l+m+2)^{\frac{1}{2}}F_{l+1,m+1}\Big\}$$

$$= F_{lm}\sigma(cf_\alpha - 1) + \frac{1}{v}Q_{lm}, \qquad |m| \le l. \tag{9}$$

THEOREM 11. *If we drop in (9) terms with indices $l = |m|$, then the remaing system will be closed.*

This phenomenon depends only on $\Omega \operatorname{grad}_x F$. This is very important and is true for other kinetic equations.
It is used very often in inverse problems. In particular, terms $\sigma(cf_\alpha - 1), l = 1, 2, \cdots$, belonging to some classes can be found if $F_0 = F\,|_{z=0}$ (Anikonov, 1988).

REFERENCES

Anikonov, Ju.E. (1978). *On Some Methods of Investigation Multidimensional Inverse Problems for Differential Equations.* Nauka, Novosibirsk (in Russian).

Anikonov, Ju.E. and Amirov, A.Kh. (1983). A uniqueness theorem for the solution of an inverse problem for kinetic equation. *Dokl. Akad. Sci. SSSR* **272**, 510–511 (in Russian).

Anikonov, Ju.E. and Bubnov, B.A. (1988). Inverse problems of transport theory. *Dokl. Akad. Nauk. SSSR* **299**, 497–500 (in Russian).

Anikonov, Ju.E. (1986). *Psevdodifferential Operators and Inverse Problems.* Preprint No. 671. Computer Center Siberian Branch Akad. Sci. USSR, Novosibirsk (in Russian).

Anikonov, Ju.E. and Pestov, L.N. (1990). *Formulas in Line and Nonlinear Problems of Tomography.* Univ. Press, Novosibirsk (in Russian).

Anikonov, Ju.E. (1987). New mathematical models and inverse problems. In: *Mathematical Model in Geophysics.* Ed. by A.S. Alekseev. Computer Center Siberian Branch Akad. Sci. USSR, Novosibirsk, 3–42 (in Russian).

Anikonov, Ju.E. (1990). Theorem uniqueness solution inverse problem for quantum kinetic equation. *Soviet Math. Sbornik.* **181**, 68–75 (in Russian).

Anikonov, Ju.E. (1988). On a linear system of equations of the transfer theory, not containing sections and scattering indicatrix. *Dokl. Akad. Nauk. SSSR.* **305**, 319–322 (in Russian).

Ill-Posed Problems in Natural Sciences, pp. 254 – 269
A. Tikhonov (Ed.)
1992 VSP/TVP

AN INVERSE PROBLEM ORIGINATING FROM MAGNETOHYDRODYNAMICS. SOME NUMERICAL EXPERIMENTS

E. BERETTA, E. FISCHER, and M. VOGELIUS*
Istituto di Analisi Globale e Applicazioni, Via S. Marta 13/A, 50139 Firenze and Department of Mathematics, Rutgers University, New Brunswick, New Jersey 08903
Ecole Polytechnique, F-91128 Palaiseau Cedex, France
Department of Mathematics, Rutgers University, New Brunswick, New Jersey 08903

INTRODUCTION

The boundary value problem

$$\Delta u = -f(u) \leq 0 \ \text{ in } \ \Omega, \quad u = 0 \ \text{ on } \ \partial\Omega \tag{1}$$

represents a very simplified model for the equilibrium state of the plasma in a magnetohydrodynamic fusion device. The function u is a magnetic flux, and in this simplified model the typical toroidal plasma configuration has been replaced by a configuration in the form of an infinite cylinder. The domain $\Omega \subset \mathbf{R}^2$ is the cross section of the cylinder. In this presentation we shall discuss an important inverse problem that arises in connection with (1). Briefly described the problem is to determine f based on measurements of $\frac{\partial u}{\partial n}|_{\partial\Omega}$. It is fairly easy to see that without further assumptions this inverse problem may not always be solved: let Ω be the unit ball, and let u_d, $0 < d \leq 1/2$, denote the function

$$u_d = (1 - r^2) + d(1 - r^2)^2.$$

A simple calculation gives that

$$\Delta u_d = -f_d(u_d) = -4 - 8d + 16dr^2 \leq 0 \ \text{ in } \ \Omega,$$

$$u_d = 0 \ \text{ on } \ \partial\Omega.$$

The function f_d is given by the formula

---

*The first author is supported by Consiglio Nazionale delle Ricerche; the third author is partially supported by NSF Grant DMS-89-02532 and by AFOSR Contract 89-NM-605

$$f_d(t) = -4 - 8d + 8\sqrt{1+4dt}.$$

Concerning the normal derivative, we have

$$\frac{\partial u_d}{\partial n} = -2 \quad \text{on} \quad \partial\Omega;$$

it is therefore not possible to distinguish among the different functions f_d by means of the normal derivative of u_d at $\partial\Omega$.

This example is a special case of a more general phenomenon:

If $\Omega = B$ is a ball and $u_i \in C^2(\Omega) \cap C^0(\overline{\Omega})$, $i = 1,2$, are nontrivial classical solutions to

$$\Delta u_i = -f_i(u_i) \leq 0 \ \text{ in } \ B, \quad u_i = 0 \ \text{ on } \ \partial B,$$

then $\partial u_1/\partial n = \partial u_2/\partial n$ on ∂B if and only if $\int_B f_1(u_1) = \int_B f_2(u_2)$.

To see this we observe that, due to the Maximum principle, u_i is positive in B. From (Gidas et al., 1979) it now follows that u_i is a radial function (provided f_i is Lipschitz); the normal derivative, $\partial u_i/\partial n$, is therefore constant on ∂B. By Green's formula the constants corresponding to two different f's are equal if and only if the total forces are equal. The degeneracy discussed above stems from the fact that the data $\partial u/\partial n$ is constant on $\partial\Omega$. From the analysis in (Serrin, 1971) it follows that a ball is the only C^2 domain for which a positive solution to $\Delta u = -f(u)$ in Ω, $u = 0$ on $\partial\Omega$, can have a constant normal derivative on $\partial\Omega$. This discussion suggests a more restricted question: "under what conditions on Ω is it possible to identify a (sufficiently regular) force $f(u)$ from measurements of $\frac{\partial u}{\partial n}|_{\partial\Omega}$"? In light of the above discussion it might even seem reasonable to ask whether balls are the only simply connected domains where such identification is not possible.

The answers to these questions are currently quite unknown, even though some progress has recently been made for domains with corners. If the domain Ω is piecewise smooth and has (at least) one corner x_0 of interior angle $\pi/2 < \theta_0 < 2\pi$, $\theta_0 \neq \pi$, $\theta_0 \neq 3\pi/2$, if $\Gamma \subset \partial\Omega$ denotes an open curve with one endpoint at x_0, and if $u_i \in C^2(\Omega) \cap C^0(\overline{\Omega})$ are nontrivial classical solutions to $\Delta u_i = -f_i(u_i) \leq 0$ in Ω, $u_i = 0$ on $\partial\Omega$, where the f_i are infinitely often differentiable in a neighborhood of zero, then

$$\frac{\partial u_1}{\partial n} = \frac{\partial u_2}{\partial n} \ \text{ on } \Gamma \ \text{ implies that } \ f_1^{(k)}(0) = f_2^{(k)}(0) \ \text{ for any integer } k \geq 0. \quad (2)$$

This result is proven in (Beretta and Vogelius, 1991 a), the proof relies on a detailed analysis of the asymptotic behaviour of the u_i at the corner x_0. A quite similar result has also been obtained for the physically much more relevant Grad-Shafranov operator, cf. (Beretta and Vogelius, 1991 b). In the latter case one must also exclude angles θ_0 that are multiples of $\pi/3$ and it is necessary to have measurements on both sides of the corner x_0; the proof is considerably more technical than the proof for the case of the Laplacian.

To give a very brief account of where singularities enter into the proof of a result like (2) and to see why smooth domains are more difficult than domains with true corners, let us discuss the formal "linearization" of this inverse problem (with data on all of $\partial\Omega$). Let u_0 denote a solution to $\Delta u_0 = -f_0(u_0) \le 0$ in Ω, $u_0 = 0$ on $\partial\Omega$. After formal "linearization" about $f = f_0$ (and $u = u_0$) the inverse problem is to determine δf of the equation

$$\Delta\delta u = -f_0'(u_0)\delta u - \delta f(u_0) \text{ in } \Omega, \quad \delta u = 0 \text{ on } \partial\Omega,$$

from knowledge of $\frac{\partial\delta u}{\partial n}|_{\partial\Omega}$. Since this problem is linear, the uniqueness question boils down to does $\frac{\partial\delta u}{\partial n}|_{\partial\Omega} = 0$ imply that $\delta f(u_0) = 0$? For simplicity let us consider the case $f_0 = 1$, so that δu solves

$$\Delta\delta u = -\delta f(u_0) \text{ in } \Omega, \quad \delta u = 0 \text{ on } \partial\Omega, \tag{3}$$

where u_0 is the solution to $\Delta u_0 = -1$ in Ω, $u_0 = 0$ on $\partial\Omega$. By integration by parts, it follows immeditaly that the condition

$$\frac{\partial\delta u}{\partial n}|_{\partial\Omega} = 0$$

is equivalent to

$$\int_\Omega \delta f(u_0)\phi \, dx = 0$$

for any $\phi \in C^2(\Omega)\cap C^0(\overline{\Omega})$, with $\Delta\phi = 0$ in Ω. The linearized uniqueness question may therefore be restated does $\int_\Omega \delta f(u_0)\phi\, dx = 0$ for any harmonic ϕ imply that $\delta f(u_0) = 0$? The answer to this question is yes, if δf is real analytic and Ω has a true corner, whose angle is larger than $\pi/2$ and is not a multiple of $\pi/2$. One may simply take different harmonic functions ϕ to prove that successive derivatives of δf vanishes at zero. We illustrate how to pick ϕ in order to show that $\delta f(0) = 0$. Assume that the corner x_0 is at the origin and assume that the domain Ω locally consists of the conical sector $\Omega_\rho = \Omega\cap B_\rho = \{(r,\theta)\colon 0 < \theta < \theta_0, 0 < r < \rho\}$. We identify $\mathbf{R}^2$ with the complex numbers and take ϕ_ε to be the harmonic function

$$\phi_\varepsilon = \Re\left(\frac{1}{(z+\varepsilon e^{i\theta_0/2})^2}\right).$$

It is not difficult to see that

$$\int_\Omega \delta f(u_0)\phi_\varepsilon\, dx = \delta f(0)\int_{\Omega_\rho} \phi_\varepsilon\, dx + O(1),$$

where $O(1)$ signifies a term that stays bounded as ε approaches zero. A simple calculation shows that if $\theta_0 \ne \pi$ and $\delta f(0) \ne 0$, then the term

$$\delta f(0)\int_{\Omega_\rho} \phi_\varepsilon\, dx \tag{4}$$

becomes unbounded (logarithmically) as ε approaches zero. If we apriori assume that $\theta_0 \ne \pi$, then the only way that the identity

$$\int_\Omega \delta f(u_0)\phi_\epsilon \, dx = 0 \tag{5}$$

can hold is therefore if $\delta f(0) = 0$. On the other hand if $\theta_0 = \pi$ (the domain Ω is smooth at x_0) then the term (4) stays bounded even if $\delta f(0) \neq 0$, and we may therefore *not* use the identity (5) to conclude that $\delta f(0) = 0$. As the example at the beginning of this section indicates it is not necessarily the case that $\delta f(0) = 0$ (that example shows that we may not determine $f(0)$ from measurements of $\frac{\partial u}{\partial n}|_{\partial\Omega}$).
At this point it may be appropriate to note that there appears to be (at least a superficial) similarity between the "linearized" question and the so called Schiffer conjecture, which states that if Ω is a simply connected C^2 domain, and not a ball, then the only solution to the overdetermined eigenvalue problem

$$\begin{aligned} \Delta v &= -\lambda v + \text{const} \quad \text{in } \Omega, \\ v &= \frac{\partial v}{\partial n} = 0 \quad \text{on } \partial\Omega, \end{aligned} \tag{6}$$

is $v = 0$ (in which case const $= 0$) (cf. (Berenstein, 1980; Garofalo and Segala, 1983)). In general, the fact that 0 represents the only solution to (6) for any λ, is known to be equivalent to the fact that the domain Ω has the so called Pompeiu property, cf. (Williams, 1976). Whereas we shall not here go into the exact definition of this property, it is striking to note that one of the first classes of domains that were proven to possess the Pompeiu property were polygonal domains, and convex domains with at least one corner (Brown et al., 1973). It was fairly early on recognized that balls do not possess the Pompeiu property.
The purpose of this paper is to describe some numerical work we have carried out, partially to illustrate the feasibility of the reconstruction of f in such cases where we have already verified the theoretical possibility of this, and partially to test whether the limitations of our theoretical results appear to be reflected in practice. We shall in particular address ourselves to the following questions:

(Q1) for a polygon with at least one angle $\pi/2 < \theta_0$, $\theta_0 \neq k\pi/2$ is it feasible to effectively reconstruct f?

(Q2) does it appear that the condition $\theta_0 \neq k\pi/2$ is essential for the ability to reconstruct f?

(Q3) does it appear that one may reconstruct truely interior features of f, and not only $f^{(k)}(0)$, say in a case where f is not analytic on the entire range of u?

(Q4) what is the situation in case of a smooth domain whose boundary is not circular, e.g., a proper ellipse?

We now briefly turn to a description of the numerical algorithm by means of which we shall try to address the questions above. We do want to mention that while we have tried to develop a reasonably efficient algorithm — at least as a partial answer to **(Q1)** — an equally important goal has been to address the questions **(Q2)**–**(Q4)**. We feel that much needs to be done in terms of the development of more novel and more efficient algorithms. This is also the case for the Grad-Shafranov operator, where some encouraging preliminary results have recently been obtained by (Blum, 1989).

THE ALGORITHM

Given h, let v denote a classical solution to

$$\begin{aligned} \Delta v &= -h(v) \le 0 \quad \text{in} \quad \Omega, \\ v &= 0 \quad \text{on} \quad \partial\Omega, \end{aligned} \tag{7}$$

if such a solution exists. Our numerical algorithm is based on the minimization of the functional

$$\int_{\partial\Omega} \left| \text{data} - \frac{\partial v}{\partial n} \right|^2 dx \tag{8}$$

over an appropriate set of $h's$, where v is a solution to (7), and where data represents the measured values of the normal derivative. The facts that the problem (7) may have no solution or may have more than one solution pose problems, which we shall only deal with in a very simplistic manner here. We consider $h's$ that are in some finite dimensional vector space generated by the basis $\{\phi_i\}_1^N$. The coordinates of h with respect to this basis are denoted x_i. The derivative of (8) with respect to x_i is now formally given by

$$y^i = -2 \int_{\partial\Omega} \left(\text{data} - \frac{\partial v}{\partial n} \right) \frac{\partial w^i}{\partial n} \, dx,$$

where v is a solution to (7), and $w^i = L\phi_i$ is the solution to the linear boundary value problem

$$\begin{aligned} \Delta w^i &= -h'(v) w^i - \phi_i(v) \quad \text{in} \quad \Omega, \\ w^i &= 0 \quad \text{on} \quad \partial\Omega, \end{aligned} \tag{9}$$

(assuming this solution exists and is unique). We shall denote the function $g = \sum_{i=1}^N y^i \phi_i$ the "gradient" of (8) with respect to h. Our algorithm uses a conjugate direction approach to minimization. A rough outline of the algorithm is as follows

(i) pick an initial guess h_0, set $p_0 = 0$,

(ii) for $k = 1, 2, \ldots, K$,

(ii.a) compute a solution to (7), v_{k-1}, corresponding to $h = h_{k-1}$,

(ii.b) compute the "gradient" of (8) with respect to h at h_{k-1} (and v_{k-1}), denote this gradient g_{k-1}.

(ii.c) introduce a new direction $p_k = -g_{k-1} + \alpha_k p_{k-1}$,

(ii.d) minimize (8) along the line $h_{k-1} + s p_k$,

(ii.e) define $h_k = h_{k-1} + s_{\min} p_k$,

(iii) solution $= h_K$.

The selection of the parameter α_k is done somewhat in the spirit of the "conjugate residual" method, cf. (Eisenstat et. al., 1983). We select α_k so that the new direction p_k and the old direction p_{k-1} are orthogonal with respect to the (semi) innerproduct

$$(\cdot,\cdot) = \int_{\partial\Omega} \frac{\partial}{\partial n} L_{k-1} \frac{\partial}{\partial n} L_{k-1}\, dx,$$

where L_{k-1} is the linear operator defined by (9), corresponding to $h = h_{k-1}$ and $v = v_{k-1}$. For $k \geq 2$ this results in the formula

$$\alpha_k = \frac{\int_{\partial\Omega} \frac{\partial}{\partial n} L_{k-1}(g_{k-1}) \frac{\partial}{\partial n} L_{k-1}(p_{k-1})\, dx}{\int_{\partial\Omega} (\frac{\partial}{\partial n} L_{k-1}(p_{k-1}))^2\, dx}, \tag{10}$$

(provided the denominator is nonzero). We take $\alpha_1 = 0$, so that the first step is a steepest descent step. Since the functions $\frac{\partial}{\partial n} L_{k-1}(\phi_i)$ have already been computed for $i = 1, \ldots, N$, in connection with the calculation of the "gradient", the calculation of (10) only requires a simple quadrature. The minimization in step **(ii.d)** is performed by means of a backtracking line search algorithm, very similar to that described in (Dennis and Schnabel, 1983). In connection with the line search one difficulty arises when the argument s accidentally becomes so large that the problem (7) (corresponding to $h = h_{k-1} + sp_k$) fails to have a solution. We deal with this problem by imposing an ad hoc bound on s if necessary. In our calculations we select h to be in the space of polynomials below a certain degree, for basis functions, ϕ_i, we select the Chebychev polynomials rescaled on the range of v_{k-1}. To be quite specific: we select

$$\phi_i(t) = T_i\left(\frac{2}{r_{k-1}}t - 1\right),$$

where $r_{k-1} = \max v_{k-1}$ and T_i is the Chebychev polynomial of degree i (relative to the interval [-1,1]). The Chebychev polynomials are selected because of their well known connection to min-max approximation. For the solution of the boundary value problems involved in **(ii.a)** and **(ii.b)** we use either a finite difference or a finite element approach. The finite difference approach is based on a uniform grid and the standard five point stencil; it employs the nonlinear solver SNSQE from the library CMLIB to solve the resulting nonlinear algebraic equations. For finite element computations we use the package PLTMG, including its adaptive mesh refinement procedure, (Bank, 1990). In the case of smooth domains (ellipses) we have only implemented a version of our algorithm which is based on finite elements. We have experimented with two kinds of domains: parallelograms and ellipses. In all cases we have by means of symmetries reduced the boundary value problems to problems on a domain a quarter the size of the original domain. When we compute a value of the least square functional it is only done corresponding to a quarter of the boundary. Figure 1.a shows a parallelogram with angle $\theta_0 = 3\pi/4$, Fig. 1.b shows a corresponding (coarse, 7×7) finite difference mesh (using 28 nodes). Figure 2 shows a typical (refined) finite element mesh generated by PLTMG (using 94 nodes); this mesh was obtained by setting a target value of 100 vertices. In the case of a parallelogram (where the normal derivative has singularities at the corners) we have slightly amended the cost functional (8), in that we do not include the contribution corresponding to the quadrature "intervals" immediately adjacent to the corners.

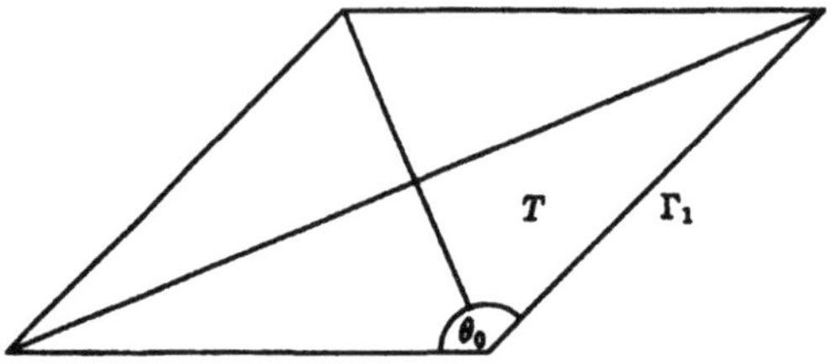

Fig. 1a

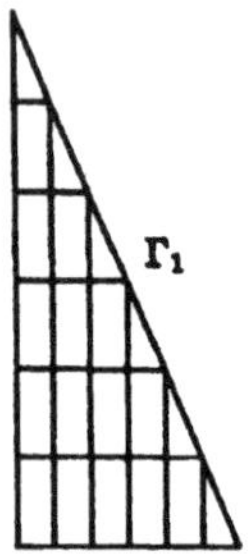

Fig. 1b

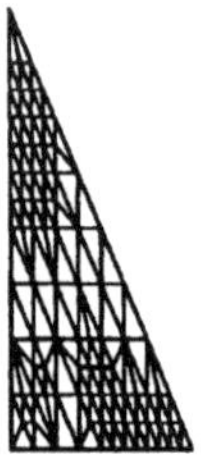

Fig. 2

In the computations reported in this paper we shall always work with functions h that have $h(0)$ bounded away from 0 (and similarly we shall take data $\frac{\partial u}{\partial n}|_{\partial\Omega}$ which comes from a function f with $f(0) > 0$). As mentioned before, we run into difficulties if the solution to (7) is not unique. As it is currently set up our software for the solution of (7) attempts to approximate the solution which is "smallest" in an appropriate sense. If $f(0) = 0$, then $h(0)$ will want to become very small, as a consequence v becomes nearly zero and our algorithm goes nowhere. Even when $f(0) > 0$ a similar difficulty arises if the given data corresponds to a solution other than the "smallest".

The least squares approach we use is one that tries to fit the normal derivatives on the boundary, maintaining homogeneous Dirichlet data. The computation of the normal derivative introduces a significant amount of numerical error (unless the solution is very accurate). It might seem much more natural to work with a least squares formulation that seeks to minimize the L^2-norm of the boundary values, maintaining the given Neumann data. Let v denote a solution to $\Delta v = -h(v)$ in Ω, $\frac{\partial v}{\partial n}|_{\partial\Omega}$ =data. By Green's formula we must now have

$$\int_{\partial\Omega} \text{data}\, dx = -\int_{\Omega} h(v)\, dx.$$

This implicit "constraint" seems to result in many difficulties for our iterative approach, most often the Neumann problem fails to have a solution and we are at a loss to continue. We have not seriously studied possible ways to overcome this problem.
Last but not least the entire "output least squares" idea upon which the algorithm is based in only very minor ways make use of the special structure of the underlying PDE.
We feel it would be very important to try to correct these shortcommings of our current numerical approach. The computations we are about to present nonetheless give important indications about the answers to the questions **(Q1)–(Q4)**, as stated previously.
It is crucial for the success of the algorithm presented here to have a reasonable initial guess. In all the computations we have performed so far a constant value which lies in the range of $f(u)$ has worked well. One possible way to produce such a guess is to take a value near the constant

$$-\int_{\partial\Omega} \text{data } dx/|\Omega|. \tag{11}$$

COMPUTATIONAL RESULTS

In the first example we consider the domain is a parallelogram with an angle $\theta_0 = 3\pi/4$ and sidelength=1 (cf. Fig. 1.a). The "true" function f from which the data is generated is a third order polynomial given by the formula

$$f(t) = 20 + 9t - 24t^2 + 16t^3.$$

The dashed curve in Fig. 3.a is the graph of f. The solid curve in Fig. 3.a represents the result after 25 iterations of our algorithm using polynomials of degree ≤ 2. The initial guess is taken to be the constant 20.5. The least-squares functional (8) has achieved a value of 1.47×10^{-5}. The interval over which these curves are plotted corresponds to the range of the solution v_K of the last step of our algorithm. Figure 3.b shows the error between f and h_K after an additional 75 iterations using polynomials of degree ≤ 3. The initial guess in this case is taken to be the quadratic polynomial displayed in Fig. 3.a. The least squares functional has now reached a value of 9.59×10^{-12}. The computations corresponding to Fig. 3.a-b were based on finite element solution of the boundary value problems — in all cases we prescribed a target value of 100 vertices in the (adaptively) generated meshes. The data was also generated with a mesh of approximately 100 vertices. We believe that the results in Fig. 3.a-b qualitatively represent the behaviour of the continuous inverse problem. To substantiate this claim we show in Fig. 4.a-b similar computations carried out with a target value of 200 vertices and with data that was also generated with approximately 200 vertices. It is evident that the results are qualitatively very similar. We are not claiming that a solution corresponding to 100 vertices is so close to a solution of the PDE that it does not matter whether we use that data or say data generated based on approximately 200 vertices for our inversion. The noise level does negatively influence the reconstruction in a significant way as illustrated in Fig. 5. In that figure we show the result of 24 iterations with our algorithm using polynomials of degree ≤ 3 and finite element meshes of approximately 100 vertices, but the data was generated by a finite element mesh of approximately 200 vertices. The value of the least squares functional (8) does in this case not want to go below 4.14×10^{-2} (the data is inconsistent with the discrete models based on 100 vertices; the size of the least squares functional is consistent with the fact that the (square of the) L^2 error between the data generated by 200 vertices and the data generated by 100 vertices is of the order 6×10^{-2}). Finally in Fig. 6 we show the error between f and h_K after 44 iterations of our algorithm using polynomials of degree ≤ 3, but this time we have based the solution of the boundary value problems on finite differences on a (15×15) uniform grid with 120 vertices. The data was generated using finite differences and the same grid. The least squares functional has achieved a value of 4.14×10^{-11}. The result is qualitatively similar to those in Fig. 3.b and 4.b.

The results shown in Fig. 3–6 indicate the feasibility of practical reconstruction of the function f in a case where it has theoretically been established that it should be possible to reconstruct a real analytic nonlinearity. Not surprisingly the error in reconstruction is largest at at the top of the range of v_K. Furthermore, Fig. 3.a and Fig. 4.a give a very strong indication that the data contains significant information about the interior behaviour of f as opposed to only the boundary information about $f^{(k)}(0)$. This would seem to indicate some hope for the reconstruction of functions that are not necessarily real analytic.

As mentioned earlier, our theorem concerning the identifiability of $f^{(k)}(0)$ makes certain assumptions about the angle θ_0. In the case of parallelograms the only domains which are excluded are rectangles (corresponding to $\theta_0 = \pi/2$). To see whether this exclusion is an artifact of our proof or it is indeed necessary in practice, we perform an experiment similar to the previous, but with Ω equal to the unit square. The true profile is

$$f(t) = 11 + 9t - 24t^2 + 16t^3.$$

We use the finite difference version of our program with a uniform grid of 120 vertices. The data was generated using finite difference and the same grid. The initial guess is the constant 11.5. Figure 7.a shows the true profile (dashed curve) and the reconstructed curve (solid curve) after 33 iterations. The reconstruction is based on using polynomials of degree ≤ 3. The least squares error has reached a value of 9.83×10^{-9}. Figure 7.b illustrates the same experiment using the finite element version of our program with about 100 vertices (and using data that was generated by a similar finite element approximation). Based on these and many other computations we feel pretty confident that rectangles are not different from other parallelograms as far as uniqueness is concerned – it does however appear that the inverse problem is less well posed (at any fixed value of the least squares functional the reconstruction is worse than for a parallelogram with angles $\neq \pi/2$).

So far we have deliberately picked "true" profiles that are smooth but with $f(0)$ so large that the maximum of u is of the order of magnitude 1. This was done to improve the sensitivity of the data to changes in f and at the same time reconstruct f over an interval where it is not approximately linear. We now provide an example with a "true" solution that is orders of magnitude smaller than one. For the domain we take the parallelogram with angle $\theta_0 = 2\pi/3$ and sidelength=1. The true profile is

$$f(t) = 1 + t + t^2 + t^3.$$

For the reconstruction we use polynomials of degree ≤ 3. Figure 8.a shows the error in our reconstruction after 111 iterations with the finite element version of our program. The reconstructed polynomial at that point is $h_K(t) = 1.00008 + .980561t + 1.95897t^2 - 10.8971t^3$, and the least squares functional is 4.26×10^{-13}. Figure 8.b shows the reconstruction error after 70 iterations of the finite difference version of our program. The reconstructed polynomial is $h_K(t) = .999992 + 1.00459t + .681276t^2 + 5.63375t^3$, and the least squares functional is of order 10^{-13}. The reconstruction is quite satisfactory but it does of course only pertain to an interval where f is already extremely well approximated by $1+t$ (in particular when compared to the accuracy of our finite element or finite difference approximations). To illustrate the point we have made earlier about multiple solutions we consider the positive solutions to

$$\Delta u_\lambda = -\lambda(1 + u_\lambda + u_\lambda^2 + u_\lambda^3) \quad \text{in } \Omega, \quad u_\lambda = 0 \quad \text{on } \partial\Omega,$$

for $\lambda \geq 0$. The domain Ω is the parallelogram with angle $\theta_0 = 2\pi/3$ (sidelength=1), and we use PLTMG with a fixed uniform mesh of 153 vertices to compute approximations to u_λ (the continuation feature of PLTMG requires a fixed mesh). To provide an idea of the set of solutions, Fig. 9.a shows the curve representing approximate points $(\lambda, \int u_\lambda^2 \, dx)$. The integral in the second coordinate is an integral over a quarter of Ω, because of our use of symmetries. Corresponding to any value $0 < \lambda < \lambda^* \approx 6.44$ there appear to be two solutions. The reconstructions performed in Fig. 8.a-b were done using data corresponding to the solution with the smallest value of $\int u^2 \, dx$. The solution corresponding to the largest value of $\int u^2 \, dx$ is clearly more sensitive to changes in f – it might be interesting to investigate whether one could benefit from this extra sensitivity when it comes to reconstruction, even though there might of course be serious questions about the physical availability of the data corresponding to this unstable solution. Our algorithm is currently unable to perform any reconstruction based on the data corresponding to the "largest" solution. Figure 9.b compares the boundary data corresponding to the "smallest" solution with the data corresponding to the "largest" solution (at $\lambda = 1$). Both data sets were computed using finite elements, the graphs shown depict $-\partial u/\partial n$ on a logarithmic scale. The normal derivatives were evaluated at 100 equidistant points, starting a distance of $1/101$ from the corner corresponding to angle $\pi/3$ and ending a distance of $1/101$ from the corner corresponding to angle $2\pi/3$. The logarithmic scale makes the data look unnecessarily constant, we chose this scale such as to be able to display both graphs in the same frame. The data from the "largest" solution is shown in dashed lines, the data from the "smallest" solution is shown in solid lines. It is evident that the two data sets differ significantly in magnitude. One should note that there is an increase in the error of the computed normal derivatives as one approaches the corners. It is this error which is the reason why we do not include the quadrature intervals nearest to the corners when we compute the least squares cost functional.

A natural question concerns the minimal regularity of f required for identifiability. In Fig. 10 and Fig. 11 we show the results of two experiments. In the first experiment f is Lipschitz but not C^1, in the second experiment f is Hölder continuous (exponent 1/2), but not Lipschitz. The computations are based on the finite element version of our algorithm using meshes with approximately 100 vertices. The domain Ω is the parallelogram with angle $\theta_0 = 3\pi/4$ and sidelength=1. The least squares functional in the Lipschitz case is about 2×10^{-6} the least squares functional in the Hölder case is about 8×10^{-5}. In the first case the reconstruction took 76 iterations and it utilized polynomials of degree ≤ 2 (there was no apparent improvement by admitting polynomials of degree 3). In the second case the least squares functional did not decrease after 24 iterations (9 iterations with polynomials of degree ≤ 2, 15 iterations with polynomials of degree ≤ 3). It is quite clear that there is fairly good reconstruction up to t=0.1, the value corresponding to the singularity of f. Beyond the singularity the reconstruction deteriorates. If anything the deterioration is worse for the Hölder continuous case – we suspect this reflects a lack of identifiability. In the case of a Lipschitz function the deterioration is less severe, but it is still strong enough that it must raise questions about identifiability. We have performed similar experiments corresponding to functions that are C^1 but not C^2; in some of those experiments the reconstructions were significantly better, even beyond the point of singularity.
Finally we come to the problem concerning reconstruction on (simply connected) smooth domains other than balls. Figure 12 shows the result of a reconstruction of the function

$$f(t) = 1.8 + 9t - 24t^2 + 16t^3$$

on an ellipse of aspect ratio 1:2. The solid line shows the results after 58 iterations of our algorithm based on finite element meshes of approximately 100 vertices. The least squares functional has reached a value of 3.36×10^{-7}. As a comparison Fig. 13 shows how a similar attempt at reconstruction completely fails on the unit ball. The least squares functional is in this case equally small (of order 10^{-6}), but there is clearly no idenfiability. The value of $\int h_K(v_K)\, dx$ is 3.05 as compared to a value for $\int f(u)\, dx$ of 3.03 (the integrals are over a quarter of the circle). These values approximately differ by an amount comparable to the square root of the least squares functional, as one would expect. Based on the contrast between these two experiments (and based on other similar experiments) we are hopeful that it may be possible to identify a sufficiently regular f even for certain types of smooth domains.

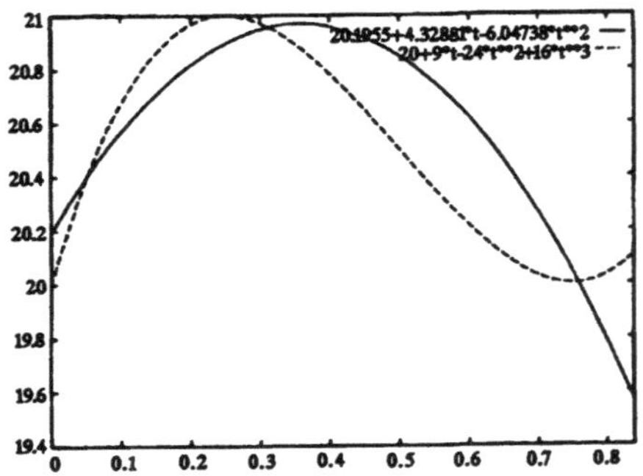

Fig. 3a

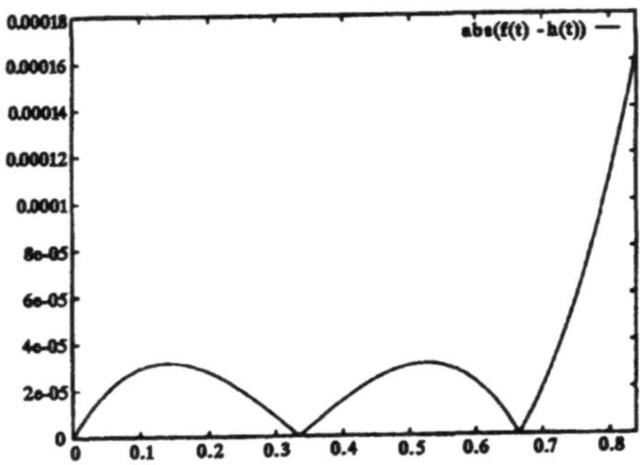

Fig. 3b

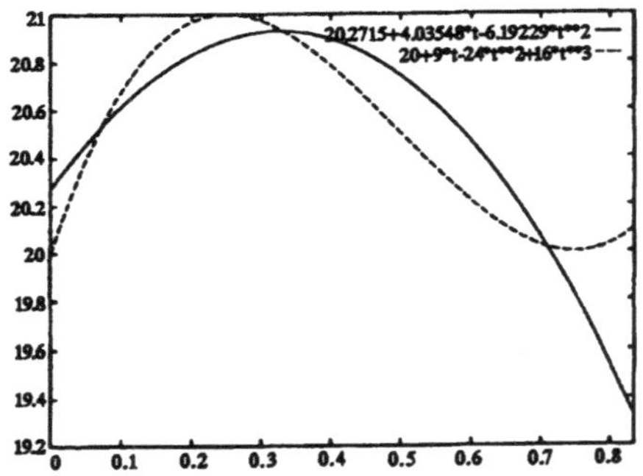

Fig. 4a

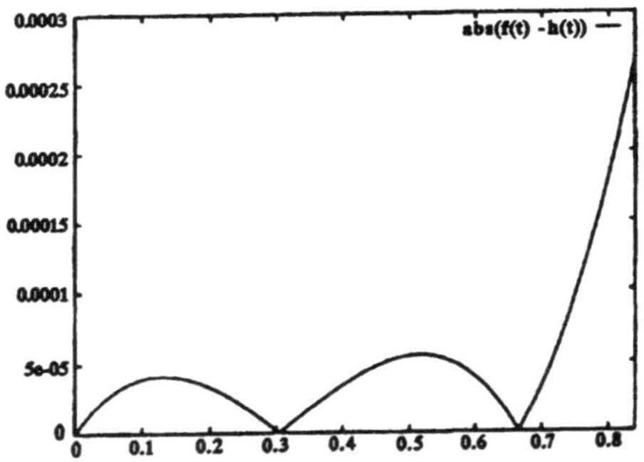

Fig. 4b

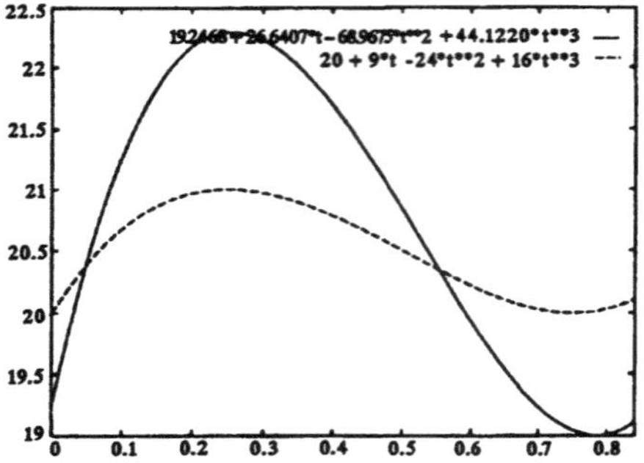

Fig. 5

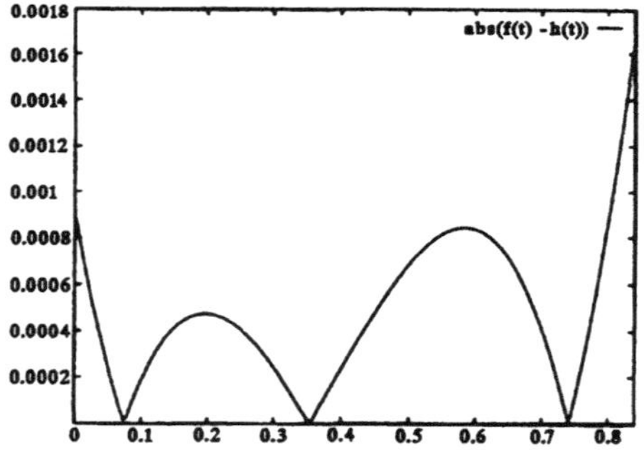

Fig. 6

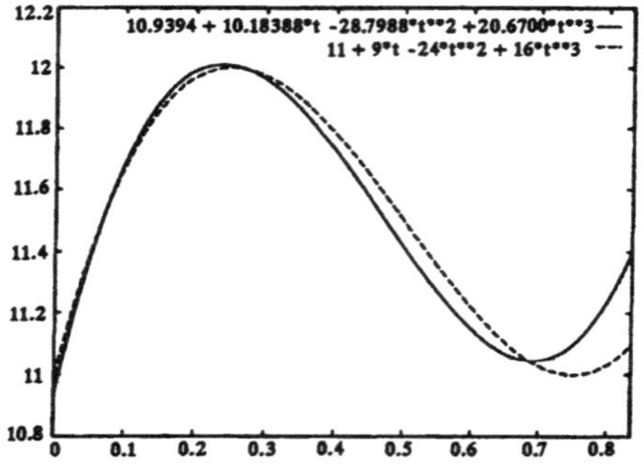

Fig. 7a

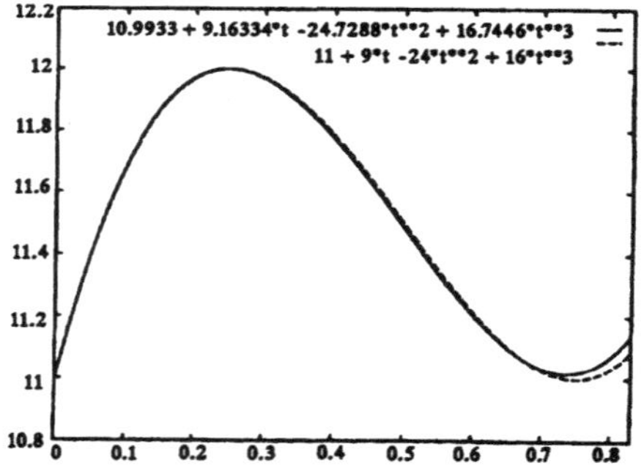

Fig. 7b

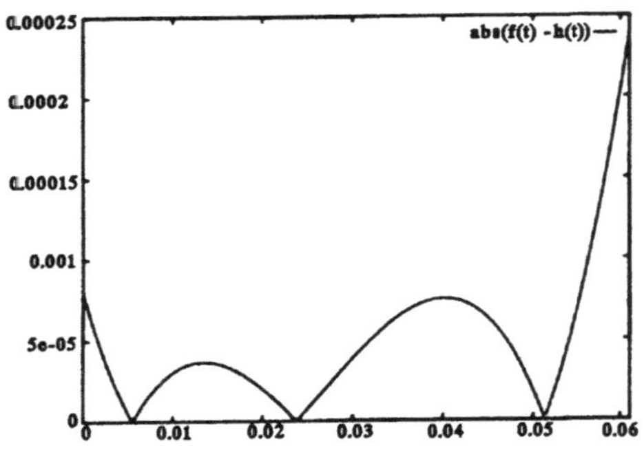

Fig. 8a

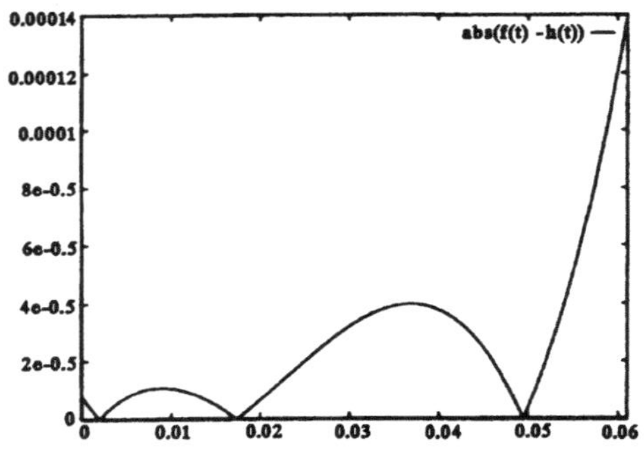

Fig. 8b

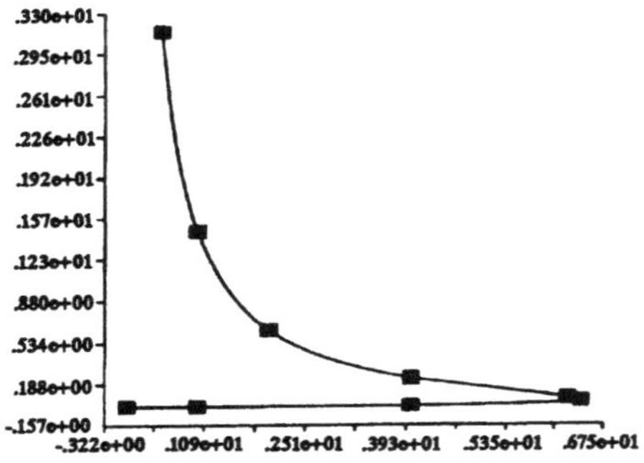

Fig. 9a

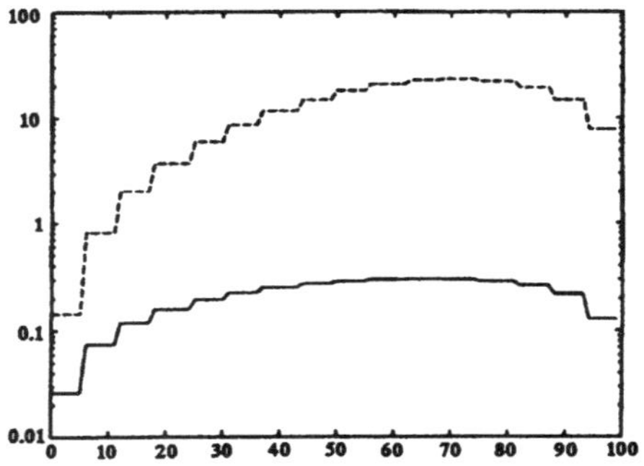

Fig. 9b

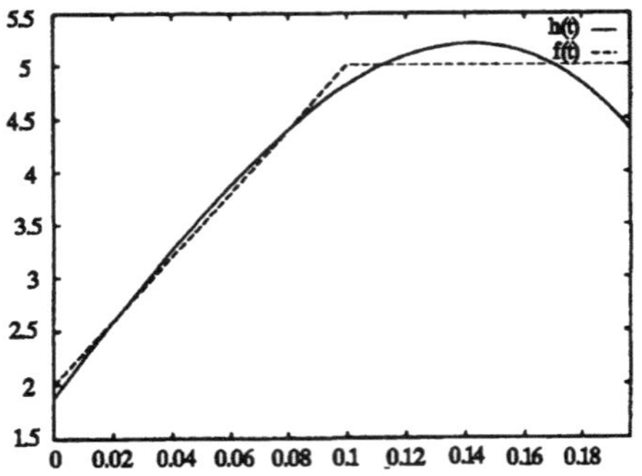

Fig. 10

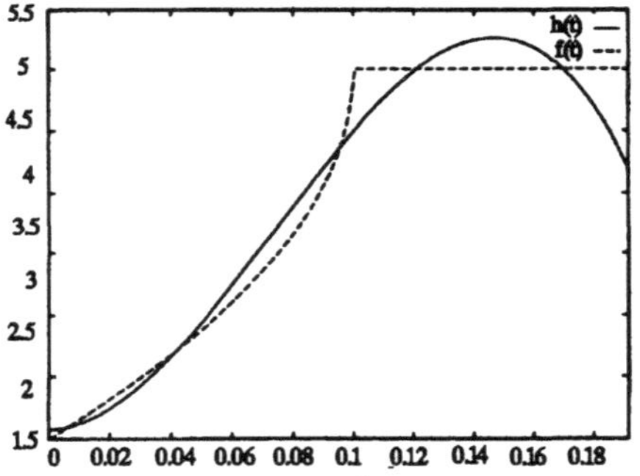

Fig. 11

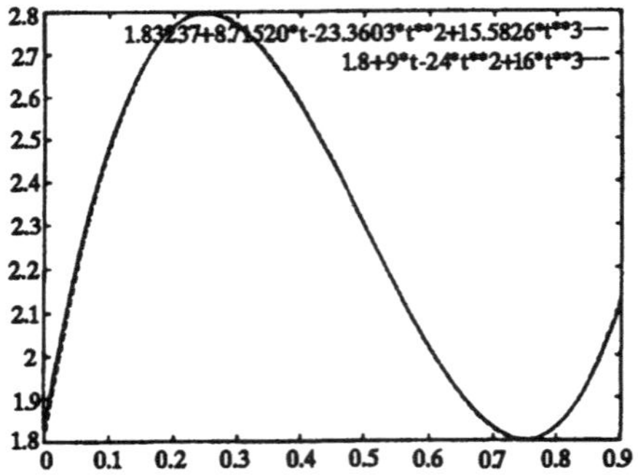

Fig. 12

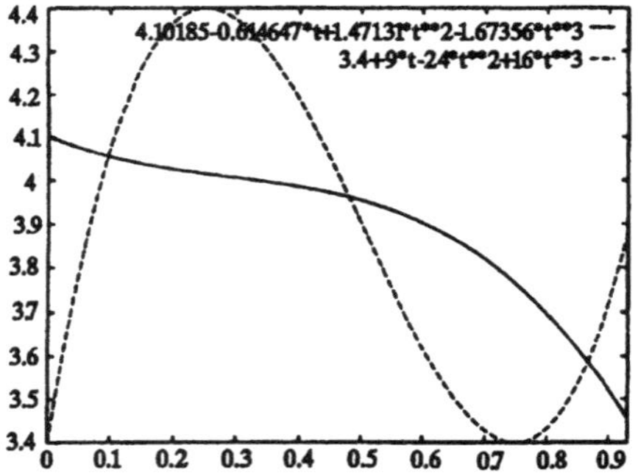

Fig. 13

REFERENCES

Bank, R.E. (1990). *PLTMG: A Software Package for Solving Elliptic Partial Differential Equations.* SIAM, Philadelphia.

Berenstein, C.A. (1980). An inverse spectral problem and its relation to the Pompeiu problem. *J. Anal. Math.* **37**, 128–144.

Beretta, E. and Vogelius, M. (1991 a). An inverse problem originating from magnetohydrodynamics. *Arch. Rat. Mech. Anal.* (to appear).

Beretta, E. and Vogelius, M. (1991 b). An inverse problem originating from magnetohydrodynamics II. The case of the Grad-Shafranov equation. *Indiana Univ. Math. J.* (to appear).

Blum, J. (1989). *Numerical Simulation and Optimal Control in Plasma Physics.* Gauthier Villars, Paris, Wiley, New York.

Brown, L., Schreiber, B.M., and Taylor, B.A. (1973). Spectral synthesis and the Pompeiu problem. *Ann. Inst. Fourier* **23**, 125–154.

Dennis, J. and Schnabel, R. (1983). *Numerical Methods for Unconstrained Optimization and Nonlinear Equations.* Prentice-Hall, Englewood Cliffs , N.J.

Eisenstat, S., Elman, H., and Schultz, M. (1983). Variational iterative methods for nonsymmetric systems of linear equations. *SIAM J. Numer. Anal.* **20**, 345–357.

Garofalo, N. and Segala, F. (1983). New results on the Pompeiu problem. *Trans. Amer. Math. Soc.* **325**, 273–286.

Gidas, B., Ni, W.-M., and Nirenberg, L. (1979). Symmetry and related properties via the maximum principle. *Comm. Math. Phys.* **68**, 209–243.

Serrin, J. (1971). A symmetry problem in potential theory. *Arch. Rat. Mech. Anal.* **43**, 304–318.

Williams, S.A. (1976). A partial solution of the Pompeiu problem. *Math. Ann.* **223**, 183–190.

Ill-Posed Problems in Natural Sciences, pp. 270 – 276
A. Tikhonov (Ed.)
1992 VSP/TVP

INVERSE CONDUCTIVITY PROBLEM IN THE TWO-DIMENSIONAL CASE

V.G. CHEREDNICHENKO and G.V. VERYOVKINA
Novosibirsk Institute of Soviet Cooperative Trade, K. Marx str., 26, 630087 Novosibirsk, Russia

ABSTRACT

A method for a general class of inverse problems with unknown boundaries is presented. These problems are met in conductivity, electrocardiology and geophysics. The local uniqueness of the solution of the inverse problem is proved using techniques of complex analysis.

1. In this paper, we give statements of several inverse problems and show the connection between them. Let us begin with a particular problem.
Let T^+, D^+, $\overline{D^+} \subset T^+$ be bounded simple connected domains on the complex plane $z = x + iy$ with smooth boundaries L, S respectively.

Direct Problem 1. *Find $v(x, y)$ in T^+ such that*

$$\operatorname{div}[1 + \chi(D^+)\nabla v] = 0 \quad \text{in } T^+, \tag{1}$$

$$(v)_n = g_1 \quad (v = h) \quad \text{on } L. \tag{2}$$

Here $\chi(D^+)$ is the indicator function of D^+; the function g_1, h are defined on L; $(\cdot)_n$ denotes the differentiation in the direction of the exterior normal to the boundary.

This is the refraction problem or the Neumann (Dirichlet) problem for elliptic equation with discontinuous coefficient.

Inverse Problem 1. *Find D^+, $\overline{D^+} \subset T^+$ and v in T^+ such that they satisfy* (1), (2) *and*

$$v = h \quad \big((v)_n = g_1\big) \quad \text{on } L. \tag{3}$$

This problem was studied in (Bellout and Friedman, 1988; Alessandrini, 1989; Friedman and Isakov, 1989; Isakov and Powell, 1990).
If we denote

$$v = \begin{cases} v^+ & \text{in } D^+, \\ v^- & \text{in } T^+ \setminus \overline{D^+}, \end{cases}$$

then the solution of the Direct Problem 1 is equivalent (Ladyzenskaya and Uralzeva, 1968) to finding the function v^+, v^- such that

$$\Delta v^+ = 0 \text{ in } D^+, \quad \Delta v^- = 0 \text{ in } T^+ \backslash \overline{D^+}, \tag{4}$$

$$\begin{cases} v^+ - v^- = 0 \\ (v^+)_n - \frac{1}{2}(v^-)_n = 0 \end{cases} \quad \text{on } S, \tag{5}$$

$$(v^-)_n = g_1 \;\; (v^- = h) \quad \text{on } L. \tag{6}$$

2. Instead of (5) we shall consider more common conditions:

$$\begin{cases} a_{11}v^+ + a_{12}v^- = b_1 \\ a_{21}(v^+)_n + a_{22}(v^-)_n = b_2 \end{cases} \quad \text{on } S. \tag{5'}$$

Here a_{kl} are constants, b_1, b_2 are given functions on S such that $b_2 = (B)_S$, $(\cdot)_S$ denotes differentiation in the tangent direction to S, B is a given function.

DIRECT PROBLEM 2. *Find functions v^+, v^- in D^+, $T^+ \backslash \overline{D^+}$ respectively, for which* (4), (5′), (6) *are satisfied.*

Relation (5′) is called the condition of conjugation of the limiting values of function and their derivatives which are given in the domains on different sides of their common boundary S.

INVERSE PROBLEM 2. *Find D^+ and functions v^+, v^- in D^+, $T^+ \backslash \overline{D^+}$ respectively satisfying conditions* (4), (5′), (6) *and*

$$v^- = h \;\; \big((v^-)_n = g_1\big) \quad \text{on } L. \tag{7}$$

3. In the two-dimensional case it is convenient to study Problem 2 using the theory of functions of complex variables. Using the harmonic functions v^+, v^- in $D^+, T^+ \backslash \overline{D^+}$ respectively, we shall construct the functions which will be analytic in the same domains

$$w_0^+(z) = v^+ + iv_c^+, \quad w^-(z) = v^- + iv_c^-, \tag{8}$$

where v_c^+, v_c^- are the harmonic conjugated functions. Let us denote also

$$\begin{gathered} f = -(2b_1a_{21} + i2Ba_{11})/d, \quad \beta = -(a_{11}a_{22} - a_{12}a_{21})/d, \\ d = a_{11}a_{22} + a_{12}a_{21}, \quad w^+ = -2a_{11}a_{21}w_0^+/d. \end{gathered} \tag{9}$$

Using these notations, the Cauchy–Riemann conditions and (5′), (6), we obtain

$$w^+(z) = w^-(z) + \beta\overline{w^-(z)} + f(z), \quad z \in S, \tag{10}$$

$$\operatorname{Im} w^-(z) = g, \;\; \big(\operatorname{Re} w^-(z) = h\big), \quad z \in L, \tag{11}$$

where $g = (g_1)_S$. Here we assume that

$$a_{11} \neq 0, \quad a_{21} \neq 0, \quad d \neq 0. \tag{12}$$

It is known (Cherednichenko, 1982) that we can go from the problem of conjugation for analytic functions (10), (11), to Direct Problem 2.

DIRECT PROBLEM 3. *Find w^+, w^- which are analytic in D^+, $T^+\backslash\overline{D^+}$ respectively satisfying conditions* (10), (11), *where the functions f, g are defined on S, L respectively and β is a constant (it is possible that β is complex).*

INVERSE PROBLEM 3. *A function $f(z)$ is given in T^+, real functions g, h are given on L. Find domain D^+ and functions w^+, w^- which are analytic in D^+, $T^+\backslash\overline{D^+}$ respectively such that* (10) *is satisfied and*

$$w^-(z) = h + ig \quad \text{on } L. \tag{13}$$

4. The given boundary L plays an auxiliary role. Let us show it.

DIRECT PROBLEM 4. *Find functions U^+, U^- which are analytic in D^+, D^- respectively, $U^-(\infty) = 0$ satisfying*

$$U^+(z) = U^-(z) + \beta\overline{U^-(z)} + F(z), \quad z \in S, \tag{14}$$

where F is function defined general on S.

This is a special case of the common problem of linear conjugation for analytic functions. When $\beta = 0$, we obtain the simplest problem the solution of which is written obviously using the help a Cauchy type integral. If $|\beta| < 1$, this problem has a unique solution. In the case $|\beta| > 1$, as V.G. Cherednichenko and G.V. Veryovkina have shown (see Cherednichenko and Veryovkina, 1987)), the homogeneous problem (4) can have a nonzero solution.

INVERSE PROBLEM 4. *The function $F(z)$ is given in T^+, an analytic function $U(z)$ is given in T^-, $U(\infty) = 0$. Find a domain D^+ and functions $U^+(z)$, $U^-(z)$ satisfying a condition (14) and*

$$U(z) = U^-(z), \quad z \in T^-. \tag{15}$$

Let us remark that in the last formulation T^+ can coincide with the whole plane. In this case T^- is any domain (any set of uniqueness for analytic function) in D^-, i.g., for example, is any neighbourhood of the point $z = \infty$.

5. We shall establish the connection between Inverse Problem 3 and Inverse Problem 4. By virtue of the integral Cauchy theorem, any function $w^-(z)$ which is analytic in double connected domain $T^+\backslash\overline{D^+}$ can be represented as

$$w^-(z) = F_1(z) + U^-(z), \tag{16}$$

where $U^-(z)$ is an analytic function in D^-, $U^-(\infty) = 0$ and $F_1(z)$ is an analytic function in T^+. This representation is unique. Substituting (16) in boundary condition (10), (13), we find that

$$U^+(z) = U^-(z) + \beta\overline{U^-(z)} + \beta\overline{F_1(z)} + f(z), \quad z \in S, \tag{17}$$

$$F_1(z) + U^-(z) = h(z) + ig(z), \quad z \in L, \tag{18}$$

where $U^+(z) = w^+(z) - F_1(z)$. Now we denote

$$-\frac{1}{2\pi i}\int_L \frac{h(\zeta)+ig(\zeta)}{\zeta - z}\,d\zeta = \begin{cases} F_2(z), & z \in T^+, \\ U(z), & z \in T^-, \end{cases} \tag{19}$$

so that $-[h(z) + ig(z)] = F_2(z) - U(z),\ z \in L$. Comparing this with (18), we obtain

$$F_1(z) + F_2(z) = U(z) - U^-(z), \quad z \in L.$$

On the left-hand side we have a function which is analytic in T^+, and on the right-hand side we have a function which is analytic in T^- and equals zero at $z = \infty$.

Using the principle of continuous analytic continuation, we conclude that

$$U(z) = U^-(z), \quad z \in T^-. \tag{20}$$

$$F_1(z) = -F_2(z), \quad z \in T^+. \tag{21}$$

Let us remark that the first of these formulas gives analytic continuation of the function $U(z)$ from T^- to D^-.

Thus if D^+ is the solution of the Inverse Problem 3, then D^+ is the solution of the Inverse Problem 4 for the function $U(z)$, which is given by (19) and for the density

$$F(z) = -\beta\overline{F_2(z)} + f(z), \tag{22}$$

where $F_2(z)$ is also given by (19). Reasoning in the opposite direction from the Problem 4 with these data, we shall come to the Problem 3 with the density $f(z)$.

6. Inverse Problem 4 includes many special problems which appear in the applied sciences. Each problem is characterized by the number β and the function (density) $F(z)$. Let us remind that the number β is constructed by the matrix $A = \begin{pmatrix} a_{11} & a_{12} \\ a_{21} & a_{22} \end{pmatrix}$, which is the matrix of coefficients of conjugation condition (5′).

The function $F(z)$ is defined by (22), by the functions b_1, b_2 which are given on the unknown boundary S, and by the functions g, h which are given on the known boundary L.

Example 1. For the inverse conductivity problem which is defined by (4), (5), (6), (7) we have

$$A = \begin{pmatrix} 1 & -1 \\ 1 & -1/2 \end{pmatrix}, \quad \beta = -\frac{1}{3}, \quad F(z) = -\beta\overline{F_2(z)},$$

where the analytic function $F_2(z)$ is defined in T^+ by (19).

Example 2. For the inverse problem of electroprospecting in the artificial magnetization we have (see (Cherednichenko, 1982))

$$A = \begin{pmatrix} 1 & -1 \\ \mu_2 & \mu_1 \end{pmatrix}, \quad \beta = \frac{\mu_1 - \mu_2}{\mu_1 + \mu_2}, \quad F = c\,\overline{\ln(z - z_0)}.$$

Here z_0 is a point in T^-, c is a constant, the positive numbers μ_1, μ_2 characterize the magnetic properties of domains D^+, $T^+\backslash\overline{D^+}$.

Example 3. Let $\beta = 0$. The solution of the Direct Problem 4 is given by Cauchy type integral, and in this case the inverse problem is called the inverse problem for Cauchy integral. If $\beta = 0$ then condition (14) is characteristic for the potential of the body with density $\mu = F_{\bar{z}}$. Therefore the Inverse Problem 4 contains the inverse problem for the potential of the body. In particular, if $F = \mu_0 \bar{z}$, $\mu_0 = \text{const}$ we obtain the constant density μ_0.
Taking into consideration the known connection between the Cauchy type integral and the potentials of simple and double layers, we deduce that Inverse Problem 4 includes both the inverse problem for potentials of simple layers and double layers (see (Cherednichenko, 1977)). It should be noted that the function $F(z)$ in (14) which can depend not only on point z, but also on the curve S, too. Such a case will be considered follow below.

Example 4 is the inverse problem of electrocardiology (see (Beretta and Vessella, 1988; Colli-Franzone et al., 1984)). Here $A = \begin{pmatrix} 1 & -1 \\ 1 & -1 \end{pmatrix}$, i.e. $\beta = 0$, the functions b_1, b_2 in (5′) have a special form and depend on the curve S. Let the function $z(t)$ maps conformally the circle $|t| < 1$ on the unknown domain and satisfies the conditions

$$z(0) = 0, \quad z'(0) > 0. \tag{23}$$

The function $F(z)$ in (14) is written as

$$F(z) = \frac{m_2 - m_1}{2} p^2(z) \frac{\overline{tz'(t)}}{tz'(t)}, \quad t = t(z), \ z \in S,$$

where m_1, m_2 are constants, the function $p(z)$ is defined in T^+, $|p(z)| = 1$.
The inverse problem is solved for the function $U(z)$, which is constructed by (19) with $g = 0$.
The given examples show the importance of our unified approach to the solution of a wide class of inverse problems.
7. Let us consider the problem about the local uniqueness of the solution of Inverse Problem 4. Let us confine ourselves to the case of (14) when

$$F = -\beta \overline{F_2(z)}, \quad |\beta| < 1, \ \beta \neq 0,$$

i.e.,

$$U^+(z) = U^-(z) + \beta \overline{U^-(z)} - \beta \overline{F_2(z)}, \quad z \in S. \tag{24}$$

Here $F_2(z)$ is an analytic function in T^+. So, the inverse conductivity problem is studied.

THEOREM. *Let D^+ be a domain with the boundary S belonging to C^1_α, $0 < \alpha < 1$, U^+, U^- the solution of boundary problem (24),*

$$(U^-(z) - F_2(z))' \neq 0, \quad z \in S. \tag{25}$$

Then there exists a constant $d > 0$, which depend on D^+, F_2, β, such that if D_1^+ is the solution of the Inverse Problem 4 with $U = U^-$ and $\|z(t) - z_1(t)\|_{C^0_\alpha} \leq d$, *where $z(t), z_1(t)$ are conformal mapping of the circle $|t| < 1$ on D^+, D_1^+ respectively normed as in* (23), *then $D^+ = D_1^+$.*

Proof. We shall look for the solution of the inverse problem in the form $z_1(t) = z(t) + \omega(t)$, where $\omega(t)$ is an analytic function when $|t| < 1$,

$$\omega(0) = 0, \quad \operatorname{Im} \omega'(0) = 0 \tag{26}$$

and $\|\omega\| \leq d_1$, the number d_1 small enough is chosen for the functions $z(t) + \omega(t)$ to be univalent in the disk $|t| \leq 1$.
Now, as in (Cherednichenko, 1982) one can prove that the finding of the domain D_1^+ is equivalent to the following nonlinear boundary problem, find analytic functions $\omega(t)$, $\varphi^+(t)$ in the circle $|t| < 1$ satisfying the condition

$$\varphi^+ = U^-[z+\omega] + \beta U^-[z+\omega] - \beta \overline{F_2[z+\omega]}, \quad |t| = 1 \tag{27}$$

Excluding the linear part, we obtain

$$\varphi^+ = (U^-)'\omega + \beta \overline{[(U^-)' - F_2']} \bar{\omega} + B\omega,$$

$$B\omega = U^-[z+\omega] - U^-[z] - (U^-)'\omega + \beta\left[\overline{U^-[z+\omega]} - \overline{U^-[z]} - \overline{(U^-)'}\bar{\omega}\right]$$
$$-\beta\left[\overline{F_2[z+\omega]} - \overline{F_2[z]} - \overline{F_2'[z]}\bar{\omega}\right]. \tag{28}$$

Let us calculate $(U^-)'$ using (24):

$$(U^+)' = (U^-)' + \beta \overline{(U^- - F_2)'} \frac{d\bar{z}}{dz}, \quad z \in S.$$

Substituting $z = z(t)$ and using the formula $\frac{d\bar{z}}{dz} = -\frac{1}{t^2} \frac{\overline{z'(t)}}{z'(t)}$, we obtain

$$\varphi_1^+ = \beta \overline{(U^- - F_2)'} (\bar{\omega} + \frac{\omega}{t^2} \frac{\overline{z'(t)}}{z'(t)}) + B\omega, \quad |t| = 1, \tag{29}$$

where $\varphi_1^+ = \varphi^+ - (U^+)'\omega$ is a new unknown function.
Assume that condition (25), is satisfied and denote

$$\varkappa = \operatorname{Ind} \overline{(U^- - F_2)'}, \quad z \in S. \tag{30}$$

LEMMA. *If condition* (25) *is satisfied then $\varkappa \leq 0$*

Proof. Denote $U_1^+ = U^+ - F_2$ and write (24) in the form $U_1^+ = U^- - F_2 + \beta\overline{(U^- - F_2)}$, $z \in S$. Combining this with the conjugate condition, we obtain

$$(U^- - F_2)' = \frac{1}{1-\beta^2}\left[\left((U^+)' + \beta(\overline{U^+})\right) \frac{d\bar{z}}{dz}\right],$$

$$(U^- - F_2)' = \frac{1}{1-\beta^2}(U^+)'(1+G), \quad G = \beta \frac{\overline{(U^+)'}}{(U^+)'} \frac{d\bar{z}}{dz}. \tag{31}$$

We have $|G| < 1$, so the index of the function $1 + G$ is zero. Using the property of the index, we conclude from (31) $\text{Ind}\,(U^- - F_2)' = \text{Ind}\,(U^+)' \geq 0$. The lemma is proved.

We return to boundary problem (29). Let l be the number of solutions of the homogeneous problem corresponding to (29) (i.e., without the term $B\omega$), with the condition $\omega(0) = 0$ and p be the number of solvability conditions on $B\omega$. Then it is known (Mikhailov, 1963) that $l = 1$, $p = |2\varkappa - 2| - 1$ if $\varkappa < 1$. Dropping the solution $\varphi_1^+ = 0$, $\omega = itz'(t)$, which does not satisfy the condition $\text{Im}\,\omega'(0) = 0$, we see that the homogeneous problem has only the trivial solution. $B\omega$ maps a sufficiently small ball $\|\omega\| \leq d \leq d_1$ into itself and is a contraction operator. Now we apply to (29) a ramification theorem for solutions of nonlinear equation in the singular case. The theorem is proved.

REFERENCES

Alessandrini, G. (1989). Remark on a paper by Bellout and Friedman. *Bull. Unione Mat. Ital.* **7**, 243–249.

Bellout, H. and Friedman, A. (1988). Identification problem in potential theory. *Arch. Rational Mech. Anal.* **101**, 143–160.

Beretta, E. and Vessela, S. (1988). Stability results for an inverse problem in potential theory. *Pubblicazioni Dell Istituto di Analisi Globale e Applicazioni.* **39**, 1–44.

Cherednichenko, V.G. (1977). Inverse problem for type Cauchy integral. *Different. Equat.* **13**, 174–176 (in Russian).

Cherednichenko, V.G. (1982). A problem of the conjugation of harmonic function and its inverse. *Different. Equat.* **18**, 682–689 (in Russian).

Cherednichenko, V.G. and Veryovkina, G.V. (1987). About the solvability of the general boundary value Riemann problem for analytic functions. *Different. Equat.* **23**, 2179–2181 (in Russian).

Colli-Franzone, L., Guerri, E., and Maganes, E. (1984). Oblique double layer potentials for and inverse problems of electrocardiology. *Math. Biosciences.* **68**, 23–55.

Friedman, A. and Isakov, V. (1989). On the uniqueness in the inverse conductivity problem with one measurement. *Indiana University Math. J.* **38**, 563–579.

Isakov, V. and Powell, J. (1990). On the inverse conductivity problem with one measurement. *Inverse Problems* **1**, 311–318.

Ladyzenskaja, O.A. and Uralceva, N.N. (1968). *Linear and Quasilinear Equations of Elliptic Type.* Academic Press, New York.

Mikhailov, L.G. (1963). The general conjugacy problem for analytic functions and its applications. *Izv. Akad. Nauk SSSR, Ser. Mat.* **27**, 969–992.

Ill-Posed Problems in Natural Sciences, pp. 277 – 283
A. Tikhonov (Ed.)
1992 VSP/TVP

INVERSE PROBLEMS FOR A SYSTEM OF NONLINEAR PARTIAL DIFFERENTIAL EQUATIONS

A.M. DENISOV
Moscow State University, 119899 Moscow, Russia

ABSTRACT

Inverse problems for a system of nonlinear partial differential equations arising in the theory of absorption are considered. Theorems of uniqueness of solutions of these inverse problems are proved.

In this paper we consider the following problem

$$u_x + a_t = 0, \quad 0 \le x \le l, \quad 0 \le t \le T, \tag{1}$$

$$a_t = \gamma(u)(\varphi(u) - a), \quad 0 \le x \le l, \quad 0 \le t \le T, \tag{2}$$

$$u(0,t) = \mu(t), \quad 0 \le t \le T, \tag{3}$$

$$a(x,0) = 0, \quad 0 \le x \le l. \tag{4}$$

This problem is a mathematical model of the dynamics absorption process. The functions $\varphi(\xi)$ and $\gamma(\xi)$ are called the isoterm of absorption and the kinetic coefficient respectively.

We consider two inverse problems for problem (1)–(4) arising in dynamic absorption experiments. The first inverse problem is: find the function $\varphi(\xi)$, if the functions $\gamma(\xi)$, $\mu(t)$ and some additional information about $u(x,t)$ are known. The second inverse problem is: find the function $\gamma(\xi)$, if the functions $\varphi(\xi)$, $\mu(t)$ and some additional information about $u(x,t)$ are known.

First we consider some properties of the solution of problem (1)–(4).

Let us assume that the functions $\gamma(\xi)$, $\varphi(\xi)$ and $\mu(t)$ satisfy the following conditions:

$$\gamma(\xi) \in \mathbf{C}^2(-\infty,\infty); \qquad 0 < \gamma_1 \le \gamma(\xi) \le \gamma_2,$$
$$0 \le \gamma'(\xi) \le \gamma_3, \qquad \xi \in (-\infty,\infty), \tag{5}$$

where $\gamma_1, \gamma_2, \gamma_3$ are constants;

$$\varphi(\xi) \in \mathbf{C}^2(-\infty,\infty); \qquad \varphi(0) = 0; \qquad |\varphi(\xi)| \le \varphi_1,$$

$$0 < \varphi'(\xi) \leq \varphi_2, \qquad \xi \in (-\infty, \infty); \qquad \varphi_1, \varphi_2 = \text{const}; \tag{6}$$

$$\mu(t) \in \mathbf{C}^1\Big[0, T\Big]; \quad \mu(0) = 0; \quad \mu'(t) > 0, \quad t \in (0, T]. \tag{7}$$

Define the set

$$Q_{l\tau} = \Big\{(x,t)\colon\ 0 \leq x \leq l,\ 0 \leq t \leq \tau\Big\}.$$

LEMMA. *If the functions $\gamma(\xi)$, $\varphi(\xi)$, and $\mu(t)$ satisfy conditions (5), (6) and (7) respectively, then there exists only one pair of functions $\{u(x,t),\ a(x,t)\}$ such that $u(x,t),\ a(x,t) \in \mathbf{C}^1[Q_{lT}]$; $u(x,t),\ a(x,t)$ satisfy (1)–(4). These functions have the following properties*

$$u_t(x,t) > 0, \quad a_t(x,t) > 0, \quad \text{for} \quad x \in \Big[0, l\Big], \quad t \in (0, T]; \tag{8}$$

$$0 \leq u(x,t) \leq \mu(\tau), \quad 0 \leq a(x,t) < \varphi\big(\mu(\tau)\big), \tag{9}$$

in $Q_{l\tau}$ for each $\tau \in (0, T]$.

Proof. If we assume that there exists a pair of functions $\{u(x,t),\ a(x,t)\}$ such that $u(x,t),\ u_x(x,t),\ a(x,t),\ a_t(x,t) \in \mathbf{C}[Q_{lT}]$ and $u(x,t),\ a(x,t)$ satisfy (1)-(4), then these functions satisfy the following equations

$$u(x,t) = \mu(t) - \int_0^x \gamma\Big(u(\xi,t)\Big)\varphi\Big(u(\xi,t)\Big)d\xi$$

$$+ \int_0^x \int_0^t \gamma\Big(u(\xi,t)\Big) \exp\Big\{ - \int_\tau^t \gamma\Big(u(\xi,\theta)\Big)d\theta\Big\}\varphi\Big(u(\xi,\tau)\Big)\gamma\Big(u(\xi,\tau)\Big)d\tau d\xi, \tag{10}$$

$$a(x,t) = \int_0^t \exp\Big\{ - \int_\tau^t \gamma\Big(u(x,\theta)\Big)d\theta\Big\}\varphi\Big(u(x,\tau)\Big)\gamma\Big(u(x,\tau)\Big)d\tau. \tag{11}$$

It is easy to prove that the nonlinear integral equation (10) has only one solution $u(x,t) \in \mathbf{C}[Q_{lT}]$. Then from (10), (11) we obtain that there exists only one couple of functions $u(x,t),\ a(x,t) \in \mathbf{C}[Q_{lT}]$ satisfying (10), (11) in Q_{lT}. From (10) and (11) we immediatelly have that $u_x(x,t),\ a_t(x,t),\ a_x(x,t) \in \mathbf{C}[Q_{lT}]$ and the functions $u(x,t),\ a(x,t)$ satisfy (1)-(4).
Substituting (11) in (10), we get

$$u(x,t) = \mu(t) - \int_0^x \gamma\Big(u(\xi,t)\Big)\varphi\Big(u(\xi,t)\Big)d\xi + \int_0^x \gamma\Big(u(\xi,t)\Big)a(\xi,t)d\xi. \tag{12}$$

Using (5)–(7), (12) we can prove that $u_t(x,t) \in \mathbf{C}[Q_{lT}]$. Consequently, $u(x,t),\ a(x,t) \in \mathbf{C}^1[Q_{lT}]$. Now from (5)–(7), (10), (11) we derive that

$$a_{tt}(x,t),\ u_{xt}(x,t),\ a_{xt}(x,t),\ u_{xx}(x,t) \in \mathbf{C}[Q_{lT}].$$

Let us consider the functions

$$z(x,t) = u_t(x,t), \qquad w(x,t) = a_t(x,t).$$

This pair of functions is a solution of the following problem

$$z_x + w_t = 0, \quad (x,t) \in Q_{lT}, \tag{13}$$

$$w_t = \Phi(x,t)z - \gamma\Big(u(x,t)\Big)w, \quad (x,t) \in Q_{lT}, \tag{14}$$

$$z(0,t) = \mu'(t), \quad 0 \le t \le T, \tag{15}$$

$$w(x,0) = 0, \quad 0 \le x \le l, \tag{16}$$

where

$$\Phi(x,t) = \gamma'\Big(u(x,t)\Big)\varphi\Big(u(x,t)\Big) + \gamma\Big(u(x,t)\Big)\varphi'\Big(u(x,t)\Big) - \gamma'\Big(u(x,t)\Big)a(x,t).$$

From (13)–(16) we easily obtain the following Volterra integral equation of the second kind for the function $z(x,t)$

$$z(x,t) = \mu'(t)\exp\Big\{ - \int_0^x \Phi(s,t)\,ds\Big\}$$

$$+\int_0^x\int_0^t \exp\Big\{ - \int_\xi^x \Phi(s,t)\,ds - \int_\tau^t \gamma\Big(u(\xi,\theta)\Big)d\theta\Big\}\Phi(\xi,\tau)\gamma\Big(u(\xi,t)\Big)z(\xi,\tau)\,d\tau d\xi. \tag{17}$$

It is easy to prove that

$$\Phi(x,t) > 0 \quad \text{in} \quad Q_{lT}. \tag{18}$$

From (5), (7), (17), and (18) we obtain that $z(x,t) > 0$ for $x \in [0,l]$, $t \in (0,T]$. Furthermore from (14), (16), (18) we deduce that $w(x,t) > 0$ for $x \in [0,l]$, $t \in (0,T]$. Thus inequalities (8) are proved. Finally, from (1)–(8) we obtain (9).

Now we will consider inverse problems for problem (1)–(4). We observe that some inverse problems arising in the theory of absorption were examined in (Denisov and Tuikina, 1984; Denisov, 1986; Denisov and Chanov, 1988; Lorenzi and Paparoni, 1988).

The explicit formulation of the first inverse problem is as follows: find a triplet of functions $\{\varphi(\xi),\ u(x,t),\ a(x,t)\}$ satisfying (1)–(4) and the additional condition

$$u(l,t) = q(t), \qquad t \in [0,T]. \tag{19}$$

Here we assume that the functions $\gamma(\xi), \mu(t)$, and $q(t)$ are known. We observe that the additional information (19) is the typical information for dynamic absorption experiments.

Definition. A triplet of functions $\{\varphi(\xi),\ u(x,t),\ a(x,t)\}$ is called a solution of the first inverse problem if: $\varphi(\xi)$ satisfies (6); $u(x,t),\ a(x,t) \in \mathbf{C}^1[Q_{lT}]; \varphi(u(x,t))$, $u(x,t),\ a(x,t)$ satisfy (1) and (2); $u(x,t)$ satisfies (3) and (19), and $a(x,t)$ satisfies (4).

THEOREM 1. *Assume that the functions $\gamma(\xi), \mu(t)$ satisfy conditions (5) and (7), respectively and that $\{\varphi_i(u_i(x,t)),\ u_i(x,t),\ a_i(x,t)\}$, $i = 1,2$, are solutions of the first inverse problem. If there exists an $\varepsilon > 0$ such that $\varphi_1(\xi) = \varphi_2(\xi)$ for $\xi \in [0,\varepsilon]$, then $\varphi_1(\xi) = \varphi_2(\xi)$ for $\xi \in [0,\mu(T)]$ and $u_1(x,t) = u_2(x,t),\ a_1(x,t) = a_2(x,t)$ in Q_{lT}.*

Proof. Define

$$v(x,t) = u_1(x,t) - u_2(x,t), \quad y(x,t) = a_1(x,t) - a_2(x,t),$$

$$\alpha(\xi) = \varphi_1(\xi) - \varphi_2(\xi).$$

Since $\{\varphi_i(u_i(x,t)),\ u_i(x,t),\ a_i(x,t)\}$, $i = 1,2$, satisfy (1)–(4), (19) the functions $v(x,t),\ y(x,t)$ are a solution of the following problem

$$v_x + y_t = 0, \qquad (x,t) \in Q_{lT}, \tag{20}$$

$$y_t = p(x,t)v + \gamma\Big(u_2(x,t)\Big)\Big(\alpha(u_2(x,t)) - y\Big), \quad (x,t) \in Q_{lT}, \tag{21}$$

$$v(0,t) = 0, \quad 0 \le t \le T, \tag{22}$$

$$y(x,0) = 0, \quad 0 \le x \le l, \tag{23}$$

$$v(l,t) = 0, \quad 0 \le t \le T. \tag{24}$$

Here

$$p(x,t) = \int_0^1 \gamma'\Big(u_2(x,t) + \theta(u_1(x,t) - u_2(x,t))\Big)d\theta\Big[\varphi_1(u_1(x,t)) - a_1(x,t)\Big]$$

$$+\gamma(u_2(x,t)) \int_0^1 \varphi_1'\Big(u_2(x,t) + \theta(u_1(x,t) - u_2(x,t))\Big)d\theta.$$

Integrating equations (20),(21) with conditions (22),(23) we obtain the Volterra integral equation of the second kind for $v(x,t)$:

$$v(x,t) = f(x,t) + \int_0^x \int_0^t K(x,\xi,t,\tau)v(\xi,\tau)\, d\tau d\xi. \tag{25}$$

The kernel $K(x,\xi,t,\tau)$ and the function $f(x,t)$ are defined by the equations

$$K(x,\xi,t,\tau) = \exp\Big\{ - \int_\xi^x p(s,t)ds - \int_\tau^t \gamma(u_2(\xi,\theta))d\theta\Big\}\gamma\Big(u_2(\xi,t)\Big)p(\xi,\tau),$$

$$f(x,t) = -\int_0^x g(x,\xi,t)\alpha\big(u_2(\xi,t)\big)d\xi + \int_0^x\int_0^t h(x,\xi,t,\tau)\alpha\big(u_2(\xi,\tau)\big)d\tau d\xi,$$

where

$$g(x,\xi,t) = \exp\Big\{-\int_\xi^x p(s,t)ds\Big\}\gamma\big(u_2(\xi,t)\big),$$

$$h(x,\xi,t,\tau) = \gamma\big(u_2(\xi,t)\big)\gamma\big(u_2(\xi,\tau)\big)\exp\Big\{-\int_\xi^x p(s,t)ds - \int_\tau^t \gamma\big(u_2(\xi,\theta)\big)d\theta\Big\}.$$

From (25) we have

$$v(x,t) = f(x,t) + \int_0^x\int_0^t R(x,\xi,t,\tau)f(\xi,\tau)d\tau d\xi, \tag{26}$$

where $R(x,\xi,t,\tau)$ is the resolvent kernel for $K(x,\xi,t,\tau)$.
Setting $x = l$ in (26) and using (24) we obtain

$$\int_0^l g(l,\xi,t)\alpha\big(u_2(\xi,t)\big)d\xi - \int_0^l\int_0^t h(l,\xi,t,\tau)\alpha\big(u_2(\xi,\tau)\big)d\tau d\xi$$

$$+\int_0^l\int_0^t R(l,\xi,t,\tau)\int_0^\xi g(\xi,s,\tau)\alpha\big(u_2(s,\tau)\big)dsd\tau d\xi$$

$$-\int_0^l\int_0^t R(l,\xi,t,\tau)\int_0^\xi\int_0^\tau h(\xi,s,\tau,\theta)\alpha\big(u_2(s,\theta)\big)d\theta dsd\tau d\xi = 0, \tag{27}$$

for $t \in [0,T]$. Let us denote

$$B(\xi,t,\tau) = -h(l,\xi,t,\tau) + \int_\xi^l R(l,s,t,\tau)g(s,\xi,\tau)\,ds$$

$$-\int_\xi^l\int_\tau^t R(l,s,t,\theta)h(s,\xi,\theta,\tau)\,d\theta ds.$$

Using this notation, we rewrite (27) in the form

$$\int_0^l g(l,\xi,t)\alpha\Big(u_2(\xi,t)\Big)d\xi$$

$$+\int_0^l\int_0^t B(\xi,t,\tau)\alpha\big(u_2(\xi,\tau)\big)d\tau d\xi=0,\qquad t\in[0,T]. \tag{28}$$

Let us denote by t_0 the root of the equation $\mu(t)=\varepsilon$. From (9) we obtain that $0\le u_2(x,t)\le\varepsilon$ in Q_{lt_0} and

$$\alpha\Big(u_2(x,t)\Big)=0\quad\text{in}\quad Q_{lt_0}. \tag{29}$$

From (28) and (29) we have

$$\int_0^l g(l,\xi,t)\alpha\big(u_2(\xi,t)\big)d\xi$$

$$+\int_0^l\int_{t_0}^t B(\xi,t,\tau)\alpha\big(u_2(\xi,\tau)\big)d\tau d\xi=0,\quad t\in[t_0,T]. \tag{30}$$

Indroducing the new variable

$$z=u_2(\xi,t),\qquad \xi=u_2^{-1}(z,t), \tag{31}$$

from (30), (31), (3), (19) we obtain

$$\int_{q(t)}^{\mu(t)} g_1(l,z,t)\alpha(z)dz+\int_{t_0}^t\int_{q(\tau)}^{\mu(\tau)} B_1(t,\tau,z)\alpha(z)dzd\tau=0,\qquad t\in[t_0,T], \tag{32}$$

where

$$g_1(l,z,t)=\frac{g\Big(l,u_2^{-1}(z,t),t\Big)}{u_{2x}\Big(u_2^{-1}(z,t),t\Big)},\quad B_1(t,\tau,z)=\frac{B(u_2^{-1}(z,\tau),t,\tau)}{u_{2x}\Big(u_2^{-1}(z,\tau),\tau\Big)}.$$

Let t_1 be a root of the equation $q(t)=\mu(t_0)$. Then from (32) we obtain for $t\in[t_0,t_1]$

$$\int_\varepsilon^{\mu(t)} g_1(l,z,t)\alpha(z)dz+\int_\varepsilon^{\mu(t)}\alpha(z)\int_{\mu^{-1}(z)}^t B_1(t,\tau,z)d\tau dz=0. \tag{33}$$

Introduce the new variable $\omega=\mu(t),\ t=\mu^{-1}(\omega)$. From (33) we obtain the Volterra integral equation of the first kind for $\alpha(z)$:

$$\int_\varepsilon^\omega H(\omega,z)\alpha(z)dz=0,\quad \omega\in[\varepsilon,\mu(t_1)], \tag{34}$$

where

$$H(\omega, z) = g_1\Big(l, z, \mu^{-1}(\omega)\Big) + \int\limits_{\mu^{-1}(z)}^{\mu^{-1}(\omega)} B_1\Big(\mu^{-1}(\omega), \tau, z\Big) d\tau.$$

The function $H(\omega, z)$ has the following properties: $H(\omega, \omega) \neq 0$ for $\omega \in [\varepsilon, \mu(t_1)]$; $H_\omega(\omega, z)$ is continuous for $\varepsilon \le z \le \omega \le \mu(t_1)$. Hence, differentiating (34), we see that $\alpha(z) = 0$ for $z \in [\varepsilon, \mu(t_1)]$. Consequently, $\alpha(z) = 0$ for $z \in [0, \mu(t_1)]$.
Let us denote by t_2 the root of the equation $q(t) = \mu(t_1)$. In the same way we can prove that $\alpha(z) = 0$ for $z \in [0, \mu(t_2)]$. If we continue this process, we obtain that $\alpha(z) = 0$ for $z \in [0, \mu(T)]$.
Hence $\varphi_1(\xi) = \varphi_2(\xi)$ for $\xi \in [0, \mu(T)]$. Then from (25) it follows that $u_1(x,t) = u_2(x,t)$ in Q_{lT}. Finally, from (21), (23) we obtain $a_1(x,t) = a_2(x,t)$ in Q_{lT}. Theorem 1 is proved.
Let us consider the second inverse problem.The explicit formulation of the second inverse problem is as follows: find a triplet of functions $\{\gamma(\xi),\ u(x,t),\ a(x,t)\}$ satisfying (1)–(4) and the additional condition (19). Here we assume that the functions $\varphi(\xi)$, $\mu(t)$, and $q(t)$ are known.

Definition. A triplet of functions $\{\gamma(\xi),\ u(x,t),\ a(x,t)\}$ is called a solution of the second inverse problem if: $\gamma(\xi)$ satisfies (5); $u(x,t),\ a(x,t) \in \mathbf{C}^1[Q_{lT}]$; $\gamma(u(x,t))$, $u(x,t)$, and $a(x,t)$ satisfy (1),(2); $u(x,t)$ satisfies (3) and (19), and $a(x,t)$ satisfies (4).

THEOREM 2. *Assume that the functions $\varphi(\xi)$, $\mu(t)$ satisfy conditions* (6), (7) *respectively and $\{\gamma_i(\xi),\ u_i(x,t),\ a_i(x,t)\}$, $i = 1,2$, are solutions of the second inverse problem. If there exists an $\varepsilon > 0$ such that $\gamma_1(\xi) = \gamma_2(\xi)$ for $\xi \in [0, \varepsilon]$, then $\gamma_1(\xi) = \gamma_2(\xi)$ for $\xi \in [0, \mu(T)]$ and $u_1(x,t) = u_2(x,t)$, $a_1(x,t) = a_2(x,t)$ in Q_{lT}.*

The proof of Theorem 2 is similar to the proof of Theorem 1.

REFERENCES

Denisov, A.M. (1986). Uniqueness of the solution for some inverse problems of nonequilibrium dynamics of absorption. *Vestnik Moscow Univ.* Ser.15, Computer Math. Cybern. **2**, 25–30.

Denisov, A.M. and Chanov, A.V. (1988). The kinetic coefficient determination for some models of nonequilibrium dynamics of absorption, *Vestnik Moscow Univ.* Ser. 15, Computer Math. Cybern. **4**, 29–33.

Denisov, A.M. and Tuikina, S.R. (1984). On some inverse problems of nonequilibrium dynamics of absorption, *Dokl. Akad. Nauk SSSR* **276**, 100–102.

Lorenzi, A. and Paparoni, E. (1988). *An Inverse Problem Arising in the Theory of Absorption.* Preprint N25, Milano Univ., Quaderno.

Ill-Posed Problems in Natural Sciences, pp. 284 – 299
A. Tikhonov (Ed.)
1992 VSP/TVP

AN INVERSE PROBLEM IN THREE-DIMENSIONAL LINEAR THERMOVISCOELASTICITY OF BOLTZMANN TYPE

M. GRASSELLI
Istituto di Analisi Numerica del C.N.R., Corso Carlo Alberto 5, 27100 Pavia, Italy

ABSTRACT

We consider the equations governing the evolution of the displacement and temperature fields in a linear, three-dimensional, thermoviscoelastic material subject to a Boltzmann damping mechanism. This model consists of a system of two coupled Volterra integro-differential equations containing some time-dependent relaxation functions accounting for the memory effects. Such effects influence the behaviour of the body, especially with respect to the wave propagation, but the relaxation coefficients are a priori unknown. Here we suggest a mathematical technique to identify them. It consists in formulating a suitable overspecified initial and boundary value problem for the system under consideration. Then the problem is reduced to a system of nonlinear second kind Volterra integrodifferential equations for the unknown relaxation functions, which can be solved (locally in time) via Contraction Principle assuming some nonvanishing conditions on the data. Local existence, uniqueness, and continuous dependence upon the data are shown.

1. INTRODUCTION

Let us consider a sample of linear, compressible, viscoelastic material of Boltzmann type located, for each time $t \in [0,T]$ $(T > 0)$, in a bounded domain $\Omega \subset \mathbf{R}^3$ having a boundary Γ of class $\mathbf{C}^2$. We denote by $\mathbf{u}(x,t)$ and $\vartheta(x,t)$ its displacement and its temperature deviation from a given constant temperature computed at a point $x \in \Omega$ at time $t \in [0,T]$, respectively. Moreover, we denote by $\sigma(x,t)$ the stress tensor field, by $\eta(x,t)$ the specific entropy field, and by $\mathbf{q}(x,t)$ the heat flux vector. We assume, for the sake of simplicity, that the body is homogeneous and isotropic with unitary density. Then a linear model describing its thermoviscoelastic behaviour is given by the following set of constitutive assumptions (cf. (Hughes and Marsden, 1983; Navarro, 1978), see also (Burns et al., 1990 and Leugering, 1985) for one-dimensional models)

$$\sigma(x,t) = \mathbf{G}(0)\varepsilon(\mathbf{u})(x,t) + \int_0^t \mathbf{G}'(\tau)\varepsilon(\mathbf{u})(x,t-\tau)\,d\tau - \delta l(0)\vartheta(x,t)$$

$$-\int_0^t \sigma l'(\tau)\vartheta(x,t-\tau)\,d\tau, \quad (x,t) \in \Omega \times [0,T], \tag{1.1}$$

$$\eta(x,t) = l(0)\nabla\cdot \mathbf{u}(x,t) + h(0)\vartheta(x,t) + \int_0^t l'(\tau)\nabla\cdot\mathbf{u}(x,t-\tau)\,d\tau$$
$$+ \int_0^t h'(\tau)\vartheta(x,t-\tau)\,d\tau, \quad (x,t) \in \Omega \times [0,T], \tag{1.2}$$

$$\mathbf{q}(x,t) = -k_0\nabla\vartheta(x,t), \qquad (x,t) \in \Omega \times [0,T], \tag{1.3}$$

where $\delta := [\delta_{ij}]$ denotes the Kronecker tensor and ∇, $\nabla\cdot$ are the gradient and the divergence operators acting on the space variables, respectively. Moreover, $\varepsilon(\mathbf{u})$ denotes the usual strain tensor $\varepsilon_{ij}(\mathbf{u}) := \frac{1}{2}(u^i_{,j} + u^j_{,i})$. Here $\mathbf{G}$ is the so-called relaxation tensor, l, h are other relaxation functions, and k_0 is a given positive constant.

We remark that the classical Fourier's law (1.3) can also be modified by introducing memory mechanism (cf., e.g., (Bercia, 1990; Bhattacharyya, 1991; Chakrabarti, 1976; Luca, 1989a, b)).

As we consider the isotropic case, the relaxation tensor $\mathbf{G}$ reduces to (cf., e.g., (Grasselli, b))

$$G_{ijlm}(t) = \frac{1}{3}\left[\psi_b(t) - \psi_s(t)\right]\delta_{ij}\delta_{lm}(t) + \frac{1}{2}\psi_s(t)[\delta_{il}\delta_{lm} + \delta_{im}\delta_{jl}], \quad t \in [0,T], \tag{1.4}$$

where ψ_b and ψ_s are named the bulk relaxation function and the shear relaxation function, respectively. We also observe that physical considerations imply

$$\psi_b(0) = 3k, \qquad \psi_s(0) = 2\mu, \tag{1.5}$$

where k, $\mu > 0$ are the elastic bulk and the elastic shear moduli. That implies $\mathbf{G}(0) = \mathbf{G}_0$, where $\mathbf{G}_0$ is a fourth-order isotropic tensor, whose components are known constants, which is hyperelastic and strongly elliptic (cf., e.g., (Hughes and Marsden, 1983)). Moreover, for simplicity, we will assume that $l(0) = l_0$, $h(0) = h_0$, where l_0, h_0 are given positive constants.

In the above constitutive equations we have supposed that the Volterra hypothesis on the memory kernels holds, that is we have neglected the terms containing the past histories of $\mathbf{u}$ and ϑ up to $t = 0$. However, it is possible to treat such terms as well, provided that the past histories satisfy some suitable conditions (see, e.g., (Grasselli, b)).

Putting the equations (1.1)–(1.3) in the balance of linear momentum and of the energy we get the Volterra integrodifferential equations (cf. (Hughes and Marsden, 1983; Navarro, 1978))

$$\ddot{\mathbf{u}}(x,t) = \nabla\cdot\left(\mathbf{G}_0\varepsilon(\mathbf{u})(x,t) + \int_0^t \mathbf{G}'(\tau)\varepsilon(\mathbf{u})(x,t-\tau)\,d\tau\right) - l_0\nabla\vartheta(x,t)$$

$$- \int_0^t l'(\tau)\nabla\vartheta(x,t-\tau)\,d\tau + \mathbf{f}(x,t), \quad (x,t) \in \Omega \times (0,T), \tag{1.6}$$

$$h_0\vartheta_t(x,t) = k_0\Delta\vartheta(x,t) + \left(\int_0^t h'(\tau)\vartheta(x,t-\tau)\,d\tau\right)_t - l_0\nabla\cdot\dot{\mathbf{u}}(x,t)$$
$$- \left(\int_0^t l'(\tau)\nabla\cdot\mathbf{u}(x,t-\tau)\,d\tau\right)_t + r(x,t), \quad (x,t) \in \Omega \times (0,T), \tag{1.7}$$

where $\dot{v} = v_t$ denotes the time derivative and Δ is the Laplace operator. Moreover $\mathbf{f}$ and r are the body forces and the heat supply, respectively.
The so-called direct problem consists in finding $\mathbf{u}$ and ϑ satisfying the equations (1.6), (1.7) and, for instance, the Cauchy–Dirichlet conditions

$$\mathbf{u}(x,0) = \mathbf{u}_0(x), \qquad \dot{\mathbf{u}}(x,0) = \mathbf{u}_1(x), \qquad \vartheta(x,0) = \vartheta_0(x), \quad x \in \Omega, \tag{1.8}$$
$$\mathbf{u}(x,t) = 0, \qquad \vartheta(\mathbf{x},\mathbf{t}) = 0, \qquad (\mathbf{x},\mathbf{t}) \in \mathbf{\Gamma} \times (\mathbf{0},\mathbf{T}). \tag{1.9}$$

This problem is well-posed in a generalized sense (cf., e.g., (Navarro, 1983)). Nevertheless, in the applications, the memory functions ψ_b, ψ_s, h, l are a priori unknown or scarcely known and, since they influence the wave propagation, their identification is of great importance for the study of the behaviour of the material under consideration (see, e.g., (Bhattacharyya, 1991; Burns et al., 1990)). These facts motivate us to consider an inverse problem, that is

$$\text{Find} \quad \psi_b,\ \psi_s,\ h,\ l \quad \text{from} \quad (1.6)\text{–}(1.9).$$

This problem is clearly ill-posed and we need other information to determine the relaxation functions. Since we deal with 4 unknown scalar time-dependent functions the overinformation shall consist of 4 equations only involving the time variable. Such equations have to be chosen by means of physical considerations as we now show.
Let us consider the relaxation tensor $\mathbf{G}$. In the isotropic case its determination reduces to finding the scalar functions ψ_b and ψ_s (cf. (1.4)). In order to do that, physically reasonable information seems to be the one consisting in the measurements of two traction components exerted over two portions Γ_1, Γ_2 of the boundary Γ for each time $t \in [0,T]$, i.e., (cf. (Grasselli, b))

$$\int_{\Gamma_i} \sigma_{ij}(y,t)n_j(y)\,d\Sigma(y) = \phi_i(t), \qquad t \in [0,T], \quad i = 1,2, \tag{1.10}$$

where $d\Sigma$ denotes the surface Lebesgue measure, $|\Gamma_i| > 0$ ($|\Gamma_i|$ being the two-dimensional Lebesgue measure of Γ_i). and $\mathbf{n} = (n^1, n^2, n^3)$ is the outgoing normal vector to the boundary Γ. Here ϕ_i: $[0,T] \to \mathbf{R}$ $(i = 1, 2)$ are given functions. We note that the two portions may coincide, i.e., $\Gamma_1 \equiv \Gamma_2$, provided that we measure two different traction components (cf. (2.26)).

Besides, for the identification of h and l, we assume that we know the temperature values for each $t \in [0,T]$ measured at two different points x_1, $x_2 \in \Omega$ by a real thermic sensor. This information can be written as follows (cf. (Grasselli, a) and the references therein)

$$\int_\Omega \vartheta(x,t)\omega_i(x)\,dx = \lambda_i(t), \qquad t \in [0,T], \quad i = 1,\, 2, \tag{1.11}$$

where λ_i: $[0,T] \to \mathbf{R}$, ω_i: $\Omega \to \mathbf{R}$, $(i = 1,\, 2)$, are known functions. Moreover, ω_i $(i = 1,\, 2)$, has compact support containing x_i and increasing as $|x - x_i|$ decreases (e.g., $\omega_i(x) = \exp\{-|x - x_i|^2\}$ in a neighbourhood of x_i).
Summing up, the equations (1.10), (1.11) form the additional information we need to give a reasonable formulation of the inverse problem under consideration. More precisely we want to study the following

PROBLEM (P1). *Find* $\mathbf{u}$, ϑ, ψ_b, ψ_s, h, l *satisfying* (1.6)–(1.9).

In the next section we will state two results whose proofs are contained in (Grasselli, c). The first one shows that, under suitable assumptions on the data, the Problem (P1) is locally (in time) well-posed. The second one is concerned with the uniqueness and the Lipschitz continuous dependence upon the data for any finite time interval $[0,T]$.
§ 3 is devoted to existence, uniqueness, and regularity results related to the direct problem (1.6)–(1.9) which will be useful to understand the final § 4. In this section the main idea underlying the proofs is presented. It consists in formulating an equivalent problem solvable by the Contraction Principle, namely a system of four nonlinear Volterra integral equations of the second kind for the unknowns ψ_b'', ψ_s'', $h^{(3)}$, $l^{(3)}$.

2. MAIN RESULTS

In order to give a precise formulation of Problem (P1) and state the main results, we need to introduce some notations.
The symbol $\mathcal{H}^r(0,T;X)$ (X being a Hilbert space, $r \in \mathbf{N}$) represents the Hilbert space consisting of all measurable functions w: $(0,T) \to X$ whose time derivatives up to the order r belong to $\mathcal{H}^0(0,T;X) := \mathbf{L}^2(0,T;X)$. If $X \equiv \mathbf{R}$, we set $\mathcal{H}^r(0,T;\mathbf{R}) := \mathcal{H}^r(0,T)$. The Hilbert space $\mathcal{H}^r(0,T;X)$ is endowed with the following norm

$$\|w\|_{r,X} = \|w\|_{\mathcal{H}^r(0,T;X)} = \left\{ \sum_{j=0}^{r} \int_0^T \|w^{(j)}(t)\|_X^2\,dt \right\}^{1/2},$$

where the integral is understood in the sense of Bochner. We simply set $\|\cdot\|_r := \|\cdot\|_{r,X}$, when $X \equiv \mathbf{R}$.
Moreover, we denote by $\mathbf{C}^r([0,T];X)$ the space of all measurable functions w: $[0,T] \to X$ continuous along with their first r time derivatives. This space is a

Banach space, endowed with the norm defined by

$$\|w\|_{r,\infty,X} = \sum_{j=0}^{r} \operatorname{ess}\sup_{t\in(0,T)} \|w^{(j)}(t)\|_X.$$

Let us set

$$\mathbf{v}_0 := \nabla\cdot\mathbf{G}_0\varepsilon(\mathbf{u}_0) - l_0\nabla\vartheta_0 + \mathbf{f}(\cdot,0), \tag{2.1}$$

$$w_0 := h_0^{-1}\Big(k_0\Delta\vartheta_0 - \nabla\cdot\big(l_0\mathbf{u}_1 + l'(0)\mathbf{u}_0\big) - h'(0)\vartheta_0 + r(\cdot,0)\Big), \tag{2.2}$$

$$\mathbf{v}_1 := \nabla\cdot\big(\mathbf{G}(0)\varepsilon(\mathbf{u}_1) + \mathbf{G}'(0)\varepsilon(\mathbf{u}_0)\big) - \nabla\big(l_0w_0 + l'(0)\vartheta_0\big) + \dot{\mathbf{f}}(\cdot,0), \tag{2.3}$$

$$z_0 := h_0^{-1}\Big(k_0\Delta w_0 - \nabla\cdot\big(l_0\mathbf{v}_0 + l'(0)\mathbf{u}_1 + l''(0)\mathbf{u}_0\big) - h'(0)w_0 - h''(0)\vartheta_0 + r_t(\cdot,0)\Big), \tag{2.4}$$

$$\xi_0 := h_0^{-1}\Big(k_0\Delta z_0 - \nabla\cdot\big(l_0\mathbf{v}_1 + l'(0)\mathbf{v}_0 + l''(0)\mathbf{u}_1 + l^{(3)}(0)\mathbf{u}_0\big) - h'(0)z_0 - h''(0)w_0 - h^{(3)}(0)\vartheta_0 + r_{tt}(\cdot,0)\Big). \tag{2.5}$$

The regularity assumptions on the data are

$$\mathbf{u}_0,\ \mathbf{u}_1 \in H_0^1(\Omega;\mathbf{R}^3)\cap H^2(\Omega;\mathbf{R}^3), \tag{2.6}$$

$$\nabla\nabla\cdot\mathbf{u}_0,\ \Delta\mathbf{u}_0 \in H_0^1(\Omega;\mathbf{R}^3), \tag{2.7}$$

$$\nabla\cdot\mathbf{u}_0 \in H_0^1(\Omega)\cap H^2(\Omega),\quad \Delta^2(\nabla\cdot\mathbf{u}_0)\in\mathbf{L}^2(\Omega),\quad \Delta(\nabla\cdot\mathbf{u}_1)\in\mathbf{L}^2(\Omega), \tag{2.8}$$

$$\vartheta_0 \in H_0^1(\Omega)\cap H^2(\Omega),\quad \nabla\vartheta_0\in H_0^1(\Omega;\mathbf{R}^3),\quad \Delta^2\vartheta_0\in\mathbf{L}^2(\Omega), \tag{2.9}$$

$$\mathbf{v}_0 \in H_0^1(\Omega;\mathbf{R}^3)\cap H^2(\Omega;\mathbf{R}^3), \tag{2.10}$$

$$\tilde{\mathbf{v}}_1 := \nabla\cdot\mathbf{G}_0\varepsilon(\mathbf{u}_1) + \dot{\mathbf{f}}(\cdot,0) \in H_0^1(\Omega;\mathbf{R}^3), \tag{2.11}$$

$$\Xi_1 := k_0\Delta\vartheta_0 - l_0\nabla\cdot\mathbf{u}_1 + r(\cdot,0) \in H_0^1(\Omega)\cap H^2(\Omega), \tag{2.12}$$

$$\Xi_2 := \nabla\big(k_0\Delta\vartheta_0 - l_0\nabla\cdot\mathbf{u}_1 + r(\cdot,0)\big) \in H_0^1(\Omega;\mathbf{R}^3), \tag{2.13}$$

$$\Xi_3 := k_0^3\Delta\Big[\Delta\big(\Delta\vartheta_0 - l_0\nabla\cdot\mathbf{u}_1 + r(\cdot,0)\big) + r_t(\cdot,0)\Big] + r_{tt}(\cdot,0) \in \mathbf{L}^2(\Omega), \tag{2.14}$$

$$\mathbf{f}\in\mathcal{H}^3(0,T;\mathbf{L}^2(\Omega;\mathbf{R}^3)), \tag{2.15}$$

$$r\in\mathcal{H}^3(0,T;H^{-1}(\Omega)),\quad r_t(\cdot,0)\in\mathbf{L}^2(\Omega), \tag{2.16}$$

$$\varphi_1,\ \varphi_2\in\mathcal{H}^2(0,T), \tag{2.17}$$

$$\lambda_1,\ \lambda_2\in\mathcal{H}^4(0,T), \tag{2.18}$$

$$\omega_1,\ \omega_2\in H_0^1(\Omega). \tag{2.19}$$

Here $H^{-1}(\Omega)$ denotes, as usual, the dual space of $H_0^1(\Omega)$.
We assume further the following consistency conditions

$$\int_{\Gamma_i} \big(G_{ijlm}(0)\varepsilon_{lm}(\mathbf{u}_0)(y) - \delta_{ij}l_0\vartheta_0(y)\big)n_j(y)\,d\Sigma(y) = \varphi_i(0),\quad i=1,2, \tag{2.20}$$

$$\int_{\Gamma_i} \big(G_{ijlm}(0)\varepsilon_{lm}(\mathbf{u}_1)(y) - \delta_{ij}l_0w_0(y)\big)n_j(y)\,d\Sigma(y)$$

$$+ \int_{\Gamma_i} \big(G'_{ijlm}(0)\varepsilon_{lm}(\mathbf{u}_0)(y) - \delta_{ij}l'(0)\vartheta_0(y)\big)n_j(y)\,d\Sigma(y) = \varphi'_i(0), \quad i = 1, 2, \tag{2.21}$$

$$\int_\Omega \vartheta_0(x)\omega_i(x)\,dx = \lambda_i(0), \qquad \int_\Omega w_0(x)\omega_i(x)\,dx = \lambda'_i(0), \quad i = 1, 2, \tag{2.22}$$

$$\int_\Omega z_0(x)\omega_i(x)\,dx = \lambda''_i(0), \qquad \int_\Omega \xi_0(x)\omega_i(x)\,dx = \lambda_i^{(3)}(0), \quad i = 1, 2. \tag{2.23}$$

Now let us introduce the 2×2 matrices

$$I(\mathbf{z}) := \begin{pmatrix} \int_{\Gamma_1} \varepsilon_{kk}(\mathbf{z})(y)n_r(y)\,d\Sigma(y) & \int_{\Gamma_1} \varepsilon_{rj}(\mathbf{z})(y)n_j(y)\,d\Sigma(y) \\ \int_{\Gamma_2} \varepsilon_{kk}(\mathbf{z})(y)n_s(y)\,d\Sigma(y) & \int_{\Gamma_2} \varepsilon_{sj}(\mathbf{z})(y)n_j(y)\,d\Sigma(y) \end{pmatrix}$$

$$\forall \mathbf{z} \in H^2(\Omega;\mathbf{R}^3),$$

$$J(w,z) := \begin{pmatrix} \int_\Omega w(x)\omega_1(x)\,dx & \int_\Omega z(x)\omega_1(x)\,dx \\ \int_\Omega w(x)\omega_2(x)\,dx & \int_\Omega z(x)\omega_2(x)\,dx \end{pmatrix}$$

$$\forall w,\ z \in \mathbf{L}^2(\Omega).$$

The local existence and uniqueness result is

THEOREM 2.1. *Under the assumptions* (2.6)–(2.23), *let the following conditions*

$$\alpha := \det I(\mathbf{u}_0) \neq 0, \tag{2.24}$$

$$\beta := \det J(\nabla \cdot \mathbf{u}_0, \vartheta_0) \neq 0, \tag{2.25}$$

hold. Then there exists $T^* \in (0,T]$ *such that Problem* (P1) *has a unique solution* $(\mathbf{u},\ \vartheta,\ \psi_b,\ \psi_s,\ h,\ l)$ *satisfying*

$$\mathbf{u} \in \mathbf{C}\big([0,T^*];\ H_0^1(\Omega;\mathbf{R}^3) \cap H^2(\Omega;\mathbf{R}^3)\big) \cap \mathbf{C}^1\big([0,T^*];\ H_0^1(\Omega;\mathbf{R}^3)\big), \tag{2.26}$$

$$\vartheta \in \mathbf{C}\big([0,T^*];\ \mathbf{L}^2(\Omega)\big) \cap \mathbf{L}^2\big(0,T^*;\ H_0^1(\Omega)\big), \tag{2.27}$$

$$\psi_b,\ \psi_s \in \mathcal{H}^2(0,T^*), \quad h,\ l \in \mathcal{H}^4(0,T^*). \tag{2.28}$$

Moreover, we have

$$\ddot{\mathbf{u}} \in \mathbf{C}\big([0,T^*];\ H_0^1(\Omega;\mathbf{R}^3) \cap H^2(\Omega;\mathbf{R}^3)\big) \cap \mathbf{C}^1\big([0,T^*];\ H_0^1(\Omega;\mathbf{R}^3)\big), \tag{2.29}$$

$$\mathbf{u}^{(4)} \in \mathbf{C}\big([0,T^*];\ \mathbf{L}^2(\Omega;\mathbf{R}^3)\big) \cap \mathbf{L}^2\big(0,T^*;\ H_0^1(\Omega;\mathbf{R}^3)\big), \tag{2.30}$$

$$\vartheta_t,\ \vartheta_{tt},\ \vartheta_{ttt} \in \mathbf{C}\big([0,T^*];\ \mathbf{L}^2(\Omega)\big) \cap \mathbf{L}^2\big(0,T^*;\ H_0^1(\Omega)\big), \tag{2.31}$$

$$\vartheta^{(4)} \in \mathbf{L}^2\big(0,T^*;\ H^{-1}(\Omega)\big). \tag{2.32}$$

The second result is concerned with the Lipschitz continuity upon the data.

THEOREM 2.2. *Let* $(\mathbf{u}_{0j},\ \mathbf{u}_{1j},\ \vartheta_{0j},\ \mathbf{f}_j,\ r_j,\ \varphi_{1j},\ \varphi_{2j},\ \lambda_{1j},\ \lambda_{2j})$, $j = 1,2$, *be two sets of data satisfying the hypotheses* (2.5)–(2.18), (2.24)–(2.25) *and the consistency conditions* (2.20)–(2.23) *and let* ω_1, ω_2 *satisfy* (2.19). *Denote by* $(\mathbf{u}_j,\ \vartheta_j,\ \psi_{bj},\ \psi_{sj},\ h_j,\ l_j)$, $j = 1,2$, *the corresponding solutions of Problem* (P1) *and let* L, M *be constants such that*

$$\max\Big\{\big\|\mathbf{u}_{0j}\big\|_{H^3(\Omega;\mathbf{R}^3)},\ \big\|\Delta^2(\nabla\cdot\mathbf{u}_{0j})\big\|_{\mathbf{L}^2(\Omega)},\ \big\|\mathbf{u}_{1j}\big\|_{H^2(\Omega;\mathbf{R}^3)},\ \big\|\Delta(\nabla\cdot\mathbf{u}_{1j})\big\|_{\mathbf{L}^2(\Omega)},$$
$$\big\|\mathbf{v}_{0j}\big\|_{H^2(\Omega;\mathbf{R}^3)},\ \big\|\tilde{\mathbf{v}}_{1j}\big\|_{H^1(\Omega;\mathbf{R}^3)},\ \big\|\vartheta_{0j}\big\|_{H^4(\Omega)},\ \big\|\Xi_{1j}\big\|_{H^2(\Omega)},$$
$$\big\|\Xi_{2j}\big\|_{H^1(\Omega;\mathbf{R}^3)},\ \big\|\Xi_{3j}\big\|_{\mathbf{L}^2(\Omega)},\ \big\|\mathbf{f}_j\big\|_{3,\mathbf{L}^2(\Omega;\mathbf{R}^3)},\ \big\|r_j\big\|_{3,H^{-1}(\Omega)},$$
$$\big\|(r_j)_t(\cdot,0)\big\|_{\mathbf{L}^2(\Omega)},\ \big\|\omega_j\big\|_{H^1(\Omega)},\ |\alpha_j|^{-1},\ |\beta_j|^{-1}\Big\} \le L,$$
$$\max\big\{\|\psi_{bj}\|_2,\ \|\psi_{sj}\|_2,\ \|h_j\|_4,\ \|l_j\|_4\big\} \le M,$$

for $j = 1, 2$. *Then there exists a positive function* $\mathbf{\Lambda} \in \mathbf{C}^0((0,+\infty)^3)$ *such that*

$$\|\mathbf{u}_1-\mathbf{u}_2\|_{2,\infty,H^2(\Omega;\mathbf{R}^3)} + \|\mathbf{u}_1-\mathbf{u}_2\|_{3,\infty,H^1(\Omega;\mathbf{R}^3)}$$
$$+\|\vartheta_1-\vartheta_2\|_{3,\infty,\mathbf{L}^2(\Omega)} + \|\vartheta_1-\vartheta_2\|_{3,H^1(\Omega)}$$
$$+\|\psi_{b1}-\psi_{b2}\|_2 + \|\psi_{s1}-\psi_{s2}\|_2 + \|h_1-h_2\|_4 + \|l_1-l_2\|_4$$
$$\le \mathbf{\Lambda}(L,\ M,\ T)\Big\{\big\|\mathbf{u}_{01}-\mathbf{u}_{02}\big\|_{H^3(\Omega;\mathbf{R}^3)} + \big\|\Delta^2(\nabla\cdot(\mathbf{u}_{01}-\mathbf{u}_{02}))\big\|_{\mathbf{L}^2(\Omega)}$$
$$+\big\|\mathbf{u}_{11}-\mathbf{u}_{12}\big\|_{H^2(\Omega;\mathbf{R}^3)} + \big\|\Delta(\nabla\cdot(\mathbf{u}_{11}-\mathbf{u}_{12}))\big\|_{\mathbf{L}^2(\Omega)}$$
$$+\big\|\mathbf{v}_{01}-\mathbf{v}_{02}\big\|_{H^2(\Omega;\mathbf{R}^3)} + \big\|\tilde{\mathbf{v}}_{11}-\tilde{\mathbf{v}}_{12}\big\|_{H^1(\Omega;\mathbf{R}^3)} + \big\|\vartheta_{01}-\vartheta_{02}\big\|_{H^4(\Omega)}$$
$$+\big\|\Xi_{11}-\Xi_{12}\big\|_{H^2(\Omega)} + \big\|\Xi_{21}-\Xi_{22}\big\|_{H^1(\Omega;\mathbf{R}^3)} + \big\|\Xi_{31}-\Xi_{32}\big\|_{\mathbf{L}^2(\Omega)}$$
$$+\big\|\mathbf{f}_1-\mathbf{f}_2\big\|_{3,\mathbf{L}^2(\Omega;\mathbf{R}^3)} + \big\|r_1-r_2\big\|_{3,H^{-1}(\Omega)} + \big\|(r_1-r_2)_t(\cdot,0)\big\|_{\mathbf{L}^2(\Omega)}\Big\}.$$

Moreover, the function $\mathbf{\Lambda}$ *(also depending on* k, μ, h_0, k_0, l_0, Ω, $|\Gamma_1|$, $|\Gamma_2|$*) is increasing with respect to each of its arguments.*

3. THE DIRECT PROBLEM

Here we present two results related to the direct problem, namely, find $\mathbf{u}$ and ϑ solving the Cauchy–Dirichlet problem (1.6)–(1.9) under the assumption that the memory kernels are given.

The first result is an existence and uniqueness theorem which has been proved in (Navarro, 1978) (see also (Grasselli, c) for an alternative and simpler proof), while the second one is a regularity result (cf. (Grasselli, c)).

In order to state our theorems it is convenient to rewrite the direct problem (1.6)–(1.9) in a Hilbert space setting.

Let us set $H^N := \mathbf{L}^2(\Omega;\mathbf{R}^N)$, $V^N := H_0^1(\Omega;\mathbf{R}^N)$, where N is a given positive integer. As is well known, H^N and V^N are Hilbert spaces. We denote their natural norms by $|\cdot|_N$ and $\|\cdot\|_N$, respectively. If $N = 1$, we simply set $H = H^1$,

$V = V^1$ and $|\cdot| = |\cdot|_1$, $\|\cdot\| = \|\cdot\|_1$. Moreover, we recall that $V^N \hookrightarrow H^N \hookrightarrow (V^N)' = H^{-1}(\Omega; \mathbf{R}^N)$ with compact and dense injections.
Consider now the symmetric bilinear forms e: $V^3 \times V^3 \to \mathbf{R}$ and a: $V \times V \to \mathbf{R}$ defined by

$$e(\mathbf{w}, \mathbf{z}) := \int_\Omega G_{ijlm}(0)\varepsilon_{ij}(\mathbf{w})(x)\varepsilon_{lm}(\mathbf{z})(x)\,dx \quad \forall \mathbf{w},\ \mathbf{z} \in V^3,$$

$$a(w, z) := \int_\Omega k_0 \nabla w(x) \cdot \nabla z(x)\,dx \quad \forall w,\ z \in V,$$

where the symbol $\cdot$ denotes the scalar product in $\mathbf{R}^3$.
It is well known that such forms are bounded and coercive and define equivalent norms in V^3 and V. Then, we denote by E and A the linear operators associated with the bilinear forms e and a, respectively. Note that

$$E := -\nabla \cdot \mathbf{G}_0\varepsilon\colon V^3 \to (V^3)', \quad A := -k_0\Delta\colon V \to V'. \tag{3.1}$$

Introducing now the set $D_E := \{\mathbf{z} \in H^3\colon E\mathbf{z} \in H^3\}$, and endowing it with the graph norm, we observe that the following identification holds (cf., e.g., (Hughes and Marsden, 1983)): $D_E \equiv H_0^1(\Omega; \mathbf{R}^3) \cap H^2(\Omega; \mathbf{R}^3)$.
Taking into account the previous considerations (cf. in particular (3.1)), the direct problem (1.6)–(1.9) can be written as

$$\ddot{\mathbf{u}}(t) + E\mathbf{u}(t) + (\mathbf{G}' * B\mathbf{u})(t) = -P(l_0\vartheta + l' * \vartheta)(t) + \mathbf{f}(t) \quad \text{for a.e. } t \in (0,T), \tag{3.2}$$

$$h_0\vartheta_t(t) + A\vartheta(t) + (h' * \vartheta)_t(t) = -Q\big(l_0\dot{\mathbf{u}} + (l' * \mathbf{u})_t\big)(t) + r(t) \quad \text{for a.e. } t \in (0,T), \tag{3.3}$$

$$\mathbf{u}(0) = \mathbf{u}_0, \quad \dot{\mathbf{u}}(0) = \mathbf{u}_1, \quad \vartheta(0) = \vartheta_0, \tag{3.4}$$

where the symbol $*$ denotes the convolution product with respect to time over the interval $(0,t)$ and B: $V^3 \to (V^3)'$, P: $V \to H^3$, Q: $V^3 \to H$ are defined by

$$B\mathbf{z} := -\nabla \cdot \hat{\mathbf{G}}\varepsilon(\mathbf{z}) \quad \text{with} \quad \hat{\mathbf{G}} := \Big[\frac{1}{2}(\delta_{il}\delta_{jm} + \delta_{im}\delta_{jl})\Big], \quad \forall \mathbf{z} \in V^3,$$

$$Pz := \nabla z, \quad \forall z \in V, \qquad Q\mathbf{z} := \nabla \cdot \mathbf{z}, \quad \forall \mathbf{z} \in V^3.$$

Note that we also have Q: $H^3 \to V'$.
Let us assume

$$\mathbf{f} \in \mathcal{H}^1(0,T;\, (V^3)') + \mathbf{L}^2(0,T;\, H^3), \quad r \in \mathbf{L}^2(0,T;\, V'), \tag{3.5}$$

$$\mathbf{u}_0 \in V^3, \quad \mathbf{u}_1 \in H^3, \quad \vartheta_0 \in H, \tag{3.6}$$

$$\psi_b,\ \psi_s,\ h,\ l \in \mathcal{H}^2(0,T). \tag{3.7}$$

Then we have

THEOREM 3.1. *Under the assumptions (3.5)–(3.7), there exists a unique solution $(\mathbf{u},\vartheta)$ to problem (3.2)–(3.4) such that*

$$\mathbf{u}\in \mathbf{C}([0,T];\,V^3)\cap \mathbf{C}^1([0,T];\,H^3),\quad \ddot{\mathbf{u}}\in \mathcal{H}^1(0,T;\,(V^3)')+\mathbf{L}^2(0,T;\,H^3),$$
$$\vartheta\in \mathbf{C}([0,T];\,H)\cap \mathbf{L}^2(0,T;\,V),\quad \vartheta_t\in \mathbf{L}^2(0,T;\,V').$$

Replacing now the hypotheses (3.5) and (3.6) by

$$\mathbf{f}\in \mathcal{H}^1(0,T;\,H^3)+\mathbf{L}^2(0,T;\,V^3),\quad r\in \mathcal{H}^1(0,T;\,V'), \tag{3.8}$$

$$\mathbf{u}_0\in D_E,\quad \mathbf{u}_1\in V^3,\quad \vartheta_0\in H,\quad -A\vartheta_0+r(0)\in H, \tag{3.9}$$

we can prove

THEOREM 3.2. *Under the assumptions (3.7)–(3.9), the unique solution $(\mathbf{u},\vartheta)$ to problem (3.2)–(3.4) has the following regularity properties:*

$$\mathbf{u}\in \mathbf{C}([0,T];\,D_E)\cap \mathbf{C}^1([0,T];\,V^3),\quad \ddot{\mathbf{u}}\in \mathcal{H}^1(0,T;\,H^3)+\mathbf{L}^2(0,T;\,V^3), \tag{3.10}$$

$$\vartheta\in \mathbf{C}^1([0,T];\,H)\cap \mathcal{H}^1(0,T;\,V),\quad \vartheta_{tt}\in \mathbf{L}^2(0,T;\,V'). \tag{3.11}$$

4. AN EQUIVALENT PROBLEM

Let us assume that the hypotheses (2.6)–(2.25) be satisfied and assume moreover that Problem (P1) has a unique solution possessing the reqularity properties (2.26)–(2.32).

We now show that this solution solves an equivalent problem consisting in a system of four nonlinear Volterra equations of the second kind.

First we observe that $\psi_b'(0)$, $\psi_s'(0)$, $h'(0)$, $h''(0)$, $h^{(3)}(0)$, $l'(0)$, $l''(0)$, and $l^{(3)}(0)$ can be expressed in terms of the data by using the consistency conditions (2.21)–(2.23) owing to the nonvanishing conditions (2.24)–(2.25).

Indeed, in view of (2.2), the right side consistency conditions (2.22) is representable as

$$\Big(\int_\Omega \nabla\cdot\mathbf{u}_0(x)\omega_i(x)\,dx\Big)l'(0)+\Big(\int_\Omega \vartheta_0(x)\omega_i(x)\,dx\Big)h'(0)=-h_0\lambda_i'(0)$$
$$+k_0\int_\Omega \Delta\vartheta_0(x)\omega_i(x)\,dx+l_0\int_\Omega \nabla\cdot\mathbf{u}_1(x)\omega_i(x)\,dx+\int_\Omega r(x,0)\omega_i(x)\,dx,$$

$$i=1,2. \tag{4.1}$$

This system is uniquely solvable owing to condition (2.25). Hence we can compute $l'(0)$ and $h'(0)$ in terms of known quantities, i.e.,

$$l'(0)=\tilde{l}_0,\qquad h'(0)=\tilde{h}_0, \tag{4.2}$$

where $\tilde{l}_0$ and $\tilde{h}_0$ are constants only depending on h_0, k_0, l_0, ϑ_0, $\mathbf{u}_0$, $\mathbf{u}_1$, ω_1, ω_2, $\lambda_1(0)$, $\lambda_2(0)$.
Then, taking into account (2.4), (2.25), and (4.1), we can, by the left side consistency conditions (2.23), express $l''(0)$ and $h''(0)$ in terms of known quantities, namely,

$$l''(0) = \bar{l}_0, \qquad h''(0) = \bar{h}_0, \tag{4.3}$$

where $\bar{l}_0$ and $\bar{h}_0$ are constants depending only upon the data.
Similarly, (2.5), (2.25), and (4.1)–(4.2) allow us to obtain, from the right side consistency conditions (2.23), the equations

$$l^{(3)}(0) = \hat{l}_0, \qquad h^{(3)}(0) = \hat{h}_0, \tag{4.4}$$

$\hat{l}_0$, $\hat{h}_0$ being constants depending only upon the data.
Consider now the consistency conditions (2.21). By (1.4) and (4.1), these conditions can be written as

$$\begin{aligned}
&\Big(\int_{\Gamma_i} \varepsilon_{kk}(\mathbf{u}_0)(y)n_i(y)\,d\Sigma(y)\Big)b(0) + 2\Big(\int_{\Gamma_i} \varepsilon_{ij}(\mathbf{u}_0)(y)n_j(y)\,d\Sigma(y)\Big)s(0) \\
=&\varphi_i'(0) - \int_{\Gamma_i} \big(G_{ijlm}(0)\varepsilon_{lm}(\mathbf{u}_1)(y) - \delta_{ij}l_0 w_0(y)\big)n_j(y)\,d\Sigma(y) \\
&+\tilde{l}_0 \int_{\Gamma_0} \vartheta_0(y)n_i(y)\,d\Sigma(y), \quad i = 1,\, 2,
\end{aligned} \tag{4.5}$$

where, for simplicity, we have set

$$b(t) := \frac{1}{3}\big[\dot{\psi}_b(t) - \dot{\psi}_s(t)\big], \qquad s(t) := \frac{1}{2}\dot{\psi}_s(t), \quad \forall t \in [0,T]. \tag{4.6}$$

Hence, taking (1.5) and (4.1) into account, the nonvanishing condition (2.24) permits to compute $b(0)$ and $s(0)$ from (4.5), that is (cf. (4.6)) $\psi_b'(0)$ and $\psi_s'(0)$. Thus we have $\mathbf{G}'(0) = \tilde{\mathbf{G}}_0$, where $\tilde{\mathbf{G}}_0$ is a fourth-order isotropic tensor whose coefficients only depend upon known quantities.
Taking equations (4.1)–(4.3), (4.6) into account, we prove, by differentiating in time, that the functions

$$\mathbf{v} := \ddot{\mathbf{u}}, \quad z := \vartheta_{tt}, \quad \tilde{b} := b', \quad \tilde{s} := s', \quad \tilde{h} := h^{(3)}, \quad \tilde{l} := l^{(3)}, \tag{4.7}$$

solve an inverse problem analogous to Problem (P1). Then, owing to conditions (2.24) and (2.25), the problem obtained will be transformed into a system of four nonlinear second kind Volterra integral equations for the unknowns $\tilde{b}$, $\tilde{s}$, $\tilde{h}$, $\tilde{l}$.
Taking the limit as $t \to 0^+$ in equations (1.6) and (1.7), we easily find that

$$\mathbf{v}(x,0) = \ddot{\mathbf{u}}(x,0) = \mathbf{v}_0(x), \qquad \vartheta_t(x,0) = w_0(x) \quad \text{for a.e.} \quad x \in \Omega. \tag{4.8}$$

Differentiating in time the equation (1.6), we obtain

$$\mathbf{u}^{(3)}(x,t) = \nabla \cdot \Big(\mathbf{G}_0\varepsilon(\dot{\mathbf{u}})(x,t) + \mathbf{G}'(t)\varepsilon(\mathbf{u}_0)(x) + \int_0^t \mathbf{G}'(\tau)\varepsilon(\dot{\mathbf{u}})(x,t-\tau)\,d\tau\Big)$$

$$- l_0\nabla\vartheta_t(x,t) - l'(t)\nabla\vartheta_0(x) - \int_0^t l'(\tau)\nabla\vartheta_t(x,t-\tau)\,d\tau + \dot{\mathbf{f}}(x,t)$$

$$\text{for a.e.} \quad (x,t) \in \Omega \times (0,T). \tag{4.9}$$

Taking now the limit as $t \to 0^+$ in the equation (4.9), we get (cf. (2.3))

$$\dot{\mathbf{v}}(x,0) = \mathbf{u}^{(3)}(x,0) = \mathbf{v}_1(x) \quad \text{for a.e.} \quad x \in \Omega. \tag{4.10}$$

Then, further differentiation in time of the equation (4.9) gives

$$\ddot{\mathbf{v}}(x,t) = \nabla \cdot \Big(\mathbf{G}_0\varepsilon(\mathbf{v})(x,t) + \int_0^t \mathbf{G}'(\tau)\varepsilon(\mathbf{v})(x,t-\tau)\,d\tau\Big) - l_0\nabla z(x,t)$$

$$- \int_0^t l'(\tau)\nabla z(x,t-\tau)\,d\tau + \ddot{\mathbf{f}}(x,t) - l''(t)\nabla\vartheta_0(x) - l'(t)\nabla w_0(x)$$

$$+ \nabla \cdot \big(\mathbf{G}''(t)\varepsilon(\mathbf{u}_0)(x) + \mathbf{G}'(t)\varepsilon(\mathbf{u}_1)(x)\big)$$

$$\text{for a.e.} \quad (x,t) \in \Omega \times (0,T). \tag{4.11}$$

Let us now consider equation (1.7) and differentiate it with respect to time. We obtain

$$h_0\vartheta_{tt}(x,t) = k_0\Delta\vartheta_t(x,t) - h''(t)\vartheta_0(x) - h'(t)w_0(x) - \int_0^t h'(\tau)\vartheta_{tt}(x,t-\tau)\,d\tau$$

$$- l_0\nabla \cdot \ddot{\mathbf{u}}(x,t) - l''(t)\nabla \cdot \mathbf{u}_0(x) - l'(t)\nabla \cdot \mathbf{u}_1(x) - \int_0^t l'(\tau)\nabla \cdot \ddot{\mathbf{u}}(x,t-\tau)\,d\tau$$

$$+ r_t(x,t), \qquad \text{for a.e.} \quad (x,t) \in \Omega \times (0,T). \tag{4.12}$$

Passing to the limit as $t \to 0^+$ in the equation (4.12), we find (cf. (2.4))

$$z(x,0) = \vartheta_{tt}(x,0) = z_0(x) \quad \text{for a.e.} \quad x \in \Omega. \tag{4.13}$$

Another differentiation in time of (4.12) yields the following equation (cf. (4.7))

$$h_0 z_t(x,t) = k_0\Delta z(x,t) - \Big(\int_0^t h'(\tau)z(x,t-\tau)\,d\tau\Big)_t$$

$$- l_0\nabla \cdot \dot{\mathbf{v}}(x,t) - \Big(\int_0^t l'(\tau)\nabla \cdot \mathbf{v}(x,t-\tau)\,d\tau\Big)_t + r_{tt}(x,t)$$

$$- h^{(3)}(t)\vartheta_0(x) - h''(t)w_0(x) - l^{(3)}(t)\nabla \cdot \mathbf{u}_0(x) - l''(t)\nabla \cdot \mathbf{u}_1(x)$$

$$\text{for a.e.} \quad (x,t) \in \Omega \times (0,T). \tag{4.14}$$

Moreover, from the boundary conditions (1.9) we infer

$$\mathbf{v}(x,t) = \mathbf{0}, \qquad z(x,t) = 0 \quad \text{for a.e.} \quad (x,t) \in \Gamma \times (0,T). \tag{4.15}$$

Summing up, using the notations introduced in § 3 (cf. (3.2)–(3.4)),we deduce from (4.8), (4.10)–(4.11), (4.13)–(4.15) (see also (1.4), (4.5)) that $\mathbf{v}$ and z satisfy the following Cauchy–Dirichlet problem

$$\ddot{\mathbf{v}}(t) + E\mathbf{v}(t) + (\mathbf{G}' * B\mathbf{v})(t) = -P(l_0 z + l' * z)(t) + \ddot{\mathbf{f}}(t) + \mathcal{F}(b,s,l)(t)$$
$$\text{for a.e.} \quad t \in (0,T), \tag{4.16}$$
$$h_0 z_t(t) + Az(t) + (h' * z)_t(t) = -Q\big(l_0\dot{\mathbf{v}} + (l' * \mathbf{v})_t\big)(t) + r_{tt}(t) + \mathcal{R}(h,l)(t)$$
$$\text{for a.e.} \quad t \in (0,T), \tag{4.17}$$
$$\mathbf{v}(0) = \mathbf{v}_0, \quad \dot{\mathbf{v}}(0) = \mathbf{v}_1, \quad z(0) = z_0, \tag{4.18}$$

where $\mathcal{F}$ and $\mathcal{R}$ are defined by

$$\mathcal{F}(b,s,l)(t) := \nabla\cdot\big(\mathbf{G}''(t)\varepsilon(\mathbf{u}_0) + \mathbf{G}'(t)\varepsilon(\mathbf{u}_1)\big) - l''(t)\nabla\vartheta_0 - l'(t)\nabla w_0, \tag{4.19}$$
$$\mathcal{R}(h,l)(t) := -h^{(3)}(t)\vartheta_0 - h''(t)w_0 - l^{(3)}(t)\nabla\cdot\mathbf{u}_0 - l''(t)\nabla\cdot\mathbf{u}_1, \tag{4.20}$$

for a.e. $t \in (0,T)$.
Let us now consider the additional information (1.10) from which we infer the equations for $\tilde{b}$ and $\tilde{s}$. Differentiating twice relations (1.10) we get (cf. (1.1), (1.8), (4.8))

$$\int_{\Gamma_i} G_{ijlm}(0)\varepsilon_{lm}(\mathbf{v})(y,t)n_j(y)\,d\Sigma(y) + b'(t)\int_{\Gamma_i}\varepsilon_{kk}(\mathbf{u}_0)(y)n_i(y)\,d\Sigma(y)$$
$$+ 2s'(t)\int_{\Gamma_i}\varepsilon_{ij}(\mathbf{u}_0)(y)n_j(y)\,d\Sigma(y) + b(t)\int_{\Gamma_i}\varepsilon_{kk}(\mathbf{u}_1)(y)n_i(y)\,d\Sigma(y)$$
$$+ 2s(t)\int_{\Gamma_i}\varepsilon_{ij}(\mathbf{u}_1)(y)n_j(y)\,d\Sigma(y) + \int_0^t G'_{ijlm}(\tau)\,d\tau\int_{\Gamma_i}\varepsilon_{lm}(\mathbf{v})(y,t-\tau)n_j(y)\,d\Sigma(y)$$
$$- l_0\int_{\Gamma_i} z(y,t)n_i(y)\,d\Sigma(y) - l''(t)\int_{\Gamma_i}\vartheta_0(y)n_i(y)\,d\Sigma(y) - l'(t)\int_{\Gamma_i} w_0(y)n_i(y)\,d\Sigma(y)$$
$$- \int_0^t l'(\tau)\,d\tau\int_{\Gamma_i} z(y,t-\tau)n_i(y)\,d\Sigma(y) = \varphi_i''(t)$$

$$\text{for a.e.} \quad t \in (0,T), \quad i = 1,2. \tag{4.21}$$

In order to write the equations for $h^{(3)}$ and $l^{(3)}$, we multiply by ω_i the both terms of equation (1.7). Then, taking (1.11) into account, integrating over Ω and differentiating three times, we obtain the following equations (cf. also (1.8), (2.19), (2.22)–(2.24), (4.8)):

$$h_0\lambda_i^{(3)}(t) = k_0 \int_\Omega \nabla z(x,t)\cdot\nabla\omega_i(x)\,dx - h^{(3)}(t)\lambda_i(0) - h''(t)\lambda_i'(0)$$
$$- h'(t)\lambda_i''(0) - \int_0^t h'(\tau)\lambda_i^{(3)}(t-\tau)\,d\tau$$
$$- l_0 \int_\Omega \nabla\cdot\dot{\mathbf{v}}(x,t)\omega_i(x)\,dx - l^{(3)}(t)\int_\Omega \nabla\cdot\mathbf{u}_0(x)\omega_i(x)\,dx$$
$$- l''(t)\int_\Omega \nabla\cdot\mathbf{u}_1(x)\omega_i(x)\,dx - l'(t)\int_\Omega \nabla\cdot\mathbf{v}_0(x)\omega_i(x)\,dx$$
$$- \int_0^t l'(\tau)\int_\Omega \dot{\mathbf{v}}(x,t-\tau)\cdot\nabla\omega_i(x)\,dx\,d\tau + \int_\Omega r_{tt}(x,t)\omega_i(x)\,dx \quad (4.22)$$

for a.e. $t \in (0,T)$, $i = 1,2$.
It is useful to rewrite the equations (4.21) and (4.22) as

$$\Big(\int_{\Gamma_i} \varepsilon_{kk}(\mathbf{u}_0)(y)n_i(y)\,d\Sigma(y)\Big)b'(t) + 2\Big(\int_{\Gamma_i}\varepsilon_{ij}(\mathbf{u}_0)(y)n_j(y)\,d\Sigma(y)\Big)s'(t)$$
$$= \mathcal{S}^i(\mathbf{v},z,b,s,l)(t) \qquad \text{for a.e.} \quad t\in(0,T) \quad i = 1,\ 2, \qquad (4.23)$$

and

$$\Big(\int_\Omega \nabla\cdot\mathbf{u}_0(x)\omega_i(x)\,dx\Big)h^{(3)}(t) + \Big(\int_\Omega \vartheta_0(x)\omega_i(x)\,dx\Big)l^{(3)}(t)$$
$$= \mathcal{T}^i(\mathbf{v},z,h,l)(t) \qquad \text{for a.e.} \quad t\in(0,T) \quad i = 1,\ 2, \qquad (4.24)$$

where

$$\mathcal{S}^i(\mathbf{v},z,b,s,l) := \varphi_i''(t) - \int_{\Gamma_i} G_{ijlm}(0)\varepsilon_{lm}(\mathbf{v})(y,t)n_j(y)\,d\Sigma(y)$$
$$- b(t)\int_{\Gamma_i}\varepsilon_{kk}(\mathbf{u}_1)(y)n_i(y)\,d\Sigma(y) - 2s(t)\int_{\Gamma_i}\varepsilon_{ij}(\mathbf{u}_1)(y)n_j(y)\,d\Sigma(y)$$
$$- \int_0^t G'_{ijlm}(\tau)\,d\tau\int_{\Gamma_i}\varepsilon_{lm}(\mathbf{v})(y,t-\tau)n_j(y)\,d\Sigma(y)$$
$$+ l_0\int_{\Gamma_i} z(y,t)n_i(y)\,d\Sigma(y) + l''(t)\int_{\Gamma_i}\vartheta_0(y)n_i(y)\,d\Sigma(y)$$

$$+\int_0^t l'(\tau)\int_{\Gamma_i} z(y,t-\tau)n_i(y)\,d\Sigma(y)$$

$$+\,l'(t)\int_{\Gamma_i} w_0(y)n_i(y)\,d\Sigma(y), \tag{4.25}$$

$$\begin{aligned}\mathcal{T}^i(\mathbf{v},z,h,l)(t) := &-h_0\lambda_i^{(3)}(t)+k_0\int_\Omega \nabla z(x,t)\cdot\nabla\omega_i(x)\,dx\\ &-h''(t)\lambda_i'(0)-h'(t)\lambda_i''(0)-\int_0^t h'(\tau)\lambda_i^{(3)}(t-\tau)\,d\tau\\ &-l_0\int_\Omega \dot{\mathbf{v}}(x,t)\cdot\nabla\omega_i(x)\,dx-l''(t)\int_\Omega \nabla\cdot\mathbf{u}_1(x)\omega_i(x)\,dx\\ &-l'(t)\int_\Omega \nabla\cdot\mathbf{v}_0(x)\omega_i(x)\,dx-\int_0^t l'(\tau)\int_\Omega \nabla\cdot\dot{\mathbf{v}}(x,t-\tau)\omega_i(x)\,dx\,d\tau\\ &+\int_\Omega r_{tt}(x,t)\omega_i(x)\,dx,\end{aligned} \tag{4.26}$$

for a.e. $t \in (0,T)$, $i = 1, 2$.
Then, owing to the nonvanishing conditions (2.24)–(2.25), from (4.23)–(4.24) we infer the equations

$$[b',\, s']^{tr} = I(\mathbf{u}_0)^{-1}\left[\mathcal{S}^1(\mathbf{v},z,b,s,l),\, \mathcal{S}^2(\mathbf{v},z,b,s,l)\right]^{tr} \tag{4.27}$$

$$[h^{(3)},\, l^{(3)}]^{tr} = J(\nabla\cdot\mathbf{u}_0,\vartheta_0)^{-1}\left[\mathcal{T}^1(\mathbf{v},z,h,l),\, \mathcal{T}^2(\mathbf{v},z,h,l)\right]^{tr} \tag{4.28}$$

holding a.e. in $(0,T)$, where tr denotes the transposition operation.
Observe now that for any set of memory functions ψ_b, ψ_s, h, l such that (cf. Theorem 2.1) $\psi_b,\ \psi_s \in \mathcal{H}^2(0,T)$ and $h,\ l \in \mathcal{H}^4(0,T)$, there exists a unique solution $(\mathbf{v},\ z)$ to problem (4.16)–(4.18) satisfying (3.10)–(3.11). This is a straightforward consequence of Theorem 3.2. Indeed, recalling equations (2.4)–(2.5), hypotheses (2.6)–(2.18), and equations (4.8), (4.10), (4.13), (4.19)–(4.20), we have (cf. (3.8), (3.9))

$$\ddot{\mathbf{f}}+\mathcal{F}(b,s,l)\in\mathcal{H}^1(0,T;\,H^3)+\mathbf{L}^2(0,T;\,V^3),\quad r_{tt}+\mathcal{R}(k,h,l)\in\mathcal{H}^1(0,T;\,V'),$$

$$\mathbf{v}_0\in D_E,\quad \mathbf{v}_1\in V^3,\quad z_0\in H,\quad -Az_0+r_{tt}(0)+\mathcal{R}(h,l)(0)\in H.$$

Thus, the mapping $\mathcal{X} := (\aleph,\mathcal{Z})$ defined by

$$\mathbf{v} := \aleph(b,s,h,l),\quad z := \mathcal{Z}(b,s,h,l), \tag{4.29}$$

$(\mathbf{v},z)$ being the unique solution to problem (4.16)–(4.18), acts from $\mathcal{H}^1(0,T)\times\mathcal{H}^1(0,T)\times\mathcal{H}^4(0,T)\times\mathcal{H}^4(0,T)$ to $[\mathbf{C}([0,T];D_E)\cap\mathbf{C}^1([0,T];V)\cap\mathcal{H}^2(0,T;H)]\times[\mathbf{C}^1([0,T];H)\cap\mathcal{H}^1(0,T;V)]$ (cf. (3.10) and (3.11)).

On the other hand, by equations (4.1)–(4.2), (4.5)–(4.7), we have

$$b(t) = b_0 + \int_0^t \tilde{b}(\tau)\,d\tau, \quad s(t) = s_0 + \int_0^t \tilde{s}(\tau)\,d\tau, \tag{4.30}$$

$$l(t) = l_0 + \tilde{l}_0 t + \frac{\tilde{l}_0}{2}t^2 + \int_0^t (t-\tau)^2 \tilde{l}(\tau)\,d\tau, \tag{4.31}$$

$$h(t) = h_0 + \tilde{h}_0 t + \frac{\tilde{h}_0}{2}t^2 + \int_0^t (t-\tau)^2 \tilde{h}(\tau)\,d\tau, \tag{4.32}$$

for any $t \in [0,T]$, where b_0, s_0 are the scalars associated to the known tensor $\tilde{\mathbf{G}}_0$ (cf. (1.4), (4.6)). Hence the system (4.27)–(4.28) can be rewritten in the fixed-point form

$$[\tilde{b}, \tilde{s}]^{tr} = I(\mathbf{u}_0)^{-1}\left[\tilde{\mathcal{S}}^1(\tilde{b},\tilde{s},\tilde{h},\tilde{l}),\ \tilde{\mathcal{S}}^2(\tilde{b},\tilde{s},\tilde{h},\tilde{l})\right]^{tr}, \tag{4.33}$$

$$[\tilde{h}, \tilde{l}]^{tr} = J(\nabla\cdot\mathbf{u}_0;\vartheta_0)^{-1}\left[\tilde{\mathcal{T}}^1(\tilde{b},\tilde{s},\tilde{h},\tilde{l}),\ \tilde{\mathcal{T}}^2(\tilde{b},\tilde{s},\tilde{h},\tilde{l})\right]^{tr}, \tag{4.34}$$

a.e. in $(0,T)$, where $\tilde{\mathcal{S}}^i$, $\tilde{\mathcal{T}}^i$, $i = 1, 2$, are the operators obtained from (4.25)–(4.26) by using (4.29)–(4.32).
Conversely, assume that $(\tilde{b},\tilde{s},\tilde{h},\tilde{l}) \in (\mathbf{L}^2(0,T))^2 \times (\mathcal{H}^1(0,T))^2$ solves system (4.33)–(4.34) and consider the functions ψ_b, ψ_s, h, l, given by (1.5), (4.6), (4.30)–(4.32) and the functions $\mathbf{u}$, ϑ defined by

$$\mathbf{u}(x,t) := \mathbf{u}_0(x) + t\mathbf{u}_1(x) + \int_0^t (t-\tau)\mathbf{v}(x,\tau)\,d\tau,$$

$$\vartheta(x,t) := \vartheta_0(x) + t\vartheta_1(x) + \int_0^t (t-\tau)z(x,\tau)\,d\tau$$

for any $(x,t) \in \Omega \times (0,T)$, $\mathbf{v}$, z given by (4.29). Then, thanks to the consistency conditions (2.20)–(2.23), it is not difficult to prove that $(\mathbf{u},\ \vartheta,\ \psi_b,\ \psi_s,\ h,\ l)$ is a solution to Problem (P1) having the regularity properties (2.26)–(2.32).
Thus, we have proved the following equivalence result.

THEOREM 4.1. *Let the assumptions of Theorem* 2.1 *hold. Problem* (*P*1) *has a unique solution* $(\mathbf{u},\ \vartheta,\ \psi_b,\ \psi_s,\ h,\ l)$ *satisfying* (2.26)–(2.32) *and only if the system* (4.33)–(4.34) *has a unique solution* $(\tilde{b},\tilde{s},\tilde{h},\tilde{l}) \in (\mathbf{L}^2(0,T))^2 \times (\mathcal{H}^1(0,T))^2$.

This theorem allows us to prove our main results by solving the system (4.33) and (4.34) via Contraction Principle, once suitable local boundedness and local Lipschitz continuity estimates for the mapping $\mathcal{X}$ (cf. (4.29)) have been obtained (see (Grasselli, c)).

REFERENCES

Bercia, R. (1990). Existence and uniqueness for linear wave thermoviscoelasticity of integral type. *Stud. Cerc. Math.* **42**, 95–102.

Bhattacharyya, R.K. (1991). Wave propagation in random thermoelastic materials with memory. In: *Mathematical and Numerical Aspects of Wave Propagation*. Ed. by G. Cohen, L. Halpern and P. Joly, SIAM, Philadelphia, pp. 626–634.

Burns, J.A., Liu, Z.Y., and Miller, R.E. (1990). Approximations of thermoelastic and viscoelastic control systems. *NASA-ICASE rep.* **90–43**, 1–66.

Chakrabarti, S. (1976). Waves in thermoelastic materials with memory. *Bull. Math. Soc. Sci. Math. R. S. Roum.* **20**, 45–57.

Grasselli, M. (a). An identification problem for a linear integrodifferential equation occurring in heat flow. *Math. Meth. Appl. Sci.* (to appear).

Grasselli, M. (b). An identification problem for an abstract linear hyperbolic integrodifferential equation with applications. *J. Math. Anal. Appl.* (to appear).

Grasselli, M. (c). Inverse problems in linear thermoviscoelasticity of integral type. (to appear).

Hughes, T.J.R. and Marsden, J.E. (1983). In: *Mathematical Foundations of Elasticity*. Prentice–Hall, Englewood, Cliffs, pp. 315–364.

Leugering, G. (1985). Boundary controllability in one-dimensional linear thermoviscoelasticity. *J. Integr. Equat.* **10**, 157–173.

Luca, I. (1989a). A uniqueness theorem in anisotropic viscothermoelasticity of integral type. *Arch. Mech.* **41**, 535–542.

Luca, I. (1989b). Domain of influence and uniqueness in viscothermoelasticity of integral type. *Continuum Mech. Thermodyn.* **1**, 213–226.

Navarro, C.B. (1978). Asymptotic stability in linear thermoviscoelasticity. *J. Math. Anal. Appl.* **65**, 399–431.

Ill-Posed Problems in Natural Sciences, pp. 300 – 306
A. Tikhonov (Ed.)
1992 VSP/TVP

AN INVERSE BOUNDARY VALUE PROBLEM FOR THE EQUATION OF THE PLATE IN THE LOVE–KIRCHHOFF THEORY

M. IKEHATA
Department of Applied Physics, School of Engineering Nagoya University, Chikusa-ku, Nagoya 464-01, Japan

INTRODUCTION

In this paper, we study an inverse boundary value problem for the equation of the flexure of the linear, inhomogeneous isotropic, elastic, thin plate in the context of the Love–Kirchhoff theory. More precisely, let Ω be a bounded domain of $\mathbf{R}^2$ with a C^∞ boundary $\partial\Omega$. In what follows, $L_{(a,b)}$ denotes the partial differential operator of order 4 acting on functions on Ω such that

$$L_{(a,b)}w = \sum_{\alpha,\beta} \frac{\partial^2}{\partial x_\alpha \partial x_\beta} M_{\alpha\beta}(w),$$

where $(a,b) = (a(x), b(x))$ is a pair of functions on Ω, $w = w(x)$ is a function on Ω, and $M_{\alpha,\beta}(w)$ is defined by

$$M_{\alpha,\beta}(w) = -\left\{ a(\Delta w)\delta_{\alpha\beta} + 2b\frac{\partial^2}{\partial x_\alpha \partial x_\beta} w \right\}.$$

Let $h > 0$. We consider $\Omega \times]-h/2, h/2[$ as a mathematical model of the bounded, linear, isotropic, inhomogeneous, elastic, thin plate with the thickness of h and the median section Ω. The material forming the plate determines the above (a,b), indeed, it is given by

$$a = \frac{h^3}{12}\frac{\nu E}{(1-\nu^2)}, \tag{1}$$

$$2b = \frac{h^3}{12}\frac{(1-\nu)E}{(1-\nu^2)}, \tag{2}$$

where $E = E(x)$ and $\nu = \nu(x)$ denote the Young's modulus and the Poisson's ratio of the material forming the plate, respectively. It should be noted that the transverse displacement at $x \in \Omega$ is represented by $w(x)$. For the details, see, e.g., (Dautray and Lions, 1990).

MAIN RESULTS

BOUNDARY VALUE PROBLEM. *For given functions ϕ_1, ϕ_2 on $\partial\Omega$, find a function w in Ω such that*

$$\begin{aligned} L_{(a,b)}w &= 0 \quad \text{in} \quad \Omega, \\ w &= \phi_1 \quad \text{on } \partial\Omega, \\ \frac{\partial}{\partial n}w &= \phi_2 \quad \text{on } \partial\Omega, \end{aligned}$$

where n denotes the unit outward normal vector field to $\partial\Omega$.

By the Lax–Milgram method, we easily obtain the following result.

PROPOSITION 1: *Let $(a,b) \in \{L^\infty(\Omega)\}^2$ and $\eta > 0$. If (a,b) satisfies that*

$$\min\left(a(x) + b(x),\ b(x)\right) > \eta^{-1}$$

a.e. *in Ω, then, for any $(\phi_1, \phi_2) \in H^{3/2}(\partial\Omega) \times H^{1/2}(\partial\Omega)$, there exists a unique $w \in H^2(\Omega)$ such that*

$$\begin{aligned} L_{(a,b)}w &= 0 \quad \text{in} \quad \Omega, \\ w &= \phi_1 \quad \text{on } \partial\Omega, \\ \frac{\partial}{\partial n}w &= \phi_2 \quad \text{on } \partial\Omega, \end{aligned} \tag{$*$}$$

Furthermore, the estimate

$$\|w\|_{H^2(\Omega)} \leq C(\Omega,\eta)\{1 + \|a\|_{L^\infty(\Omega)} + \|b\|_{L^\infty(\Omega)}\} \times \{\|\phi_1\|_{H^{3/2}(\partial\Omega)} + \|\phi_2\|_{H^{1/2}(\partial\Omega)}\}$$

holds for a positive constant $C(\Omega,\eta)$.

By this proposition, there exists a unique bounded operator $\Lambda_{(a,b)}$ from $H^{3/2}(\partial\Omega)\times H^{1/2}(\partial\Omega))$ to its dual such that

$$\begin{aligned} \langle \Lambda_{(a,b)}(\phi_1^1, \phi_2^1), (\phi_1^2, \phi_2^2)\rangle &= -\sum_{\alpha,\beta} \int_\Omega M_{\alpha,\beta}(w_1)\frac{\partial^2}{\partial x_\alpha \partial x_\beta} w_2\, dx \\ &= \int_\Omega a\Delta w_1 \Delta w_2 + 2b\nabla^2 w_1 \cdot \nabla w_2\, dx, \end{aligned} \tag{3}$$

where $(\phi_1^j, \phi_2^j) \in H^{3/2}(\partial\Omega) \times H^{1/2}(\partial\Omega)$, w_j satisfies $(*)$ with $(\phi_1, \phi_2) = (\phi_1^j, \phi_2^j)$ for $j = 1, 2$. The operator $\Lambda_{(a,b)}$ is called the Dirichlet to Neumann map for $L_{(a,b)}$. It follows that

$$\langle \Lambda_{(a,b)}(\phi_1^1, \phi_2^1), (\phi_1^2, \phi_2^2)\rangle = \langle \Lambda_{(a,b)}(\phi_1^2, \phi_2^2), (\phi_1^1, \phi_2^1)\rangle. \tag{4}$$

Our inverse problem is the following.

INVERSE BOUNDARY VALUE PROBLEM. *Determine the value of* (a,b) *in* Ω *by* $\Lambda_{(a,b)}$.

In this paper, we are interested in the uniqueness of recovering (a,b) in Ω by $\Lambda_{(a,b)}$. The following theorem gives us a partial solution of the problem.

THEOREM 1. *Let* $\eta > 0$. *There exists a positive constant* $\varepsilon(\Omega,\eta)$ *such that, if* $(a_\infty, b_\infty) \in \mathbf{R}\times\mathbf{R}$ *and* $(a_j,b_j) \in C^\infty(\overline{\Omega})\times C^\infty(\overline{\Omega})$ $(j=1,2)$ *satisfy*

$$\sum_j \|\nabla a_j\|_{W^{13,\infty}(\Omega)} + \|\nabla b_j\|_{W^{13,\infty}(\Omega)} < \eta, \tag{5}$$

$$\min_j \inf_\Omega \min(a_j+b_j, b_j) > \eta^{-1}, \tag{6}$$

$$\sum_j \|a_j - a_\infty\|_{H^3(\Omega)} + \|b_j - b_\infty\|_{H^4(\Omega)} < \varepsilon(\Omega,\eta), \tag{7}$$

$$\Lambda_{(a_1,b_1)} = \Lambda_{(a_2,b_2)}, \tag{8}$$

and

$$D^\alpha a_1\big|_{\partial\Omega} = D^\alpha a_2\big|_{\partial\Omega} \quad for\ |\alpha| = 0,1, \tag{9}$$

$$D^\beta b_1\big|_{\partial\Omega} = D^\beta b_2\big|_{\partial\Omega} \quad for\ |\beta| = 0,1,2,3, \tag{10}$$

then

$$(a_1,b_1) = (a_2,b_2) \quad in\ \Omega.$$

This type of problem was proposed in (Calderon, 1980) for an elliptic scalar partial differential operator of order 2 related to electrical impedance tomography and several results were obtained by many authors, for example (Kohn and Vogelius, 1984, 1985, 1987; Sylvester and Uhlmann, 1986, 1987, 1988) and see the references in (Sylvester and Uhlmann, 1990). For the second order elliptic system appearing in the linear elasticity, see (Ikehata, 1990; Akamatsu, 1991), and (Nakamura and Uhlmann, 1990).

The proof of Theorem 1 proceeds as follows. First, (3) and (4) imply that the following elementary but important identity holds:

$$\begin{aligned}&\langle\{\Lambda_{(a_1,b_1)} - \Lambda_{(a_2,b_2)}\}(\phi_1^1,\phi_2^1),(\phi_1^2,\phi_2^2)\rangle\\ &= \int_\Omega (a_1-a_2)\Delta w_1 \Delta w_2 + 2(b_1-b_2)\nabla^2 w_1 \nabla^2 w_2\, dx,\end{aligned} \tag{11}$$

where w_j satisfies $(*)$ in Proposition 1 with $(a,b)=(a_j,b_j)$ and (ϕ_1,ϕ_2), (ϕ_1^j,ϕ_2^j). Put $f = a_1 - a_2$ and $g = b_1 - b_2$. As an immediate consequence, we obtain that

$$\Lambda_{(a_1,b_1)} = \Lambda_{(a_2,b_2)}$$

if and only if

$$\int_\Omega f\Delta w_1 \Delta w_2 + 2g\nabla^2 w_1 \nabla^2 w_2\, dx = 0 \tag{12}$$

for any $w_j \in H^2(\Omega)$ such that $L_{(a_j,b_j)}w_j = 0$ in Ω.

Remark 1. G. Alessandrini and A.G. Ramm (see (Alessandrini, 1988; Ramm, 1988)) used the identities corresponding to (11) and (12) in the second order case. Therefore, our problem reduces to showing that $f = g = 0$ in Ω under (12) and

$$D^\alpha f\big|_{\partial\Omega} = 0 \quad \text{for } |\alpha| = 0,\, 1, \tag{13}$$

$$D^\beta g\big|_{\partial\Omega} = 0 \quad \text{for } |\beta| = 0,\, 1,\, 2,\, 3. \tag{14}$$

A well-known effective method in two dimensions proposed by J. Sylvester and G. Uhlman is studied in (Sylvester and Uhlmann, 1986; Sun, 1989). In those papers, the second order case is considered, but the method is applicable to the fourth order case. We basically follow the argument of (Sylvester and Uhlmann, 1986; Sun, 1989). In the method, two types of solutions to $L_{(a_j,b_j)}w = 0$ in Ω are needed, one of which is called the low frequency solutions, and the other is called the high frequency solutions. We describe those more precisely. Let $k \in \mathbf{R}^2\backslash\{0\}$ and put

$$u_0^\pm(x,k) = e^{ixz_\pm},$$
$$v_0^\pm(x,k) = e^{ixz_\pm}\frac{x\overline{z}_\pm}{2i|z_\pm|^2}$$

with

$$z_\pm = \frac{1}{2}(-k \pm iJk),$$
$$J = \begin{pmatrix} 0 & 1 \\ -1 & 0 \end{pmatrix}. \tag{15}$$

$u_0^\pm$ and $v_0^\pm$ are solutions of $L_{(a_\infty,b_\infty)} = 0$ in $\mathbf{R}^2$ since

$$L_{(a_\infty,b_\infty)} = (a_\infty + 2b_\infty)\Delta^2,$$
$$\Delta u_0^\pm = 0,$$
$$\Delta v_0^\pm = u_0^\pm.$$

Then, the solutions of $L_{(a_j,b_j)}w = 0$ in Ω with

$$\left(w\Big|_{\partial\Omega}, \frac{\partial}{\partial n}w\Big|_{\partial\Omega}\right) = \left(u_0^\pm\Big|_{\partial\Omega}, \frac{\partial}{\partial n}u_0^\pm\Big|_{\partial\Omega}\right)$$
$$= \left(v_0^\pm\Big|_{\partial\Omega}, \frac{\partial}{\partial n}v_0^\pm\Big|_{\partial\Omega}\right)$$

are called the low frequency solutions.
Inserting these into (12), we obtain the following basic lemma.

LEMMA A. *Let $R > 0$. Then, there exists a positive constant $C(\Omega,\eta,R)$ such that*

$$\|\{\Delta^2 g\}^\wedge\|_{L^2(|k|\le R)} \le C(\Omega,\eta,R)(\|f\|_{H^2(\Omega)} + \|g\|_{H^2(\Omega)})$$
$$\times\left(\sum_j \|b_j - b_\infty\|_{L^2(\Omega)} + \prod_j \|b_j - b_\infty\|_{L^2(\Omega)}\right) \tag{16}$$

and

$$
\begin{aligned}
&\left\| \left\{ \Delta f + \frac{1}{8}\Delta^2(|x|^2 g) + \Delta g \right\}^{\wedge} \right\|_{L^2(|k|\le R)} \\
&\le C(\Omega,\eta,R)\left(\|f\|_{H^2(\Omega)} + \|g\|_{H^2(\Omega)}\right) \\
&\times \left\{ \sum_j \left(\|a_j - a_\infty\|_{L^2(\Omega)} + \|b_j - b_\infty\|_{L^2(\Omega)}\right) \right. \\
&\left. + \prod_j \left(\|a_j - a_\infty\|_{L^2(\Omega)} + \|b_j - b_\infty\|_{L^2(\Omega)}\right) \right\},
\end{aligned}
\tag{17}
$$

where $\hat{c}(k) = \int_\Omega e^{-ixk} c(x)\, dx$ for any $c \in L^1(\Omega)$.

Remark 2. $C(\Omega, \eta, R)$ is independent of $(a_\infty, b_\infty) \in \mathbf{R}^2$.

Remark 3. These estimates are called the low frequency estimates.
The construction of the high frequency solutions is very complicated. For any $c \in C^\infty(\overline{\Omega})$, $\tilde{c}$ denotes a C^∞-extension of c to $\mathbf{R}^2$ with a compact support. Put

$$
\begin{aligned}
\gamma_j^1 &= 2(a_j + 2b_j)^{-1}\nabla(a_j + 2b_j), \\
\gamma_j^2 &= (a_j + 2b_j)^{-1}\Delta a_j, \\
\gamma_j^3 &= 2(a_j + 2b_j)^{-1}\nabla^2 b_j.
\end{aligned}
$$

These functions belong to $C^\infty(\overline{\Omega})$ and we can take $\tilde{\gamma}_j^p$ for $p = 1, 2, 3$.
Then we will construct the special solutions of $L_{(a_j,b_j)}w = 0$ in Ω of the form $w = e^{ixz_\pm} \times u(\cdot,k)\big|_\Omega$, where $u(\cdot,k)$ solves an equation with coefficients given by $\tilde{\gamma}_j^p$'s in the whole space. More precisely, we define $\Delta_z^j = e^{-ixz}\Delta^j(e^{ixz}\cdot)$, $\nabla_z^j = e^{-ixz}\nabla^j(e^{ixz}\cdot)$ for $j = 1, 2$ and $z \in C^2$ with $z \cdot z = 0$. Note that $z_\pm \cdot z_\pm = 0$.
Furthermore, put $V_j(x, \nabla_z) = \tilde{\gamma}_j^1 \nabla_z \Delta_z + \tilde{\gamma}_j^2 \Delta_z + \tilde{\gamma}_j^3 \nabla_z^2$. Then, we see that the following identity holds: $L_{(a_j,b_j)} = (a_j + 2b_j)\{\Delta^2 + \gamma_j^1 \nabla\Delta + \gamma_j^2 \Delta + \gamma_j^3 \nabla^2\}$ in Ω and if u solves

$$
\Delta_z^2(u) + V_j(x, \nabla_z)u = 0 \quad \text{in } \mathbf{R}^2 \tag{18}
$$

then $w = e^{ix\cdot z_u}\big|_\Omega$ solves $L_{(a_j,b_j)}w = 0 \quad \text{in } \Omega$.
The solutions to (18) with $z = z_\pm$ are constructed in the following form:

$$
u_j^\pm(x,k) = 1 + \sum_{l=0}^{5} (-i\omega_\pm)^{-l} A_{l,j}^\pm(x) + \dots, \tag{19}
$$

$$
v_j^\pm(x,k) = \frac{x\bar{z}_\pm}{2i|z_\pm|^2} + (-i\omega_\pm)^{-1} \sum_{l=0}^{5} (-i\omega_\pm)^{-l} B_{l,j}^\pm(x) + \dots, \tag{20}
$$

where $\omega_\pm = k_1 \pm (-i)k_2$, $A_{l,j}^\pm = A_l$ and $B_{l,j}^\pm = B_l$ are satisfying

$$
\Gamma_{j,2}^\pm A_0 + \frac{1}{4}(\tilde{\gamma}_j^3 \nu_\pm \nu_\pm) = 0, \tag{21}
$$

$$\Gamma_{j,2}^{\pm} A_1 + \Gamma_{j,1}^{\pm} A_0 = 0, \tag{22}$$

$$\Gamma_{j,2}^{\pm} A_{l+2} + \Gamma_{j,1}^{\pm} A_{l+1} + \Gamma_{j,0}^{\pm} A_l = 0 \quad \text{for} \quad l = 0, 1, \ldots, 3, \tag{23}$$

$$\Gamma_{j,2}^{\pm} B_0 + \frac{1}{2}(\tilde{\gamma}_j^1 \nu_\pm) + \frac{1}{8}(\tilde{\gamma}_j^3 \nu_\pm \nu_\pm)\bar{x}_\pm = 0, \tag{24}$$

$$\Gamma_{j,2}^{\pm} B_1 + \Gamma_{j,1}^{\pm} B_0 + \tilde{\gamma}_j^2 + \frac{1}{2} \operatorname{tr} \tilde{\gamma}^3 = 0, \tag{25}$$

$$\Gamma_{j,2}^{\pm} B_{l+2} + \Gamma_{j,1}^{\pm} B_{l+1} + \Gamma_{j,0}^{\pm} B_l = 0 \quad \text{for} \quad l = 0, 1, \ldots, 3, \tag{26}$$

where

$$\Gamma_{j,0}^{\pm} = \Delta^2 + \tilde{\gamma}_j^1 \nabla \Delta + \tilde{\gamma}_j^2 \Delta + \tilde{\gamma}^3 \nabla^2,$$

$$\Gamma_{j,1}^{\pm} = 2\partial_\pm \Delta + \tilde{\gamma}_j^1 \nabla \partial_\pm + \frac{1}{2}(\tilde{\gamma}_j^1 \nu_\pm)\Delta + \tilde{\gamma}_j^2 \partial_\pm + \nu_\pm \tilde{\gamma}_j^3 \nabla,$$

$$\Gamma_{j,2}^{\pm} = \partial_\pm^2 + \frac{1}{2}(\tilde{\gamma}_j^1 \nu_\pm)\partial_\pm + \frac{1}{4}(\tilde{\gamma}_j^3 \nu_\pm \nu_\pm),$$

$$\partial_\pm = \frac{\partial}{\partial x_1} \pm i \frac{\partial}{\partial x_2},$$

$$x_\pm = x_1 \pm i x_2,$$

$$\nu_\pm = {}^t(1, \pm i).$$

The remainders of (19) and (20) are constructed and estimated by the estimates of $g_z^{(2)}(x)$, the fundamental solution $\Delta_z^{\pm}$ being constructed in (Ikehata, 1991a). For convenience, we describe it.

THEOREM 2. *Let $a > 0$ and $0 < \delta < 1$. Then, there exists a positive constant $C(a, \delta)$ such that, for $|\alpha| \le 4$,*

$$\|D^\alpha g_z^{(2)} * f\|_{-(2-\delta)} \le C(a, \delta)|z|^{|\alpha|-2}\|f\|_{\delta+1}, \tag{27}$$

where $z \in C^2$, $z \cdot z = 0$, $|z| \ge a$, $\| \cdot \|_S = \left\|(1 + |x|^2)^{s/2}\right\|_{L^2(\mathbf{R}^2)}$ $(s \in \mathbf{R})$ and f is a rapidly decreasing function.

Substituting $e^{ix \cdot z_\pm} \times u_j^{\pm}(x, k)$ and $e^{ix \cdot z_\pm} \times v_j^{\pm}(x, k)$ into (12), we obtain the following lemma complementary to Lemma A.

LEMMA B. *There exist positive constants $C(\Omega, \eta), \varepsilon(\Omega, \eta)$ and $K(\eta)$ such that*

$$\|\{\Delta g\}^\wedge\|_{L^2(|k| \ge R)} \le C(\Omega, \eta)(\varepsilon + R^{-1})(\|f\|_{H^2(\Omega)} + \|g\|_{H^4(\Omega)}), \tag{28}$$

$$\left\|\left\{\Delta f + \Delta g + \tfrac{1}{8}\Delta^2(|x|^2 g)\right\}^\wedge\right\|_{L^2(|k| \ge R)} \le C(\Omega, \eta)(\varepsilon + R^{-1})(\|f\|_{H^2(\Omega)} + \|g\|_{H^4(\Omega)}), \tag{29}$$

where $R \ge K(\eta)$ and $0 < \varepsilon \equiv \sum_j \|a_j - a_\infty\|_{H^3(\Omega)} + \|b_j - b_\infty\|_{H^4(\Omega)} < \varepsilon(\Omega, \eta)$.

For the details of the proof of Lemma A and B, see (Ikehata, 1991b). Hereafter, by the same argument as in the proof of Theorem 1 in (Sun, 1989), for the corresponding inverse boundary value problem that corresponds to the operator $\nabla \gamma \nabla$ in two dimensions, we conclude that

$$\|f\|_{H^2(\Omega)} + \|g\|_{H^4(\Omega)} = 0$$

if ε is sufficiently small.

REFERENCES

Akamatsu, M. Nakamura, G. and Steinberg, S. (1991). Identification of Lame coefficients from boundary observations. *Inverse Problems* (to appear).

Alessandrini, G. (1988). Stable determination of conductivity by boundary measurements. *Appl. Anal.* **27**, 153–172.

Calderon, A.P. (1980). On an inverse boundary value problem. In: *Seminar on Numerical Analysis and Applications to Continuum Physics.*, Soc. Brasileira de Mathematica, Rio de Janeiro pp. 65–73.

Ikehata, M. (1990). Inversion formulas for the linearized problem for an inverse boundary value problem in elastic prospection. *SIAM J. Appl. Math.* **50**, 1635–1644.

Ikehata, M. (1991a). A special Green's function for the biharmonic operator and its application to an inverse boundary value problem. *Comp. Math. Appl.* (to appear).

Ikehata, M. (1991b). A local uniqueness theorem for an inverse boundary value problem for the equation of the flexure of the plate in the Love–Kirchhoff theory. *SIAM J. Appl. Math.* (to appear).

Kohn, R. and Vogelius, M. (1984). Determining conductivity by boundary measurements. *Comm. Pure Appl. Math.* **37**, 289–298.

Kohn, R. and Vogelius, M. (1985). Determining conductivity by boundary measurements II, Interior results. *Comm. Pure Appl. Math.* **38**, 643–667.

Kohn, R. and Vogelius, M. (1987). Relaxation of a variational method for impedance computed tomography. *Comm. Pure Appl. Math.* **40**, 745–777.

Nakamura, G. and Uhlmann, G. (1990). Identification of Lame parameters by boundary measurements. Preprint.

Ramm, A.G. (1988). Multidimensional inverse problems and completeness of the products of solutions to PDE. *J. Math. Anal. Appl.* **134**, 211–253.

Sun, Z. (1989). On an inverse boundary value problem in two dimensions. *Comm. PDE* **14**, 1101–1113.

Sylvester, J. and Uhlmann, G. (1986). A uniqueness theorem for an inverse boundary value problem in electrical prospection. *Comm. Pure Appl. Math.* **39**, 91–112.

Sylvester, J. and Uhlmann, G. (1987). A global uniqueness theorem for an inverse boundary value problem. *Ann. Math.* **125**, 153–169.

Sylvester, J. and Uhlmann, G. (1988). Inverse boundary value problems at the boundary-continuous dependence. *Comm. Pure Appl. Math.* **41**, 197–221.

Sylvester, J. and Uhlmann, G. (1990). The Dirichlet to Neumann map and applications. In: *Inverse Problems in Partial Differential Equations*, SIAM, Philadelphia, pp. 99–139.

Dautray, R. and Lions, J.L. (1990). *Mathematical Analysis and Numerical Methods for Science and Technology.* Vol. 2, Springer-Verlag, Berlin.

Ill-Posed Problems in Natural Sciences, pp. 307 - 311
A. Tikhonov (Ed.)
1992 VSP/TVP

INVERSE PROBLEMS FOR THE NONSTATIONARY KINETIC TRANSPORT EQUATION

A.L. IVANKOV
Moscow Institute of Engineering and Physics, Kashirskoye Shosse, 31, 115409 Moscow, Russia

ABSTRACT

Two kinds of inverse problems for the nonstationary kinetic transport equation are considered. The first group includes the inverse problems of searching stationary parameters by time-dependent overdetermination. The theorems on global uniqueness and stability are proved for this case. The second group consists of the inverse problems with no initial condition. In these problems it is necessary to find time-dependent unknown parameters by the overdetermination on unbounded time interval.

INTRODUCTION

The nonstationary kinetic transport equation, which we consider, has the form

$$(\frac{\partial}{\partial t} v\nabla)u(x,v,t,) + \Sigma(x,v,t)u(x,v,t) = \int\limits_{\nabla} J(x,v,v',t)u(x,v',t)\,dv' + F(x,v,t),$$

$$(x,v,t) \in D = G \times V \times (0,T). \tag{0.1}$$

Here $V, G \subset \mathbf{R}^n, I = (0,T), G$ is a bounded strictly convex domain with a C^1 boundary, $V = \{v \in \mathbf{R}^n\colon\ 0 < \nu_- \leq \mid v \mid \leq \nu_+ < \infty\}$, $\varkappa < n$. We consider below, inverse problems of two kinds and divide the text into two sections. The first of them contains the inverse problems whose formulations include the boundary condition

$$u\mid_{\Gamma_-} = \psi \tag{0.2}$$

and the initial condition

$$u\mid_{t=0} = \psi, \tag{0.3}$$

(see (Ivankov, 1990)) while the second one considers the inverse problems with no initial condition (see (Ivankov, 1991)). These results continue the investigations, begun in the article (Ivankov and Prilepko, 1989; Prilepko and Ivankov, 1984; Prilepko and Ivankov, 1985).
Here and below $\Gamma_\pm = \gamma_\pm \times I$, $\gamma_\pm = \{(x,v) \in \partial G \times I\colon\ \pm(n_x, v) < 0\}$, $a(x,v) = \max\{t\colon\ x + vt \in \partial G\}$, $d = \sup a$.

1. THE SEARCH OF THE NONHOMOGENEOUS COEFFICIENTS OF KINETIC EQUATION BY READING A POINT MEASURER

In this section the searching parameters Σ, J, F have the form

$$\Sigma(x,v,t) = \sigma(x)Q_1(x,v,t) + H_1(x,v,t) \tag{1.1}$$

$$J(x,v,v',t) = g(x)Q_2(x,v,v',t) + H_2(x,v,v',t), \tag{1.2}$$

$$F(x,v,t) = f(x)Q_3(x,v,t) + H_3(x,v,t), \tag{1.3}$$

where $\sigma = \sigma(x)$, $g = g(x)$, and $f = f(x)$ are the stationary parameters we seek for and the remaining functions Q_i, H_i $(i = 1,2,3)$ are known a priori.
We formulate below inverse problems and uniqueness and stability theorems.

INVERSE PROBLEM 1.1. *Find functions u_1, u_2, u_3, f, σ, and g, such that, for $i = 1,2,3$, the functions $u = u_i$, f, σ, and g satisfy equation (0.1) with coefficients Σ, J, F of form (1.1)–(1.3), boundary condition (0.2) with $\psi = \psi_i$, initial condition (0.3) with $\phi = \phi_i$, and the condition of overdetermination*

$$u(x,v,t)|_{\Pi_\nu} = \chi(x,v) \tag{1.4}$$

with $\chi = \chi_i$. Here $\Pi_\nu = \{(x,v,t) \in D\colon\ x = x_0 \in \overline{G},\ (v,t) \in V \times I\colon\ |v| = \nu_+, 0 < t < \alpha(x_0,-v)\}$.

Remark. Inverse Problem 1.1 is a mathematical model of the series of three experiments, in each of them the initial conditions ϕ_i, the boundary conditions ψ_i and the conditions of overdetermination are measured. Condition of overdetermination (1.4) is interpreted as a mathematical model of the point measure, which registers the particles at a point x_0, flying with velocity v whose absolute value equals ν_+ on the time interval $(0, \alpha(x,-v,))$.

THEOREM 1.2. *Let Q_i, H_i, ψ_i, $\chi_i \in \overline{C}^1_t$, $\phi_i \in \overline{C}^1_{v\nabla}$, $i = 1,2,3$; $T \geq d$ and the determinant of the matrix*

$$A = \begin{pmatrix} (\phi_1 Q_1)(x,v,t) & \int\limits_V (\phi_1 Q_2)(x,v,v',t)dv' & Q_3(x,v,t) \\ (\phi_2 Q_1)(x,v,t) & \int\limits_V (\phi_2 Q_2)(x,v,v',t)dv' & Q_3(x,v,t) \\ (\phi_3 Q_1)(x,v,t) & \int\limits_V (\phi_3 Q_2)(x,v,v',t)dv' & Q_3(x,v,t) \end{pmatrix} \tag{1.5}$$

on the set of points $G_{x_0} = \{(x,v,t) = (x, \frac{x_0 - x}{|x-x_0|}\nu_+, 0)\colon\ x \in G\backslash\{x_0\}\}$ is bounded away from zero by a number $q > 0$. Under these conditions the solution of Inverse Problem 1.1 (u_1,u_2,u_3,f,s,g) is unique in the class of functions $K^3 = [C^1_{tv\nabla}(\overline{D})]^3 \times [C(\overline{G})]^3$ and is conditionally stable. An estimate of conditional stability has the form

$$\sum_{i=1}^{3} \|u_{1i} - u_{2i}\|_{tv\nabla} + \|\sigma_1 - \sigma_2\| + \|g_1 - g_2\| + \|f_1 - f_2\|$$

$$\leq C\sum_{i=1}^{3}(\|\phi_{i1}-\phi_{i2}\|_{v\nabla}+\|\psi_{i1}-\psi_{i2}\|_{t}+\|\chi_{i1}-\chi_{i2}\|_{t}+\|H_{3i1}-H_{3i2}\|) \quad (1.6)$$

We explain now the notation, used in the formulation of the theorem. Class C consists of continuous functions, $C^1_t=\{u\in C\colon u_t\in C\}$, $C^1_{v\nabla}=\{u\in C\colon \frac{\partial}{\partial s}u(x+vs,t+s,v)\mid_{s=0}\in C\}$, $C^1_{tv\nabla}=C^1_t\cap C^1_{v\nabla}$. Class $\overline{C}$ consists of the functions, which are continuous on the closures of their domains of definition. Classes $\overline{C}^1_t$, $\overline{C}^1_{v\nabla}$, $\overline{C}^1_{tv\nabla}$ are defined analogously. The norms used in estimate (1.6) are the norms of the corresponding classes, for example, $\|u\|_{tv\nabla}=\|u\|+\|u_t\|+\|v\nabla u\|$.

INVERSE PROBLEM 1.3. *Find functions u_1, u_2, f, σ, and g, such that, for $i=1,2$, the functions $u=u_i$, f, σ, and g satisfy equation* (0.1) *whose coefficients Σ, J have the forms* (1.1) *and* (1.2), *the boundary condition* (0.2) *with $\psi=\psi_i$, the initial condition* (0.3) *with $\phi=\phi_i$, and the the condition of overdetermination* (1.4) *with $\chi=\chi_i$.*

Remark. Inverse Problem 1.2 is a mathematical model of the series of two experiments the same as for Inverse Problem 1.1. The inverse problem of determination of the pairs of functions (σ,f), (g,f) are formulated in a similar way. The inverse problem of determination of the pair of functions (σ,f) will be considered in more details since the problem of searching all coefficients in the absence of radiation sources is more interesting from a practical point of view.

THEOREM 1.4. *Let Q_i,H_i,ψ_i,χ_i, $F\in\overline{C}^1_t$, $\phi_i\in\overline{C}^1_{v\nabla}$, $i=1,2$; $T\geq d$ and*

$$\left|\det\begin{pmatrix}(\phi_1Q_1)(x,v,t) & \int_V(\phi_1Q_2)(x,v,v',t)\,dv'\\ (\phi_1Q_1)(x,v,t) & \int_V(\phi_2Q_2)(x,v,v',t)\,dv'\end{pmatrix}\right|\geq q>0$$

on G_{x_0}. Then the solution of Inverse Problem 1.3 is unique in the class of functions $K^2=[C^1_{tv\nabla}(\overline{D})]^2\times[C(\overline{G})]^2$ and conditionally stable. An estimate of stability has the form

$$\sum_{i=1}^{2}\|u_{1i}-u_{1i}\|_{tv\nabla}+\|\sigma_1-\sigma_2\|+\|g_1-g_2\|+\|f_1-f_2\|\leq C\sum_{i=1}^{2}(\|\phi_{i1}-\phi_{i2}\|_{v\nabla}$$
$$+\|\psi_{i1}-\psi_{i2}\|_t+\|\chi_{i1}-\chi_{i2}\|_t+\|F_{1i}-F_{1i}\|),$$

where C depends on $\|Q_i\|$, $\|u_{i2}\|_{tv\nabla}$, $\|\sigma_1\|$, $\|g_1\|$, $\|\phi_{i2}\|_{v\nabla}$, q, $\varkappa$, $\|H_i\|$, $i=1,2$.

INVERSE PROBLEM 1.5. *Find functions u and f, satisfying equation* (0.1) *whose right-hand side F has the form* (1.3) *and conditions* (0.2), (0.3), *and* (1.4).

THEOREM 1.6. *Let Q_3, H_3, ψ, χ, J, $\Sigma\in\overline{C}^1_t$, $\phi_i\in\overline{C}^1_{v\nabla}$, $T\geq d$ and*

$$\left|Q(x,\ \frac{x_0-x}{|x_0-x|}\nu_+,\ v',\ 0)\right|\geq q>0 \quad \text{on } G_0.$$

Then the solution of Inverse Problem 1.5 is unique in the class of functions $K=C^1_{tv\nabla}(\overline{D})\times C(\overline{G})$ and stable. An estimate of stability has the form

$$\|u_1-u_2\|_{tv\nabla}+\|f_1-f_2\|\leq C(\|\phi_1-\phi_2\|_{v\nabla}+\|\psi_1-\psi_i\|_t+\|\chi_1-\chi_2\|_t+\|H_{31}-H_{32}\|,$$

where C depends on $\|Q_3\|_t$, $\|\Sigma\|_t$, $\|J\|_t$, q, $\varkappa$, and on the domain D.

Remark. The inverse problems of determination of the pairs of functions (u, σ) and (u, g) are formulated in the same way. Analogous theorems concerning the uniqueness and conditional stability are valid for these problems.

2. INVERSE PROBLEMS WITHOUT INITIAL CONDITION

The inverse problems of determinating the homogeneous part of the full scatering or the density of sources of a nonstationary kinetic transport equation with overdetermination on the boundary without initial condition are considered. These inverse problems are mathematical models of experiments, in which an unknown homogeneous isotropic influence set on the known evolution of properties of a medium. The medium is subject to an external radiation, for example, Roentgen or neutron radiation, the outgoing flux is registered at one point, and it is necessary to determine by these data the dependence of properties of the medium on time.

We introduce the class of functions $B^1_{tv\nabla\lambda} = \{u \in \overline{C}\colon \sup u < \infty\}$, $B^1_t = \{u \in B\colon u_t \in B\}$, $B^1_{v\nabla} = \{u \in B\colon v\nabla u \in B\}$, $B^1_{tv\nabla} = B^1_t \cap B^1_{v\nabla}$, $B_\lambda = \{u \in B\colon \exp(\lambda t)u \in B\}$, $B^1_{t\lambda} = \{u \in B_\lambda\colon \exp(\lambda t)u_t \in B\}$, $B^1_{v\nabla\lambda} = \{u \in B_\lambda\colon v\nabla u \in B_\lambda$, $B^1_{tv\nabla\lambda} = B^1_{t\lambda} \cap B^1_{v\nabla\lambda}$, $X_\lambda(D) = \{u \in B_\lambda(D)\colon \frac{\partial}{\partial s}u(x+vs,\ v,\ t+s)|_{s=0} \in B_\lambda(D)\}$ and norms $\|u\|_\lambda = \|\exp(\lambda t)u\|$, $\|u\|_{t\lambda} = \|u\|_\lambda + \|u_t\|_\lambda$, $\|u\|_{v\nabla\lambda} = \|u\|_\lambda + \|v\nabla u\|$, $\|u\|_{tv\nabla\lambda} = \|u\|_{v\nabla\lambda} + \|u_t\|_\lambda$. Let $T = \infty$. We proceed now to the statement of inverse problems.

INVERSE PROBLEM 2.1. *Find a pair of functions $u = u(x, v, t)$, $f = f(t)$ satisfying equation* (0.1) *with the right-hand side F of the form*

$$F(x, v, t) = f(t)Q(x, v, t) + H(x, v, t), \tag{2.1}$$

the boundary condition (0.2), *and the condition of overdetermination*

$$u|_{x=x_0, v=v_0} = \chi, \quad (x_0, v_0) \in \gamma_-. \tag{2.2}$$

Two results concerning Inverse Problem 2.1 are obtained. They are the global theorem of well-posedness in the class of functions $(u, f) \in B^1_{tv\nabla\lambda}(\overline{D}) \times B_\lambda([0, \infty))$ and the theorem of uniqueness and stability in the class of functions $(u, f) \in X_\lambda(\overline{D}) \times B_\lambda([0, \infty))$ for sufficiently large values of λ. We formulate the theorems below.

THEOREM 2.2. *Let $\Sigma = \Sigma^1 + \Sigma^2$, $H = H^1 + H^2$; Σ^1, $J \in B^1_t$; Σ^2, $Q \in B^1_{v\nabla}$ and for sufficiently large values of λ depending on Σ^2, Q, J, $\varkappa$, H^1, $\eta \in B^1_{t\lambda}$; $H^2 \in B^1_{v\nabla\lambda}$. Let the function $Q(x_0, v_0, t)$ be bounded away from zero on $[0, \infty]$.*

Under these conditions a solution of Inverse Problem 2.1 exists, it is unique and stable in the class of functions $(u, f) \in B^1_{tv\nabla\lambda}(\overline{D}) \times B_\lambda([0, \infty))$. An estimate of stability has the form

$$\|u_1 - u_2\|_{tv\nabla\lambda} + \|f_1 - f_2\|_\lambda$$
$$\le C(\|\eta_1 - \eta_2\|_{t\lambda} + \|\chi_1 - \chi_2\|_{t\lambda} + \|H^1_1 - H^1_2\|_{t\lambda} + \|H^1_1 - H^1_2\|_{v\nabla\lambda}),$$

where the constant C depends only on Σ^1, Σ^2, J, Q, $\varkappa$.

THEOREM 2.3. *If Σ, J, $H \in B_\lambda(\overline{D})$; $Q \in X_\lambda(\overline{D})$ and the function $Q(x_0, v_0, t)$ is bounded away from zero on $[0, \infty]$, then the solution of Inverse Problem 2.1 is unique in the class of functions $(u, f) \in X_\lambda(\overline{D}) \times B_\lambda([0, \infty))$ for sufficiently large values of λ. This solution is stable, and an estimate of its stability has the form*

$$\|u_1 - u_2\|_{tv\nabla\lambda} + \| \int_0^t \partial\tau f_1(\tau) - \int_0^t \partial\tau f_2(\tau)\| \le C(\|\eta_1 - \eta_2\|_\lambda$$

$$+\|\chi_1 - \chi_2\|_\lambda + \|H_1 - H_2\|_\lambda),$$

where the constant C depends only on Q, J, Σ, $\varkappa$.

REFERENCES

Ivankov, A.L. (1990). Some inverse problems of searching the nonhomogeneous environment parameters (coefficients of the kinetic equation) by reading a point measurer. In: *Numerical Methods of the Transport Equation Solution.* Acad. Nauk Est. SSR Publ., Tartu, pp. 76–78.

Ivankov, A.L. (1991). The inverse problems of searching the homogeneous isotropic parameters for the nonstationary kinetic transport equation without initial condition. In: *Inverse Problems for the Mathematical Models of the Physical Processes.* Energoatomizdat, Moscow, pp. 25–39 (in Russian).

Ivankov, A.L. and Prilepko, A.I. (1989). The uniqueness and stability of the nonhomogeneous parameters searching of inverse problems in the nonstationary transport theory. In: *Mathematical Modeling. Modern Problems of the Mathematical Physics and Computational Mathematics.* Nauka, Moscow, pp. 185–206 (in Russian).

Prilepko, A.I. and Ivankov, A.L. (1984). Inverse problems for the nonstationary transport equation. *Dokl. Acad. Nauk SSSR* **276**, 555–559 (in Russian).

Prilepko, A.I. and Ivankov, A.L. (1985). Inverse problems of searching the coefficient, scattering indicatriss and right-hand part of the nonstationary many-speeds transport equation. *Different. Equat.* **21**, 870–885.

Ill-Posed Problems in Natural Sciences, pp. 312 – 325
A. Tikhonov (Ed.)
1992 VSP/TVP

OPTIMIZATION METHODS OF SOLVING INVERSE PROBLEMS OF GEOELECTRICS

S.I. KABANIKHIN and A.L. KARCHEVSKY

Institute of Mathematics, Universitetskii prospekt, 4, 630090 Novosibirsk, Russia

ABSTRACT

Direct and inverse problems for the system of Maxwell equations are investigated. A general algorithm for determining the conductivity coefficient, dielectric and magnetic permeability is based on the conjugate gradient method. A formula for the gradient of the inverse problem functional is presented.

INTRODUCTION

In this paper we consider the dynamic case of the direct and inverse problems for the system of Maxwell equations (a source of impulse type, additional information is an electromagnetic field (depending on time) on the Earth surface).

In § 1 we formulate the general statement of direct and inverse problems.

In § 2 we give a short survey of the preceding works.

In § 3 we give a general algorithm for finding coefficients of Maxwell equations. It is based on the conjugate gradient algorithm. This algorithm is used in § 4 to solve the one-dimensional undersurface radiolocation inverse problem, and in § 5 to obtain quasi-stationary approximation of Maxwell equations.

We notice, that this algorithm one can be used for many other inverse problems of electrodynamics.

1. A GENERAL STATEMENT OF THE DIRECT AND INVERSE PROBLEMS

Geophysical methods are a major part in the investigation of the inner Earth structure. They are based on the study of some physical fields on the Earth surface (or in a borehole). These fields gives us an information about the Earth structure. Electromagnetic field is a field of that type. Components of this field depend upon the distribution of dielectric and magnetic penetrations and conductivity of the medium. The process of electromagnetic waves spreading in the medium is described by Maxwell equations. We give the statement of inverse problem for the nonstationary system of Maxwell equations (Romanov and Kabanikhin, 1989a,b).

$$\begin{aligned} &\varepsilon\frac{\partial \mathbf{E}}{\partial t} - \operatorname{rot}\mathbf{H} + \sigma\mathbf{E} + \mathbf{j} = \mathbf{0}, \quad (x_1, x_2, x_3) \in \mathbf{R}^3, \\ &\mu\frac{\partial \mathbf{H}}{\partial t} + \operatorname{rot}\mathbf{E} = \mathbf{0}, \qquad x_3 \neq 0, \quad t \in \mathbf{R}_+ = \{t \in \mathbf{R}\colon t > 0\}. \end{aligned} \tag{1.1}$$

Here $\mathbf{E} = (E_1, E_2, E_3)^*$ and $\mathbf{H} = (H_1, H_2, H_3)^*$ are the vectors of electric and magnetic strength, ε and μ are dielectric and magnetic penetrations, σ is conductivity, $\mathbf{j}$ is the external electric current density.
The Earth surface is thought to be plane in our geophysical model. In this model the whole physical space $\mathbf{R}^3$ of variables $x = (x_1, x_2, x_3)$ is divided by plane $x_3 = 0$ into two semispaces (air-earth) $\mathbf{R}^2 \times \mathbf{R}_- = \{x \in \mathbf{R}^3\colon x_3 < 0\}$, $\mathbf{R}^2 \times \mathbf{R}_+ = \{x \in \mathbf{R}^3\colon x_3 > 0\}$, in one of which (namely, $\mathbf{R}^2 \times \mathbf{R}_-$) the parameters ε, μ, σ are constant while in the second one they are smooth functions with respect to $x \in \mathbf{R}^2 \times \mathbf{R}_+$ up to the boundary of the semispace. The continuity condition of tangential components of this field must be implemented on the media separation boundary:

$$[E_j]_{x_3=0} = 0 \qquad [H_j]_{x_3=0} = 0 \quad j = 1, 2. \tag{1.2}$$

When studying the nonstationary problem it is assumed that there are no electromagnetic oscillations are absent up to the moment $t = 0$:

$$\mathbf{H}|_{t<0} = \mathbf{0}, \qquad \mathbf{E}|_{t<0} = \mathbf{0}, \qquad \mathbf{j}|_{t<0} = \mathbf{0}, \tag{1.3}$$

and then they are induced by the external current $\mathbf{j} = \mathbf{j}(x, t)$.

DIRECT PROBLEM. *To find* $\mathbf{E}$, $\mathbf{H}$ *as the solution of the generalized Cauchy problem* (1.1)–(1.3), *using the known coefficients* ε, μ, σ.

Let us assume that the following additional information concerning the solution of direct problem (1.1)–(1.3) is known:

$$E_j|_{x_3=0} = G_j(x_1, x_2, t), \qquad H_j|_{x_3=0} = Q_j(x_1, x_2, t), \quad j = 1, 2. \tag{1.4}$$

The functions G_j, Q_j, $j = 1, 2$, can not be arbitrary, because they are connected by two linear equations. For example, if we know functions G_j, $j = 1, 2$, then we state initial-boundary problem (1.1), (1.3) in the domain $\mathbf{R}^2 \times \mathbf{R}_+$ with the boundary conditions $E_j|_{x_3=0} = G_j(x_1, x_2, t)$, $j = 1, 2$, and we can determine uniquely all the components of dielectric and magnetic strength vectors in this domain and, therefore, calculate functions Q_j, $j = 1, 2$, as linear combinations of G_j, $j = 1, 2$. Hence, we can choose only G_1, G_2 or Q_1, Q_2. It's also possible to consider the pair G_1, Q_1 or G_2, Q_2.
Thus, we can consider direct problem (1.1), (1.3) with the boundary conditions:

$$E_1|_{x_3=0} = G_1(x_1, x_2, t) \qquad H_1|_{x_3=0} = Q_1(x_1, x_2, t). \tag{1.5}$$

INVERSE PROBLEM. *To find* σ, ε, μ, $\mathbf{E}$, $\mathbf{H}$ *in* $\mathbf{R}^2 \times \mathbf{R}_+$ *using relations* (1.1), (1.3), (1.5) *and the additional information:*

$$E_2|_{x_3=0} = G(x_1, x_2, t), \qquad H_2|_{x_3=0} = Q_2(x_1, x_2, t). \tag{1.6}$$

The theory of direct and inverse problems such kind was published in (Romanov and Kabanikhin, 1989a,b). You can find there the uniqueness theorems and stability estimates for one-, two-, and three-dimensional cases (see, for example, Romanov's paper in thes volume).

2. SURVEY

Let's look through the works for the direct and inverse problems of Maxwell equations.

Tikhonov A.N. gave the important investment in the development of the electromagnetic methods for the study of the Earth structure. V.I. Tikhonov and D.N. Shakhsuvarov (see (Tikhonov and Shakhsuvarov, 1956, 1961)) worked out a numerical algorithm for the horizontal electric dipole field. Dipole stayed on the surface of an arbitrary n-stratum semispace.

In (Dmitriev, 1968,1976; Van'yan 1965) V.I. Dmitriev, A.L. Van'yan and others developed this algorithm. In 1986 V.I. Dmitriev (see (Dmitriev, 1968)) extended this method to the case of gradient medium and for arbitrary sources. The detailed survey and analysis of direct and inverse problems for electrodynamics were given in (Dmitriev et al., 1976).

M.M. Lavrent'ev and other authors (see (Lavrent'ev and Romanov, 1966; Blagoveshensky, 1966; Alekseev, 1967; Romanov, 1984)) gave the first statements of dynamic inverse problems for hyperbolic systems.

We must mention the analogous statements in (Prilepko et al., 1973, 1984).

V.G. Romanov, S.I. Kabanikhin and T.P Pukhnacheva in (Romanov et al., 1984) stated the general approach for the investigation of dynamic inverse problems for Maxwell equations. S.I. Kabanikhin and others (see (Kabanikhin and Abdiev, 1982, 1986; Kabanikhin and Martakov, 1988; Kabanikhin et al.,1989; Kabanikhin et al., 1990; Iskakov and Kabanikhin, 1991)) presents the investigation of the numerical methods of the direct and inverse problems of the geoelectrics and show the results of the numerical experiments.

M.S. Zhdanov (see (Zhdanov et al., 1981, 1982a,b, 1984, 1986)) gave the methods of modelling and numerical solution for the quasi-stationary electromagnetic fields in inhomogeneous media.

The selection method is one of the most popular methods of inverse problems solving. A possible structure of medium is taken and the physical field is calculated. The problem solution is chosen when the calculated field differs a little from the observing physical field.

V.I. Tikhonov and others in (Tikhonov, 1963; Marchuk, 1964; Alekseev, 1967) suggested solving the inverse and ill-posed problems as the problems of optimal control.

In (Dmitriev and Rudneva, 1971, 1973) the inverse problem was stated as the problem for the search of the functional minimum, which depends on the large number of the parameters, which characterize the geoelectrics section. This method was developed in (Dmitriev and Fedorova, 1983).

A. Bamberger, G. Chavent, Ch. Hemon and P. Lailly in (Bamberger et al., 1979) investigated the inverse problem for the one-dimensional wave equation and suggested the stable algorithm for its solution. The conclusions and results for the

work of method on the real models were presented in (Bamberger et al., 1982).
D.A. Cook and W.A. Scheider in (Cook and Scheider, 1983) used the least squares method for the inverse problem solution with the real seismic data.
P.W. Marquart and other authors (see (Marquart, 1963; Chen and Seinfeldt, 1982; Kravaris and Seinfeldt, 1985)) considered the method for the estimation of the different parameters in the geophysical models.
The survey for the wide spectrum of this investigation were presented in (Ursin and Berteussen, 1986).
B.M. Budak and A.D. Iskenderov (see (Budak and Iskenderov, 1966)) considered two extreme problems (hyperbolic and parabolic). They proved the existence and uniqueness of problems solution and worked out the numerical method for the search of unknown coefficient depending time only. Then A.D. Iskenderov (Iskenderov, 1968) developed this method. Inverse problem was considered for the definition of two unknown coefficients for the hyperbolic equation and the ordinary differential equation with quasi-linear right part. He considered the questions of solution stability and investigated the algorithm convergence.
A.D. Iskenderov and R.K. Tagiev in (Iskenderov and Tagiev, 1983) investigated the optimal control problem for parabolic equation with control as the coefficients of this equation. They investigated the questions of the problem's statement correctness, proved the existence and the uniqueness theorem for the solution, got the necessary condition of the optimality. They gave necessary and sufficient condition of optimal control.
A.M. Denisov (see (Denisov, 1982a,b)) investigated the uniqueness of inverse problem solution for the heat conduction equation with piece-wise coefficients.
O.M. Alifanov, E.A. Artyukhin, and S.V. Rumyantsev (see (Alifanov et al., 1988)) considered many types of linear and nonlinear inverse problems and the numerical solutions for these problems on the basis of the minimizing functional by the different gradient methods.
The optimal control method was used in (Nyambaa and Cheverda, 1988) for the inverse problem solution of geoelectrics for the constant current and the vertically unhomogeneous medium. They used the of the incomplete division method of the variables. This method simplifies the three-dimensional problem to the one-parameter system of the ordinary differential equations. The minimum is searched in the temporary frequencies domain. Then Alekseev and other authors (see (Alekseev et al., 1991)) developed this algorithm.
F.P. Vasil'ev in (Vasil'ev 1981, 1988) stated different questions of the optimal control problems and applications in the investigation of the inverse problems.

3. THE OPTIMIZATION METHOD FOR FINDING COEFFICIENTS OF THE SYSTEM OF MAXWELL EQUATIONS

We suppose that the additional information concerning inverse problem (1.6) has some δ-error:

$$\|G_2 - G_2^\delta\|_{L_2(S)} < \delta \qquad \|Q_2 - Q_2^\delta\|_{L_2(S)} < \delta, \tag{3.1}$$
$$S = \{(x_1, x_2, t)\colon (x_1, x_2) \in \mathbf{R}^2,\ t \in \mathbf{R}_+\}.$$

The system of Maxwell equations is of hyperbolic type. We suppose that the external current source has a finite support. Hence, at each moment $t \in (0,T)$ the direct problem solution will have a finite support with respect to the space variables. Thus, we can pose the following boundary condition:

$$(\mathbf{E},\mathbf{H})\Big|_{\substack{x_1=\pm L\\ x_2=\pm L\\ x_3=\ L}} = \mathbf{0}. \tag{3.2}$$

Here $L = L(T,\varepsilon,\mu)$ (in the parabolic case we suppose that σ = const while $x_3 > L$, and we get another boundary condition on $x_3 = L$, see § 5).

We formulate the following problem of optimal control. We search on a set $\mathcal{M}$ of admissible solutions for a vector-function $V = (\sigma,\varepsilon,\mu)$ minimizing to the functional:

$$\begin{aligned} J(\sigma,\varepsilon,\mu) = &\int\limits_{S_1} \Big(\big[E_2(x_1,x_2,0,t) - G_2^\delta(x_1,x_2,t)\big]^2 \\ &+ \big[H_2(x_1,x_2,0,t) - Q_2^\delta(x_1,x_2,t)\big]^2\Big)\, dt\, dx_1\, dx_2 \\ &+ \int\limits_{\mathcal{D}} \Big(\alpha_1(\sigma-\sigma_0)^2 + \alpha_2(\varepsilon-\varepsilon_0)^2 + \alpha_3(\mu-\mu_0)^2\Big)\, dx_1\, dx_2\, dx_3, \qquad (3.3)\\ \mathcal{D} =& \big\{(x_1,x_2,x_3)\colon x_1 \in [-L,L],\ x_2 \in [-L,L],\ x_3 \in [0,L]\big\}, \\ S_1 =& \big\{(x_1,x_2,t)\colon x_1 \in [-L,L],\ x_2 \in [-L,L],\ t \in [0,T]\big\}. \end{aligned}$$

Here σ_0, ε_0, μ_0 are the known in advance approximation of σ, ε, μ.

Generally speaking, as it follows from inverse problem theory (Romanov and Kabanikhin, 1989a,b) the inverse problem can have several solutions. In fact, we have only two independent functions as the additional information on three variables, but we want to find three functions on three variables. However, in many practically interesting cases we know one of the three functions σ, ε, μ. (For example, in the undersurface radiolocation problem we can consider μ as the known constant, see § 4; in the quasi-stationary approximation inverse problem we put $\varepsilon = 0$, see § 5).

At each step of an iteration process we must know the functional gradient formula for the minimization of functional (3.3) by some gradient method.

For example, using the conjugate gradient method we choose the $(k+1)$th approximation $V^{k+1} = (\sigma^{k+1},\varepsilon^{k+1},\mu^{k+1})$ by means of the formulas:

$$V^{k+1} = V^k - \nu_k p_k, \qquad p_k = J'(V^k) - \beta_k p_{k-1}, \quad p_0 = J'(V^0).$$

Here, ν_k and β_k are special constants of the conjugate gradient's method, $J'(V)$ is the gradient of functional (3.3). It is easy to show that for a wide class of models on $\mathcal{M}$ the gradient $J'(V)$ looks as follows

$$J'(V) = (J'_\sigma, J'_\varepsilon, J'_\mu),$$
$$J'_\sigma = \int_0^T (\mathbf{E}^*, \mathbf{\Phi}^*)\, dt + 2\alpha_1(\sigma - \sigma_0),$$
$$J'_\varepsilon = \int_0^T (\mathbf{E}^*_t, \mathbf{\Phi}^*)\, dt + 2\alpha_2(\varepsilon - \varepsilon_0),$$
$$J'_\mu = \int_0^T (\mathbf{H}^*_t, \mathbf{\Psi}^*)\, dt + 2\alpha_3(\mu - \mu_0).$$

Here the vector-functions $\mathbf{\Phi} = (\Phi_1, \Phi_2, \Phi_3)^*$, $\mathbf{\Psi} = (\Psi_1, \Psi_2, \Psi_3)^*$ are the solution of the conjugate problem:

$$\varepsilon\frac{\partial \mathbf{\Phi}}{\partial t} - \mathrm{rot}\mathbf{\Psi} - \sigma\mathbf{\Phi} = \mathbf{0},$$
$$\mu\frac{\partial \mathbf{\Psi}}{\partial t} + \mathrm{rot}\mathbf{\Phi} = \mathbf{0},$$
$$(\mathbf{\Phi}, \mathbf{\Psi})|_{t=T} \equiv \mathbf{0}, \qquad (\mathbf{\Phi}, \mathbf{\Psi})|_{x_3=L} \equiv \mathbf{0},$$

$$\Psi_1(x_1, x_2, 0, t) = 2\big[G_2^\delta(x_1, x_2, t) - E_2(x_1, x_2, 0, t)\big],$$
$$\Phi_1(x_1, x_2, 0, t) = 2\big[H_2(x_1, x_2, 0, t) - Q_2^\delta(x_1, x_2, t)\big].$$

In the following two paragraphs we shall consider the examples of application of this algorithm to the various concrete problems.

4. UNDERSURFACE RADIOLOCATION PROBLEM

Let the external current have the form

$$\mathbf{j} = (0, 1, 0)^* p(x_1) q(x_3) r(t). \tag{4.1}$$

At the beginning we transform Maxwell equations and obtain second order equations.

Let constant μ be known. We suppose that functions ε and σ are independent from variables x_1, x_2 and positive. There is no electromagnetic oscillation up to the moment $t = 0$.

Then the Maxwell equations have three terms E_2, H_1, H_3 and looks as follows:

$$\begin{aligned} &\varepsilon\frac{\partial E_2}{\partial t} + \frac{\partial H_3}{\partial x_1} - \frac{\partial H_1}{\partial x_3} + \sigma E_2 + p(x_1)q(x_3)r(t) = 0,\\ &\mu\frac{\partial H_1}{\partial t} + \frac{\partial E_2}{\partial x_3} = 0, \qquad \mu\frac{\partial H_3}{\partial t} - \frac{\partial E_2}{\partial x_1} = 0. \end{aligned} \tag{4.2}$$

Excluding H_1 and H_3 from (4.2) we can easily arrive to the second order equation with respect to E_2:

$$\varepsilon\frac{\partial^2 E_2}{\partial t^2}+\sigma\frac{\partial E_2}{\partial t}=\frac{1}{\mu}\Delta_{x_1x_3}E_2-p(x_1)q(x_3)r'(t), \tag{4.3}$$

We must add the initial condition to (4.3):

$$E_2|_{t<0}\equiv 0 \tag{4.4}$$

and after all transformations we obtain the continuity conditions:

$$[E_2]_{x_3=0}=0, \qquad [\tfrac{\partial}{\partial x_3}E_2]_{x_3=0}=0. \tag{4.5}$$

Here $\varepsilon=\varepsilon_0\cdot\varepsilon_{\text{rel}}$, $\varepsilon_0=8.854\cdot 10^{-12}\,F/m$, $\mu=\mu_0\cdot\mu_{\text{rel}}$, $\mu_0=1.257\cdot 10^{-6}\,H/m$, $\mu_{\text{rel}}=1$, $p(x_1)$ and $q(x_3)$ are functions describing the source.
Now we give a physical interpretation of the undersurface radiolocation problem. We must determine functions $\varepsilon(x_3)$ and $\sigma(x_3)$ about 2–3 m depth. (Such works are useful for the investigations of the road and airdrome covers and other objects.) The external current source $\mathbf{j}$ is switched on the surface. The duration of the switches is nearly 2 ns. We measure the electromagnetic field on the surface during nearly 50 ns. The external current source is a long cable in the centre and along the road cover (on the axis x_2). In this case we neglect the influence of the cable ends and road boundaries.
We can transform problem (4.3)–(4.5). Let $F_{x_1}[\cdot]$ denote the operator of the Fourier transformation with respect to variable x_1.
Then we arrive to the problem:

$$\varepsilon U_{tt}+\sigma U_t=\frac{1}{\mu}(U_{x_1x_3}-\lambda^2 U)-P(\lambda)q(x_3)r'(t)=0, \tag{4.6}$$

$$U|_{t<0}\equiv 0. \tag{4.7}$$

Here $U=F_{x_1}[E_2(x_1,x_3,t)]$, $P(\lambda)=F_{x_1}[p(x_1)]$ and λ is the parameter of the Fourier transformation.
The continuity condition has the form

$$[U]_{x_3=0}=0, \qquad [U_{x_3}]_{x_3=0}=0. \tag{4.8}$$

We can add the boundary conditions, since equation (4.6) is of hyperbolic type:

$$U|_{x_3=-l_1}=0, \qquad U|_{x_3=l_2}=0. \tag{4.9}$$

Here l_1 and l_2 are sufficiently large numbers.

DIRECT PROBLEM. *Using the known functions $\sigma(x_3)$, $\varepsilon(x_3)$, constant μ and relations (4.6)–(4.8) we must find the function U.*

After all transformation we know the additional information concerning the direct problem solution at the point $x_3=0$:

$$U(0,t,\lambda_j)=f(\lambda_j,t), \quad j=1,2. \tag{4.10}$$

INVERSE PROBLEM. *Using the known constant μ, additional information (4.10) and correlations (4.6)–(4.9) we must find the functions $\varepsilon(x_3)$, $\sigma(x_3)$, $U(x_3,t,\lambda_1)$, $U(x_3,t,\lambda_2)$.*

Remark. We use additional information with two different parameters λ_j of Fourier transformation for the uniqueness of the inverse problem to have a unique solution (Romanov, 1984).

We shall solve inverse problem (4.6)–(4.10) as the problem of optimal control. On the set $\mathcal{M}$ consisting of the pairs (ε, σ) we shall find the control $V = (\varepsilon, \sigma)$ which minimizes the functional:

$$J(\sigma, \varepsilon) = \sum_{j=1}^{2} \int_0^T \left[U(0, t, \lambda_j, \sigma, \varepsilon) - f(\lambda_j, t)\right]^2 dt + \int_0^L \left(\alpha_1(\sigma - \sigma_0)^2 + \alpha_2(\varepsilon - \varepsilon_0)^2\right) dx_3 \tag{4.11}$$

We suppose that functions $\sigma(x_3)$ and $\mu(x_3)$ are smooth functions. Then gradient of this functional has the following form

$$J'(V) = (J'_\sigma, J'_\varepsilon),$$

$$J'_\sigma = \sum_{j=1}^{2} \int_0^T U_t(x_3, t, \lambda_j)\psi(x_3, t, \lambda_j)\, dt + 2\alpha_1(\sigma - \sigma_0),$$

$$J'_\varepsilon = \sum_{j=1}^{2} \int_0^T U_{tt}(x_3, t, \lambda_j)\psi(x_3, t, \lambda_j)\, dt + 2\alpha_2(\varepsilon - \varepsilon_0).$$

If we suppose that functions $\varepsilon(x_3)$ and $\sigma(x_3)$ are piece-wise functions, then we can describe these functions as vectors, for example:

$$\sigma(x_3) = (\sigma_{x_3^1}, \ldots, \sigma_{x_3^k}, \ldots, \sigma_{x_3^M}),$$

$\sigma_{x_3^k}$ is the value of function σ on the interval $[x_3^k, x_3^{k+1}]$ (k a gap number, $k = 1, \ldots, M$, M is the number of the gaps). We must add the continuity conditions at points $x_3 = x_3^k$:

$$[U]_{x_3 = x_3^k} = 0, \qquad [U_{x_3}]_{x_3 = x_3^k} = 0. \tag{4.12}$$

Then we can get the other type of gradient formula:

$$J'(V) = (J'_\sigma, J'_\varepsilon),$$

$$J'_\sigma = (\overline{J}'_{x_3^1}, \ldots, \overline{J}'_{x_3^k}, \ldots, \overline{J}'_{x_3^M}),$$
$$J'_\varepsilon = (J'_{x_3^1}, \ldots, J'_{x_3^k}, \ldots, J'_{x_3^M}),$$

$$\vec{J}_{x_3^k} = \sum_{j=1}^{2} \int_{x_3^k}^{x_3^{k+1}} \Big(\int_0^T \psi(x_3,t,\lambda_j) U_t(x_3,t,\lambda_j)\, dt \Big)\, dx_3$$

$$+ 2\alpha_1(\sigma_{x_3^k} - \sigma^0_{x_3^k}) \cdot \Delta_{x_3^k},$$

$$J'_{x_3^k} = \sum_{j=1}^{2} \int_{x_3^k}^{x_3^{k+1}} \Big(\int_0^T \psi(x_3,t,\lambda_j) U_{tt}(x_3,t,\lambda_j)\, dt \Big)\, dx_3$$

$$+ 2\alpha_2(\varepsilon_{x_3^k} - \varepsilon^0_{x_3^k}) \cdot \Delta_{x_3^k},$$

$$\Delta_{x_3^k} = x_3^{k+1} - x_3^k.$$

Here ψ is the solution of the conjugate problem:

$$\varepsilon\psi_{tt} - \sigma\psi_t = \frac{1}{\mu}(\psi_{x_3x_3} - \lambda_j^2\psi),$$

$$[\psi]_{x_3=0} = 2\big(f(\lambda_j,t) - U(0,t,\lambda_j)\big), \qquad [\psi_{x_3}]_{x_3=0} = 0,$$

$$\psi|_{x_3=-l_1} = 0, \qquad \psi|_{x_3=l_2} = 0.$$

When σ and ε are piece-wise functions we must add the continuity conditions:

$$[\psi]_{x_3=x_3^k} = 0, \qquad [\psi_{x_3}]_{x_3=x_3^k} = 0.$$

In (Kabanikhin et al., 1990) the results of numerical calculations for this problem are presented.

5. QUASI-STATIONARY APPROXIMATION PROBLEM

Physical interpretation of the quasi-stationary approximation problem has the form analogous to the undersurface radiolocation problem, but the various properties of medium and physical assumption allows us to put $\varepsilon = 0$ (we neglect the displacement currents).
We obtain the following problem:

$$\sigma\frac{\partial E_2}{\partial t} = \frac{1}{\mu}\Delta_{x_1x_3}E_2 - p(x_1)q(x_3)r'(t), \tag{5.1}$$

with the initial condition:

$$E_2|_{t<0} \equiv 0, \tag{5.2}$$

and the continuity condition:

$$[E_2]_{x_3=0} = 0, \qquad [\tfrac{\partial}{\partial x_3}E_2]_{x_3=0} = 0. \tag{5.3}$$

The conductivity σ in the air is equal to 0, so we consider the equation of elliptical type in the semispace $x_3 \in \mathbf{R}_-$ (air)

$$\Delta_{x_1 x_3} E_2 = 0, \quad x_3 < 0. \tag{5.4}$$

We put the condition of periodicity on x_1 at sufficiently large distance from axis x_3:

$$E_2|_{x_1=+D} = E_2|_{x_1=-D}. \tag{5.5}$$

We apply Fourier-series expansion with respect to variable x_1 and reduce problem (5.4)–(5.5) to the series of problems:

$$U_{x_3 x_3} - \lambda^2 U = 0, \quad x_3 \in \mathbf{R}_-, \tag{5.6}$$

$$[U]_{x_3=0} = 0, \qquad [U_{x_3}]_{x_3=0} = P(\lambda)\mu r'(t), \tag{5.7}$$

$$U|_{t=0} = 0, \tag{5.8}$$

$$\sigma U_t = \frac{1}{\mu}(U_{x_3 x_3} - \lambda^2 U), \quad x_3 \in \mathbf{R}_+. \tag{5.9}$$

Here we put $q(x_3) = \delta(x_3)$ where $\delta(x_3)$ is the Dirac δ-function.
Following (Oristaglio and Hohmann, 1984) we shall consider (5.6)–(5.9) only when $x_3 > 0$, writing down the boundary condition when $x_3 = 0$ taking into account the influence of the air:

$$\sigma U_t = \frac{1}{\mu}(U_{x_3 x_3} - \lambda^2 U), \quad x_3 \in (0, L) \tag{5.10}$$

$$(U_{x_3} - |\lambda|U)|_{x_3=0} = P(\lambda) \cdot \mu \cdot r'(t) \quad t \in [0, T] \tag{5.11}$$

$$U|_{t=0} = 0, \quad x_3 \in [0, L] \tag{5.12}$$

The quantity L is a sufficiently large number. If $\sigma = \text{const} \neq 0 \ (x_3 > L)$, then we can put the boundary condition:

$$U|_{x_3=L} = 0. \tag{5.13}$$

If function $\sigma = 0 \ (x_3 > L)$, then we can put the boundary condition which is analogous (5.11):

$$(U_{x_3} + |\lambda|U)|_{x_3=0} = 0, \quad t \in [0, T]. \tag{5.14}$$

DIRECT PROBLEM. *To find function U using relations* (5.10)–(5.14), *the known function σ and constant μ.*

We know the additional information concerning the direct problem solution:

$$U(0, t, \lambda) = f(\lambda, t). \tag{5.15}$$

INVERSE PROBLEM. *To find functions $\sigma(x_3)$ and U using relations* (5.10)–(5.14), *the known constant μ and additional information* (5.15).

We solve inverse problem (5.9)–(5.15) as problem of optimal control. On the set $\mathcal{M}$ we find the control σ minimizing the functional:

$$J(\sigma) = \int_0^T \left[U(0,t,\lambda,\sigma) - f(\lambda,t)\right]^2 dt + \int_0^T \alpha_1(\sigma - \sigma_0)^2\, dx_3. \tag{5.16}$$

As in § 4 we obtain the gradient for functional (4.16). In the case, when σ – is a smooth function, the gradient will look like:

$$J'_\sigma = \int_0^T U_t(x_3,t,\lambda)\psi(x_3,t,\lambda)\, dt + 2\alpha_1(\sigma - \sigma_0).$$

In the case, when σ is a piece-wise function, the gradient will be as follows

$$J'_\sigma = (\overline{J}'_{x_3^1}, \dots, \overline{J}'_{x_3^k}, \dots, \overline{J}'_{x_3^M});$$

$$\overline{J}'_{x_3^k} = \int_{x_3^k}^{x_3^{k+1}} \left(\int_0^T \psi(x_3,t,\lambda) U_t(x_3,t,\lambda)\, dt \right) dx_3$$
$$+2\alpha_1(\sigma_{x_3^k} - \sigma^0_{x_3^k})\Delta_{x_3^k}.$$

Here ψ is the solution of the conjugate problem:

$$-\sigma\psi_t = \frac{1}{\mu}(\psi_{x_3x_3} - \lambda^2\psi),$$

$$(\psi_{x_3} - |\lambda|\psi)|_{x_3=0} = 2\big(f(\lambda,t) - U(0,t,\lambda)\big),$$
$$\psi|_{t=0} = 0,$$
$$\psi|_{x_3=L} = 0 \quad \big(\text{ or } (\psi_{x_3} + |\lambda|\psi)|_{x_3=L} = 0\big).$$

We must remember that if the function σ is a piece-wise function, it is necessary to add the continuity condition for problem (5.10)–(5.14):

$$[U]_{x_3=x_3^k} = 0, \qquad [U_{x_3}]_{x_3=x_3^k} = 0$$

and for the conjugate problem:

$$[\psi]_{x_3=x_3^k} = 0, \qquad [\psi_{x_3}]_{x_3=x_3^k} = 0.$$

K.T. Iskakov and S.I. Kabanikhin in (Iskakov and Kabanikhin, 1991) present the results of numerical calculations for this problem.

REFERENCES

Alekseev, A.S. (1967). Inverse dynamical problems of seysmics. In: *Some Methods and Algorithms for Interpretation of Geophysics Data.* Nauka, Moscow, pp. 9–84 (in Russian).

Alekseev, A.S., Avdeev, A.V., Fat'yanov, A.G., and Cheverda, V.A. (1991). *Wave Process in the Vertically Unhomogeneous Medium.* Preprint No. 924. Computer Center, Siberian Branch of USSR Academy Sci., Novosibirsk (in Russian).

Alifanov, O.M., Artyukhin, E.A., and Rumyantsev, S.V. (1988). *Extremal Methods of Solution for Ill-Posed Problems.* Nauka, Moscow (in Russian).

Bamberger, A., Chavent, G., and Lailly, P. (1979). About the stability of the inverse problem in one-dimensional wave equations. Application to the interpretation of seismic profiles. *Appl. Math. Optim.* **5**, 1–47.

Bamberger, A., Chavent, G., Hemon, Ch., and Lailly, P. (1982). Inversion of normal incidence seismograms. *Geophys.* **47**, 757–770.

Blagoveshensky, A.S. (1966). About inverse problem of seismic wave spreading theory. *Problems Math. Phys.* **1**, 68–81.

Budak, B.M. and Iskenderov, A.D. (1966). Numerical methods of solution for some coefficient boundary problem. *Dokl. Akad. Nauk SSSR* **171**, 1054–1057 (in Russian).

Chen, W.N. and Seifeld, J.N. (1972). Estimation of spatially varging partial differential equations. *Int. J. Control.* **15**, 487–495.

Cook, D.A. and Shneider, W.A. (1983). Generalized linear inversion of reflection seismic data. *Geophys.* **48**, 665–676.

Denisov, A.M. (1982). About inverse problem for heat conduction equation with piece-wise coefficients. *Calc. Methods Progr.* **37**, 93–97.

Denisov, A.M. (1982). Solution uniqueness in some inverse problems for heat conduction equation with piece-wise coefficients. *J. Calc. Math. and Math. Phys.* **22**, 858–864.

Dmitriev, V.I. and Rudneva, T.L. (1971). *Algorithm for Inverse Problem Solution of Magnetotelluric Sounding.* Moscow Univ. Press, Moscow (in Russian).

Dmitriev, V.I. and Rudneva, T.L. (1973). About inverse problem of magnetotelluric sounding. *Calc. Methods Progr.* **20**, 421–430 (in Russian).

Dmitriev, V.I., Il'insky, A.S., and Sveshnikov, A.G. (1976). Development of mathematical investigation methods for direct and inverse problems of the electrodynamics. *Uspekhi Math. Nauk* **31**, 123–142 (in Russian).

Dmitriev, V.I. (1968). General calculation method for electromagnetic field in the layer medium. *Calc. Methods Progr.* **10**, 55–65 (in Russian).

Dmitriev,V.I. and Fedorova, E.A. (1983). About inverse problem solution in the frequency testing method for layer medium. In: *Biblioteka Program dlya Geophysiki,* Moscow Univ. Press, Moscow, pp. 11–18 (in Russian).

Iskakov, K.T. and Kabanikhin, S.I. (1991). The solution of one-dimensional inverse problem of geoelectrics by the method of conjugate gradients *J. Theor. Appl. Mech.* **3**, 78–88.

Iskenderov, A.D. (1968). About inverse boundary problems with unknown coefficients for some quasi-linear equations. *Dokl. Akad. Nauk SSSR* **178**, 999–1002 (in Russian).

Iskenderov, A.D. and Tagiev, R.K. (1983). Optimal problem with control in the coefficients of the parabolic equations. *Different. Equat.* **19**, 1324–1334.

Kabanikhin, S.I. and Abdiev, K.S. (1982). Modeling the initial stage of electromagnetic current establishment and using this in the problem of defining the conductivity tensor. In: *Questions of Correctness of Mathematical Physics and Analysis Problem.* Computer Center, Siberian Branch of USSR Academy Sci., Novosibirsk, pp. 85–94 (in Russian).

Kabanikhin, S.I. and Abdiev, K.S. (1986). Projection-difference method for solution of three-dimensional problem of geoelectrics. In: *Questions of Correctness of Mathematical Physics and Analysis Problem.* Computer Center, Siberian Branch of USSR Academy Sci., Novosibirsk, pp. 61–72 (in Russian).

Kabanikhin, S.I. and Martakov, S.V. (1988). *Investigation of the Projection-Difference Methods for Geoelectrics.* Preprint No. 13. Inst. of Math., Siberian Branch of USSR Academy Sci., Novosibirsk (in Russian).

Kabanikhin, S.I., Martakov, S.V., Tabarovskii, L.A., and Shiphon, N.Y. (1989). Studying the dielectric permeability of medium near borehole in the early stady process. In: *Investigations on Conditional Correctness of the Problems of Mathematical Physics,* Inst. of Math., Siberian Branch of USSR Academy Sci., Novosibirsk, pp. 40–59 (in Russian).

Kabanikhin, S.I., Martakov, S.V., Morgunova, T.Ya., and Podgornaya, L.E. (1990). Undersurface radiolocation problem for the underground investigations on the small depths. In: *Solution Methods of Inverse Problems.* Inst. of Math., Siberian Branch of USSR Academy Sci., Novosibirsk, pp. 61–72 (in Russian).

Kravaris, C. and Seifeld, J.N. (1985). Indentification of parameters in distributed parameter systems by regularization. *SIAM J. Control Opt.* **23**, 217–241.

Lavrent'ev, M.M. and Romanov, V.G. (1966). About three linearized inverse problems for hyperbolic equations. *Dokl. Akad. Nauk SSSR* **171**, 1279–1281 (in Russian).

Marchuk, G.I. (1964). About statement of ill-posed inverse problems. *Dokl. Akad. Nauk SSSR* **156**, 503–506 (in Russian).

Marquart, P.W. (1963). An algorithm for least squares estimation of nonlinear parameters. *J. SIAM* **11**, 413–441.

Nyambaa, Sh. and Cheverda, V.A. (1988). *Optimal Method of Solution for Inverse Problem of Electromagnetics on Constant Current for Vertically-Unhomogeneous Medium.* Preprint No. 794. Computer Center, Siberian Branch of USSR Academy Sci., Novosibirsk (in Russian).

Oristaglio, M.L. and Hohmann, G.W. (1984). Diffusion of electromagnetic field into a two-dimensional Earth. A finite-difference approach. *Geophysics* **49**, 870–894.

Prilepko, A.I. (1973). Inverse problems of potential theory. *Math. Zametki* **14**, 755-767 (in Russian).

Prilepko, A.I., Ivankov, A.L., and Orlovsky, D.G. (1984). Inverse problem for transport equation and system of hyperbolic type equations. In: *Ill-Posed Problems of Mathematical Physics and Analysis.* Nauka, Novosibirsk, pp. 240–241.

Prilepko, A.I. and Orlovsky, D.G. (1984). Inverse problem for the evolutionary half-linear equation. *Dokl. Akad. Nauk SSSR* **277**, 799–803 (in Russian).

Romanov, V.G. (1984). *Inverse Problems of Mathematical Physics.* Nauka, Moscow (in Russian).

Romanov, V.G., Kabanikhin, S.I., and Pukhnacheva, T.P. (1984). *Inverse Problems of Electrodynamics.* Computer Center, Siberian Branch of USSR Academy Sci., Novosibirsk (in Russian)

Romanov, V.G. and Kabanikhin, S.I. (1989). *Inverse Problems of Geoelectrics (Uniqueness and Existence Theorems).* Preprint No. 25. Inst. of Math. Siberian Branch of USSR Academy Sci., Novosibirsk (in Russian).

Romanov, V.G. and Kabanikhin, S.I. (1989). *Inverse Problems of Geoelectrics (Numerical Methods of Solution)*. Preprint No. 32. Inst. of Math., Siberian Branch of USSR Academy Sci., Novosibirsk (in Russian).

Tikhonov, V.I. and Shakhsuvarov, D.N. (1956). Calculation method of electromagnetic field excited by the variable current in the layer medium. *Novosti AN SSSR. Ser. Geogr. and geoph.* **20**, 245–251 (in Russian).

Tikhonov, V.I. and Shakhsuvarov, D.N. (1961). About unevenness of electromagnetic field asymptotic excited by the variable current dipole in the layer medium. *Novosti AN SSSR. Ser. Geophys.* **1**, 107–110 (in Russian).

Tikhonov, A.N. (1963). About regularization of ill-posed problem and regularization methods. *Dokl. Akad. Nauk SSSR* **151**, 501–504 (in Russian).

Ursin, B. and Berteussen, K.-A. (1986). Comparison of some inverse methods for wave propagation in layered media. *Proc. IEEI* **74**, 7–19.

Vasil'ev, F.P. (1981). *Methods for Solution of Extreme Problems*. Nauka, Moscow (in Russian).

Vasil'ev, F.P. (1988). *Numerical Methods for Solution of Extreme Problems*. Nauka, Moscow (in Russian).

Van'yan, A.L. (1965). *Principles of Electromagnetic Testing*. Nedra, Moscow (in Russian).

Zhdanov, M.S., Varentsov, I.M., Golubev, N.G., and Spichak, V.V. (1981). *Problems of Development for Modeling Calculation Methods of Electromagnetic Anomalies*. IZMIRAN, Moscow (in Russian).

Zhdanov, M.S. and Spichak, V.V. (1982). Conditions and developments of the modeling for the electromagnetic field in the three-dimensional medium. In: *Algorithms and Problems for the Direct and Inverse Problem Solutions of Electromagnetic Induction in the Earth*. IZMIRAN, Moscow, pp. 3–10 (in Russian).

Zhdanov, M.S., Golubev, N.G., Spichak, V.V. and Varentsov, I.M. (1982). The construction of elective methods for electromagnetic modeling. *Geophys.* **68**, 589–607.

Zhdanov, M.S. and Spichak, V.V. (1984). *Modern Modeling Methods for Quasi-Stationary Electromagnetic Field in Three-Dimensional Unhomogeneous Medium*. Preprint No. 45. IZMIRAN, Moscow (in Russian).

Zhdanov, M.S., Spichak, V.V., and Zaslavsky, L.Y. (1986). *Numerical Modeling of the Quasi-Stationary Electromagnetic Fields in Three-Dimensional Axis-Symmetrical Medium by Balance Method*. Preprint No. 68. IZMIRAN, Moscow (in Russian).

Ill-Posed Problems in Natural Sciences, pp. 326 – 332
A. Tikhonov (Ed.)
1992 VSP/TVP

ON PASSAGE TO THE LIMIT IN THE INVERSE PROBLEMS FOR NONDIVERGENCE PARABOLIC EQUATIONS

V.L. KAMYNIN
The Computer Centre of the Russia Academy of Sciences, Vavilova str., 40, 117967 Moscow, Russia

ABSTRACT

Conditions are established under which the solutions of the sequences of the inverse problems for parabolic equations with weakly converging coefficients tend to the solutions of the corresponding limit problems. An example revealing the essentiality of these conditions, is given.

The subject of this paper is passage to the limit in a sequence of inverse problems for equations with weakly converging coefficients. Problems of this type arise in mathematical simulation of physical processes in strongly inhomogeneous media in the case where due to certain reasons, some characteristic of the media are unknown (even in the average), but some additional information about the process under study is available.

In earlier papers, various aspects of passage to the limit in differential equations were examined when studying direct problems. One of the methods which can be used here is based on the G-convergence theory and the theory of averaging differential operators, see (De Giorgi and Spagnolo, 1973; Bensoussan et al., 1978; Zhikov et al., 1981; Bakhvalov and Panasenko, 1984).

Another method is based on the study of convergence of solutions of differential equations with weakly converging coefficients. Using this, one also can provide the transition from a "microscopic" description of strongly inhomogeneous media to a "macroscopic" one. In (Kruzhkov and Kamynin, 1983; Kamynin, 1988; 1990) this method was developed for various classes of elliptic and parabolic equations. The main additional restriction on the weakly convergent coefficients in the equations under consideration is their equicontinuity in a certain weak sense.

This paper presents a similar approach to the study of the passage to the limit for two types of inverse problems in a cylinder $Q \equiv [0,T] \times \Omega$ (where Ω is a bounded domain in $\mathbf{R}^n$ with a smooth boundary) for the equations

$$L^m u(t,x) = f(x)h^m(t,x) + g^m(t,x) \tag{1}$$

with the boundary condition

$$u(0,x)=0; \qquad u(t,x)=0, \qquad x\in\partial\Omega; \tag{2}$$

and the condition of final overdetermination

$$u(T,x)=\Psi^m(x); \tag{3}$$

and also for the equations

$$L^m u(t,x)+f(x)u(t,x)=g^m(t,x) \tag{4}$$

with the boundary condition

$$u(0,x)=\theta^m; \qquad u(t,x)=\theta^m, \quad x\in\partial\Omega; \tag{5}$$

and with the condition of final overdetermination (3). Here $m = 1,2,...,\infty$; the pair $(u(t,x), f(x))$ is unknown, and

$$L^m u \equiv -d^m(x)u_t + a_{ij}^m(x)u_{x_ix_j} + b_i^m(x)u_{x_i} + c^m(x)u.$$

Throughout this paper we employ the usual convention of summation over repeated indices from 1 to n.

Note that the solvability of the inverse problems was studied in (Prilepko and Solov'ev, 1987 a,b).

We are interested in the conditions on the data (1), (2), (3) and (4), (5), (3) under which the solutions $(u^m(t,x), f^m(x))$ of the problems under consideration converge as $m\to\infty$ to the solutions of the corresponding limit problems (1), (2), (3) or (4), (5), (3), (for $m=\infty$).

Let problems (1)–(3) satisfy the following assumptions.

A. The coefficients of equations (1) are measurable, $a_{ij}^m(x) = a_{ji}^m(x)$, and the following inequalities are true for any $(t,x)\in Q$ uniformly with respect to $m = 1,2,...,\infty$:

$$\Lambda_0|\zeta|^2 \le a_{ij}^m(x)\zeta_i\zeta_j \le \Lambda_1|\zeta|^2, \quad \zeta\equiv(\zeta_1,...,\zeta_n)\in\mathbf{R}^n; \tag{6}$$

$$d_0 \le d^m(x) \le d_1; \tag{7}$$

$$\left|b_i^m(x)\right|, \quad \left|g^m(t,x)\right|, \quad \left|g_t^m(t,x)\right|, \quad \left|\Psi^m(x)\right|, \quad \left|\Psi_{x_i}^m(x)\right|,$$

$$\left|\Psi_{x_ix_j}^m(x)\right| \le M_1; \quad -M_1 \le c^m(x) \le -c_0; \tag{8}$$

$$\left|h^m(t,x)\right|, \quad \left|h_t^m(t,x)\right| \le M_1, \quad \left|h^m(T,x)\right| \ge h_0;$$

where the positive constants Λ_0, Λ_1, d_0, d_1, M_1, c_0, h_0 do not depend on m.

B. The compatibility condition $\Psi^m(x)=0$, $x\in\partial\Omega$, $m=1,2,...,\infty$, holds.

C. The constant h_0 of assumption A satisfies the inequality

$$h_0 \ge \frac{M_1 d_1}{d_0}\Big[\frac{d_1}{c_0} + \exp\big(-\frac{c_0 T}{d_1}\big)\big(1-\frac{d_1}{c_0}\big)\Big].$$

D. For $n \geq 2$ the Cordes type condition (see e.g., (Alkhutov and Mamedov, 1985)) is valid:

$$\sup_{m=1,2,\ldots,\infty} \sup_{x\in\Omega} \frac{\sum_{i,j=1}^{n} \left(a_{ij}^{m}(x)\right)^2}{\left(\sum_{i=1}^{n} a_{ii}^{m}(x)\right)^2} < \frac{1}{n-\mu^2},$$

where

$$\mu = \inf_{m=1,2,\ldots,\infty} \frac{\inf_{x\in\Omega} \sum_{i=1}^{n} A_{ii}^{m}(x)}{\sup_{x\in\Omega} \sum_{i=1}^{n} A_{ii}^{m}(x)}, \quad A_{ii}^{m}(x) = \frac{a_{ii}^{m}(x)}{d^{m}(x)}.$$

E. The following convergence conditions are satisfied as $m \to \infty$:

$$d^m(x) \rightharpoonup d^\infty(x),\ a_{ij}^m(x) \rightharpoonup a_{ij}^\infty(x),\ b_i^m(x) \rightharpoonup b_i^\infty(x),\ c^m(x) \rightharpoonup c^\infty(x) \tag{9}$$

weakly in $\mathbf{L}_2(\Omega)$;

$$g^m(t,x) \rightharpoonup g^\infty(t,x) \tag{10}$$

weakly in $\mathbf{L}_2(Q)$;

$$\Psi^m(x) \rightrightarrows \Psi^\infty(x) \tag{11}$$

uniformly on Ω; $h^m(t,x) \to h^\infty(t,x)$ in the $\mathbf{L}_2(Q)$-norm.
As far as problems (3)–(5) are concerned, we suppose the following assumptions to hold.

A$'$. The coefficients of equations (4) are measurable, $a_{ij}^m(x) = a_{ji}^m(x)$, inequalities (6)–(8) and the inequalities $\Psi^m(x) \geq \Psi_0$, $\Theta_0 \leq \theta^m \leq \theta_1$ are true uniformly with respect to m for any $(t,x) \in Q$. Here, the positive constants Λ_0, Λ_1, d_0, d_1, M_1, c_0, Ψ_0, Θ_0, Θ_1 are supposed to be independent of m.

B$'$. The compatibility condition $\Psi^m(x) = \Theta^m$, $x \in \partial\Omega$, $m = 1,2,\ldots,\infty$, holds.

C$'$. The constant T satisfies the inequality

$$T > \frac{1}{c_0} \log \frac{\theta_1}{\Psi_0}.$$

D$'$. Condition D holds.

E$'$. Convergence conditions (9)–(11) and the condition $\theta^m \to \theta^\infty$ hold as $m \to \infty$.

Definition 1. A pair of functions $(u^m(t,x), f^m(x))$ is said to be a generalized solution of problem (1)–(3) if $u^m(t,x) \in C^{0,\gamma}(Q) \cap W_2^{1,2}(Q)$, $f^m(x) \in \mathbf{L}_\infty(\Omega)$; these functions satisfy equation (1) almost everywhere in Q and the function $u^m(t,x)$ satisfies conditions (2) and (3).

Definition 2. A pair of functions $(u^m(t,x), f^m(x))$ is said to be a generalized solution of problem (3)–(5) if $u^m(t,x) \in C^{0,\gamma}(Q) \cap W_2^{1,2}(Q)$, $f^m(x) \in \mathbf{L}_\infty(\Omega)$, $f^m(x) \leq 0$, $x \in \Omega$; these functions satisfy equation (4) almost everywhere in Q and the function $u^m(t,x)$ satisfies conditions (5) and (3).
We assume that for any fixed $m \neq \infty$ all smoothness and compatibility conditions on the data of problems (1)–(3) and (3)–(5) are fulfilled which ensure the existence of solutions of the corresponding inverse problems (see (Prilepko and Solov'ev, 1987 a,b)).

We emphasize that the uniformity (with respect to m) of the restrictions mentioned is not required.
By virtue of the above assumptions and basing on the results due to (Prilepko and Solov'ev, 1987 a,b; Kruzhkov, 1966; Krylov and Safonov, 1980; Alkhutov and Mamedov, 1985) it is not difficult to obtain the following uniform with respect to m estimates for the solutions (u^m, f^m) of problems (1)–(3) (or for these of problems (3)–(5)):

$$|f^m|_{\mathbf{L}_\infty(\Omega)} \le K_1, \quad |u^m|_{C^{0,\gamma}(Q)} \le K_2, \quad \|u^m\|_{W_2^{1,2}(Q)} \le K_3; \tag{12}$$

here the constants $K_1, K_2, K_3 > 0$ do not depend on m and on the smoothness of the coefficients in (1) or in (4).
By virtue of estimates (12) there exist such a subsequence $m_\nu \to \infty$ and a pair of functions $(\hat{u}^\infty(t,x), \hat{f}^\infty(x))$ that

$$u^{m_\nu}(t,x) \rightrightarrows \hat{u}^\infty(t,x) \text{ uniformly on } Q,$$

$$u^{m_\nu}_{x_i}(t,x) \to \hat{u}^\infty_{x_i}(t,x) \text{ in the } \mathbf{L}_2(Q)\text{–norm}$$

,

$$u^{m_\nu}_t(t,x),\ u^{m_\nu}_{x_i x_j}(t,x) \rightharpoonup \hat{u}^\infty_t(t,x),\ \hat{u}^\infty_{x_i x_j}(t,x)$$

weakly in $\mathbf{L}_2(Q)$, $f^{m_\nu}(x) \rightharpoonup \hat{f}^\infty(x)$ weakly in $\mathbf{L}_2(\Omega)$ as $\nu \to \infty$.
The question is whether the pair $(\hat{u}^\infty, \hat{f}^\infty)$ is a generalized solution of problem (1)–(3) (or, respectively, of problem (3)–(5) $m = \infty$). This statement will be true if for we require the following additional condition similar to the one used in (Kruzhkov and Kamynin, 1983; Kamynin 1988, 1990): there exists a function $\alpha(m) \to 0$ as $m \to \infty$ such that for all $g(t,x) \in \overset{\circ}{C}{}^\infty(Q)$ and $m = 1, 2, \ldots$

$$\Big| \int_Q [a^m_{ij}(x) - a^\infty_{ij}(x)] g_{x_i x_j}\, dx\, dt \Big| \le \alpha(m) \{ \|g\|_{W_2^{1,2}Q} + |g|_{C^{0,\gamma}(Q)} \}. \tag{13}$$

THEOREM 1. *If conditions $A - E$ and (13) are satisfied, then the pair $(\hat{u}^\infty, \hat{f}^\infty)$ is a generalized solution of problem (1) – (3) $m = \infty$.*

THEOREM 2. *If conditions $A' - E'$ and (13) are satisfied, then the pair $(\hat{u}^\infty, \hat{f}^\infty)$ is a generalized solution of problem (3)–(5) $m = \infty$.*

The proofs of Theorems 1 and 2 are similar. In what follows we give a brief schetch of these proofs. Let us consider (for the sake of definiteness) the case of Theorem 1. We show that the pair $(\hat{u}^\infty, \hat{f}^\infty)$ satisfies equation (1) for $m = \infty$ almost everywhere in Q. Let $\varphi(t,x)$ be an arbitrary but fixed function in $\overset{\circ}{C}{}^\infty(Q)$. The quantity J defined by the ratio

$$J \equiv \Big| \int_Q \big[L^\infty \hat{u}^\infty(t,x) - \hat{f}^\infty(x) h^\infty(t,x) - g^\infty(t,x) \big] \varphi(t,x)\, dx\, dt \Big|,$$

satisfies the inequality $J \le J_1^\nu + J_2^\nu$, where

$$J_1^\nu \equiv \left| \int\limits_Q \left[a_{ij}^\infty \hat{u}_{x_i x_j}^\infty - a_{ij}^{m_\nu} u_{x_i x_j}^{m_\nu} \right] \varphi \, dx \, dt \right|,$$

$$J_2^\nu \equiv \left| \int\limits_Q \left\{ \left[-d^\infty \hat{u}_t^\infty + d^{m_\nu} u_t^{m_\nu} \right] + \left[b_i^\infty \hat{u}_{x_i}^\infty - b_i^{m_\nu} u_{x_i}^{m_\nu} \right] \right. \right.$$

$$\left. \left. + \left[c^\infty \hat{u}^\infty - c^{m_\nu} u^{m_\nu} \right] - \left[\hat{f}^\infty h^\infty - f^{m_\nu} h^{m_\nu} \right] - \left[g^\infty - g^{m_\nu} \right] \right\} \varphi \, dx \, dt \right|.$$

To prove the required statement it is sufficient to show that for any $\varepsilon > 0$ there exists such $\nu_0(\varepsilon) > 0$ that for all $\nu \ge \nu_0(\varepsilon)$

$$J_1^\nu < \varepsilon, \quad J_2^\nu < \varepsilon. \tag{14}$$

We prove here only the first estimate as the second one is get easier. We have $J_1^\nu \le \sum_{k=1}^6 I_k$, where

$$I_1 \equiv \left| \int\limits_Q a_{ij}^\infty \left[u_{x_i x_j}^{m_\nu} - \hat{u}_{x_i x_j}^\infty \right] \varphi \, dx \, dt \right|,$$

$$I_2 \equiv \left| \int\limits_Q \left[a_{ij}^{m_\nu} - a_{ij}^\infty \right] (u^{m_\nu} \varphi)_{x_i x_j} \, dx \, dt \right|,$$

$$I_3 \equiv 2 \left| \int\limits_Q \left[a_{ij}^{m_\nu} - a_{ij}^\infty \right] \left[u_{x_i}^{m_\nu} - \hat{u}_{x_i}^\infty \right] \varphi_{x_j} \, dx \, dt \right|,$$

$$I_4 \equiv \left| \int\limits_Q \left[a_{ij}^{m_\nu} - a_{ij}^\infty \right] \left[u^{m_\nu} - \hat{u}^\infty \right] \varphi_{x_i x_j} \, dx \, dt \right|,$$

$$I_5 \equiv \left| \int\limits_Q \left[a_{ij}^{m_\nu} - a_{ij}^\infty \right] \hat{u}_{x_i}^\infty \varphi_{x_j} \, dx \, dt \right|,$$

$$I_6 \equiv \left| \int\limits_Q \left[a_{ij}^{m_\nu} - a_{ij}^\infty \right] \hat{u}^\infty \varphi_{x_i x_j} \, dx \, dt \right|.$$

The integrals I_1, I_3, I_4 tend to zero as $\nu \to \infty$ by virtue of the definition of the function $\hat{u}^\infty(t,x)$, while the integrals I_5 and I_6 are small for all sufficiently large ν in accordance with assumption E.
Finally, condition (13) and estimates (12) imply the estimate

$$I_2 \le \alpha(m_\nu) \left\{ \| u^{m_\nu} \varphi \|_{W_2^{1,2}(Q)} + \left| u^{m_\nu} \varphi \right|_{C^{0,\alpha}(Q)} \right\} \le C(K_2, K_3, \varphi) \alpha(m_\nu).$$

Therefore $I_2 \to 0$ as $\nu \to \infty$, and hence estimate (14) for J_1^ν is obtained. Thus, the pair $(\hat{u}^\infty, \hat{f}^\infty)$ satisfies equation (1) (for $m = \infty$) almost everywhere in Q. Evidently, $\hat{u}(t,x) \in C^{0,\gamma}(Q) \cap W_2^{1,2}(Q)$, $\hat{f}^\infty(x) \in \mathbf{L}_\infty(\Omega)$. The proposition that $\hat{u}^\infty(t,x)$ satisfies conditions (2), (3) (for $m = \infty$) is also obvious. Consequently, the pair $(\hat{u}^\infty, \hat{f}^\infty)$ is a generalized solution of problem (1)–(3) for $m = \infty$. The theorem is proved.

Remark. Condition (13) in Theorems 1 and 2 is essential. This can be seen from the following example.

Example. Consider in the cylinder $P = \{(t,x,y) : 0 \le t \le 1,\ r^2 \equiv (x^2+y^2) \le 1\}$ the inverse problems of determination of the pair (u,f) with boundary condition (2) and the condition of overdetermination

$$u(T,x) = 0 \tag{15}$$

for the equations

$$\begin{aligned} u_t - \Big[(10 - \sin m\pi r^2)u_{xx} + 10u_{yy} - u\Big] &= f(x,y) \\ +\frac{\pi}{m^2}\cos mt \sin m\pi r^2 &- \sin \pi t(10 - \sin m\pi r^2) \\ \times\left[\frac{2\pi}{m}\cos m\pi r^2 - 4\pi^2x^2 \sin m\pi r^2\right] &- \frac{20\pi}{m}\sin \pi t \cos m\pi r^2 \\ +\sin m\pi r^2\left[40\pi^2y^2 \sin \pi t + \frac{1}{m^2}\sin \pi t - 1\right], &\quad m = 1,2,..., \end{aligned} \tag{16}$$

and for the equation

$$u_t - \Big[10u_{xx} + 10u_{yy} - u\Big] = f(x,y) - 2\pi^2x^2 \sin \pi t. \tag{17}$$

One can check easily that all the conditions of Theorem 1 except assumption (13) are satisfied for these problems. The solutions of problems (16), (2), (15) with $m \ne \infty$ are the pairs of functions $(\frac{1}{m^2}\sin \pi t \sin m\pi r^2,\ \sin m\pi r^2)$ which converge, as $m \to \infty$, to the trivial pair $(0,0)$. The latter trivial pair is not, however, a solution of the limit problem (17), (2), (15).

REFERENCES

Alkhutov, Yu.A. and Mamedov, I.T. (1985). Some properties of the solutions of the first boundary value problem for parabolic equation with discontinuous coefficients. *Soviet Math. Dokl.* **32**, 343–347.

Bakhvalov, N.S. and Panasenko, G.P. (1984). *Homogenization of Processes in Periodic Media.* Nauka, Moscow (in Russian).

Bensoussan, A., Lions, J.L., and Papanicolaou, P. (1978). *Asymptotic Analysis for Periodic Structures*. North-Holland/Elsevier, Amsterdam.

De Giorgi, E. and Spagnolo, S. (1973). Sulla convergenza degli integrali dell energia per operatori ellitici del secondo ordine *Bull. Un. Mat. Ital. Ser.* 4. **8**, 391–411.

Kamynin, V.L. (1988). On passage to the limit in quasilinear elliptic equations with several independent variables. *Math. USSR Sbornik.* **60**, 47–66.

Kamynin, V.L. (1990). Passage to the limit in quasilinear parabolic equations with weakly converging coefficients and asymptotic behaviour of the Cauchy problem solutions. *Math. USSR Sbornik.* **181**, 1031–1047.

Kruzhkov, S.N. and Kamynin, V.L. (1983). Convergence of the solutions of quasilinear parabolic equations with weakly converging coefficients. *Soviet Math. Dokl.* **27**, 628–632.

Kruzhkov, S.N. (1966). A priori estimate for the derivative of a solution to a parabolic equation and some of its applications. *Dokl. Acad. Nauk SSSR.* **170**, 501–504 (in Russian).

Krylov, N.V. and Safonov, M. V. (1980). Certain property of the solutions of the parabolic equations with measurable coefficients. *Izv. Akad. Nauk SSSR. Ser. Math.* **44**, 161–175 (in Russian).

Prilepko, A.I. and Solov'ev, V.V. (1987). Existence theorems and Rothe's method in the inverse problems for an equation of parabolic type, II. *Different, Equat.* **23**, 1971–1980 (in Russian).

Prilepko, A.I. and Solov'ev, V.V. (1987 b). On the solvability of inverse boundary problems of determination of the coefficient before the minor derivative in a parabolic equation. *Different, Equat.* **23**, 136–141(in Russian).

Zhikov, V.V., Kozlov, S.M. and Oleinik, O.A. (1981). G-convergence of parabolic operators. *Russian Math. Surveys.* **36**, 11–15 (in Russian).

Ill-Posed Problems in Natural Sciences, pp. 333 – 341
A. Tikhonov (Ed.)
1992 VSP/TVP

ONE METHOD OF SOLUTION OF THE INVERSE PROBLEMS FOR LAYERED MEDIA

M.M. LAVRENT'EV, Jr.
Institute of Mathematics, Siberian Branch of Russia Academy of Sciences, Maltseva str. 14, 630090 Novosibirsk, Russia

ABSTRACT

One method for the solution of a number of identification problems of piecewise constant coefficient for wave equation and for heat equation is investigated. All the problems considered are reduced to the Helmholtz equation with a parameter.
To solve the inverse problems some modifications of the algorithm which was proposed by the author are used. The algorithm consists simply in calculating the integral moments related to the over information and algebraic transformations.
It is important that the information is used only on bounded intervals of frequency parameters which can be chosen away from zero.

1. INTRODUCTION

We consider some identification problems of piecewise constant coefficients for wave equation and for heat equation. One can associate these problems with the problem of finding the parameters of isotropic layered medium or with the problem of identifying the structure of an infinite slab which consists of a number of different heat conductors.
The inverse problems with piecewise constant coefficients were investigated by many authors, cf. (Baev, 1988; Gerver, 1974; Belishev and Kupriyanova, 1983; Denisov, 1982; Sgheri, to appear).
To recall the general theory and some applications of the inverse problems we refer the reader to (Lavrentiev et al, 1986; Aki and Richards, 1980; Yakhno, 1990).
We consider the following problem for wave equation

$$\frac{\partial^2 u}{\partial t^2} = c^{-2}(x)\frac{\partial^2 u}{\partial x^2}, \quad x > 0,$$

$$u\,|_{t<0} = 0, \qquad u_x\,|_{x=+0} = H(t), \tag{1.1}$$

here $c(x)$ is a positive piecewise constant function having discontinuities at points $a_1 < a_2 < \ldots < a_k$, i.e.,

$$c(x) = \begin{cases} c_m, & x \in (a_{m-1}, a_m), \quad m = 1, \ldots, k; \\ c_{k+1}, & x > a_k, \end{cases} \tag{1.2}$$

where c_m are constants, $a_0 = 0$.
Let us add to (1.1) the interface conditions at the discontinuity points

$$[u]_{x=a_m} = [c^{-2}u_x]_{x=a_m} = 0 \quad (m = 1, \dots, k). \tag{1.3}$$

We will consider the following inverse problem.

INVERSE PROBLEM 1 (I.P.1). *Given $\Phi(\omega)$ for $\omega \in [\omega_1, \omega_2]$, find the coefficient $c(x)$ (i.e., a set of $2k+1$ numbers $a_1, \dots, a_k; c_1, \dots, c_{k+1}$) such that*

$$\Phi(\omega) = \int_{-\infty}^{\infty} e^{-i\omega t} u_t(0,t)\, dt, \quad \omega \in [\omega_1, \omega_2], \tag{1.4}$$

where u: $(0, +\infty) \times (0, +\infty) \to \mathbf{R}$ solves equations (1.1)–(1.3).

Remark 1.1. One can consider either $F_1(t) = u(0,t)$ or $F_2(t) = u_t(0,t)$ as the overinformation. We use here (1.4) because this form is exploited in the sequel. A comparatively simple algorithm for solving I.P.1 was proposed by the author (Lavrent'ev, Jr., 1991). It consists of algebraic transformations related to the integral moments of overinformation (1.4).

From now on we will assume that

$$\tau = c_m(a_m - a_{m-1}), \quad m = 1, \dots, k, \tag{1.5}$$

the quantity τ is known and is such that $\tau > \pi/(\omega_2 - \omega_1)$. This is one of the necessary conditions to use our algorithm.
Let us note that if one associate (1.1) with the process of seismic wave propagation through the Earth crust then (1.5) means that the times of wave propagation are the same for all the layers. This is the so-called Goupillaud hypothesis and it is often used in geophysics.

Remark 1.2. Due to the k equalities (1.5) we have to determine only the $k+1$ numbers $c_1, \dots, c_{k+1}$.
In this paper we investigate how one can use the above mentioned algorithm to solve some other inverse problems.
In §2 our algorithm is described. In §§ 3 and 4 some inverse problems for the wave equation and for the heat equation respectively are discussed.

2. AN ALGORITHM FOR SOLVING I.P.1

It is well known, see (Yakhno, 1990; Aki and Richards, 1980), that the following problem for the Helmholtz equation

$$\frac{\partial^2 v}{\partial x^2} + \omega^2 c^2(x) v = 0, \tag{2.1}$$

$$v_x(0, \omega) = h(\omega), \tag{2.2}$$

$$v_x - i\omega cv \to 0 \quad \text{as } x \to +\infty, \tag{2.3}$$

$$[v]_{x=a_m} = [c^{-2}v_x]_{x=a_m} = 0 \quad (m = 1, \dots, k), \tag{2.4}$$

is connected with problem (1.1). Here $v(x,\omega), h(\omega)$ are the Fourier transforms of $u(x,t), H(t)$, i.e.,

$$v(x,\omega) = \int_{-\infty}^{\infty} e^{-i\omega t} u(x,t)\, dt, \qquad h(\omega) = \int_{-\infty}^{\infty} e^{-i\omega t} H(t)\, dt.$$

Thus to solve I.P.1 one can investigate the following problem:

INVERSE PROBLEM 2 (I.P.2). *Given $\Phi(\omega)$ for $\omega \in [\omega_1, \omega_2]$, find a coefficient $c(x)$ (see (1.2)) such that*

$$\Phi(\omega) = i\omega v(0,\omega), \quad \omega \in [\omega_1, \omega_2], \tag{2.5}$$

where v: $(0, +\infty) \times (\omega_1, \omega_2) \to \mathbf{R}$ solves equations (2.1)–(2.4).

According to (Lavrent'ev, Jr, 1991) one can solve I.P.2 as follows.
Let us write the general solution of (2.1) on intervals $a_{m-1} \le x \le a_m$ as

$$v(x,\omega) = B_1^m(\omega) \exp\{ic_m\omega(x - a_m)\} + B_2^m(\omega)\exp\{-ic_m\omega(x - a_m)\}. \tag{2.6}$$

One can rewrite the interface conditions (2.4) at $a_1, \dots, a_{k-1}$ in the following way

$$B_1^m + B_2^m = B_1^{m+1}e^q + B_2^{m+1}e^{-q},$$

$$\frac{i\omega}{c_m}(B_1^m - B_2^m) = \frac{i\omega}{c_{m+1}}(B_1^{m+1}e^q - B_2^{m+1}e^{-q}), \tag{2.7}$$

where $q = i\omega c_{m+1}(a_m - a_{m+1})$.
Using (1.5), it is easy to transform (2.7) into the recursive formulas

$$B_1^m(\omega) = \frac{1}{2}(\lambda_m^+ B_1^{m+1} e^{-i\omega\tau} + \lambda_m^- B_2^{m+1} e^{i\omega\tau}),$$

$$B_2^m(\omega) = \frac{1}{2}(\lambda_m^- B_1^{m+1} e^{-i\omega\tau} + \lambda_m^+ B_2^{m+1} e^{i\omega\tau}), \tag{2.8}$$

where

$$\lambda_m^\pm = 1 \pm c_m/c_{m+1}, \quad m = 1, \dots, k. \tag{2.9}$$

Let us represent B_p^k using one unknown function (see (2.3), (2.4)).

$$B_1^k = \frac{1}{2}B(\omega)\lambda_k^+, \qquad B_2^k = \frac{1}{2}B(\omega)\lambda_k^-. \tag{2.10}$$

Because of (2.8) all the functions B_p^m also depend only on $B(\omega)$. So in order to calculate $B(\omega)$ one can use (2.2):

$$i\omega c_1(B_1^1(\omega)e^{-i\omega\tau} - B_2^1(\omega)e^{i\omega\tau}) = h(\omega). \tag{2.2'}$$

From (2.7)–(2.10), and (2.2') we conclude that one can represent the overinformation (2.5) $\Phi(\omega) = i\omega v(0,\omega) = i\omega(B_1^1 e^{-i\omega\tau} + B_2^1 e^{i\omega\tau})$ as follows

$$\Phi(\omega) = \frac{h(\omega)}{c_1} R(e^{-i\omega\tau}, e^{i\omega\tau}). \tag{2.11}$$

Here $R(\xi,\eta)$ is a ratio of homogeneous polynomials of degree k. Their coefficients are also homogeneous polynomials of degree k and they consist of the summands of the form $\lambda_1^{\pm}\lambda_2^{\pm}\dots\lambda_k^{\pm}$.
Let us consider the variable $z = e^{2i\omega\tau}$ and define the function $F(z)$ by $F(z) = R(z^{-1/2}, z^{1/2})$ (see (2.11)). Obviously

$$F(z) = \frac{f_0 + f_1 z + \dots + f_k z^k}{g_0 + g_1 z + \dots + g_k z^k}, \tag{2.12}$$

where, as it was already mentioned above, f_m, g_m are homogeneous polynomials of degree k with the summands of the form $\pm\lambda_1^{\pm}\lambda_2^{\pm}\dots\lambda_k^{\pm}$. The precise formulae are out of our interest.
Let us define the following quantities

$$\varphi_m = \frac{1}{2ic_1\tau} \int\limits_{|z|=1} F(z)\frac{dz}{z^{m+1}}, \quad m = 0,\dots,k; \tag{2.13}$$

$$\gamma_m = \frac{\lambda_m^-}{\lambda_m^+} = c_{m+1} - c_m/(c_{m+1} + c_m), \quad m = 1,\dots,k; \tag{2.14}$$

$$h_p^m = h_{m-1}^m h_{m-p-1}^{m-1} + h_p^{m-1}, \quad m = 2\dots k, \quad p = 0,\dots,m-2;$$

$$h_{m-1}^m = -\gamma_{m-1}, \qquad m = 2,\dots,k; \qquad h_0^1 = h_m^m = 0. \tag{2.15}$$

One can see easily that integrals (2.13) are connected with the overinformation (2.5) of I.P.2 by (see (2.11), (2.12))

$$\varphi_m = \int\limits_{\omega_0}^{\omega_0+\pi/\tau} \Phi(\omega)h^{-1}(\omega)e^{-2im\omega\tau}d\omega. \tag{2.16}$$

Here ω_0 is such that $[\omega_0, \omega_0 + \pi/\tau] \subset [\omega_1, \omega_2]$.
We suppose that the following conditions are satisfied:

i) equalities (1.5) hold;

ii) the function $h(\omega)$ in (2.2) is such that $h \neq 0$ for $\omega \in [\omega_0, \omega_0 + \pi/\tau]$;

iii) function $F(z)$ (2.12) which is constructed with the help of the solution of (2.1)–(2.4) has no poles in the disk $|z| \leq 1$.

Then one can solve I.P.2 as follows.
1. Compute the numbers φ_m (2.16), $m = 0,\dots,k$.
2. Find the numbers $\gamma_m (m = 1,\dots,k)$ recursively in the following way (cf. (2.15))

$$\gamma_1 = \varphi_1/2\varphi_0,$$

$$\gamma_m = \frac{1}{2\varphi_0}\left(\varphi_m + \sum_{p=1}^{m-1} h_p^m \varphi_{m-p}\right) \Big/ \prod_{p=1}^{m-1}(1-\gamma_p^2). \tag{2.17}$$

3. Solve I.P.2 as

$$c_1 = \pi/(\tau\varphi_0),$$

$$c_{m+1} = \frac{1+\gamma_m}{1-\gamma_m} c_m, \quad m = 1,\dots,k. \tag{2.18}$$

One can find some conditions sufficient for ii), iii) to be true in the above mentioned paper (Lavrent'ev, Jr, 1991).

3. INVERSE PROBLEM FOR LAME SYSTEM

Some problems for the Lamé system can be reduced to the problem like (1.1), (1.3) (see Yakhno, 1990). Let as consider one of these statements:

$$\rho\frac{\partial^2 u_i}{\partial t^2} = \sum_{j=1}^{3} \frac{\partial}{\partial x_j}\sigma_{ij} \quad (i = 1,\dots,3),$$

$$u_i|_{t<0} = 0 \quad (i = 1,\dots,3),$$

$$\sigma_{3i}|_{x_3=+0} = -e_3\delta(x_1,x_2)\delta(t),$$

$$[u_i]_{x_3=a_m} = [\sigma_{3i}]_{x_3=a_m} = 0, \quad m = 1,\dots,k. \tag{3.1}$$

Here $(x_1,x_2) \in \mathbf{R}^2, x_3 \in \mathbf{R}_+\backslash\{a_1,\dots,a_k\}, t \in \mathbf{R}, U = (u_1,u_2,u_3)$ is the displacement vector, $\sigma_{ij} = \mu(\partial u_i/\partial x_j + \partial u_j/\partial x_i) + \delta_{ij}\lambda \mathrm{div} U$ is the stress tensor, $\lambda(x_3), \mu(x_3)$ are the Lamé parameters, $\rho(x_3)$ is the density, e_3 is the unit vector of Ox_3 axis. At the points $x_3 = a_m$ the piecewise constant functions λ, μ, ρ can have jumps.

The seismic wave propagation through isotropic elastic medium due to the instantaneous vertical (in the direction of e_3) point force is simulated by system (3.1), cf. (Yakhno, 1990).

Suppose that U can be represented as $U = \partial^2 w/\partial t^2$, where $w|_{t=0} = 0, w_t|_{t=0} = 0$ and $w \in W_2^1(\mathbf{R}^2 \times \mathbf{R}_+ \times (0,T))$. Then one can reduce (3.1) to the following problem

$$\partial^2 v/\partial x^2 + \omega^2 c^2(x) v = 0,$$

$$v_x(0,\omega) = -\left(\lambda(0) + 2\mu(0)\right)^{-1},$$

$$v_x - i\omega c v \to 0 \quad \text{as} \quad x \to +\infty,$$

$$[v]_{x=a_m} = [\rho c^{-2} v_x]_{x=a_m} = 0, \quad m = 1,\dots,k, \tag{3.2}$$

for the function $v = F_t[F_{x_1,x_2}[u_3]], x = x_3$, where $\nu_1 = \nu_2 = 0$. Here F_t, F_{x_1,x_2} represent the Fourier transform operators with respect to t and x_1, x_2 respectively, ν_1, ν_2 are the variables dual to $x_1, x_2, c = (\rho/(\lambda + 2\mu))^{1/2}$ is the reciprocal to the longitudinal wave velocity.

System (3.2) is different from (2.1)–(2.4) only by the interface conditions. Applying the argument given in §2 to (3.2), we get

$$\frac{\rho_m}{c_m}(B_1^m - B_2^m) = \frac{\rho_{m+1}}{c_{m+1}}(B_1^{m+1}e^{-i\omega\tau} - B_2^{m+1}e^{i\omega\tau})$$

instead of the second formula in (2.7). Thus formulae (2.8), (2.10)–(2.13), and (2.15) remain true if we put (see (2.9) and (2.14))

$$\lambda_m^{\pm} = 1 \pm \frac{c_m}{\rho_m}\frac{\rho_{m+1}}{c_{m+1}}, \quad \gamma_m = \frac{\lambda_m^-}{\lambda_m^+} = \frac{c_{m+1}\rho_m - c_m\rho_{m+1}}{c_{m+1}\rho_m + c_m\rho_{m+1}}.$$

Thus under the assumption that the density in the above layer ρ_1 is known one can find the values of $c/\rho = (\rho(\lambda + 2\mu))^{1/2}$. In order to do this we have to use the algorithm from §2 replacing (2.18) by

$$\frac{c_1}{\rho_1} = \frac{\pi}{\tau\varphi_0\rho_1},$$

$$\frac{c_{m+1}}{\rho_{m+1}} = \frac{1+\gamma_m}{1-\gamma_m}\frac{c_m}{\rho_m} \ (m = 1, \ldots, k).$$

4. INVERSE PROBLEMS FOR HEAT EQUATION

Let us consider the following problem

$$\frac{\partial u}{\partial t} = c^{-2}(x)\frac{\partial^2 u}{\partial x^2}, \qquad u|_{t<0} = 0,$$

$$u_x(0,t) = H(t), \qquad u(a_k, t) = 0,$$

$$[u]_{x=a_m} = [c^{-2}u_x]_{x=a_m} = 0 \quad (m = 1, \ldots, k-1), \tag{4.1}$$

where $(x,t) \in [0, a_k] \times \mathbf{R}_+$.
Without going into details about physical processes simulated by (4.1) we refer the reader to (Denisov, 1982 and Sgheri, to appear).

INVERSE PROBLEM 4 (I.P.4). *Given $G(t)$ for $t \in \mathbf{R}_+$, find the coefficient $c(x)$ (see (1.2)) such that*

$$G(t) = u(0,t), \tag{4.2}$$

where u: $(0, a_k) \times (0, +\infty) \to \mathbf{R}$ solves equations (4.1).

As it was already mentioned above we assume that equalities (1.5) hold.
Using the Laplace transform with respect to t one can easily get from (4.1) the following problem

$$\partial^2 v/\partial x^2 - pc^2(x)v = 0, \tag{4.3}$$

$$v_x(0,p) = h_\alpha(p), \tag{4.4}$$

$$v(a_k, p) = 0, \tag{4.5}$$

$$[v]_{x=a_m} = [c^{-2}v_x]_{x=a_m} = 0 \quad (m = 1, \dots, k-1). \tag{4.6}$$

Here v, h_α are the Laplace transforms of u, H, i.e.,

$$v(x,p) = \int_0^{+\infty} e^{-pt}u(x,t)\,dt, \qquad h_\alpha(p) = \int_0^{+\infty} e^{-pt}H(t)\,dt. \tag{4.7}$$

Suppose that $v(x,p)$ exists for $p = p_0$ such that

$$p_0 < -\pi^2/\tau^2. \tag{4.8}$$

INVERSE PROBLEM 4.1 (I.P.4.1). *Given $\Psi_\alpha(p)$ for $p \in [p_0, 0]$, find the coefficient $c(x)$ (see (1.2)) such that*

$$\Psi_\alpha(p) = i\sqrt{|p|}v(0,p), \tag{4.9}$$

where $v(x,p)$ solves equations (4.3) – (4.6).

Let us denote $\omega = \sqrt{-p}$. By the reasons given in §2 (see (2.6)–(2.15)) I.P.4.1 is distinct from I.P.2 only by the definition of B_p^k:

$$B_1^k(\omega) + B_2^k(\omega) = 0 \tag{4.10}$$

and, consequently, by the values $\lambda_k^\pm$. In order for (2.10) to remain true let us set (see (4.10))

$$\lambda_k^\pm = \pm 2. \tag{4.11}$$

Thus one can solve I.P.4.1 using the algorithm from §2. However, it is necessary to exclude the calculation of $\varphi_k, \gamma_k, c_{k+1}$.

Finally let us suppose that the function $v(x,p)$ (see (4.7)) does not exist for all p_0 satisfying (4.8).

After using the Fourier transform we see that the function

$$W(x,\omega) = \int_{-\infty}^{\infty} e^{-i\omega t}u(x,t)\,dt \tag{4.12}$$

is a solution of the following problem:

$$\partial^2 W/\partial x^2 - i\omega c^2(x)W = 0, \tag{4.13}$$

$$W_x(0,\omega) = h_F(\omega), \tag{4.14}$$

$$W(a_k,\omega) = 0, \tag{4.15}$$

$$[W]_{x=a_m} = [c^{-2}W_x]_{x=a_m} = 0 \quad (m = 1, \dots, k-1). \tag{4.16}$$

Here $h_F(\omega)$ is the Fourier transform of $H(t)$ see (4.1).

Let $p, \omega > 0$. Then the general solutions of (4.3, 13) for $a_{m-1} \le x \le a_m$ are

$$v(x,p) = B_1^m(p)\exp\{p^{1/2}c_m(x-a_m)\} + B_2^m(p)\exp\{-p^{1/2}c_m(x-a_m)\},$$

$$W(x,\omega) = D_1^m(\omega)\exp\{\frac{i+1}{\sqrt{2}}\sqrt{\omega}c_m(x-a_m)\} + D_2^m(\omega)\exp\{-\frac{i+1}{\sqrt{2}}\sqrt{\omega}c_m(x-a_m)\}.$$

By reasons similar to the ones given in §2 one can write the following representation formulae for $v(0,p), W(0,\omega)$ (see (4.10), (4.11))

$$F_v = \frac{f_0 + f_1\xi + \ldots + f_k\xi^k}{g_o + g_1\xi + \ldots + g_k\xi^k},$$

$$F_W = \frac{f_0 + f_1\eta + \ldots + f_k\eta^k}{g_0 + g_1\eta + \ldots + g_k\eta^k}, \tag{4.17}$$

where $\xi = e^{2\sqrt{p}\tau}, \eta = \exp\{(1+i)\sqrt{2}\sqrt{\omega}\tau\}$ and f_m, g_m are the same.
Let us put $\lambda = \sqrt{p} = \sqrt{\omega}$ and use the notation (see I.P.4)

$$\Psi_\alpha(\lambda) = \int_0^\infty e^{-\lambda^2 t}G(t)\,dt, \qquad \Psi_F(\lambda) = \int_{-\infty}^\infty e^{-i\lambda^2 t}G(t)\,dt; \tag{4.18}$$

$$\psi_m = 2\int_{\lambda_0/\sqrt{2}}^{\sqrt{2}\lambda_0} \lambda\Psi_\alpha(\lambda)h_\alpha^{-1}(\lambda^2)e^{-2m\lambda\tau}d\lambda +$$

$$+i\int_{\lambda_0}^{2\lambda_0} \lambda\Psi_F(\lambda)h_F^{-1}(\lambda^2)\exp\{-(1+i)\sqrt{2}m\lambda\tau\}d\lambda, \tag{4.19}$$

$$m = 0,\ldots,k-1, \qquad \lambda_0 = \sqrt{2}\pi/\tau.$$

Observe that due to (4.17)–(4.19) and (2.12) the equality

$$\psi_m = \frac{1}{\tau c_1}\int_{c(\tau)} F(z)\frac{dz}{z^{m+1}} \tag{4.20}$$

holds. Here the contour $c(\tau)$ of the complex plane consists of the interval $[e^{2\pi}, e^{4\pi}]$ of the real axis and of the spiral part $z = \exp\{\sqrt{2}(1+i)\lambda\tau\}, \lambda \in [\lambda_0, 2\lambda_0]$.
Assume that:

i) equalities (1.5) hold;

ii) the Fourier and Laplace transforms of $H(t)$ (4.1) are such that

$$h_\alpha(p) \neq 0, \quad p \in [\pi^2/\tau^2,\ 4\pi^2/\tau^2]; \qquad h_F(\omega) \neq 0, \quad \omega \in [2\pi^2/\tau^2,\ 8\pi^2/\tau^2];$$

iii) the function $F(z)$ (2.12) has no poles in the domain bounded by $c(\tau)$ (see (4.20)).

Summarizing stated above one can solve I.P.4 as follows:

1. Compute the numbers ψ_m (4.19), $m = 0, \ldots, k-1$.
2. Find the numbers γ_m ($m = 1, \ldots, k-1$) recursively by (2.17) with φ_m replaced by ψ_m.
3. Solve the I.P.4 by taking:

$$c_1 = 2\pi i / \tau\psi_0,$$

$$c_{m+1} = \frac{1+\gamma_m}{1-\gamma_m} c_m \quad (m = 1, \ldots, k-1).$$

Acknowledgement
The author would like to thank V.G.Yakhno for helpful discussions.

REFERENCES

Aki, K. and Richards, P. (1980). *Quantitative Seismology.* Freeman, San Francisco.

Baev, A. (1988). About solution of an inverse boundary value problem for wave equation with coefficient with breaks. *J. Vych. Mekh. Mat. Fis.* **28**, 1619–1633 (in Russian).

Belishev, M. and Kupriyanova, N. (1983). On a refection of inclined plane wave from periodic fayered nalfspace. *Akust. J.* **29**, 733–735 (in Russian).

Denisov, A. (1982). On an inverse problem for the heat conduction equation with piecewise constant coefficients. In: *Metody Vychislenii i Programmirovanie,* Vol. 37, Moscow Univ. Press, Moscow, **37**, 93–97 (in Russian).

Gerver, M. (1974). *An Inverse Problem for One-Dimensional Wave Equation with Unknown Source of Ossilation.* Nauka, Moscow (in Russian).

Lavrentiev, M., Romanov, V., and Shishat'skii, S. (1986). *Ill-Posed Problems of Mathematical Physics and Analysis.* American Math. Society, Providence, RI.

Lavrent'ev, M., Jr. (1991). On the reconstruction of piecewise constant coefficient of wave equation. *Dokl. Akad. Nauk SSSR* **317**, 1108–1112 (in Russian).

Sgheri, L. Identification of piecewise constant heat diffusion coefficients (to appear).

Yakhno, V. (1990). *An Inverse Problems for Didderential Equations of Elasticity.* Nauka, Novosibirsk (in Russian).

Ill-Posed Problems in Natural Sciences, pp. 342 – 366
A. Tikhonov (Ed.)
1992 VSP/TVP

IDENTIFICATION PROBLEMS FOR INTEGRODIFFERENTIAL EQUATIONS

A. LORENZI
Dipartimento di Matematica F. Enriques, Università degli Studi di Milano, via Saldini 50, 20133 Milano, Italy.

ABSTRACT

Linear and nonlinear identification problems related to parabolic and hyperbolic integrodifferential equations are considered. Existence and uniqueness results in the small and some (conditional) uniqueness and stability results in the large are given.

INTRODUCTION

This paper reports the results on identification of unknown terms in (possibly) nonlinear partial integrodifferential equations obtained in the last five years initially by a group of Italian reseachers from Milan, Pavia and Rome and, later, in cooperation with Soviet mathematicians from Moscow and Novosibirsk.
The preliminary work consists in reducing the problems under consideration to an abstract form concerning first-order or second-order integrodifferential equations, in order to be able to deal with the parabolic and hyperbolic cases.

1. PARABOLIC IDENTIFICATION PROBLEMS: GENERALITIES

We assume that we are given a Banach space X, two (possibly) nonlinear operators A: $\mathcal{D}(A) \subset X \to X$, B: $\mathcal{D}(B) \subset X \to X$, and two (possibly) nonlinear functionals Φ: $\mathcal{D}(\Phi) \subset X \to \mathbf{R}$, Ψ: $\mathcal{D}(\Psi) \subset X \to \mathbf{R}$. We pose the following identification problem: determine a pair of functions u: $[0,T] \to X$ and h: $[0,T] \to \mathbf{R}$ such that

$$u'(t) - A(u(t)) - \int_0^t h(t-s)B(u(s))ds = f(t), \quad 0 < t < T, \tag{1.1}$$

$$u(0) = u_0, \tag{1.2}$$

$$\Phi[u(t)] + \int_0^t h(t-s)\Psi[u(s)]\, ds = g(t), \quad 0 \le t \le T. \tag{1.3}$$

Here T is a fixed positive real, f: $[0,T] \to X$ and g: $[0,T] \to \mathbf{R}$ are prescribed functions and $u_0 \in X$ is an assigned vector.
We observe that, when h (the relaxation kernel) is known, problem (1.1–2) is a usual Cauchy integrodifferential (abstract) problem.
On the contrary, when h is unknown, for its determination we need some additional information. In our case, such information is represented by functionals Φ and Ψ. To this purpose we observe that in concrete problems condition (3) stands for either a physical measurement or a quantity explicitly derivable from it. To fix ideas, when u stands for the temperature of a physical body Ω, Φ may denote the temperature at a fixed point $x_0 \in \Omega$ or the action of a thermic sensor located in a neighbourhood of $x_0 \in \Omega$. In addition condition (3) may express a flux condition at $x_0 \in \Omega$ or in a neighbourhood of x_0.
We conclude this section by introducing some terminology convenient for the sequel. We will call problem (1.1–3) linear in the case when operators A and B are linear. Note that, strictly speaking, this terminology is rough, because our problem is always nonlinear with respect to the pair (u, h).

2. PARABOLIC IDENTIFICATION PROBLEMS: THE LINEAR CASE

We assume here that X is a Banach space, $\Psi = 0$ and operators A: $\mathcal{D}(A) \subset X \to X$, B: $\mathcal{D}(B) \subset X \to X$ and Φ: $\mathcal{D}(\Phi) \subset X \to \mathbf{R}$ appearing in equations (1.1) and (1.3) satisfy the following assumptions:

i) A and B are linear and closed such that $Y := \mathcal{D}(A) \subset \mathcal{D}(B) \cap \mathcal{D}(\Phi)$;

ii) there exist $M > 0$ and $\theta \in (\pi/2, \pi)$ such that $(z - A)^{-1} \in \mathcal{L}(X)$ and $\|(z - A)^{-1}\|_{\mathcal{L}(X)} \le M|z|^{-1}$ for any $z \in \mathbf{C}$ with $|\arg z| \le \theta$;

iii) $\Phi \in X^*$.

Remark 2.1. If X_1 and X_2 are Banach spaces, $\mathcal{L}(X_1, X_2)$ denotes the Banach space consisting of all linear bounded operators from X_1 into X_2. In particular, if $X_1 = X_2 = X$, we set $\mathcal{L}(X, X) = \mathcal{L}(X)$.

Remark 2.2. According to **ii)** A turns out to be a generator of a analytic semigroup $\{e^{tA}\}_{t>0}$ of linear bounded operators in $\mathcal{L}(X)$. The semigroup may be not continuous at $t = 0$, since the domain Y of A may be not dense in X.

Remark 2.3. We can assume that A admits an inverse $A^{-1} \in \mathcal{L}(X)$. In fact, if A has not such a property, replace the unknown (u, h) and the datum (f, g) in problem (2.2–4) by $(u, h) = e^{\lambda t}(\bar{u}, \bar{h})$ and $(f, g) = e^{\lambda t}(\bar{f}, \bar{g})$ respectively, λ being a fixed (arbitrary) positive number. Then $(\bar{u}, \bar{h})$ turns out to be a solution to problem (2.2–4) with (f, g) and A replaced by $(\bar{f}, \bar{g})$ and $A - \lambda$ respectively. It is then an easy matter to check that $A - \lambda$ satisfies properties **i)**, **ii)**.
Concerning problem (1.1–3) we can state that it is ill-posed in the sense of Hadamard. In fact, assume that $f = 0$ and $u_0 = 0$. Then, under general assumptions on kernel h, such as its membership in $L^p(0,T)$ $(1 < p \le +\infty)$, we deduce that the direct problem (1.1–2) admits only the solution $u = 0$. Hence equation (1.3)

admits either no solution or an infinity of them, according as function g is different or coincides with the null function.
This fact is not strange, because our identification problem is nonlinear with respect to the pair (u, h).
Consequently, one of the aims of our investigation consists in finding out sufficient conditions on data in order problem (1.1–3) to be (at least locally in time) uniquely solvable.
A further aspect of our investigation is concerned with uniqueness and stability in the large of the solution (u, h). This question is, as is well known, basic from the point of view of practical applications. Generally speaking, we cannot prove stability results in the large without assuming that our unknown kernel h satisfies some apriori estimate, such as its membership in a fixed ball of a suitable Banach space.
We note, however, that our locally in time solution (u, h) depends continuously on data with respect to suitable metrics. Moreover, we can, generally speaking, prove the existence of a maximal in time solution (u, h). The existence and the stability in the large are equivalent to the possibility of proving the quoted apriori estimate for h.

3. THE FORMAL PROCEDURE FOR SOLVING LINEAR IDENTIFICATION PROBLEMS

We sketch here the basic ideas we follow to solve problem (1.1–3) under assumptions **i–iii**). Roughly speaking, we use a differentiation procedure (similar to the one used to solve, e.g., D'Alembert ordinary differential equations). It forces us to suppose some smoothness about data and unknowns.
First we introduce the auxiliary unknown

$$v(t) = u'(t), \quad 0 \le t \le T \quad \Leftrightarrow \quad u(t) = u_0 + \int_0^t v(s)\,ds, \quad 0 \le t \le T \tag{3.1}$$

The pair (v, h) turns out to be a solution to the problem

$$v'(t) - Av(t) = \int_0^t h(t-s)Bv(s)\,ds + h(t)Bu_0 + f'(t), \quad 0 < t < T. \tag{3.2}$$

$$v(0) = Au_0 + f(0) := v_0, \tag{3.3}$$

$$\Phi[v(t)] = g'(t), \quad 0 \le t \le T. \tag{3.4}$$

Use then the semigroup representation for the solution of the differential Cauchy problem and apply linear operator Φ to both members in (3.2), we get the following integrodifferential system for (v, h):

$$v(t) = e^{tA}v_0 + \int_0^t e^{(t-s)A} f'(s)\,ds + \int_0^t e^{(t-\sigma)A}\,ds \int_0^\sigma h(\sigma - s)Bv(s)\,ds$$

$$+\int_0^t e^{(t-s)A}h(s)Bu_0\,ds, \quad 0 \le t \le T, \tag{3.5}$$

$$g''(t) = \Phi\Big[Av(t) + \int_0^t h(t-s)Bv(s)ds + f'(t)\Big] + h(t)\Phi[Bu_0], \quad 0 \le t \le T. \tag{3.6}$$

Introduce now the new unknown $w = Av$ and assume that $\chi := \Phi[Bu_0] \neq 0$. You easily derive the following fixed point system for (w, h):

$$w(t) = Ae^{tA}v_0 + A\int_0^t e^{(t-s)A}f'(s)\,ds + A\int_0^t e^{(t-\sigma)A}\,ds\int_0^\sigma h(\sigma - s)$$

$$\times BA^{-1}w(s)\,ds + \int_0^t e^{(t-s)A}h(s)ABu_0\,ds, \quad 0 \le t \le T, \tag{3.7}$$

$$h(t) = \chi^{-1}g''(t) - \chi^{-1}\Phi[w(t) + \int_0^t h(t-s)BA^{-1}w(s)ds + f'(t)], \quad 0 \le t \le T. \tag{3.8}$$

Remark 3.1. Operator BA^{-1} belongs to $\mathcal{L}(X)$ according to hypothesis **i)** and the closed graph theorem.
System (3.7–8) can be solved using the generalized contraction principle and the following basic formula valid, e.g., for functions in $C^\gamma([0,T];X)$ or $W^{\gamma,p}((0,T);X)$ with $\gamma \in (0,1)$ and $p \in (1,+\infty)$:

$$A\int_0^t e^{(t-s)A}f(s)\,ds = \int_0^t Ae^{(t-s)A}[f(s) - f(t)]\,ds + (e^{tA} - I)f(t).$$

Once we have solved system (3.7–8), function v can be expressed in terms of w by formula (2.6) with the function Bv, appearing in the second integral, replaced by $BA^{-1}w$. Hence the required function u is assigned by formula (3.1).

Remark 3.2. When condition $\Phi[Bu_0] \neq 0$ is not satisfied, problem (3.2–4) may have an infinity of solutions. To this purpose we assume: **i)** $A = B$; **ii)** $Az = \lambda z$ for some $z \in X\backslash\{0\}$ and $\lambda \in \mathbf{R}\backslash\{0\}$; **iii)** $f = \varphi z$ with φ: $[0,T] \to \mathbf{R}$; **iv)** $u_0 = z$; **v)** $\Phi[z] = 0$; **vi)** $g = 0$. Then the pair $(\psi z, h)$, where ψ: $[0,T] \to \mathbf{R}$, solves problem (3.2–4) if and only if the pair (ψ, h) solves the scalar Cauchy problem

$$\psi'(t) - \lambda\psi(t) - \lambda\int_0^t h(t-s)\psi(s)\,ds = \varphi(t) \quad 0 < t < T, \quad \psi(0) = 1.$$

The main results concerning local existence and uniqueness for problem (1.1–3) corresponding to linear operators A and B are reported in (Lorenzi and Sinestrari, 1988; Lorenzi and Paparoni, 1990). The former is concerned with relaxation kernels of the weakly singular form $h(t) = t^{\alpha-1}k(t)$ ($\alpha \in (0,1)$), k being a continuous function. Consequently the functional framework is that of continuous functions. On the contrary, the latter paper is concerned with the L^p-theory ($1 < p < +\infty$). Before stating the corresponding results, we need to introduce the following intermediate spaces $D_A(\sigma, p)$ ($\sigma \in (0,1)$, $p \in (1, +\infty]$) between X and Y, where $\|\cdot\|$ denotes the norm in X:

$$D_A(\sigma,p) = \left\{ x \in X\colon |x|_{\sigma,p} = \left(\int_0^{+\infty} t^{(1-\sigma)p-1} \|Ae^{tA}x\|^p \, dt \right)^{1/p} < +\infty \right\},$$

$$D_A(\sigma,+\infty) = \left\{ x \in X\colon |x|_{\sigma,+\infty} = \sup_{t>0} t^{1-\sigma} \|Ae^{tA}x\| < +\infty \right\}$$

according as $p \in (1, +\infty)$ and $p = +\infty$.

THEOREM 3.1. *Let $\alpha \in (0,1)$ and $\beta \in (0, \min(\alpha, 1-\alpha))$ and let assumptions* **i)–iii)** *be satisfied. Let data f, u_0, g enjoy the following properties:* **i)** $f \in C^{1+\beta}([0,T_0]; X)$; **ii)** $u_0, Bu_0, Au_0 + f(0) \in Y$ *and* $A[Au_0 + f(0)] + f'(0) \in D_A(\beta, +\infty)$; **iii)** $g \in C^1([0,T_0]) \cap C^2((0,T_0]; X)$ *and* $t^{1-\alpha}g''C^\beta([0,T_0])$; **iv)** $g(0) = \Phi[u_0]$; **v)** $g'(0) = \Phi[Au_0 + f(0)]$; **vi)** $\Phi[Bu_0] \neq 0$. *Then there exists a positive $T \in (0, T_0]$ for which problem (1.1–3) has a unique solution (u,h) such that $u \in C^1([0,T_0];\ Y) \cap C^2([0,T_0]; X)$ and $t^{1-\alpha}h \in C^\beta([0,T_0])$.*

THEOREM 3.2. *Let Y be dense in X and let assumptions* **i)– iii)** *be satisfied. Let $(f, u_0, g) \in W^{1+\sigma,p}((0,T_0); X) \times Y \times W^{2+\sigma,p}((0,T_0))$ be a triplet satisfying the following properties:* **i)** $Au_0 + f(0) \in D_A(\sigma + 1/p', p)$; **ii)** $Bu_0 \in D_A(\theta, p)$ *for some $\theta \in (0,1)$, when $\sigma \in (0, 1/p)$;* **iii)** $Bu_0 \in D_A(\sigma - 1/p, p)$ *when $\sigma \in (1/p, 1)$;* **iv)** $A[Au_0 + f(0)] + f'(0) + \Phi[Bu_0]^{-1}\{g''(0) - \Phi[A[Au_0 + f(0)] + f'(0)]Bu_0\} \in D_A(\sigma - 1/p, p)$; **v)** $g(0) = \Phi[u_0]$; **vi)** $g'(0) = \Phi[Au_0 + f(0)]$; **vii)** $\Phi[Bu_0] \neq 0$. *Then there exists a positive $T \in (0, T_0]$ for which problem (1.1–3) has an unique solution $(u,h) \in [W^{2+\sigma,p}((0,T); Y) \cap W^{1+\sigma,p}((0,T); X)] \times W^{\sigma,p}((0,T))$.*

Remark 3.3. In (Lorenzi and Paparoni, 1990) we actually consider the following generalization of equation (3.2):

$$u'(t) - Au(t) - \int_0^t h(t-s)[Bu(s) + b(s)] \, ds = f(t), \quad 0 < t < T.$$

The addition of the term b allows, in concrete cases, to deal with nonhomogeneous boundary value conditions.

Remark 3.4. Assume that data in theorems 2.1 and 2.2 belong to a fixed ball in the Banach spaces pointed out and satisfy an estimate of the form $|\Phi[Bu_0]| \geq \chi_0$ with a fixed $\chi_0 > 0$ and h is apriori known to satisfy a suitable estimate. Then

the mapping $(f, u, g) \to (u, h)$ is Lipschitz continuous with respect to the norms of the Banach spaces appearing in Theorems 3.1, 3.2 (cf. (Lorenzi and Sinestrari, 1989; Lorenzi and Paparoni, 1990)).
We have thus restored the stability of the solution to linear problem (1.1–3).
We conclude this section dealing with a case of interest in applications. It corresponds to a homogeneous initial datum, i.e. $u_0 = 0$. In this case we have, of course, $\Phi[Bu_0] = 0$. In order to overcome this difficulty, we introduce the new unknown

$$\hat{u}(t) = u'(t), \quad 0 \le t \le T \quad \Leftrightarrow \quad u(t) = \int_0^t \hat{u}(s)\, ds, \quad 0 \le t \le T.$$

The corresponding identification problem for $(\hat{u}, h)$ is the following:

$$\hat{u}'(t) - A\hat{u}(t) - \int_0^t h(t-s)B\hat{u}(s)ds = f'(t), \quad 0 < t < T, \tag{3.9}$$

$$\hat{u}(0) = f(0), \tag{3.10}$$

$$\Phi[\hat{u}(t)] = g'(t), \quad 0 \le t \le T. \tag{3.11}$$

Under the assumption $\Phi[Bf(0)] \neq 0$ we can apply Theorems 3.1 and 3.2 to problem (3.9–11) and obtain existence and uniqueness results.

4. PARABOLIC IDENTIFICATION PROBLEMS: THE NONLINEAR CASE

We list here our basic assumptions concerning operators A, B and functionals Φ, Ψ, X and Y denoting Banach spaces with $Y \to X$:

i) $A, B\colon Y \to X$ and $\Phi, \Psi\colon Y \to \mathbf{R}$ are twice Fréchet differentiable;

ii) the Fréchet derivatives $A', B'\colon Y \to \mathcal{L}(Y, X)$ and Φ', $\Psi'\colon Y \to Y^*$ are locally bounded;

iii) the Fréchet derivatives $A'', B''\colon Y \to \mathcal{L}(Y \times Y, X)$ and $\Phi'', \Psi''\colon Y \to (Y \times Y)^*$ satisfy the estimates:

$$\|A''(u)(z_1, z_2)\|_X + \|B''(u)(z_1, z_2)\|_X \le \Lambda_1(\rho)\|z_1\|_Y\, \|z_2\|_Y,$$

$$\|A''(u)(z_1, z_2) - A''(w)(z_1, z_2)\|_X + \|B''(u)(z_1, z_2) - B''(w)(z_1, z_2)\|_X$$
$$\le \Lambda_2(\rho)\|u - w\|_Y\, \|z_1\|_Y\, \|z_2\|_Y,$$

$$|\Phi''(u)(z_1, z_2) + \Psi''(u)(z_1, z_2)| \le \Lambda_3(\rho)\|z_1\|_Y\, \|z_2\|_Y,$$

$$|\Phi''(u)(z_1, z_2) - \Phi''(w)(z_1, z_2)| + |\Psi''(u)(z_1, z_2) - \Psi''(w)(z_1, z_2)|$$
$$\le \Lambda_4(\rho)\|u - w\|_Y\, \|z_1\|_Y\, \|z_2\|_Y,$$

for any $z_1, z_2 \in Y$ and any u, $w \in Y$ such that $\|u\|_Y \le \rho$, $\|w\|_Y \le \rho$, where Λ_j: $[0,+\infty) \to (0,+\infty)$ $(j = 1,2,3,4)$ are continuous nondecreasing functions;

iv) $u_0 \in Y$ and $A_0 := A'(u_0)$ generates an analytic semigroup $S(t)$ of linear bounded operators.

As far as the data f, u_0, g are concerned, we make the following assumptions, where $p \in (1,+\infty)$:

v) $f \in W^{1+\theta,p}((0,T_0);X)$ for some $\theta \in (0,\gamma(p))$ with $\gamma(p) = \min(1/p, 1/p')$;

vi) $u_0 \in Y$, $v_0 := -A(u_0)+f(0)$, $B(u_0) \in D_{A_0}(\theta'+1/p,p)$ for some $\theta' \in (0,\gamma(p))$;

vii) $g \in W^{2,p}((0,T_0))$;

viii) $g(0) = \Phi[u_0]$, $g'(0) = \Phi[v_0]$;

ix) $\Psi(u_0) \neq 0$.

THEOREM 4.1. *Under assumptions* **i)–ix)** *there exists* $T \in (0,T_0]$ *for which problem* (1.1–3) *has an unique solution* $(u,h) \in W^{1+\theta,p}((0,T);Y) \times L^p((0,T))$ *such that*

x) $u \in C^1([0,T];D_{A_0}(\theta+1/p',p)) \cap W^{1+\alpha,p}((0,T);D_{A_0}(\beta,p))$ *for any* $\alpha, \beta \in (0,1)$ *with* $\alpha+\beta = 1$;

xi) $u'' \in L^p((0,T);D_{A_0}(\theta',p)) + W^{\theta,p}((0,T);X)$.

Proof (sketch). First we introduce the auxiliary unknown

$$v(t) = u'(t), \quad 0 \le t \le T \quad \Leftrightarrow \quad u(t) = u_0 + \int_0^t v(\tau)\,d\tau, \quad 0 \le t \le T. \tag{4.1}$$

The pair (v,h) turns out to be a solution to the problem

$$v'(t) - A'\left(u_0 + \int_0^t v(s)\,ds\right) v(t) - \int_0^t h(t-s)B'\left(u_0 + \int_0^t v(\tau)\,d\tau\right) v(s)\,ds$$

$$+h(t)Bu_0 + f'(t), \quad \text{for a.e. } t \in (0,T), \tag{4.2}$$

$$v(0) = Au_0 + f(0) := v_0, \tag{4.3}$$

$$\Phi'\left[u_0 + \int_0^t v(s)\,ds\right] v(t) + \int_0^t h(t-s)\Psi'\left[u_0 + \int_0^t v(\tau)\,d\tau\right] v(s)\,ds$$

$$+h(t)\Psi(u_0) = g'(t), \quad \text{for a.e. } t \in (0,T). \tag{4.4}$$

We rewrite equation (4.2) in the following operator form

$$v'(t) - A'(u_0)v(t) = P(v,h)(t) + h(t)\Psi(u_0) + f'(t), \quad \text{for a.e. } t \in (0,T), \quad (4.5)$$

where operator P is defined by

$$P(v,h) = \left[A'\left(u_0 + \int_0^t v(s)\,ds\right) - A'(u_0)\right]v(t) + \int_0^t h(t-s)$$

$$\times B'\left(u_0 + \int_0^t v(\tau)d\tau\right)v(s)\,ds, \quad t \in (0,T). \quad (4.6)$$

Consequently, using the semigroup $\{S(t)\}$, we easily realize that identification problem (4.2–4) is equivalent to the following fixed point system:

$$v(t) = S(t)v_0 + \{S * [P(v,h) + h(t)\Psi(u_0) + f'\}(t), \quad \text{for a.e. } t \in (0,T), \quad (4.7)$$

$$h(t) = \Psi(u_0)^{-1}\left\{g'(t) - \Phi'\left[u_0 + \int_0^t v(s)\,ds\right]v(t) + \int_0^t h(t-s)\right.$$

$$\left.\times\Psi'\left[u_0 + \int_0^t v(\tau)\,d\tau\right]v(s)\,ds\right\}, \quad \text{for a.e. } t \in (0,T). \quad (4.8)$$

System (4.7–8) can be solved using the generalized contraction principle. Once we have solved such a system, function u is assigned by formula (4.1).

Remark 4.1. Problem (4.2–4) can be generalized (cf. (Grasselli and Lorenzi, 1990)) to the case of two unknown relaxation kernels h, k appearing in the equation

$$u'(t) + \int_0^t k(t-s)u'(s)\,ds - A(u(t)) - \int_0^t h(t-s)B(u(s))\,ds = f(t).$$

Of course, we have to add a further condition to determine the new unknown k. In the quoted paper it has the form

$$\Xi[u(t)] = \xi(t), \quad t \in [0,T]. \quad (4.9)$$

Here $\Xi \in X^*$ and ξ: $[0, T_0] \to \mathbf{R}$ is a prescribed smooth function.

Remark 4.2. Obviously for problem (4.2–4) it is possible to prove stability results analogous to those proved in §3. They are not reported in (Grasselli and Lorenzi, 1990) for lack of space only.

Remark 4.3. Identifications problems can be solved in the framework of Hilbert spaces using apriori estimates as in (Lions and Magenes, 1972; Lions, 1969) instead of semigroup theory. To this purpose cf. (Grasselli, 1989).
To conclude this section, we want to spend some words about some open problems.

PROBLEM 1. *Solve the identification problem when operators A, B, M, N depend explicitly on time.*

This problem seems be not very hard to solve (at least under standard regularity assumptions).

PROBLEM 2. *Solve the identification problem when operators A, B, M, N depend, in addition to u, also on an unknown element $z \in Z, Z$ being a prescribed Banach space. In this case we have to give an additional information, e.g., of form (4.9), where $\Psi \in X^*$ and ψ: $[0, T_0] \to Z$ is an assigned (smooth) function. Our identification problem consists in determining the triplet (u, z, h).*

This problem is of interest for applications, but hard to solve.

PROBLEM 3. *Solve the identification problem when the relaxation kernel h is an operator instead of a scalar function.*

This problem is pratically new in the case when h is a space dependent function too and it is as well of interest in the case when h is a four-dimensional tensor depending on time only, as in the anisotropic theory of viscoelasticity.

PROBLEM 4. *Solve the identification problem when the domain $\mathcal{D}(A)$ of operator A strictly contains the domain $\mathcal{D}(B)$ of operator B.*

This problem seems be hard to solve. To our knowledge it was dealt in the linear case only under the assumption $A = 0$ in (Lorenzi, 1989a). In such a case it is strictly related to the construction in a bounded time interval of a so called resolvent family for linear Cauchy problem (1.1–2) (cf. (Da Prato et al., 1983, 1985)).

5. FREDHOLM TYPE RESULTS FOR LINEAR PARABOLIC IDENTIFICATION PROBLEMS

In this section we pose the following identification problem: determine a function u: $[0, T] \to X$ and an element $z \in X$ such that

$$u'(t) - Au(t) - \int_0^t B(t-s)u(s)\,ds = E(t)z + f(t), \quad 0 < t < T, \tag{5.1}$$

$$u(0) = u_0, \tag{5.2}$$

$$\Psi[u] = g, \quad 0 \le t \le T. \tag{5.3}$$

Here Ψ and g denote a linear operator from $C((0, T]; X)$ into X and an element in X respectively. In concrete applications Ψ is often prescribed by the formula $\Psi[u] = u(T)$.

The objective of the present section consists, unlike the previous ones, in establishing results of Fredholm type (i.e., uniqueness implies existence and stability) for problem (5.1–3). Consequently, an ill-posed identification problem of Fredholm type is nothing else but a problem without uniqueness.
We explicitly note that our problem is linear with respect to the pair (u,z). This allows us to use Fredholm theory for linear compact operators.
In order to state our basic result (cf. (Lorenzi and Prilepko, 1991)) we need to introduce the following functional spaces, where $\sigma \in (0,2)$ and $p \in (1,+\infty)$:

$$Y^{j,p} = \mathcal{D}(A^j) \quad (j = 0,1), \quad Y^{\sigma,p} = \mathcal{D}_A(\sigma,p) \quad \sigma \in (0,1),$$

$$Y^{\sigma,p} = \{x \in \mathcal{D}(A)\colon\ Ax \in Y^{\sigma-1,p}\}, \quad \sigma \in (1,2),$$

$$U^{s,p}_{\sigma,T} = W^{s,p}((0,T);Y^{\sigma,p}) \cap W^{s-1,p}((0,T);Y^{1+\sigma,p}), \quad \sigma \in (0,1).$$

We list now our assumptions concerning operators A, B, E, Ψ, where σ is a fixed real number in $(0,1/p)$ and $\mathcal{C}(X_1,X_2)$ denotes the Banach space of all compact linear operators from X_1 into X_2:

i) $A\colon Y \subset X \to X$ is a linear closed operator admitting an inverse $A^{-1} \in \mathcal{L}(X;Y) \cap \mathcal{L}\left(Y^{\sigma,p};Y^{1+\sigma,p}\right)$;

ii) there exist $M > 0$ and $\theta \in (\pi/2,\pi)$ such that $(z-A)^{-1} \in \mathcal{L}(X)$ and $\|(z-A)^{-1}\|_{\mathcal{L}(X)} \le M|z|^{-1}$ for any $z \in \mathbf{C}$ with $|\arg z| \le \theta$;

iii) $(\lambda_0 - A)^{-1} \in \mathcal{C}\left(Y^{\sigma,p};Y^{\sigma+1/p',p}\right)$ for some λ_0 in the resolvent set of A;

iv) $B \in L^p((0,T);\mathcal{L}\left(Y^{1+\sigma,p};Y^{\sigma,p}\right) \cap W^{\epsilon,q}\left((0,T);\mathcal{L}\left(Y^{1+\sigma,p};Y^{\sigma,p}\right)\right)$ for some $q \in (1,+\infty)$ and $\epsilon \in (0,1)\backslash\{1/q\}$;

v) $E \in W^{1,p}((0,T);\mathcal{L}\left(Y^{\sigma+1/p',p};Y^{\sigma,p}\right)$ and $E(0) \in \mathcal{L}(Y^{\sigma+1/p'})$;

vi) $\Psi \in \mathcal{L}\left(C\left((0,T];\mathcal{C}\left(Y^{\sigma+1/p',p}\right)\right);\ \mathcal{C}\left(Y^{\sigma+1/p',p}\right)\right)$ and $A\Psi[u] = \Psi[Au]\ \forall u \in U^{2,p}_{\sigma,T}$;

vii) the linear operator $z \to \Psi[Ez] \in \mathcal{L}\left(Y^{\sigma+1/p',p}\right)$ admits an inverse $\Psi_1 \in \mathcal{L}\left(Y^{\sigma+1/p',p}\right)$.

As far as data are concerned, we suppose that they enjoy the properties:

viii) $f \in W^{1,p}\left((0,T);Y^{\sigma,p}\right)$ and $f(0) \in Y^{\sigma+1/p',p}$;

ix) $u_0 \in Y^{2+\sigma-1/p,p}$;

x) $g \in Y^{1+\sigma,p}$.

THEOREM 5.1. *Let operators A, B, E, Ψ and data f, u_0, g satisfy properties* **i)–vii)** *and* **viii)–x)** *respectively. Assume further that problem (5.1–3) with $(f,u_0,g) = (0,0,0)$ admits in $U^{2,p}_{\sigma,T} \times Y^{\sigma+1/p',p}$ only the trivial solution $(u,z) =$*

$(0,0)$. *Then problem* (5.1–3) *admits an unique solution* $(u,z) \in U^{2,p}_{\sigma,T} \times Y^{\sigma+1/p',p}$ *depending continuously on* (f, u_0, g) *with respect to the metrics pointed out.*

Consider now the following (nonlinear) generalization of problem (5.1–3): determine a function u: $[0,T] \to X$ and an element $z \in X$ such that

$$u'(t) - A_0 u(t) - A_1[u(t), z] - \int_0^t B(t-s)u(s)\,ds = f(t), \quad 0 < t < T, \tag{5.4}$$

$$u(0) = u_0, \tag{5.5}$$

$$\Psi[u] = g, \quad 0 \le t \le T. \tag{5.6}$$

Here A_0: $Y \subset X \to X$ and A_1: $Y \times Y^{\sigma+1/p',p} \subset X \times Y^{\sigma+1/p',p}$ are a linear and a bilinear operator respectively.
We introduce now the linear operators $A(z)$: $Y \subset X \to X$ and $E(z)$: $(0,T) \to \mathcal{L}\left(Y^{\sigma+1/p',p}\right)$ defined by

$$A(z)x = A_0 x + A_1(x,z), \quad E(u(t))z = A_1[u(t), z]. \tag{5.7}$$

Taking advantage of Theorem 5.1, we easily derive the following global (uniqueness and) stability result for problem (5.4–6):

THEOREM 5.2. *Let* $(u_j, z_j) \in U^{2,p}_{\sigma,T} \times Y^{\sigma+1/p',p}$ $(j = 1,2)$ *be solutions to problems* (5.1–3) *corresponding, respectively, to data* $(f_j, u_{0,j}, g_j)$ $(j = 1,2)$ *enjoying properties* **viii)–x)**. *Assume that operators* $A(z_1), B, E(u_2), \Psi$ *satisfy properties* **i)–vii)**. *Suppose further that problem* (5.1–3) *with* (A, E) *and* (f, u_0, g) *replaced by* $(A(z_1), E(u_2))$ *and* $(0,0,0)$ *respectively admits in* $U^{2,p}_{\sigma,T} \times Y^{\sigma+1/p',p}$ *only the trivial solution* $(u,z) = (0,0)$. *Then there exists a positive constant* C *depending on* (u_2, z_1) *only such that the following stability estimate holds true:*

$$\|u_2 - u_1\|_{U^{2,p}_{\sigma,T}} + \|z_2 - z_1\|_{Y^{\sigma+1/p',p}} \le C\Big\{\|f_2 - f_1\|_{W^{1,p}((0,T),Y^{\sigma,p})}$$

$$+\|f_2(0) - f_1(0)\|_{Y^{\sigma+1/p',p}} + \|u_{0,2} - u_{0,1}\|_{Y^{2+\sigma-1/p,p}} + \|g_2 - g_1\|_{Y^{1+\sigma,p}}\Big\}.$$

Remark 5.1. It would be of interest to establish sufficient conditions on a general nonlinear operator A_1 appearing in equation (5.4) in order uniqueness to imply (global or local) existence, as in the linear case.

Remark 5.2. The same question, of course, arises also in the cases of the identification problems solved, under specific assumptions, in §§2, 3, 4.

6. HYPERBOLIC IDENTIFICATION PROBLEMS: GENERALITIES

We assume that we are given a so called Hilbert triplet W, H, W^*, i.e., a triplet of Hilbert spaces and a functional a: $W \times W \to \mathbf{R}$ such that

i) H is separable;

ii) $W \to H \to W^*$ with dense embeddings;

iii) a is a bounded, coercive, quadratic and symmetric form.

Assumptions **i)–iii)** guarantee the existence of a linear operator $A \in \mathcal{L}(W, W^*)$ enjoying the following properties:

iv) A: $\mathcal{D}(A) \subset H \to H$, where $\mathcal{D}(A) = \{u \in H: Au \in H\}$, is closed, self-adjoint and positive-definite;

v) $\mathcal{D}(A^{1/2}) = W$ and $W^\beta := \mathcal{D}(A^{\beta/2})$ $(\beta \geq 0)$ is dense in W^α if $\beta \geq \alpha \geq 0$;

vi) $A^\alpha \in \mathcal{L}(W^\beta, W^{\beta-2\alpha})$ if $\beta \geq 2\alpha$, W^γ $(\gamma \geq 0)$ being endowed with the graph-norm $\|\cdot\|_\gamma$.

We pose now the following identification problem: determine a pair of functions u: $[0,T] \to X$ and h: $[0,T] \to \mathbf{R}$ such that

$$u''(t) + A\left\{u(t) + \int_0^t h(t-s)B(u(s))\,ds\right\} = f(t), \quad 0 < t < T, \tag{6.1}$$

$$u(0) = u_0, \tag{6.2}$$

$$u'(0) = u_1, \tag{6.3}$$

$$\Phi[u(t)] = G(h, f, u_0, u_1, g)(t), \quad 0 \leq t \leq T. \tag{6.4}$$

Here T is a fixed real, B: $X \to X$ is a nonlinear operator, f: $[0,T] \to X$ and g: $[0,T] \to \mathbf{R}$ are prescribed functions and $u_0, u_1 \in X$ are assigned vectors.
We observe that, when h (the relaxation kernel) is known , problem (6.1–3) is a usual Cauchy (abstract) problem. On the contrary, when h is unknown, for its determination we need some additional information. It, in our case, it is represented by operator Φ. To this purpose we observe that in concrete problems Φ denotes either a physical measurement or a quantity explicitly derivable from it. To fix ideas, when u stands for the displacement from a reference position of a physical body Ω, Φ may denote the traction at $x_0 \in \partial\Omega$ or (more physically) the mean traction in a (surface) neighbourhood of x_0 on $\partial\Omega$.
We conclude this section by introducing a terminology similar to the one introduced in §1. We agree to call problem (6.1–4) linear in the case when operator B is linear. Obviously our problem is always nonlinear with respect to the pair (u, h) .

7. HYPERBOLIC IDENTIFICATION PROBLEMS: THE LINEAR CASE

We assume here that $B = I$ and $G(h, f, u_0, g) = g$. We sketch the formal procedure we follow to reduce problem (6.1–4) to a system of fixed point equations (cf. (Grasselli et al., 1988, 1990)).

First we rewrite equation (6.1) (with $B = I$) in the form

$$Au(t) + \int_0^t h(t-s)Au(s)\, ds = -u''(t) + f(t), \quad 0 < t < T. \tag{7.1}$$

Then we recall that the solution to the Volterra integral equation

$$y(t) + \int_0^t h(t-s)y(s)\, ds = z(t), \quad 0 < t < T \tag{7.2}$$

is given by

$$y(t) = z(t) + \int_0^t k(t-s)z(s)\, ds, \quad 0 < t < T. \tag{7.3}$$

The resolvent kernel k solves, in turn, the scalar integral equation

$$k(t) + \int_0^t h(t-s)k(s)ds = -k(t), \quad 0 < t < T. \tag{7.4}$$

Consequently, equation (7.1) is equivalent to the following

$$u''(t) + Au(t) + \int_0^t k(t-s)u''(s)\, ds = f(t) + \int_0^t k(t-s)f(s)\, ds, \quad 0 < t < T. \tag{7.5}$$

We introduce now the new unknown

$$v(t) = Au''(t), \quad 0 \le t \le T \quad \Leftrightarrow \quad u(t) = u_0 + tu_1 + \int_0^t v(s)\, ds, \quad 0 \le t \le T. \tag{7.6}$$

Under the assumption that k is differentiable, the pair (v, k) turns out to be a solution to the identification problem

$$v''(t) + Av(t) = \int_0^t k'(t-s)[Af'(s) - v'(s)]\, ds + \alpha v'(t)$$

$$+k'(t)A^2u_0 + F(t), \quad 0 < t < T, \tag{7.7}$$

$$v(0) = -A^2u_0 + Af(0) := v_0, \tag{7.8}$$

$$v'(0) = -A^2u_1 - \alpha A^2u_0 + Af'(0) := v_1, \tag{7.9}$$

$$\Phi[v(t)] = \int_0^t k'(t-s)\left\{\Phi[f'(s)] - g^{(3)}(s)\right\} ds + \alpha g^{(3)}(t) - g^{(4)}(t)$$

$$+\Phi[f''(t)] - \alpha f'(t) + \chi k'(t), \quad 0 \le t \le T. \tag{7.10}$$

We have set here

$$F = Af'' - \alpha Af', \quad \alpha := -k(0), \quad \chi := \Phi[Au_0]. \tag{7.11}$$

Remark 7.1. To derive equation (7.7) we have applied operator AD_t^2 to both members in (7.5) and we have made use of the following basic formula

$$D_t^2 \int_0^t k(t-s)z(s)ds = D_t \left\{ k(0)z(t) + \int_0^t k'(s)z(t-s)\, ds \right\}$$

$$= k(0)z'(t) + k'(t)z(0) + \int_0^t k'(s)z'(t-s)\, ds.$$

Remark 7.2. To derive equation (7.10) we have applied operator ΦD_t^2 to equation (7.7) and have used condition (6.4) (with $G(h, f, u_0, g) = g$).
Our next task consists in computing the initial value $k(0)$ in terms of data. To this purpose we apply operator ΦD_t to both members in (7.5) and, then, we compute the trace at $t = 0$. Using conditions (6.2–4), from (7.11) and the formula $u''(0) = f(0) - Au_0$ (under assumption $\chi \neq 0$) we easily derive

$$k(0) = \chi^{-1}\left\{g^{(3)}(0) + \Phi[Au_1] - \Phi[f'(0)]\right\} := \beta. \tag{7.12}$$

The next step to solve problem (7.7–10) consists in deriving a (functional) representation formula and the corresponding energy estimates for the solution v to the direct problem

$$v''(t) + Av(t) = f(t), \quad 0 < t < T, \tag{7.13}$$

$$v(0) = v_0, \quad v'(0) = v_1. \tag{7.14}$$

To do that, we need to introduce the following Banach spaces $U_t^{r,s}$ ($r \in \mathbf{N}$, $s \in (-1, +\infty)$, $t \in (0, T]$) normed in the natural way, where we set $W^{-s} = (W^s)^*$ for $s \in (0, 1]$:

$$U_t^{r,s} = \left\{u \in C([0,t]; W^{r+s}):\ u^{(j)} \in C([0,t]; W^{r+s-j})\ j = 1, ..., r\right\}.$$

THEOREM 7.1. *Let $(f, v_0, v_1) \in U_T^{0,m-1} \times W^m \times W^{m-1}$. Then problem (7.13–14) admits an unique solution $u = L_0(v_0, v_1) + L_1 f \in U_T^{2,m-2}$, where $L_0 \in \mathcal{L}(W^m \times W^{m-1}; U_T^{2,m-2})$ and $L_1 \in \mathcal{L}(U_T^{0,m-2}; U_T^{2,m-2})$. Moreover L_1 satisfies the estimate*

$$\|D_t L_1 f(t)\|_{m-1}^2 + \|L_1 f(t)\|_m^2 \le C \int_0^t \|f(s)\|_{m-1}^2 \, ds, \quad \forall t \in (0,T].$$

By virtue of Theorem 7.1 and formula (7.12) problem (7.7– 10) is equivalent to the following, where $*$ denotes convolution:

$$v = L_0(v_0, v_1) + L_1 \left[k' * (Af' - v') + \alpha v' + k' A^2 u_0 + F\right], \tag{7.15}$$

$$k' = \chi^{-1}\Phi[v] - \chi^{-1} k' * \left(\Phi[f'] - g^{(3)}\right) + k_0, \tag{7.16}$$

$$k(0) = \beta. \tag{7.17}$$

Introduce then the new pair of unknown (w, p) defined by

$$w(t) = v'(t), \quad 0 \le t \le T \quad \Leftrightarrow \quad v(t) = v_0 + \int_0^t w(s)\, ds, \quad 0 \le t \le T, \tag{7.18}$$

$$p(t) = k'(t), \quad 0 \le t \le T \quad \Leftrightarrow \quad k(t) = \beta + \int_0^t p(s)\, ds, \quad 0 \le t \le T. \tag{7.19}$$

Then problem (7.15–17) is equivalent to the following system of fixed-point equations:

$$w = D_t L_0(v_0, v_1) + D_t L_1 \left[p * (Af' - w) + \alpha w + p A^2 u_0 + F\right], \tag{7.20}$$

$$p = \chi^{-1}\Phi\left[D_t L_0(v_0, v_1)\right] + \chi^{-1}\Phi\left[D_t L_1[p * (Af' - w)]\right.$$

$$\left. - \chi^{-1} p * \left(\Phi[f'] - g^{(3)}\right) + k_0. \tag{7.21}$$

THEOREM 7.2. *Let Φ, f, u_0, u_1, g satisfy the properties:* **i)** $\Phi \in W^{m/2}$; **ii)** $f \in U_{T_0}^{2,m+1}$; **iii)** $u_j \in W^{m+3+j}$ $(j = 1,2)$; **iv)** $g \in C^4([0, T_0])$ *and the consistency conditions:* **v)** $g^{(j)}(0) = \Phi[u_j]$ $(j = 0,1)$; **vi)** $g''(0) = \Phi[-Au_0 + f(0)]$; **vii)** $\Phi[Au_0] = 0$. *Then there exists $T \in (0, T_0]$ for which identification problem (6.1–4) with $B = I$ and $G(h, f, u_0, u_1, g) = g$ admits an unique solution $(u, h) \in U_T^{4,m} \times C^1([0,T])$ depending continuously on data.*

THEOREM 7.3. *Let $f_j, u_{0,j}, u_{1,j}, g_j$ $(j = 1,2)$ satisfy the hypotheses listed in Theorem 7.2. If $(u_j, h_j) \in U_{T_0}^{4,m} \times C^1([0, T_0])$ $(j = 1,2)$ are solutions to problems*

(6.1–4) with $B = I$ and $G(h, f, u_0, u_1, g) = g$ corresponding to data $(f_j, u_{0,j}, u_{1,j}, g_j)$, then the following estimate holds true:

$$\|u_2 - u_1\|_{U^{4,m}_{T_0}} + \|h_2 - h_1\|_{C^1([0,T_0])} \le C(\gamma^{-1}, M_1, M_2)\Big\{\|f_2 - f_1\|_{U^{2,m+1}_{T_0}}$$

$$+\|u_{0,2} - u_{0,1}\|_{m+4} + \|u_{1,2} - u_{1,1}\|_{m+3} + \|g_2 - g_1\|_{C^1([0,T_0])}\Big\}.$$

We have set:

$$\gamma = \min_{1\le j\le 2} \Phi[Au_{0,j}],\ M_1 = \max_{1\le j\le 2} \|h_j\|_{C^1([0,T_0])},$$

$$M_2 = \max_{1\le j\le 2} \left(\|f_j\|_{U^{2,m+1}_{T_0}} + \|u_{0,j}\|_{m+4} + \|u_{1,j}\|_{m+3} + \|g_j\|_{C^1([0,T_0])}\right).$$

The function C is continuous from $(\mathbf{R}_+)^3$ to $\mathbf{R}_+$ and increasing in each arguments of its. C depends also on T_0, A, Φ, m.

Remark 7.3. The abstract formulation (6.1–4) with $B = I$ and $G(h, f, u_0, g) = g$ was derived from the concrete case dealt with in (Grasselli et al., 1988), where $\Omega = \mathbf{R}^n$ and $A = B = -\Delta$.

Remark 7.4. An important case in applications is the one corresponding to homogeneous initial data, i.e., $u_0 = u_1 = 0$. In this case we have, of course, $\Phi[Au_0] = 0$. In order to overcome this difficulty, we introduce the new unknown

$$\hat{u}(t) = u''(t), \quad 0 \le t \le T \quad \Leftrightarrow \quad u(t) = \int_0^t (t-s)\hat{u}(s)\, ds, \quad 0 \le t \le T.$$

The corresponding identification problem for $\hat{u}$ is the following:

$$\hat{u}''(t) + A\left\{\hat{u}(t) + \int_0^t h(t-s)\hat{u}(s)\, ds\right\} = f''(t), \quad 0 < t < T, \tag{7.22}$$

$$\hat{u}(0) = f(0), \tag{7.23}$$

$$\hat{u}'(0) = f'(0), \tag{7.24}$$

$$\Phi[\hat{u}(t)] = g''(t), \quad 0 \le t \le T. \tag{7.25}$$

Under the assumption $\Phi[Af(0)] \ne 0$ we can apply Theorems 7.2 and 7.3 to problem (7.22–25). Thus we obtain existence, uniqueness and stability results.

We conclude this section by noting that some identification problems related to parabolic linear integrodifferential equations can be reduced to hyperbolic ones (under smoothness assumptions). Consider, in fact, the following problem:

$$u'(t) - \int_0^t h(t-s)Bu(s)\, ds = f(t), \quad 0 < t < T, \tag{7.26}$$

$$u(0) = u_0, \tag{7.27}$$

$$\Phi[u(t)] = g(t), \quad 0 \le t \le T. \tag{7.28}$$

Differentiate once both members in equations (7.26) and (7.28). You easily obtain the (equivalent) identification problem

$$u''(t) - h(0)Bu(t) - \int_0^t h'(t-s)Bu(s)\,ds = f'(t), \quad 0 < t < T, \tag{7.29}$$

$$u(0) = u_0, \tag{7.30}$$

$$u'(0) = f(0), \tag{7.31}$$

$$\Phi[u(t)] = g(t), \quad 0 \le t \le T. \tag{7.32}$$

Assume then $\Phi[Bu_0] \neq 0$, apply operator Φ to both members in (7.29) and compute the trace at $t = 0$. You easily derive the equation $h(0) = \Phi[Bu_0]^{-1}\{g''(0) - \Phi[f'(0)]\} := h_0$. If we assume $h_0 < 0$, we can apply Theorems 7.2 and 7.3 (under assumptions of smoothness of data).

Remark 7.5. The previous procedure does not work in the case dealt with in §3, when $h(t) = t^{\alpha-1}k(t)$, where $\alpha \in (0,1)$ and k is a smooth function not necessarily vanishing at $t = 0$.

Remark 7.6. The cases $h_0 > 0$ and $h_0 = 0$ are open problems, the former being of elliptic type, the latter being related to a third-order integrodifferential equation.

We conclude this section by considering a Cauchy problem, related to the identification of an unknown coefficient in Maxwell's equations (cf. (Lorenzi and Priymenko, 1991)), with distributional data. To this purpose with spaces W^r $(0 \le r \le N)$ we associate their duals

$$W^{-r} = (W^r)^* \quad (0 \le r \le N)$$

endowed with their usual norms and we denote by $< \cdot, \cdot >_r$ the pairing between W^{-r} and W^r $(r = 1, ..., N)$. Then we define by duality the operator A on W^{-r} $(r = 1, ..., N)$. This implies $A \in \mathcal{L}(W^{1-r}; W^{-1-r})$ $(r = 0, ..., N)$.

Finally we introduce the following Banach spaces $V^{m,-r}$, basic to our investigations:

$$V^{m,-r} = \left\{u \in C([0,T]; W^{-r}) \colon u^{(j)} \in C([0,T]; W^{-r-j}), \quad 0 \le j \le m\right\} \tag{7.33}$$

normed in the natural way, with norm denoted by $\|\cdot\|_{m,r,T}$.

Assume then that r is a fixed integer in $[2, N]$. Our identification problem is the following: determine a pair of functions u: $(-\infty, T) \to X$ with $u \in V^{2,2-r}$, $h \in C([0,T])$ and a real number β such that

$$u''(t) + Au(t) + \beta u'(t) + h * u'(t) = \delta(t) * u_1 + Y_+(t)f(t), \quad t \in (-\infty, T), \tag{7.34}$$

$$u(t) = u_0, \quad t \in (-\infty, 0], \tag{7.35}$$

$$\langle u(t), \varphi \rangle_r = g(t), \quad t \in (0, T], \tag{7.36}$$

where $*$ denotes convolution and $\varphi \in W^r$.
As far as data f, u_0, u_1, g are concerned, we assume that they enjoy the following properties:

$$(f, u_0, u_1, g) \in C^1([0,T]; W^{1-r}) \times W^{2-r} \times W^{1-r} \times C^3([0,T]), \tag{7.37}$$

$$Au_0 = 0. \tag{7.38}$$

Using standard arguments (cf. (Dautray and Lions, 1985)), we easily realize that problem (7.34–36) is equivalent to the following: determine a pair of functions $u \in V^{2,2-r}, h \in C([0,T])$ and a real number β such that

$$u''(t) + Au(t) + \beta u'(t) + h * u'(t) = f(t), \quad t \in (0, T), \tag{7.39}$$

$$u(0) = u_0, \tag{7.40}$$

$$u'(0) = u_1, \tag{7.41}$$

$$\langle u(t), \varphi \rangle_r = g(t), \quad t \in [0, T]. \tag{7.42}$$

We observe now that β can be apriori determined in terms of data, if the following condition is satisfied:

$$\chi = \langle u_1, \varphi \rangle_r \neq 0. \tag{7.43}$$

In fact, if we compute the crochet of both members in (7.39) with φ at $t = 0$ and use equations (7.41–42), we easily get the required value β :

$$\beta = \chi^{-1} \left\{ \langle f(0), \varphi \rangle_r - g''(0) \right\}. \tag{7.44}$$

Assume then that $u \in V^{3,3-r}$ solves problem (7.39– 42) and introduce the function $v \in V^{2,2-r}$ defined by

$$v(t) = u'(t), \quad t \in [0,T] \quad \Leftrightarrow \quad u(t) = u(0) + \int_0^t v(s)\, ds, \quad t \in [0,T].$$

By differentiation we easily deduce that the pair $(v,h) \in V^{2,2-r} \times C([0,T])$ solves the following problem, equivalent to (7.39–42):

$$v''(t) + Av(t) + \beta v'(t) + h * v'(t) + h(t)u_1 = f'(t), \quad t \in (0,T), \tag{7.45}$$

$$v(0) = u_1, \tag{7.46}$$

$$v'(0) = f(0) - \beta u_1 := u_2, \tag{7.47}$$

$$\langle v(t), \varphi \rangle_r = g'(t), \quad t \in [0,T]. \tag{7.48}$$

Hence, from (7.46–48) we deduce that our data have to satisfy the following consistency conditions

$$\langle u_j, \varphi \rangle_r = g^{(j)}(0), \quad j = 0,1,2. \tag{7.49}$$

Take now the crochets of both members in (7.45) by φ. From assumption (7.43) we deduce the following equation for h :

$$h(t) = \chi^{-1}\{\langle f'(t), \varphi \rangle_r - g^{(3)}(t) - \langle Av(t), \varphi \rangle_r$$

$$-\langle \beta v'(t) + h * v'(t), \varphi \rangle_r\}, \quad t \in [0,T]. \tag{7.50}$$

We note now that, as in the case of regular data, we can prove by the means of energy estimates the following existence theorem for direct problem (7.45–47):

THEOREM 7.4. *For any quadruplet $(h, f, u_0, u_1) \in C([0,T]) \times C^1([0,T]; W^{1-r}) \times W^{2-r} \times W^{1-r}$ problem (7.45–47) admits an unique solution $v \in V^{2,2-r}$ satisfying the estimate*

$$\|v\|^2_{1,2-r,t} \le C_1(r)\left[\|v_0\|^2_{2-r} + \|v_1\|^2_{1-r}\right] \exp\left[\int_0^t m_1(\|h\|_{0,s})\, ds\right]$$

$$+C_1(r)\int_0^t \exp\left[\int_s^t m_1(\|h\|_{0,\sigma}) d\sigma\right] \|f(s)\|^2_{1-r}\, ds. \tag{7.51}$$

Moreover, if $v, v^* \in V^{2,2-r}$ are solutions to problems (7.45–47) corresponding to data $(h, f, u_0, u_1), (h^*, f^*, u_0^*, u_1^*) \in C([0,T]) \times C^1([0,T]; W^{1-r}) \times W^{2-r} \times W^{1-r}$, then $v - v^*$ satisfies the estimate

$$\|v - v^*\|^2_{1,2-r,t} \le C_1(\rho)\left[\|v_0 - v_0^*\|^2_{2-r} + \|v_1 - v_1^*\|^2_{1-r}\right] \exp(\rho t)$$

$$+C_1(r)\int_0^t \exp[\rho(t-s)]\left[\|f - f^*\|^2_{C^1([0,s];W^{1-r})} + \|u_0 - u_0^*\|^2_{2-r}\right.$$

$$+\|u_1 - u_1^*\|_{1-r}^2 + \|h - h^*\|_{0,s}^2]ds, \tag{7.52}$$

$t \in (0,T]$ and $\|h\|_{0,T} \leq \rho$, $\|h^*\|_{0,T} \leq \rho$.
Associate then with any $h \in C([0,T])$ the unique solution $v = M(h) \in V^{2,2-r}$ to problem (7.45–47). According to (7.52), we deduce that $M \in C(C([0,T]); V^{2,2-r})$ and satisfies the estimate

$$\|M(h_2) - M(h_1)\|_{1,2-r,t}^2 \leq C_1(\rho) \int_0^t \exp[\rho(t-s)]\|h_2 - h_1\|_{0,s}^2\, ds \tag{7.53}$$

$t \in (0,T]$, $h_j \in C([0,T])$ and $\|h_j\|_{0,T} \leq \rho$ $(j = 1,2)$.
Observe now that identification problem (7.45–47), (7.50) is equivalent to the following fixed-point equation in $C([0,T])$

$$h(t) = \chi^{-1}\{\langle f'(t), \varphi\rangle_r - g^{(3)}(t) - \langle AM(h)(t), \varphi\rangle_r$$

$$+\langle \beta D_t M(h)(t) + h * D_t M(h)(t), \varphi\rangle_r\}, \quad t \in [0,T]. \tag{7.54}$$

Such an equation can be solved by the generalized contraction principle (cf. (Lorenzi and Priymenko, 1991)). Consequently, the solution h is locally Lipschitz continuous with respect to data.

Remark 7.7. In this section we have outlined two different procedures to solve our linear identification problem. The former consists in deducing a fixed-point system for the pair (u, h), while the latter in solving first the linear integrodifferential equation for u in dependence of the function h and in investigating then the stability of the mapping $h \to u$ with respect to suitable norms.
We observe, however, that such procedures are substantially equivalent. In fact, in the cases under consideration, we can always solve the direct problem for u in terms of h and deduce a fixed-point equation for h, which, in turn, can be solved by the contraction principle.
Yet, the latter procedure allows to derive explicit stability results for the solution u of the direct problem in terms of the functional parameter h.
This technique is used, e.g., in (Grasselli, 1990).

Remark 7.8. As far as identification problems related to linear viscoelasticity and thermoviscoelasticity are concerned, the reader is referred to (Grasselli, 1991).

Remark 7.9. The linear identification problems dealt with in this section can be reduced to the unifying formulation exposed in (Grasselli, 1990).

8. HYPERBOLIC IDENTIFICATION PROBLEMS: THE NONLINEAR CASE

We list here the basic assumptions on operators B and G and functional Φ, where $\alpha \in (0,1)$:

i) $B \in C(W^{3+\alpha}; W^{3+\alpha})$ and $B' \in C(W^{3+\alpha}; \mathcal{L}(W^{2+\alpha}; W^{2+\alpha}))$;

ii) $\|B(u_2)-B(u_1)\|_{3+\alpha}+\|B'(u_2)-B'(u_1)\|_{2+\alpha,2+\alpha} \le \Lambda_1(\rho)\|u_2-u_1\|_{3+\alpha}$ for any $u_1, u_2 \in W^{3+\alpha}$ such that $\|u_1\|_{3+\alpha} \le \rho$, $\|w_2\|_{3+\alpha} \le \rho$, where Λ_1: $[0,+\infty) \to (0,+\infty)$ $(j = 1,2)$ is a continuous nondecreasing function;

iii) $\Phi \in (W^{1+\alpha})^*$;

iv) $G \in C(C^1([0,T_0]) \times W^{3+\alpha} \times W^{2+\alpha} \times U_T^{0,2+\alpha} \times C^4([0,T_0]); C^2([0,T_0]))$;

v) $D_t^2 G(h, u_0, u_1, f, g) = q(u_0, u_1, f, g)h' + G_2(h, u_0, u_1, f, g)$;

vi) $q \in C(W^{3+\alpha} \times W^{2+\alpha} \times U_T^{0,2+\alpha} \times C^4([0,T_0]); C([0,T_0]))$ and $q \in C^{0,1}(\mathcal{E}(c))$ for some $c > 0$, where $\mathcal{E}(c) = \{(u_0, u_1, f, g) \in W^{3+\alpha} \times W^{2+\alpha} \times U_T^{0,2+\alpha} \times C^4([0,T_0])$: $q(u_0, u_1, f, g)(t) \ge c$ for any $t \in [0,T_0]\}$;

vii) $G_2 \in C(C^1([0,T_0]) \times W^{3+\alpha} \times W^{2+\alpha} \times U_T^{0,2+\alpha} \times C^4([0,T_0]); C([0,T_0]))$;

viii)

$$\|G_2(h_2, u_{0,2}, u_{1,2}, f_2, g_2) - G_2(h_1, u_{0,1}, u_{1,1}, f_1, g_1)\|_{C([0,t])}$$

$$\le \Lambda_2(\rho)\Big\{ \int_0^t \|h_2 - h_1\|_{C^1([0,s])} ds$$

$$+\|(u_{0,2}, u_{1,2}, f_2, g_2) - (u_{0,1}, u_{1,1}, f_1, g_1)\|\Big\}$$

for any $t \in (0,T_0]$ and for any $h_1, h_2 \in C^1([0,T_0])$ such that $\|h_1\|_{3+\alpha} \le \rho$, $\|h_2\|_{3+\alpha} \le \rho$ and for any $(u_{0,1}, u_{1,1}, f_1, g_1)$, $(u_{0,2}, u_{1,2}, f_2, g_2) \in \mathcal{E}(c)$. Here Λ_2: $[0,+\infty) \to (0,+\infty)$ is a continuous nondecreasing function and $\|\cdot\|$ denotes the norm in the space of data.

We list now the consistency conditions for problem (6.1–3):

ix) $D_t^j G(h, u_0, u_1, f, g)(0) = \Phi(u_j)$, $j = 1,2$;

x) $D_t^2 G(h, u_0, u_1, f, g)(0) = \Phi(Au_0 + f(0))$.

Finally we assume that the following properties hold:

xi) $G(h, u_0, u_1, f, g)(0)$ is independent of h;

xii) $D_t G(h, u_0, u_1, f, g)(0)$ depends only on $h(0)$ and can be solved for it in terms of data $(u_0, u_1, f, g) \in \mathcal{E}(c)$, i.e., $h(0) = H(u_0, u_1, f, g)$. Moreover $H \in C^{0,1}(\mathcal{E}(c); \mathbf{R})$.

xiii) $G_2(h, u_0, u_1, f, g)(0)$ is independent of $h(0)$.

We state now the main results concerning identification problem (6.1–4) (cf. (Lorenzi, 1989b)).

THEOREM 8.1. *Let $\alpha \in (1/2, 1)$ and let assumptions* **i)–xiii)** *be satisfied. Then there exists $T \in (0,T_0]$ for which problem (6.1–4) admits an unique solution $(u,h) \in U_T^{2,1+\alpha} \times C^1([0,T])$ depending continuously on data in the norm pointed out.*

THEOREM 8.2. *Let $\alpha \in (1/2, 1)$ and let assumptions* **i)–xiii)** *be satisfied with the quadruplet (u_0, u_1, f, g) replaced by $(u_{0,j}, u_{1,j}, f_j, g_j) \in \mathcal{E}(c)$ $(j = 1,2)$. Let*

$(u_j, h_j) \in U_T^{2,1+\alpha} \times C^1([0,T])$ $(j = 1,2)$ be solutions to problem (6.1–4) corresponding to data $(u_{0,j}, u_{1,j}, f_j, g_j)$ such that $\|h_j\|_{C^1([0,T])} \le \rho$ $(j = 1,2)$ for some given $\rho > 0$. Then the following estimate holds true:

$$\|u_2 - u_1\|_{U_T^{2,1+\alpha}} + \|h_2 - h_1\|_{C^1([0,T])} \le C\|(u_{0,2}, u_{1,2}, f_2, g_2)$$

$$-(u_{0,1}, u_{1,1}, f_1, g_1)\|, \quad (u_{0,j}, u_{1,j}, f_j, g_j) \in \mathcal{E}(c) \ (j = 1,2),$$

C being a constant depending on c, ρ and the norm of data.

9. IDENTIFICATION PROBLEMS RELATED TO MAXWELL'S EQUATIONS

We pose the following identification problem: determine a quadruplet of functions E: $[0,T] \to \mathcal{H}_1$, H: $[0,T] \to \mathcal{H}_2$ and a, b: $[0,T] \to \mathbf{R}$ such that

$$A_1 H(t) = E'(t) + \sigma_0 E(t) + \int_0^t a(s)E'(t-s)\,ds + f(t), \quad 0 < t < T, \tag{9.1}$$

$$A_2 E(t) = -H'(t) - \int_0^t b(s)H'(t-s)\,ds, \quad 0 < t < T, \tag{9.2}$$

$$E(0) = 0, \tag{9.3}$$

$$H(0) = 0, \tag{9.4}$$

$$\Phi[E(t), H(t)] = h(t), \quad 0 \le t \le T. \tag{9.5}$$

To derive the formulation suitable for solving problem (9.1–4) we differentiate three times both members in (9.1– 2), (9.5). To this purpose, according to the following formula

$$D_t^3 \int_0^t \gamma(s)Z(t-s)ds = \gamma''(t)Z(0) + \gamma'(t)Z'(0) + \gamma(0)Z''(t)$$

$$+ \int_0^t \gamma'(s)Z''(t-s)\,ds,$$

and equations (9.1–2), we need to compute the initial values $E^{(j)}(0)$ and $H^{(j)}(0)$ $(j = 1,2)$. From equations (9.1–2) we immediately derive the formulas

$$E'(0) = -f(0) := E_1, \quad H'(0) = 0,$$

$$E''(0) = -[\sigma + a(0)]E_1 - f'(0) := E_2, \quad H''(0) = -A_2 E_1 := H_2,$$

$$E^{(3)}(0) = A_1 H_2 - \sigma_0 E_2 - a'(0)E_1 - f''(0) := E_3,$$

$$H^{(3)}(0) = -A_2 E_2 - b(0)H_2 := H_3.$$

Introduce then the auxiliary unknowns

$$\widehat{E}(t) = E^{(3)}(t) \quad \Leftrightarrow \quad E(t) = tE_1 + \frac{1}{2}t^2 E_2 + \frac{1}{2}\int_0^t (t-s)^2 \widehat{E}(s)\, ds,$$

$$\widehat{H}(t) = H^{(3)}(t) \quad \Leftrightarrow \quad H(t) = \frac{1}{2}t^2 H_2 + \frac{1}{2}\int_0^t (t-s)^2 \widehat{H}(s)\, ds.$$

It is an easy task to check that the quadruplet $(\widehat{E}, \widehat{H}, a, b)$ solves the following identification problem, equivalent to (9.1–5):

$$A_1 \widehat{H}(t) = \widehat{E}'(t) + [\sigma_0 + a(0)]\widehat{E}(t) + a''(t)E_1 + a'(t)E_2$$

$$+ \int_0^t a'(s)\widehat{E}(t-s)\, ds + f^{(3)}(t), \quad 0 < t < T, \tag{9.6}$$

$$A_2 \widehat{E}(t) = -\widehat{H}'(t) - b(0)\widehat{H}(t) - b'(t)H_2 - \int_0^t b(s)\widehat{H}(t-s)\, ds, \quad 0 < t < T, \tag{9.7}$$

$$\widehat{E}(0) = E_3, \tag{9.8}$$

$$\widehat{H}(0) = H_3, \tag{9.9}$$

$$\Phi[\widehat{E}(t), \widehat{H}(t)] = h^{(3)}(t), \quad 0 \le t \le T. \tag{9.10}$$

We list now our assumptions concerning spaces $\mathcal{H}_j$, operators A_j and Φ:

i) W_j, X_j, W_j^* $(j = 1, 2)$ are two Hilbert triplets (cf. **i)**–**ii)** in §6);

ii) A_j: $W_j \to X_j$ $(j = 1, 2)$ are linear closed operators;

iii) $\mathcal{A} \subset \mathcal{A}^*$, where $\mathcal{A} = \begin{pmatrix} 0 & A_2 \\ -A_1 & 0 \end{pmatrix}$: $\mathcal{D}(\mathcal{A}) \subset \mathcal{H} \to \mathcal{H}$, $\mathcal{D}(\mathcal{A}) = W_1 \times W_2$ and $\mathcal{H} = \mathcal{H}_1 \times \mathcal{H}_2$;

iv) $(\mathcal{A}U, U) = 0$ for any $U \in \mathcal{H}$, $(\cdot,\cdot)$ denoting the scalar product in $\mathcal{H}$;

v) the resolvent of $\mathcal{A}$ contains a point $\lambda_0 \in \mathbf{R}_-$;

vi) $\Phi \in \mathcal{D}(\mathcal{A})^*$.

As far as data f, h are concerned, we assume that they enjoy the following properties:

vii) $f \in W^{2,2}((0,T_0);\mathcal{D}(\mathcal{A}))$;

viii) $h \in C^3([0,T_0];\mathbf{R}^2)$;

ix) $(E_j, H_j) \in \mathcal{D}(\mathcal{A})^3 \quad (j = 1,2,3)$.

Finally, we assume that the following consistency conditions are satisfied:

x) $h(0) = 0$;

xi) $h^{(j)}(0) = \Phi(U_j)$, where $U_j = (E_j, H_j)^T \quad (j = 1,2,3)$, T denoting transposition;

Remark 8.1. According to the definitions of vectors E_1 and H_1 , the condition in **xi)** with $j = 1$ is an actual consistency one, since the previous vectors are independent of functions a and b. On the contrary,the conditions related to $j = 1,2$ supply the consistency condition

xii) $\det(-\Phi(E_1,0), h''(0) + \Phi(-\sigma_0 E_1 + f'(0), 0)\Phi(0, A_2 E_1)) = 0$ and the initial values $(a(0), a'(0), b(0))$ in terms of E_2, H_2, E_3, H_3, provided the following inequalities hold:

xiii) $m_1 := |\Phi(E_1,0)| > 0, \; m_2 = |\det(-\Phi(E_1,0), \; \Phi(0, A_2 E_1))| > 0$.

Before stating our main results (cf. (Kabanikhin and Lorenzi, 1991)) we need to introduce some functional spaces

$$\mathcal{U}_T^s = W^{s,2}((0,T);\mathcal{H}) \cap C^{s-1}([0,T];\mathcal{D}(\mathcal{A})), \quad s \in \mathbf{N}\backslash\{0\},$$

$$C^{(q,s)}([0,T],\mathbf{R}^2) = C^q([0,T],\mathbf{R}) \times C^s([0,T],\mathbf{R}), \; q, s \in \mathbf{N},$$

$$\mathcal{G}_T^m = L^2((0,T);\mathcal{D}(\mathcal{A}^m)) \times C^4([0,T],\mathbf{R}^2), \quad m \in \mathbf{N}.$$

THEOREM 9.1. *Let properties* **i)–xiii)** *be satisfied. Then there exists* $T \in (0, T_0]$ *such that identification problem* (9.1–5) *admits an unique solution* $(E, H, a, b) \in \mathcal{U}_T^{4,2} \times C^{(2,1)}([0,T],\mathbf{R}^2)$ *continuously depending on data with respect to the norms pointed out.*

THEOREM 9.2. *Let properties* **i)–xiii)** *be satisfied with* (f,h) *replaced everywhere by* $(f_j, h_j) \quad (j = 1,2)$. *Let* $(E_j, H_j, a_j, b_j) \in \mathcal{U}_T^{4,2} \times C^{(2,1)}([0,T],\mathbf{R}^2) \quad (j = 1,2)$ *be solutions to problems* (9.1–5) *corresponding to data* (f_j, h_j) . *Then the following estimate holds true:*

$$\|(E_2 - E_1, H_2 - H_1)\|^2_{\mathcal{U}_T^{4,2}} + \|(a_2 - a_1, b_2 - b_1)\|^2_{C^{(2,1)}([0,T],\mathbf{R}^2)}$$

$$\leq C(\rho, r, T, m_1^{-1}, m_2^{-1})\|(f_2 - f_1, 0, h_2 - h_1)\|^2_{\mathcal{G}_T^2},$$

where

$$\rho = \max(\|(a_1, b_1)\|_{C^{(2,1)}([0,T],\mathbf{R}^2)}, \; \|(a_2, b_2)\|_{C^{(2,1)}([0,T],\mathbf{R}^2)}),$$

$$r = \max(\|(f_1, 0, h_1)\|_{\mathcal{C}^2_T}, \ \|(f_2, 0, h_2)\|_{\mathcal{C}^2_T}).$$

The positive constant C depends increasingly and continuously on the arguments pointed out.

REFERENCES

Da Prato, G., Iannelli, M., and Sinestrari, E. (1983). Temporal regularity for a class of integro-differential equations in Banach spaces. *Bull. Un. Mat. Ital. An. Funz. Appl.* **2C**, 171–185.

Da Prato, G., Iannelli, M., and Sinestrari, E. (1985). Regularity of solutions of a class of integrodifferential equations in Banach spaces. *J. Integr. Equat.* **8**, 27–40.

Dautray, R. and Lions, J.L. (1985). *Analyse Mathématique et Calcul Numérique Pour les Sciences et les Techniques*. Vol. III. Masson, Paris.

Grasselli, M. (1989). An identification problem for a linear integrodifferential equation occurring in heat flow. *J. Math. Anal.* (to appear).

Grasselli, M. (1990). An identification problem for an abstract linear hyperbolic integrodifferential equations with applications. *J. Math. Anal.* (to appear).

Grasselli, M. (1991). An inverse problem in three-dimensional linear thermoviscoelasticity of Boltzmann type. In: *Ill-Posed Problems in Natural Sciences,* Ed. by A. Tikhonov. TVP Sci. Publ., Moscow, pp. 284–299.

Grasselli, M., Kabanikhin, S.I., and Lorenzi, A. (1989). An inverse hyperbolic integrodifferential problem arising in Geophysics I. *Siberian Math. J.* (to appear).

Grasselli, M., Kabanikhin, S.I., and Lorenzi, A. (1990). An inverse hyperbolic integrodifferential problem arising in Geophysics II. *Nonlinear Anal. T.M.A.* **15**, 283–298.

Grasselli, M. and Lorenzi, A. (1991). An inverse problem for an abstract nonlinear parabolic integrodifferential equation. (to appear).

Kabanikhin, S.I. and Lorenzi, A. (1991). An identification problem related to integrodifferential Maxwell's equations. (to appear).

Lions, J.L. (1969). *Quelques Méthodes de Résolution des problèmes aux Limites non Linéaires.* Dunod, Paris.

Lions, J.L. and Magenes, E. (1972). *Nonhomogeneous Boundary Value Problems and Applications.* Vol. I. Springer-Verlag, Berlin.

Lorenzi, A. (1989a). *An Inverse Problem in the Theory of Materials with Memory.* Vol. II. (Ser. Pure and Appl. Math., **116**,) M. Dekker, New York.

Lorenzi, A. (1989b). An identification problem related to a nonlinear hyperbolic integrodifferential equations. (to appear).

Lorenzi, A. and Paparoni, E. (1990). Direct and inverse problems in the theory of materials with memory. *Rend. Sem. Univ. Padova* (to appear).

Lorenzi, A. and Prilepko, A.I. (1991). Fredholm type results for integrodifferential parabolic problems. (to appear).

Lorenzi, A. and Priymenko, V. (1991). A duality approach for solving identification problems related to integrodifferential equations. (to appear).

Lorenzi, A. and Sinestrari, E. (1988). An inverse problem in the theory of materials with memory. *Nonlinear Anal. T.M.A.* **12**, 1317–1335.

Lorenzi, A. and Sinestrari, E. (1989). Stability results for a partial integrodifferential inverse problem. *Pitman Research Notes Math.* **190**, 271–294.

Ill-Posed Problems in Natural Sciences, pp. 367 – 374
A. Tikhonov (Ed.)
1992 VSP/TVP

THE FREDHOLM SOLVABILITY OF INVERSE PROBLEMS FOR ABSTRACT DIFFERENTIAL EQUATIONS

D.G. ORLOVSKII
Moscow Engineering-Physics Institute, Kashirskoe shosse, 31, 115409 Moscow, Russia

ABSTRACT

This paper deals with inverse problems for evolution equations of the first and second order in a Banach space. For the equations of the second order we consider the hyperbolic and the elliptic cases separately. Under certain conditions the Fredholm solvability of inverse problems is established.

The inverse problems for partial differential equations form an important part of the theory of ill-posed problems. Various formulations of the problems and surveys of the results can be found in (Tikhonov and Arsenin, 1977; Lavrent'ev et al., 1986; Prilepko, 1973). One aspect of the investigation of these problems is the interpretation of partial differential equations as ordinary differential equations in a Banach space. From the viewpoint of this approach the inverse problems are considered in (Iskenderov, 1976; Rundell, 1980; Eidelman, 1983; Prilepko and Orlovskii, 1989; Orlovskii, 1989, 1990; Eidelman, 1990), where the problem of existence and uniqueness of the solution is studied. The main subject of this paper is the Fredholm solvability of the inverse problems for abstract differential equations of the first and second order in a Banach space.

1. EQUATION OF THE FIRST ORDER

Let X be a Banach space with the norm $\|\cdot\|$ and let $\mathcal{L}(X)$ denote the set of all bounded linear operators from X to X. Let A be a closed linear operator in X. We consider its domain $\mathcal{D}(A)$ as a Banach space with the norm of the graph of A. We suppose that $\mathcal{D}(A)$ is dense in X. The identity operator is denoted by I.
We consider the following problem

$$u'(t) = Au(t) + f(t), \quad 0 \leq t \leq T, \tag{1}$$

$$u(0) = u_0, \tag{2}$$

$$f(t) = \Phi(t)p + F(t), \quad 0 \leq t \leq T, \tag{3}$$

$$u(T) = u_1, \tag{4}$$

where Φ: $[0,T] \to \mathcal{L}(X)$, F: $[0,T] \to X$, $u_0, u_1 \in X$. A solution of (1)–(4) is a pair (u,p), where $u \in \mathbf{C}^1([0,T];X) \cap \mathbf{C}([0,T];\mathcal{D}(A))$ and $p \in X$. Our fundamental assumption is that A is the infinitesimal generator of a C_0-semigroup $V(t)$. In this case the Cauchy problem (1)–(2) is well-posed and the solution of (1)–(2) is given by

$$u(t) = V(t)u_0 + \int_0^t V(t-s)f(s)\,ds \tag{5}$$

(if $u_0 \in \mathcal{D}(A)$ and $f \in \mathbf{C}^1([0,T];X) + \mathbf{C}([0,T]);\ \mathcal{D}(A))$.

We now introduce the notion of admissible data. The data u_0, u_1, F are admissible if

$$u_0,\ u_1 \in \mathcal{D}(A), \tag{6}$$

$$F \in \mathbf{C}^1([0,T];X) + \mathbf{C}([0,T];\mathcal{D}(A)). \tag{7}$$

Condition (6) is necessary for the solvability of problem (1)–(4). Condition (7) is a common restriction for evolution equations.

In the general case the solution of problem (1)–(4) is not unique. Suppose that the operator A has an eigenvalue λ with the corresponding eigenvector f. Let $\Phi \in \mathbf{C}^1([0,T]:\ \mathbf{R}^1)$ and

$$\int_0^T \Phi(s)\exp(-\lambda s)\,ds = 0$$

(obviously such function exists). Let us choose $u_0 = 0,\ u_1 = 0,\ F = 0$. These data are admissible. Besides the trivial solution $u = 0,\ p = 0$ problem (1)–(4) has a non-trivial one:

$$u(t) = \Big(\int_0^t \Phi(s)\exp(\lambda(t-s))\,ds\Big)f, \quad p = f.$$

Let us choose $\lambda \in \rho(A)$, where $\rho(A)$ is the resolvent set of A. Define

$$h = (A - \lambda I)(u_2 - V(T)u_1 - \int_0^T V(T-s)F(s)\,ds).$$

THEOREM 1. *Let the semigroup $V(t)$ be compact (i.e., the operator $V(t)$ is compact for every $t > 0$), $\Phi \in \mathbf{C}^1([0,T];\ \mathcal{L}(X))$, $\Phi(T)^{-1} \in \mathcal{L}(X)$. Then the Fredholm property holds:*

1) problem (1)-(4) is solvable for every admissible data if and only if only the trivial solution exists for $u_0 = 0,\ u_1 = 0,\ F = 0$;

2) the space of all solutions of problem (1)-(4) for $u_0 = 0,\ u_1 = 0,\ F = 0$ is finite-dimensional in $\mathbf{C}^1([0,T];\ X) \times X$;

3) There exist $l_1, \dots,\ l_n \in X^*$ such that problem (1)-(4) is solvable if and only if

$$l_1(h) = \ \dots\ = l_n(h) = 0.$$

The functionals $l_1, \dots, l_n$ depend on A and Φ only.

Proof. From (5), by using (3) and (4), we deduce an equation for p:

$$Bp = g, \tag{8}$$

where

$$Bp = \int_0^T V(T-s)\Phi(s)p\, ds, \tag{9}$$

$$g = u_1 - V(T)u_0 - \int_0^T V(T-s)F(s)\, ds. \tag{10}$$

We have $g \in \mathcal{D}(A)$ for any admissible data. Convesely, for every $g \in \mathcal{D}(A)$ there are admissible data such that equality (10) is valid (for example, we can choose $u_0 = 0$, $u_1 = g$, $F = 0$). Therefore, the inverse problem is reduced to equation (8) with $g \in \mathcal{D}(A)$. Since $(A - \lambda I)g = h$ and $\lambda \in \rho(A)$, equation (8) is equivalent to

$$(A - \lambda I)Bp = h. \tag{11}$$

It is important to note that h is an arbitrary element of X.
It is well-known that, for every $\varphi \in \mathbf{C}^1([0,T];\ X)$, there holds

$$A\int_0^t V(t-s)\varphi(s)\, ds = \int_0^t V(t-s)\varphi'(s)\, ds + V(t)\varphi(0) - \varphi(t). \tag{12}$$

Using formula (12) for $\varphi(t) = \Phi(t)p$, we deduce from (11)

$$\int_0^T V(T-s)\Phi'(s)p\, ds + V(T)\Phi(0)p - \Phi(T)p - \lambda\int_0^T V(T-s)\Phi(s)p\, ds = h.$$

Taking into account the fact that $\Phi(T)^{-1} \in \mathcal{L}(X)$, we obtain the equation

$$p - B_1 p = h_1, \tag{13}$$

where

$$B_1 = \Phi(T)^{-1}\Big(V(T)\Phi(0) + \int_0^T V(T-s)[\Phi'(s) - \lambda\Phi(s)]\, ds\Big),$$

$$h_1 = -\Phi(T)^{-1}h.$$

Since the semigroup $V(t)$ is compact, the operator B_1 is compact too. Now from the Fredholm theorem for equations of the second kind we obtain the conclusions of our theorem.

2. EQUATION OF THE SECOND ORDER. HYPERBOLIC CASE

Let us consider the inverse problem for the equation of the second order:

$$u''(t) = Au(t) + f(t), \quad 0 \le t \le T, \tag{14}$$

$$u(0) = u_0, \quad u'(0) = u_1, \tag{15}$$

$$f(t) = \Phi(t)p + F(t), \quad 0 \le t \le T, \tag{16}$$

$$u(T) = u_2. \tag{17}$$

A solution of (14)–(17) is a pair (u,p), where $u \in \mathbf{C}^2([0,T];X) \cap \mathbf{C}([0,T];\mathcal{D}(A))$ and $p \in X$. Our fundamental assumption is that A is the infinitesimal generator of a strongly continuous cosine function $C(t)$. The associated sine function is defined by

$$S(t)f = \int_0^t C(s)f\, ds.$$

We shall also use the set

$$E = \{f \in X:\ C(t)f \in \mathbf{C}^1(\mathbf{R}^1;X)\}.$$

The set E with the norm

$$\|f\|_E = \|f\| + \sup_{0 \le t \le 1} \|C'(t)f\|$$

is a Banach space.

It is well-known that the operator A is the infinitesimal generator of a strongly continuous cosine function if and only if the Cauchy problem (14)–(15) is uniformly well-posed. The solution of (14)–(15) is given by

$$u(t) = C(t)u_0 + S(t)u_1 + \int_0^t S(t-s)f(s)\, ds \tag{18}$$

(if $u_0 \in \mathcal{D}(A)$, $u_1 \in E$ and $f \in \mathbf{C}^1([0,T];X) + \mathbf{C}([0,T];\mathcal{D}(A))$).

In this case the admissible data is defined as

$$u_0, u_2 \in \mathcal{D}(A), \quad u_1 \in E, \tag{19}$$

$$F \in \mathbf{C}^1([0,T];X) + \mathbf{C}([0,T];\mathcal{D}(A)). \tag{20}$$

In general, the solution of problem (14)–(17) is not unique. We define the operator

$$\mathcal{D} = \Phi(T) - C(T)\Phi(0).$$

As distinct from the inverse problem for the equation of the first order, we have to use the supplementary restriction

$$\mathcal{D}^{-1} \in \mathcal{L}(X). \tag{21}$$

Let us choose $\lambda \in \rho(A)$ and define

$$h = (A - \lambda I)(u_2 - C(T)u_0 - S(T)u_1 - \int_0^T S(T-s)F(s)\,ds).$$

THEOREM 2. *Let the sine function $S(t)$ be compact (i.e., the operator $S(t)$ is compact for every t), $\Phi \in \mathbf{C}^2([0,T];\mathcal{L}(X))$ and condition (21) be satisfied. Then the Fredholm property holds:*
1) problem (14)-(17) is solvable for every admissible data if and only if only the trivial solution exists for $u_0 = 0,\ u_1 = 0, u_2 = 0,\ F = 0$;
2) the space of all solutions of problem (14)–(17) for $u_0 = 0,\ u_1 = 0,\ u_2 = 0,\ F = 0$ is finite dimenshional;
3) there exist $l_1, \dots, l_n \in X^*$ such that problem (14)–(17) is solvable if and only if

$$l_1(h) = \ldots = l_n(h) = 0.$$

Proof. From (18), by using (16) and (17), we derive the equation

$$Bp = g, \tag{22}$$

where

$$Bp = \int_0^T S(T-s)\Phi(s)p\,ds,$$

$$g = u_2 - C(T)u_0 - S(T)u_1 - \int_0^T S(T-s)F(s)\,ds.$$

The element g is an arbitrary element of $\mathcal{D}(A)$. Since $(A - \lambda I)g = h$, equation (22) is equivalent to

$$(A - \lambda I)Bp = h. \tag{23}$$

To reduce (23) to an equation of the second kind, we use the formula

$$A\int_0^t S(t-s)\varphi(s)\,ds = \int_0^t S(t-s)\varphi''(s)\,ds$$

$$+C(t)\varphi(0) + S(t)\varphi'(0) - \varphi(t), \tag{24}$$

which is valid for every $\varphi \in \mathbf{C}^2([0,T];X)$.
Using formula (24) for $\varphi(t) = \Phi(t)p$ and taking into account the fact that $\mathcal{D}^{-1} \in \mathcal{L}(X)$, we deduce the equation

$$p - B_1 p = h_1,$$

where

$$B_1 = \mathcal{D}^{-1}(S(T)\Phi'(0) + \int_0^T S(T-s)[\Phi''(s) - \lambda\Phi(s)]\, ds),$$

$$h_1 = -\mathcal{D}^{-1}h.$$

Since the sine function $S(t)$ is compact, the operator B_1 is compact. From the Fredholm theorem we obtain the conclusions of Theorem 2.

3. EQUATION OF THE SECOND ORDER. ELLIPTIC CASE

Let us consider the following inverse problem

$$u''(t) = Au(t) + f(t), \quad 0 \le t \le T, \tag{25}$$

$$u(0) = u_0, \quad u(T) = u_2, \tag{26}$$

$$f(t) = \Phi(t)p + F(t), \quad 0 \le t \le T, \tag{27}$$

$$u'(0) = u_1. \tag{28}$$

A solution of this problem is a pair (u,p), where $u \in \mathbf{C}^2([0,T];X) \cap \mathbf{C}([0,T];\mathcal{D}(A))$ and $p \in X$.

Suppose that A is weakly positive, i.e., there exists $c > 0$ such that for every $\lambda \ge 0$ we have $(A+\lambda I)^{-1} \in \mathcal{L}(X)$ and

$$\|(A+\lambda I)^{-1}\| \le c/(1+\lambda).$$

It is well known that if the operator A is weakly positive, then problem (25)–(26) is well-posed. In this case the fractional powers of A are well defined and $-A^{1/2}$ is the infinitesimal generator of a contracting C_0-semigroup $V(t)$. The Green's function of problem (25)–(26) is given by

$$G(t,s) = \frac{1}{2}A^{-\frac{1}{2}}(I - V(2T))^{-1}[V(t+s) + V(2T-t-s) - V(|t-s|) - V(2T-|t-s|)].$$

If $u_0, u_2 \in \mathcal{D}(A)$ and $f \in \mathbf{C}^\alpha([0,T];X) + \mathbf{C}([0,T];\mathcal{D}(A^{1/2}))$ $(0 < \alpha < 1)$,then the solution of problem (25)–(26) is defined by the formula

$$u(t) = V(t)a + V(T-t)b + \int_0^T G(t,s)f(s)\, ds, \tag{29}$$

where

$$a = (I - V(2T))^{-1}(u_0 - V(T)u_2),$$
$$b = (I - V(2T))^{-1}(u_2 - V(T)u_0).$$

Let us consider inverse problem (25)–(28). Now the admissible data are defined as

$$u_0, u_2 \in \mathcal{D}(A), \quad u_1 \in \mathcal{D}(A^{1/2}),$$

$$F \in \mathbf{C}^{\alpha}([0,T];X) + \mathbf{C}([0,T];\mathcal{D}(A^{1/2})), \quad 0 < \alpha < 1.$$

We define

$$g = u_1 - A^{1/2}a - V(T)A^{1/2}b - \int_0^T G_t(0,s)F(s)\,ds, \tag{30}$$

$$h = A^{1/2}g \tag{31}$$

(g is an arbitrary element of $\mathcal{D}(A^{1/2})$ and h is an arbitrary element of X).

THEOREM 3. *Let the semigroup $V(t)$ be compact, $\Phi \in \mathbf{C}^1([0,T];\ \mathcal{L}(X))$, $\Phi(0)^{-1} \in \mathcal{L}(X)$. Then the Fredholm property holds:*
1) *problem (25)–(28) is solvable for every admissible data if and only if only the trivial solution exists for the trivial admissible data;*
2) *the space of all solutions of problem (25)–(28) for the trivial data is finite dimensional;*
3) *three exist $l_1, \ldots, l_n \in X^*$ such that problem (25)–(28) is solvable if and only if*

$$l_1(h) = \ldots = l_n(h) = 0.$$

Proof. Carrying out computations similar to those used to prove the previous theorems, we can derive the equation

$$p - B_1 p = h_1,$$

where

$$B_1 = \Phi(0)^{-1}(I + V(2T))^{-1}(2V(T)\Phi(T) - \int_0^T [V(2T-s) + V(s)]\Phi'(s)\,ds).$$

The operator B_1 is compact and the conclusions of Theorem 3 follow from the Fredholm theorem.

REFERENCES

Eidelman, J.S. (1983). Two-points bundary value problem for differential equations with a parameter. *Dokl. Akad. Nauk Ukr.SSR, Ser. A.* **4**, 15 18 (in Russian).

Eidelman, J.S. (1990). Solvability conditions of the inverse problems for evolution equations. *Dokl. Akad. Nauk Ukr.SSR, Ser. A.* **7**, 28–31 (in Russian).

Iskenderov, A.D. (1976). Some inverse problems of determining the right sides for differential equations. *Dokl. Akad. Nauk Azerb.SSR.* **2**, 35–44 (in Russian).

Lavrent'ev, M.M., Romanov, V.G., and Shishatskii, G.P. (1980). *Ill-Posed Problems of Mathematical Physics and Analysis*. Nauka, Moscow (in Russian).

Orlovskii, D.G. (1989). On one inverse problem for differential equation of the second order in Banach space. *Different. Equat.* **25**, 1000–1009 (in Russian).

Orlovskii, D.G. (1990). One the problem of determining a parameter of evolution equation. *Different. Equat.* **26**, 1614–1621 (in Russian).

Prilepko, A.I. (1973). Inverse problems of the potential theory. *Math. Notes.* **14**, 755–767 (in Russian).

Prilepko, A.T. and Orlovskii, D.G. (1989). On a semigroup approach to the problem of determining the nonhomogeneous term in evolution equation. *Dokl. Akad. Nauk. SSSR.* **305**, 1045–1049 (in Russian).

Rundell, W. (1980). Determination of an unknown nonhomogeneous term in a linear partial differential equation from overspecified boundary data. *Appl. Anal.* **10**, 231–242.

Tikhonov, A.N. and Arsenin, V.Y. (1979). *Methods of Solving Ill-Posed Problems.* Nauka, Moscow (in Russian).

Ill-Posed Problems in Natural Sciences, pp. 375 – 378
A. Tikhonov (Ed.)
1992 VSP/TVP

REGULARIZED TRACES OF SINGULAR DIFFERENTIAL OPERATORS OF HIGHER ORDERS

A.S. PECHENTSOV
Moscow State University, Len. Gory, 119899 Moscow, Russia

ABSTRACT

In space $L_2[o,\infty)$ we consider a class of self-adjoint operators generated by the expression

$$l(y) \equiv (-1)^n d^{2n}y(x)/dx^{2n} + xy(x).$$

For such operators regularized traces of all orders, i.e., the series

$$\sum_{k=1}^{\infty}[\lambda_k^m - A_m(k)]$$

are calculated, where λ_k are eigenvalues of the operator, $A_m(k)$ are fixed numbers depending on k and providing convergency of series, $m \in \mathbb{N}$. Regularized traces can be used for approximate calculation of first eigenvalues of the operator in question.

Consider the regular Sturm–Liouville problem:

$$\frac{d^2y}{dx^2} + g(x)y = \lambda y,$$

$$y(0) = y(\pi) = 0.$$

If the function $g(x) \in C^1[0,\pi]$, then the eigenvalues λ_k, $k = 1,2,3,\ldots$, of this problem satisfy the following:

$$\lambda_k = k^2 + c_0 + O\left(\frac{1}{k^2}\right), \quad k \to \infty, \quad c_0 = \frac{1}{\pi}\int_0^{\pi} g(x)\,dx.$$

Consequently, the series $\sum_{k=1}^{\infty}[\lambda_k - k^2 - c_0]$ converges.
Firstly (Gel'fand and Levitan, 1953) calculated the sum of this series:

$$\sum_{k=1}^{\infty}[\lambda_k - k^2 - c_0] = \frac{1}{2}c_0 - \frac{g(0)+g(\pi)}{4}.$$

In a large number of papers that have been published since 1953 regularized traces for particular differential operators (specially of the second order) have been calculated.

Then (Lidskii and Sadovnichii, 1967) suggested a method of obtaining regularized sums of zeros for the class K of entire functions. Eigenvalues of general boundary problems generated by ordinary differential expression on a finite interval are zeros of an entire function of the class K. Therefore the theory of regularized traces of eigenvalue problems on a finite interval is practically complete at the present time.

Let us consider in space $\mathbf{L}_2[0,\infty)$ the operator T generated by the expression

$$l(y) \equiv (-1)^n d^{2n} y(x)/dx^{2n} + xy(x),$$

and boundary conditions at the point $x = 0$ making it self-adjoint.

The characteristic determinant for operator T is not a function of class K: $\Delta(\lambda) \notin K$, $\Delta(\lambda_k) = 0 \Leftrightarrow \lambda_k$ are eigenvalues of T. In the present paper we extend the method of (Lidskii and Sadovnichii, 1967) on a class of entire functions associated with the operator T.

Without loss of generality, consider a homogeneous boundary conditions at $x = 0$:

$$y(0) = y'(0) = \ldots = y^{(n-1)} = 0.$$

Then in the sector $S = \{\lambda \in \mathrm{C}\colon -3\pi/2(2n+1) \le \arg\lambda \le \pi/2(2n+1)\}$ the characteristic determinant $\Delta(\lambda)$ has the asymptotic expansion for $\lambda \to \infty$ (see (Pechentsov, 1991)):

$$\Delta(\lambda) \sim \lambda^{-n/4} \exp\left(\alpha\lambda^{(2n+1)/2n}\right) \sum_{s=0}^{\infty} \gamma_{1s} \lambda^{-(2n+1)s/2n}$$

$$+\lambda^{-n/4} \exp\left(\bar{\alpha}\lambda^{(2n+1)/2n}\right) \sum_{s=0}^{\infty} \gamma_{2s} \lambda^{-(2n+1)s/2n}, \quad (1)$$

where the coefficients γ_{is}, $i = 1, 2$, are calculated successively by recurrent formulas.

Substituting the above asymptotic expansion in the equation $\Delta(\lambda) = 0$ and applying the method of iteration (see (Olver, 1974)) we get asymptotic series for eigenvalues:

$$\lambda_k \sim ((2n+1)\pi k/2n)^{2n/(2n+1)} \left(1 + \sum_{s=1}^{\infty} \chi_s k^{-s}\right), \quad k \to \infty.$$

The coefficients χ_s are calculated successively by recurrent formulas and $\chi_1 = n(n-2)/2(2n+1)$.

Raising the asymptotic expansion for eigenvalues λ_k to the power $(-\sigma)$, where σ is an arbitrary complex number, we get

$$\lambda_k^{-\sigma} \sim \sum_{s=0}^{\infty} \xi_s k^{-2n\sigma/(2n+1)-s}. \quad (2)$$

In the plane λ we consider the ray l from the origin that forms the angle φ: $0 < \varphi < \pi/2(2n+1)$ with the positive part of the real axis. The ray l is chosen so, that on l there are no zeros of $\Delta(\lambda)$.
One adden in the asymptotic expression (1) of $\Delta(\lambda)$ is exponentially small for $\lambda \in l$, $\lambda \to \infty$, in comparison with another one.
Therefore, we have the following asymptotic expansion for $\lambda \in l$, $\lambda \to \infty$:

$$\frac{\Delta'(\lambda)}{\Delta(\lambda)} \sim \lambda^{1/2n} \sum_{s=0}^{\infty} \gamma_s \lambda^{-(2n+1)s/2n}, \tag{3}$$

where the coefficients γ_s can be expressed in terms of the parameters $\gamma_{1s}, \gamma_{2s}, s = 0, 1, \ldots$, of the asymptotic expression for $\Delta(\lambda)$ and $\gamma_0 = 2i/\big(\exp(\pi i/n) - 1\big)$.
Introduce the Zeta function associated with $\Delta(\lambda)$ by the integral

$$Z(\sigma) = \frac{1}{2\pi i} \int_{\Gamma} \lambda^{-\sigma} \frac{\Delta'(\lambda)}{\Delta(\lambda)}\, d\lambda, \quad \operatorname{Re} \sigma > 1 + \frac{1}{2n}, \tag{4}$$

where the contour Γ is chosen on the Riemann surface $\lambda^{1/2n}$ with a slit along the ray l. The contour Γ is composed by: a) the ray l lying on the first sheet of the Riemann surface; b) a curve γ consisting of $2n$-fold circuit around zero with sufficiently small radius; c) ray l lying on the $2n$-th sheet and directed in the opposite direction (see Fig. 1 for the case n=1).

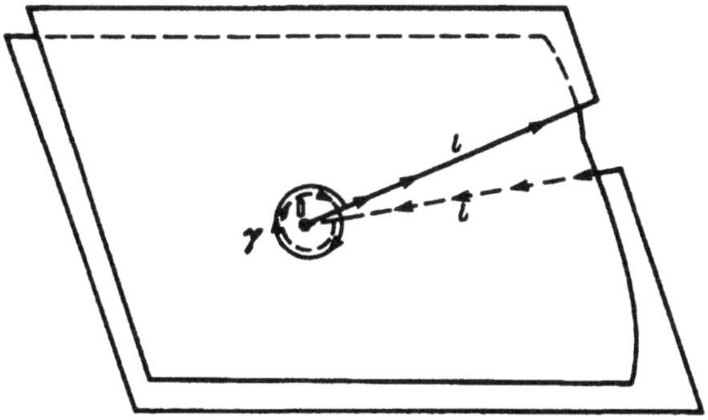

Fig. 1. The contour Γ on Riemann surface $\sqrt{\lambda}$

Function $Z(\sigma)$ is entire function and for $\operatorname{Re} \sigma > 1 + 1/2n$ the following correlation is valid (see (Pechentsov, 1991)):

$$\sum_{k=1}^{\infty} \lambda_k^{-\sigma} = \frac{Z(\sigma)}{\Theta(\sigma)}, \qquad \text{where } \Theta(\sigma) = \frac{\exp(-4n\pi\sigma i) - 1}{\exp(-2\pi\sigma i) - 1}. \tag{5}$$

Let τ be a sufficiently large integer. We introduce the function

$$\Psi_\tau(\sigma) = \sum_{k=1}^{\infty}{}' \left[\lambda_k^{-\sigma} - \sum_{s=0}^{\tau} \xi_s k^{-2n\sigma/(2n+1)-s}\right]$$

and the prime on the summation sign is connected with the regularization defect (see (Lidskii and Sadovnichii, 1967)).
From (2) it follows that the function $\Psi_\tau(\sigma)$ admit analytic extension into half-plane Re$\sigma > -\tau - \tau/2n$. On the other hand, applying (5) for Re$\sigma > 1 + 1/2n$, we have

$$\Psi_\tau(\sigma) = \frac{Z(\sigma)}{\Theta(\sigma)} - \sum_{s=0}^{\tau} \xi_s \zeta\Big(\frac{2n\sigma}{2n+1} + s\Big),$$

where $\zeta(t)$ is the Riemann zeta function. The values of the analytic extension of $Z(\sigma)$ at points $\sigma = -1, -2, \ldots$ are obtained in (Pechentsov, 1991). Thus we have the following theorem.

THEOREM . *For any integer τ and any integer m such that $0 \le m < (2n+1)\tau/2n$, the following equalities hold*

$$\sum_{k=1}^{\infty}{}' \Big[\lambda_k^m - \big((2n+1)\pi k/2n\big)^{2nm/(2n+1)} - \sum_{s=1}^{\tau} \xi_s k^{2nm/(2n+1)-s}\Big]$$

$$= \gamma_{1+2nm/(2n+1)} - \sum_{s=0}^{\tau} \xi_s \zeta\Big(s - \frac{2nm}{2n+1}\Big),$$

where $\gamma_{1+2nm/(2n+1)} = \gamma_{1+2nl}$ if $m = (2n+1)l$, $l \in \mathbf{N}$ (see (3)), and $\gamma_{1+2nm/(2n+1)} = 0$ if $m \ne (2n+1)l$, the coefficients ξ_s are calculated successively by recurrent formulas (see (2)) and $\xi_0 = \big((2n+1)\pi/2n\big)^{2nm/(2n+1)}$.

Remark. Calculating regularized sums $\lambda_k^{p/2n}$, $k = 1, 2, 3, \ldots$, where $p \in \mathbf{N}$ we must know values of the analytic extension of $Z'(\sigma)$ at points $\sigma = -p/2n$. In this case regularized sums of $\lambda_k^{p/2n}$ can be expressed in terms of integrals of

$$\mathrm{Sp}\Big[T^{(2n+1)/2n}\big(\omega - T^{1/2n}\big)\Big]^{-1}.$$

Acknowledgement
The author expresses his thanks to V.A. Sadovnichii for the discussion of the work.

REFERENCES

Gel'fand, I. and Levitan, B. (1953). On a simple identity for the characteristic values a differential operator of the second order. *Dokl. Akad. Nauk SSSR* **88**, 593–596 (in Russian).

Lidskii, V. and Sadovnichii, V. (1967). Regularized sums of roots of a certain class of an entire function. *Funct. Anal. Appl.* **1**, 52–59 (in Russian).

Olver, F. (1974). *Asymptotics and Special Functions.* Academic Press, New York, London.

Pechentsov, A. (1991). The traces of a certain class of singular differential operators. *Trudy seminara im I.G. Petrovskogo* **15**, 221–231 (in Russian).

Ill-Posed Problems in Natural Sciences, pp. 379 – 389
A. Tikhonov (Ed.)
1992 VSP/TVP

INVERSE PROBLEMS FOR EVOLUTION EQUATIONS

A.I. PRILEPKO, A.B. KOSTIN, and I.V. TIKHONOV
Department of Mathematics Moscow Engineering Physics Institute, Kashirskoe Shosse, 31, 115409 Moscow, Russia

ABSTRACT

This paper consists of two parts. The first part contains a brief review of results obtained by A.I. Prilepko and A.B. Kostin, while the second part describes the results of the joint research made by A.I. Prilepko and I.V. Tikhonov.
The numeration of paragraphs, items, theorems and formulae in the first part is independent on that in the second.

PART I. PARABOLIC INVERSE PROBLEMS

§1. *Formulation of inverse problems. Basic results*
Let Ω be a bounded simply connected domain in $\mathbf{R}^n$ with the boundary $\partial\Omega$ of the class C^2.
In the cylinder $Q = \Omega\times(0,T)$ (where $T > 0$) with the lateral surface $S = \partial\Omega\times[0,T]$ we consider the following inverse problems.

PROBLEM I (linear case): *Find a pair of functions* $\{u(x,t), f(x)\}$ *satisfying the following system*

$$u_t - Lu = h\cdot f + g \quad \text{in } Q, \tag{1}$$

$$u(x,0) = a(x) \quad \text{on } \Omega, \tag{2}$$

$$\mathcal{B}u = b(x,t) \quad \text{on } S, \tag{3}$$

$$lu = \chi(x) \quad \text{on } \Omega \tag{4}$$

if all the functions (besides u and f) and operators of (1) – (4) *are known.*

PROBLEM II (nonlinear case): *Find a pair of functions* $\{u(x,t), c(x)\}$ *satisfying the equation*

$$u_t - Lu = c\cdot u + g \qquad \text{in } Q, \tag{5}$$

initial condition (2) *with* $a(x) = 0$ *and conditions* (3)–(4), *if all the other functions (besides u and c) and operators are known.*

Here, the operator L is supposed to be uniformly eliptic:

$$Lu = \sum_{i,j=1}^{n} \frac{\partial}{\partial x_i}(a_{ij}(x)\frac{\partial}{\partial x_j}u) + \sum_{i=1}^{n} b_i(x)\frac{\partial u}{\partial x_i} + c_0(x)\cdot u,$$

with real-valued coefficients $a_{ij} \in C^1(\overline{\Omega})$; $b_i, \frac{\partial}{\partial x_i} b_i, c_0 \in L_\infty(\Omega)$ so that $a_{ij}(x) = a_{ji}(x)$ and

$$0 < \nu \leq \sum_{i,j=1}^{n} a_{ij}(x)\xi_i\xi_j \leq \mu$$

for any $x \in \Omega$ and $\xi \in \mathbf{R}^n, |\xi| = 1\left(|\xi|^2 = (\xi,\xi) = \sum_{i=1}^n \xi_i^2\right)$.
The operator $\mathcal{B}$ in (3) can be introduced in one of the two possible forms: either $\mathcal{B}u = u$ or $\mathcal{B}u = \frac{\partial u}{\partial N} + \sigma(x) \cdot u$.
Here, N is a co-normal vector, $\frac{\partial u}{\partial N} \equiv \sum_{i,j=1}^n a_{ij}\frac{\partial u}{\partial x_j}\cos(\widehat{\overrightarrow{n}, \overrightarrow{ox_i}})$, $\overrightarrow{n}$ is the external normal with respect to $\partial\Omega, \sigma \in C(\partial\Omega)$.
The expression lu in (4) has one of the following two forms: either $lu = u(x,t_1)$, $x \in \Omega$, $0 < t_1 \leq T$, or $lu = \int_0^T u(x,\tau)\omega(\tau)\,d\tau$, $x \in \Omega$, where t_1 is fixed, ω is a given function belonging to $L_2(0,T)$.
We shall call these restrictions on the function $u(x,t)$ as conditions of observation of the final and integral type respectively. The norm in the Banach space $L_p(\Omega)(1 \leq p \leq \infty)$ is denoted by $\|\cdot\|_{p,\Omega}$. The Sobolev spaces $W_p^l(\Omega)$, $W_p^{2,1}(Q)(l \in \mathcal{N}, 1 \leq p < \infty)$ are introduced in the customary fashion and the norms in these spaces are defined as

$$\|u\|_{p,\Omega}^{(l)} = \Big(\sum_{|\alpha|\leq l} \|\mathbf{D}^\alpha u\|_{p,\Omega}^p\Big)^{1/p},$$

$$\|v\|_{p,Q}^{(2,1)} = \left(\|v_t\|_{p,Q}^p + \int_0^T (\|v(\cdot,t)\|_{p,\Omega}^{(2)})^p dt\right)^{1/p},$$

Here, $\alpha = (\alpha_1, \ldots, \alpha_n)$ is a multi-index,

$$|\alpha| = \alpha_1 + \ldots + \alpha_n, \qquad \alpha_i \in \mathbf{Z}_+, \qquad i = 1, \ldots, n,$$

$$\mathbf{D}^\alpha u = \partial^{|\alpha|}u/\partial x_1^{\alpha_1}\ldots\partial x_n^{\alpha_n}, \qquad \mathbf{D}^0 u = u.$$

By $V_2^{1,0}(Q)$ we denote the Banach space of functions whose norm $\|v\| = \max_{[0,T]}\|v(\cdot,t)\|_{2,\Omega} + \||v_x|\|_{2,Q}$ is finite, where $v_x = (v_{x_1}, \ldots, v_{x_n})$.
The set of measurable functions $u(x,t)$ with finite integral $\int_0^T \|u(\cdot,t)\|_{q,\Omega}^r\,dt$ $(1 \leq q \leq \infty, 1 \leq r < \infty)$ is denoted by $L_{q,r}(Q)$.
All the equalities and inequalities in the text of **Part I** are understood to be valid almost everywhere with respect to the Lebesque measure.
We will use also the sets $E \equiv L_\infty(\Omega)$ and

$$E_- = \{c \in E\colon\ c(x) \leq 0 \text{ in } \Omega\}.$$

Definition 1. A pair of functions $\{u(x,t), f(x)\}$ is a generalized solution of Inverse Problem I if $u \in W_2^{2,1}(Q), f \in L_2(\Omega)$ and (1)–(4) are satisfied.

Definition 2. A pair of functions $\{u(x,t), c(x)\}$ is a generalized solution of Inverse Problem II, if $u \in W_2^{2,1}(Q), c \in E_-$ and equalities (5), (2) with $a(x) \equiv 0$, (3) and (4) are valid.
Any extension of the function $b(x,t)$ in (3) to the cylinder Q is denoted by the same symbol.
The following conditions play an important role in **Part I**:

$$c_0(x) \leq 0 \text{ on } \Omega, \qquad \sigma(x) \geq 0 \text{ on } \partial\Omega,$$

$$\omega(t) \geq 0 \text{ on } [0,T], \tag{6}$$

$$g(x,t), g_t(x,t) \geq 0 \text{ in } Q; \qquad b(x,t) \geq 0,$$

$$b_t(x,t) \geq 0 \text{ on } S \tag{7}$$

Problem I for $g \equiv 0, a \equiv 0$ and $b \equiv 0$ will be denoted as Problem I_0.

Remark 1. Suppose that $g, g_t \in L_2(Q); a, \chi \in W_2^2(\Omega); b, b_t \in W_2^{2,1}(Q)$. Let the compatibility conditions $\mathcal{B}a = b(x,0), \mathcal{B}\chi = lb$ on $\partial\Omega$ be satisfied. Without loss of generality we can assume for Problem I that $g \equiv 0, a \equiv 0, b \equiv 0$ and $\chi \in W_2^2(\Omega)$ is such that $\mathcal{B}\chi = 0$ on $\partial\Omega$. This conclusion is based on the fact that $u_0(x,t)$ exists and is unique as a solution of (1)–(3) (with $f \equiv 0$) and has the necessary differential properties.
And so a pair of functions $\{u(x,t), f(x)\}$ is a solution of Problem I if and only if $\{u - u_0, f\}$ is a solution of Problem I_0 where $\chi - lu_0$ is substituted for χ .
Therefore, to analize Problem I it is sufficient to investigate Problem I_0.
The following theorems are true.

THEOREM 1. *If $h, h_t \in L_{\infty,2}(Q); |lh| \geq \delta > 0$ in Ω and $\chi \in W_2^2(\Omega)$ is such that $\mathcal{B}\chi = 0$ on $\partial\Omega$. Then Problem I_0 has a Fredholm character.*

Let $r = 2 \cdot p/(p-2)$ correspond to $p > 2$ and $r = \infty$ correspond to $p = 2$.

THEOREM 2. *Let $h, h_t \in L_{r,2}(Q); h(x,t) \geq 0, h_t(x,t) \geq 0$ in $Q, l(h) > 0$ in Ω and (6) hold.*
If functions $u \in W_2^{2,1}(Q)$ and $f \in L_p(\Omega)$, $p \geq 2$ satisfy (1) – (4) with $g \equiv 0, a \equiv 0, b \equiv 0, \chi \equiv 0$, then $u \equiv 0, f \equiv 0$.

Theorems 1 and 2 imply the following basic result.

THEOREM 3. *Let $h, h_t \in L_{\infty,2}(Q)$; $g, g_t \in L_2(Q)$; $a, \chi \in W_2^2(\Omega)$; $b, b_t \in W_2^{2,1}(Q)$ and (6) be valid. Let the compatibility conditions $\mathcal{B}a = b(x,0), \mathcal{B}\chi = lb$ be true when $x \in \partial\Omega$.*
In addition let us assume that $h(x,t) \geq 0$, $h_t(x,t) \geq 0$ in $Q, l(h) > \delta > 0$ in Ω.
Then Problem I is well-posed, i.e., its has a unique solution and the following stability estimate holds:

$$\|u - u_0\|_{2,Q}^{(2,1)} + \|f\|_{2,\Omega} \leq C \cdot \|\chi - lu_0\|_{2,\Omega}^{(2)}.$$

The proofs of Theorems 1, 2 and 3 for the heat equation and overdetermination $lu = u(x,t_1)$ can be found in (Prilepko and Kostin, 1991*a*). Theorems 1, 2 and 3 formulated in the present paper can be obtained by the same method and their proofs are qiven in (Prilepko and Kostin, 1991*b*). Section 2 deals with Problem II and contains proofs of the following theorems.

THEOREM 4. *Let $g, g_t \in L_2(Q)$; $b, b_t \in W_2^{2,1}(Q)$; $\chi(x) > 0$ on Ω and conditions (6), (7) be satisfied. Then a solution of Problem II is unique.*

We denote by $u^0(x,t)$ the solution of problem (5), (2), (3) with $c(x) \equiv 0, a(x) \equiv 0$ in Ω.

THEOREM 5. *Let $g, b, b_t \in W_p^{2,1}(Q), p > n; L\chi \in E$, $\chi(x) \geq \delta > 0$ on Ω and conditions (6), (7) be satisfied. Let the compatibility conditions $b(x,0) = 0, \mathcal{B}\chi = lb$, $\mathcal{B}g(x,0) = b_t(x,0)$ be valid on $\partial\Omega$. Suppose that $L[lu^0 - \chi] \leq 0$ in Ω. If the relations $\mathcal{B}u = \frac{\partial u}{\partial N}$, $c_0(x) \equiv 0$ and $L[lu^0 - \chi] \equiv 0$ in Ω are valid simultaneously than we require in addition that the equality $(lu^0 - \chi)(x_0) = 0$ holds at least at one point $x_0 \in \overline{\Omega}$. Then there exists a unique solution $u \in W_p^{2,1}(Q)$, $c \in E_-$ of Problem II.*

We study inverse Problems I and II because coefficients of the equations describing properties and characteristics of the medium under consideration are actually unknown.

We give in addition two interpretations of Problems I and II which have various applications.

We suggest to consider these problems as special problems of control. Namely, we have to find a control function $f(x)$ (or $c(x)$) in order to transform the system from the initial state $u(x,0)$ to the given state lu.

On the other hand, these problems can be interpreted as problems of behaviour prediction when one knows only a part of information about the properties of the system with distributed parameters (it means for Problem I, for example, that one can represent a source function in the form $h(x,t)f(x)$ where $h(x,t)$ is given and $f(x)$ is unknown). In this case prediction means that using additional information about the solution we seek at $t = t_1$ we find the value of the solution at any $t \in (0,T]$.

All the interpretations given above can be applied also to the problems considered in **Part II.**

Now, we dwell upon a brief review of research articles connected with the subject of the present paper.

Cauchy problem and boundary value problems for parabolic type equations with inverse time directions were considered by many authors (see(Tikhonov, 1935; Krein, 1957; Agmon and Nirenberg, 1967; Ghidaglia, 1986)). In some cases these problems are ill-posed (see(Tikhonov and Arsenin, 1979; Lavrent'ev et al., 1980; Ivanov et al., 1978)).

We note that I.G. Petrovskii and V.P. Michailov (see(Petrovskii, 1934; Michailov, 1963)) considered the Dirichlet problem for parabolic equation, when right-part function $F(x,t)$ and Dirichlet data on the boundary ∂Q of the domain Q are given. In particular, it was shown that this problem is ill-posed if Q is a cylinder.

For the first time the right-part function of the form $h(x,t) \cdot f(x)$ can be found in (Prilepko 1966, 1967, 1973), where elliptic equations were examined. The uniqueness results were obtained under the conditions $h > 0, h_t \geq 0$.

The same restrictions on $h(x,t)$ and overdetermination $lu = u(x,T)$ were used in parabolic equation inverse problems by V.M. Isakov (see (Isakov, 1982*a*)), where uniqueness results for Problems I and II were announced.
In (Isakov, 1982*b*) a theorem of classical solvability for Problem II was formulated. Problems I and II with Dirichlet data on the boundary ∂Q were examined in (Prilepko and Soloviev, 1987*a*, *b*), where theorems of existence and uniqueness of classical solutions were proved. In this work for the first time one can find a Fredholm type result for inverse Problem I. For some evolution equations a Fredholm character of Problem I with $lu = u(x,T)$ was proved in (Orlovskii, 1988, 1990).
In (Isakov, 1991) one can find a proof of existence and uniqueness theorem of classical solution for Problem I with $lu = u(x,T)$. This proof uses the method elaborated in (Prilepko and Soloviev, 1987*a*). In (Isakov, 1991) an existence theorem for Problem II is proved by the method of monotone sequences.
In (Prilepko and Kostin, 1991*a*, *b*) theorems are proved for generalized solutions. The inverse problem with integral overdetermination considered in this paper has not been investigated earlier. To show the uniqueness of linear problem generalized solutions we use the positivity method (see(Prilepko and Kostin, 1991*a*)). Earlier this method of positivity was offered in (Prilepko and Soloviev, 1987*a*) for classical solutions of inverse problems.
The results of V.V. Soloviev (see (Soloviev, 1989)) are also devoted to the inverse problems in question.
It should be noted that the inverse problem of simultaneous identification of the temperature u and coefficient c has a nonlinear character. That is why it is dificult to get sufficient conditions of solvability without restrictions on norms of the given functions and diameter of the domain.
The investigation of solvability becomes easier when one takes into account the property of positivity of the solution under the positive input data and uses the property of the solution monotonicity with respect to the coefficient under identification.
Developing method of monotonicity (see (Isakov, 1982; Prilepko and Kostin, 1991*a*)) we prove the existence theorem of a generalized solution for Problem II applying the theory of isotonic operators and Birkhoff-Tarskii theorem.

§2. *Proofs of Theotrems 4 and 5*

2.1. *Proof of Theorem* 4.

Let two distinct pairs of functions (u, c) and (u_1, c_1) be solutions of Problem II. Then the functions $w = u_1 - u$ and $f = c_1 - c$ satisfy the equalities:

$$w_t - Lw - c_1 w = u \cdot f \qquad \text{in } Q,$$

$$w(x,0) = 0 \qquad \text{on } \Omega,$$

$$\mathcal{B}w = 0 \qquad \text{on } S,$$

$$lw = 0 \qquad \text{on } \Omega.$$

By the virtue of Theorem 4 we can write that $u \geq 0, u_t \geq 0$ in Q and $lu = \chi > 0$ in Ω.

Applying Theorem 2 we obtain $w \equiv 0, f \equiv 0$. This proves the theorem.
2.2. *Prof of Theorem* 5.
Let $\tilde{c}(x) = c(x)\chi(x)$. Since $\chi(x) \geq \delta > 0$ in $\Omega, \chi \in E$, then $\tilde{c} \in E$ only if $c \in E$.
Let $\tilde{c} \in E$ be an arbitrary fixed function. We introduce $u(x,t;\tilde{c})$ as a solution of the following problem:

$$u_t - Lu = \tilde{c} \cdot \chi^{-1} u + g \qquad \text{in } Q, \tag{8}$$

$$u(x,0) = 0 \qquad \text{on } \Omega, \tag{9}$$

$$\mathcal{B}u = b(x,t) \qquad \text{on } S. \tag{10}$$

According to the information given in the theorem we derive that $v = u_t$ is a solution of the problem:

$$v_t - Lv = \tilde{c}\chi^{-1} v + g_t \qquad \text{in } Q,$$

$$v(x,0) = g(x,0) \qquad \text{on } \Omega,$$

$$\mathcal{B}v = b_t(x,t) \qquad \text{on } S.$$

Therefore, $u_t \in W_p^{2,1}(Q),\ lu_t \in E,\ l(Lu) = L(lu) \in E$.
Consequently, we can define a single-valued nonlinear operator A: $E \to E$ as follows $A\tilde{c} = l(u_t) - L\chi - lg$. The range of definition for A is $\mathbf{D}(A) = E$.
Consider the operator equation

$$A\tilde{c} = \tilde{c}. \tag{11}$$

Now we show that the solvability of Problem II follows from the solvability of equation (11) on E_-.
Let $\tilde{c} \in E_-$ be a solution of (11). Considering $u(x,t;\tilde{c})$ we see that this function satisfies the equality

$$l(u_t) - L(lu) = \tilde{c}\chi^{-1} lu + lu + lg \qquad \text{in } \Omega \tag{12}$$

To show that $l(u) = \chi(x)$ on Ω we subtract (11) from (12) and use the definition of the operator A. According to the compatibility conditions

$$L[lu - \chi] + \tilde{c}\chi^{-1}[lu - \chi] = 0 \qquad \text{in } \Omega,$$

$$\mathcal{B}[lu - \chi] = 0 \qquad \text{on } \partial\Omega,$$

and the inequalities $\tilde{c}\chi^{-1} \leq 0, c_0 \leq 0$ for $x \in \Omega$ we find that $lu = \chi(x), x \in \Omega$.
Therefore, the pair $(u, \tilde{c}\chi^{-1})$ satisfies (5), (2)–(4) with $a(x) \equiv 0$.
Now, we show that under the assumptions of Theorem 5 a solution of (11) exists and belongs to E_-.
To prove this we begin with demonstrating that A is an isotonic on E operator, i.e., $Ac_1 \leq Ac_2$ for any $c_1, c_2 \in E$ satisfying the inequality $c_1 \leq c_2$.
Note that $E(E_-)$ is a partially ordered set which is complete with respect to the operations sup and inf, i.e., each bounded with respect to the order set in $E(E_-)$ posesses the least upper and greatest lower bounds in $E(E_-)$. Sometimes this type of sets is defined as conditionally complete set S (see (Birkhoff, 1967)).

Let us denote $u^i(x,t) = u(x,t;c_i)$ $i = 1,2$; $c = c_2 - c_1$, $u = u^2 - u^1$. If $c_1 \le c_2$, then $c \ge 0$ in Ω and

$$u_t - Lu - c_2\chi^{-1}u = c\chi^{-1}u^1 \qquad \text{in } Q,$$

$$u(x,0) = 0 \qquad \text{on } \Omega, \qquad \mathcal{B}u = 0 \qquad \text{on } S.$$

Using the assumptions of the present theorem we see that $c\chi^{-1}u^1 \ge 0, c\chi^{-1}u^1_t \ge 0$ in Q. Consequently $Ac_2 - Ac_1 = l(u_t) \ge 0$. This proves that the operator A is isotonic.

The next step is made to show that there exists an ordinal segment which belongs to E_- and is maped by A into itself.

In fact, $A0 \equiv L[lu^0 - \chi] \le 0$ in Ω by assumptions. Therefore, A: $E_- \to E_-$. Since $A(-M) = l(u_t) - L\chi - lg, L\chi + lg \in E$, and $l(u_t) \ge 0$ for any $M \in \mathbf{R}^1$ then there exists $M > 0$ satisfying the inequality $A(-M) \ge -M$.

With this in view we find that the isotonic operator A maps the ordinal segment $[-M;0] = \{c \in E$: $-M \le c \le 0\}$ of conditionally complete, partially ordered set E_- into itself. The Birkhoff-Tarskii theorem (see (Birkhoff, 1967)) can therefore be invoked to conclude that A has on $[-M;0] \subset E_-$ at least one fixed point.

And so Problem II has a solution. The uniqueness of this solution follows from Theorem 4. This proves Theorem 5.

PART II. INVERSE PROBLEM FOR AN ABSTRACT EVOLUTION EQUATION IN A BANACH SPACE

§1. *Formulation of the problem. Basic results*

1.1. *Formulation of the problem.*

Consider the nonhomogeneous evolution equation in Banach space E.

$$u'(t) = Au(t) + \Phi(t)f \ (0 \le t \le T). \tag{1}$$

Here $T > 0$ is a real number, A is a closed linear operator from E to E with the range of definition $\mathbf{D}(A)$ which is dense in E. Assume, that A is a generator of C_0-semigroup $S(t), \Phi(t) \in \mathcal{L}(F;E)$ for $0 \le t \le T$ and F is a Banach space. In this part we investigate the problem of finding a function $u(t)$ and an element $f (f \in F)$ from equation (1) satisfying the following conditions

$$u(0) = \psi_0, \tag{2}$$

$$u(T) = \psi_1. \tag{3}$$

Here ψ_0, ψ_1 are fixed elements from $\mathbf{D}(A)$.

Definition. A pair $\{u(t), f\}$ is a solution of problem (1)–(3), if $u \in C^1([0,T];E)$ $u(t) \in \mathbf{D}(A)$ for $0 \le t \le T, f \in F$, and equalities (1)–(3) are true.

1.2. *Notations and basic assumptions.*

Let $(E, \ge)$ be a real partially ordered vector space with cone E_+, and let it be a vector lattice, i.e., for any pair $f, g \in E$ there exists a $\sup(f,g) \in E$. Denote

$f^+ = \sup(f, 0)$, $f^- = \sup(-f, 0)$, $|f| = f^+ + f^-$, then $f = f^+ - f^-$. Elements f and g are said to be disjoint if $\inf(|f|, |g|) = 0$. We write $f \perp g$ to denote that elements f and g are disjoint. Let E be a Banach space with a norm $\|\cdot\|$. If the inequality $|f| \leq |g|$ implies the inequality $\|f\| \leq \|g\|$, then we say that E is a Banach lattice. Let us consider two Banach lattices E and F. An operator $M \in \mathcal{L}(F; E)$ such that $MF_+ \subset E_+$ is called a positive operator and is denoted as $M \geq 0$. It is said that an operator $M \in \mathcal{L}(F; E)$ saves disjointness if $f \perp g$ implies $Mf \perp Mg$ $(f, g \in F)$. We assume that A is a generator of C_0-semigroup $S(t)$ of positive operators, i.e., $S(t)E_+ \subset E_+$ for any $t \geq 0$. Such a semigroup is called a positive semigroup. As usual $\rho(A)$ is the resolvent set of A, $\sigma(A) = \mathbf{C} \setminus \rho(A)$ is its spectrum. We denote by $s(A) = \sup\{Re\lambda\colon \lambda \in \sigma(A)\}$ the spectral boundary of A. It is known that positivity of $S(t)$ implies the inclusion

$$(\lambda I - A)^{-1} E_+ \subset E_+ \qquad \text{if } \lambda > s(A) \tag{4}$$

Note that $s(A) \in \sigma(A)$ if $\sigma(A)$ is nonempty. (See (Greiner et al.,1981)).

1.3. *Basic results.*

Section 2 contains the proofs of the following theorems.

THEOREM 1. *Let us assume, that A is a generator of a positive C_0-semigroup $S(t)$, and let $s(A) < 0$, $\Phi \in C^1([0, T]; \mathcal{L}(F, E))$ be an operator conserving the disjointness at any t and in addition, $\Phi(t) \geq 0$, $\Phi'(t) \geq 0$, $t \in [0, T]$ and* Ker $\Phi(T) =$ *$\{0\}$. Then a solution of problem (1) – (3) is unique.*

THEOREM 2. *Let us assume, that A is a generator of a positive compact (at $t > 0$) C_0-semigroup $S(t)$ and $s(A) < 0$.*
Let $\Phi(t) \in C^1([0, T]; \mathcal{L}(E))$, $\Phi(t) \geq 0$, $\Phi'(t) \geq 0$ at $t \in [0, T]$, $[\Phi(T)]^{-1} \in \mathcal{L}(E)$ and $[\Phi(T)]^{-1} \geq 0$.
Then for any $\psi_0, \psi_1 \in \mathbf{D}(A)$ problem (1) – (3) has a unique solution $(u(t), f)$ and the following stability estimates are true:

$$\|f\| \leq C(|||\psi_0||| + |||\psi_1|||), \qquad \max_{[0,T]} |||u(t)||| \leq C(|||\psi_0||| + |||\psi_1|||),$$

(here $|||\psi||| = \|\psi\| + \|A\psi\|$).

§2. *Proofs of Theorems 1 and 2*

2.1. *Proof of Theorem 1 (uniqueness).*

Let $\{u(t), f\}$ be a solution of (1)-(3) with $\psi_0 = 0, \psi_1 = 0$. Let us show that $u(t) \equiv 0, f = 0$. We represent f as $f = f^+ - f^-$ and consider the two Cauchy problems

$$\begin{cases} u_1'(t) &= Au_1 + \Phi(t) f^+ \\ u_1(0) &= 0, \end{cases} \qquad \begin{cases} u_2'(t) &= Au_2 + \Phi(t) f^- \\ u_2(0) &= 0 \end{cases} \tag{5}$$

A solution of problems (5) exists and is unique and the equality $f = f^+ - f^-$ implies $u(t) = u_1(t) - u_2(t)$ for $t \in [0, T]$. Since $u(T) = \psi_1 = 0$ then $u_1(T) = u_2(T) = \varphi$ where $\varphi \in \mathbf{D}(A)$.

Taking $t = T$ in (5) we have

$$\begin{cases} u_1'(T) &= A\varphi + \Phi(T)f^+ \\ u_2'(T) &= A\varphi + \Phi(T)f^- \end{cases} \tag{6}$$

Now it will be shown that (6) implies

$$A\varphi \geq 0. \tag{7}$$

The following formula is true for $u_1'(T)$

$$u_1'(T) = S(T)\Phi(0)f^+ + \int_0^T S(T-s)\Phi'(s)f^+ ds$$

By the assumptions of the present theorem we conclude that $u_1'(T) \geq 0$ and $u_2'(T) \geq 0$. Therefore, (6) implies $A\varphi + \Phi(T)f^+ \geq 0$, $A\varphi + \Phi(T)f^- \geq 0$ and besides this $-A\varphi \leq \inf(\Phi(T)f^+, \Phi(T)f^-)$. Since the elements $\Phi(T)f^+$ and $\Phi(T)f^-$ are non-negative and disjoint then $-A\varphi \leq 0$ and $A\varphi \geq 0$.
Since $s(A) < 0$ then $(-A)^{-1} \geq 0$ (see (4)). Therefore, $\varphi = A^{-1}(A\varphi) = -(-A^{-1})(A\varphi) \leq 0$. On the other hand, $\varphi = u_1(T) = \int_0^T p(s)\,ds$ where $p(s) = S(T-s)\Phi(s)f^+$ for $0 \leq s \leq T, p \in C([0,T]; E)$ and $p(s) \geq 0$ if $s \in [0,T]$.
By the virtue of the inequality $\varphi \leq 0$ we conclude that the integral of $p(s)$ is nonpositive, i.e., $p(s) \equiv 0$ on $[0,T]$.
But $p(T) = \Phi(T)f^+$ and Ker $\Phi(T) = \{0\}$. Consequently, $f^+ = 0$. By analogy we have $f^- = 0$. Finaly we obtain that $f = 0, u(t) \equiv 0$. This proves the theorem.
2.2. *Proof of Theorem* 2.
From the initial assumptions it is clear that inverse problem (1)–(3) is equivalent to the operator equation

$$f = Bf + h, \tag{8}$$

where $B \in \mathcal{L}(E), h \in E$ look as follows

$$Bf = [\Phi(T)]^{-1}\Big(S(T)\Phi(0)f + \int_0^T S(T-s)\Phi'(s)f\,ds\Big),$$

$$h = [\Phi(T)]^{-1}A\big(S(T)\psi_0 - \psi_1\big).$$

Since $S(t)$ is a compact semigroup for $t > 0$, then taking into account assumptions of the theorem we conclude that the operator B is compact. Equation (8) is a Fredholm type equation of the second kind (Orlovskii, 1990). In particular, uniqueness the inverse problem solution (Theorem 1) implies existence and stability of solution (8). Stability estimate follow from (8) and (1)–(2). This completes the proof.

References and remarks.
A set of the inverse problems formulation of the type concerned can be found in (Prilepko, 1973).
Problem (1)–(3) was investigated in (Iskenderov and Tagiev, 1979; Eidelman 1987, 1990; Rundell, 1980) for the case $\Phi(t) \equiv I$.
In (Orlovskii, 1990; Eidelman, 1990) one can find the case when $\Phi(t)$ is a scalar-valued function.
The general case of an operator function $\Phi(t)$ was considered in (Orlovskii, 1988, 1990), where Fredholm character of (1)–(3) was proved assuming that the semigroup is compact.
Under consideration the authors note in conclusion that in parallel with our research D.G. Orlovskii has got a result similar to Theorem 1. This result of D.G. Orlovskii was proved on the basis of other techniques and under additional restrictions on the operator A.

Remark 1. The results of Theorem 1 can be applied in particular, to the inverse problems for kinetic transport equation with final overdetermination (see (Prilepko and Tikhonov, 1991)).

Remark 2. Inverse problem (1), (2), (3) allows similar consideration if condition (3) has the form $l(u) = \psi_1$ where l is a definite operator.

REFERENCES

Agmon, S. and Nirenberg, L. (1967). Lower bounds and uniqueness theorems for solutions of differential equation in a Hilbert space. *Comm. Pure Appl. Math.* **20**, 207–229.

Beznotsenko, N.Y. (1974). On a determination of a coefficient in parabolic equation. *Different. Equat.* **10**, 24–35.

Birkhoff, G. (1967). *Lattice Theory.* Providence.

Cannon, J.R. (1963). Determination of an unknown coefficients in a parabolic differential equation. *Duke Math. J.* **30**, 313–323.

Cannon, J.R. and Duchateau, P. (1973). Determining unknowm coefficients in a nonlinear heat conduction problem. *SIAM J. Appl. Math.* **24**, 298–314.

Eidelman, Y.S. (1987). Uniqueness of inverse problem solution for a differential equation in a Banach space. *Different. Equat.* **23**, 1647–1649.

Eidelman, Y.S. (1990). Conditions of solvabily for inverse problems for evolution equation. *Dokl. Akad. Nauk Ukr. SSR.* **7**, 28–31 (in Russian).

Ghidaglia, J.-M. (1986). Some backward uniqueness results. *Nonlinear Anal. Theory. Method. Appl.* **10**, 777–790.

Greiner, G., Voight, J., and Wolff, M.P. (1981). On the spectral bound of the generator of semigroups of positive operators. *J. Oper. Theory.* **5**, 245–256.

Ivanov, V.K., Vasin, V.V., and Tanana, V.P. (1978). *Theory of Linear Ill-Possed Problems and Its Applications.* Nauka, Moscow (in Russian).

Isakov, V.M. (1982a). On a class of inverse problems for parabolic equations. *Dokl. Akad. Nauk SSSR.* **263**, 1296–1299 (in Russian).

Isakov, V.M. (1982b). An inverse problem for a parabolic equation. *Uspekhi Mat. Nauk.* **37**, 108–109 (in Russian).

Isakov, V.M. (1991). Inverse parabolic problems with the final overdetermination. *Comm. Pure Appl. Math.* **44**, 185–209.

Iskenderov, A.D. and Tagiev, R.G. (1979). Inverse source problem for evolution equations in a Banach space. *Voprosi Prikladn. Mat. i Kiber.* **1**, 51–56 (in Russian).

Krein, S.G. (1957). Classes of correctness for boundary value problems. *Dokl. Akad. Nauk SSSR.* **114**, 1162–1165 (in Russian).

Lavrentiev, M.M., Romanov, V.G., and Shishatskii, S.P. (1980). *Ill-Possed Problems of Mathematical Physics and Analysis.* Nauka, Moscow (in Russian).

Michailov, V.P. (1963). On a Dirichlet problem for parabolic equation I, II. *Mat. Sbornik* **61**, 40–64; **62**, 140–159 (in Russian).

Orlovskii, D.G. (1988). Inverse Cauchy problem for evolution equations in a Banach space. In: *Analysis of Mathematical Models of Physical Processes.*, MIFI, Moscow, pp. 84–90 (in Russian).

Orlovskii, D.G. (1990). On a determination problem of evolution equation parameter. *Different. Equat.* **26**, 1614–1621.

Petrovskii, I.G. (1934). Zur ersten Randwertaufgabe der warmeleitungleichung. *Compositio Mathem.* **1**, 383–419.

Prilepko, A.I. (1966). On the uniqueness of an inverse problem represented by the integral equation of the first kind. *Dokl. Akad. Nauk SSSR* **167**, 751–756 (in Russian).

Prilepko, A.I. (1967). On inverse problems in potential theory. *Different. Equat.* **3**, 30–44.

Prilepko, A.I. (1973). Inverse problems of potential theory. *Mat. Zametki.* **14**, 755–767 (in Russian).

Prilepko, A.I. and Soloviev, V.V. (1987a). Solvability theorems and the Rothe method in inverse problems for an equation of parabolic type II. *Different. Equat.* **23**, 1971–1980.

Prilepko, A.I. and Soloviev, V.V. (1987 b). A solvability of inverse coefficients problems for parabolic equation. *Different. Equat.* **23**, 136–143.

Prilepko, A.I. and Kostin, A.B. (1991 a). On a solvability of inverse problem for heat equation. In: *Inverse Problems for Mathematical Models of Physical Processes.*, MIFI, Moscow, pp. 52–58 (in Russian).

Prilepko, A.I. and Kostin, A.B. (1991 b). Inverse problems for parabolic equations with final and integral observations (to appear).

Prilepko, A.I. and Tikhonov, I.V. (1991). Uniqueness of inverse problem solution for evolution euqation and applications to transport equation (to appear).

Rundell, W. (1980). Determination of an unknown nonhomogeneous term in a linear partial differential equation from overspecified boundary data. *Appl. Anal.* **10**, 231–242.

Rundell, W. (1987). The determination of a parabolic equation from initial and final data. *Proc. Amer. Math. Soc.* **99**, 637–642.

Soloviev, V.V. (1989). On a solvability of inverse source problem for parabolic equation with final overdetermination. *Different. Equat.* **25**, 1577–1583.

Suzuki, T. and Murayama, R. (1980). A uniqueness theorem in an identification problem for coefficients of parabolic equations. *Proc. Japan Acad.* **56**, 259–263.

Tikhonov, A.N. (1935). Uniqueness theorems for heat equation. *Mat. Sbornik.* **42**, 199–216 (in Russian).

Tikhonov, A.N. and Arsenin, V.V. (1979). *Methods of Solving Ill-Possed Problems.* Nauka, Moscow (in Russian).

Ill-Posed Problems in Natural Sciences, pp. 390 – 407
A. Tikhonov (Ed.)
1992 VSP/TVP

INVERSE PROBLEMS IN MATHEMATICAL PHYSICS

A.I. PRILEPKO, D.G. ORLOVSKII, and I.A. VASIN
Department of Mathematics, Moscow Engineering Physics Institute,
Kashirskoe Shosse, 31, Moscow 115409, Russia

ABSTRACT

The paper deals with the inverse problems for evolution equations. The first section is devoted to the parabolic equation, the Boltzmann equation, the Maxwell system, the transport equation and the elasticity system. The second section contains the investigation on inverse problems for the Navier–Stokes system. In Section 3, inverse problems in an abstract formulation are considered.

The inverse problems for partial differential equations form an important part of the theory of ill-posed problems. One can find various approaches and a survey of the results which are related to them in (Tikhonov and Arsenin, 1977; Lavrentiev et al., 1986; Ivanov et al., 1978; Prilepko, 1973).

Many works are devoted to the heat equation. The investigation of inverse problems for the heat equation was started by A.N. Tikhonov (see (Tikhonov, 1935)). One of the main objectives of the study of inverse problems is the determination of unknown coefficients in a parabolic equation from an overspecification which is the result of measuring a temperature at a fixed point: $u(x_0, t) = \varphi(t)$. We note the papers (Jones, 1963; Beznotchenko, 1974; Uzlov, 1979; Rundell, 1980; Prilepko and Orlovskii, 1985; Prilepko and Solovyev, 1987). The case of a medium with memory was studied by A. Lorenzi and E. Sinestrary (see (Lorenzi and Sinestrary, 1988)). Another approach is based on measuring a temperature at some moment: $u(x, T) = \varphi(x)$. This problem was studied in (Iskenderov, 1976; Rundell, 1980; Eidelman, 1983; Prilepko and Orlovskii, 1989; Solovyev, 1988; Orlovskii, 1990).

Analogous problems were studied for some other equations of mathematical physics. The Maxwell system and the system of elasticity equations were considered in (Prilepko and Orlovskii, 1989). A.I. Prilepko and D.G. Orlovskii examined also the Boltzmann equation. The same problems for the transport equation were discussed in (Prilepko and Volkov, 1988; Orlovskii, 1991c).

Section 2 is devoted to the investigation of inverse problems for linearized and nonlinear Navier–Stokes systems arising in viscous incompressible fluid dynamics. The results formulated in this section were obtained by A.I. Prilepko and I.A. Vasin in 1989–1991.

One aspect of the investigation of inverse problems is the interpretation of partial differential equations as ordinary differential equations in a Banach space. Such approach was used in (Iskenderov, 1976; Rundell, 1980; Eidelman, 1983; Prilepko and Orlovskii, 1985, 1991; Orlovskii, 1990, 1991). Some of these articles study the existence and uniqueness of a solution. The Fredholm solvability in an abstract formulation was proved by D.G. Orlovskii (see (Orlovskii, 1990)). Section 3 describes the results on the aspects concerned which were obtained by A.I. Prilepko and D.G. Orlovskii.

1. THE INVERSE PROBLEMS FOR EVOLUTION EQUATIONS

First we consider the following problem for the parabolic equation

$$u_t = \sum_{i,j=1}^{n} \frac{\partial}{\partial x_i}\left(a_{ij}(x)\frac{\partial u}{\partial x_j}\right) + c(x)u + \Phi(x,t)p(t), \tag{1}$$

$$u\big|_{t=0} = u_0(x), \qquad u\big|_{\partial\Omega} = 0, \tag{2}$$

$$\int_{\Omega} u(x,t)\,\omega(x)\,dx = \varphi(t), \tag{3}$$

where $x \in \Omega$, $t \in [0,T]$, Ω is a bounded domain in $\mathbf{R}^n$ having a sufficiently smooth boundary $\partial\Omega$. A solution of (1)–(3) is a pair $u(x,t)$, $p(t)$. So, the nonhomogeneous term in equation (1) is unknown since it involves the function $p(t)$.
In order to determine the unknown function $p(t)$ one gives the additional information or overspecification (3). This additional information can be treated as the result of measuring by a real gauge which produces some mollification. For a perfect gauge $\omega(x) = \delta(x - x_0)$ where δ is Dirac's function.
Let $1 < p < \infty$ and $X = L_p(\Omega)$. We define an operator A in X by the relations

$$A(u) = \sum_{i,j=1}^{n} \frac{\partial}{\partial x_i}\left(a_{ij}(x)\frac{\partial u}{\partial x_j}\right) + c(x)u,$$
$$D(A) = W_p^2(\Omega) \cap \mathring{W}_p^1(\Omega).$$

The coefficients a_{ij} satisfy the conditions of symmetry and positive definiteness

$$a_{ij} = a_{ji}, \qquad \sum_{i,j=1}^{n} a_{ij}(x)\xi_i\xi_j \geq \alpha \sum_{i=1}^{n} \xi_i^2, \qquad \alpha = \text{const} > 0.$$

THEOREM 1. *Let* $a_{ij} \in C^1(\overline{\Omega})$, $c \in C(\overline{\Omega})$, $\Phi, A\Phi \in C([0,T]; L_p(\Omega))$, $\omega \in L_q(\Omega)$, $q = p/(p-1)$, $u_0 \in W_p^2(\Omega) \cap \mathring{W}_p^1(\Omega)$, $\varphi \in C^1[0,T]$, $\int_\Omega u_0(x)\omega(x)\,dx = \varphi(0)$, $\int_\Omega \Phi(x,t)\omega(x)\,dx \neq 0$. *Then problem* (1)–(3) *has a unique solution* $u \in C^1([0,T]; L_p(\Omega))$, $p \in C[0,T]$.

Let us consider the system of Maxwell equations in a bounded domain Ω with a perfectly conducting boundary:

$$\frac{\partial D}{\partial t} = \operatorname{rot}\frac{B}{\mu} - J, \qquad \frac{\partial B}{\partial t} = -\operatorname{rot}\frac{D}{\varepsilon},$$

$$J = \sigma\frac{D}{\varepsilon} + I, \qquad [n \times D]\big|_{\partial\Omega} = 0, \tag{4}$$

$$D(x,0) = D_0(x), \qquad B(x,0) = B_0(x),$$

where $D(x,t)$ is the electric displacement vector, $B(x,t)$ is the magnetic induction vector, ε and μ are respectively the dielectric and magnetic media permittivity, J is the current density, I is the extraneous current density, and σ is the conductivity of the medium. Let us assume that the extraneous current density is unknown and can be represented in the form

$$I(x,t) = \Phi(x,t)p(t) + F(x,t), \tag{5}$$

where the 3×3 matrix $\Phi(x,t)$ and the vector $F(x,t)$ are known and $p(t)$ has to be found. For that we introduce the additional information

$$B(x_0,t) = \varphi(t). \tag{6}$$

THEOREM 2. *Let* $\varepsilon, \mu \in C^3(\overline{\Omega})$, $\varepsilon > 0$ *and* $\mu > 0$ *in* $\overline{\Omega}$, $\sigma \in C^4(\overline{\Omega})$, $D_0, B_0 \in \mathring{W}_2^3(\Omega)$, $\operatorname{div} B_0 = 0$, $\Phi, F \in C([0,T]; \mathring{W}_2^3(\Omega))$, $\det\Phi(x_0,t) \neq 0$, $\varphi \in C^1([0,T]; \mathbf{R}^3)$, $B_0(x_0) = \varphi(0)$. *Then problem* (4)–(6) *has a unique solution* $D, B \in C^1([0,T]; L_2(\Omega))$, $p \in C([0,T]; \mathbf{R}^3)$.

Now we consider the elasticity system. Let Ω be a bounded domain in $\mathbf{R}^3$ with a sufficiently smooth boundary. We denote by $u = (u_1, u_2, u_3)$ the displacement vector; by $f = (f_1, f_2, f_3)$, the density of external forces; by σ_{ij}, the stress tensor; by ρ, the density of the medium. The system of elasticity equations has the form

$$\rho\frac{\partial^2 u_i}{\partial t^2} = \frac{\partial \sigma_{ij}}{\partial x_j} + \rho f_i, \qquad 1 \le i \le 3,$$

$$u(x,0) = u_0(x), \qquad u_t(x,0) = u_1(x), \tag{7}$$

$$u\big|_{\partial\Omega} = 0,$$

with the summation over repeated indices. The equation of a state is expressed by the law

$$\sigma_{ij} = a_{ijmn}e_{mn}(u). \tag{8}$$

The coefficients a_{ijmn} satisfy the conditions of symmetry and positive definiteness

$$a_{ijmn} = a_{jimn} = a_{ijnm} = a_{mnij}, \quad a_{ijmn}(x)\xi_{ij}\xi_{mn} \ge \alpha\xi_{ij}\xi_{ij}, \quad \alpha = \text{const} > 0,$$

and $e_{mn}(u)$ is the strain tensor,

$$e_{mn}(u) = \frac{1}{2}\left(\frac{\partial u_m}{\partial x_n} + \frac{\partial u_n}{\partial x_m}\right).$$

Let us assume that the unknown density of external forces in (7) can be represented as follows:

$$f(x,t) = \Phi(x,t)p(t) + F(x,t), \tag{9}$$

where the 3×3 matrix $\Phi(x,t)$ and the vector $F(x,t)$ are given and the vector $p(t)$ has to be found. Additional information is given in the form

$$u(x_0,t) = \varphi(t). \tag{10}$$

THEOREM 3. *Let $\rho, a_{ijmn} \in C^3(\overline{\Omega})$, $\rho(x) \geq \rho_0 > 0$; Φ, $F \in C([0,T]; \mathring{W}_2^3(\Omega))$, $\det\Phi(x_0,t) \neq 0$, $u_0 \in W_2^4(\Omega) \cap \mathring{W}_2^3(\Omega)$, $u_1 \in \mathring{W}_2^3(\Omega)$, $\varphi \in C^2([0,T]; \mathbf{R}^3)$, $u_0(x_0) = \varphi(0)$, $u_1(x_0) = \varphi'(0)$. Then problem (7)–(10) has a unique solution $u \in C^2([0,T]; L_{2,\rho}(\Omega))$, $p \in C([0,T]; \mathbf{R}^3)$.*

A similar result can be proved for the system of thermoelasticity equations. In this case, law (8) must be changed into

$$\sigma_{ij} = a_{ijmn}e_{mn}(u) - \gamma_{ij}\vartheta, \tag{11}$$

where γ_{ij} is a symmetric tensor and ϑ is the deviation of the temperature from the steady-state temperature T_0. System (7) must be supplemented by the heat equation

$$c\rho\frac{\partial\vartheta}{\partial t} = \frac{\partial}{\partial x_i}\left(k_{ij}\frac{\partial\vartheta}{\partial x_j}\right) - T_0\gamma_{ij}\frac{\partial}{\partial x_j}\frac{\partial u_i}{\partial t} + G, \tag{12}$$

where we denote by c the specific heat of the medium and by k_{ij}, the thermal conductivity tensor satisfying the conditions

$$k_{ij} = k_{ji}, \qquad k_{ij}(x)\xi_i\xi_j \geq k_0\xi_i\xi_i, \qquad k_0 = \mathrm{const} > 0.$$

We consider for equation (12) the Dirichlet initial-boundary data

$$\vartheta(x,0) = \vartheta_0(x), \qquad \vartheta\big|_{\partial\Omega} = 0. \tag{13}$$

THEOREM 4. *Let $\rho, c, \gamma_{ij} \in C^4(\overline{\Omega})$, $\rho(x) \geq \rho_0 > 0$, $a_{ijmn} \in C^3(\overline{\Omega})$, $k_{ij} \in C^5(\overline{\Omega})$; $\Phi, F \in C([0,T]; W_2^5(\Omega) \cap \mathring{W}_2^4(\Omega))$, $G \in C([0,T]; W_2^6(\Omega) \cap \mathring{W}_2^5(\Omega))$, $\det\Phi(x_0,t) \neq 0$, $u_0 \in W_2^4(\Omega) \cap \mathring{W}_2^3(\Omega)$, $u_1 \in W_2^5(\Omega) \cap \mathring{W}_2^4(\Omega)$, $\vartheta_0 \in W_2^6(\Omega) \cap \mathring{W}_2^5(\Omega)$, $u_0(x_0) = \varphi(0)$, $u_1(x_0) = \varphi'(0)$, $\varphi \in C^2([0,T]; \mathbf{R}^3)$. Then problem (7), (9)–(13) has a unique solution $u \in C^2([0,T]; W_2^2(\Omega))$, $\vartheta \in C^1([0,T]; W_2^4(\Omega))$, $p \in C([0,T]; \mathbf{R}^3)$.*

The next objects of the paper are the kinetic equations. First we consider the linearized Boltzmann equation. Let Ω be a strictly convex bounded closed domain in $\mathbf{R}^3$ with the Lyapunov boundary; let $B = \partial\Omega \times \mathbf{R}^3$, n_x be the unit vector of the inward normal to $\partial\Omega$ at the point x, $B^+ = \{(x,v) \in B\colon vn_x > 0\}$, $B^- = \{(x,v) \in B\colon vn_x < 0\}$; let u^+ and u^- be the traces of u on B^+ and B^- respectively,

$$H^{\pm} = L_{2,M(v)|vn_x|}(B^{\pm}), \qquad \mathcal{H} = L_{2,M(v)}(\mathbf{R}^3),$$

where $M(v)$ is the Maxwell distribution

$$M(v) = \rho(2\pi kT)^{-3/2}\exp(-v^2/(2kT)).$$

The Maxwell distribution describes the equilibrium state of a gas with density ρ and temperature T.
The linearized Boltzmann equation is

$$\frac{\partial u}{\partial t} + v \operatorname{grad}_x u = \Gamma(u) + F(x, v, t), \tag{14}$$

where $\Gamma(u) = Lu - \nu(v)u$.
We shall indicate the properties of the operator L and the function ν which are assumed to be satisfied. The operator L is compact and self-adjoint in $\mathcal{H}$; if $u, \nu u \in \mathcal{H}$, then $(\Gamma(u), u) \leq 0$, and $\nu(v)$ is a nonnegative measurable function. Besides equation (14) the initial and boundary conditions

$$u(x, v, 0) = u_0(x, v), \qquad u^+ = Ru^- \tag{15}$$

must be given, where R is a bounded linear operator from H^- to H^+. This operator satisfies one of the following two conditions:

a) $\|R\| < 1$;

b) $R1^- = 1^+$; for some $\varepsilon > 0$ and all $u^- \in H^-$ the dissipative inequality $\|Ru^-\|^2 + \varepsilon\|Pu^-\|^2 \leq \|u^-\|^2$ is satisfied, where P is the orthogonal projection in H^- onto the orthogonal complement of the subspace of functions depending on the variable x only.

We consider the question of finding a solution u and recovery of the inhomogeneous term F in (14) provided that F can be represented in one of the following two forms:

$$F(x, v, t) = \Phi(x, v, t)p(t) + G(x, v, t), \tag{16}$$

$$F(x, v, t) = \Phi(x, v, t)p(v, t) + G(x, v, t). \tag{17}$$

The functions $p(t)$ and $p(v, t)$ are unknown. To determine them, the additional information is specified whether in the form

$$\int\limits_{\Omega \times \mathbf{R}^3} u(x, v, t)\omega(x, v)\, dx\, dv = \varphi(t) \tag{18}$$

or in the form

$$\int\limits_{\Omega} u(x, v, t)\omega(x)\, dx = \varphi(v, t), \tag{19}$$

respectively.

THEOREM 5. *Let Φ, $v \operatorname{grad}_x \Phi + \nu\Phi$, G, $v \operatorname{grad}_x G + \nu G \in C([0, T]; \mathcal{H})$, u_0, $v \operatorname{grad}_x u_0 + \nu u_0 \in \mathcal{H}$; Φ, G, u_0 satisfy the boundary condition, $\omega \in L_{2,1/M(v)}(\Omega \times \mathbf{R}^3)$, $\varphi \in C^1[0, T]$, the compatibility condition*

$$\int\limits_{\Omega \times \mathbf{R}^3} u_0(x, v)\omega(x, v)\, dx\, dv = \varphi(0)$$

be satisfied, and for each $t \in [0,T]$

$$\int\limits_{\Omega\times\mathbf{R}^3} \Phi(x,v,t)\omega(x,v)\,dx\,dv \neq 0.$$

Then problem (14)–(16) has a unique solution $u \in C^1([0,T];\mathcal{H})$, $p \in C[0,T]$.

THEOREM 6. *Let Φ, $v\,\mathrm{grad}_x \Phi + \nu\Phi$, G, $v\,\mathrm{grad}_x G + \nu G \in C([0,T];\mathcal{H})$, u_0, $v\,\mathrm{grad}_x u_0 + \nu u_0 \in \mathcal{H}$; Φ, G, u_0 satisfy the boundary condition, $\omega \in L_2(\Omega)$, $\varphi \in C^1([0,T];\mathcal{H})$,*

$$\int\limits_{\Omega} u_0(x,v)\omega(x)\,dx = \varphi(v,0),$$

and for each $v \in \mathbf{R}^3$, $t \in [0,T]$

$$\left|\int\limits_{\Omega} \Phi(x,v,t)\omega(x)\,dx\right| \geq \gamma > 0.$$

Then problem (14), (15), (17), (19) has a unique solution $u \in C^1([0,T];\mathcal{H})$, $p \in C([0,T];\mathcal{H})$.

Now we consider the initial-boundary value problem for the transport equation

$$u_t + v\,\mathrm{grad}_x u + \sigma u = \int\limits_{\mathcal{D}} K(x,v,v',t)u(x,v',t)\,dv' + f, \tag{20}$$

$$u(x,v,0) = u_0(x,v), \qquad x \in \Omega, \qquad v \in \mathcal{D}, \tag{21}$$

$$u(x,v,t) = 0, \quad x \in \partial\Omega, \qquad vn_x < 0, \quad t \in [0,T], \tag{22}$$

where Ω is a strictly convex bounded closed domain with a smooth boundary, $\mathcal{D}$ is a closed bounded domain, $\Omega \subset \mathbf{R}^3$, $\mathcal{D} \subset \mathbf{R}^3$, n_x is the unit vector of the outward normal to $\partial\Omega$ at the point x. Suppose that

$$f(x,v,t) = \Phi(x,v,t)p(t) + F(x,v,t), \tag{23}$$

and the function p is unknown. To determine this function, we consider the overspecification having the form

$$\int\limits_{\Omega\times\mathcal{D}} u(x,v,t)\omega(x,v)\,dx\,dv = \varphi(t), \qquad 0 \leq t \leq T. \tag{24}$$

We assume that the following conditions are fulfilled:

1°) $\sigma \in C([0,T];L_\infty(\Omega\times\mathcal{D}))$.

2°) The operator

$$K(t)\colon\ u(x,v) \mapsto \int\limits_{\mathcal{D}} K(x,v,v',t)u(x,v')\,dv'$$

satisfies the condition

$$K \in C([0,T]; \mathcal{L}(L_s(\Omega \times \mathcal{D}))), \qquad s > 1.$$

3°) $\Phi \in C([0,T]; L_s(\Omega \times \mathcal{D}))$ and for any $t \in [0,T]$

$$\int\limits_{\Omega\times\mathcal{D}} \Phi(x,v,t)\omega(x,v)\,dx\,dv \neq 0.$$

4°) $F \in C([0,T]; L_s(\Omega \times \mathcal{D}))$.
5°) $u_0 \in L_s(\Omega \times \mathcal{D})$.
6°) $\omega, v\,\mathrm{grad}_x\,\omega \in L_r(\Omega \times \mathcal{D})$, $r = s/(s-1)$, $\omega(x,v) = 0$, $x \in \partial\Omega$, $v \in \mathcal{D}$, $n_x v > 0$.
7°) $\varphi \in C^1[0,T]$, $\qquad \varphi_0 = \int\limits_{\Omega\times\mathcal{D}} u_0(x,v)\omega(x,v)\,dx\,dv.$

THEOREM 7. *Let conditions* **1°)–7°)** *be satisfied. Then problem* (20)–(24) *has a unique solution* $u \in C([0,T]; L_s(\Omega \times \mathcal{D}))$, $p \in C[0,T]$.

2. THE INVERSE PROBLEMS IN HYDRODYNAMICS

This section deals with a brief survey of the results on inverse problems for non-stationary Navier–Stokes equations both linear

$$\vec{V}_t - \nu\Delta\vec{V} = -\nabla p + \vec{F}(x,t), \qquad \mathrm{div}\,\vec{V} = 0 \tag{25}$$

and nonlinear

$$\vec{V}_t - \nu\Delta\vec{V} + (\vec{V},\nabla)\vec{V} = -\nabla p + \vec{F}(x,t), \qquad \mathrm{div}\,\vec{V} = 0. \tag{26}$$

While solving these problems we try to determine a set of functions which describe the viscous incompressible fluid flow, i.e., the velocity of the fluid, the pressure gradient and the vector of the external force causing the motion.
We begin with some notations. Everywhere throughout Section 2 the symbol Ω denotes a bounded time-invariable domain in $\mathbf{R}^n$ ($n = 2$ or $n = 3$) with C^2-smooth boundary.
Let $\overset{\circ}{J}(\Omega)$ be a closure of the set of smooth finite solenoidal vectors with respect to the $L_2(\Omega)$-norm. The orthogonal complement of $\overset{\circ}{J}(\Omega)$ in $L_2(\Omega)$ is denoted by $G(\Omega)$, and $G(\Omega)$ is the set of all the vectors of the form $\nabla\psi$, where ψ is single-valued, square-summable and has the first derivatives in $L_2(\Omega)$.
Therefore, we have $L_2(\Omega) = \overset{\circ}{J}(\Omega) \oplus G(\Omega)$.
The Sobolev spaces $L_p(\Omega)$, $W_p^l(\Omega)$, $L_{q,r}(Q_T)$ and $W_p^{l,m}(Q_T)$ and their norms are understood as usual.
We note that all the definitions of the function spaces used in the research can be found in (Ladyzhenskaya, 1970).
To formulate the inverse problem for system (25), we suppose that the function $\vec{F}$ allows the representation

$$\vec{F}(x,t) = \vec{f}(x)g(x,t), \tag{27}$$

where the vector-function $\vec{f}(x)$ has to be found, and the scalar coefficient $g(x,t)$ is given.
One can pose the problem of finding a triple of vector-valued functions $\{\vec{V}, \nabla p, \vec{f}\}$ satisfying in $Q_T = \Omega \times [0,T]$ the system of linearized Navier–Stokes equations

$$\vec{V}_t - \nu\Delta\vec{V} = -\nabla p + \vec{f}(x)g(x,t), \qquad \operatorname{div}\vec{V} = 0, \tag{28}$$

the initial condition

$$\vec{V}(x,0) = 0, \qquad x \in \Omega, \tag{29}$$

the boundary condition

$$\vec{V}(x,t) = 0, \qquad (x,t) \in S_T = \partial\Omega \times [0,T] \tag{30}$$

and the conditions of final overdetermination

$$\vec{V}(x,T) = \vec{\varphi}(x), \quad \nabla p(x,T) = \nabla\psi(x), \quad x \in \Omega, \tag{31}$$

if g, $\vec{\varphi}$, $\nabla\psi$ and ν are given.
Inverse problem (28)–(31) was investigated in (Prilepko and Vasin, 1989a).
It should be noted that (29) and (30) are taken homogeneous only to simplify calculations.
A generalized solution of inverse problem (28)–(31) is a vector-function set $\{\vec{V}, \nabla p, \vec{f}\}$ such that (28)–(31) are satisfied and $\vec{V} \in \overset{\circ}{W}{}_2^{2,1}(Q_T) \cap \overset{\circ}{J}(Q_T)$, $\vec{f} \in L_2(\Omega)$, $\nabla p(\cdot,t)$ belongs to $G(\Omega)$ for each t from $[0,T]$ and depends continuously on t in the $L_2(\Omega)$-norm on $[0,T]$. We suppose that $g, g_t \in C(\overline{Q}_T)$ and $|g(x,T)| \geq g_T > 0$ ($g_T \equiv \text{const}$).
Let $\vec{f}$ be an arbitrary chosen function from $L_2(\Omega)$. Then we substitute this element $\vec{f}$ in (28) and consider (28)–(30) as the direct problem of finding a pair of functions $\{\vec{V}, \nabla p\}$.
By virtue of the results formulated in (Ladyzhenskaya, 1970; Temam, 1979) the existence and uniqueness of $\{\vec{V}, \nabla p\}$ is guaranteed.
Consequently, one can introduce a linear operator A acting from $L_2(\Omega)$ to $L_2(\Omega)$ by the relation

$$(A\vec{f})(x) = \frac{1}{g(x,T)}\vec{V}_t(x,T), \qquad x \in \Omega, \tag{32}$$

where $\vec{V}$ belongs to the set $\{\vec{V}, \nabla p\}$ founded above.
Our further considerations will be connected with the following equation of the second kind:

$$\vec{f} = A\vec{f} + \vec{\chi}, \tag{33}$$

where $\vec{\chi}$ belongs to $L_2(\Omega)$.
The conditions of solvability equivalence between inverse problem (28)–(31) and equation (33) are formulated in the following theorem.

THEOREM 8. *Let $g, g_t \in C(\overline{Q}_T)$, $|g(x,T)| \geq g_T > 0$, $\vec{\varphi} \in W_2^2(\Omega) \cap H(\Omega)$, $\nabla\psi \in G(\Omega)$, where $H(\Omega) = \overset{\circ}{J}(\Omega) \cap \overset{\circ}{W}{}_2^1(\Omega)$. Then the following two statements are true.*

1. *If inverse problem* (28)–(31) *has a unique solution, then equation* (33) *has a unique solution under*

$$\vec{\chi} = \frac{1}{g(x,T)}(-\nu\,\Delta\vec{\varphi} + \nabla\psi), \qquad x \in \Omega. \tag{34}$$

2. *If* $\vec{\chi}$ *satisfies* (34) *and equation* (33) *has a unique solution, then inverse problem* (28)–(31) *has a unique solution.*

Using the corresponding energy equalities and estimates and applying the Gronwall lemma, we can show that under certain restrictions imposed on input data the linear operator A possesses a contraction property. Then due to Theorem 8 (item **2**) we have

THEOREM 9. *Let* $g, g_t \in C(\overline{Q}_T)$, $|g(x,T)| \geq g_T > 0$, $\vec{\varphi} \in W_2^2(\Omega) \cap H(\Omega)$, $\nabla\psi \in G(\Omega)$. *Suppose that*

$$m_1 < 1, \tag{35}$$

where

$$m_1 = \frac{1}{\inf\limits_{\Omega} |g(x,T)|}\Big(\sup_{\Omega} |g(x,0)| \exp\left(-\nu T / C_1^2(\Omega)\right) + \int_0^T \sup_{\Omega} |g_t(x,t)| \exp\left(-\nu(T-t)/C_1^2(\Omega)\right) dt\Big)$$

and $C_1(\Omega)$ *is the constant from the Poincaré inequality.*
Then inverse problem (28)–(31) *has a unique solution, and the corresponding stability estimates are true.*

If we continue to investigate properties of the operator, then by applying the corresponding embedding theorems the following result can be obtained.

THEOREM 10. *Let* $g, g_t \in C(\overline{Q}_T)$, $|g(x,T)| \geq g_T > 0$. *Then the linear operator* A *is completely continuous on* $L_2(\Omega)$.

By virtue of Theorem 10 we see that equation (8) is a Fredholm equation of the second kind.
Therefore, Theorems 8 and 10 yield that the solution of inverse problem (28)–(31) possesses the Fredholm property, i.e., its uniqueness implies existence and stability.
We note that Theorems 8, 9, 10 and the Fredholm-type result for system (28) were proved in (Prilepko and Vasin, 1989a).
The next inverse problem to be considered here is an inverse problem for the nonlinear Navier–Stokes system (26).
Suppose that $\vec{F}$ in (26) can be represented in form (27).
We formulate a nonlinear inverse problem of finding a triplet of functions $\{\vec{V}, \nabla p, \vec{f}\}$ satisfying the system

$$\vec{V}_t - \nu\Delta\vec{V} + (\vec{V},\nabla)\vec{V} = -\nabla p + \vec{f}(x)g(x,t),$$

$$\operatorname{div}\vec{V} = 0, \qquad (x,t) \in Q_T = \Omega \times [0,T], \tag{36}$$

the initial condition

$$\vec{V}(x,0) = \vec{a}(x), \qquad x \in \Omega, \tag{37}$$

the boundary condition

$$\vec{V}(x,t) = 0, \qquad (x,t) \in S_T = \partial\Omega \times [0,T] \tag{38}$$

and the condition of final overdetermination

$$\vec{V}(x,T) = \vec{\varphi}(x), \quad \nabla p(x,T) = \nabla\psi(x), \quad x \in \Omega, \tag{39}$$

where $\vec{a}$, $\vec{\varphi}$, $\nabla\psi$, g and ν are given.
If one begins to investigate inverse problem (36)–(39), then one should take into account that the uniqueness and existence theorem for the corresponding direct problem in three spatial dimensions has a local character.
Using energy estimates, embedding theorems, the Schauder fixed-point theorem and arguing in a way similar to the linear case, we prove the following solvability results.
For two spatial dimensions we obtain

THEOREM 11 (Prilepko and Vasin, 1989b). *Let $\Omega \subset \mathbf{R}^2$, $g, g_t \in C(\overline{Q}_T)$, $|g(x,T)| \geq g_T > 0$, $\vec{a}, \vec{\varphi} \in W_2^2(\Omega) \cap H(\Omega)$, $\nabla\psi \in G(\Omega)$, and let the following estimates be valid:*

$$\sqrt[4]{2}\,\sqrt{C_1(\Omega)}\,\|\vec{\varphi}\|_{4,\Omega} < \nu, \qquad m_2 + \|\vec{\chi}\|_{2,\Omega} < 1,$$

where

$$m_2 = \frac{1}{\inf_{\Omega} |g(x,T)|} \Big\{ \big[\|\nu\Delta\vec{a} - (\vec{a},\nabla)\vec{a}\|_{2,\Omega} + \sup_{\Omega} |g(x,0)|\big] \exp\big[-\nu T/C_1^2(\Omega)\big]$$

$$+ \int_0^T \exp\big[-\nu(T-t)/(2C_1^2(\Omega))\big] \sup_{\Omega} |g_t(x,t)|\,dt \Big\}$$

$$\times \exp\Big\{\frac{1}{\nu^2}\Big[\|\vec{a}\|_{2,\Omega}^2 + \frac{3}{2}\Big(\int_0^T \sup_{\Omega} |g(x,t)|\,dt\Big)^2\Big]\Big\}$$

$$\vec{\chi} = (-\nu\Delta\vec{\varphi} + (\vec{\varphi},\nabla)\vec{\varphi} + \nabla\psi)/g(x,T),$$

$C_1(\Omega)$ is the constant from the Poincaré inequality.
Then there exists a solution of inverse problem (36)–(39).

In the case of three spatial dimensions we have

THEOREM 12 (Prilepko and Vasin, 1990b). *Let $\Omega \subset \mathbf{R}^3$, $g, g_t \in C(\overline{Q}_T)$, $|g(x,T)| \geq g_T > 0$, $\vec{a}, \vec{\varphi} \in W_2^2(\Omega) \cap H(\Omega)$, $\nabla\psi \in G(\Omega)$, and let the following estimates be true:*

$$\sqrt{C_2(\Omega)}\,\|\vec{\varphi}\|_{4,\Omega} < \nu, \quad m_3 m_4 < \nu^3/(4C_2(\Omega))^2, \quad m_5 + \|\vec{\chi}\|_{2,\Omega} < 1,$$

where

$$m_3 = \|\vec{a}\|_{2,\Omega} + \int_0^T \sup_\Omega |g(x,t)|\, dt,$$

$$m_4 = \|\nu\Delta\vec{a} - (\vec{a},\nabla)\vec{a}\|_{2,\Omega} + \sup_{Q_T} |g(x,t)| + \sup_\Omega |g(x,0)| + \int_0^T \sup_\Omega |g_t(x,t)|\, dt,$$

$$m_5 = \frac{1}{\inf_\Omega |g(x,T)|}\Big\{ \big[\|\nu\Delta\vec{a} - (\vec{a},\nabla)\vec{a}\|_{2,\Omega} + \sup_\Omega |g(x,0)|\big] \exp\big[-\nu T/(2C_1^2(\Omega))\big]$$

$$+ \int_0^T \sup_\Omega |g_t(x,t)| \exp\big[-\nu(T-t)/(2C_1^2(\Omega))\big]\, dt\Big\}$$

$$\vec{\chi} = (-\nu\Delta\vec{\varphi} + (\vec{\varphi},\nabla)\vec{\varphi} + \nabla\psi)/g(x,T), \qquad C_2(\Omega) = (4/3)^{3/4}\sqrt{C_1(\Omega)}.$$

Then there exists a solution of inverse problem (36)–(39).

Let us suppose that the coefficient g is independent of x. In this case we can change the formulation of the inverse problems concerned and drop the overdetermination relation which prescribes the value of $\nabla p(\cdot,T)$ in (28) and (29).
Then the vector-valued function $\vec{f}$ should be looked for in the space $\overset{\circ}{J}(\Omega)$.
Results dealing with the case $g = g(t)$ can be found in (Prilepko and Vasin, 1990a) for the linear system and in (Prilepko and Vasin, 1991) for the nonlinear one.
Now we are going to discuss an inverse problem where overdetermination is given in an integral form.
Assume that $\vec{F}$ in (25) is expressed as

$$\vec{F} = f(t)\vec{g}(x,t), \tag{40}$$

where the coefficient $f(t)$ is unknown and the vector-valued function $\vec{g}(x,t)$ is given.
We consider the inverse problem of finding a set of functions $\{\vec{V}, \nabla p, f\}$ satisfying the system

$$\vec{V}_t - \nu\Delta\vec{V} = -\nabla p + f(t)\vec{g}(x,t), \quad \operatorname{div}\vec{V} = 0, \quad (x,t) \in Q_T, \tag{41}$$

$$\vec{V}(x,0) = 0, \qquad x \in \Omega, \tag{42}$$

$$\vec{V}(x,t) = 0, \qquad (x,t) \in S_T \tag{43}$$

and the integral overdetermination

$$\int_\Omega \vec{V}(x,t)\vec{\omega}(x)\, dx = \varphi(t), \qquad t \in [0,T], \tag{44}$$

where $\vec{g}$, $\vec{\omega}$, φ and ν are given.

To solve the formulated inverse problem, we derive a second kind operator equation with respect to $f(t)$ and invoke the principle of contracting mapping.
Under certain restrictions on input data we obtain the global theorem showing that inverse problem (41)–(44) has a unique solution

$$\vec{V} \in \overset{\circ}{W}{}_2^{2,1}(Q_T) \cap \overset{\circ}{J}(Q_T), \quad \nabla p \in G(Q_T), \quad f \in C[0,T]$$

(see (Prilepko and Vasin, 1989a)).
Using a similar procedure we can formulate and examine the inverse problem for nonlinear Navier–Stokes system (26) with integral overdetermination (44) if the external force function $\vec{F}$ from (26) is represented in form (40).
It was found under what sufficient conditions the solution of this nonlinear inverse problem can be shown to exist and to be unique (see (Vasin, 1992b)).
If a mathematical model of viscous incompressible fluid motion contains the system

$$\vec{V}_t - \nu\Delta\vec{V} + \sum_{k=1}^{n} B_k(x,t)\vec{V}_{x_k} + A(x,t)\vec{V} = -\nabla p + \vec{F},$$

$$\operatorname{div}\vec{V} = 0, \qquad (x,t) \in Q_T, \tag{45}$$

then we can offer similar formulations of inverse problems with final and integral overdetermination.
Here B_k and A are square matrices.
Applying the techniques described above, one can generalize the results concerning problems (28)–(31) and (41)–(42) to the case of system (45) (see (Prilepko and Vasin, 1990a)).
Note that integral overdetermination can be given in the form

$$\int_{\Omega} \nabla p(x,t)\vec{\chi}(x)\,dx = \psi(t)$$

instead of (44) (see (Vasin, 1992a)).
To conclude Section 2, we describe an approach to the coefficient identification inverse problems of hydrodynamics.
Let us assume that $B_k(x,t) = \beta_k(x)I$ in (45), where I is the unitary matrix and $\beta_k(x)$ is a scalar function. Therefore

$$\sum_{k=1}^{n} B_k(x,t)\vec{V}_{x_k} = (\vec{\beta}(x), \nabla)\vec{V}, \tag{46}$$

where the symbol $\vec{\beta}$ stands for a vector-valued function with components $\beta_k(x)$.
We pose the coefficient inverse problem of finding a function set $\{\vec{V}, \nabla p, \vec{\beta}\}$ which satisfies system (45)–(46), the initial condition $\vec{V}(x,0) = \vec{a}(x)$ $(x \in \Omega)$, the boundary condition $\vec{V}(x,t) = \vec{b}(x,t)$ $((x,t) \in S_T)$ and conditions of final overdetermination (31), if $\vec{a}$, $\vec{b}$, φ, $\nabla\psi$, A, ν and the external force function $\vec{F}$ are given.
The uniqueness theorem for the formulated coefficient inverse problem is proved in (Prilepko and Vasin, 1990c).

3. THE INVERSE PROBLEMS FOR ABSTRACT DIFFERENTIAL EQUATIONS

In this section we consider several problems for differential equations in a Banach space. Let X be a Banach space and $A(t)$ be a family of closed linear operators in X with dense domains. Let Y be a Banach space also, and

$$L_1 \in C([0,T];\mathcal{L}(X)), \qquad L_2 \in C([0,T];\mathcal{L}(Y,X)).$$

Let us consider the problem of determining the functions $u \in C^1([0,T];X)$ and $p \in C([0,T];Y)$ which satisfy the conditions

$$\begin{gathered} u'(t) + A(t)u(t) = L_1(t)u(t) + L_2(t)p(t), \qquad 0 \le t \le T, \\ u(0) = u_0, \\ Bu(t) = \varphi(t), \qquad 0 \le t \le T, \end{gathered} \tag{47}$$

where B is a linear operator from $D(B) \subset X$ to Y.
The parabolic case is characterized by the conditions:
(P1) The domain of $A(t)$ is constant: $D(A(t)) = D_1$.
(P2) For every λ with $\operatorname{Re}\lambda \ge 0$ and every $t \in [0,T]$

$$\|(A(t) + \lambda I)^{-1}\| \le c/(1 + |\lambda|).$$

(P3) For every $t, s, \tau \in [0,T]$

$$\|[A(t) - A(s)]A^{-1}(\tau)\| \le c|t-s|^\alpha, \qquad 0 < \alpha \le 1.$$

We suppose that the operator B satisfies the condition
(P4) For some $\beta < \alpha$

$$B \in \mathcal{L}(D(A^\beta(0)), Y),$$

where $D(A^\beta(0))$ is endowed with the norm $\|u\|_\beta = \|A^\beta(0)u\|$.

THEOREM 13. *Let $u_0 \in D(A^{1+\beta}(0))$, $\varphi \in C^1([0,T];Y)$, $Bu_0 = \varphi(0)$, L_1, AL_1A^{-1}, $A^{1+\beta}L_1A^{-1-\beta} \in C([0,T];\mathcal{L}(X))$, $A^{1+\beta}L_2 \in C([0,T];\mathcal{L}(Y,X))$, $(BL_2)^{-1} \in C([0,T];\mathcal{L}(Y))$.
Then problem (47) has a unique solution $u \in C^1([0,T];X)$, $p \in C([0,T];Y)$.*

Let us consider now the hyperbolic case which is characterized by the following conditions:
(H1) For every $t \in [0,T]$ the operator $-A(t)$ is a generator of a strongly continuous semigroup in X.
(H2) There exist M and β such that for every $\lambda > \beta$ and every $t_1, \dots, t_k$ $(0 \le t_1 \le \dots \le t_k \le T)$

$$\|(A(t_k) + \lambda I)^{-1} \dots (A(t_1) + \lambda I)^{-1}\| \le M/|\lambda - \beta|^k.$$

(H3) There exist a Banach space X_0, continuously and densely embedded into X, and

$$S(t) \in C([0,T];\mathcal{L}(X_0, X))$$

possessing the following properties: $S(t)$ is strongly continuously differentiable, for every $t \in [0,T]$ $S(t)^{-1} \in \mathcal{L}(X, X_0)$ and

$$S(t)A(t)S(t)^{-1} = A(t) + R(t)$$

where $R(t) \in \mathcal{L}(X)$, the function $R(t)$ is a strongly measurable and bounded in $\mathcal{L}(X)$ one.
(H4) For every $t \in [0,T]$ $X_0 \subset D(A(t))$ and $A(t) \in C([0,T]; \mathcal{L}(X_0, X))$.
The operator B satisfies the condition
(H5) $B \in \mathcal{L}(X,Y)$, $\overline{BA}(t) \in C([0,T]; \mathcal{L}(X,Y))$.

THEOREM 14. *Let* $u_0 \in X$, $\varphi \in C^1([0,T];Y)$, $Bu_0 = \varphi(0)$, $L_1 \in C([0,T]; \mathcal{L}(X))$, $L_2 \in C([0,T]; \mathcal{L}(Y,X))$, $(BL_2)^{-1} \in C([0,T]; \mathcal{L}(Y))$.
Then problem (47) *has a unique solution*

$$u \in C([0,T]; X), \qquad p \in C([0,T]; Y).$$

A similar result can be proved for the second order hyperbolic equation

$$\begin{aligned} u''(t) = Au(t) + L_1(t)u + L_2(t)u' + L_3(t)p, \qquad 0 \le t \le T, \\ u(0) = u_0, \qquad u'(0) = u_1, \\ Bu(t) = \varphi(t), \qquad 0 \le t \le T. \end{aligned} \tag{48}$$

Our fundamental assumption is that the operator A is an infinitesimal generator of a strongly continuous cosine function $C(t)$. The associated sine function is defined as

$$S(t) = \int_0^t C(s)\, ds.$$

We shall also make use of the set

$$E = \{f \in X\colon\ C(t)f \in C^1\}.$$

The set E with the norm

$$\|f\|_E = \|f\| + \sup_{0 \le t \le 1} \|C'(t)f\|$$

is a Banach space.
We suppose that for some natural number m

$$B \in \mathcal{L}(D(A^m), Y),$$

where $D(A^m)$ is endowed with the graph norm.

THEOREM 15. *Let* $L_1 \in C([0,T]; \mathcal{L}(E,X))$, $L_2 \in C([0,T]; \mathcal{L}(X) \cap \mathcal{L}(E))$,

$$L_1 A^{-1} \in C([0,T]; \mathcal{L}(X,E)), \quad A^m L_1 A^{-m-1} \in C([0,T]; \mathcal{L}(X,E)),$$

$$A^m L_2 A^{-m} \in C([0,T];\mathcal{L}(E)), \quad A^m L_3 \in C([0,T];\mathcal{L}(Y,E)),$$

$$(BL_3)^{-1} \in C([0,T];\mathcal{L}(Y)), \quad u_0 \in D(A^{m+1}), \quad u_1 \in D(A^m), \quad A^m u_1 \in E,$$

$$\varphi \in C^2([0,T];Y), \quad Bu_0 = \varphi(0), \quad Bu_1 = \varphi'(0).$$

Then problem (48) *has a unique solution*

$$u \in C^2([0,T];X), \qquad p \in C([0,T];Y).$$

Now we consider a boundary value problem for abstract evolution equations. The problem is to find a pair (u,p) satisfying the conditions

$$u'(t) = Au(t) + \Phi(t)p, \qquad 0 \le t \le T,$$

$$u(0) = u_0, \qquad u(T) = u_1, \tag{49}$$

where A is the generator of a strongly continuous semigroup,

$$\Phi \in C([0,T];\mathcal{L}(X)), \quad u \in C^1([0,T];X), \quad p \in X.$$

In the general case, problem (49) may have more than one solution; however, if the semigroup $\exp(At)$ is compact (i.e., the operator $\exp(At)$ is compact for every $t > 0$) and

$$\Phi \in C^1([0,T];\mathcal{L}(X)), \qquad (\Phi(T))^{-1} \in \mathcal{L}(X),$$

then problem (49) possesses the property of Fredholm's solvability. Let X be a Hilbert space, and let the operator A be self-adjoint and negative. Suppose that X is a lattice and the semigroup $\exp(At)$ is positive.

THEOREM 16. *Let $A^{-1} \in \mathcal{L}(X)$ and be compact; let A have a strictly positive eigenvector, $\Phi \in C^1([0,T];\mathcal{L}(X))$, $\Phi^{-1}(T) \in \mathcal{L}(X)$, for every $t \in [0,T]$ $\Phi(t)$ and $\Phi'(t)$ be positive; let $\Phi^{-1}(T)$ be strictly positive, and let the operator $\Phi'(t)$ be strictly positive for some $t \in [0,T)$. Then problem* (49) *has a unique solution*

$$u \in C^1([0,T];X), \qquad p \in X$$

for every $u_0, u_1 \in D(A)$.

In most cases these results can be generalized from the linear case onto the semi-linear one. We formulate one of such generalizations. Let us consider the following inverse problem for the evolution equation of the first order

$$u'(t) = Au(t) + f(t,u(t),p(t)), \qquad 0 \le t \le T,$$

$$u(0) = u_0, \tag{50}$$

$$Bu(t) = \varphi(t), \qquad 0 \le t \le T,$$

where A is the generator of a strongly continuous semigroup in a Banach space X and B is a linear bounded operator acting from X to Y. The solution of problem (50) is a pair (u,p). We introduce on the space $D(A)$ the graph norm

$$\|u\|_\Gamma = \|u\| + \|Au\|$$

and set

$$S_{R,T} = \{(t,u,p): \quad t \in [0,T], \quad \|u\| \le R, \quad \|p\| \le R\},$$
$$\tilde{S}_{R,T} = \{(t,u,p): \quad t \in [0,T], \quad \|u\|_\Gamma \le R, \quad \|p\| \le R\}.$$

THEOREM 17. *Suppose that $f(t,u,p) \in C([0,T];H)$ for each fixed $u \in H$ and $p \in Y$, $f(t,u,p)$ is bounded on the set $S_{R,T}$ for every R and satisfies the Lipschitz condition*

$$\|f(t,u_2,p_2) - f(t,u_1,p_1)\| \le \mathcal{L}(\|u_2 - u_1\| + \|p_2 - p_1\|); \tag{51}$$

suppose further that $Af(t,u,p)$ is bounded on the set $\tilde{S}_{R,T}$ for every $R > 0$ and satisfies the Lipschitz condition

$$\|Af(t,u_2,p_2) - Af(t,u_1,p_1)\| \le \mathcal{L}(\|u_2 - u_1\| + \|p_2 - p_1\|). \tag{52}$$

Let $f(t,u,p) \in C([0,T];D(A))$ for each fixed $u \in D(A)$ and $p \in Y$, and let

$$\varphi f(t,u,p) = \tilde{f}(t,\varphi u,p), \tag{53}$$

where $\tilde{f}$ is a function from $\mathbf{R}^1_+ \times Y \times Y$ to Y such that the mapping $w = \tilde{f}(t,\varphi(t),p)$ is invertible with respect to p, and for each fixed $w \in Y$ the inverse mapping $p = \Phi(t,w)$ is of the class $C([0,T];Y)$ and satisfies with respect to w the Lipschitz condition

$$\|\Phi(t,w_2) - \Phi(t,w_1)\| \le \mathcal{L}\|w_2 - w_1\|. \tag{54}$$

Under these assumptions, if the initial data satisfy the necessary solvability conditions

$$Bu_0 = \varphi(0), \qquad \varphi \in C^1([0,T];Y),$$

then there exists a $T^* > 0$ such that for $0 \le t \le T^*$ inverse problem (50) has a unique solution.

REFERENCES

Beznotchenko, N.Y. (1974). Determination of a coefficient in a parabolic equation *Different. Equat.* **10**, 24–35.

Beznotchenko, N.Y. (1983). The Cauchy problem for the equation $u_t = \Delta u + uAu = f$ *Different. Equat.* **19**, 991–1000.

Cannon, J.R. and Lin, Y. (1998). Determination of a parameter $p(t)$ in some quasi-linear parabolic differential equations *Inverse Problems* **4**, 35–45.

Eidelman, Y.S. (1983). Two-points boundary value problem for a differential equation with a parameter *Dokl. Akad. Nauk SSSR* **4**, 15–18 (in Russian).

Isakov, V. (1982). On a class of inverse problems for parabolic equations *Dokl. Akad. Nauk SSSR* **25**, 519–521 (in Russian).

Iskenderov, A.D. (1976). Some inverse problems of determining the right sides of differential equations *Izv. Akad. Nauk Azerb. SSR* **2**, 35–44 (in Russian).

Ivanov, V.K., Vasin, V.V., and Tanana, V.P. (1978). *Theory of Linear Ill-Posed Problems and Its Applications.* Nauka, Moscow (in Russian).

Jones, F. (1963). Various methods for finding unknown coefficient in parabolic differential equations *Comm. Pure Appl. Math.* **16**, 33–44.

Ladyzhenskaya, O.A. (1967). Uniqueness and smoothness of generalized solutions of Navier–Stokes equations *Zapiski Nauchn. Seminarov LOMI* **5**, 169–185 (in Russian).

Ladyzhenskaya, O.A. (1970). *Mathematical Theory of Viscous Incompressible Flow.* Nauka, Moscow (in Russian).

Lavrentiev, M.M., Romanov, V.G., and Shyshatsky, S.P. (1980). *Ill-Posed Problems of Mathematical Physics and Analysis.* Nauka, Moscow (in Russian).

Lorenzi, A. and Sinestrary, E. (1988). An inverse problem in the theory of material with memory *Nonlinear Anal.* **12**, 1317–1335.

Orlovskii, D.G. (1988). *Cauchy Inverse Problem for the Evolution Equations in a Banach Space.* Energoatomizdat, Moscow, pp. 84–90 (in Russian).

Orlovskii, D.G. (1989). Inverse problem for the differential equation of the second order in a Banach space *Different. Equat.* **25**, 1000–1009.

Orlovskii, D.G. (1990). The problem of determining a parameter of the evolution equation *Different. Equat.* **26**, 1614–1621.

Orlovskii, D.G. (1991a). Weak and strong solution of inverse problems for differential equations in a Banach space *Different. Equat.* **27**, 867–874.

Orlovskii, D.G. (1991b) Determining the evolution of a parameter in the abstract quasilinear parabolic equation *Mat. Zametki* **50**, 111–119 (in Russian).

Orlovskii, D.G. (1991c). The solution of an inverse problem for the transport equation with an integral overspecification In: *Obratnye Zadachi dlya Matematicheskikh Modeley Fizicheskikh Protsessov.* MIFI, Moscow, pp. 71–76 (in Russian).

Prilepko, A.I. (1973). Inverse problems in potential theory (elliptic, parabolic, hyperbolic equations and transport equation) *Mat. Zametki* **14**, 755–767 (in Russian).

Prilepko, A.I. and Kostin, A.B. (1991). About the solvability of an inverse problem for the heat equation In: *Obratnye Zadachi dlya Matematicheskikh Modeley Fizicheskikh Protsessov,* MIFI, Moscow, pp. 51–58 (in Russian).

Prilepko, A.I. and Orlovskii, D.G. (1985a). The determination of parameter of the evolution equation and inverse problems of mathematical physics. 1 *Different. Equat.* **21**, 119–129.

Prilepko, A.I. and Orlovskii, D.G. (1985b). The determination of parameter of the evolution equation and inverse problems of mathematical physics. 2 *Different. Equat.* **21**, 694–701.

Prilepko, A.I. and Orlovskii, D.G. (1987a). The determination of parameter of the evolution equation and inverse problems of mathematical physics. 3 *Different. Equat.* **23**, 1343–1353.

Prilepko, A.I. and Orlovskii, D.G. (1987b). Some inverse problems of the kinetic theory of gas for the near equilibrium states *J. Math. Phys. Comput. Math.* **27**, 1690–1700.

Prilepko, A.I. and Orlovskii, D.G. (1989). About the semigroup approach to the problem of determining the nonhomogeneous term in the evolution equations *Dokl. Akad. Nauk SSSR* **305**, 1045–1049 (in Russian).

Prilepko, A.I. and Orlovskii, D.G. (1991). The determination of the parameter's evolution in an abstract parabolic equation *Different. Equat.* **27**, 114–120.

Prilepko, A.I. and Solovyev, V.V. (1987). Solvability theorems and Rothe's method for the inverse problems for an equation of the parabolic type *Different. Equat.* **23**, 1971–1980.

Prilepko, A.I. and Vaśin, I.A. (1989a). Inverse initial-boundary-value problems for linearized nonstationary Navier–Stokes equations *Different. Equat.* **25**, 106–117.

Prilepko, A.I. and Vasin, I.A. (1989b). Inverse boundary-value problems for nonlinear Navier–Stokes equations with final over-determination *Different. Equat.* **25**, 2164–2177.

Prilepko, A.I. and Vasin, I.A. (1990a). Nonstationary inverse problems of hydrodynamics with final over-determination *Dokl. Akad. Nauk SSSR* **314**, 1075–1078 (in Russian).

Prilepko, A.I. and Vasin, I.A. (1990b). Solvability of a three-dimensional inverse problem for nonlinear Navier–Stokes equations *Zh. Vychisl. Mat. i Mat. Fiz.* **29**, 1540–1552 (in Russian).

Prilepko, A.I. and Vasin, I.A. (1990c). The investigation of uniqueness in nonlinear inverse problems of hydrodynamics *Different. Equat.* **26**, 109–120.

Prilepko, A.I. and Vasin, I.A. (1991). On a nonlinear nonstationary inverse problem of hydrodynamics *Inverse Problems* **7**, 13–16.

Prilepko, A.I. and Volkov, N.P. (1987). Inverse problems of the determination of parameters in a nonstationary transport equation from overspecification of the integral type *Different. Equat.* **23**, 124-136.

Rundell, W. (1980). Determination of an unknown nonhomogeneous term in a linear partial differential equations from overspecified boundary data *Appl. Anal.* **10**, 231–242.

Solonnikov, V.A. (1973). Estimates of solutions of nonstationary Navier–Stokes systems *Zapiski Nauchn. Seminarov LOMI* **38**, 153–231 (in Russian).

Solovyev, V.V. (1988). Fredholm's property of the inverse problem of determination of the right part in a parabolic equation In: *Analysis of the Mathematical Models of Physical Processes*, Energoatomizdat, Moscow, pp. 90–95.

Temam, R. (1979). *Navier–Stokes Equations. Theory and Numerical Analysis.* North-Holland, Amsterdam.

Tikhonov, A.N. (1935). Uniqueness theorem for a heat equation *Mat. Sbornik* **42**, 199–216 (in Russian).

Tikhonov, A.N. and Arsenin, V.Ya. (1977). *Solutions of Ill-Posed Problems.* Winston, Washington

Uzlov, A.E. (1979). The existence and uniqueness theorem for inverse problems for equations of the second order of the parabolic type *Different. Equat.* **15**, 370–372.

Vasin, I.A. (1992a). Inverse problems of viscous incompressible fluid dynamics with integral overdetermination (to appear).

Vasin, I.A. (1992b). The existence and uniqueness of generalized solution for a nonlinear nonstationary Navier–Stokes system (to appear).

Ill-Posed Problems in Natural Sciences, pp. 408 – 416
A. Tikhonov (Ed.)
1992 VSP/TVP

STABILITY ESTIMATES FOR INVERSE PROBLEMS OF GEOELECTRICS

V.G. ROMANOV

Institute of Mathematics of the Siberian Branch of the Russian Academy of Sciences, Universitetskii prospekt, 4, 630090 Novosibirsk, Russia

ABSTRACT

Results of investigations of some inverse problems related to the electrodynamics system of equation are described. A geophysical model is chosen for which all the space $\mathbf{R}^3$ of variables $(x_1, x_2, x_3) = x$ is divided by the plane $S = \{x \mid x_3 = 0\}$ in to two half-spaces $\mathbf{R}^3_- = \{x \mid x_3 < 0\}$, $\mathbf{R}^3_+ = \{x \mid x_3 > 0\}$. The coefficients ε, μ, σ of the Maxwell system of equations are known in the domain $\mathbf{R}^3_-$ as constants but they are unknown smooth functions of x in the domain $\mathbf{R}^3_+$. The problems of finding the coefficients ε, μ, σ in $\mathbf{R}^3_+$ from the observations of some components of the electromagnetic field on the interface S are considered. Stability estimates of one- and three-dimensional (linear and nonlinear) inverse problems are given. The quasi-stationary approximation of the Maxwell system of equations and the issues of inverse problems related to this system are discussed.

INTRODUCTION

Beginning with the classical works of A.N.Tikhonov (see (Tikhonov, 1949, 1950)), the inverse problems of electrodynamics were studied by many mathematicians and geophysicians. There is extensive scientific literature related to these problems. Our short review presents some basic results of studying stability issues for inverse problems of geoelectrics.

We will consider the following geophysical model. Let the physical space $\mathbf{R}^3$ of variables $(x_1, x_2, x_3) = x$ be divided by the plane $S = \{x \mid x_3 = 0\}$ into two parts $\mathbf{R}^3_+ = \{x \mid x_3 > 0\}$ and $\mathbf{R}^3_- = \{x \mid x_3 < 0\}$. We will assume that the halfspace $\mathbf{R}^3_-$ is filled with a homogeneous electromagnetic medium with given constant parameters: the dielectrical permeability $\varepsilon = \varepsilon^- > 0$, the magnetic permeability $\mu = \mu^- > 0$, and the conductivity $\sigma = \sigma^- \geq 0$. However the halfspace $\mathbf{R}^3_+$ is filled with a nonhomogeneous medium with the variable electromagnetic parameters ε, μ, and σ, which are functions of $x \in \mathbf{R}^3_+$. The latter parameters are unknown and have to be found. To this end, the information is used about the electromagnetic field caused by an external current source and observed on the plane S. Namely, let H, E be vector functions, $H = (H_1, H_2, H_3)$, $E = (E_1, E_2, E_3)$, which are the solution of the following problem for the Maxwell equations

$$\text{rot } H = \varepsilon \frac{\partial E}{\partial t} + \sigma E,$$

$$\text{rot } E = -\mu \frac{\partial H}{\partial t}, \quad (x,t) \in (\mathbf{R}^3_- \cup \mathbf{R}^3_+) \times \mathbf{R}^1, \tag{1}$$

$$(H,E)_{t<0} \equiv 0, \quad [H_k]_{x_3=0} = g_k(x_1,x_2,t), \quad [E_k]_{x_3=0} = 0, \; k = 1,2 \tag{2}$$

Here the symbol $[\cdot]_{x_3=0}$ means the difference of the limiting values of a function computed at points of S from the outside of the domains $\mathbf{R}^3_+$, $\mathbf{R}^3_-$, respectively. The conjugation conditions in the form (2) correspond to the external current j as follows: $j = (-g_2, g_1, 0)\delta(x_3)$. Thus, the functions g_1, g_2 determine the value and the direction of the external current on the plane S.

The general formulation of the inverse problem.
Given $\varepsilon^-, \mu^-, \sigma^-, g_1, g_2$ find ε, μ, σ within the domain $\mathbf{R}^3_+$ if the following information about the solution of problem (1), (2), is available:

$$E_k\big|_{x_3=0} = f_k(x_1,x_2,t), \quad k = 1,2. \tag{3}$$

Some remarks related to this formulation of the problem are needed.

Remark 1. All components of the electromagnetic field on the Earth surface are usually measured in the geophysical prospecting. Thus it may seem to be necessary to include all the components in (3). However, from the mathematical point of view, relations (3) determine uniquely the vectors H, E within the domain $\mathbf{R}^3_- \times \mathbf{R}^1$ and therefore their limiting values on the plane S. Indeed, the initial-boundary value problem for equations (1) with the initial data (2) and the boundary conditions (3) is well posed. Hence, H and E are constructed uniquely within the domain $\mathbf{R}^3_- \times \mathbf{R}^1$. Taking this into account, we can change the formulation of the inverse problem. Namely, the system of equations (1) can be considered within the domain $\mathbf{R}^3_+ \times \mathbf{R}^1$ only and all the tangential components of the field can be given on the boundary of this domain. It should be remembered, however, that they are functionally dependent and can be computed from data (3) only.

Remark 2. We can replace the components E_1, E_2 in (3) by one of the following pairs (E_1, H_1^-), (E_2, H_2^-), (H_1^-, H_2^-). The sign (-) corresponds to the external limiting values to the domain $\mathbf{R}^3_- \times \mathbf{R}^1$. Of course, it is possible to use the same pairs for the external limiting values to the domain $\mathbf{R}^3_+ \times \mathbf{R}^1$.

Remark 3. It is easy to show that the information in the form

$$E_k\big|_{x_3=x_3^0} = f_k(x_1,x_2,t), \quad k = 1,2, \; (x_1,x_2) \in Q_\rho, \; t \in \mathbf{R}^1,$$

where $x_3^0 < 0$, $Q_\rho = \{(x_1,x_2) |\; x_1^2 + x_2^2 < \rho^2\}$, $\rho > 0$, determines H and E within the domain $\mathbf{R}^3_- \times \mathbf{R}^1$ uniquely, since this information can be given instead of (3). Having the above remarks in mind, we now give a review of the results related to the of estimates stability of solutions of inverse problems.

1. THE ONE-DIMENSIONAL INVERSE PROBLEM

Let $\varepsilon = \varepsilon(x_3), \mu = \mu(x_3), \sigma = \sigma(x_3)$. Assume that in equalities (2) $g_1 = 0,\ g_2 = \delta(x_1, x_2, t)$. The fourier transform of any function $u(x,t),\ x \in \mathbf{R}^3$ with respect to the first two variables $x_1,\ x_2$ will be denoted by $\tilde{u}(\nu, x_3, t)$, where $\nu = (\nu_1, \nu_2)$ is the parameter of the transform. We consider only a small part of information to find the unknown coefficients ε, μ, σ. Namely, let us assume that we know the Fourier transform of the function $E_1|_{x_3=0}$ for three different values of the parameter ν:

$$\widetilde{E}_1(\nu^{(k)}, 0, t) = f_k(t), \quad k = 1,2,3, \quad t \in \mathbf{R}^1, \quad \nu^{(k)} = (\nu_{1k}, \nu_{2k}), \tag{4}$$

The problem is to determine the coefficients ε, μ, σ in the domain $\mathbf{R}^1_+ = \{x_3 |\ x_3 > 0\}$ in terms of f_1, f_2, f_3.
Let $M = M(\varepsilon^-, \mu^-, \sigma^-, m_1, m_2)$ be a set of the triples ε, μ, σ. The triple $(\varepsilon, \mu, \sigma) \in M(\varepsilon^-, \mu^-, \sigma^-, m_1, m_2)$, if

1) $\varepsilon = \varepsilon^- > 0,\ \mu = \mu^- > 0,\ \sigma = \sigma^- \geq 0$, if $x_3 < 0$,

2) $(\varepsilon, \mu) \in C^3[0, \infty),\ \sigma \in C^2[0, \infty)$,

3) $\varepsilon(x_3) \geq m_1 > 0,\ \mu(x_3) \geq m_1$,

4) $\max\left(\|\varepsilon\|_{C^3[0,\infty)}, \|\mu\|_{C^3[0,\infty)}, \|\sigma\|_{C^2[0,\infty)}\right) \leq m_2 < \infty$.

Analysis of the direct problem (1), (2) shows that under the condition $(\varepsilon, \mu, \sigma) \in M$ the functions $f_k(t),\ k = 1,2,3$, can be represented as $f_k(t) = a\delta(t) + f_{0k}(t)\theta(t),\ k = 1,2,3$, where $\theta(t)$ is the Heaviside function, $f_{0k} \in C^2[0, \infty)$.

THEOREM 1. (See (Romanov et al. 1984).) *Let the condition*

$$\begin{vmatrix} 1 & \nu_{11}^2 & \nu_{21}^2 \\ 1 & \nu_{12}^2 & \nu_{22}^2 \\ 1 & \nu_{13}^2 & \nu_{23}^2 \end{vmatrix} \neq 0 \tag{5}$$

for the parameters $\nu^{(k)} = (\nu_{1k}, \nu_{2k}),\ k = 1,2,3$, be satisfied and let the functions $f_k(t) = a\delta(t) + f_{0k}(t)\theta(t),\ k = 1,2,3$, defined by (4) correspond to the triple $(\varepsilon, \mu, \sigma) \in M$, and the functions $f_k(t) = \bar{a}\delta(t) + \bar{f}_{0k}(t)\theta(t),\ k = 1,2,3$, correspond to the triple $(\bar{\varepsilon}, \bar{\mu}, \bar{\sigma}) \in M$. Then for any $L > 0$ there exists a positive number $K = K(\varepsilon^-, \mu^-, \sigma^-, m_1, m_2, \nu^{(1)}, \nu^{(2)}, \nu^{(3)})$ such that

$$\|\varepsilon - \bar{\varepsilon}\|_{C^1[0,L]} + \|\mu - \bar{\mu}\|_{C^1[0,L]} + \|\sigma - \bar{\sigma}\|_{C[0,L]}$$

$$\leq K \left[|a - \bar{a}| + \sum_{j=1}^{3} \|f_{0j}(t) - \bar{f}_{0j}(t)\|_{C^2[0,T]} \right],$$

$$T \geq 2m_2 L.$$

Similar results were also obtained for special models of anisotropic media (see (Romanov, 1984; Romanov et al., 1984; Romanov and Savin, 1984; Romanov and Yakhno, 1984)).

Another version of the inverse problem of determination of the coefficient $\sigma(x_3)$ (the parameters ε, μ are constants) in which one component of the electromagnetic field is given at a fixed point $x \in S$ has been studied in (Priimenko, 1989).

2. THE LINEARIZED THREE-DIMENSIONAL INVERSE PROBLEM

In this case it is supposed that for $x \in \mathbf{R}^3_+$ the functions ε, μ, σ are represented as $\varepsilon = \varepsilon_0(x_3) + \varepsilon_1(x)$, $\mu = \mu_0(x_3) + \mu_1(x)$, $\sigma = \sigma_0(x_3) + \sigma_1(x)$, where the functions $\varepsilon_0(x_3) > 0$, $\mu_0(x_3) > 0$, $\sigma_0(x_3)$ are given, $(\varepsilon_0, \mu_0) \in C^3[0,\infty)$, $\sigma_0 \in C^2[0,\infty)$, and $\varepsilon_1, \mu_1, \sigma_1$ are unknown small functions finite in $\mathbf{R}^3_+$, $(\varepsilon_1, \mu_1) \in C^3(\mathbf{R}^3_+)$, $\sigma_1 \in C^2(\mathbf{R}^3_+)$. The assumption concerning the smallness of $\varepsilon_1, \mu_1, \sigma_1$ enables us to study this problem by using a linear approximation (see (Romanov et al., 1984)).
Assume that in (2), $g_1 = 0$, $g_2 = \delta(x_1 - x_1^0, x_2 - x_2^0, t)$ and let $x^0 = (x_1^0, x_2^0, 0)$ run through all the points of the plane $S = \{x|\ x_3 = 0\}$. In this case the solution of the direct problem (1), (2) depends on the coordinates x_1^0, x_2^0 of this point which should be treated as parameters. Denote by $E_1(\nu, x_3, t, \lambda)$ the following Fourier transform of the function $E_1(x, t, x^0)$ with respect to the variables x_1, x_2, x_1^0, x_2^0:

$$\widetilde{E}_1(\nu, x_3, t, \lambda) = \int\limits_{\mathbf{R}^4} E_1(x, t, x^0) \exp\{i[\nu_1(x_1 - x_1^0) + \nu_2(x_2 - x_2^0)$$

$$+\lambda_1 x_1^0 + \lambda_2 x_2^0]\} dx_1 dx_2 dx_1^0 dx_2^0,$$

$$\nu = (\nu_1, \nu_2), \quad \lambda = (\lambda_1, \lambda_2).$$

Let the following information be given:

$$\widetilde{E}_1(\nu^{(k)}, 0, t, \lambda) = f_k(t, \lambda), \quad k = 1, 2, 3,$$

$$0 < |\lambda| < \delta, \quad \delta > 0, \quad \lambda \in \mathbf{R}^2, \quad t > 0, \quad \nu^{(k)} = (\nu_{1k}, \nu_{2k}). \tag{6}$$

The problem consists in determining $\varepsilon_1, \mu_1, \sigma_1$ in terms of the data (6).
It is not difficult to show that, for $t > 0$, the functions $f_k(t, \lambda)$, $k = 1, 2, 3$, are representable as $f_k(t, \lambda) = a_k(t)\delta(\lambda) + f_{0k}(t, \lambda)$, where $a_k(t) \in C^2[0, \infty)$, $f_{0k} \in C^3[0, \infty)$ for any fixed $\lambda \in \mathbf{R}^2$ and $f_{0k}(t, \lambda)$ are analytic functions in $\lambda \in \mathbf{R}^2$ and can be continued to the complex space C^2 as entire analytic functions.
Let $q_1 = (\varepsilon_1, \mu_1, \sigma_1)$ be the vector of the unknown functions such that $q_1 \equiv 0$ if $x_1^2 + x_2^2 \geq r_0^2$, $r_0 > 0$, and let $f = (f_1, f_2, f_3)$ be the vector of data of the inverse problem. Define

$$\|q_1\|_L = \sup_{0 \leq x_3 \leq L} \left(\iint\limits_{x_1^2 + x_2^2 \leq r_0^2} \left[\varepsilon_1^2 + \mu_1^2 + \sigma_1^2 + \left(\frac{\partial \varepsilon_1}{\partial x_3}\right)^2 + \left(\frac{\partial \mu_1}{\partial x_3}\right)^2 \right] dx_1 dx_2 \right)^{1/2},$$

$$\|f\|_T = \max_{k=1,2,3} \sup_{\substack{0\le t\le T\\ 0<|\lambda|<\delta}} \left|\frac{\partial^2 f(t,\lambda)}{\partial t^2}\right|, \quad T>0.$$

THEOREM 2. *Under condition (5) for any $L>0$ there exists a positive constant K depending on $L, r_0, \delta, \nu^{(1)}, \nu^{(2)}, \nu^{(3)}$, the norms of the functions $\varepsilon_0, \mu_0, \varepsilon_1, \mu_1$ in the space $C^3(D_{r_0L})$, $D_{r_0L} = \{x|\ 0\le x_3\le L,\ x_1^2+x_2^2\le r_0^2\}$, and the norms of the functions σ_0, σ_1 in the space $C^2(D_{r_0L})$, such that*

$$\|q_1\|_L \le K\,(1+|\ln\|f\|_T|)^{1/2}, \quad T = 2\int\limits_0^L \sqrt{\varepsilon_0(s)\mu_0(s)}\,ds.$$

This estimate follows from the analysis contained in (Romanov et al., 1984)) and from the well-known estimates of the analytical continuation in the complex space C^2.

3. THREE-DIMENSIONAL NONLINEAR PROBLEM

Unfortunately, the simultaneous determination of ε, μ, σ as functions of $x\in\mathbf{R}^3_+$ has not been studied. There are results showing how to find only one of these functions under the condition that the other two are known. We will give a stability estimate for the problem of determination of $\varepsilon(x)$ in $\mathbf{R}^3_+$ under the assumption that $\mu=\mu^+>0$, $\sigma=\sigma^+$ are known constants in $\mathbf{R}^3_+$. Let us refine more the formulation of the inverse problem. Assume that, in (2), $g_k = g_k^0(\bar{x})\delta(t)$, $k=1,2$, $\bar{x}=(x_1,x_2)$, and the data have the form (3). When investigating the inverse problem it is convenient to pass to a parametrical representation of the function $\varepsilon(x)$. To this end, we consider solution $z=z(x)$ of the following Cauchy problem $|\nabla z|^2=\varepsilon\mu$, $z|_{x_3=0}=0$, $z_{x_3}>0$. Let $x_3=h(\bar{x},z)$, $\bar{x}=(x_1,x_2)$ be the inverse function for $z=z(x)$ with respect to the variable x_3. Define $\hat{\varepsilon}(\bar{x},z)=\varepsilon(\bar{x},h(\bar{x},z))$. The pair $\hat{\varepsilon}(\bar{x},z)$, $h(\bar{x},z)$ provides a parametrical representation of the function $\varepsilon=\varepsilon(x)$. We will consider the problem of finding the pair $\hat{\varepsilon}$, h in the class of analytic functions with respect to $\bar{x}$ and continuously differentiable with respect to z.

Let a scale of Banach spaces, A_s, $s>0$, be introduced as follows. The space A_s is the set of analytic functions $\varphi(\bar{x})$ whose s norm $\|y\|_s$ defined as

$$\|\varphi\|_s = \sum_{n,m=0}^{\infty} \frac{s^{n+m}}{n!m!}\left|\frac{\partial^{n+m}}{\partial x_1^n\partial x_2^m}\varphi(\bar{x})\right|_{\bar{x}=0}$$

is finite. If a function $u(\bar{x},z,t)$ is an element of A_s for fixed then we use the notation $\|u\|_s(z,t)$ for its s norm. Finally, if a function $u(\bar{x},z,t)$ is a continuous element of A_s with $(z,t)\in D$ and

$$\sup_{(z,t)\in D}\|u\|_s(z,t)\equiv\|u\|_{C(A_s,D)}<\infty,$$

then we say that $u\in C(A_s,D)$.

Let $\mathcal{E}(s_0, T_0, m)$ be the set of functions $\hat{\varepsilon}(\bar{x}, z)$ which satisfy the conditions

$$\|\hat{\varepsilon}\|_{C(A_{s_0},[0,T_0])} \le m, \quad \|1/\hat{\varepsilon}\|_{C(A_{s_0},[0,T_0])} \le m,$$

$$\|\frac{\partial}{\partial z}\hat{\varepsilon}\|_{C(A_{s_0},[0,T_0])} \le m$$

with some positive s_0, T_0, m, and let $G(s_0, M)$ be the set of functions $g(\bar{x})$ such that $\max(\|g\|_{s_0}, \|1/g\|_{s_0}) \le M$.

THEOREM 3 (Romanov and Pukhnacheva, 1990). *Let the data* (3) $f_k(\bar{x}, t)$, $k = 1, 2$, *correspond to the functions* $\hat{\varepsilon} \in \mathcal{E}(s_0, T_0, m)$, $g_k^0 \in G(s_0, M)$, $k = 1, 2$, *and let the data* $f'_k(\bar{x}, t)$, $k = 1, 2$, *correspond to the functions* $\hat{\varepsilon}'_k \in \mathcal{E}(s_0, T_0, m)$, $(g_k^0)' \in G(s_0, M)$, $k = 1, 2$. *Then, for any* $p > 1$, $s < s_1 < s_0$, *there exist positive* T, K, $T \in (0, T_0)$, $K = K(s_0, T_0, m, M, p, s_1)$, *such that*

$$\|\hat{\varepsilon} - \hat{\varepsilon}'\|_s(z) + \|h - h'\|_s(z)$$

$$\le K \sum_{j=1}^{2} \left[\|g_j^0 - (g_j^0)'\|_{s_0} + \max_{0 \le t \le T(1 - s_1/s_0)} \|f_j - f'_j\|_{s_1}(t) \right],$$

$$z \in [0, T(1 - s_1/s_0)/(1+p)].$$

Here, h *and* h' *correspond to* $\hat{\varepsilon}$ *and* $\hat{\varepsilon}'$, *respectively.*

Similar estimates were obtained for the problems of determination of the coefficients μ, σ.

4. STATIONARY CASE: THE ONE-DIMENSIONAL INVERSE PROBLEM FOR ELLIPTIC EQUATION

If the coefficients ε, μ, σ and the functions g_1, g_2 do not depend on x_2, then problem (1), (2) reduces two independent problems. The first them is the problem for the components (H_1, H_3, E_2), the second concerns the components (H_2, E_1, E_3). The first problem can easily be rewritten in terms of the function E_2 only and has the form

$$\varepsilon \frac{\partial^2 E_2}{\partial t^2} + \sigma \frac{\partial E_2}{\partial t} = \frac{\partial}{\partial x_2}\left(\frac{1}{\mu}\frac{\partial E_2}{\partial x_1}\right) + \frac{\partial}{\partial x_3}\left(\frac{1}{\mu}\frac{\partial E_2}{\partial x_3}\right),$$

$$E_2\big|_{t<0} \equiv 0, \quad [E_2]_{x_3=0} = 0, \quad \left[\frac{1}{\mu}\frac{\partial E_2}{\partial x_3}\right]_{x_3=0} = \frac{\partial}{\partial t} g_2(x_1, t).$$

We will consider the case $\mu \equiv 1$, $g_2(x_1, t) = e^{i\omega t}\delta(x_1)$, $\omega > 0$. If $t \to \infty$, then periodical oscillations are established and the complex amplitude $u(x_1, x_3)$ of these oscillations satisfies the relations

$$\left(\frac{\partial^2}{\partial x_1^2} + \frac{\partial^2}{\partial x_3^2}\right) u - qu = 0, \tag{7}$$

$$[u]_{x_3=0} = 0, \quad [u_{x_3}]_{x_3=0} = g_0\delta(x_1), \tag{8}$$

where $q = -\omega^2\varepsilon + i\omega\sigma$, $g_0 = i\omega$.
In the domain $x_3 < 0$ the coefficient q is constant, $q = q^- = -\omega^2\varepsilon^- + i\omega\sigma^-$. Assume that in the domain $x_3 \geq 0$, the coefficients ε, σ depend on x_3 only. Then the following formulation of the inverse problem looks reasonable: for a fixed $\omega > 0$, given the trace of the solution of problem (7), (8) on the straight line $x_3 = 0$:

$$u|_{x_3=0} = f(x_1), \quad x_1 \in \mathbf{R}^1, \tag{9}$$

find $q(x_3)$ in the domain $x_3 \geq 0$.
It is clear that the complex coefficient q determines both the unknown functions ε, σ. A stability estimate for the solution of inverse problem (7)–(9) has been obtained in (Romanov and Mamayusupov, 1985).
Consider the set $Q(q^-, \alpha, m, M)$ of complex functions $q(x_3)$ which satisfy the following conditions:

1) $q = q^-$, if $x_3 < 0$,
2) the coefficient $q(x_3)$ is representable in the domain $x_3 > 0$ as $q = q_0 + \hat{q}(x_3)/(1 + x_3)^\alpha$, $\alpha > 1$,
3) $\operatorname{Im} q \geq 0$, $\operatorname{Im} q^- \geq m > 0$, $\operatorname{Im} q_0 \geq m$,
4) $\|\hat{q}\|_{C^2[0,\infty)} \leq M$, $|q^-| \leq M$, $|q^0| \leq M$.

It can be shown that any function $f(x_1)$ which is the trace of the solution of direct problem (7), (8) with a coefficient $q \in Q(q^-, \alpha, m, M)$ on the line $x_3 = 0$ is representable as

$$f(x_1) = -\frac{1}{2\pi} g_0 K_0(x_1\sqrt{q^-}) + f_0(x_1),$$

where K_0 is the modified Bessel function and $f_0(x_1)$ is a continuous function which decreases exponentially as $|x_1| \to \infty$.

THEOREM 4. *Let $(q_1, q_2) \in Q(q^-, \alpha, m, M)$ and let f_1, f_2 be the traces of the solutions of problem (7), (8) on the line $x_3 = 0$, which correspond to the functions $q = q_1$ and $q = q_2$ respectively. Then, for any $L > 0$, there exists a positive constant $K = K(L, q^-, \alpha, m, M)$ such that*

$$\|q_1 - q_2\|_{C[0,L]} \leq K\left(1 + \left|\ln \|f_1 - f_2\|_{C(-\infty,\infty)}\right|\right)^{-1/8}.$$

5. THE QUASI-STATIONARY SYSTEM OF EQUATIONS

In the case where the displacement current $\varepsilon\frac{\partial}{\partial t}E$ is substantially smaller than the conductivity current σE, use can be made of the following quasi-stationary system of equations for the description of the process of waves propagation:

$$\operatorname{rot} H = \sigma E, \quad \operatorname{rot} E = -\mu\frac{\partial H}{\partial t},$$

$$(x,t) \in (\mathbf{R}^3_- \cup \mathbf{R}^3_+) \times \mathbf{R}^1. \tag{10}$$

It is reasonable to assume that $\sigma(x) > 0$. Under this assumption system (10) is a system of parabolic equations describing process of diffusion of electromagnetic waves. We will consider system (10) under the conditions

$$(H,E)_{t<0} \equiv 0;$$

$$[H_k]_{x_3=0} = g_k^0(x_1,x_2)\delta(t), \quad [E_k]_{x_3=0} = 0, \quad k = 1,2. \tag{11}$$

To investigate the inverse problems of determination of the coefficients σ, μ in terms of the data (3), we can reduce them to the inverse problems for the auxiliary hyperbolic system

$$\mathrm{rot}\hat{H} = \sigma\frac{\partial\hat{E}}{\partial t}, \quad \mathrm{rot}\hat{E} = -\mu\frac{\partial\hat{H}}{\partial t} \tag{12}$$

under the conditions

$$(\hat{H},\hat{E})_{t<0} \equiv 0; \quad [\hat{H}_k]_{x_3=0} = g_k^0(x_1,x_2)\delta(t),$$
$$[\hat{E}_k]_{x_3=0} = 0, \quad k = 1,2. \tag{13}$$

The solutions of problems (10), (11) and (12), (13) are connected by the relations

$$H(x,t) = \int_{-\infty}^{\infty} \hat{H}(x,\tau)G(t,\tau)\,d\tau,$$

$$E(x,t) = \int_{-\infty}^{\infty} \hat{E}_\tau(x,\tau)G(t,\tau)\,d\tau, \tag{14}$$

where

$$G(t,\tau) = \theta(t)\frac{\tau}{2\sqrt{\pi t^3}}\exp\left(-\frac{\tau^2}{4t}\right).$$

Formula (14) has been established in (Priimenko, 1982).
Define

$$\hat{f}_k(x_1,x_2,t) = \hat{E}_k(x_1,x_2,0,t), \quad k = 1,2. \tag{15}$$

According to (14), there is the following relation between the functions $f_k, \hat{f}_k$:

$$f_k(x_1,x_2,t) = \int_{-\infty}^{\infty} \left[\frac{\partial}{\partial\tau}\hat{f}_k(x_1,x_2,\tau)\right] G(t,\tau)\,d\tau$$

$$= \frac{\theta(t)}{4\sqrt{\pi t^3}}\int_{-\infty}^{\infty} \left[\frac{\partial}{\partial\tau}\hat{f}_k(x_1,x_2,\tau)\right]_{\tau=\sqrt{z}} \exp\left(-\frac{z}{4t}\right) dz \tag{16}$$

In fact, (16) is the Fourier transformation with the parameter $p = 1/4t$. Inverting this transformation, we can find $f_k(x_1, x_2, t)$, $k = 1, 2$, uniquely and the inverse problem of determination of the coefficients σ, μ reduces to the corresponding problem (12), (13), (15). To obtain a stability estimate of the problem of determination of $\sigma(x)$, for example, one can use the estimates from Theorems 1–3. To do this, it is necessary to obtain a stability estimate of the solution of equation (16) only. In fact, the problem reduces to that of estimation of the analytical continuation of the function f_k, as a function of $p = 1/4t$, from the real axis to the complex half-plane $\mathrm{Re}\, p \geq \alpha > 0$. Such a problem has been considered in (Romanov and Mamayusupov, 1985) as a part of a more complex problem. The results given above have been used in (Priimenko, 1990) to obtain a stability estimate for the one-dimensional problem of determination of the coefficient $\sigma = \sigma(x_3)$.

REFERENCES

Priimenko, V.I. (1982). On the uniqueness of finding the conductivity tensor for a one-dimensional medium. In: *Correctness Questions for Inverse Problems of Mathematical Physics*, Computing Centre, Novosibirsk, pp. 120–130 (in Russian).

Priimenko, V.I. (1989). A problem of conductivity finding from the Maxwell system of equations. In: *Correctness Questions of Analysis Problems*, Institute of Mathematics, Novosibirsk, pp. 125–141 (in Russian).

Priimenko, V.I. (1990). A stability estimete of the solution of a one-simensional inverse problem for a mixed type equation. In: *Methods of Inverse Problems Solution*, Institute Mathematics, Novosibirsk, pp. 121–143 (in Russian).

Romanov, V.G. (1984). On the uniqueness of finding the dielectric and magnetic permeabilities for an anisotropic one-dimensional non-homogeneous medium. *Different. Equat.* **20**, 325–332.

Romanov, V.G. (1989). On the solvability of inverse problems for hyperbolic equations in a class of functions analytic in some of their variables. *Soviet Math. Dokl.* **39**, 160–164.

Romanov, V.G. (1984). *Inverse Problems of Electrodynamics.* Computing Centre, Siberian Branch of USSR Academy Sci., Novosibirsk (in Russian).

Romanov, V.G. and Mamayusupov, O.Sh. (1985). *A Stability Estimate of a Solution of a One-Dimensional Inverse Problem for a Elliptic Type Equation.* Preprint No 583. Computing Centre Siberian Branch of USSR Academy Sci., Novosibirsk, pp. 1–48 (in Russian).

Romanov, V.G. and Pukhnacheva, T.P. (1990). Determination of the coefficients of a system of equations of electrodynamics in a class of functions analytic in some of their variables. *Soviet. Math. Dokl.* **41**, 417–420.

Romanov, V.G. and Savin, M.G. (1984). A problem of finding the conductivity tensor for a non-homogeneous with respect to depth anisotropic medium. *J. Phys. Earth*, 84–94.

Romanov, V.G. and Yakhno, V.G. (1984). A problem of finding the tensor dielectric permeability for a one-dimensional non-homogeneous medium. In: *Ill-Posed Problems of Mathematical Physics*, Nauka, Novosibirsk, pp. 123–128 (in Russian).

Tikhonov, A.N. (1949). On the uniqueness of solution for a problem of electric prospecting. *Dokl. Acad. Nauk SSSR* **69**, 797–800 (in Russian).

Tikhonov, A.N. (1950). On the finding of electrical properties of deep layers of the Earth crust. *Dokl. Acad. Nauk SSSR* **73**, 295–298 (in Russian).

Ill-Posed Problems in Natural Sciences, pp. 417 – 422
A. Tikhonov (Ed.)
1992 VSP/TVP

INVERSE PROBLEMS FOR ELLIPTIC DIFFERENTIAL OPERATORS

V.A. SADOVNICHII, V.V. DUBROVSKII, and A.V. NAGORNY
Moscow State University, Len. Gory, 118899 Moscow, Russia

ABSTRACT

There is solved inverse problem for power of operator of Laplace with potential, on which authors restrict bounds. This method admits using for others elliptic operators. There is investigated stability of solution of inverse problems with potential from some class functions.

In this article we investigate the problem of restoring of potential using one spectrum of the perturbed operator. The basic operators are powers of the laplacian and Sturm–Liouville operators. In addition we need restriction on the potential. We investigate the stability of the solution of inverse problem for a self-adjoint discrete operator. Some of these results were published in (Sadovnichii and Dubrovskii, 1979; Dubrovskii and Nagorny, 1990).
Let $\Pi = \{0 \le x \le a,\ 0 \le y \le b\}$ and $\Pi_{\frac{1}{4}} = \{0 \le x \le a/2,\ 0 \le y \le b/2\}$ be rectangles, where a^2/b^2 is an algebraic number of power $n \ge 2$. We investigate in $\mathbf{L}^2(\Pi)$ the boundary problem L

$$-\Delta u = \lambda u, \qquad u|_{\partial\Pi} = 0,$$

where $\partial\Pi$ is the bound of any of the rectangle Π. We introduce the operator $T = \int \lambda^\beta dE(\lambda)$, where $E(\lambda)$ is the decomposition of unity of the operator L, $\beta > 0$. Let $v_{m,n}(x,y) = 2/\sqrt{ab}\sin\frac{\pi}{a}mx\sin\frac{\pi}{b}ny$ be the orthonormal eigen-functions of T corresponding to the eigenvalues $\lambda_{m,n} = ((\pi^2/a^2)m^2 + (\pi^2/b^2)n^2)^\beta$; $d_{m,n} = \rho(\sigma(T)\backslash\lambda_{m,n})$, $m,n = 1,2\dots$. If a^2/b^2 is an irrational number, then the spectrum of operator T is single (Dubrovskii and Nagorny, 1990). We will numerate $\lambda_{m,n}$ by a single index t in the increasing order of λ. Let P be the operator of multiplication by a function $p(x,y) \in \mathbf{L}^2(\Pi)$, μ_t be the eigenvalues of the operator $T + P$ numerated in the order of increasing Re μ_t taking into account multiplicity. The function $p(x,y)$ satisfies the condition

$$p(a-x,y) = p(x,y) = p(x,b-y), \qquad (x,y) \in \Pi, \tag{1}$$

which is necessary for restoring the potential with single spectrum. Let conditions

$$\iint\limits_{\Pi} p(x,y)\cos\frac{2\pi}{a}mx\,dx\,dy$$

$$= \iint_{\Pi} p(x,y) \cos \frac{2\pi}{b} ny dx dy = 0, \tag{2}$$

hold where $m, n = 0, 1, 2, \ldots$.

THEOREM 1. *If $\beta > 2\frac{1}{2}$ in $\mathbf{L}^2(\Pi_{\frac{1}{4}})$, then three exists an ball $u(0,\varepsilon) = \{p(x,y)\colon \|p(x,y)\|_2 \leq \varepsilon\}$ such that for a sequence of numbers $\{\mu_t\}_{t=1}^{\infty}$ satisfying the condition $(\sum_{t=1}^{\infty} |\mu_t - \lambda_t|^2)^{\frac{1}{2}} < \delta$, where $\delta = \delta(\varepsilon, a, b)$, there exists one potential $p(x,y) \in u(0,\varepsilon)$ satisfying conditions (1), (2) and such that the operator $T + P$ have the spectrum $\{\mu_t\}_{t=1}^{\infty}$.*

Proof. If a^2/b^2 is an algebraic number of power $n \geq 2$, then (Dubrovskii and Nagorny, 1990) $d_t \geq$ const $t^{\beta-2-\gamma}$, where γ is arbitrary small. Using the inequalities

$$\|P(T - \lambda E)^{-1}\| \leq \frac{\text{const}\|P\|}{d_t}, \tag{3}$$

$$\|(T + P - \lambda E)^{-1}\| \leq \frac{\text{const}}{d_t},$$

which are valid if $\lambda \in c_t$, $c_t = \{\lambda\colon |\lambda_t - \lambda| = d_t/2\}$, we can prove that when $\|p\|_2$ is small the spectrum of $T + P$ is single and $|\mu_t - \lambda_t| \leq$ const $\|p\|_2$.
We use the identity

$$\mu_t - \lambda_t = \frac{(Pu_t, v_t)}{(u_t, v_t)}. \tag{4}$$

The following equality is valid

$$(u_t, v_t) =$$

$$1 - \frac{1}{2\pi i}\Big(\int_{c_t} (T + P - \lambda E)^{-1} \left[P(T - \lambda E)^{-1}\right]^2 v_t d\lambda, v_t\Big).$$

The system of functions $\psi_{m,n}(x,y) = (4/\sqrt{ab}) \cos \frac{2\pi}{a} mx \cos \frac{2\pi}{b} ny$, $m, n = 0, 1, 2, \ldots$ is an orthonormal base in $\mathbf{L}^2(\Pi_{\frac{1}{4}})$. If $p_{m,n}$ or p_t is the Fourier coefficient for the function $p(x,y)$ in $\mathbf{L}^2(\Pi_{\frac{1}{4}})$ with respect to $\psi_{m,n}(x,y)$, then

$$(Pv_{m,n}, v_{m,n}) = \frac{p_{m,n}}{\sqrt{ab}}.$$

From (4) we obtain the following equality

$$p_t = \sqrt{ab}(\mu_t - \lambda_t) + \sqrt{ab}(\mu_t - \lambda_t)\Phi_t'(p)$$

$$+ \sqrt{ab}\Phi_t(p), \tag{5}$$

where

$$\Phi_t(p) = \frac{1}{2\pi i}\Big(\int_{c_t} P(T + P - \lambda E)^{-1} P(T - \lambda E)^{-1} v_t d\lambda, v_t\Big),$$

$$\Phi_t'(p) = -\frac{1}{2\pi i}\Big(\int_{c_t} (T + P - \lambda E)^{-1} \left[P(T - \lambda E)^{-1}\right]^2 v_t d\lambda, v_t\Big).$$

We multiply equality (5) by the systems of functions $\{\psi_{m,n}(x,y)\}_{m,n=1}^{\infty}$ and thus obtain the operator equation in the space $\mathbf{L}^2(\Pi_{\frac{1}{4}})$

$$p = x_0 + \Phi'(p) + \Phi(p), \tag{6}$$

where

$$\Phi'(p) = \sqrt{ab}\sum_{t=1}^{\infty}(\mu_t - \lambda_t)\Phi'_t(p)\psi_t,$$

$$\Phi(p) = \sqrt{ab}\sum_{t=1}^{\infty}\Phi_t(p)\psi_t,$$

$$x_0 = \sqrt{ab}\sum_{t=1}^{\infty}(\mu_t - \lambda_t)\psi_t.$$

We shall find the solution for operator equation (6) in the space $\mathbf{L}^2(\Pi_{\frac{1}{4}})$ in the class of functions satisfying conditions (1), (2). The spectrum of the operator $T+P$ coincides with the sequense $\{\mu_t\}_{t=1}^{\infty}$ if and only if, when $p(x,y)$ is a solution of equation (6).
We prove that the operator $\Phi(p)$ when $\|p\|_2$ is small is a contractive operator with the coefficient of contraction $\alpha < 1/4$. Using inequalities (3), we obtain

$$\|\Phi(p) - \Phi(\tilde{p})\|_2 \le C\left(\sum_{t=1}^{\infty}\frac{(\|p\|_2 + \|\tilde{p}\|_2 + \|p\|_2\|\tilde{p}\|_2)^2}{d_t^2}\right)^{\frac{1}{2}}$$

$$\times\|p - \tilde{p}\|_2 = \alpha\|p - \tilde{p}\|_2,$$

where

$$C = C(a,b),$$

$$\alpha = C\left(\sum_{t=1}^{\infty}\frac{(\|p\|_2 + \|\tilde{p}\|_2 + \|p\|_2\|\tilde{p}\|_2)^2}{d_t^2}\right)^{\frac{1}{2}}.$$

The series $\sum_{t=1}^{\infty} 1/d_t^2$ is convergent if $\beta > 2(1/2)$. Hence if $\|p\|_2$ and $\|\tilde{p}\|_2$ are sufficiently small we have $\alpha < 1/4$.
Similarily we can prove, that the operator $\Phi'(p)$ is contractive with the coefficient of contraction $\alpha' < 1/4$. Consequently the operator $\Phi'(p) + \Phi(p)$ is contractive with the coefficient of contraction $\alpha' + \alpha$ in some ball $u(0,\varepsilon) = \{p(x,y) | \|p\|_2 \le \varepsilon\}$.
We shall prove that the operator $x_0 + \Phi'(p) + \Phi(p)$ maps the ball $u(0,\varepsilon)$ into itself. Choosing δ appropriately we can assume that $\|x_0\|_2 < (1/2)\varepsilon$ and under this assumption we obtain

$$\|x_0 + \Phi'(p) + \Phi(p)\|_2 \le \|x_0\|_2 + \|\Phi'(p) + \Phi(p)\|_2$$

$$\le \frac{1}{2}\varepsilon + \|\Phi'(p) + \Phi(p) - \Phi'(0) - \Phi(0)\|_2 \le \frac{1}{2}\varepsilon + \|\Phi'(p) - \Phi(0)\|_2 + \|\Phi(p) - \Phi(0)\|_2$$

$$\le \frac{1}{2}\varepsilon + \alpha'\varepsilon + \alpha\varepsilon < \varepsilon.$$

Hence operator equation (6) has a unique solution in the ball $u(0,\varepsilon)$, which can be found by the method of successive approximations.

Theorem 1 is proved.

If a^2/b^2 is a rational number, then the spectrum of the operator T consists of single and multiple eigenvalues. Let $\{\lambda_{t_k}\}_{k=1}^{N}$, $N \leq \infty$ be the sequence of single eigenvalues of the operator T. This sequence may be finite or infinite, or coincide with the whole spectrum, if a^2/b^2 is an irrational number. We suppose that $p(x,y) \in C(\Pi)$ and satisfies conditions (1), (2).

THEOREM 2. *If* $\sum_{k=1}^{N} d_{t_k}^{-1} < \infty$, *then in* $C(\Pi_{\frac{1}{4}})$ *there exists a ball* $u(0,\varepsilon) = \{p(x,y) | \|p\|_\infty \leq \varepsilon\}$ *such that for a set* $\{\xi_{t_k}\}_{k=1}^{N}$ *satisfying the condition* $\sum_{k=1}^{N} |\xi_{t_k} - \lambda_{t_k}| < \delta$, *where* $\delta = \delta(\varepsilon, a, b)$, *there exists one and only one potential* $p(x,y) \in u(0,\varepsilon)$ *satisfying conditions (1), (2), with the following properties*

a) $\mu_{t_k} = \xi_{t_k}, \qquad k = 1, 2, \ldots, N$

b) $\iint_\Pi p(x,y) \cos \frac{2\pi}{a} mx \cos \frac{2\pi}{b} ny dx dy = 0$, *if* $\lambda_{m,n} \notin \{\lambda_{t_k}\}_{k=1}^{N}$,

c) *Fourier series of the function* $p(x,y)$ *converges uniformly and absolutely on* $\Pi_{\frac{1}{4}}$.

Proof of Theorem 2. is analogous to the proof of Theorem 1. Operator equation (6) is investigated in the space $C(\Pi_{\frac{1}{4}})$. If the sequence $\{\lambda_{t_k}\}_{k=1}^{N}$ coincideds with the whole spectrum of the operator T, then the series $\sum_{t=1}^{\infty} d_t^{-1}$ converges if $\beta > 3$. If $\beta > 2$, then the potential may be restored by an infinite subsequence of the spectrum in the class of the corresponding functions, when a^2/b^2 is an algebraic number and the series $\sum_{k=1}^{\infty} d_{t_k}^{-1} < \infty$. The restoring of potentials, which are finite trigonometric polynomials may be done by a finite subsequence of the spectrum if $\beta > 0$.

Example 1. Let $b = a/\sqrt{2}$. Then the eigenvalues of the operator will have the following form $\lambda_{m,n} = (\pi^{2\beta}/a^{2\beta})(m^2 + 2n^2)^\beta$. The operator T have the infinite subsequence of single eigenvalues $\lambda_{m,n} = (\pi^{2\beta}/a^{2\beta})p^\beta$, where p is a prime number of the form $p = 8l + 1$ or $p = 8l + 3$, which is represented uniquely in the form $p = m^2 + 2n^2$. Under this condition $d_{t_k} \geq \text{const } t_k^{\beta-1}$, and hence $\sum_{k=1}^{\infty} d_{t_k}^{-1} < \infty$ for $\beta > 2$.

Example 2. Put $a = 1, b = 1/M$, where M is a large number. The eigenvalues of the operator T have the following form $\lambda_{m,n} = \pi^{2\beta}(m^2 + M^2 n^2)^\beta$.

If $n \geq 2$, then $\lambda_{m,n} \geq \pi^{2\beta}(1 + 4M^2)^\beta$. The finite subsequence of the spectrum $\lambda_{m,n} < \pi^{2\beta}(1 + 4M^2)^\beta$, where $m < [\sqrt{3M^2 + 1}]$ consists of single eigenvalues.

The suggested method of restoring of potentials by one spectrum does not depend on the form of a particular operator and admites extention to other operators, in particular to a power of the Sturm–Liouville operator.

Consider in $\mathbf{L}^2[0,\pi]$ the boundary problem L

$$-\frac{d^2y}{dx^2} = \lambda y, \qquad y(0) = y(\pi) = 0.$$

We introduce the operator $T = \int \lambda^\beta dE(\lambda)$, where $E(\lambda)$ is a decomposition of unity of the operator L, $\beta > 0$. Let P be the operator of multiplication by a function

$q(x) \in \mathbf{L}^2[0,\pi]$ with the following properties

$$q(x) = q(\pi - x), \qquad x \in [0,\pi], \qquad \int_0^\pi q(x)dx = 0. \tag{7}$$

Theorem 1 is valid for a power of the Sturm–Liouville operator if $\beta > 3/4$. This method makes it possible to calculate ε. If $q(x) \in C(\Pi)$, then the main statement of Theorem 2 is valid for $\beta > 1$. Let T be a selfadjoint discrete operator in the Hilbert space $H = \mathbf{L}^2(d)$, where D is a bounded domain in $\mathbf{R}^n$. Denote by λ_t the eigenvalues of the operator T; $v_t(x), x \in D$ are orthonormal eigenfunctions. Furthermore let P be the operator of multiplication by a function $p(x) \in \mathbf{L}^2(D)$. Denote by μ_t the eigenvalues of the operator $T + P$ numbered in the increasing order of $\text{Re}\mu_t$ with multiplicity taken into account and by $u_t(x)$ the corresponding normed eigenfunctions.

We suppose that the spectrum of the operator T is single and the following conditions are fulfilled

1) $\sum\limits_{t=1}^{\infty} d_t^{-2} < \infty$,

2) $\sup\limits_{x\in D} |v_t(x)| \le C, \qquad t = 1, 2, \ldots.$

We shall suppose that $p(x)$ belongs to some class of functions $B(C_1, C_2) \subset \mathbf{L}^2(D)$ for which the following condition is fullfilled

$$C_1 \int_D |p(x)|^2 dx \le \sum_{t=1}^{\infty} |(Pv_t, v_t)|^2$$
$$\le C_2 \int_D |p(x)|^2 dx.$$

For functions $p(x,y) \in \mathbf{L}^2(\Pi)$ satisfying conditions (1), (2) we have the equation

$$\sum_{t=1}^{\infty} |(Pv_t, v_t)|^2 = \frac{1}{4ab} \iint_{\Pi} |p(x,y)|^2 dxdy.$$

Thus functions satisfying conditions (1), (2) belong to the class $B(1/4ab, 1/4ab)$. Hence $\sum_{t=1}^{\infty} d_t^{-2} < \infty$ when $\beta > 2.5$. A function $q(x) \in \mathbf{L}^2[0,\pi]$ with property (7) satisfies the equation

$$\sum_{t=1}^{\infty} |(Pv_t, v_t)|^2 = \frac{1}{4\pi} \int_0^\pi |q(x)|^2 dx.$$

The series $\sum_{t=1}^{\infty} d_t^{-2} < \infty$ if $\beta > 3/4$. Hence functions $q(x)$ with property (7) belong to the class $B(1/4\pi, 1/4\pi)$.

THEOREM 3. *If a selfadjoint descrete operator T has single spectrum and conditions 1) and 2) are satisfied, then there exist $\varepsilon > 0$ such that if $\|p\|_2, \|\tilde{p}\|_2 < \varepsilon$ and $p - \tilde{p} \in B(C_1, C_2)$, then*

$$C_1' \sqrt{\sum_{t=1}^{\infty} |\mu_t - \tilde{\mu}_t|^2} \leq \|p - \tilde{p}\|_2 \leq C_2' \sqrt{\sum_{t=1}^{\infty} |\mu_t - \tilde{\mu}_t|^2},$$

where $C_1', C_2' > 0$.

LEMMA. *If conditions 1) and 2) are fulfilled, then the normed eigenfunctions $\{u_t(x)\}_{t=1}^{\infty}$ of the operator $T+P$ are uniformly bounded in $\mathbf{L}_{\infty}(D)$ for small $\|p\|_2$.*

Proof of Theorem 3. We have

$$\frac{1}{(u_t, v_t)} = 1 + \sum_{l=1}^{\infty} \left(\frac{1}{2\pi i}\right)^l \times \left(\int_{c_t} (T + P - \lambda E)^{-1} \left[P(T - \lambda E)^{-1}\right]^2 v_t d\lambda, v_t \right)^l.$$

Hence in the well known identity

$$\mu_t - \lambda_t = \frac{(Pu_t, v_t)}{(u_t, v_t)}$$

we can restrict ourself to the investigation of the first term in the right side. We have $\mu_t - \lambda_t = p_t + (P(u_t - v_t), v_t)$, where $p_t = (Pv_t, v_t)$. We get the inequality

$$|\mu_t - \tilde{\mu}_t - p_t + \tilde{p}_t| \leq C_3 \frac{\|p(x) - \tilde{p}(x)\|_2}{d_t},$$

where $C_3 = C_3(\|p\|_2, \|\tilde{p}\|_2) \to 0$ as $\|p\|_2, \|\tilde{p}\|_2 \to 0$. Summing with respect to t we have

$$\left| \sqrt{\sum_{t=1}^{\infty} |\mu_t - \tilde{\mu}_t|^2} - \sqrt{\sum_{t=1}^{\infty} |((p - \tilde{p})v_t, v_t)|^2} \right| \leq \text{const} \sqrt{\sum_{t=1}^{\infty} \frac{\|p - \tilde{p}\|_2}{d_t^2}}.$$

But $p(x) - \tilde{p}(x) \in B(C_1, C_2)$. From this the assertion of the theorem follows.

Remark. The theorem implies the uniqueness of solution of the inverse problem for the operators $T + P$ and $T + \widetilde{P}$ when $p(x) - \tilde{p}(x) \in B(C_1, C_2)$ and $\|p\|_2$ and $\|\tilde{p}\|_2$ are small.

REFERENCES

Dubrovskii, V.V. and Nagorny, A.V. (1990). On inverse problem for the Laplace operator with continuous potential. *Different. Equat.* **26**, 1563–1567 (in Russian).

Sadovnichii, V.A. and Dubrovskii, V.V. (1979). On some properties of operators with disctrete spectrum. *Different. Equat.* **15**, 1206–1211 (in Russian).

Ill-Posed Problems in Natural Sciences, pp. 423 – 430
A. Tikhonov (Ed.)
1992 VSP/TVP

INVERSE BOUNDARY VALUE PROBLEMS IN VISCOUS FLUID DYNAMICS

I.A. VASIN
Department of Mathematics, Moscow Engineering Physics Institute, Kashirskoe Highway, 31, 115409 Moscow, Russia

ABSTRACT

The subject of this paper are the inverse boundary value problems based on a model of viscous incompressible fluid. Inverse problems with final and integral overdetermination are considered. Existence and uniqueness theorems are proved.

We consider the linear Navier–Stokes equations which describe the fluid motion in a bounded domain Ω:

$$\begin{gathered}\vec{V} - \nu\Delta\vec{V} + \sum_{k=1}^{n} B_k(x,t)\vec{V}_{x_k} = -\nabla p + \vec{F}(x,t), \\ \operatorname{div} \vec{V} = 0, \quad (x,t) \in Q_T \equiv \Omega \times [0,T].\end{gathered} \tag{1}$$

Here $\vec{V} = \vec{V}(x,t)$ is the velocity at a point (x,t), $p = p(x,t)$ is the pressure at this point, $\vec{F} = \vec{F}(x,t)$ is an external force function, B_k are known square $(n \times n)$-matrices with elements b_k^{ij}.
The symbol ∇ denotes the gradient with the respect to the space variables and ν denotes the kinematic viscousity. The density is supposed to be constant and equal 1.
Let Ω be a bounded time-invariant domain in $\mathbf{R}^n$ $(n = 2; 3)$ with a $\mathbf{C}^2$-smooth boundary.
In some cases the external force vector is actually unknown and subject has to be modelled on the basis of indirect measurements. Thus there this calls for studying inverse problems for the Navier–Stokes equations.
One of the main purposes of the study is to find a formulation and a set of restrictions on the input data for the inverse problem to be well-posed.
The inverse problems concerned can be interpreted as problems of the fluid motion control. That is why it is important to find out when the solution can be shown to exist and to be unique.
Various aspects of the general theory of ill-posed and inverse problems in mathematical physics were investigated in detail in (Tikhonov and Arsenin, 1977; Lavrent'ev et al., 1980; Prilepko, 1973), etc.

A detailed review of investigations of direct problems for the Navier–Stokes equations was given in (Ladyzhenskaya,1970; Temam, 1979). A study of properties of the solutions of direct problems for (1) can be found in (Ladyzhenskaya, 1967 and Solonnikov, 1973).

1. INVERSE PROBLEM 1

We assume that the external force $\vec{F}$ can be represented in the form

$$\vec{F} = \vec{f}(x)g(x,t), \tag{2}$$

where the coeficient $g(x,t)$ is known and the vector-valued function $\vec{f}(x)$ has to be found.

To simplify our calculations, we suppose in what follows that $B_k(x,t) = \beta_k(x)I$, where I is the unit matrix and $\beta_k(x)$ is a scalar function.

Let us denote by $\vec{\beta}(x)$ the vector-valued function whose components are $\beta_k(x)$ $(k = 1, \dots, n)$.

We examine the inverse problem of determination of the set of functions which consists of the velocity vector, the pressure gradient and the unknown vector-valued external force function.

In order to have an appropriate mathematical formulation of the inverse problem of the type concerned one should complement the initial and boundary data with some additional information about the solution (the so-called overdetermination condition).

In view of that it is necessary to investigate various types of overdetermination.

In particular, the additional information can be introduced in the form of final overdetermination, i.e., by assuming that the value of the velocity and the value of the pressure gradient are known at the final time moment T.

INVERSE PROBLEM 1. *Find a set of vector-valued functions* $\{\vec{V}, \nabla p, \vec{f}\}$, *that satisfy the system of equations*

$$\begin{gathered}\vec{V}_t - \nu\Delta\vec{V} + (\vec{\beta}(x), \nabla)\vec{V} = -\nabla p + \vec{f}(x)g(x,t),\\ \operatorname{div}\vec{V} = 0, \quad (x,t) \in Q_T,\end{gathered} \tag{3}$$

the initial condition

$$\vec{V}(x,0) = 0, \quad x \in \Omega, \tag{4}$$

the boundary condition

$$\vec{V}(x,t) = 0, \quad (x,t) \in S_T \equiv \partial\Omega \times [0,T], \tag{5}$$

and the conditions of final overdetermination

$$\vec{V}(x,T) = \vec{\varphi}(x), \quad \nabla p(x,T) = \nabla\psi(x), \quad x \in \Omega, \tag{6}$$

assuming that the vector-functions $\vec{\beta}$, $\vec{\varphi}$, $\nabla\psi$ *and the coefficient* g *are given. Here,* $(\vec{\beta}(x), \nabla)\vec{V} = \sum_{k=1}^{n} \beta_k(x)\vec{V}_{x_k}$.

We shall look for a solution in the following function spaces: $\vec{V} \in \overset{\circ}{W}{}^2_{2,1}(Q_T) \cap \overset{\circ}{J}(Q_T)$, $\vec{f} \in \mathbf{L}_2(\Omega)$, $\nabla p(\cdot, t)$ belongs to $G(\Omega)$ for any t and continuously depends on t in the norm of the space $\mathbf{L}_2(\Omega)$ for $t \in [0, T]$.
All the definitions of the function spaces we use in this paper can be found in (Ladyzhenskaya, 1970). Here we give a short review of some of them.
$\overset{\circ}{W}{}^2_{2,1}(Q_T)$ is a subspace of the Sobolev function space $W_2^{2,1}(Q_T)$ consisting of the vector-valued functions which are equal to zero on S_T.
The symbol $\overset{\circ}{J}(\Omega)$ denotes the closure of the set of smooth finite solenoidal vectors with respect to the norm in $\mathbf{L}_2(\Omega)$.
$\overset{\circ}{J}(\Omega)$ is a subspace of $\mathbf{L}_2(\Omega)$. It is known that $\mathbf{L}_2(\Omega) = \overset{\circ}{J}(\Omega) \oplus G(\Omega)$, where $G(\Omega)$ is the orthogonal complement of $\overset{\circ}{J}(\Omega)$ in $\mathbf{L}_2(\Omega)$. $G(\Omega)$ consists of all vectors of the form $\nabla\psi$, where ψ is a single-valued, square summable function having first derivatives which belong to $\mathbf{L}_2(\Omega)$.
The norms in $\mathbf{L}_q(\Omega)$, $\mathbf{L}_{q,r}(Q_T)$ are introduced in the usual way.
We begin with deriving a linear second kind operator equation for the function $\vec{f}$. In what follows it is supposed that $g,\ g_t \in \mathbf{C}(\bar{Q}_T)$,

$$|g(x,T)| \geq g_T > 0, \quad \vec{\beta} \in \overset{\circ}{J}(\Omega) \cap \mathbf{L}_4(\Omega), \quad \vec{\varphi} \in W_2^2(\Omega) \cap H(\Omega), \quad \nabla\psi \in G(\Omega).$$

If we select an arbitrary vector-function $\vec{f}$ in $\mathbf{L}_2(\Omega)$ and keep it fixed, we may condider (3)–(5) as the direct problem examined by Ladyzhenskaya and Solonnikov (Ladyzhenskaya, 1967; Solonnikov, 1973). It was proved that the solution $\{\vec{V}, \nabla p\}$ of such a direct problem exists and is unique. That is why we can define the linear operator A: $\mathbf{L}_2(\Omega) \to \mathbf{L}_2(\Omega)$ by

$$(A\vec{f})(x) = \frac{1}{g(x,T)}\vec{V}_t(x,T), \quad x \in \Omega, \tag{7}$$

where the function $\vec{V}$ is the solution of the direct problem (3)–(5) corresponding to $\vec{f}$.
Now we consider the following linear operator equation of the second kind

$$\vec{f} = A\vec{f} + \vec{\chi}, \tag{8}$$

where $\vec{\chi}$ is given and belongs to $\mathbf{L}_2(\Omega)$.
The first result shows how the solvability of equation (8) implies the solvability of inverse problem (3)–(6) and conversly.

THEOREM 1. *Let* $g,\ g_t \in \mathbf{C}(\bar{Q}_T)$, $|g(x,T)| \geq g_T > 0$, $\vec{\beta} \in \overset{\circ}{J}(\Omega) \cap \mathbf{L}_4(\Omega)$, $\vec{\varphi} \in W_2^2(\Omega) \cap H(\Omega)$, $\nabla\psi \in G(\Omega)$.

1) *If the relation*

$$\vec{\chi} = (-\nu\Delta\vec{\varphi} + (\vec{\beta},\nabla)\vec{\varphi} + \nabla\psi)/g(x,T), \tag{9}$$

holds and linear equation (8) *has a unique solution, then there exists a unique solution of Inverse Problem* 1.

2) *If there exists a solution of Inverse Problem* 1 *and this solution is unique, then the linear equation* (8) *has a unique solution when* (9) *is true.*

Proof. We first prove 1). Let (9) hold and let equation (8) have a unique solution. Let this solution be denoted by $\vec{f}$.

If we substitute $\vec{f}$ to the Navier–Stokes equations (3), then system (3)–(5) can be considered as the system of a direct problem.

By the corresponding results of Ladyzhenskaya and Solonnikov (Ladyzhenskaya, 1967; Solonnikov, 1973) there exists a unique pair $\{\vec{V},\nabla p\}$ which belongs to the reguired function classes and satisfies (3)–(5).

Now we show that the over-determination conditions (6) are true for this pair $\{\vec{V},\nabla p\}$. Suppose that $\vec{V}(x,T) = \vec{\varphi}_1$ and $\nabla p(x,T) = \nabla\psi_1$. Then $\vec{\varphi}_1 \in H(\Omega)$ and

$$\vec{V}_t(x,T) - \nu\Delta\vec{\varphi}_1 + (\vec{\beta},\nabla)\vec{\varphi}_1 = -\nabla\psi_1 + \vec{f}(x)g(x,T), \qquad \operatorname{div}\vec{\varphi}_1 = 0. \tag{10}$$

On the other hand,

$$A\vec{f} + (-\nu\Delta\vec{\varphi} + (\vec{\beta},\nabla)\vec{\varphi} + \nabla\psi)/g(x,T) = \vec{f}. \tag{11}$$

Using relations (7), (10), and (11), we observe that the following stationary system

$$\begin{aligned} -\nu\Delta(\vec{\varphi} - \vec{\varphi}_1) + (\vec{\beta},\nabla)(\vec{\varphi} - \vec{\varphi}_1) &= -\nabla(\psi - \psi_1), \\ \operatorname{div}(\vec{\varphi} - \vec{\varphi}_1) &= 0, \quad x \in \Omega, \end{aligned} \tag{12}$$

with the boundary condition

$$(\vec{\varphi} - \vec{\varphi}_1)(x) = 0, \quad x \in \partial\Omega \tag{13}$$

is satisfied.

Let us take the scalar product in $\mathbf{L}_2(\Omega)$ of $(\vec{\varphi} - \vec{\varphi}_1)$ and both sides of the first equation in (12). We obtain $\nu\|(\vec{\varphi} - \vec{\varphi}_1)_x\|^2_{2,\Omega} = 0$. Here, we took into account that $\vec{\beta} \in \overset{\circ}{J}(\Omega)$ and $\int_\Omega (\vec{\beta},\nabla)(\vec{\varphi} - \vec{\varphi}_1)\cdot(\vec{\varphi} - \vec{\varphi}_1)\,dx = 0$. Hence, $\vec{\varphi} = \vec{\varphi}_1$, $\nabla\psi = \nabla\psi_1$ and thus Inverse Problem 1 has a solution.

The assumption about nonuniqueness of the solution of inverse problem (3)–(6) implies the nonuniqueness of the solution of equation (8) when $\vec{\chi}$ is represented by (9). This conclusion is in contradiction with the original assumption.

Now we prove 2). Let Inverse Problem 1 have a unique solution. This solution will be denoted by $\{\vec{V},\nabla p,\vec{f}\}$. It is clear that $\vec{F} = \vec{f}g$ and $\vec{F}_t = \vec{f}g_t$ belong

to $\mathbf{L}_2(Q_T)$. Using this fact and the smoothness properties of the solution of the Navier–Stokes system, we conclude that $\vec{V}_t$, $\Delta\vec{V}$ and ∇p are continuous in $\mathbf{L}_2(\Omega)$-norm with respect to $t \in [0,T]$.
Hence, the solution of system (3) at $t = T$ satisfies the equality

$$\vec{V}_t(x,T) - \nu\Delta\vec{V}(x,T) + (\vec{\beta},\nabla)\vec{V}(x,T) = -\nabla p(x,T) + \vec{f}(x)g(x,T),$$

which, by (6), implies

$$\vec{V}_t(x,T) - \nu\Delta\vec{\varphi} + (\vec{\beta},\nabla)\vec{\varphi} + \nabla\psi = \vec{f}(x)g(x,T).$$

Then, taking into account definition (7), we obtain that the relation $A\vec{f} + \vec{\chi} = \vec{f}$ holds and thus equation (8) has a solution.
Now suppose that (8) has more than one solution. Then, using the procedure described in the above proof of 1), we can conclude that Inverse Problem 1 has two different solutions. But this conclusion contradicts our assumption. The theorem is thereby proved.
The following result shows under what restrictions on the input data there exists a unique solution of Inverse Problem 1.

THEOREM 2. *Let* $g,\ g_t \in \mathbf{C}(\bar{Q}_T)$, $|g(x,T)| \geq g_T > 0$, $\vec{\beta} \in \overset{\circ}{J}(\Omega) \cap \mathbf{L}_4(\Omega)$, $\vec{\varphi} \in W_2^2(\Omega) \cap H(\Omega)$, $\nabla\psi \in G(\Omega)$ *and let the inequality*

$$m < 1 \tag{14}$$

hold, where

$$m = \frac{1}{g_T}\left(\sup_{\Omega}|g(x,0)|\exp\left(-\nu T/c(\Omega)\right) + \int_0^T \exp\left(-\nu(T-t)/c(\Omega)\right)\sup_{\Omega}|g_t(x,t)|\,dt\right)$$

and $c(\Omega)$ *is the constant from the Poincaré inequality.*
Then a solution of Inverse Problem 1 exists and is unique and the following estimate is true:

$$\|\vec{f}\|_{2,\Omega} \leq \frac{1}{(1-m)g_T}\left\| -\nu\Delta\vec{\varphi} + (\vec{\beta},\nabla)\vec{\varphi} + \nabla\psi\right\|_{2,\Omega}. \tag{15}$$

Proof. Consider system (3) with an arbitrary vector-function $\vec{f}$ from $\mathbf{L}_2(\Omega)$. Using the smoothness properties of the solutions of (3) we can obtain the following energy equality:

$$\begin{aligned} &\frac{1}{2}\frac{d}{dt}\|\vec{V}_t(\cdot,t)\|_{2,\Omega}^2 + \nu\|\vec{V}_{tx}(\cdot,t)\|_{2,\Omega}^2 + \int_\Omega (\vec{\beta},\nabla)\vec{V}_t\cdot\vec{V}_t\,dx = \\ &\int_\Omega \vec{f}_t(x)g(x,t)\cdot\vec{V}_t(x,t)\,dx, \qquad 0 < \varepsilon \leq t \leq T, \end{aligned} \tag{16}$$

where ε is any strictly positive constant belonging to $(0, T)$. Since $\vec{\beta} \in \overset{\circ}{J}(\Omega)$ and $\int_\Omega (\vec{\beta}, \nabla)\vec{V}_t \cdot \vec{V}_t \, dx = 0$, relation (16) implies the estimate

$$\frac{d}{dt}\|\vec{V}_t(\cdot,t)\|_{2,\Omega} + \frac{\nu}{c(\Omega)}\|\vec{V}_t(\cdot,t)\|_{2,\Omega} \leq \|\vec{f}(x)g_t(\cdot,t)\|_{2,\Omega}, \quad 0 < \varepsilon \leq t \leq T, \tag{17}$$

where $c(\Omega)$ is the constant from the Poincaré inequality. Now we have to use the Gronwall lemma and operator definition (7).
By virtue of (17), the following inequality is true:

$$\|A\vec{f}\|_{2,\Omega} \leq m\|\vec{f}\|_{2,\Omega}, \quad \vec{f} \in \mathbf{L}_2(\Omega), \tag{18}$$

where the constant m is specified in (14). That is why the linear operator A is bounded and its norm is less than 1. Hence, we conclude that the linear equation (8) has a unique solution and estimate (15) holds. Now the existence and the uniqueness of the solution of inverse problem (3)–(6) follow from Theorem 1. This completes the proof.
The following result deals with further examination of properties of the linear operator A introduced in (7).

THEOREM 3. *If* $g,\ g_t \in \mathbf{C}(\bar{Q}_T)$, $|g(x,T)| \geq g_T > 0$, $\vec{\beta} \in \overset{\circ}{J}(\Omega) \cap \mathbf{L}_4(\Omega)$, *then the linear operator* A *is completely continuous on* $\mathbf{L}_2(\Omega)$.

Theorems 1 and 3 show that inverse problem (3)–(6) is of the Fredholm type which means that the uniqueness implies the existence and the stability.

THEOREM 4. *Let* $g,\ g_t \in \mathbf{C}(\bar{Q}_T)$, $|g(x,T)| \geq g_T > 0$, $\vec{\beta} \in \overset{\circ}{J}(\Omega) \cap \mathbf{L}_4(\Omega)$, $\vec{\varphi} \in W_2^2(\Omega) \cap H(\Omega)$, $\nabla\psi \in G(\Omega)$.

1) *If the linear homogeneous equation* $A\vec{f} = \vec{f}$ *has only trivial solution* $\vec{f} = 0$, *then Inverse Problem 1 has a solution (which is unique).*

2) *If the uniqueness theorem for Inverse Problem 1 is valid, then Inverse Problem 1 has a solution (which is unique).*

To prove Theorems 3 and 4, one can use the procedure elaborated in (Prilepko and Vasin, 1989a; 1990a).
We illustrate the research by

Example. Suppose that $g = g(t)$, $g \in \mathbf{C}^1([0,T])$, $g \geq 0$, $g' \geq 0$, $g(T) \neq 0$ and $\vec{\beta}$, $\vec{\varphi}$, $\nabla\psi$ are taken from the required classes. Then inequality (14) is true for any $T > 0$ and Theorem 2 has a global character.

2. INVERSE PROBLEM 2

In this section we consider the case where the overdetermination has an integral form.

Suppose that the function $\vec{F}$ (see (1)) can be written as

$$\vec{F} = f(t)\vec{g}(x,t), \tag{19}$$

where $f(t)$ is an unknown element of $\mathbf{C}([0,T])$ and the vector-valued function $\vec{g}$ is given.

INVERSE PROBLEM 2. *Find a set $\{\vec{V}, \nabla p, f\}$ of functions satisfying system* (1) *(with $\vec{F}$ represented by* (19)*), the initial condition* (4)*, the boundary condition* (5) *and the condition of integral over-determination*

$$\int_\Omega \vec{V}(x,t)\cdot\vec{\omega}(x)\,dx = \varphi(t), \quad 0 \le t \le T,$$

where the matrices B_k, the functions $\vec{g}$, $\vec{\omega}$, and φ are given.

In what follows we suppose that $B_k(x,t) = \beta_k(x,t)I$, where I is the unit matrix and $\vec{\beta}$ denotes a vector-valued function with components $\beta_k(x,t)$.
To solve the above inverse problem, the method developed in (Prilepko and Vasin, 1989a) can be applied. Then we obtain the following global result.

THEOREM 5. *Let $\vec{g} \in \mathbf{C}([0,T],\mathbf{L}_2(\Omega))$, $\vec{\beta} \in \mathbf{C}([0,T],\mathbf{L}_4(\Omega))$, $\vec{\omega} \in W_2^2(\Omega)\cap H(\Omega)$, $\varphi \in \mathbf{C}^1([0,T])$, $\varphi(0)=0$, and*

$$\left|\int_\Omega \vec{g}(x,t)\cdot\vec{\omega}(x)\,dx\right| \ge g_0 > 0, \quad 0 \le t \le T.$$

Then Inverse Problem 2 has a unique solution

$$\vec{V} \in \overset{\circ}{W}{}^2_{2,1}(Q_T)\cap \overset{\circ}{J}(Q_T), \qquad \nabla p \in G(Q_T), \qquad f \in \mathbf{C}([0,T])$$

and f satisfies the inequality $\|f\|_{\mathbf{C}([0,T])} \le m\|\varphi'\|_{\mathbf{C}([0,T])}$.

To conclude, we note that similar formulation of inverse problems can be examined for nonlinear nonstationary Navier–Stokes equations both in two and three dimensions (see (Prilepko and Vasin, 1989b; 1990b; 1991)).

Acknowledgement
The author is grateful to A.I. Prilepko for his help and encouragement during the research.

REFERENCES

Ladyzhenskaya, O.A. (1967). Uniqueness and smoothness of generalized solutions of Navier–Stokes equations. *Zap. Nauchn. Semin. LOMI* **5**, 169–185 (in Russian).

Ladyzhenskaya, O.A. (1970). *The Mathematical Theory of Viscous Incompressible Flows.* Nauka, Moscow (in Russian).

Lavrent'ev, M.M., Romanov, V.G., and Shyshatskii, S.P. (1980). *Ill-posed Problems of Mathematical Physics and Analysis.* Nauka, Moscow (in Russian).

Prilepko, A.I. (1973). Inverse problems in potential theory (elliptic, parabolic, hyperbolic equations. *Math. Notes* **14**, 755–767 (in Russian).

Prilepko, A.I. and Vasin, I.A. (1989a). Inverse initial-boundary-value problems for linearized nonstationary Navier–Stokes equations. *Different. Equat.* **25**, 106–117 (in Russian).

Prilepko, A.I. and Vasin, I.A. (1989b). Inverse boundary-value problems for nonlinear Navier–Stokes equations with final over-determination. *Different. Equat.* **25**, 2164–2177 (in Russian).

Prilepko, A.I. and Vasin, I.A. (1990a). Nonstationary inverse problems of hydrodynamics with final over-determination. *Dokl. Akad. Nauk SSSR* **314**, 1075–1078 (in Russian).

Prilepko, A.I. and Vasin, I.A. (1990b). Solvability of a three-dimensional inverse problem for nonlinear Navier–Stokes equations. *USSR Computer Math. Math. Phys. J.* **29**, 1540–1552 (in Russian).

Prilepko, A.I. and Vasin, I.A. (1991). On a nonlinear nonstationary inverse problem of hydrodynamics. *Inverse Problems* **7**, L13–L16.

Solonnikov, V.A. (1973). Estimates of solutions of nonstationary Navier–Stokes system. *Zapiski Nauchn. Semin. LOMI* **38**, 153–231 (in Russian).

Temam, R. (1979). *Navier–Stokes Equations. Theory and Numerical Analysis.* North-Holland/ Elsevier, Amsterdam.

Tikhonov, A.N. and Arsenin, V.Y. (1977). *Solutions of Ill-Posed Problems.* Winston, Washington, DC.

Ill-Posed Problems in Natural Sciences, pp. 431 – 438
A. Tikhonov (Ed.)
1992 VSP/TVP

ON SOME INVERSE PROBLEMS FOR THE TIME-DEPENDENT TRANSPORT EQUATION

N.P. VOLKOV

Moscow Engineering Physics Institute, Kashirskoye Shosse, 31, 115409 Moscow, Russia

ABSTRACT

Inverse problems for the time-dependent multi-speed anisotropic kinetic transport equation are considered. Sufficient conditions for existence and uniqueness of generalized solutions to these inverse problems are obtained. The proofs are based on the method of successive approximations. Recommendations for constructing solutions of the problems under study are given. The formulations of some inverse problems considered in this paper are close to those of control observation and diagnosis problems for physical processes described by the transport equation.

In the past years the methods based on the theory of inverse problems of mathematical physics gained wide acceptance in mathematical modelling of physical processes and in processing results of diagnostics of diverse media.

This paper is concerned with the research of mass transfer processes which are described by the time-dependent multi-speed anisotropic kinetic transport equation

$$u_t + (v, \nabla_x)u + \Sigma(x,v,t)u(x,v,t) = \int\limits_V J(x,v',t,v)u(x,v',t)\,dv' \\ +F(x,v,t), \quad (x,v,t) \in \mathcal{D} = G \times V \times (0,T). \tag{1}$$

Here, the function $u(x,v,t)$ characterizes the density of particle distribution in a domain $G \times V$ of the space $\mathbf{R}^n \times \mathbf{R}^n$ at time $t \in (0,T)$. The functions $\Sigma(x,v,t)$, $J(x,v',t,v)$ and $F(x,v,t)$ characterize properties of the medium in which the process occurs. These functions are called the absorbtion coefficient, the scattering function and the function of interior sources, respectively. We assume in what follows that G is a bounded strictly convex domain with a boundary ∂G belonging to the class C^1 and a bounded closed set V is contained in a sphere layer $\{v \in \mathbf{R}^n\colon\ 0 < v_0 \le \|v\| \le v_* < \infty\}$.
Let all the characteristics Σ, J, F of the medium be specified, and let the initial condition

$$u(x,v,0) = \varphi(x,v), \quad (x,v) \in \bar{G} \times V, \tag{2}$$

and the incoming flux

$$u(x,v,t)=\mu(x,v,t), \quad (x,v,t)\in\gamma_-\times[0,T] \tag{3}$$

be given, where $\gamma_\pm=\{(x,v)\in\partial G\times V\colon \pm(v,n_x)>0\}$, n_x is the outer normal to the boundary ∂G of the domain G at a point x. Then it is possible to determine uniquely the state of the process for all points $(x,v)\in G\times V$ and for all moments $t\in[0,T]$.

To analyse inverse problems we introduce the following function spaces:
1) $L_\infty(\mathcal{D};L_p(V))$ is the space of functions u from $L_\infty(\mathcal{D})$ to $L_p(V)$, $1<p<\infty$;
2) $H_p(\mathcal{D})$ is the space of functions u which belong to $L_p(\mathcal{D})$ together with their generalized derivatives u_t, $(v,\nabla)u$ on $\mathcal{D}$ and whose traces $u|_{\Gamma_-}$ on $\Gamma_-=\gamma_-\times(0,T)$ belong to $L_p(\Gamma_-)$, i.e.,

$$H_p(\mathcal{D})=\{u\in L_p(\mathcal{D}):u_t,\ (v,\nabla)u\in L_p(\mathcal{D}),\ u|_{\Gamma_-}\in L_p(\Gamma_-)\},\quad 1<p<\infty;$$

this space is a Banach space with respect to the norm

$$\|u\|_{H_p}=\left[\|u\|^p_{p,\mathcal{D}}+\|u_t\|^p_{p,\mathcal{D}}+\|(v,\nabla)u\|^p_{p,\mathcal{D}}+\|u|_{\Gamma_-}\|^p_{p,\Gamma_-}\right]^{1/p}<\infty;$$

3) $H_\infty(\mathcal{D})$ is the space of functions u from $L_\infty(\mathcal{D})$ whose generalized derivatives u_t, $(v,\nabla)u$ belong to $L_\infty(\mathcal{D})$ and whose traces $u|_{\Gamma_-}$ on Γ_- belong to $L_\infty(\Gamma_-)$, i.e.,

$$H_\infty(\mathcal{D})=\{u\in L_\infty(\mathcal{D}):\ u_t,(v,\nabla)u\in L_\infty(\mathcal{D}),u|_{\Gamma_-}\in L_\infty(\Gamma_-)\},$$

this space is a Banach space with respect to the norm

$$\|u\|_{H_\infty}=\left[\|u\|_{\infty,\mathcal{D}}+\|u_t\|_{\infty,\mathcal{D}}+\|(v,\nabla)u\|_{\infty,\mathcal{D}}+\|u|_{\Gamma_-}\|_{\infty,\Gamma_-}\right]<\infty,$$

where $\|\cdot\|_{p,\mathcal{D}},\|\cdot\|_{\infty,\mathcal{D}}$ are norms in the spaces $L_p(\mathcal{D})$, $L_\infty(\mathcal{D})$;
4) $H_\infty((0,T);L_p(G\times V))$ is the space of functions u from the set $\{u:\ u,u_t,(v,\nabla)u\in L_\infty((0,T);L_p(G\times V)),\ u|_{\Gamma_-}\in L_\infty((0,T);L_p(\gamma_-))\}$ with the norm

$$|\!|\!|u|\!|\!|=\|u\|_{\infty;p,G\times V}+\|u_t\|_{\infty;p,G\times V}+\|(v,\nabla)u\|_{\infty;p,G\times V}+\|u|_{\Gamma_-}\|_{\infty;p,\gamma_-}<\infty,$$

where $\|\cdot\|_{\infty;p,G\times V}$ is the norm of the space $L_\infty((0,T);L_p(G\times V))$ and $\|\cdot\|_{\infty;p,\gamma_-}$ is the norm of the space $L_\infty((0,T);L_p(\gamma_-))$. We shall consider and study a number of inverse problems.

The following condition of overdetermination supplements the formulation of direct problem (1)–(3):

$$u(x,v,t_1)=\psi(x,v),\quad (x,v)\in\bar G\times V,\quad t_1\in(0,T]. \tag{4}$$

INVERSE PROBLEM 1(F). *Find a generalized solution of* (1)–(4), *i.e., a pair of functions* $u(x,v,t)$ *and* $F(x,v,t)=f(x,v)g(x,v,t)+h(x,v,t)$ *satisfying almost everywhere conditions* (1)–(4); *here,* f *is to be found while* g *and* h *are given.*

THEOREM 1. *Let* $\Sigma, \Sigma_t \in L_\infty(\mathcal{D})$; $J, J_t \in L_\infty(\mathcal{D}; L_p(V))$;

$$g, g_t \in L_\infty((G \times V); L_p(0,T)), \quad |g(x,v,t_1)| \geq g_1 > 0; \quad h, h_t \in L_p(\mathcal{D});$$

$$\mu, \mu_t \in L_p(\Gamma_-); \quad \varphi, (v,\nabla)\varphi, \psi, (v,\nabla)\psi \in L_p(G \times V), \quad \varphi\big|_{\gamma_-}, \psi\big|_{\gamma_-} \in L_p(\gamma_-);$$

$$v_0^{-1} \operatorname{diam} G < a = \min\{t_1, \left(\|\Sigma\| + \sqrt[q]{\operatorname{mes} V}\|J\|\right)^{-1}, (\sqrt{C^{-1}+1} - 1)\},$$

where C is a constant depending on p, q, V, and on the norms of the functions Σ, J, g in the corresponding spaces; $q = p/(p-1)$. Let the compatibility conditions

$$\varphi(x,v) = \mu(x,v,0), \quad (x,v) \in \gamma_-, \tag{5}$$

$$\psi(x,v) = \mu(x,v,t_1), \quad (x,v) \in \gamma_-, \tag{6}$$

be satisfied. Then there exists a unique pair of functions $u \in H_p(\mathcal{D})$ and $f \in L_p(G \times V)$ which is a generalized solution of Inverse Problem 1(F).

Example. Let $\{u, f\}$ be a generalized solution of Inverse Problem 1(F) in the function class $H_p(\mathcal{D}) \times L_p(G \times V)$. Let us introduce a local coordinate system, where any point $x \in G$ has the form $x = y + v\xi$ and y belongs to Π_v, the projection of G on an not intersecting G and orthogonal to the vector v, and hyperplane and $\xi \in [\zeta_-, \zeta_+]$, $\zeta_- : (y + v\zeta_-, v) \in \gamma_-$, $\zeta_+ : (y + v\zeta_+, v) \in \gamma_+$.
Integrating equation (1) along the characteristic $t - \xi =$const and taking into account conditions (2) and (3) we obtain the integral equation

$$\begin{aligned} u(y+v\xi, v, t) = \Phi(\xi, y, v, t) + \int_\eta^\xi &[(Pu + h)(y + v\theta, v, \theta - \xi + t) \\ &+ f(y+v\theta, v) g(y + v\theta, v, \theta - \xi + t)]\, d\theta, \end{aligned} \tag{7}$$

where $\Phi(\xi, y, v, t) = \varphi(y + v(\xi - t), v)$, $\eta = \xi - t$, for $0 \le t \le \xi - \zeta_-$, $\zeta_- \le \xi \le \zeta_+$ and $\Phi(\xi, y, v, t) = \mu(y + v\zeta_-, v, t - \xi + \zeta_-)$, $\eta = \zeta_-$ for $\xi - \zeta_- \le t \le T$, $\zeta_- \le \xi \le \zeta_+$.
Differentiating (7) with respect to ξ and using simple transformations based on (4),we obtain, for $v_0^{-1} \operatorname{diam} G \le t_1$, the following equality

$$\begin{aligned} f(y + v\xi, v) = [g(y+v\xi, v, t_1)]^{-1} \{ &\mu_t(y + v\zeta_-, v, t_1 - \xi + \zeta_-) \\ &+ (v,\nabla)\psi(y+v\xi, v) - (P\psi)(y + v\xi, v, t_1) \\ &+ \int_{\zeta_-}^\xi \Big[\frac{\partial}{\partial t}(Pu + h)(y + v\theta, v, \theta - \xi + t_1) \\ &+ f(y+v\theta, v) g_t(y+v\theta, v, \theta - \xi + t_1)\Big]\, d\theta\}, \end{aligned} \tag{8}$$

where

$$\begin{aligned} (Pu)(y+v\xi, v, t) \equiv &-\Sigma(y + v\xi, v, t) u(y + v\xi, v, t) \\ &+ \int_V J(y + v\xi, v', t, v) u(y + v\xi, v', t)\, dv'. \end{aligned}$$

Inverse Problem 1(F) is equivalent to the system of integral equations (7) and (8) in the function class $\{u, f\} \in \mathcal{H}_p = H_p(\mathcal{D}) \times L_p(G \times V)$ for $v_0^{-1}\operatorname{diam} G < a_1 = \min\{t_1, \left(\|\Sigma\| + \sqrt[q]{\operatorname{mes} V}\|J\|\right)^{-1}\}$ (here $\|\cdot\|$ is the norm of indicated function in the corresponding space).
Therefore, we can study further the system of integral equations (7) and (8). We rewrite this system in the operator form

$$\{u, f\} = \mathcal{A}\{u, f\} \equiv \{\mathcal{A}_1\{u, f\}, \mathcal{A}_2\{u, f\}\},$$

where $\mathcal{A}_1$ and $\mathcal{A}_2$ are the operators of the right-hand sides of equations (7) and (8) with the domain of definition $\mathcal{H}_p$. The operator $\mathcal{A}$ maps continuously the space $\mathcal{H}_p$ into itself (it is a Banach space with respect to the norm $\|\{u, f\}\|_{\mathcal{H}_p} = \|u\|_{H_p} + \|f\|_{p,G\times V}$).
Let us take any two elements $\{u_1, f_1\}$ and $\{u_2, f_2\}$ from $\mathcal{H}_p$ and estimate the difference $\mathcal{A}^2\{u_1, f_1\} - \mathcal{A}^2\{u_2, f_2\}$, where $\mathcal{A}^2$ is the second degree of the operator $\mathcal{A}$. By the Hölder and Minkowski inequalities we obtain the estimate

$$\|\mathcal{A}^2\{u_1, f_1\} - \mathcal{A}^2\{u_2, f_2\}\|_{\mathcal{H}_p} < C(k^2 + 2k)\|\{u_1, f_1\} - \{u_2, f_2\}\|_{\mathcal{H}_p},$$

where C is a constant depending on p, q, V and in the norms of the functions Σ, J, g in the corresponding spaces; $q = p/(p-1)$ and $k = v_0^{-1}\operatorname{diam} G$. From this estimate it follows that the operator $\mathcal{A}^2$ is contracting if $k < a_2 = \sqrt{C^{-1} + 1} - 1$. Since all the conditions of the Banach principle on a fixed point are satisfied, there exists a unique element of the space $\mathcal{H}_p$ being a solution of the operator equation. Moreover, if $v_0^{-1}\operatorname{diam} G < \min\{a_1, a_2\}$, there exists a unique generalized solution of the Inverse Problem 1(F). The theorem is proved.
The inverse problem just considered is linear while the following two inverse problems of finding coefficients of transport equation (1) are nonlinear.

INVERSE PROBLEM 1(Σ). *Find a pair of functions $u(x, v, t)$ and $\Sigma(x, v, t) = \sigma(x, v)g(x, v, t) + h(x, v, t)$ (a generalized solution) satisfying almost everywhere conditions (1)–(4), where σ is an unknown function and g, h are given functions.*

Sufficient conditions for the solution of Inverse Problem 1(Σ) to exist and be unique are formulated in the following theorem which was proved for $p = 2$ in (Prilepko and Volkov, 1988).

THEOREM 2. *If $J, J_t \in L_\infty(\mathcal{D}; L_p(V))$; $F, F_t \in L_\infty(\mathcal{D})$;*

$$g, g_t \in L_\infty(\mathcal{D}), \qquad |(g\psi)(x, v, t_1)| \geq g_2 > 0; \qquad h, h_t \in L_\infty(\mathcal{D});$$

$$\varphi, (v, \nabla)\varphi, \psi, (v, \nabla)\psi \in L_\infty(G \times V), \qquad \varphi\big|_{\gamma_-}, \psi\big|_{\gamma_-} \in L_\infty(\gamma_-);$$

$\mu, \mu_t \in L_\infty(\Gamma_-)$; $v_0^{-1}\operatorname{diam} G < b$ (the constant b has the same properties as the constant a from Theorem 1) and compatibility conditions (5) and (6) be satisfied, then there exists a unique pair of functions $u(x, v, t)$ and $\sigma(x, v)$ (a generalized solution of Inverse Problem 1(Σ)) in the set

$$\mathcal{K} = \{\{u, \sigma\} : \ u \in H_\infty(\mathcal{D}), \sigma \in L_\infty(G \times V), \|u\|_{\infty,\mathcal{D}} \leq k_1,$$

$$\|\sigma\|_{\infty,G\times V} \leq k_2, \|u_t\|_{\infty,\mathcal{D}} \leq k_3, \|(v, \nabla)u\|_{\infty,\mathcal{D}} \leq k_3\},$$

where k_1, k_2, k_3 are constants depending only on the initial data of the problem.

Remark 1. The assertion of Theorem 2 remains true if we replace the conditions $F, F_t \in L_\infty(\mathcal{D})$; $h, h_t \in L_\infty(\mathcal{D})$ by the conditions $F, (v, \nabla)F \in L_\infty(\mathcal{D})$; $h, (v, \nabla)h \in L_\infty(\mathcal{D})$ of Theorem 2.
Now let us consider the inverse problem of finding the kernel J of the scattering integral.

INVERSE PROBLEM 1(J). *Find a pair of functions $u(x, v, t)$ and $J(x, v', t, v) = j(x, v)\tilde{g}(x, v', t, v) + \tilde{h}(x, v', t, v)$ (a generalized solution) satisfying almost everywhere conditions (1)–(4), where j is to be found and $\tilde{g}, \tilde{h}$ are known.*

THEOREM 3. *If the functions F, φ, ψ, μ satisfy the conditions of Theorem 2, $\Sigma, \Sigma_t \in L_\infty(\mathcal{D})$; $\tilde{h}, \tilde{h}_t \in L_\infty(\mathcal{D}; L_p(V))$,*

$$\tilde{g}, \tilde{g}_t \in L_\infty(\mathcal{D}; L_p(V)), \qquad \left| \int\limits_V \tilde{g}(x, v', t, v)\psi(xv')\, dv' \right| \geq g_3 > 0,$$

and compatibility conditions (5) and (6) hold, then, in the set $\mathcal{K}$, there exists a unique generalized solution $\{u, j\}$ of Inverse Problem 1(J).

Remark 2. In Theorem 3, the conditions $F, F_t \in L_\infty(\mathcal{D})$; $\Sigma, \Sigma_t \in L_\infty(\mathcal{D})$ can be replaced by the conditions $F, (v, \nabla)F \in L_\infty(\mathcal{D}); \Sigma, (v, \nabla)\Sigma \in L_\infty(\mathcal{D})$. Such being the case the assertion of Theorem 3 remains true.
Let us consider an other type of inverse problems, where the overdetermination condition (4) is replaced by the condition

$$u(x, v, t) = \nu(x, v, t), \quad (x, v, t) \in \Gamma'_+ = \gamma_+ \times [0, \zeta_+ - \zeta_-]. \tag{9}$$

Same as in the previous case, we shall consider some formulations of inverse problems and sufficient conditions for their solvability.

INVERSE PROBLEM 2(F). *Find a pair of functions $u(x, v, t)$ and $F(x, v, t) = f(x, v)g(x, v, t) + h(x, v, t)$ (a generalized solution) satisfying almost everywhere conditions (1)–(4) and (9), where f is to be found and g, h are given functions.*

The following theorem guarantees that the solution of this problem exists and is unique.

THEOREM 4. *Let the functions $\Sigma, J, h, \varphi, \mu$ satisfy conditions of Theorem 1 and let $\nu, \nu_t \in L_2(\Gamma'_+)$; $g, g_t \in L_\infty(G \times V); L_p(0, T))$, $|g(x, v, 0)| \geq g_4 > 0$ for $(x, v) \in G \times V$; $Tv_0^{-1} \operatorname{diam} G < d_1$. Furthermore, let the compatibility conditions (5) and the condition*

$$\nu(x, v, 0) = \varphi(x, v), \quad (x, v) \in \gamma_+, \tag{10}$$

hold.
Then there exists a unique generalized solution

$$\{u, f\} : \ u \in H_p(\mathcal{D}), \ f \in L_p(G \times V)$$

of *Inverse Problem* 2(F).

Now we consider nonlinear inverse problems of finding the coefficient Σ and the kernel J of the scattering integral of equation (1).

INVERSE PROBLEM 2(Σ). *Find a pair of functions*

$$u(x,v,t), \qquad \Sigma(x,v,t) = \sigma(x,v)g(x,v,t) + h(x,v,t)$$

(a generalized solution) satisfying almost everywhere conditions (1)–(3) *and* (9), *where* σ *is unknown and* g, h *are given functions.*

THEOREM 5. *Let the functions* J, F, h, φ, μ, *satisfy conditions of Theorem* 1 *and let* $\nu, \nu_t \in L_\infty(\Gamma'_+)$, $g, g_t \in L_\infty(\mathcal{D})$, $|g(x,v,0)\varphi(x,v)| \geq g_5 > 0$, $Tv_0^{-1}\operatorname{diam} G < d_2$. *Assume that compatibility conditions* (5) *and* (10) *hold.*
Then there exists a unique pair of functions $\{u, \sigma\}$ *in the set* $\mathcal{K}$ *which is a generalized solution of Inverse Problem* 2(Σ).

Remark 3. The results of Theorem 5 remains valid if we replace the conditions $F, F_t \in L_\infty(\mathcal{D})$ and $h, h_t \in L_\infty(\mathcal{D})$ by the conditions $F, (v,\nabla)F \in L_\infty(\mathcal{D})$ and $h, (v,\nabla)h \in L_\infty(\mathcal{D})$.

INVERSE PROBLEM 2(J). *Find a generalized solution, i.e., a pair of functions* $u(x,v,t)$, $J(x,v',t,v) = j(x,v)\tilde{g}(x,v',t,v) + \tilde{h}(x,v',t,v)$ *satisfying almost everywhere conditions* (1)–(3) *and* (9), *where* j *is to be found and* $\tilde{g}, \tilde{h}$ *are given functions.*

THEOREM 6. *If the functions* F, φ, μ *satisfy the conditions of Theorem* 2,

$$\Sigma, \Sigma_t \in L_\infty(\mathcal{D}); \qquad \tilde{h}, \tilde{h}_t \in L_\infty(\mathcal{D}; L_p(V)); \qquad \nu, \nu_t \in L_\infty(\Gamma'_+);$$

$$\tilde{g}, \tilde{g}_t \in L_\infty(\mathcal{D}; L_p(V)), \qquad \left| \int_V \tilde{g}(x,v',0,v)\varphi(x,v')\,dv' \right| \geq g_6 > 0;$$

$Tv_0^{-1}\operatorname{diam} G < d_3$ *and the compatibility conditions* (5) *and* (10) *are fulfilled, then in the set* $\mathcal{K}$ *there exists a unique generalized solution* $\{u, j\}$ *of Inverse Problem* 2(J).

Remark 4. If we use the conditions $\Sigma, (v,\nabla)\Sigma \in L_\infty(\mathcal{D})$; $F, (v,\nabla)F \in L_\infty(\mathcal{D})$ instead of the conditions $\Sigma, \Sigma_t \in L_\infty(\mathcal{D})$; $F, F_t \in L_\infty(\mathcal{D})$ in Theorem 6, then its assertion remains true.
Now we will use the overdetermination condition in the integral form

$$\int_{G\times V} \omega(x,v)u(x,v,t)\,dx\,dv = \chi(t), \quad t \in [0,T], \tag{11}$$

where $\omega \in C(G \times V)$, $\operatorname{supp}\omega \subset G \times V$, $\operatorname{dist}(\operatorname{supp}\omega, \partial G \times V) > 0$.

INVERSE PROBLEM 3(F). *Find a generalized solution, i.e., a pair of functions $u(x,v,t)$, $F(x,v,t) = f(t)g(x,v,t) + h(x,v,t)$ satisfying almost everywhere conditions (1)–(3) and (11), where f is unknown and g, h are given functions.*

THEOREM 7. *Let $\Sigma, \Sigma_t \in L_\infty(\mathcal{D})$, $J, J_t \in L_\infty(\mathcal{D}; L_p(V))$,*

$$g, (v,\nabla)g \in L_\infty((0,T),\ L_p(G\times V)), \qquad \left|\int\limits_{G\times V} \omega(x,v)g(x,v,t)\,dx\,dv\right| \geq g_7 > 0,$$

$$h, h_t \in L_p(\mathcal{D}), \quad \varphi, (v,\nabla)\varphi \in L_p(G\times V),$$

$\varphi\big|_{\gamma_-} \in L_p(\gamma_-)$, $\mu, \mu_t \in L_p(\Gamma_-)$, $\chi, \chi' \in L_p(0,T)$, *and let compatibility conditions* (5) *and*

$$\int\limits_{G\times V} \omega(x,v)\varphi(x,v)\,dx\,dv = \chi(0) \tag{12}$$

be satisfied.
Then there exists a unique generalized solution $\{u,f\}$: $u \in H_p(\mathcal{D})$; $f \in L_p(0,T)$, of Inverse Problem 3(F).

Remark 5. The assertion of Theorem 5 remains true if the conditions $\Sigma, \Sigma_t \in L_\infty(\mathcal{D})$ and $h, h_t \in L_p(\mathcal{D})$ are replaced by the conditions $\Sigma, (v,\nabla)\Sigma \in L_\infty(\mathcal{D})$ and $h, (v,\nabla)h \in L_p(\mathcal{D})$.
Now we study some nonlinear inverse problem of finding the coefficient Σ.

INVERSE PROBLEM 3(Σ). *Find a generalized solution, i.e., a pair of functions $u(x,v,t)$, $\Sigma(x,v,t) = \sigma(t) + h(x,v,t)$ satisfying almost everywhere conditions (1)–(3) and (11), where σ is to be found and h is a given function.*

THEOREM 8. *If the functions J, φ satisfy the conditions of Theorem 7,*

$$F, F_t \in L_\infty((0,T),\ L_p(G\times V)), \qquad \chi, \chi' \in L_\infty(0,T),$$

$$|\chi(t)| \geq k_1 > 0, \qquad \mu, \mu_t \in L_\infty((0,T, L_p(\gamma_-))$$

and compatibility conditions (5) *and* (12) *hold, then there is a number a_1 (depending on the initial data of this problem) such that for $Tv_0^{-1}\operatorname{diam} G < a_1$ there exists a pair of functions $\{u,\sigma\}$: $u \in H_\infty((0,T); L_p(G\times V))$, $\sigma \in L_\infty(0,T)$, which is a generalized solution of Inverse Problem 3(Σ). For all T, v_0 and $\operatorname{diam} G$ this problem has at most one such solution.*

For the cases $p = 2$ and $p = \infty$ Inverse Problems 3(F) and 3(Σ) have been studied earlier in (Prilepko and Volkov, 1987).
We supplement condition (12) by the overdetermination condition

$$\int\limits_{G\times V} \omega_1(x,v)u(x,v,t)\,dx\,dv = \chi_1(t), \quad t \in [0,T], \tag{13}$$

and study the following inverse problem.

Inverse Problem 3(F,Σ). *Find a triple of functions* $u(x,v,t)$, $F(x,v,t) = f(t)g(x,v,t) + h_1(x,v,t)$, $\Sigma(x,v,t) = \sigma(t) + h_2(x,v,t)$ *(a generalized solution) satisfying almost everywhere conditions* (1)–(3), (11) *and* (13), *where* f, σ *are unknown and* g, h_1, h_2 *are given functions.*

We shall formulate sufficient conditions for the existence and uniqueness of a generalized solution to Inverse Problem 4(F,Σ).

Theorem 9. *Let* $J, J_t \in L_\infty(\mathcal{D}; L_p(V))$; $h_2, h_{2t} \in L_\infty(\mathcal{D})$;

$$h_1, h_{1t} \in L_\infty((0,T); L_p(G \times V)); \qquad \mu, \mu_t \in L_\infty((0,T); L_p(\gamma_-));$$

$$\varphi, (v,\nabla)\varphi \in L_p(G \times V), \varphi|_{\gamma_-} \in L_p(\gamma_-); \chi, \chi', \chi_1, \chi_1' \in L_\infty(0,T);$$

$$\left| \chi(t) \int\limits_{G\times V} \omega_1(x,v) g(x,v,t)\, dx\, dv - \chi_1(t) \int\limits_{G\times V} \omega(x,v) g(x,v,t)\, dx\, dv \right| \geq k_2 > 0;$$

$g, (v,\nabla)g \in L_\infty((0,T); L_p(G \times V))$, *and let the compatibility conditions* (5) *and* (12) *and the condition*

$$\int\limits_{G\times V} \omega_1(x,v)\varphi(x,v)\, dx\, dv = \chi_1(0) \tag{14}$$

satisfied.
Then it is possible to indicate a number a_2 *such that for* $Tv_0^{-1} \operatorname{diam} G < a_2$ *there exists a triple of functions* $\{u, f, \sigma\}$ *(a generalized solution of Inverse Problem* 3(F, Σ)) *from the function class* $u \in H_\infty((0,T); L_p(G \times V))$, $f \in L_\infty(0,T)$, $\sigma \in L_\infty(0,T)$. *For all* T, v_0 *and* $\operatorname{diam} G$ *this inverse problem has no more than one generalized solution.*

Similar results are obtained for inverse problems of finding the functions $u(x,v,t)$, $F(x,v,t) = f(v,t)g(x,v,t) + h(x,v,t)$, $\Sigma(x,v,t) = \sigma(v,t) + h_1(x,v,t)$, where f, σ are to be found and g, h, h_1 are functions given a priori, with the overdetermination condition having the form

$$\int\limits_G \omega_2(x) u(x,v,t)\, dx = \chi_2(v,t), \quad (v,t) \in V \times [0,T]. \tag{15}$$

REFERENCES

Prilepko, A.I. and Volkov, N.P. (1987). Inverse problems for determinating parameters of non-stationary transport equation by overdeterminations of integral type. *Different. Equat.* **23**, 124–136 (in Russian).

Prilepko, A.I. and Volkov, N.P. (1988). Inverse problems for determinating parameters of nonstationary kinetic transport equation by additional information on tracks of searched function. *Different. Equat.* **24**, 136–146 (in Russian).

Ill-Posed Problems in Natural Sciences, pp. 439 – 445
A. Tikhonov (Ed.)
1992 VSP/TVP

DETERMINATION OF CONSTANT PARAMETERS IN SOME SEMILINEAR PARABOLIC EQUATIONS

M. YAMAMOTO
Department of Mathematics, College of Arts and Sciences University of Tokyo, Komaba, Meguro, Tokyo 153, Japan

ABSTRACT

For $\alpha > 0$, let $u(x,t;\alpha)$ be a positive classical solution to $u_t = \Delta u + u^\alpha$, $u(x,0) = f(x)$ $(x \in \Omega,\ t > 0)$ and $\partial_N u(x,t) = 0$ $(x \in \partial\Omega,\ t > 0)$. For given $x_0 \in \overline{\Omega}$ and $t_0 > 0$, we prove the following:

(i) If $f(x) \geq 1$ $(x \in \overline{\Omega})$, then $u(x_0,t_0;\alpha)$ determines α uniquely;

(ii) If $f(x) > 1$ $(x \in \overline{\Omega})$, then the map $u(x_0,t_0;\alpha) \to \alpha$ is Lipschitz continuous.

1. INTRODUCTION

In this paper, we consider an inverse problem, where, from an observation of solutions at a single point and a fixed time, we determine a single constant parameter $\alpha > 0$ in the semilinear parabolic equation

$$u_t = \Delta u + u^\alpha \quad (x \in \Omega,\ t > 0),$$

$$\partial_N u(x,t) = 0 \quad (x \in \partial\Omega,\ t > 0),$$

$$u(x,0) = f(x) > 0 \quad (x \in \Omega),$$

where Ω is a bounded domain in $\mathbf{R}^n$ with a smooth boundary $\partial\Omega$ and ∂_N denotes the normal derivative. Let the positive initial value f be fixed and let a positive solution to this initial value problem be denoted by $u(x,t;\alpha)$. Then, for fixed $x_0 \in \overline{\Omega}$ and $t_0 > 0$, our inverse problem is to study

(i) whether a map G: $u(x_0,t_0;\alpha) \to \alpha$ is well defined.

(ii) whether G is continuous.

If G is well defined, then we can uniquely determine α from the single observation $u(x_0,t_0;\alpha)$, and moreover if G is continuous, then we can expect a stability estimate such as $|\alpha - \beta| \leq C|u(x_0,t_0;\alpha) - u(x_0,t_0;\beta)\,|$.

There are many papers dealing with inverse problems for nonlinear parabolic equations. For example, we can refer to (Cannon and DuChateau, 1973, 1980a, 1980b;

DuChateau and Rundell, 1985; Lorenzi, 1982; Cannon and Yin, 1991). Those papers focus on the unique determination of the unknown functions in equations from functions in t (i.e., from data givenon a time interval).
In contrast, this paper treats the unique determination of the unknown constants in nonlinear equations from a scalar-type observation. For similar problems in linear parabolic equations, we can refer to (Cannon, 1964; Kitamura and Nakagiri, 1977). Our method is based on the monotonicity of semilinear terms u^α, and hence is applicable to other semilinear parabolic equations such as $u_t = \Delta u + e^{\alpha u}$, where α is unknown constant.
This paper consists of three sections. In §2 we state two main results on the unique determination and stability estimates. The proofs of these results are given in §3.

2. MAIN RESULTS

Let

$$\alpha > 0 \tag{2.1}$$

and let f be smooth and positive on $\overline{\Omega}$. Let $u(x,t;\alpha)$ be a positive classical solution to (2.2):

$$\begin{aligned} &u_t = \Delta u + u^\alpha \quad (x \in \Omega,\ 0 < t < T), \\ &\partial_N u(x,t) = 0 \quad (x \in \partial\Omega,\ 0 < t < T), \\ &u(x,0) = f(x) \quad (x \in \Omega), \end{aligned} \tag{2.2}$$

where T is fixed. Here by a classical solution to (2.2) we mean that

$$u, u_t, u_x, u_{xx} \in C^0(\overline{\Omega} \times [0,T]).$$

Remark. We always consider positive solutions to (2.2). Throughout this paper, assuming the existence of classical solutions, we will concentrate on inverse problems. For the existence, see, e.g., (Henry, 1981).
Throughout this paper, the initial value $f = f(x)$ is assumed to be known and fixed.
Our conclusions heavily depend on whether the range of the initial value f is above l or not. Roughly speaking, in the former case we can get a "global" estimate of the stability.
Henceforth let $G(x,y,t)$ denote the fundamental solution of $\frac{\partial}{\partial t} - \Delta$ in Ω with the boundary condition $\partial_N u(x,t) = 0$. Then, by the positivity of G (see, e. g., (Itô, 1977)), we have

$$l(\delta) = l(\delta; x_0, t_0) = \inf\{G(x_0, y, t);\ \delta \le t \le t_0,\ y \in \overline{\Omega}\} > 0 \tag{2.3}$$

for $x_0 \in \overline{\Omega}$ and $\delta \in (0, t_0]$.

THEOREM 1 (Uniqueness). *Let $\alpha > 0$ and $\beta > 0$. Assume that*

$$f(x) \ge 1 \quad (x \in \overline{\Omega}). \tag{2.4}$$

Let $x_0 \in \overline{\Omega}$ and $t_0 > 0$ be arbitrarily fixed. Then $u(x_0, t_0; \alpha) = u(x_0, t_0; \beta)$ implies $\alpha = \beta$.

THEOREM 2 (Stability estimate). *For $\epsilon > 0$, assume that*

$$f(x) \geq 1 + \epsilon \quad (x \in \overline{\Omega}). \tag{2.5}$$

Then, for every $\delta \in (0, t_0)$, we have an estimate depending only on δ:

$$|\alpha - \beta| \leq \big(l(\delta)(t_0 - \delta)|\Omega| \cdot \log(1 + \epsilon)\big)^{-1} |u(x_0, t_0; \alpha) - u(x_0, t_0; \beta)|. \tag{2.6}$$

In the case $\Omega = (0, a)$, we can give an explicit estimate for $l(\delta)$. Namely, we have the following

COROLLARY. *Let $\Omega = (0, a)$ $(a > 0)$ and let (2.5) be satisfied. Then*

$$l(\delta) \geq \frac{1}{2\sqrt{\pi}} \min\Big\{\frac{1}{\sqrt{\delta}} \exp(-\frac{x_1^2}{4\delta}), \frac{1}{\sqrt{t_0}} \exp(-\frac{x_1^2}{4t_0})\Big\}, \tag{2.7}$$

where

$$x_1 = \max\{x_0,\ a - x_0\}. \tag{2.8}$$

That is, for every $\delta \in (0, t_0)$, estimate (2.6) can be written as

$$|\alpha - \beta| \leq 2\sqrt{\pi}\{a(t_0 - \delta)\log(1 + \epsilon)\}^{-1} \max\Big\{\sqrt{\delta}\exp(\frac{x_1^2}{4\delta}), \sqrt{t_0}\exp(\frac{x_1^2}{4t_0})\Big\}$$
$$\times |u(x_0, t_0; \alpha) - u(x_0, t_0; \beta)|. \tag{2.9}$$

In particular,

$$|\alpha - \beta| \leq 4\sqrt{\pi}\{a\sqrt{t_0}\log(1 + \epsilon)\}^{-1} \exp(\frac{a^2}{2t_0})|u(x_0, t_0; \alpha) - u(x_0, t_0; \beta)|.$$

For the uniqueness, *the assumption $f(x) \geq 1$ is essential.* More precisely, if $f(x) \geq 1$ $(x \in \overline{\Omega})$ does not hold, then, generally we can not determine α from $u(x_0, t_0; \alpha)$ at any $t_0 > 0$. This is seen from the following example. Consider $f(x) = e^{-\gamma}$ – const $(0 \leq x \leq 1;\ \gamma > 0)$ as an initial value in (2.2). Then (2.2) is written as

$$\frac{du(t)}{dt} = u(t)^{\alpha}\ (0 < t < T), \quad u(0) = e^{-\gamma}. \tag{2.10}$$

Here, we choose $\gamma > 0$ such that

$$1 < \frac{1}{\gamma} \log \frac{e^{\gamma}}{4\gamma} < 2. \tag{2.11}$$

Since $1 - e^{-1/4} \le \frac{1}{s}\log\frac{e^s}{4s}$ $(s > 0)$, such a γ exists. For $\alpha \ge 1$, we denote the solution to (2.10) by $u(t;\alpha)$. Moreover, we set $S(\alpha) = \left((1-\alpha)e^\gamma/4 + e^{\gamma(\alpha-1)}\right)^{1/(1-\alpha)}$ $(2 \le \alpha \le 3)$. By (2.11), S is monotone decreasing in $(2, 1 + \frac{1}{\gamma}\log e^\gamma/4\gamma)$ and increasing in $(1 + \frac{1}{\gamma}\log e^\gamma/4\gamma, 3)$. Therefore, there exist α_1, α_2 such that

$$2 < \alpha_1 < \alpha_2 < 3 \text{ and } S(\alpha_1) = S(\alpha_2). \tag{2.12}$$

Then we can prove that

$$u(\frac{e^\gamma}{4};\alpha_1) = u(\frac{e^\gamma}{4};\alpha_2) \text{ in spite of } \alpha_1 \ne \alpha_2.$$

In fact, we can solve (2.10) to get $u(t;\alpha_i) = ((1-\alpha_i)t + e^{\gamma(\alpha_i-1)})^{1/(1-\alpha_i)}$ $(i = 1, 2)$. Since $2 < \alpha_1, \alpha_2 < 3$, the inequality $(1-\alpha_i)e^\gamma/4 + e^{\gamma(\alpha_i-1)} > 0$ holds, so $u(t;\alpha_i)$ exists for $t \in [0, e^\gamma/4]$. We have $u(e^\gamma/4;\alpha_i) = S(\alpha_i)$ $(i = 1, 2)$ and therefore $u(e^\gamma/4;\alpha_1) = u(e^\gamma/4;\alpha_2)$ by (2.12).

3. PROOF OF THEOREMS 1 AND 2

First we will prove a comparison lemma.

LEMMA. *Let $u > 0$ and $v > 0$ be positive and continuous on $\overline{\Omega} \times [0,T]$ and twice continuously differentiable in x and once continuously differentiable in t on $\overline{\Omega} \times [0,T]$. If*

$$u_t - \Delta u \ge u^\alpha \quad (x \in \Omega,\ 0 < t < T),$$

$$v_t - \Delta v \le v^\alpha \quad (x \in \Omega,\ 0 < t < T),$$

$$\partial_N u(x,t) = \partial_N v(x,t) = 0 \quad (x \in \partial\Omega,\ 0 < t < T),$$

$$u(x,0) \ge v(x,0) \quad (x \in \Omega),$$

then

$$u(x,t) \ge v(x,t) \quad (x \in \overline{\Omega},\ 0 \le t \le T).$$

Proof. Although the proof of this lemma is a routine in maximum principles for parabolic equations, we will give its proof for convenience. We may assume that there exists some $m_1 > 0$ such that $m_1 \le u(x,t), v(x,t)$ $(x \in \overline{\Omega},\ 0 \le t \le T)$. Let us set $w(x,t) = u(x,t) - v(x,t)$ and $c(x,t) = (u^\alpha(x,t) - v^\alpha(x,t))/(u(x,t) - v(x,t))$ $(x \in \overline{\Omega},\ 0 \le t \le T)$. Then $c(x,t)$ is continuous on $\overline{\Omega} \times [0,T]$ and there exists an $M > 0$ such that

$$|c(x,t)| < M \quad (x \in \overline{\Omega},\ 0 \le t \le T). \tag{3.1}$$

Moreover, w satisfies $w_t - \Delta w \ge c(x,t)w$ $(x \in \Omega,\ 0 < t < T)$. Set $w(x,t) = e^{Mt}W(x,t)$ $(x \in \overline{\Omega},\ 0 \le t \le T)$. Then W is continuous on $\overline{\Omega} \times [0,T]$ and we have

$$W_t - \Delta W \ge (c(x,t) - M)W \quad (x \in \Omega,\ 0 < t < T), \tag{3.2}$$

$$\partial_N W(x,t) = 0 \quad (x \in \partial\Omega,\ 0 < t < T), \tag{3.3}$$

and

$$W(x,0) \geq 0 \quad (x \in \Omega). \tag{3.4}$$

We have only to prove that $W(x,t) \geq 0$ $(x \in \overline{\Omega}, 0 \leq t \leq T)$. Contrarily assume that $\min\{W(x,t);\ x \in \overline{\Omega},\ 0 \leq t \leq T\} = W(x_0,t_0) < 0$. By (3.4), we have $t_0 > 0$ and so $W_t(x_0,t_0) \leq 0$. If $x_0 \in \partial\Omega$, then $\Delta W(x_0,t_0) \geq 0$ by (3.3) and a well-known property (e.g. (Itô, 1977)).
On the other hand, if $x_0 \in \Omega$, then we directly get $\Delta W(x_0,t_0) \geq 0$. Therefore by (3.2) we obtain $0 \geq W_t(x_0,t_0) - \Delta W(x_0,t_0) \geq (c(x_0,t_0) - M)W(x_0,t_0)$, namely, $(c(x_0,t_0) - M)W(x_0,t_0) \leq 0$. Since $c(x_0,t_0) - M < 0$ by (3.1) and $W(x_0,t_0) < 0$, this inequality is a contradiction. Thus the lemma is proved.

Proof of Theorem 1. Putting $u(x,t) = u(x,t;\alpha)$ or $= u(x,t;\beta)$ and $v(x,t) = 1$ in Lemma, we can get

$$u(x,t;\alpha) \geq 1, \quad u(x,t;\beta) \geq 1 \quad (x \in \overline{\Omega},\ 0 \leq t \leq T). \tag{3.5}$$

In order to prove $\alpha = \beta$, we contrarily assume that $\alpha \neq \beta$. It is sufficient to derive a contradiction from $\alpha > \beta$. By (3.5) we have $u(x,t;\beta)^\beta \leq u(x,t;\beta)^\alpha$ and hence $u(x,t) = u(x,t;\alpha)$ and $v(x,t) = u(x,t;\beta)$ satisfies $u_t - \Delta u = u^\alpha$, $v_t - \Delta v \leq v^\alpha$ $(x \in \Omega,\ 0 < t < T)$, $\partial_N u(x,t) = \partial_N v(x,t) = 0$ $(x \in \partial\Omega,\ 0 < t < T)$ and $u(x,0) = v(x,0)$ $(x \in \Omega)$. Therefore Lemma implies

$$u(x,t;\beta) \leq u(x,t;\alpha) \quad (x \in \overline{\Omega},\ 0 \leq t \leq T). \tag{3.6}$$

In view of that $u(x,t;\alpha)$ and $u(x,t;\beta)$ are representable as

$$\begin{aligned} u(x,t;\alpha) &= \int_\Omega G(x,y,t)f(y)dy + \int_0^t \int_\Omega G(x,y,t-s)u(y,s;\alpha)^\alpha dyds \\ u(x,t;\beta) &= \int_\Omega G(x,y,t)f(y)dy + \int_0^t \int_\Omega G(x,y,t-s)u(y,s;\beta)^\beta dyds \end{aligned} \tag{3.7}$$

$$(x \in \overline{\Omega},\ t > 0).$$

Therefore $u(x_0,t_0;\alpha) = u(x_0,t_0;\beta)$ implies

$$\int_0^t \int_\Omega G(x_0,y,t_0-s)(u(y,s;\alpha)^\alpha - u(y,s;\beta)^\beta)dyds = 0. \tag{3.8}$$

By $\alpha > \beta$, (3.5), and (3.6), we have

$$u(y,s;\alpha)^\alpha \geq u(y,s;\alpha)^\beta \geq u(y,s;\beta)^\beta \quad (y \in \overline{\Omega},\ 0 \leq s \leq T), \tag{3.9}$$

whence

$$u(y,s;\alpha)^\alpha - u(y,s;\beta)^\beta \geq 0 \quad (y \in \overline{\Omega},\ 0 \leq s \leq T). \tag{3.10}$$

Therefore relations (3.8), (3.10) and (2.3) imply $G(x_0,y,t_0 - s)(u(y,s;\alpha)^\alpha - u(y,s;\beta)^\beta) = 0$ $(y \in \overline{\Omega},\ 0 \leq s < t_0)$, i.e., $u(x,t;\alpha)^\alpha = u(x,t;\beta)^\beta$ $(x \in \overline{\Omega},\ 0 \leq t < t_0)$. Combining this equality with (3.9), we obtain $u(x,t;\alpha)^\alpha = u(x,t;\alpha)^\beta$ $(x \in \overline{\Omega},\ 0 \leq t < t_0)$. Hence, by $\alpha > \beta$, we are led to $u(x,t;\alpha) = 1$ $(x \in \overline{\Omega},\ 0 \leq t < t_0)$, which contradicts the condition that $u(x,t;\alpha)$ satisfies $u_t - \Delta u = u^\alpha$. Theorem 1 is there by proved.

Proof of Theorem 2. From Lemma, putting $u(x,t) = u(x,t;\alpha)$ or $= u(x,t;\beta)$ and $v(x,t) = 1+\epsilon$, we obtain

$$u(x,t;\alpha) \geq 1+\epsilon, \quad u(x,t;\beta) \geq 1+\epsilon. \tag{3.11}$$

Without loss of generality, we may assume $\alpha \geq \beta$. Then, as is seen from the proof of Theorem 1, we have (3.6) and (3.10). For simplicity, we set

$$d = |u(x_0,t_0;\alpha) - u(x_0,t_0;\beta)|,$$

and

$$g(y,s) = G(x_0,y,t_0-s)(u(y,s;\alpha)^\alpha - u(y,s;\beta)^\beta) \quad (0 \leq s < t_0,\ y \in \overline{\Omega}). \tag{3.12}$$

By (3.7), (3.10) and (2.3), we have $d = \int_0^{t_0}\int_\Omega g(y,s)dyds \geq \int_0^{t_0-\delta}\int_\Omega g(y,s)dyds$ for each $\delta \in [0,t_0)$. If $g(y,s) > d/(t_0-\delta)|\Omega|$ for each $y \in \overline{\Omega}$ and each $s \in [0,\ t_0-\delta]$, then $d \geq \int_0^{t_0-\delta}\int_\Omega g(y,s)dyds > (d/(t_0-\delta)|\Omega|)\int_0^{t_0-\delta}\int_\Omega dyds = d$, which is a contradiction. Therefore there exist $y_0 \in \overline{\Omega}$ and $s_0 \in [0, t_0-\delta]$ such that $g(y_0,s_0) \leq d/(t_0-\delta)|\Omega|$, i.e., $0 \leq (u(y_0,s_0;\alpha^\alpha - u(y_0,s_0;\beta)^\beta)G(x_0,y_0,t_0-s_0) \leq d/(t_0-\delta)|\Omega|$. Consequently, by (2.3), we obtain

$$0 \leq u(y_0,s_0;\alpha)^\alpha - u(y_0,s_0;\beta)^\beta \leq d/l(\delta)(t_0-\delta)|\Omega|. \tag{3.13}$$

Now, applying the mean value theorem to the function $\varphi(x) = u(y_0,s_0;\alpha)^x$, by (3.9) we have

$$\begin{aligned} u(y_0,s_0;\alpha)^\alpha - u(y_0,s_0;\beta)^\beta &\geq u(y_0,s_0;\alpha)^\alpha - u(y_0,s_0;\alpha)^\beta \\ &= \log u(y_0,s_0;\alpha)u(y_0,s_0;\alpha)^{\theta\alpha+(1-\theta)\beta}(\alpha-\beta) \end{aligned}$$

for some $\theta \in (0,1)$. By (3.11), we obtain

$$\begin{aligned} &u(y_0,s_0;\alpha)^\alpha - u(y_0,s_0;\beta)^\beta \\ &\geq \log(1+\epsilon)\ (1+\epsilon)^{\theta\alpha+(1-\theta)\beta}(\alpha-\beta) \geq \log(1+\epsilon)(\alpha-\beta). \end{aligned}$$

Substituting this inequality into (3.13), we obtain

$$|\alpha-\beta| \leq d/l(\delta)(t_0-\delta)|\Omega|\log(1+\epsilon).$$

The proof of (2.6) is there by completed.

Proof of Corollary. In the case of $\Omega = (0,\alpha)$, we have

$$G(x,y,t) = \frac{1}{2\sqrt{\pi t}}\sum_{n=-\infty}^{\infty}\left\{\exp\left(-\frac{(x-y+2na)^2}{4t}\right) + \exp\left(-\frac{(x+y+2na)^2}{4t}\right)\right\},$$

(e.g., (Itô, 1977)). Therefore for $\delta \leq t \leq t_0$ and $y \in [0,a]$, we obtain

$$G(x_0, y, t) \geq \frac{1}{2\sqrt{\pi t}} \exp(-\frac{(x_0 - y)^2}{4t}) \geq \frac{1}{2\sqrt{\pi t}} \exp(-\frac{x_1^2}{4t})$$

$$\geq \frac{1}{2\sqrt{\pi}} \min \{ \frac{1}{\sqrt{\delta}} \exp(-\frac{x_1^2}{4\delta}), \frac{1}{\sqrt{t_0}} \exp(-\frac{x_1^2}{4t_0}) \},$$

which implies (2.7).

REFERENCES

Cannon, J.R. (1964). Determination of certain parameters in heatconduction problems. *J. Math. Anal. Appl.* **8**, 181–201.

Cannon, J.R. and DuChateau, P. (1973). Determining unknown coefficients in a nonlinear heat conduction problem. *SIAM J. Appl. Math.* **24**, 298–314.

Cannon, J.R. and DuChateau, P. (1980a). An inverse problem for a nonlinear diffusion equation. *SIAM J. Math.* **39**, 272–289.

Cannon, J.R. and DuChateau, P. (1980b). An inverse problem for an unknown source in a heat equation. *J. Math. Anal. Appl.* **75**, 465–485.

Cannon, J.R. and Yin, H-M. (1991). On a class of nonlinear parabolic equations with nonlinear trace type functionals. *Inverse Problems.* **7**, 149–161.

DuChateau, P. and Rundell, W. (1985). Unicity in an inverse problem for an unknown reaction term in a reaction diffusion equation. *J. Differential Equations* **59**, 155–164.

Henry, D. (1981). *Geometric Theory of Semilinear Parabolic Equations.* Springer–Verlag, Berlin–Heidelberg–New York

Itô, S. (1977). *Partial Differential Equations.* Baifû-kan, Tokyo.

Kitamura, S. and Nakagiri, S. (1977). Identifiability of spatially varying and constant parameters in distributed systems of parabolic type. *SIAM J. Control Optim.* **15**, 785–802.

Lorenzi, A. (1982). An inverse problem for a semilinear parabolic equation. *Ann. Mat. Pura Appl.* **82**, 145–166.

Part Three

APPLICATIONS

Ill-Posed Problems in Natural Sciences, pp. 449 – 453
A. Tikhonov (Ed.)
1992 VSP/TVP

A NUMERICAL METHOD FOR SOLVING THE INVERSE PROBLEM OF DEEP MAGNETOTELLURIC QUASI-LAYERED MEDIA SOUNDING

I.S. BARASHKOV and V.I. DMITRIEV
Moscow State University, 119899 Moscow, Russia

ABSTRACT

The subject of this paper are the inverse problems of magnetotelluric sounding (MTS) in the case where the electric conductivity changes rather slowly along the horizontal directions but yet it is difficult to apply Tikhonov–Cagniard model with one-dimensional electric conductivity distribution in the vertical direction, first of all for the problems connected with deep MTS and restoration of conducting layers under salt thickness.

The peculiarities of the problems under consideration are that the horizontal heterogeneity takes place mainly in a surface layer $0 < z < h_1(y)$ formed by the sediments with an electrical conductivity σ_1 which may be characterized by the integral conductivity $S_1(y) = h_1(y)\sigma_1$. Our model contains also a homogeneous intermediate layer $(h_1(y) < z < h_2)$ of higher resistivity with an electrical conductivity $\sigma_2(\sigma_2 \ll \sigma_1)$ and $n-2$ homogeneous layers $(z > h_2)$ with different electrical conductivities:

$$\sigma(y,z) = \begin{cases} \sigma_1, & \text{if } 0 < z < h_1(y), \\ \sigma_k, & \text{if } h_{k-1} < z < h_k, \quad k = 2, 3, \cdots, n-1, \\ \sigma_n, & \text{if } z > h_{n-1}. \end{cases}$$

We suppose that $h_1(y)$ changes slowly in the horizontal direction and the integral conductivity of the near-surface layer $S_1(y)$ is known.

Our method permits to reduce the two-dimensional inverse problem to a one-dimensional problem. The method takes into account the influence of the non-uniform near-surface layer on the impedance.

Our model and parameters of this model are shown in Fig. 1. For numerical example $n = 5$.

It is difficult to apply Tikhonov–Cagniard model to this problem. It is well-known that the lower the frequency, the deeper the wave penetration, so it provides information about deeper layers of the Earth. The problem is that the surface layer distorts significantly the impedance for low frequencies (the so-called S-effect) and this distortion must be taken into account when interpreting MTS data.

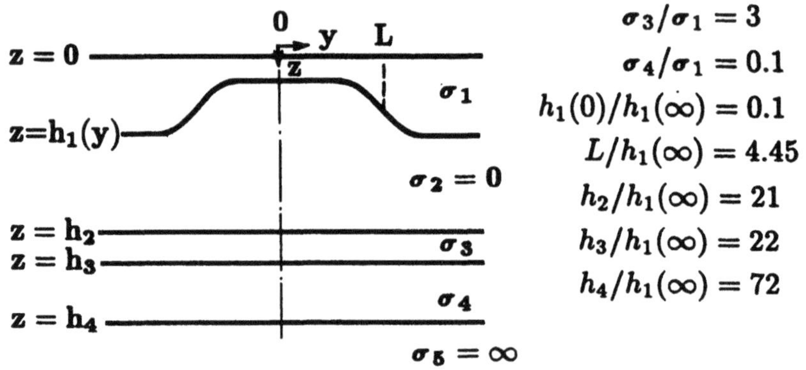

Fig. 1. Model of environment

Our purpose is to investigate the media that cannot be described adequately by the Tikhonov–Cagniard model. Those media are called quasi-layered media.
Interpretation of the data for deep MTS shows that the surface layer distorts tangibly the impedance even in the regions where the electric conductivity changes slowly along the earth surface. The explanation of this fact is as follows. The surface layer is separated from the deep conducting halfspace with a powerful layer of a poor conductor. As a result, the current from the first layer to the conducting halfspace is slow.
In the three-dimensional case the problem becomes more difficult because it is necessary to take into consideration not only the current from one layer to another but also the propagation along the surface.
Special asymptotic representations of the Maxwell system of equations are applied to solve this problem. We use a parameter ε,

$$\varepsilon = \frac{1}{\sqrt{\min S_1(y)}} \max \left(\left| \frac{dS_1}{dy} \right| \right),$$

to represent not only the slowly changing electric conductivity distribution but also the negligible distribution of electric conductivity in the second layer. We use also a parameter q:

$$q = h_2 \varepsilon^2 / \sigma_2.$$

It has been proved in (Barashkov and Dmitriev, 1987) that if the parameter q tends to zero as ε tends to zero, then the two-dimensional problem of deep magnetotelluric sounding amounts to a one-dimensional problem if ε is sufficiently small. If the parameter q does not tend to zero as ε tends to zero, then the two-dimensional problem of deep magnetotelluric sounding does not amount to a one-dimensional problem, even if ε is very small. In practice, if

$$\frac{h_2}{\sigma_2 \min S_1(y)} \max \left(\left| \frac{dS_1}{dy} \right|^2 \right) \gg 1,$$

then it is not possible to solve the inverse problem of deep magnetotelluric sounding as a one-dimensional problem.

The asymptoty for solving the problem at ε tends to zero is being investigated. As this takes place, the following formula for asymptoty of impedance is derived for two-dimensional case

$$Z = i\omega\mu t - \frac{d\theta}{dy} r + \frac{dh_1}{dy}\theta g, \tag{1}$$

where the function θ is a solution of the following differential equation

$$\frac{d}{dy}\left(\left(1 - \varkappa\right)\frac{d\theta}{dy}\right) + \frac{d}{dy}\left(\Psi\frac{dh_1}{dy}\theta\right) + i\omega\mu\sigma_2\theta = -i\omega\mu\frac{d\chi}{dy}, \tag{2}$$

with the boundary conditions

$$\lim_{y\to\infty}\frac{d\theta}{dy} = 0, \qquad \lim_{y\to-\infty}\frac{d\theta}{dy} = 0,$$

where

$$\chi = \frac{dt}{dz}\,|_{z=h_1}, \quad \kappa = \frac{dr}{dz}\,|_{z=h_1+0}, \quad \Psi = \frac{dg}{dz}\,|_{z=h_1},$$

and the functions t, r and g are solutions of equations

$$\frac{d^2u}{dz^2} + i\omega\mu\sigma_1 u = 0 \quad \text{if } 0 < z < h_1(y),$$

$$\frac{d^2u}{dz^2} = 0 \quad \text{if } h_1(y) < z < h_2,$$

with the following boundary conditions

$$\frac{dt}{dz}\,|_{z=0} = 1, \quad [t]\,|_{z=h_1} = 0, \quad \left[\frac{dt}{dz}\right]|_{z=h_1} = 0, \quad \left(\frac{dt}{dz} - \frac{i\omega\mu}{\nu}t\right)|_{z=h_2} = 0,$$

$$\frac{dr}{dz}\,|_{z=0} = 0, \quad [r]\,|_{z=h_1} = 0, \quad \left[\frac{dr}{dz}\right]|_{z=h_1} = 1, \quad \left(\frac{dr}{dz} - \frac{i\omega\mu}{\nu}r\right)|_{z=h_2} = 1,$$

$$\frac{dg}{dz}\,|_{z=0} = 0, \quad [g]\,|_{z=h_1} - 1, \quad \left[\frac{dg}{dz}\right]|_{z=h_1} = 0, \left(\frac{dq}{dz} - \frac{i\omega\mu}{\nu}g\right)|_{z=h_2} = 0.$$

The parameter ν is the impedance for environment, which lies when $z > h_2$. We solve the differential equation (2) for the function θ by difference method.

The curve $y = 0$ in Fig. 2 is the curve of seeming resistivity for our model, which corresponds to $y = 0$. The curve ρ_3 in Fig. 2 characterizes environment, which lies when $z > h_2$. The impedance ν corresponds to the curve ρ_3. The parameters of this curve are indicated in Fig. 2.

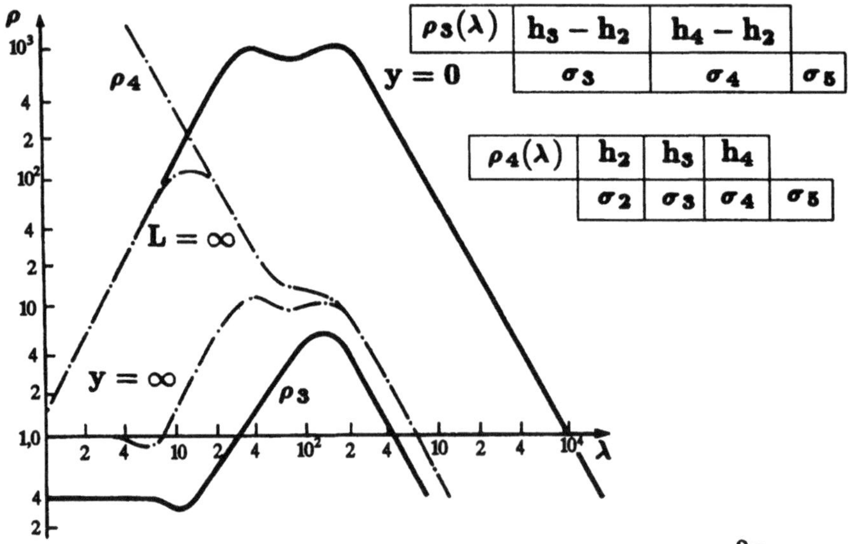

Fig. 2. Curves of seeming resistivity $\rho = \rho(\lambda)$, where $\lambda = \dfrac{2\pi}{h_1(\infty)\mathrm{Re}\,k_1}$

The inverse problem is the problem of revealing the electrical conductivity profile of the layered earth from the impedance $Z(\omega)$. It is necessary to measure the impedance at one point on the surface of the earth for different frequencies. In the numerical example this point is $y = 0$.

The following formula describes an iterative method of finding the parameter ν when σ_2 and h_2 are known:

$$\nu_{j+1} = \nu_j + \frac{Z_{\text{measured}} - Z\,|_{\nu=\nu_j}}{Z'_\nu\,|_{\nu=\nu_j}}, \tag{3}$$

where

$$Z'_\nu = i\omega\mu t'_\nu - \frac{dU}{dy}r - \frac{d\theta}{dy}r'_\nu + \frac{dh_1}{dy}(Ug + \theta g'_\nu),$$

and the function U is a solution of the equation

$$\frac{d}{dy}\Big((1-\varkappa)\frac{dU}{dy}\Big) + \frac{d}{dy}\Big(\Psi\frac{dh_1}{dy}U\Big) + i\omega\mu\sigma_2 U = -i\omega\mu\frac{d^2\chi}{d\nu dy}$$

$$+\frac{d}{dy}\Big(\varkappa'_\nu\frac{d\theta}{dy}\Big) - \frac{d}{dy}\Big(\Psi'_\nu\frac{dh_1}{dy}\theta\Big),$$

with the boundary conditions

$$\lim_{y\to\infty}\frac{dU}{dy} = 0, \qquad \lim_{y\to-\infty}\frac{dU}{dy} = 0.$$

If the parameters σ_2 and h_2 are unknown, it is not possible to use formula (3). For the sake of simplicity, we will consider the case when the parameter σ_2 equals zero and the parameter h_2 is unknown. Let N be a parameter, which we calculate by the formula

$$N = \nu - i\omega\mu h_2.$$

The parameter N is the impedance for our model if

$$\sigma_1 = \sigma_2 = 0.$$

The curve ρ_4 in Fig. 2 corresponds to the impedance N. Parameters of this curve are indicated in Fig. 2. Using the parameter N, we can solve the differential equation of our problem analytically. The inverse problem in this case reduces to the quadratic equation

$$Z = \frac{N + i\omega\mu(h_1 - \tau/k_1)}{1 + k_1\tau\big[h_1 - iN/(\omega\mu)\big]}(1 - c_1 v) + i\omega\mu c_2 v, \tag{4}$$

where

$$k_1 = \sqrt{i\omega\mu\sigma_1}, \qquad \mathrm{Im}(k_1) > 0, \qquad \tau = \tan(k_1 h_1),$$

$$c_1 = k_1 \sin(k_1 h_1), \qquad c_2 = \cos(k_1 h_1), \qquad h_1^\infty = \lim_{|y|\to\infty} h_1(y),$$

$$v = \frac{1}{k_1 \sin k_1 h_1}\left[1 - \frac{(N + i\omega\mu h_1)k_1 \sin k_1 h_1 + i\omega\mu \cos k_1 h_1}{(N + i\omega\mu h_1^\infty)k_1 \sin k_1 h_1^\infty + i\omega\mu \cos k_1 h_1^\infty}\right].$$

It is easy to solve equation (4) in order to find $N(\omega)$ for any ω.

REFERENCES

Barashkov, A. and Dmitriev, V. (1987). On the inverse problem of deep magnetotelluric sounding. *Dokl. Akad. Nauk SSSR* **295**, 83–86.

Ill-Posed Problems in Natural Sciences, pp. 454 – 462
A. Tikhonov (Ed.)
1992 VSP/TVP

ON NUMERICAL METHODS OF SOLVING INVERSE PROBLEMS OF ELECTRICAL PROSPECTING

S.M. BERSENEV
Computer Center, Siberian Branch of Russia Academy of Sciences, Akademgorodok, 660036 Krasnoyarsk, Russia

ABSTRACT

The paper describes a construction of numerical methods for solving inverse problems of electrical prospecting. The methods are based on minimization of Tikhonov's regularizing functional. Minimization and selecting of a parameter of regularization can be provided simultaneously by a steady-state scheme for a special system of differential equations. The problems of parametrization of a medium, numerical stability and complexification of geophysical methods are considered and numerical examples are given.

1. INTRODUCTION

Electrical prospecting as a geophysical technique is used to prospect for water, geothermal waters, gas, oil, and other natural resources. Most efficiently this technique can be used in combination with seismic prospecting. Various kinds of electrical prospecting techniques are generated by different types of electrical and magnetic field sources, their location and measurement techniques. The diversity of electrical prospecting techniques is connected with the properties of conductive zones in the media under investigation, the depth of exploration, availability of mathematical methods for interpreting the data measured, ecological situation, the cost and other factors.

In hydrological exploration, for example, it is customary to use the direct current sources, while the presence of zones with small conductivity makes use of the alternate current. In deep exploration the sources are the plane electromagnetic waves of the natural electromagnetic field of the earth.

The main purpose of every electrical prospecting technique is to determine the conductivity function X by the measurements of the field U at the surface of the earth or in the well, i.e., to solve the operator equation

$$F(X) = U. \tag{1}$$

The operator F determines the theoretical electromagnetic field depending on the sources and the medium at the points of measurement. The values of the operator

F are assigned by solutions of the Maxwell equations as applied to a particular form of electrical prospecting.
The problem of solving equation (1) with respect to X is not always a correctly posed problem, since X can be nonunique and unstable, with respect to the perturbations of U.

2. THE MEDIUM PARAMETRIZATION

The prospecting level of geophysical works is specified by a concept of the medium as a set of structures with a similar geophysical property. Therefore, the successful parametrization of conductivity X as a step function is of prime importance for the solution of problem (1).
The electromagnetic exploration usually, starts with assuming the medium to be a horizontally-layered structure (Fig. 1).
Parametrization of electric cross-section in this case is rather simple and can be assigned by one list

$$\sigma_1(z_0, z_1);\ \sigma_2(z_1, z_2);\ \ldots;\ \sigma_{n+1}(z_n, \infty).$$

The two-dimensional step electrical cross-section (Fig. 2) can be assigned by three lists. First we assign the list of specific points determining the electrical cross-section

$$1(y_1, z_1);\ 2(y_2, z_2);\ \ldots;\ 11(y_{11}, z_{11}).$$

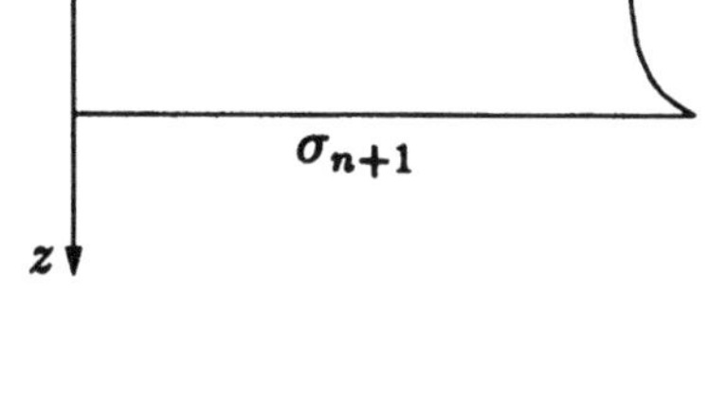

Fig. 1.

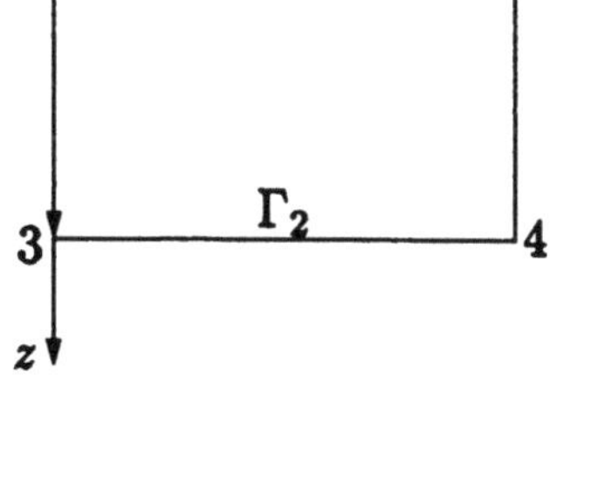

Fig. 2.

Next we assign the list of lines, connecting these points:

$$\Gamma_1(1,2);\ \Gamma_2(2,3,4,5);\ \Gamma_3(5,6,1);\ \Gamma_4(2,7);\ \Gamma_5(7,8);\ \Gamma_6(5,7);\ \Gamma_7(8,11,10,9,8).$$

The desired parametrization can be assigned then by the list of domains with similar conductivity

$$\sigma_1(\Gamma_1,\Gamma_4,-\Gamma_6,\Gamma_3);\ \sigma_2(\Gamma_7);\ \sigma_3(\Gamma_2,\Gamma_6,\Gamma_5,-\Gamma_7,-\Gamma_5,-\Gamma_4).$$

It is not difficult to see that the homogeneity domains are assigned by a set of lists, enumerated in such an order, that the domain remains on the left when by-passed. The cross-section parameters in this case are the coordinates of specific points and the values of the conductivities σ_i in the domains.
The generalization of this approach to the three-dimensional case can result in too cumbersome lists even for the simplest bodies. A better solution is therefore to form two lists.
First we assign a set of parametrized basic bodies (balls $S_i(\cdot,\cdot,\cdot,\cdot)$, parallelepipeds $P_i(\cdot,\cdot,\cdot,\cdot)$, cones $K_i(\cdot,\cdot,\cdot,\cdot)$, ellipsoids $E_i(\cdot,\cdot,\cdot,\cdot)$, and other). Then we assign a list of inhomogeneities which can be formed by the functions of mathematical logic. For example,

$$\sigma_1(S_1\cap S_2\backslash K_1);\ \sigma_2(P_1\cup P_2\cap S_1);\ \sigma_3\big((S_2\cup K_1)\cap P_3\big).$$

3. STABILITY OF THE PROBLEM

A parametrization of the medium allows us to reduce the original problem (1) to solving a system of nonlinear equations

$$F(X)=U,\qquad F\colon D\subset\mathbf{R}^n\to\mathbf{R}^m,$$

where D is a closed convex set.
Let the nonlinear mapping F satisfy the conditions

$$\|F''(X_1)-F''(X_2)\|\leqslant L\|X_1-X_2\|,\quad \forall X_1,\ X_2\in D,$$

where L is the Lipshitz constant. Assume that the functional $g(X)\equiv\|F(X)-U\|_2^2$ is convex on D. Hence the nonuniqueness of the mapping F follows. The above conditions result in the instability of the element X with respect to perturbation of the element U in equation (1).
From problem (1) we pass over to stating

$$X=\mathop{\arg\min}_{X\colon F(X)=U}\|X-\mathring{X}\|_2^2,\quad \mathring{X}\in D, \tag{2}$$

where $\mathring{X}$ is an a priori given vector. The element X is unique, but it still has no stability property. To reconstruct the stability of problem (2), we shall use the general approach connected with the minimization of the Tikhonov's regularizing functional

$$X^\alpha=\mathop{\arg\min}_{X}\Phi(\alpha,X), \tag{3}$$

where $\Phi(\alpha, X) \equiv \|F(X) - U\|_2^2 + \alpha\|X - \overset{\circ}{X}\|_2^2, \quad \alpha > 0$.
To choose the regularization parameter it is possible to use various criteria, such as the discord principle (Morozov, 1966), generalized discord (Tikhonov et al., 1983) and so on. Let the criterion for choosing the regularization parameter be assigned by an equation

$$\alpha\colon \varphi(X^\alpha, \delta, h) = 0, \tag{4}$$

where δ, h are the parameters, specifying the perturbations of the measurement data and the errors of the model respectively.

4. NUMERICAL METHOD

Problems (3) and (4) are nonlinear and must be solved simultaneously. There are several approaches to solve this problem. For example, in (Tikhonov et al., 1973) the solution of equation (4) is obtained from the solutions of equation (3). This approach is complicated for solving equation (3). Minimization of $\Phi(\alpha, X)$ performed simultaneously with the variation of α by the iterative regularization principle was considered in (Bakushinskii, 1977). This approach, however, does not solve problems (3) and (4) simultaneously. We shall consider a different approach, connected with the simultaneous solving equations (3) and (4) by transition to a steady-state scheme

$$B(Z)\frac{dZ}{dt} = -\Psi(Z), \qquad Z(0) = Z^0, \tag{5}$$

where

$$Z = \begin{pmatrix} \alpha \\ X \end{pmatrix}, \qquad \Psi(Z) = \begin{pmatrix} \varphi(X, \delta, h) \\ \Phi'_X(\alpha, X) \end{pmatrix}.$$

To make the transition to a steady-state scheme, any matrix $B(Z)$, satisfying conditions (5)

$$\det B(Z) \neq 0, \quad \mathrm{Re}\lambda\Big\{\big[B(Z)\big]^{-1}\Psi'(Z)\Big\} > 0$$

in a neighbourhood of the solution $\widetilde{Z}$ of equation

$$\Psi(Z) = 0 \tag{6}$$

can be chosen. The speed of convergence of system (5) to a steady state depends on the choice of the matrix $B(Z)$ from this class and on the vector Z^0. Let us consider the numerical integration of system (5) by Euler's method (Bakhvalov, 1973):

$$B(Z_{i-1})\frac{Z_i - Z_{i-1}}{\tau_i} = \Psi(Z_{i-1}), \quad i = 1, 2, \ldots, \tag{7}$$

where τ_i is the time step and Z_i is the approximate solution, obtained by scheme (7) at the moment $t_i = t_{i-1} + \tau_i$.
One can obtain the greatest length of the integration step under stable calculation in a neighbourhood of the solution of (6) putting

$$B(Z) = \Psi'(Z) = \begin{pmatrix} 0 & \varphi'_X(X, \delta, h) \\ X - \overset{\circ}{X} & \Phi''_X(\alpha, X) \end{pmatrix}.$$

This choice of the matrix $B(Z)$ requires, however, the computation of $F''(X)$ which might result in unjustified losses. Moreover, $\det B(Z) = 0$ for $X = \overset{\circ}{X}$. Thus $B(Z)$ should be chosen close to $\Psi''(Z)$ and such in order to ensure the steady-state mode. Besides, it is useful to preserve the properties of the algorithms, known for the case, where F is a linear mapping. Taking all this into consideration, we choose $B(Z)$ as follows:

$$B(Z) = \begin{pmatrix} 0 & \varphi'_X(X,\delta,h) \\ X - \overset{\circ}{X} & C(Z) \end{pmatrix},$$

where $C(Z) = F'^T(X)F'(X) + \alpha E$. We introduce notation $a = \varphi'_X(X,\delta,h)$, $b = X - \overset{\circ}{X}$, $Q = C(Z)$. Then,

$$B^{-1}(Z) = \begin{pmatrix} 0 & a^T \\ b & Q \end{pmatrix}^{-1} = \begin{pmatrix} -\frac{1}{a^TQ^{-1}b} & \frac{a^TQ^{-1}}{a^TQ^{-1}b} \\ \frac{Q^{-1}b}{a^TQ^{-1}b} & Q^{-1} - \frac{Q^{-1}ba^TQ^{-1}}{a^TQ^{-1}b} \end{pmatrix}$$

and

$$B^{-1}(Z)\Psi'(Z) = E + G.$$

The norm of the matrix G is small in a neighbourhood of solution $\widetilde{Z}$ and the solution is asymptotically stable. It is not difficult to see that the main difficulty in implementation of scheme (7) is connected with the computation of the vectors of the form $f = Q^{-1}b$. We shall consider this problem as the problem of solving the system

$$Qf = b.$$

Put

$$Q = Q_m = F'^T(X)F'(X) + \alpha E = \sum_{i=1}^{m} a_i^T a_i + \alpha E,$$

where

$$F'(X) = \begin{pmatrix} a_1 \\ a_2 \\ \dots \\ a_m \end{pmatrix}.$$

Since matrix $F'(x)$ may have values of different scale, the procedure of convolution $F'^T(X) \cdot F'(X)$ can result in the loss of correct digits. Therefore, it is expedient to use the Cholesky factorization recalculation technique (Bersenev, 1984), which does not need the convolution procedure. The method consists in successive transition from the matrix L_k to the matrix L_{k+1} in Cholesky expansions

$$Q_k = L_k L_k^T, \quad k = 0, 1, \dots, m; \qquad L_0 = \sqrt{\alpha}\, E.$$

In geophysical problems the domain D is the set of non-negative solutions. Getting beyond this domain can be accounted by a too large value of the time step τ_i. In this case its value can be diminished.

5. ON SOLVING DIRECT PROBLEMS

When computing $F(X)$ it is possible to use an application package. The main requirement for these packages is to ensure the prescribed accuracy over a broad range of values of X. If the source is a plane wave, the medium being layered, then the problem of computing $F(X)$ has an analytic solution. When computing the fields in horizontally-layered media with a point source, we have to deal with the quadrature formulas (Zhogolev et al., 1962). In the two-dimensional or three-dimensional media the boundary problems must be solved numerically.

For example, using the magnetotelluric sounding technique for the two-dimensional medium in the case of E-polarization it is necessary to solve the first boundary problem for the equation

$$\frac{\partial^2 E_X}{\partial z^2} + \frac{\partial^2 E_X}{\partial y^2} + i\omega\mu_0\sigma E_X = 0,$$

where i is the imaginary unit, ω is the frequency, σ is the conductivity of the medium, μ_0 is the permeability of vacuum.

The parametrization of multidimensional cross-sections described above can be implemented for this problem most flexibly by the finite elements technique, since this method distorts the media less than the method of finite differences in coarse grids. Besides, there is no difficulty in coordinating the boundary conditions at the discontinuity boundary. Sharp conductivity jumps, however, can reduce the accuracy of this method. It has already been mentioned in (Dmitriev et al., 1987) that the greatest gradients in the solution arise at the boundaries of the conductivities. Hence in the two-dimensional case we propose to condense the triangles cyclically along the discontinuity lines, preserving the coordination of the general triangulation. The triangles can be condensed by the rule shown in Fig. 3.

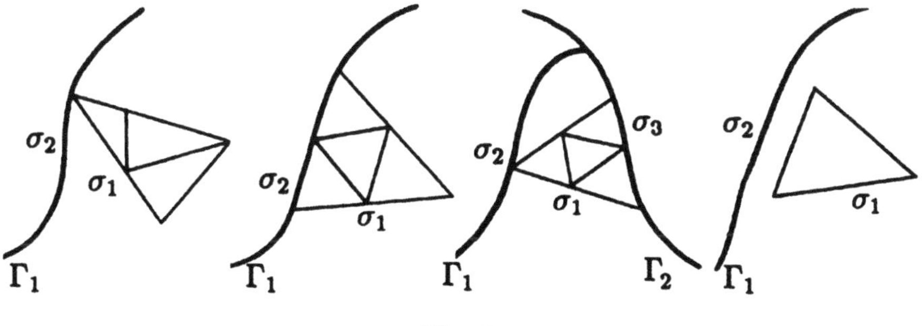

Fig. 3.

After several condensation cycles we get a sequence of compatible grids, enabling us to use multigrid methods. The total number of grid nodes is not large in this case and depends on the complexity of the cross-section. The matrices emerging in the finite element method, can be used not only to compute the fields, but also to calculate the partial derivatives of the fields with respect to the electrical cross-section parameters. It is not necessary in this case to recalculate the electrical cross-section triangulation to compute the partial derivatives with respect to geometrical parameters of the cross-section.

6. COMPLEXIFICATION OF GEOPHYSICAL METHODS

The electrical prospecting techniques provide different amount of information with respect to different parts of the electrical cross-section. The methods with point sources, for example, make better determination of the upper part of the cross-section, than the methods with plane electromagnetic waves. Besides, the electrical prospecting techniques complemented by seismic methods yield more accurate geophysical parameters of the geological structures.
We shall consider a problem of simultaneous solution of several inverse problems for different geophysical methods by studying the horizontally-layered medium (Fig. 4).

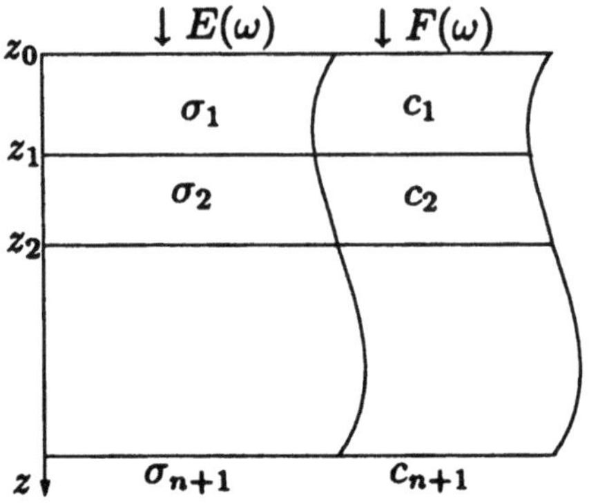

Fig. 4.

We introduce notation for the parameters, specifying the geophysical cross-section

$$H = (z_1, z_2, \dots, z_n); \quad \Sigma = (\sigma_1, \sigma_2, \dots, \sigma_{n+1}); \quad C = (c_1, c_2, \dots, c_{n+1}),$$

where z_i are the depths, σ_i are the conductivities, c_i are the velocities. If it is known that the boundaries of changes of the velocities and conductivities do not coincide, then the common boundaries can be introduced by imaginary layers, which will reconstruct the generality of the boundaries. Thus the statement of the following combined invcrsc problem

$$\begin{cases} F_1\ (H,\ \Sigma) \ = U_1 \\ \dots\dots\dots\dots\dots\dots \\ F_k\ (H,\ \Sigma) \ = U_k \\ F_{k+1}(H,\ C) = U_{k+1} \\ \dots\dots\dots\dots\dots\dots \\ F_{k+l}(H,\ C) = U_{k+l} \end{cases} \tag{8}$$

for k electrical prospecting methods and l seismic prospecting methods is of interest. The models are combined due to the common geological structure of the layers. Note that when we pass to continuous functions of conductivity and velocity, i.e., as $n \to \infty$, the combination of the electrical and seismic models vanishes.

Making the right and the left parts of relation (8) dimensionless, we get a complex geophysical model.

We will present now the results of a computing experiment on the combined interpretation of geophysical methods with the plane normally incident electromagnetic and seismic waves.

For a geophysical cross-section with parameters, given in Table 1, we have calculated, in the frequency range the electrical field corresponding to the boundary value problem (3)

$$\begin{aligned} &E'' - k(z)E = 0, \\ &E = E_0, \qquad z = 0, \\ &|E(z)| \to 0, \quad z \to \infty, \qquad k(z) = i\omega\mu_0\sigma(z), \qquad \sigma(z) = \sigma_i, \; z \in [z_i, z_{i+1}] \end{aligned} \tag{9}$$

and the seismic field, corresponding to the boundary value problem (Kiss,1986)

$$\begin{aligned} &u'' + k^2(z)u = 0, \\ &u' = F(\omega), \quad z = 0, \\ &u' - i\frac{\omega}{c_{n+1}}u = 0, \quad z = z_n; \end{aligned} \tag{10}$$

$$k(z) = \frac{\omega}{c(z)}, \qquad c(z) = c_i, \quad z \in [z_i, z_{i+1}].$$

Using one per cent perturbations of the impedance and the amplitude of the seismic field the inverse problems refining the cross-section parameters were solved. The results obtained are given in Table 2.

IBM PC/AT was used and the calculations were made for twenty values of the frequency ω. To choose the regularization parameter the discord principle was used. The number of time steps to reach the steady state circuit was not great (about 8–10). One can see from the table that the combined interpretation improves the accuracy of the geophysical parameters. It appears, besides this, that in the case where one combines inverse problems with point sources the results of the interpretation are improved also.

Table 1.

Layer number i	$d_i = z_i - z_{i-1}$ (km)	σ_i^{-1} (ohm·m)	c_i (km/sec)
1	0.4	10.	2.5
2	2.	500.	6.
3	0.2	8.	5.
4	2.	500.	8.
5		1000.	8.

Table 2.

d_2	d_3	d_4	σ_3^{-1}	c_3	Note
2.	0.2	2.0	8.	5.	real value
2.	0.23	2.3	20.	5.4	initial estimate
1.964	0.307	2.469	12.438	–	electric prospecting
1.883	0.305	2.311	–	5.113	seismic prospecting
2.014	0.198	2.462	8.076	5.202	combined interpretation

REFERENCES

Bakhvalov, N.S. (1973). *Numerical Methods.* Nauka, Moscow (in Russian).

Bakushinskii, A.B. (1977). Methods of solving monotonic variation inequalities, based on iterative regularization principles. *USSR Comput. Math. Math. Phys.* **17**, 1350–1362 (in Russian).

Bersenev, S.M. (1984). On computing schemes of regularization method. *USSR Comput. Math. Math. Phys.* **24**, 1402–1405 (in Russian).

Dmitriev, V.I., Barashkov, I.S., and Valeeva, Ye.A. (1987). On solving the two-dimensional inverse problem of magnetotelluric sounding by finite-difference method. In: *Mathematical Models and Computing Methods,* Moscow Univ. Press, Moscow, pp. 5–18 (in Russian).

Kis, R.G. (1986). The solving of an inverse problem of seismic prospecting by Marquardt's method. *TIIER.* **74**, 103–113.

Morozov, V.A. (1966). On regularization of ill-posed problems and the choice of a regularization parameters. *USSR Comput. Math. Math. Phys.* **6**, 170–175 (in Russian).

Pontryagin, N.S. (1974). *Ordinary Differential Equations.* Nauka, Moscow (in Russian).

Tikhonov, A.N., Glasko, V.B., and Kulik, N.I. (1973). Regularizing algorithms for nonlinear problems and inverse problem of magnetotelluric sounding. In: *Computing Methods and Programming,* Iss. 20, Moscow Univ. Press, Moscow, pp. 158–174 (in Russian).

Tikhonov, A.N., Goncharskii, A.V., Stcpanov, V.V., and Yagola, A.G. (1983). *Regularizing Algorithms and a priori Information.* Nauka, Moscow (in Russian).

Zhogolev, Ye.A., Trifonov, N.P., and Shakhsuvarov, D.N. (1962). Calculus of electromagnetic fields in layered media. In: *Computing Methods and Programming,* Moscow Univ. Press, Moscow, pp. 209–233 (in Russian).

Ill-Posed Problems in Natural Sciences, pp. 463 – 471
A. Tikhonov (Ed.)
1992 VSP/TVP

THE SOLUTION STABILITY AND RESTRICTIONS ON THE SPACE SCATTERER SPECTRUM IN THE TWO-DIMENSIONAL MONOCHROMATIC INVERSE SCATTERING PROBLEM

V.A. BUROV and O.D. RUMIANTSEVA
Moscow State University, Len.Gory, 119899 Moscow, Russia

ABSTRACT

This paper analyzes two reconstruction algorithms for two-dimensional inhomogeneities on the base of the scattered field data obtained by means of radiation of an inhomogeneity being looked for by a set of plane scalar waves. The observation regime is monochromatic. The foundation for the reconstruction algorithms are the functional-analytical methods. Both the scatterer reconstruction algorithms are based on the formal transition from monochromatic plane waves to fictitious inhomogeneous waves.

The first algorithm has an approximate character, because it requires the absence of the back scattering of the wave field. This requirement leads to restrictions on the space scatterer spectrum. The stronger the scatterer, the more rigid the restrictions. The approximate algorithm is of great importance for solving applied physical problems owing to the following facts: firstly, comparatively not large amount of calculating resources is needed; secondly, the algorithmic scheme suggested is stable.

The second algorithm does not impose any restrictions on the scatterer spectrum and is exact in this sense. However, the exactness of the algorithm is provided by the operation of the analytical extension. However, this operation is strong unstability. A similar unstability is inherent in the two-dimensional inverse scalar monochromatic problems.

1. THE PROBLEM STATEMENT AND FOUNDATIONS FOR THE RECONSTRUCTION ALGORITHMS

This paper considers the solution of the two-dimensional scalar inverse scattering problem by using functional-analytical methods. Functional-analytical methods appeared in the fifties. Recently, in (Grinevich and Manakov, 1986; Novikov, 1986 and 1988; Novikov and Henkin, 1987) they got further development. This paper is based on the ideas of the latter works, in which monochromatic wave fields are under study.

The statement of the inverse scattering problem is as follows. There is a finite domain within a boundless homogeneous space. This domain contains a finite inhomogeneity described by a scalar function $v(r)$, where r is a coordinate vector. According to its physical sense, in the acoustical formulation the function $v(r)$ characterizes the variation of the phase velocity in the domain of the inhomogeneity localization. The problem is to find the unknown function $v(r)$, which, by definition, is smooth and diminishes to zero on the boundary of the scattering domain. To be reconstructed, the scatterer is successively radiated from different directions by a plane monochromatic wave $\exp(ikr)$, $r \in \mathbf{R}^2$. The scattering amplitude $f(k,l)$ is measured for all directions of wave vectors l of the scattered field, where

$$k^2 = l^2 = k_0^2, \quad k \in \mathbf{R}^2, \quad l \in \mathbf{R}^2, \tag{1}$$

and k_0 is the wave number for the background homogeneous medium.

From the mathematical point of view the problem reduces to analyzing the equation

$$\Delta\Psi + k^2\Psi = v(r)\Psi. \tag{2}$$

In the case of real wave vectors k, equation (2) is supplemented by a radiation condition at the infinity. The functional-analytical methods treat equation (2), generally, for complex wave vectors $k \in \mathbf{C}^2$, i.e., $k = k_R + ik_I$, k_R and $k_I \in \mathbf{R}^2$; the same is true for vectors l of the scattered wave field. Analysis of solutions $\Psi = \Psi(r,k)$ $(k \in \mathbf{C}^2)$ of equation (2) under lies the two-dimensional scatterer reconstruction algorithms at the monochromatic observation regime. It is sufficient to consider the solutions only for the vectors k such that the condition (compare with (1))

$$k^2 = l^2 = k_0^2, \quad k \in \mathbf{C}^2, \quad l \in \mathbf{C}^2, \tag{3}$$

holds. In the case of complex wave vectors condition is equivalent to:

$$k_R \perp k_I, \quad k_R^2 - k_I^2 = k_0^2;$$

$$l_R \perp l_I, \quad l_R^2 - l_I^2 = k_0^2. \tag{4}$$

where $|k_R| = |l_R|, |k_I| = |l_I|$. In fact, the used method supposes the consideration of inhomogeneous waves scattering.

To consider directly the reconstruction algorithms one should introduce the concept of the generalized scattering amplitude $h(k,l)$ $(k,l \in \mathbf{C}^2)$, which is an analog of the classical scattering amplitude $f(k,l)$ $(k,l \in \mathbf{R}^2)$:

$$h(k,l) = -(2\pi)^{-2} \int\limits_{r \in \mathbf{R}^2} e^{-ilr}\Psi(r,k)v(r)\, dr. \tag{5}$$

Then the function $h_1(k) \equiv h\ (k, l = -\bar{k})$ (the bar over a vector denotes the complex conjugation) is responsible for the back scattering of the inhomogeneous waves.

Finally, another feature of the algorithms is the transition from the variable $r = \{x,y\}$ (x and y are the components of the vector $r \in \mathbf{R}^2$ in Cartesian coordinate system) to new complex variables $z = x + iy$ and $\bar{z}$, from the vector $k = k_R + ik_I = \{k_x, k_y\}$ to complex values $\lambda = (k_x + ik_y)/k_0$ and $\bar{\lambda}$ and from the vector

$l = l_R + il_I = \{l_x, l_y\}$ to values $\lambda' = (l_x + il_y)/k_0$ and $\bar{\lambda}'$. It is necessary to make the corresponding substitutions in the arguments of the abovementioned functions $v(r) = v(z, \bar{z})$; $f(k,l) = f(\lambda, \lambda')$, where $|\lambda| = |\lambda'| = 1$, because k and $l \in R^2$; $h_1(k) = h_1(\lambda, \lambda')$, where $\lambda \in \mathbf{C}^1$ because $k \in \mathbf{C}^2$.

2. THE APPROXIMATE RECONSTRUCTION ALGORITHM FOR A TWO-DIMENSIONAL MEAN POWER SCATTERER

This algorithm requires the absence of back scattering the inhomogeneous waves:

$$h_1(k) \equiv 0 \quad \forall k \in \mathbf{C}^2 \quad \Big(h_1(\lambda, \bar{\lambda}) \equiv 0 \quad \forall \lambda \in \mathbf{C}^1\Big), \tag{6}$$

here the vector k meets requirement (3) (or (4)). Representing the function $\Psi(r,k)$ in expression (5) as the Born–Neumann series with respect to powers of the function $v(r)$ leads to the conclusion that, if the m-fold process of wave scattering within the inhomogeneity is taken into account, then (6) can be satisfied by imposing the following restrictions on the space scatterer spectrum $\tilde{v}(\xi)$:

$$\tilde{v}(\xi) \equiv 0 \quad \text{for} \quad |\xi| \geq 2k_0/m. \tag{7}$$

On this case, the two-dimensional scatterer reconstruction can be performed using the scheme described below (Novikov, 1986; 1989). The validity of this scheme and the physical sense of the functions are discussed in detail in (Burov and Rumiantseva, 1992).

Functions $h^{\pm}(\lambda, \lambda')$ are found from the measured scattering amplitude $f(\lambda, \lambda')$, $(|\lambda| = |\lambda'| = 1)$:

$$h^{\pm}(\lambda, \lambda') - \pi i \int\limits_{|\lambda''|=1} h^{\pm}(\lambda, \lambda'')\Theta\Big[\pm\frac{1}{i}\big(\frac{\lambda''}{\lambda} - \frac{\lambda}{\lambda''}\big)\Big]$$

$$\times f(\lambda'', \lambda')\, |d\lambda''| = f(\lambda, \lambda'), \tag{8}$$

where Θ is Heaviside's function. Then the function $f_1(\lambda, \lambda')$ $(|\lambda| = |\lambda'| = 1)$ is calculated:

$$\begin{aligned} f_1(\lambda, \lambda') = {} & \Theta\Big[-\frac{1}{i}\big(\frac{\lambda'}{\lambda} - \frac{\lambda}{\lambda'}\big)\Big]h^{+}(\lambda, \lambda') \\ & - \Theta\Big[\frac{1}{i}\big(\frac{\lambda'}{\lambda} - \frac{\lambda}{\lambda'}\big)\Big]h^{-}(\lambda, \lambda'). \end{aligned} \tag{9}$$

After that the function $\rho(\lambda, \lambda')$ $(|\lambda| = |\lambda'| = 1)$ is determined from

$$\rho(\lambda, \lambda') + \pi i \int\limits_{|\lambda''|=1} \rho(\lambda, \lambda'')\Theta\Big[\frac{1}{i}\big(\frac{\lambda'}{\lambda''} - \frac{\lambda''}{\lambda'}\big)\Big]$$

$$\times f_1(\lambda'', \lambda')\, |d\lambda''| = -\pi i f_1(\lambda, \lambda'). \tag{10}$$

It is convenient to solve equation (11) below for the function $\mathcal{K}(z,\bar{z},\lambda)$ ($|\lambda| = 1$) taking its Fourier transform with respect to λ for fixed values of z and $\bar{z}$:

$$\mathcal{K}(z,\bar{z},\lambda) = \int\limits_{|\lambda'|=1} \rho(\lambda,\lambda')\exp\left[i\frac{k_0}{2}\left((\lambda'-\lambda)\bar{z} + (\frac{1}{\lambda'} - \frac{1}{\lambda})z\right)\right]$$

$$\times\left(1 + \frac{1}{2\pi i}\int\limits_{|\lambda''|=1} \frac{\mathcal{K}(z,\bar{z},\lambda'')}{\lambda'' - \lambda'(1+0)}d\lambda''\right)|d\lambda'|. \qquad (11)$$

Then the desired scatterer function is reconstructed:

$$v(z,\bar{z}) = -\frac{k_0}{\pi}\frac{\partial}{\partial z}\int\limits_{|\lambda|=1} \mathcal{K}(z,\bar{z},\lambda)d\lambda. \qquad (12)$$

The present algorithm would be exact, if condition (7) were valid. But this condition is an approximate one. First, condition (7) cannot be valid, even approximately, for full power scatterers because of the necessity to retain an infinite number of terms in the Born–Neumann series. Nevertheless, even for weak and mean power finite scatterers, when m can be treated as a finite number, condition (7) holds only approximately at a sufficiently good localization of the space scatterer spectrum within the circle of radius $2k_0/m$.

Reconstruction errors can be estimated. Having this in mind, we should take into account that both equation (11) and equation (12) become inexact. Indeed, the function $\mathcal{K}(z,\bar{z},\lambda)$ on the right side of (12) can be represented as the Born–Neumann series. Then a comparison of the terms on the left and right sides of equation (12) which are linear with respect to the scatterer results in the following conclusion: the approximate algorithm does not enable us to reconstruct the space spectral components $\tilde{v}(\xi)$ at $|\xi| \geq 2k_0$. Moreover, for non-Born scatterers there are summands nonlinear with respect to the scatterer on the right side of equation (12). These summands vanish if condition (7) holds exactly; otherwise, they introduce an error in the reconstruction of the space spectral components $\tilde{v}(\xi)$ even at $|\xi| < 2k_0$ (Burov and Rumiantseva, 1992).

However, owing to the following two reasons the approximate algorithm is highly perspective for reconstructing the scatterers with a narrow space spectrum. First, the algorithm is sufficiently stable to noises. And the scatterer reconstruction quality may be considered acceptable if the reconstructed scatterer spectrum is concentrated mainly within a domain lying inside the $2k_0$-radius circle. At the same time, the intensity of space spectral components outside the domain of concentration permits to estimate the reconstruction errors.

Secondly, let us assume that the scattering amplitude $f(k,l)$ is measured for N directions of the vector l at each of N directions of the vector k, i.e., there are N^2 data. Then the discretization of equations (8)–(12) allows us to affirm that the total number of calculating operations for the scatterer reconstruction is of order N^4 instead of $N^5 - N^6$ operations for iterative methods of solving similar problems (Johnson et al., 1984; Burov et al., 1982; Burov et al., 1989).

3. THE EXACT ALGORITHM AND THE INSTABILITY PROBLEM

Now we analyze the case of an arbitrary two-dimensional scatterer creating a back scattered field, whose intensity is not small in comparison with the global intensity of the field scattered over all other directions. Hence equations (8)–(10) for finding the function $\rho(\lambda, \lambda')$ from the scattering amplitude $f(\lambda, \lambda')$ remain valid, but equations (11) and (12) become somewhat more complicated (Novikov, 1989):

$$\mathcal{K}(z, \bar{z}, \lambda) = \int\limits_{|\lambda'|=1} \rho(\lambda, \lambda') \exp\left[i\frac{k_0}{2}\left((\lambda' - \lambda)\bar{z} + (\frac{1}{\lambda'} - \frac{1}{\lambda})z\right)\right]$$

$$\times\left[e(z, \bar{z}, \lambda') + \frac{1}{2\pi i} \int\limits_{|\lambda''|=1} \left\{\mathcal{K}(z, \bar{z}, \lambda'')\Omega^+(z, \bar{z}, \lambda'(1+0), \lambda'')\, d\lambda''\right.\right.$$

$$\left.\left. +\Omega^-(z, \bar{z}, \lambda', \lambda'')\bar{\mathcal{K}}(z, \bar{z}, \lambda'')d\bar{\lambda}''\right\}\right] |d\lambda'|, \tag{13}$$

where the auxiliary functions e, Ω^+ and Ω^- are found by solving integral equations based either on the function $h_1(\lambda, \bar{\lambda})$ responsible for back scattering the inhomogeneous waves or on the function $r_1(\lambda, \bar{\lambda})$:

$$r_1(\lambda, \bar{\lambda}) = h_1(\lambda, \bar{\lambda})\Big(- \frac{\mathrm{sgn}(|\lambda|^2 - 1)}{4\pi\bar{\lambda}}\Big);$$

$$r_1(z, \bar{z}, \lambda) = r_1(\lambda, \bar{\lambda}) \exp\left[- i\frac{k_0}{2}\left((\bar{\lambda} + \frac{1}{\lambda})z + (\lambda + \frac{1}{\bar{\lambda}})\bar{z}\right)\right];$$

$$e(z, \bar{z}, \lambda') = 1$$

$$-\frac{1}{\pi}\int\int\limits_{\eta\in\mathbf{C}^1} \frac{r_1(z, \bar{z}, \eta)\bar{e}(z, \bar{z}, \eta)}{\eta - \lambda'}\, d\eta_R\, d\eta_I, \tag{14}$$

where $\eta = \eta_R + i\eta_I$, $\eta \in \mathbf{C}^1$, η_R and $\eta_I \in \mathbf{R}^1$. The functions Ω^+ and Ω^- are given by

$$\Omega^\pm(z, \bar{z}, \lambda', \lambda'') = X_1(z, \bar{z}, \lambda', \lambda'') \pm iX_2(z, \bar{z}, \lambda', \lambda''),$$

and the functions X_1 and X_2 satisfy the integral equations

$$X_1(z, \bar{z}, \lambda', \lambda'') - \frac{1}{2(\lambda'' - \lambda')}$$

$$= -\frac{1}{\pi}\int\int\limits_{\eta\in\mathbf{C}^1} \frac{r_1(z, \bar{z}, \eta)\bar{X}_1(z, \bar{z}, \eta, \lambda'')}{\eta - \lambda'}\, d\eta_R\, d\eta_I; \tag{15}$$

$$X_1(z,\bar{z},\lambda',\lambda'') - \frac{1}{2i(\lambda''-\lambda')}$$
$$= -\frac{1}{\pi}\int\int_{\eta\in\mathbf{C}^1} \frac{r_1(z,\bar{z},\eta)\bar{X}_2(z,\bar{z},\eta,\lambda'')}{\eta-\lambda'}\, d\eta_R\, d\eta_I. \quad (16)$$

As a result, the scatterer function is expressed by the function $\mathcal{K}(z,\bar{z},\lambda)$ and the auxiliary functions:

$$v(z,\bar{z}) = \frac{2ik_0}{\pi}\frac{\partial}{\partial z}\Bigg[\int\int_{\eta\in\mathbf{C}^1} r_1(z,\bar{z},\eta)\bar{e}(z,\bar{z},\eta)\, d\eta_R\, d\eta_I$$
$$+\frac{1}{2i}\int_{|\lambda|=1}\Big\{\overline{\mathcal{K}}(z,\bar{z},\lambda)\Big[-1+\frac{1}{\pi}\int\int_{\eta\in\mathbf{C}^1} r_1(z,\bar{z},\eta)\bar{\Omega}^-(z,\bar{z},\eta,\lambda)\, d\eta_R\, d\eta_I\Big]d\lambda$$
$$+\frac{1}{\pi}\overline{\mathcal{K}}(z,\bar{z},\lambda)$$
$$\times\int\int_{\eta\in\mathbf{C}^1} r_1(z,\bar{z},\eta)\bar{\Omega}^+(z,\bar{z},\eta,\lambda)\, d\eta_R\, d\eta_I d\,\bar{\lambda}\Big\}\Bigg]. \quad (17)$$

For $h_1(\lambda,\bar{\lambda})\equiv 0$, equations (13) and (17) reduce to (11) and (12), respectively. As to equations (14)–(17), in the summands containing the function $r_1(z,\bar{z},\eta)$ the integration is carried out over the whole complex plane $\eta\in\mathbf{C}^1$. However, for practical purposes it is necessary to know a minimal domain of the complex plane η such that the knowledge of the function $r_1(\eta,\bar{\eta})$ on this minimal domain is sufficient for finding an adequate solution of integral equations (14)–(17). Let us assume that the whole space spectrum $\tilde{v}(\xi)$ is concentrated inside a circle of radius s. The value s is greater than $2k_0/m$ and so the back scattering is possible. Then, as it can be obtained by equation (5) at $l=-\bar{k}$, $h_1(k)=0$ for all vectors $k=k_R+ik_I$ such that $|k_R|\geq ms/2$. Therefore, for the corresponding complex numbers η, we have $r_1(\eta,\bar{\eta})=0$ for the order m. Then $r_1(\eta,\bar{\eta})\neq 0$ only within the ring

$$|\eta|_- < |\eta| < |\eta|_+, \quad (18)$$

where

$$|\eta|^2_\mp = 1 - \frac{2}{k_0^2}\sqrt{\left(\frac{ms}{2}\right)^2 - k_0^2}$$
$$\times\left(\sqrt{\left(\frac{ms}{2}\right)^2 - k_0^2} \mp \frac{ms}{2}\right).$$

Consequently, the corresponding integrations are made over the domain (18) only. The algorithm (8)–(10), (13)–(17) is exact. So, in the Born approximation the function $\rho(\lambda,\lambda')$ describes the Fourier-harmonics $\tilde{v}(\xi)$ with $|\xi|\leq 2k_0$, and the function $r_1(\lambda,\bar{\lambda})$ with $|\xi|>2k_0$ (see (Grinevich and Manakov, 1986)). Nevertheless, even the exact algorithm imposes a restriction on the scatterer power:

two-dimensional scatterers having the power greater than a certain value are reconstructed in nonunique way. On the contrary, three-dimensional scatterers have a unique reconstruction under the monochromatic observation regime owing to the dimensional redundancy of the scattering data for three-dimensional problem (Devaney, 1978; Novikov, 1988).

A practical implementation of the exact algorithm encounters some difficulty caused primarily by the apparent instability of the algorithm. This means that the function $h_1(k)(k \in \mathbf{C}^2)$ cannot be directly measured in an experiment: it is found from the scattering amplitude $f(k,l)$ (k and $l \in \mathbf{R}^2$) (by solving the equations given, e.g., in (Novikov, 1988)). The idea of calculating the function $h_1(k)$ is as follows. Using integral equations containing the function $f_1(k,l)$ we found a function $(\Phi - \Phi_0)(r, r', k^2)$ whose arguments r and r' belong to $\partial\mathcal{D}$, where the curve $\partial\mathcal{D}$, encircles the scatterer localization domain. By virtue of (3), the value of $(\Phi - \Phi_0)$ remains un changed when passing to complex vectors $k \in \mathbf{C}^2$. Therefore we can calculate the wave function $\Psi(r,k)$ $(r \in \partial\mathcal{D},\ k \in \mathbf{C}^2)$ using the function $(\Phi - \Phi_0)$ and then obtain the function $h_1(k)$:

$$h_1(k) = \frac{1}{(2\pi)^2} \int \int_{r,r' \in \partial\mathcal{D}} \exp(i\bar{k}r)\Psi(r',k)$$

$$\times(\Phi - \Phi_0)(r, r', k^2)drdr'. \tag{19}$$

Equation (19) is obviously instable, its instability growing as the imaginary part $|k_I|$ of the absolute value of the wave vector k increases. This instability is generated by the fact that the integrand is a fast oscillating function of the coordinate vector r, because $\exp(i\bar{k}r) = \exp(ik_R r + k_I r)$. As a result, in the Born approximation, for example, the root-mean-square error for the function $h_1(k)$ calculated from equation (19) proves to be proportional to the quantity

$$\Big[\exp(|k_I|L) - \exp(-|k_I|L)\Big] / (|k_I|L) \tag{20}$$

for in the white noise case. Here, L is the linear size of the scatterer. Therefore, it is possible to calculate $h_1(k)$ up to an acceptable accuracy for small values $|k_I|$ only: $|k_I| \lesssim L^{-1}$. On the other hand $h_1(k) \neq 0$ for $|k_R| < ms/2$, i.e., for $|k_I|^2 < (ms/2)^2 - k_0^2$. Comparing these two conditions on $|k_I|$, we see that it is possible calculation of the function $h_1(k)$ in the Born approximation ($m = 1$) and, consequently, reconstruction of the scatterer according to (8)–(10), (13)–(17) up to an acceptable accuracy is possible only under the condition

$$\left(\frac{s}{2}\right)^2 - k_0^2 \lesssim L^{-2}, \quad \text{i.e.,} \quad s \lesssim 2\sqrt{k_0^2 + L^{-2}}. \tag{21}$$

In other words, the reconstruction stability is provided only for such weak power scatterers, whose space spectrum is mainly concentrated within the circle of radius $(2(k_0^2 + L^{-2})^{1/2})$. For the correction, we have $L^{-2} \ll k_0^2$, so condition (21) allows us to widen the reconstructed inhomogeneity spectrum only slightly in comparison with the case of the absence of back scattering ($s < 2k_0$). There is an evident analogy with the analytical continuation of the space spectrum of the functions having finite support in the coordinate space. This operation has a very strong instability. For this reason extension of the space scatterer spectrum is in practice possible only for one or two sampling points in the frequency space. A similar situation occurs for the non-Born scatterers also, though estimations like (20) and (21) are, of course, different and depending on the number m defined by the scatterer power.

As the power of the scattering inhomogeneity having a wide space spectrum, increases, the instability of the inhomogeneity reconstruction increases also. This fact can be explained as follows. Consider the processes of wave rescattering within the inhomogeneity. Every spectral scatterer constituent corresponding to a fixed space frequency can generate spectral constituents of the wave field on other space frequencies, which can be both higher and lower. This results in the appearance of a fine scale structure of the wave field inside the scattering domain. On the one hand, the structure's fine scale is of great importance for a formation of the wave field outside the scattering domain. On the other hand, it amplifies the instability of the process of the inside wave field reconstruction. The reconstruction of the inside wave field in an explicit or implicit form (as an intermediate stage of the scatterer reconstruction) is necessary regardless of the way in which the inverse scattering problem is solved and is omitted only in the Born approximation.

The higher the order of the rescattering, the more complicated the space structure of the inside wave field and, consequently, the stronger the instability. Hence, the stable reconstruction is provided for the scatterers with narrow space spectrum only. While increasing the scatterer power, the restrictions imposed on the space scatterer spectrum become more rigid.

A similar situation occurs for the iterative method of solution (Johnson, 1984; Burov et al., 1989). The instability appears when the space spectrum of the secondary sources begins to spread over the circle of radius $2k_0$ (Burov and Rumiantseva, 1992). The latter circumstance reflects the appearance of back scattered field. Then the number of the equations solved happens to exceed the number of the data known from an experiment, which results in the appearance of instability effects.

We note in conclusion that the above inferences have been obtained by studying particular equations and algorithms. Nevertheless, they are common for the two-dimensional scalar inverse problems under the monochromatic observation regime. Thus, the functional-analytical methods both provide a deeper insight into other methods having been used so far and, at the same time, give answers to many questions which are of much importance for solving applied problems.

Acknowledgments

The authors wish to thank R.G. Novikov and P.G. Grinevich for numerous discussions.

REFERENCES

Burov, V., Goryunov, A. and Saskovets, S. (1982). Iterative algorithm for the solution of the inverse scattering problem. *Vestnik Moscow Univ.* Ser. Phys.. **37**, 102–104.

Burov, V., Rychagov, M. and Saskovets, A. (1989). Effect of multiple scattering in diffraction tomography: the T-matrix approach. *Vestnik Moscow Univ.* Ser. Phys.. **44**, 48–52.

Burov, V. and Rumiantseva, O. (1992). The solving of two-dimensional acoustical inverse scattering problem basing on functional-analytical methods. *Sov. Phys. Acoust.* **38**.

Devaney, A. (1978). Nonuniqueness in the inverse scattering problem. *Math. Phys.* **19**, 1526–1531.

Grinevich, P. and Manakov, S. (1986). The inverse problem of the scattering theory for two-dimensional Schrödinger operator, $\bar{\partial}$-method and nonlinear equations. *Func. Anal. Appl.* **20**, 14–24.

Johnson, S., Zhou, Y., Tracy, M., Berrgren, M. and Stanger, F. (1984). Inverse scattering solutions by a sinc basis, multiple source, moment method. Part III: Fast algorithms. Ultrason. Imag. **6**, 103–106.

Novikov, R. (1986). Reconstruction of two-dimensional Schrödinger operator from the given scattering amplitude at fixed energy. *Theor. Math. Phys.* **66**, 154–157.

Novikov, R. (1988). Multi-dimensional inverse spectral problem for the equation $-\Delta\Psi+(v(x)-Eu(x))\Psi=0$. *Func. Anal. Appl.* **22**, 263–272.

Novikov, R. (1989). *The Inverse Scattering Problem for Two-dimensional Schrödinger Equation at Fixed Energy and Nonlinear Equations.* Doctor Thesis, Moscow State Univ., Moscow (in Russian).

Novikov, R. and Henkin, G. (1987). $\bar{\partial}$-equation in multi-dimensional inverse scattering problem. *Russian Math. Surveys* **42**, 109–180.

Ill-Posed Problems in Natural Sciences, pp. 472 – 481
A. Tikhonov (Ed.)
1992 VSP/TVP

INVERSE PROBLEMS OF ASTROPHYSICS

A.M. CHEREPASHCHUK, A.V. GONCHARSKII, and A.G. YAGOLA
Moscow State University, 119899 Moscow, Russia

ABSTRACT

Some inverse problems of astrophysics are considered. Applications of the theory of ill-posed problems to solution of the problems under consideration are presented.

1. INTRODUCTION

Astrophysics deals with very far objects, such as stars and galaxies, whose properties can only be investigated by those indirect observational appearences which are observable from the Earth or from a spacecraft. To interpret the observational data, solving inverse problems of astrophysics is required most of which are ill-posed.

At present, the theory of ill-posed problems is widely used for solving inverse problems arising in optics, spectroscopy, electrodynamics, plasma diagnostics, geophysics, and radioastronomy (Tikhonov and Arsenin, 1979; Tikhonov et al.,1983; Ivanov et al., 1978; Goncharskii et al. 1978, 1985, 1991).

In this survey we present results of applying modern mathematical methods based on Tikhonov's regularization algorithms to the inverse problems arising in astrophysics (see (Goncharskii et al., 1978, 1985, 1991) for details). Up to now a number of different directions has been developed in astrophysics which are connected with the solution of inverse, ill-posed problems (Goncharskii et al., 1978, 1985, 1991; Khokhlova, 1983). We should mention also that the consideration of concrete astrophysical problems has led to the development of new mathematical methods. For example, in order to restore the brightness distributions of stellar disks on the basis of the natural assumption that the functions of interest monotonic, we have developed effective numerical methods for solving ill-posed problems on the sets of monotonic (convex, convex monotonic) functions (Yagola, 1980; Tikhonov et al., 1983; Goncharskii and Yagola, 1969; Goncharskii et al., 1972, 1973). The restoration of the strip-distributions of radio brightness over a source, with errors in the antenna directivity diagrams being taken into account, is also an inverse problem we have studied. This research, gave an impetus to the development of the theory of ill-posed problems with an approximately specified operator (Tikhonov et al., 1983; Goncharskii et al., 1972; Yagola, 1980). The corresponding methods for solving ill-posed problems were incorporated into a library of FORTRAN codes (Tikhonov et al., 1983; Goncharskii et al.,1985).

A wide range of important inverse problems in astrophysics are associated with the interpretation of the observations of close binary systems. The most important parameters of stars such as their masses, radii, temperatures, and structure of atmospheres can be estimated by studing the motions and various interaction effects of gravitationally coupled binary star systems. Such a complete set of parameters cannot be obtained for a single star. Hence, it is very fortunate for the astrophysicist when a peculiar object (e.g. Wolf–Rayet star, neutron star, black hole, etc.) is a component of a close binary system.

Almost half of the stars in Galaxy are binaries. Up to now more than 4000 eclipsing binary systems are known, i.e., systems whose orbital planes are oriented in such a way that the components eclipse each other as seen from the Earth twice each orbital period. A number of inverse problems can be formulated on the basis of the light curves and radial velocity curves of binary systems and, when solved by the regularization methods, various important parameters of stars and peculiar objects can be determined. In this survey we describe some inverse problems related to the interpretation of observational data for eclipsing close binary systems (see (Goncharskii et al., 1987) for more detail).

2. INVESTIGATION OF WOLF–RAYET STARS IN ECLIPSING BINARY SYSTEMS

There are about 10^3 Wolf–Rayet stars in Galaxy. It may be suggested that these stars are the helium remnants formed as a result of intensive mass loss by most massive stars. The mechanism of this mass loss is strong stellar wind of single O-B stars or mass exchange in close binary system. Optical spectra of Wolf–Rayet (WR) stars are very peculiar because they contain strong emission lines of hydrogen, helium, nitrogen, carbon, and oxygen in various ionization states. For many of these lines to be exited, the temperature of matter of radiation must reach several hundred thousands degrees, while the visible continuum of WR stars can be roughly described by the temperature $\approx 10000 - 20000^\circ K$. This means that the matter in WR stars is far from being in thermodynamic equilibrium. Very extended atmosphere exists around a WR star with velocity of expansion of about 1000 km/s and mass loss rate $\dot{M} \approx 10^{-5} M_\odot$/year. Moreover, the observations unambiguously indicate that WR stars have anomalous chemical composition: WR stars contain far more helium than normal stars. Therefore, WR stars are observed at late evolutional stage.

It is very difficult to do any decisive conclusions about WR stars whilst the observations only involve single stars. A WR star itself is hidden inside thick expanding atmosphere. In 1964 we started studing WR stars in eclipsing binary systems. We suggested a new approach to the interpretation of the light curves of eclipsing WR binary systems on the basis of the modern regularization methods for ill-posed problems. Efficient computer programs have been developpped to solve those problems (Goncharskii et al., 1978, 1985). A natural a priori physical assumption is that the functions of interest are monotonic and nonnegative. In addition an extensive programs of observations of all known eclipsing binary systems containing WR components were undertaken (Cherepashchuk, 1982; 1991).

Let us consider briefly the mathematical problem of interpreting the light curve of an eclipsing binary system containing the WR star and a normal star with thin atmosphere of spectral class O or B. Both stars are spherical, travelling around their mass center in circular orbits. Total luminosity of the system is supposed to be equal to unit. Light curve of eclipsing binary system can be described by the following set of equations:

$$1 - l_1(\theta) = \int_0^{R_1} K_1(\xi, \theta, i)\, I_a(\xi)\, d\xi, \tag{1}$$

$$1 - l_2(\theta) = \int_0^{R_2} K_2(\xi, \theta, i)\, I_C(\xi)\, d\xi, \tag{2}$$

$$I_a(0) = \frac{\left(1 - \int_0^{R_2} I_C(\xi) 2\pi\xi\, d\xi\right)}{\pi r_{06}^2 (1 - x/3)}, \tag{3}$$

where $1 - l_{1,2}(\theta)$ is the observed light loss during eclipses in the primary and secondary minima (θ stands for the phase angle of the binary system), $K_{1,2}$ are the kernels of the integral equations (1), (2) describing the shape of the eclipsing area (ξ is the polar distance on the disk of WR-star, i is the inclination of the orbit plane), $I_a(\xi)$, $I_C(\xi)$ are the functions of interest describing the distributions of the absorbtion properties of WR disk and brightness distribution over the WR disk, r_{06} is the radius of an 0-star expressed in terms of the relative distance between the components of the binary system, x is the value of the limb darkening for an 06 star (≈ 0.3). $R_{1,2}$ are the upper limits for the full radii of the absorbtion and emitting disk of WR component which can be estimated from the width of the primary and secondary eclipses.

The algebraic equation (3) is the normalization of the total luminosity of the binary system to unit.

There are two unknown parameters in our problem, r_{06} and i, as well as two functions of interest, $I_a(\xi)$ and $I_C(\xi)$. For solving the Fregholm integral equations of the first kind we used the algorithm of solving ill-posed problems on the set of monotonic nonnegative functions (Goncharskii et al., 1987). Solving two integral equations (1), (2) for various values of r_{06}, i and using the equation of normalization of the total luminosity of the system (3), minimizing the discrepancies between theoretical and observed light curves we can find the values r_{06} and i and the functions $I_a(\xi)$ and $I_C(\xi)$. The radius of the "core" of WR star r_0^{WR} can be obtained by using the function $I_a(\xi)$. One can find the brightness distribution for different wavelengths over the WR star disk and determine the values of the brightness and colour temperatures of the "core" of a WR star by making use of the function $I_C(\xi)$. The obtained solution of our inverse problem for the eclipsing WR binary V 444 Cyg is

$$r_{06} = 0.25 \pm 0.02 = (10 \pm 0.8)\, R_\odot, \qquad i = 78^\circ \pm 1^\circ,$$

$$r_0^{WR} = 2.6\, R_\odot, \qquad T_0^{WR} \gtrsim 70000\, K.$$

Fig. 1 shows the brightness distribution $I_C(\xi)$ over the disk of WR star restored from the analysis of light curves of V 444 Cyg at different wavelength. A large increase in the volume of the extended atmosphere of a WR star is observed in the infrared range. This increase cannot be explained in the model of homogenous stellar wind of a WR star (e.g., (Hillier, 1987; Hamman et al., 1988)). As it was pointed out in (see (Cherepashchuk et al., 1984)) infrared data suggest a nonhomogenous, clumping, blob structure of the WR wind. The theory of emission line profile in the spectra of WR stars based on the idea of clumping WR wind, has been developed (Antokhin et al.,1992).

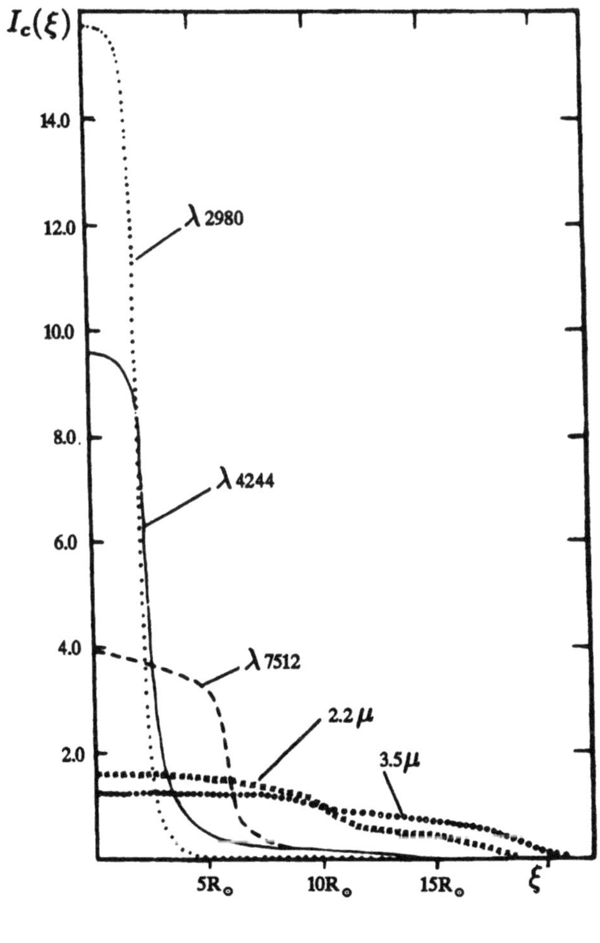

Fig. 1. The distribution of brigtness $I_C(\xi)$ over the disk of WR star restored on the set of monotonic non-negative functions analysing multiwavelength light curves of the V 444 Cyg binary system. A strong increasing of the volume of the WR extended atmosphere in infrared range is observed. It is an evidence that the extended atmosphere of WR star has a clumping, blob structure

Recently many evidences supporting the idea of the clumping structure of the WR wind have been obtained. Among them, the following are of prime importance.

1. The deficiency of X-ray radiation from the shock waves in binary WR + OB systems (Cherepashchuk, 1990).

2. The microvariability of emission lines in the spectra of WR stars (Moffat et al., 1988).

3. The discrepancy between the value of the mass loss rate $\dot{M}$ determined for WR stars from the period chages in WR + OB eclipsing binaries and that found from the radio and infrared observations (Khaliullin et al., 1984; Willis, 1991).

4. The irregular microvariability of the intensity and linear polarization of the light of WR stars (Moffat, 1988).

5. Theoretical investigations of instability of stellar wind driven by radiation (Owocki et al., 1988).

Two important conclusions can be drawn from our investigations of the clumping structure of the WR wind:

a). Due to the blob structure of the WR wind, the new three-dimensional spectral classification of the WR stars, based on the advanced multi-level non-LTE model of the homogeneous extended atmosphere (Schmutz et al., 1989) can be non-unique.

b). Since the blob structure of WR stars yields a lower value of mass loss rate $\dot{M}$, the observed number frequency of WR stars should be revised in view of the dependence between the local metallicity Z and the mass loss rate (Maeder, 1991). In particular, the binary evolution may be more important for the formation of the WR stars.

All our results obtained from the analysis of binary WR systems support the basic idea that the WR stars are bare cores of massive stars being at a late evolutionary stage (Conti, 1976; Maeder, 1991; Tutukov and Yungelson, 1973; Van den Hevel, 1976).

3. DETERMINATIONS OF MASSES OF NEUTRON STARS AND BLACK HOLE CANDIDATES IN X-RAY BINARY SYSTEMS

About 10^3 compact X-ray sources in Galaxy has been discovered up to now by means of special X-ray satellites. Most of them are binary systems consisting of a normal star and a relativistic object (neutron star, black hole) surrounded by the accretion disk. The normal star is a donor of matter. Due to presence in the binary system of a strong X-ray source, the investigation of motion, deformation and heating of the normal star enables one to estimate the mass of the relativistic object, which is very important for the relativistic astrophysics and the theory of stellar evolution. According to the modern understanding (Zeldovich and Novikov, 1971), a star at the end of its evolution can become either a white dwarf (if its mass is less than $1.4\,M_\odot$), or a neutron star (if $M > 1.4\,M_\odot$), or a black hole (if $M > 2-3\,M_\odot$). The knowledge of the mass of the relativistic object is very important for identifying this object as a neutron star or a black hole.

It has been shown (Luytyi et al., 1973), that the optical variability of X-ray binary systems is caused mainly by the effect of ellipsoidal shape of the optical star in the systems with OB supergiants and by the reflection effect-heating of the surface of the optical star by the strong X-ray radiation of the accreting relativistic object.

The amplitude of optical variability connected with the ellipsoidal shape of the normal star due to its tidal deformation in the gravitational field of the relativistic object depends on the mass ratio $q = M_x/M_v$ (M_x and M_v are the masses of the X-ray source and the normal star, respectively), on the angle i between the normal to the orbital plane and the line of site, and on the Roche lobe filling parameter μ.

Additional effects in optical variability of X-ray binary systems are connected with the eclipses of the accretion disk by normal star and normal star by accretion disk as well as with the precission of the accretion disk which may be tilted relatively to orbit plane due to slightly asymmetrical supernova explosion in a binary system.

We have developed algorithms to solve the inverse problems related to the interpretation of optical light curves of X-ray binary systems in the framework of a parametric model. These algorithms determine values of parameters and indicate confidence intervals for parameters (Goncharskii et al., 1985; 1991).

Various kinds of X-ray binary systems were considered in an extensive program (e.g., (Cherepashchuk, 1981)). Significant results were obtained by the new method (Goncharskii et al., 1987; 1991). The problem of interpretation of optical light curves of X-ray binary system is discribed in detail by A.V. Goncharskii (see (Goncharskii et al., 1987, 1991)).

Let us consider a close binary system of two stars in circular orbits. One star is a relativistic object surrounded with a thick accretion disk inclined at the angle θ_0 with respect to the orbital plane. The other component is a massive OB star whose shape coincides with the equipotential surface of the Roche model. The axial and orbital rotations of the normal star are synchronized. The tidally deformed surface of the normal star is heated by a radiation emanating from the relativistic object and the accretion disk.

Basically, the equations describing the shape and the temperature distribution over the normal star are as follows.

The shape of the normal star:

$$\Omega = \frac{1}{r} + q\left(\frac{1}{\sqrt{1 - r^2 - 2r\lambda}} - \lambda r\right) + \frac{1+q}{2} r^2 (1 - \nu^2), \tag{4}$$

where

$$r = \sqrt{x^2 + y^2 + z^2}, \qquad q = Mx/Mv, \qquad \lambda = \cos\eta, \qquad \nu = \sin\eta \, \sin\varphi.$$

The temperature distribution on the surface of the normal star:

$$T(\eta, \varphi) = \sqrt[4]{T_0^4(\eta, \varphi) + \frac{\varkappa k L_v \cos\alpha}{4\pi\sigma\rho^2}}. \tag{5}$$

An elementary surface area:

$$dS = \frac{r^2(\eta, \varphi) \sin\eta \, d\eta \, d\varphi}{\lambda l + \mu m}, \tag{6}$$

where l, m, n are the components of the normal vector to the surface area:

$$\vec{n}_1 = -\frac{\mathrm{grad}\Omega}{|\mathrm{grad}\Omega|} = (l, m, n). \tag{7}$$

The brightness of the elementary surface:

$$dI_\lambda = B_\lambda(T)\left[1 - x(\lambda, T)(1 - \cos\gamma)\cos\gamma\right] dS. \tag{8}$$

The theoretical light curve :

$$l(t) = \iint_S dI_\lambda = l(t, q, \mu, i), \tag{9}$$

where S is the surface of a visible area.

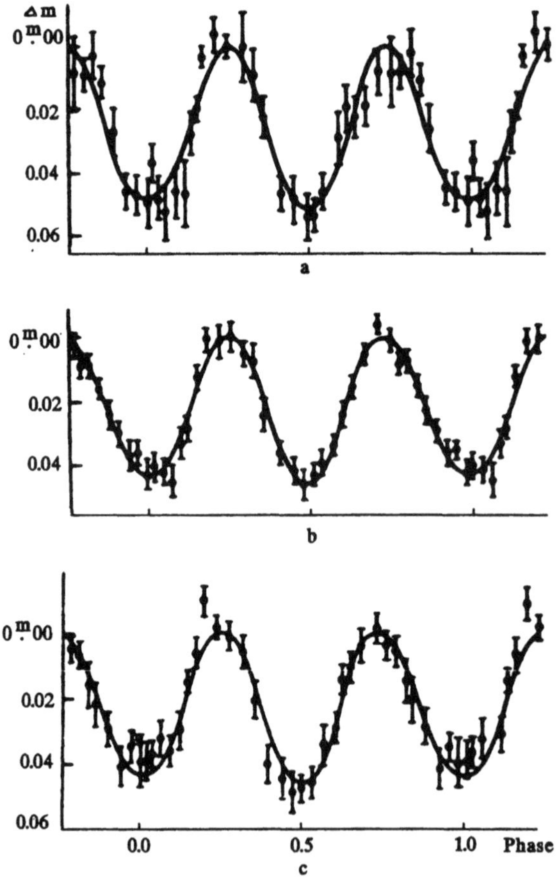

Fig. 2. Observed (dots) and theoretical light curves for the X-ray binary system Cyg X-1. The lower limit for the mass of the relativistic component at a distance more than 2 kiloparsecs from the system is $7\,M_\odot$. Hence, the object Cyg X-1 may be considered as the first candidate for black hole

Minimizing discrepancies between the observed and theoretical light curves, we obtain the values of the parameters q, μ, i and their confidence intervals. If we take into account the duration of X-ray eclipse and spectroscopic observations of X-ray binary system, we obtain the values of masses of the normal star and the relativistic object M_v, M_x.
Fig. 2 presents observed and theoretical light curves of the X-ray binary system Cyg X-1, the most probable black hole candidate. The results of mass determination for relativistic objects in X-ray binary systems are summarized in Fig. 3.

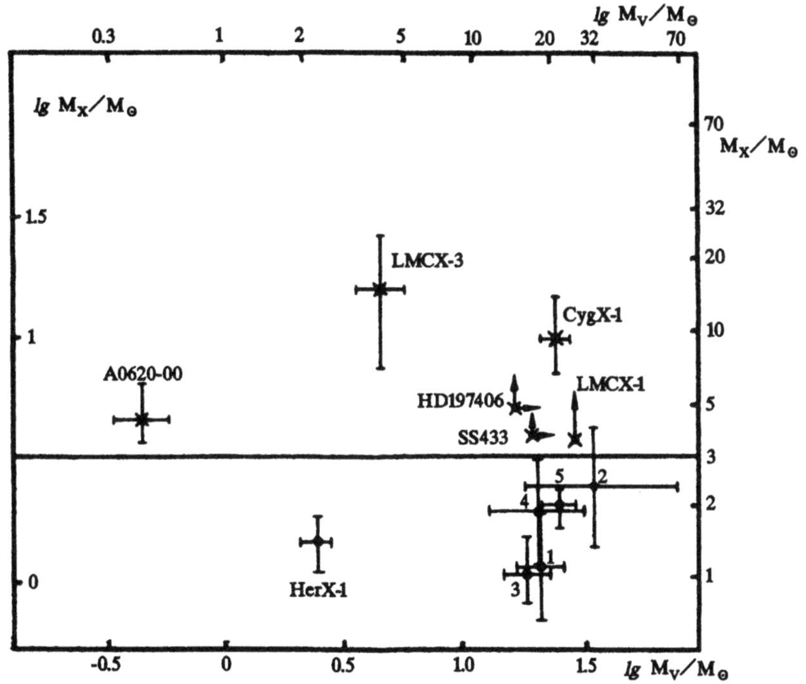

Fig. 3. Masses M_x of the neutron stars (dots) and black hole candidates (crosses) as a function of masses of normal stars in X-ray binary systems. 1 — Cen X-3, 2 — LMC-4, 3 — SNCX-1, 4 — 4 U1538-52, 5 — 4 U0900-40. The values of confidence intervales are shown. All X-ray pulsars have masses less than $3\,M_\odot$ — theoretical upper limit for the mass of neutron star predicted by Einstein General Relativity. There are no pulsars among the massive X-ray sourses ($M_x > 3\,M_\odot$). Therefore the X-ray sourses in the binary systems differ by masses as well as by observational appearences

It should be stressed that masses of X-ray pulsars do not exceed the value $3\,M_\odot$ which is theoretical upper limit predicted by Einstein General Relativity. On the other hand, there are no pulsars among the massive ($M_x > 3\,M_\odot$) X-ray sources. Therefore, the X-ray sources in binary systems differ not only by the masses but also by observational appearence.
This result is very significant for the testing of Einstein General Relativity in the

strong gravitational fields.

4. CONCLUSION

We have described some new results concerning the application of modern regularization algorithms to solving the inverse problems of astrophysics. In more detail, these results are published in the books of the authors (Goncharskii et al., 1978; 1985; 1987; 1991). In these books we have discribed also other inverse problems of astrophysics: determinations of structure and dynamic characteristics of accretion disks in Nova and Nova-like stars, mapping of the distribution of chemical elements over the surface of the peculiar A-stars, determination, of the stellar angular diameters from the diffractional light curves observed during lunar occultations of the stars, problems of automatic astronomical observations. Application of the modern regularization algorithms to solving these problems appears to be very fruitful.

REFERENCES

Antokhin, I.I., Nugis, T., and Cherepashchuk, A.M. (1992). He I and He II emission line profiles in cloud Wolf–Rayet atmosphere model (to appear).

Cherepashchuk, A.M. (1981). Periodic variability of HD50896-a Wolf–Rayet star associated with a ring nebula. *Astron. Soc.* **194**, 761.

Cherepashchuk, A.M. (1982). Properties of Wolf–Rayet stars in eclipsing lineary systems. *Atrophys. Sp. Sci.* **86**, 299.

Cherepashchuk, A.M. (1990). X-ray radiation and ragged stellar wind in Wolf–Rayet binary systems. *Astron. Zh.* **67**, 955.

Cherepashchuk, A.M. (1991). Wolf–Rayet binaries: observational aspects. In: *Wolf–Rayet Stars and Interrelation with Other Massive Stars in Galaxies.* Ed. by K.A. Van der Hucht and B. Hidayat. IAU Symp. **143**, pp. 187.

Cherepashchuk, A.M., Eaton, J.A., and Khaliullin, Kh.F. (1984). Ultraviolet photometry from the ORBITING ASTRONOMICAL OBSERVATORY. XXXIX. *Astrophys. J.* **281**, 774.

Goncharskii, A.V., Cherepashchuk, A.M., and Yagola, A.G. (1978). *Numerical Methods for Inverse Problems in Astrophysics.* Nauka, Moscow (in Russian).

Goncharskii, A.V., Cherepashchuk, A.M., and Yagola, A.G. (1982). *Ill-Posed Problems in the Natural Sciences.* Mir, Moscow (in Russian).

Goncharskii, A.V., Cherepashchuk, A.M., and Yagola, A.G. (1985). *Ill-Posed Problems in Astrophysics.* Nauka, Moscow (in Russian).

Goncharskii, A.V., Leonov, A.S., and Yagola, A.G. (1972). On a regularizing algorithm for ill-posed problems with approximate operator. *Zh. Vychisl. Matem. i Matem. Fiz.* **12**, 1592 (in Russian).

Goncharskii, A.V., Leonov, A.S., and Yagola, A.G. (1973). Generalized discrepancy principle. *Zh. Vychisl. Matem. i Matem.Fiz.* **13**, 294 (in Russian).

Goncharskii, A.V., Romanov, S.Yu., and Cherepashchuk, A.M. (1991). *Limit-Number Parametric Inverse Problems of Astrophysics.* Moscow Univ. Press., Moscow (in Russian).

Goncharskii, A.V., Yagola, A.G. (1969). On uniform approximation of monotonic solutions for ill-posed problems. *Dokl. Akad. Nauk SSSR* **184**, 771 (in Russian).

Hamman, W.R., Schmutz, W., and Wessolowski, V. (1988). Spectral analysis of the Wolf–Rayet star HD50896. *Astron. Astrophys.* **194**, 190.

Hillier, D.J. (1987). An empirical model for the Wolf–Rayet star HD50896. *Astrophys. J. Suppl.* **63**, 965.

Ivanov, V.K., Vasin, V.V., and Tanana, V.P. (1978). *Theory of Linear Ill-Posed Problems and Its Applications.* Nauka, Moscow (in Russian).

Khaliullin, Kh.F., Khaliullina, A.I., and Cherepashchuk, A.M. (1984). Increase of the period of the eclipsing binary V 444 Cyg and the mass loss by the Wolf–Rayet star. *Pis'ma Astron. Zh.* **10**, 600.

Khokhlova, V.L. (1983). Magnetic stars. In: *Itogi Nauki i Tekhniki, Astronomy.* Ed. by R.A. Sunyaev. VINITI, Moscow, pp. 233 (in Russian).

Luytyi, V.M., Sunyaev, R.A., and Cherepashchuk, A.M. (1973). On the nature of optical variability of HZ Her=Her, X1 and BD+34°3815=Cyg X1. *Astron. Zh.* **50**, 3.

Maeder, A. (1991). The relative numbers of Wolf–Rayet stars in galaxies with active star formation: the metallicity effect. *Astron. Astrophys.* **242**, 93.

Moffat, A.F.J. (1988). Wolf–Rayet stars: polarization variability and deduced parameters. In: *Polarized Radiation of CS.* Origin, Vatican, pp. 607.

Moffat, A.F.J., Drissen, L., Lamontagne, R., and Robert, C. (1988). Spectroscopic evidence for rapid blob Ejection in Wolf–Rayet stars. *Astrophys. J.* **334**, 1038.

Owocki, S.P. and Rybicki, G.B. (1988). Time dependent models of radiatively driven stellar winds I. Nonlinear evolution of instabilities for a pure absorption model. *Astrophys. J.* **355**, 914.

Schmutz, W., Hamman, W.R., and Wessolowski, V. (1989). Spectral analysis of 30 Wolf–Rayet stars. *Astron. Astrophys.* **210**, 236.

Tikhonov, A.N. and Arsenin, V. Ya. (1979). *Methods for Solving Ill-Posed Problems.* Nauka, Moscow (in Russian).

Tikhonov, A.N., Goncharskii, A.V., Stepanov, V.V., and Yagola, A.G. (1983). *Regularizing Algorithms and a Priori Information.* Nauka, Moscow (in Russian).

Tutukov, A.V. and Yungelson, L.R. (1973). Evolution of close binaries and Wolf–Rayet stars. *Nauchn. Inform. Astrosov AN USSR* **27**, 58.

Van den Heuvel, E.P.J. (1976). Late stages of close binary systems. In: *Structure and Evolution of Close Binary Systems.* Ed. by P.P. Eggleton Reidel, Dodrecht, pp. 35.

Willis, A.J. (1991). Observation of WR mass loss. In: *Wolf–Rayet Stars and Interrelation with Other Massive Stars in Galaxies.* Ed. by K.A. Van der Hucht and B. Hidayat. IAU Symp. **143**, pp. 265.

Yagola, A.G. (1980). On choosing regularization parameter while solving ill-posed problems in reflexive spaces. *Zh. Vychisl. Matem. i Matem. Fiz.* **20**, 586 (in Russian).

Zeldovich, Ya.B. and Novikov, I.D. (1971). *Gravity and Evolution of Stars.* Nauka, Moscow (in Russian).

Ill-Posed Problems in Natural Sciences, pp. 482 – 486
A. Tikhonov (Ed.)
1992 VSP/TVP

METHODS OF SOLVING THE ION EXCHANGE INVERSE PROBLEM IN COUNTER-CURRENT COLUMNS

A.M. DENISOV, S.R. TUIKINA, and A.V. CHANOV
Moscow State University, Len. Gory, 119899 Moscow, Russia

ABSTRACT

Stable methods for the determination of the isotherm function from dynamic experiment data are considered. The uniqueness theorem for the isotherm is proved. The numerical algorithms based on minimization of the discrepancy function by the gradient method and the random search method are developed.

In this paper, we propose and study stable methods for the determination of the isotherm function from data obtained in dynamic experiments in counter-current column. This inverse problem for models of ion exchange in a fixed-bed column was examined in (Denisov and Tuikina, 1983; Denisov and Chanov, 1988).
Assume that the process of ion exchange is isothermal and consider the one-dimensional model of interdiffusion nonequilibrium sorption dynamics taking into account the effect of axial dispersion in balance equations:

$$\varkappa x_t = -Lx_z + Ex_{zz} - k\big(f(x) - y\big), \quad 0 < z < l, \quad 0 < t \le T, \tag{1}$$

$$(1-\varkappa)y_t = Sy_z + k\big(f(x) - y\big), \quad 0 \le z < l, \quad 0 < t \le T, \tag{2}$$

where x and y are the average mole fractions of ion B in the solution and resign phases, respectively, L and S are the averege flow rate in the solution and resign phases, respectively, E is the coefficient of axial dispersion, k is the mass transfer coefficient, $\varkappa$ is the fraction of bed free space, z is the axial distance, and $f(x)$ is the isotherm function. The quantities E, k and $\varkappa$ are assumed to be positive constants. Let us complement equations (1) and (2) with the boundary and initial conditions

$$x(0,t) = \mu(t), \quad x(l,t) + \lambda x_z(l,t) = 0, \quad y(l,t) = 0, \quad 0 < t \le T, \quad \lambda = \text{const} > 0 \tag{3}$$

$$x(z,0) = 0, \quad y(z,0) = 0, \quad 0 \le z \le l. \tag{4}$$

These models were discussed in (Gorshkov et al. 1981).

Definition. A pair of functions $x(z,t)$, $y(z,t)$ such that

$$x,\ x_t,\ y,\ y_t \in C[\bar{Q}_T] \quad x_z,\ x_{zt} \in C[Q_{Tl}], \quad y_z,\ y_{zt} \in C[Q_{0T}],$$
$$y_{tt},\ y_{zzt},\ y_{zz},\ x_{tt},\ x_{zzt},\ x_{zz} \in C[Q_T], \tag{5}$$

is called a solution of problem (1)–(4) if these functions satisfy (1)–(4). Here $Q_T = \{(z,t),\ 0 < z < l,\ 0 < t \le T\}$, $Q_{Tl} = \{(z,t),\ 0 < z \le l,\ 0 < t \le T\}$, $Q_{0T} = \{(z,t),\ 0 \le z < l,\ 0 < t \le T\}$, and $\bar{Q}_T$ is the closure of Q_T.
Let us formulate the existence theorem for a solution of problem (1)–(4).

THEOREM 1. *Let the functions $\mu(t)$ and $f(\xi)$ satisfy the conditions*

$$\mu(t) \in C^2[0,T], \quad \mu(0) = \mu'(0) = \mu''(0) = 0,$$
$$|\mu''(t_1) - \mu''(t_2)| < L_1|t_1 - t_2| \quad \text{for} \quad t_1, t_2 \in [0,T], \tag{6}$$
$$f(\xi) \in C^1(-\infty,\infty), \quad f(0) = 0, \quad f(\infty) \le \text{const},$$
$$|f'(\xi_1) - f'(\xi_2)| < L_2|\xi_1 - \xi_2|, \quad K > f'(\xi) > 0 \quad \text{for} \quad \xi \in (-\infty,\infty), \tag{7}$$

where L_1, L_2, K are constants. Then there exists a unique solution of problem (1)–(4).

The proof of this theorem is based on Schauder's theorem and the maximum principle.
We consider the following inverse problem. It is required to find functions $x(z,t)$, $y(z,t)$, $f(\xi)$ which satisfy (1)–(4) and the supplementary condition

$$x(z^*,t) = q(t), \quad 0 < t \le T, \quad 0 < z^* \le l. \tag{8}$$

It is assumed that the functions $\mu(t)$ and $q(t)$ and the coefficients E, k, $\varkappa$, L, S and λ, in problem (1)–(4) are known.

Definition. A triplet of functions $f(\xi)$, $x(z,t)$, $y(z,t)$ is called a solution of inverse problem if $f(\xi)$ satisfies (7) and is an analytical function on an open interval containing $[0,\mu(T)]$, while $x(z,t)$, $y(z,t)$ satisfy (1)–(5).

THEOREM 2. *Assume that the function $\mu(t)$ satisfies condition (6) and $f_i(\xi)$, $x_i(z,t)$, $y_i(z,t)$, $i = 1,2$, are solutions of the inverse problem. If there exists $z^* \in (0,l]$ such that $x_1(z^*,t) = x_2(z^*,t)$ for $t \in [0,T]$, then $f_1(\xi) = f_2(\xi)$ for $\xi \in [0,\mu(T)]$ and $x_1(z,t) = x_2(z,t)$, $y_1(z,t) = y_2(z,t)$ in $\bar{Q}_T$.*

The proof of this theorem is based on the method proposed in (Denisov and Tuikina, 1983).
Now let us consider a numerical method for the determination of the isotherm function.
Let us assume that the isotherm is a function $f(\xi,d)$ depending of a real vector $d = (d_1,\ldots,d_N)$. This function can represent the Lengmurs isotherm which is commonly used in physico-chemical studies, or a polynomial of regular approximation, a spline, etc.
Let $d \in Q^N \subset \mathbf{R}^N$ (where Q^N is a compact set in $\mathbf{R}^N$) and let $f(\xi,d)$ satisfy the assumptions of Theorem 2 and the conditions $f(\xi,d) \in C^{2,2}[G]$ $(G = [0,\mu(T)] *$

Q^N) and $0 \le f'_\xi(\xi,d) \le C_3$ for $(\xi,d) \in G$. The relation between $f(\xi,d)$ and $d \in Q^N$ is unique.
The elution curve $x(z^*,t,d)$, $0 < t \le T$ is set into correspondence with the isotherme $f(\xi,d)$ by the continuous operator $A: C[0,\mu(T)] \to C[0,T]$ which is induced by the boundary-value problem (1)–(4). We may consider the problem of evaluating the isotherm function from the data of dynamic experiment $q_\delta(t)$ (which are measured with defect δ at an axial point z^*) as the problem of solving the operator equation $Af(\xi,d) = q_\delta(t)$, where $\|q_\delta(t) - \bar{q}(t)\|_{L_2[0,T]} \le \delta$.
Assume that $\bar{f}$ is the unique solution of the operator equation $A\bar{f}(\xi,d) = \bar{q}(t)$. Note that if a vector sequence $d_\delta \in Q^N$ satisfies an unequality

$$S(d) = \|Af(\xi,d) - q_\delta(t)\|_{L_2[0,T]} \le C\delta, \quad C = \text{const} \ge 1, \tag{9}$$

then $d_\delta \to \bar{d}$ for $\delta \to 0$ in $\mathbf{R}^N$ and

$$\|f(\xi,d_\delta) - f(\xi,\bar{d})\|_{C[0,T]} \to 0 \quad \text{as} \quad \delta \to 0.$$

Hence, we may consider a function $f(\xi,d_\delta)$ $(d_\delta \in Q^N)$ satisfying unequality (9) approximately. This approximate solution of (9) is sought as a minimizer of the functional

$$S(d) = \int_0^T \left(x(z^*,t,d) - q_\delta(t)\right)^2 dt$$

on the set Q^N. We apply the gradient methods to solve this problem. The gradient of the function $S(d)$ has the form

$$S_{d_i} = \sum_{k=1}^{2} \int_0^T \int_{z_{k-1}}^{z_k} \Big(f_{d_i}\big(x(z,t,d);\, d\big)\left(\gamma_2\eta_k - \gamma_1\alpha_k\right)\Big)\, dz\, dt, \quad i = 1,\dots,N.$$

Here, α_k, η_k is a solution of the conjugate problem

$$\begin{aligned}
&\alpha_{kt} = -D\alpha_{kzz} - \nu_1\alpha_{kz} + f_x\big(x(z,t,d);\, d\big)(\gamma_1\alpha_k - \gamma_2\eta_k),\\
&\qquad\qquad\qquad\qquad\qquad\qquad z_{k-1} < z < z_k, \quad 0 \le t < T,\\
&\eta_{kt} = \nu_2\eta_{kz} + (\gamma_2\alpha_k - \gamma_1\eta_k), \quad z_{k-1} \le z \le z_k, \quad 0 \le t < T,\\
&\alpha_1(0,t) = 0, \quad \eta_1(0,t) = 0, \quad 0 \le t < T,\\
&\alpha(z,T) = 0, \quad \eta(z,T) = 0, \quad z_{k-1} \le z \le z_k,\\
&(\nu_1 + D/\lambda)\alpha_2(l,t) + D\alpha_{2z}(l,t) = 0, \quad 0 \le t < T,\\
&\alpha_1(z_1,t) = \alpha_2(z_1,t), \quad \eta_1(z_1,t) = \eta_2(z_1 t), \quad 0 \le t < T,\\
&\alpha_{1z}(z_1,t) - \alpha_{2z}(z_1,t) = 2\big(x(z_1,t,d) - q_\delta(t)\big)/D, \quad 0 \le t < T,
\end{aligned}$$

where $k = 1,2$, $z_0 = 0$, $z_1 = z^*$, $z_2 = l$, $D = E/\varkappa > 0$, $\nu = L/\varkappa > 0$, $\gamma_1 = k/\varkappa > 0$, $\gamma_2 = k/(1-\varkappa) > 0$.
In the case $z^* = l$, instead of the last three conditions we consider the equation

$$(\nu_1 + D/\lambda)\alpha(l,t) + D\alpha_z(l,t) = 2\big(x(l,t,d) - q_\delta(t)\big).$$

It is convenient to search the isotherm function in the form of Bernshtein's polynomial:

$$f(\xi, d) = \sum_{i=0}^{N} d_i C_N^i \xi^i (1-\xi)^{N-i},$$

where $d_0 = 0$, and the parameters d_i, $i = 0, \dots, N$ satisfy the conditions $0 \le d_{i+1} - d_i \le C_1/N$ for $i = 0, \dots, N$, where $d_N \le C_2$, and C_1, C_2 are constants. It is evident that the vector $\bar{d}$ belongs to a compact set in $\mathbf{R}^n$. To minimize $S(d)$, we apply the method of conditional gradient and the random search method.
The same approach was used for the model with outerdiffusion kinetic, where the following equations were considered instead of equations (1) and (2):

$$\varkappa x_t = -Lx_z + Ex_{zz} - k\big(x - \psi(y)\big), \quad 0 < z < l, \quad 0 < t \le T, \tag{1'}$$

$$(1-\varkappa)y_t = Sy_z + k\big(x - \psi(y)\big), \quad 0 \le z < l, \quad 0 < t \le T, \tag{2'}$$

Here, $\psi(\xi)$ is the inverse of the isotherm function $f(x)$, k is the mass transfer coefficient.
Basing on these considerations, computer programs in FORTRAN were developed. The calculation proved the high efficiency of these methods. The results are presented in the Fig.1, where Curve 1 is the exact isotherm and Curves 2 and 3 are the isotherms calculated by the gradient method and the random search method, respectively.

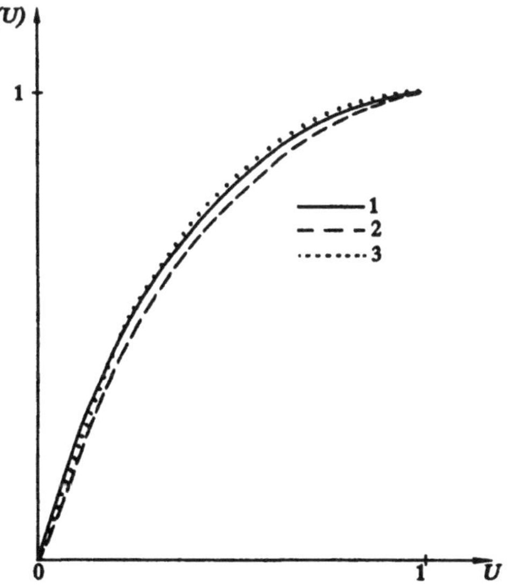

Fig. 1

REFERENCES

Denisov, A.M. and Chanov, A.V. (1988). Evaluating the kinetic coefficients for models of nonequilibrium sorption dynamics. *Vestnik Moskow. Univ.* Ser. 15, *Computer Math. Cybern.* **3**, 36–40 (in Russian).

Denisov, A.M. and Tuikina, S.R. (1983). Approximate solution of an inverse problem of sorption dynamics. *Vestnik Moskow. Univ.* Ser. 15, *Computer Math. Cybern.* **3**, 34–39 (in Russian).

Gorshkov, V.I., Safonov, M.S., and Voskresenskii, N.M. (1981). *Ion Exchange in Counter-Current Column.* Nauka, Moscow (in Russian).

Ill-Posed Problems in Natural Sciences, pp. 487 – 503
A. Tikhonov (Ed.)
1992 VSP/TVP

NUMERICAL METHODS FOR OPTICAL THIN FILMS AT THE NATIONAL RESEARCH COUNCIL OF CANADA

J.A. DOBROWOLSKI
Institute for Microstructural Sciences, National Research Council of Canada, Ottawa, Ontario, K1A 0R6 Canada

ABSTRACT

This article reviews various numerical tools for optical thin films that have been developed or implemented at the National Research Council of Canada. Different ways of specifying quantities of interest for analysis and numerical design of optical multilayer coatings are discussed. Merit functions useful for this purpose are also reviewed. The results obtained by ten different optimization algorithms as applied to one and the same problem are presented. Five different programs for numerical thin film synthesis are also briefly described as well as the results of their application to solution of another problem. The use of numerical methods for the determination of construction parameters of experimentally produced multilayers is discussed. Ways in which computer programs can be used to monitor the deposition of optical thin film systems without and with real-time reoptimization are also addressed.

1. INTRODUCTION

Numerous applications exist today for optical multilayer coatings. They are being constructed for a wavelength region that extends from about 100 Å to 200 μm. In addition to traditional uses in scientific instruments, multilayer coatings are used in such widely diverse disciplines as communications, computers, architecture and automobile engineering, energy conversion and conservation, and for various consumer products. Some of these applications require coatings of extreme precision, where cost is no issue and where only a few coated items are needed. However, in mass applications a low cost is essential if the product is to be commercially successful.

Theoretical and practical work on optical coatings has been an ongoing research topic at the NRCC since 1955. Many new devices and products have been developed in the Thin Film Laboratory. Important factors in this work have been the numerous numerical methods that have been developed or implemented for the solution of the nonlinear problems that arise during the design and the manufacture of optical multilayer coatings. This article attempts to summarize these techniques. It is not intended to be an exhaustive review of all the efforts in this field around the world.

2. CONSTRUCTION PARAMETERS

To achieve the required performance, the thin film designer has at his disposal the following construction parameters (p): the number of layers N of which the multilayer is composed and the thicknesses t, refractive indices n and extinction coefficients k of the two media that surround it and of the individual layers. Although in theory the number of layers could be very large, in practice it must be reasonable if the system is to be successfully produced. The values that the thicknesses of the layers may assume must, of course, be positive and, for mechanical reasons, they cannot by excessively large. The optical constants must lie within the limits imposed by existing real substrate and coating materials. Mathematical solutions that violate these constraints are of no practical interest.

Solutions to a particular problem may exist, but will not be found unless the multilayer is allowed to assume an adequate overall thickness that is subdivided into a sufficient number of layers. In general, with an increasing number of parameters better solutions to a problem will be found.

3. SPECIFICATIONS

Often the performance of a multilayer is defined in terms of the transmittance T or reflectance R as a function of wavelength λ and angle of incidence θ of the incident radiation. If any one of the materials used in the multilayer is absorbing, the absorptance A and the reflectance R' and absorptance A' for radiation incident from the opposite direction will also be of interest. For other applications the products TR, TR' and the differences $(1 - R)$, $(1 - R')$, $(1 - A)$, $(1 - A')$, the phase changes on transmission Δ_T and reflection Δ_R, $\Delta_{R'}$ are needed. At times one is also interested in the derivatives of these quantities with respect to λ and θ. For non-normal incidence of the radiation, additional quantities of interest (Q) are $(T_p + T_s)/2$, $(R_p + R_s)/2$, $(R'_p + R'_s)/2$, $(T_pR_p + T_sR_s)/2$, $(T_pR'_p + T_sR'_s)/2$ and so on. Here the subscripts p and s denote radiation polarized parallel and perpendicular to the plane of incidence.

In addition to the above, a number of more complicated quantities of interest are of practical importance (Dobrowolski et al., 1989). These may be defined in terms of integrals I_j of the type

$$I_j = \int_{\lambda_{L,j}}^{\lambda_{U,j}} F_j(\lambda) q_j(\lambda)\, d\lambda, \tag{1}$$

where $\lambda_{L,j} \leq \lambda \leq \lambda_{U,j}$ is the range of integration. For example, the CIE colour coordinates $\bar{x}_\lambda$, $\bar{y}_\lambda$ and the luminous transmittance LT, reflectance LR of a multilayer are defined as follows

$$\bar{x}_\lambda = \frac{I_1}{I_1 + I_2 + I_3},$$

$$\bar{y}_\lambda = \frac{I_2}{I_1 + I_2 + I_3}, \tag{2}$$

$$LT,\, LR = I_2 \Bigg/ \int_{\lambda_{L,2}}^{\lambda_{U,2}} F_2(\lambda)\, d\lambda,$$

where $F_1(\lambda)$, $F_2(\lambda)$, and $F_3(\lambda)$ are the colorimetric functions and $q_j(\lambda)$ is either $T(\lambda)$ or $R(\lambda)$. Another integral quantity of interest is the solar absorptance SA of a coated surface, given by

$$SA = I_{SA} \Bigg/ \int_{\lambda_L}^{\lambda_U} F_{SA}(\lambda)\, d\lambda, \tag{3}$$

where $F_{SA}(\lambda)$ is the solar spectral irradiance and $q_{SA}(\lambda) = 1.0 - T(\lambda) - R(\lambda)$. The normal emissivity EM of a surface is given by

$$EM = I_{EM} \Bigg/ \int_{\lambda_L}^{\lambda_U} F_{EM}(\lambda)\, d\lambda, \tag{4}$$

where $F_{EM}(\lambda)$ is the Plank black body energy distribution and $q_{EM}(\lambda) = 1.0 - R(\lambda)$. There are additional applications which require the definition of still other quantities. The ability to specify and operate on the ultimate quantities of interest is one of the strengths of numerical thin film design methods.

4. ANALYSIS

If the construction parameters of a multilayer are given, it is a relatively straightforward task to calculate any of the above properties. The calculations are based on a matrix formulation of the boundary conditions at a film interface derived from Maxwell's equations (Born and Wolfe, 1959), (Macleod, 1986). All the quantities except the integral quantities of interest can be evaluated in terms of the elements of a product matrix **M** given by the expression

$$\mathbf{M} = M_l M_{l-1} \ldots M_i \ldots M_2 M_1, \tag{5}$$

where M_i represents the matrix of the i-th layer and which is based on the construction parameters of that layer only. The expressions for the integral quantities of interest can be evaluated only after the basic quantities of interest on which they are based have been calculated for the appropriate range of wavelengths. Analysis programs, coupled with experience and intuition, can be powerful tools for thin film design. They are integral parts of the refinement and thin film synthesis programs to be described later.

5. ANALYTICAL DESIGN METHODS

It is possible to write down the general expression for the average reflectance of one, two or even three nonabsorbing films on a nonabsorbing substrate, for light of one wavelength and a nonzero angle of incidence (Pohlak, 1962). However, these expressions are exceedingly complicated and highly nonlinear in terms of the construction parameters of the layers. For practical problems the required performance must be specified at very many wavelengths and frequently at several angles of incidence. Furthermore, for most problems many layers are required to achieve this performance. Clearly, the use of explicit expressions is not a viable approach for the solution of the general thin film problem.

In the past, analytical methods for the design of multilayer coatings with a specified spectral transmittance or reflectance have been suggested that were based on series expansions or on the use of Fourier transform methods (Delano and Pegis, 1969), (Knittl, 1976). In these methods the materials are assumed to be nonabsorbing and nondispersive. Analytical methods cannot therefore be applied to the class of problems that requires a strong absorption of the incident light. In the solutions that they yield all the optical thicknesses nt of the layers are integral multiples of a basic thickness and the refractive indices n_i are treated as variables. The values of n_i that are selected are frequently outside of the range of values that correspond to those of real coating materials. Nevertheless, analytical methods can frequently provide excellent primary solutions for further calculations with numerical design programs in which the problem of out-of-range parameters can be more easily handled.

The inverse Fourier transform computer program for the design of filters with given spectral reflectance or transmittance curves implemented at the NRCC (Dobrowolski and Lowe, 1978) is based on the work of (Sossi, 1974, 1976). He has shown that it is possible to write down an integral expression which relates the refractive index profile of an inhomogeneous layer to an even function $Q(k)$ which, in turn, is a function of the required transmittance $T(\lambda)$:

$$\int_{-\infty}^{\infty} \frac{dn}{dx}\frac{1}{2n}\exp(ikx)\,dx = Q(k)\exp\left[i\phi(k)\right] = f(k). \tag{6}$$

In the above, $k = 1/\lambda$ and x is twice the optical path within the inhomogeneous layer. $\phi(k)$ is an arbitrary phase. By taking the inverse Fourier transform of the above equation, one obtains an expression for the inhomogeneous refractive index profile:

$$n(x) = \exp\left[\frac{2}{\pi}\int_{-\infty}^{\infty}\frac{Q(k)}{k}\sin\left[i\phi(k) - kx\right]dk\right]. \tag{7}$$

A homogeneous multilayer solution can also be obtained:

$$n(x) = n_0\exp\left[\frac{2}{p}\sum_{m=-\infty}^{l} F\left(\frac{m}{p}\right)\right], \quad \left(\frac{l}{p} \le x \le \frac{l+1}{p}\right). \tag{8}$$

Here $F(x)$ is the Fourier transform of $f(k)$.

The method assumes that the media on either side of the inhomogeneous layer are the same. Unfortunately it has not so far been possible to obtain an exact expression for the quantity Q. The one suggested by Sossi and used at first at the NRCC is

$$Q(k) = \sqrt{\frac{1}{2}\left[\frac{1}{T(k)} - T(k)\right]}. \tag{9}$$

This expression breaks down for values of T close to 0.0. However, other expressions have been found that permit the design of reflection filters with very high reflection coefficients (Verly et al., 1989).

6. MERIT FUNCTIONS FOR THIN FILM DESIGN

Almost all numerical methods for the design of optical multilayer coatings are based on the use of merit functions MF. These are single valued functions which, at any one stage of the design, manufacture or evaluation, express the current performance of the system and direct the flow of future calculations. A proper definition of the merit function is crucial to the success of the process. At the NRCC the form of merit function used most frequently is

$$MF = \left[\frac{1}{m}\sum_{i=1}^{m}\left(\frac{|Q_i^T - Q_i|}{\delta Q_i}\right)^k\right]^{\frac{1}{k}}, \quad k = 1, 2, 4, \ldots, \tag{10}$$

where Q_i^T, Q_i and δQ_i are the target, and present values and acceptable tolerances on the m quantities of interest that are used to define the merit function (Dobrowolski et al., 1989). The quantities of interest can include any number of the many different properties listed in § 3 above. This is possible because the division by a tolerance factor causes the quantity within the summation sign to be dimensionless. Another advantage of this formulation of the merit function is that its value is equal to unity when all quantities of interest are just within the acceptable tolerance of the target values. This provides a useful test for the convergence of the process. The choice of higher values of k $(= 4, 8, 16, \ldots)$ forces a smaller variance between the values of Q_i^T, and Q_i.

A number of specialized merit functions exist. For example, for the design of filters in which the *relative* rather than the absolute spectral reflectance is important, a merit function of the form

$$MF = \left[\frac{1}{m}\sum_{i=1}^{m}\left(\frac{|CQ_i^T - Q_i|}{\delta Q_i}\right)^k\right]^{\frac{1}{k}}, \quad C_{\min} \le C \le C_{\max} \tag{11}$$

can be used (Dobrowolski, 1970). Here C is a scaling factor which yields the lowest value of the merit function. In some NRCC programs merit functions are provided which permit the calculation of the performance of J layer systems placed in series (Dobrowolski, 1981):

$$MF = \left[\frac{1}{m}\sum_{i=1}^{m}\left(\frac{|Q_i^T - Q_i|}{\delta Q_i}\right)^2\right]^{\frac{1}{2}}, \quad \text{where} \quad Q_i^T = T_i^T \text{ and } Q_i = \prod_{j=1}^{J}(T_j). \tag{12}$$

This simplifies solutions for many types of problems.
For the design of achromatic coatings and cut-off filters the following two merit functions are useful (Pelletier et al., 1971), (Wild and Buhay, 1986):

$$MF = W_1\left[\frac{1}{m}\sum_{i=1}^{m}(|Q_i - \overline{Q}|)^k\right]^{\frac{1}{k}} + W_2|Q^T - \overline{Q}| \tag{13}$$

and

$$MF = \frac{1}{m_1}\sum_{i=1}^{m_1} W_i(R_i)^2 + \frac{1}{m_2}\sum_{j=1}^{m_2} W_j/(R_j)^2. \tag{14}$$

In (13), W_1, W_2 are weight factors that govern the relative importance of the two terms and $\overline{Q}$ is the mean value of the quantity Q. In (14), W_i, W_j are weight factors associated with the individual quantities of interest and R_i, R_j are the reflectances in the transmission and rejection regions.
For the design of multilayers for use at oblique angles of incidence (Kimura, 1962, 1963) suggested a merit function of the form

$$MF = \left(\frac{1}{2m}\sum_{i=1}^{m} W_i\left[(R_{p,i}^T - R_{p,i})^2 + (R_{s,i}^T - R_{s,i})^2\right]\right)^{1/2}. \tag{15}$$

The ability to combine in one merit function many different quantities of interest and to define merit functions of many specialized types is another key to the success of numerical thin film design methods.

7. REFINEMENT METHODS

The distinction between refinement and synthesis of thin films is becoming increasingly blurred. For the purpose of this paper, a refinement method will be any numerical method which, given a reasonable starting design, gradually approaches the closest minimum of the merit function in the multidimensional parameter space by modifying the construction parameters of the multilayer without increasing the number of layers in the system. It follows that for refinement to yield a good result, the starting design must have a satisfactory overall thickness and must be divided into a sufficient number of layers. The choice of the variables, n_i, k_i, t_i, depends on manufacturing considerations.
In the past many different optimization techniques have been applied to the solution of thin film problems. Recently at the NRCC the suitability of the following ten different procedures for the refinement of optical multilayers has been investigated: Adaptive Random Search (ARS), Damped Least Squares (DLS), Modified Gradient Method (GRAD), Golden Section Method (GOLD),Hooke and Jeeves Pattern Search (H&J), Basic Powell's Conjugate Search Method (BPCS), Rosenbrock's Rotating Coordinates Method (ROSE), Generalized Simulated Annealing (GSA), Monte Carlo Simulated Annealing (MCSA) and the Revised Nelder-Mead Simplex Method (SIMP) (Dobrowolski and Kemp, 1990). In that paper the methods are briefly described and references are given to the original literature. The refinement methods were applied to three different problems. In each case the

same starting design was used. Because the results obtained depend to a certain extent on method-specific parameters such as step sizes, convergence rates, acceleration factors, etc., the calculations were repeated for ten different sets of such constants. Here the results for one of the problems – a wide-band infrared antireflection coating for a germanium substrate based on the use of ZnS and Ge films – are given (Fig. 1). For each method, plots are presented of the best spectral reflectance and the corresponding refractive index profile as a function of optical thickness, together with the ten MF vs. CPU time curves for these calculations on a HP 1000 model 700 minicomputer. A comparison of the best M_B and average M_A values of the merit function gives an indication of the robustness of the method. It should be noted that other starting designs can lead to lower values of the merit function.

8. NUMERICAL SYNTHESIS METHODS

Synthesis methods generate their own starting designs or, if provided with one, drastically modify it. In general, the number of layers in the multilayer system will increase in the process. Many different numerical synthesis methods exist. Here the techniques available at the NRCC are described.

8.1. *Comprehensive Search Method*

In the Comprehensive Search Method (COMP), the optical thicknesses and refractive indices of each of the l layers of a system may assume i, j different values, respectively (Dobrowolski, 1965, 1981). There are therefore $(ij)^l$ different ways of selecting the layers of the multilayer. After the merit functions have been evaluated for all of these combinations of layers, the combination with the lowest value of the merit function is further refined. This method will find the best solution to a problem in the construction parameter hyperspace providing that a l layer solution exists and that the refractive index and thickness intervals are small enough. With present day desktop computers, calculations with $i = j = 3$, $l = 6$, corresponding to $(ij)^l = 5 \times 10^5$, are quite feasible. The method handles very complicated merit functions and can be used with dispersive and absorbing coating materials.

8.2. *Gradual Evolution Method*

The Gradual Evolution Method (GRAD) is an adaptation of the Comprehensive Search Method (Dobrowolski, 1965, 1981). It can be applied whenever the number of layers l required for the solution of a problem is very large. A multilayer system is gradually built up by repeatedly adding up to m ($m < l$) layers to the substrate. The layers selected from the $(ij)^m$ different combinations of thicknesses and refractive indices correspond to the set which reduces the merit function the most. After each such addition, all the layers of the system are refined. Usually several such iterations are needed to evolve a design with a satisfactory performance. The calculations can be orders of magnitude faster because only a small fraction of the $(ij)^l$ possible permutations required for a comprehensive search is examined. There is no guarantee that the solution obtained with the Gradual Evolution Method corresponds to a global minimum. However, this is not of great practical importance.

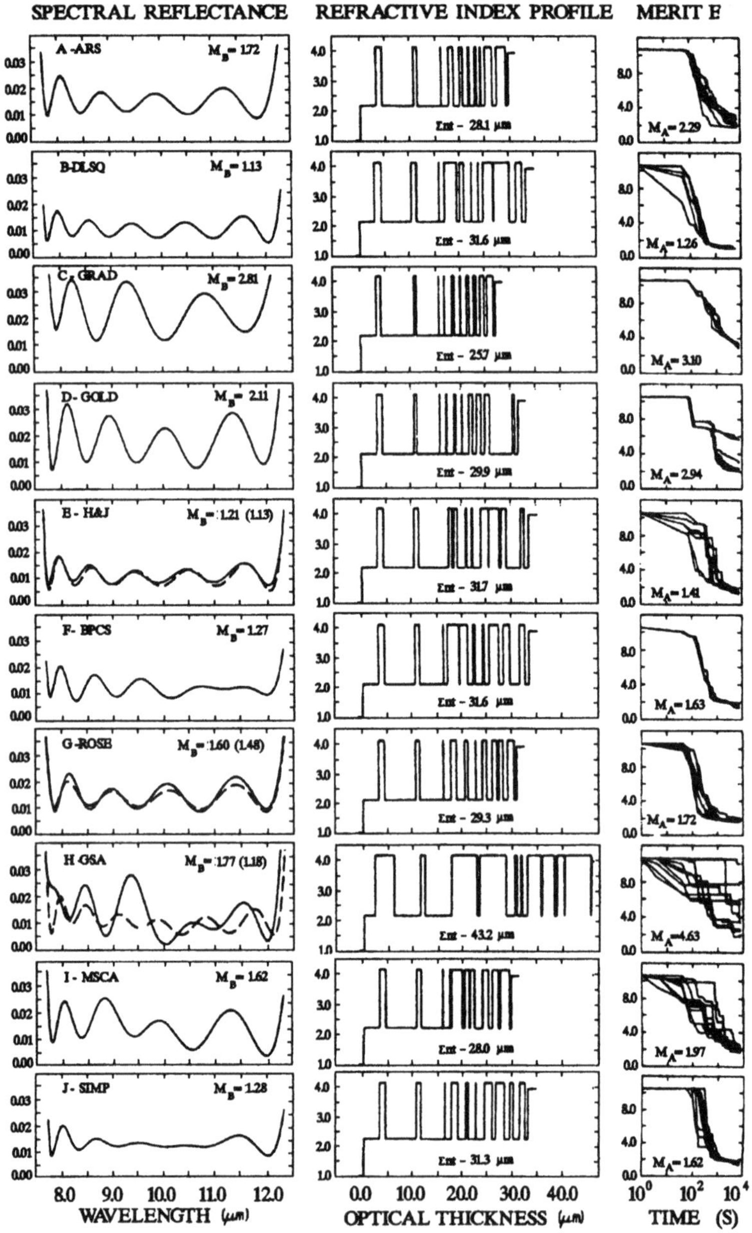

Fig. 1. Wide-band infrared antireflection coating on a Ge substrate refinement of a 21 layer starting design by ten different procedures

8.3. Minus Filter Method

Thelen (1971) has shown that it is possible to design minus filters that transmit all the light incident upon them except in a certain narrow spectral region. The minus filters are composed of dielectric layers of quarterwave optical thicknesses and many different refractive indices. The position, width and attenuation of their rejection bands can be freely controlled. It is therefore possible, by placing suitably selected minus filters in series, to obtain any desired spectral transmission curve. A Minus Filter Synthesis program (MINUS) has been written which performs the selection of the components automatically (Dobrowolski, 1973).

8.4. Flip-Flop Method

The Flip-Flop method (FLIP), first proposed by (Southwell, 1985a), makes use of a starting design composed of many thin dielectric films of equal optical thicknesses. The refractive indices of all the layers of this starting design are then "flipped" one by one in a systematic manner between a high and a low value. Of these, that index is retained which results in a lower value of the merit function. After several such flipping sequences the solution becomes stable and the performance can be improved further by refinement (Dobrowolski, 1986a). An enhanced version of the program exists which is capable of designing multilayers containing absorbing materials and which limits the "flipping" operation to parts of the layer system (Dobrowolski and Kemp, 1991).

8.5. Inverse Fourier Transform Method

The analytical basis for the Inverse Fourier Transform Method (Fourier) has been given in Section 5 above. When the method is combined with various numerical procedures, it becomes a very powerful tool for the design of filters with specified spectral transmittance or reflectance curves (Verly and Dobrowolski, 1990). It has also been possible to adapt the method for the design of antireflection coatings effective over broad spectral regions (Verly et al., 1991).

8.6. Intercomparison of the various synthesis methods

The reason for the development of various thin film synthesis methods is that, when applied to the same problems, they yield different types of solutions. This may be of importance during the practical implementation of the designs. The properties of these *primary* solutions are summarized in Table 1. In an intercomparison at the NRCC the methods were applied to several different problems (Li Li and Dobrowolski, 1991). Every time a solution was sought consisting of the same materials. This necessitated a modification and/or refinement of some of the primary solutions. The results obtained by the methods for one such problem are once again presented with sets of three graphs (Fig. 2). In the first of these graphs the desired and final reflectance curves are represented by heavy and light continuous lines. Intermediate results are represented by broken and dotted lines. The variation of the merit function with CPU time shown in the third graph was obtained on a 386 desktop computer with a 387 numeric co-processor. The properties of the five synthesis methods are compared in Table 2. In general, the

Comprehensive Search and Gradual Evolution Methods are the slowest, but they yield the most practical solutions. The Inverse Fourier Transform and Flip-Flop Methods yield primary solutions quickly, but they consist of many layers and time is lost on the transformations and additional refinement.

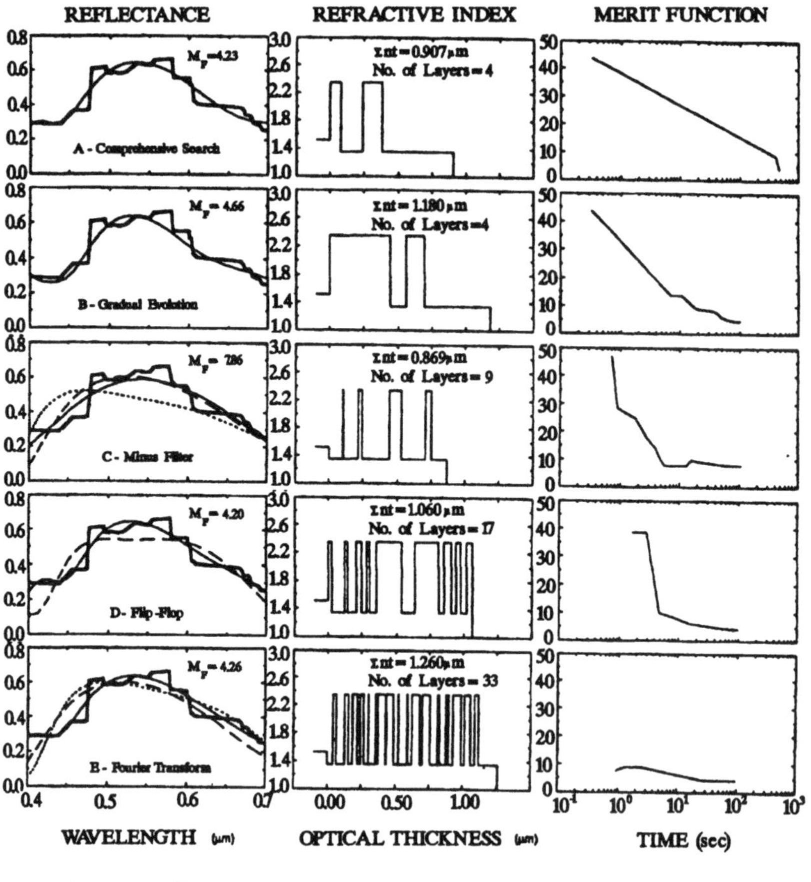

Fig. 2. A filter with a spectral reflectance curve corresponding to the silhouette of the NRCC library solutions obtained by five different thin film synthethsis methods

9. TRANSFORMATIONS OF THE PRIMARY DESIGNS

Before a solution can be implemented in practice, it may have to be modified. For example, the Inverse Fourier Transform program can yield an inhomogeneous layer solution. Routines exist which approximate such a layer by a large number of homogeneous thin films. The primary solutions may consist of many different coating materials and this may be inconvenient for the manufacturer. Computer programs based on the Herpin equivalent index concept (Epstein, 1952), (Ohmer, 1978) can be used to automatically translate such systems into two material solutions of equivalent performance (Dobrowolski and Piotrowski, 1982). W.H. Southwell (see (Southwell, 1985b)) has shown how an absorbing layer with hypothetical optical constants $n - ik$ can be replaced by three thin films of real coating materials with optical constants n_H, n_L, and $n_M - ik_M$. Some thin film synthesis programs do not allow for the dispersion of the refractive index or the finite absorption of real coating materials. Such solutions must be tuned to compensate for these effects. Some primary solutions call for the deposition of very many, very thin layers. T. Skettrup (see (Skettrup, 1989)) has shown how to reduce the number of layers in such systems. A manufacturer may prefer to deposit an inhomogeneous layer rather than a multilayer consisting of discrete optical constants. A routine to accomplish such a transformation also exists at the NRCC.

Table 1. Properties of Primary Solutions

Property	Method	A COMP	B GRAD	C MINUS	D FLIP	E FOURIER
No. of systems	One	Yes	Yes	Yes	Yes	Yes
	Several	No	Yes	Yes	No	No
System	Inhomogeneous Layer	No	No	No	No	Yes
	Two materials	Yes	Yes	No	Yes	No
	Many materials	Yes	Yes	Yes	No*	Yes
Material	Non-absorbing	Yes	Yes	Yes	Yes	Yes
	Absorbing	Yes	Yes	No	No*	No
	Non-dispersive	Yes	Yes	Yes	Yes	Yes
	Dispersive	Yes	Yes	No	No*	No
No. of layers	Few	Yes	Yes	No	No	One
	Medium	No	Yes	Yes	Yes	Yes
	Many	No	Yes	Yes	Yes	Yes
Thickness	Quarter-waves	Yes	Yes	Yes	Yes	No
	Odd	Yes	Yes	No	No	No

* With enhanced Flip-flop method, Yes.

10. CONSTRUCTION PARAMETER DETERMINATION

Depending on the deposition method and process parameters used, the thin films may be quite porous or very dense. The porosity, of course, affects the optical constants of the layers and gives rise to ageing. For example, TiO_2 films with refractive indices that varied between 2.10 and 2.59 have been recently reported (Bennett et al., 1989). It is important to know the optical constants of all the materials used at all stages of the construction of a thin film multilayer system. Numerical thin film methods can assist in the determination of these constants. One technique that has been used with some success at the NRCC is the inverse synthesis method (Dobrowolski et al., 1983). This is a multiwavelength curve fitting method in which the construction parameters of the layer system are adjusted to obtain a better agreement between the theoretical and experimental curves. The number of layers and their approximate thicknesses are known. The optical constants can be represented by any one of a number of different dispersion equations or by an effective medium theory (Sullivan and Dobrowolski, 1991). To reduce the sensitivity of the determination to random measurement errors, an rms fit over a large number of data points is used. To enhance the sensitivity of the determination and to reduce the chances of a spurious solution, one or more of the following measures are taken: a) an optimum combination of measurement quantities is chosen; b) angles of incidence are carefully selected; c) the films of the unknown material are incorporated in a specially selected multilayer; and d) measurements on J separate layers or systems containing the unknown material are simultaneously fitted with a merit function of the type (Dobrowolski, 1981, 1986b)

$$MF = \left\{ \frac{1}{(m_1 + m_2 + \ldots + m_J)} \left[\sum_{i=1}^{m_1} \left(\frac{|Q_{1,i}^T - Q_{1,i}|}{\delta Q_{1,i}} \right)^2 + \sum_{i=1}^{m_2} \left(\frac{|Q_{2,i}^T - Q_{2,i}|}{\delta Q_{2,i}} \right)^2 + \ldots \right] \right\}^{1/2} \quad (16)$$

11. MONITORING

The deposition of optical multilayer coatings may be controlled in many different ways. Optical monitoring measures the quantity of interest directly and is therefore frequently chosen. The same computer programs that are used for the analysis and design of the multilayers can provide useful information for the monitoring. At the NRCC programs are used to simulate the sensitivity of designs to random errors. A repeated perturbation of all the layers can yield statistical data on the probable merit function of the completed filter.

When a multiwavelength monitoring method is used to control the deposition of the i-th layer, the measured spectral reflectance or transmittance is displayed together with the calculated curve for the partial system composed of the first i layers. A merit function is evaluated which represents the difference between these two curves. The deposition process is terminated automatically when the merit function reaches a minimum value or when the operator feels that the best agreement between the two curves has been achieved (Vidal, 1979).

Equipment for single wavelength monitoring is much simpler, but the method is still capable of good results if for each layer the monitoring wavelength is selected for which the derivative of the transmittance with respect to thickness is maximum at the completion of the layer. This most sensitive wavelength monitoring method has been used at the NRCC for many years for the construction of layer systems with non-quarterwave optical thicknesses (Dobrowolski and Waldorf, 1970).

Table 2. Properties of Different Methods

Property / Method		A COMP	B GRAD	C MINUS	D FLIP	E FOURIER
Starting design		None	Optional	Optional	Optional	None
Overall thickness		Specified	Not specified	Not specified	Specified	Specified
Quantities of interest	T, R vs. λ	Yes	Yes	Yes	Yes	Yes
	T, R, A vs. λ, θ	Yes	Yes	No	No*	No
	Other quantities	Yes	Yes	No	No*	No
Necessity to modify primary solution		No	No	Yes	No	Yes
Execution speed	Primary solution	–	–	Medium	Fast	Very fast
	Final solution	Very slow	Medium	Medium	Medium	Medium

* With enhanced Flip-flop method, Yes.

12. DESENSITIZATION

Should an error analysis indicate that a multilayer is too sensitive to construction errors, it is sometimes possible to desensitize the design. Thus usually occurs at the expense of additional layers or a loss in performance. Two different compound merit functions have been proposed for this purpose. Both consist of two terms, the first of which describes the performance MF of the multilayer in the usual way. In the first merit function $(MF)^*$, the second term is a summation of the derivatives of MF with respect to the construction parameters p_i, calculated numerically (Dobrowolski, 1981):

$$(MF)^* = \Omega_1 MF + \Omega_2 \sum_{i=1}^{l} W_i \left| \frac{d(MF)}{dp_i} \right| . \tag{17}$$

In the other merit function the second term contains a factor P. This is the probability of producing a satisfactory coating, obtained from statistical calculations with randomly perturbed construction parameters. The perturbations result from errors in the monitoring and from fluctuations in the deposition process parameters (Zheng and Tang, 1987):

$$(MF)^* = \Omega_1 MF + \Omega_2(1 - P). \tag{18}$$

In the above equations Ω_1, Ω_2, W_i are suitably chosen weight factors. Desensitization entails considerable computation, especially when the number of construction parameters is large. In effect, the optimum solution is somewhat de-tuned to find a set of parameters that correspond to a nearby plateau with somewhat worse, but still acceptable performance.

13. REAL-TIME REOPTIMIZATION

A number of laboratories are working on the implementation of real-time reoptimization in order to significantly improve the experimental yield. Consider the manufacture of a multilayer system. After the deposition of a layer and the measurement of the spectral characteristics of the partial system that has been deposited so far, the construction parameters of the layer are determined. In general, they will differ from the intended values. It is possible at this stage to adjust the thicknesses of the remaining layers to reoptimize the performance of the system and allow for the deposition errors made thus far.

It has been shown theoretically that the effect of even large errors can be compensated in this manner (Holm, 1978). This has been confirmed in a much more thorough study in which the effectiveness of this approach has been investigated for different deposition processes and for different monitoring strategies (Sullivan and Dobrowolski, 1991). Li Li and Yen (see (Li Li and Yen, 1989)) were the first to publish a description of experimental work along these lines. In the future real-time reoptimization should make possible the manufacture of very complicated multilayer systems.

14. CONCLUSIONS

This article has reviewed many numerical procedures available at the NRCC for the design, evaluation and manufacture of optical multilayer coatings. It is true that it is possible today to design filters with almost any specified spectral transmittance or reflectance using nonabsorbing coating materials with optical constants that can be achieved in practice (Fig. 3). The advantage of numerical procedures over purely analytical methods is that they permit the specification (and hence, solution) of much more complicated problems. Different methods give rise to different types of solutions for the same problem. Solutions obtained by numerical methods take into account the dispersion of the optical constants of the coating materials and permit corrections to be made for the defects of experimentally produced coatings, such as porosity and finite extinction coefficients. When interfaced with deposition equipment, they make possible the complete automation of the manufacture of optical thin film systems.

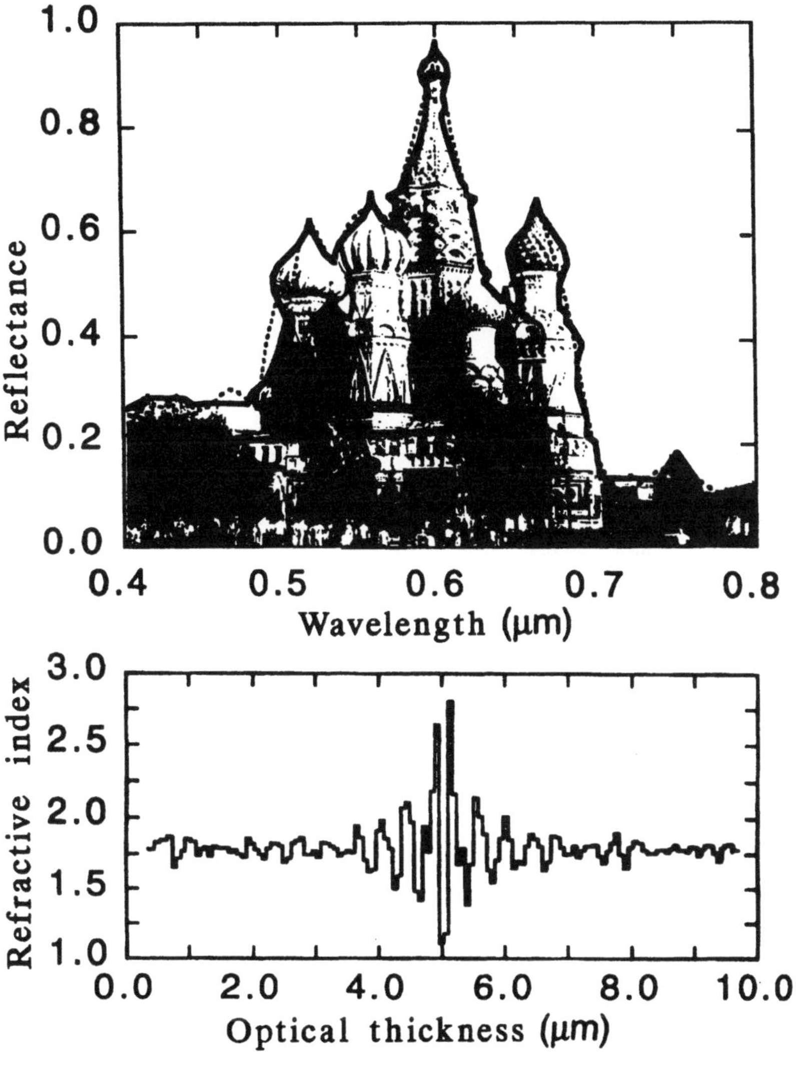

Fig. 3. A filter with a special reflectance curve corresponding to the silhouette of St. Basil's cathedral, Moscow. The light continious line in the upper diagram is the calculated reflectance of multilayer whose variation in index with optical thickness is given on the lower diagram

REFERENCES

Bennett et al. (1989). Comparison of the properties of titanium dioxide films prepared by various techniques. *Appl. Opt.* **28**, 3303–3317.

Born, M. and Wolf, E. (1959). *Principles of Optics*. Pergamon Press, London.

Delano, E. and Pegis, R.J. (1969). Methods of synthesis for dielectric multilayer filters. In: *Progress in Optics*, 7. Ed. by E. Wolf. North-Holland, Amsterdam, pp. 67–137.

Dobrowolski, J.A. (1965). Completely automatic synthesis of optical thin film systems. *Appl. Opt.* **4**, 937–946.

Dobrowolski, J.A. (1970). Optical interference filters for the adjustment of spectral response and spectral power distribution. *Appl. Opt.* **9**, 1396–1402.

Dobrowolski, J.A. (1973). Subtractive method of optical thin-film interference filter design. *Appl. Opt.* **12**, 1885–1893.

Dobrowolski, J.A. (1981). Versatile computer program for absorbing optical thin film systems. *Appl. Opt.* **20**, 74–81.

Dobrowolski, J.A. (1986a). Comparison of the Fourier transform and flip-flop thin-film synthesis methods. *Appl. Opt.* **25**, 1966–1972.

Dobrowolski, J.A. (1986b). On the determination of optical constants of films in multilayers. *Proc. S.P.I.E.* **652**, 206–208.

Dobrowolski, J.A., Ho, F.C., Belkind, A., and Koss, V.A. (1989). Merit functions for more effective thin film calculations. *Appl. Opt.* **28**, 2824–2831.

Dobrowolski, J.A., Ho, F.C., and Waldorf, A. (1983). Determination of optical constants of thin film coating materials based on inverse synthesis. *Appl. Opt.* **22**, 3191–3200.

Dobrowolski, J.A. and Kemp, R.A. (1990). Refinement of optical multilayer systems with different optimization procedures. *Appl. Opt.* **29**, 2876–2893.

Dobrowolski, J.A. and Kemp, R.A. (1991). A flip-flop thin film design program with enhanced capabilities (to appear).

Dobrowolski, J.A. and Lowe, D. (1978). Optical thin film synthesis program based on the use of Fourier transforms. *Appl. Opt.* **17**, 3039–3050.

Dobrowolski, J.A. and Piotrowski, S.H.C. (1982). Refractive index as a variable in the numerical design of optical thin films. *Appl. Opt.* **21**, 1502–1511.

Dobrowolski, J.A. and Waldorf, A. (1970). Manufacture of all-dielectric interference filters with layers of arbitrary thickness and refractive index. *J. Opt. Soc. Amer.* **60**, 725.

Epstein, I.I. (1952). The design of optical filters. *J. Opt. Soc. Amer.* **42**, 806–810.

Holm, C. (1978). Optical thin film production with continuous reoptimization of layer thicknesses. *Appl. Opt.* **18**, 1978–1982.

Kimura, N. (1962; 1963). Automatic design of multilayer dielectric films I, II. *Japan J. Appl. Phys.* **31**, 739–745; **32**, 199–212.

Knittl, Z. (1976). *Optics of Thin Films*. John Willey, London.

Li Li and Dobrowolski, J.A. (1991). Computation speeds of different optical thin film synthesis methods (to appear).

Li Li and Yen, Y.-H. (1989). Wideband monitoring and measuring system for optical coatings. *Appl. Opt.* **28**, 2889–2894.

Macleod, H.A. (1986). *Thin Film Optical Filters*. Second Ed. McGraw-Hill, New York.

Ohmer, M.C. (1978). Design of three layer equivalent films. *J. Opt. Soc. Amer.* **68**, 137–139.

Pelletier, E., Klapisch, M., and Giacomo, P. (1971). Synthesis of thin film multilayer systems. *Nouv. Rev. Opt. Appl.* **2**, 247–254.

Pohlak, H. (1962). Optik Duenner Schichten – Ein Ueberblick ueber den Stand der Entwicklung in Theorie und Praxis. In: *Optik und Spektroskopie aller Wellenlaengen,* Ed. by P. Goerlich. Akademie-Verlag, Berlin, pp. 93–125.

Skettrup, T. (1989). Three-layer approximation of dielectric thin film systems. *Appl. Opt.* **28**, 2860–2863.

Sossi, L. (1974). A method for the synthesis of multilayer dielectric interference coatings. *Eesti NSV Tead. Akad. Toim. Fuus. Mat.* **23**, 229–237.

Sossi, L. (1976). On the theory of the synthesis of multilayer dielectric light filters. *Eesti NSV Tead. Akad. Toim. Fuus. Mat.* **25**, 171–176.

Southwell, W.H. (1985a). Coating design using very thin high- and low index layers. *Appl. Opt.* **24**, 457–460.

Southwell, W.H. (1985b). Thin layer equivalence for absorbing films. *J. Opt. Soc. Amer.* **A2**, 85.

Sullivan, B.T. and Dobrowolski, J.A. (1991). Deposition error compensation for optical multilayer coatings: I. Theoretical description (to appear).

Thelen, A. (1971). Design of optical minus filters. *J. Opt. Soc. Amer.* **61**, 365–369.

Verly, P.G. and Dobrowolski, J.A. (1990). Iterative correction process for optical thin film synthesis with the Fourier transform method. *Appl. Opt.* **29**, 3672–3684.

Verly, P.G., Dobrowolski, J.A., Wild, W.J., and Burton, R.L. (1989). Synthesis of high rejection filters with the Fourier transform methods. *Appl. Opt.* **28**, 2864–2875.

Verly, P.G., Dobrowolski, J.A., and Willey, R.R. (1991). Fourier transform method for the design of wideband AR coatings. Submitted to *Appl. Opt.*

Vidal, B., Fornier, A., and Pelletier, E. (1979). Wideband optical monitoring of nonquarterwave multilayer filters. *Appl. Opt.* **18**, 3851–3856.

Wild, W.J. and Buhay, H. (1986). Thin film multilayer design optimization using a Monte Carlo approach. *Opt. Letters.* **11**, 745–747.

Zheng, Y.-F. and Tang, J.-F. (1987). New automatic design technique for optical coatings. *Appl. Opt.* **26**, 1546–1549.

Ill-Posed Problems in Natural Sciences, pp. 504 – 514
A. Tikhonov (Ed.)
1992 VSP/TVP

INVERSE SCATTERING AND SYNTHESIS PROBLEMS IN THE DIFFRACTION THEORY

Ju.A. EREMIN and A.G. SVESHNIKOV

Moscow State University, Computational Mathematics and Cybernetics Department, 119899 Moscow, Russia

ABSTRACT

This paper focuses on inverse scattering problems of determining the shape or permittivity of scattering obstacle from the knowledge of the far field pattern of the scattered field. The cases of a perfectly conducting scatterer, of nonvanishing surface impedance and of a penetrable obstacle are investigated. Examples of identification and synthesis problems are given.

INTRODUCTION

The diffraction theory (DT) is an important part of the modern mathematical physics. DT problems are not only of great scientific interest but they find multiple practical applications. They also form a theoretical foundation for the investigation of numerous problems of electrodynamics, acoustics, geophysics, computer tomography, etc. Along with the development of new methods for solution of direct problems (DP) in recent years the inverse scattering problems (ISP) involving reconstruction and synthesis problems have received considerable attention. When solving the reconstruction problems it is required to determine the obstacle from a known far field pattern (FFP). Synthesis problems consist in the construction of sources and objects realizing the given diffracted field with the required accuracy. The methods for solving these problems not only have many common features, but also have some essential distinction taking into account the specifics of their formulation. Of course the greatest practical interest present three-dimensional vector problems. Typically, implementation of algorithms for numerical solution of such problems involves many difficulties, often of a technical character. But the fundamental ideas of the main methods for investigation of the ISP of the DT are similar for two-dimensional problems.

1. DIFFRACTION PROBLEM FOR AN IMPENETRABLE OBSTACLE

1.1 Direct problem

Let D denote a bounded connected domain in $\mathbf{R}^2$ with a smooth boundary ∂D $C^{(2,\alpha)}$ and let U^i be an entire function satisfying the Helmholtz equation in $\mathbf{R}^2$ e.g., the plane wave $U^i = e^{ikx}$. We denote the total field by $U = U^i + U^s$, where U^s is the scattered field (SF). Then the formulation of the diffraction problem is

$$\Delta U + k^2 U = 0 \quad \text{in } D_e = \mathbf{R}^2 \backslash D$$

$$U(P) = 0 \quad P \in \partial D \tag{1.1.1}$$

$$\frac{\partial U^s}{\partial r} - ikU^s = o(r^{-1/2}), \quad r \to \infty.$$

The asymptotic behavior of the SF U^s is described by

$$U^s(r,\varphi) = e^{ikr}/r^{1/2}\; F(\varphi) + o(r^{-1}),$$

where $F(\varphi)$ is the FFP, and k is the wave number.

THEOREM 1.1.1 (Colton, 1984). *For any fixed $k > 0$ the* FFP *is an entire function of the independent variable φ; if U^i is an entire function, then for $k > 0$ the* FFP *is also an analytic function of k also.*

Let us consider an operator T that transforms the given input data (ID) $\{U^i, D\}$ into the FFP F. We suppose that the data $\{U^i, D\}$ completely characterize the ID of DP (1.1.1) assuming that the boundary condition and the value k are known. Thus

$$T\{U^i, D\} = F.$$

Assume that D is fixed. We shall examine $T\{U^i, D\}$, where U^i is the set of all entire solutions of the Helmholtz equation in $\mathbf{R}^2$.

THEOREM 1.1.2. *Let $\mathcal{F}_k = R(T)$ denote the range of T.*
Then
1) *if the contour ∂D is nonresonance for a given k, then $\overline{\mathcal{F}}_k = \mathbf{L}_2[0, 2\pi]$;*
2) *if the contour D is resonance, then $\overline{\mathcal{F}}_k = \mathbf{L}_2$ if and only if there exists an internal eigenfunction*

$$G(x) = \int_0^{2\pi} g(\varphi)\; e^{ikx} d\varphi, \tag{1.1.2}$$

where G is a Gerglotz function and g is its Gerglotz kernel (Colton, 1984).

Denote the class of Gerglotz functions by G. The above result means that there exists the nonzero orthogonal supplement ary span $\{g\} = R^{\perp}(T)$ corresponding to the existence of nonradiating currents. These facts are used essentially for solution of synthesis problems. They define the class of the FFP realized .

1.2. Inverse problem

We shall discuss briefly the existence of the inverse operator T^{-1},

$$T^{-1}\{U^i, F\} = D.$$

Suppose that $F \in R(T)$ and, therefore, for given U^i there exists a closed smooth contour ∂D for which F is an FFP.

Definition 1.2.1. Two scatterers D^1 and D^2 are called equivalent scatterers (ES) if for given k and U^i there holds $F_1 = F_2$, but $\text{mes}(D^1 \triangle D^2) \neq 0$.
An example of existence of ES is given in (Sveshnikov et al., 1979). It is proved that for a bounded domain only a finite number of ES exists.

THEOREM 1.2.1. (A necesstiy and sufficient condition of existence of ES (Sveshnikov et al., 1979)). *Let k and U^i be known.*
Then
1) *(Necessity) if there exist* ES $D^{1,2}$, *then the domain* $D^1 \triangle D^2$ *is resonance for a given* k;
2) *(Sufficiency)* $D^{1,2}$ are ES *if in* $D^1 \triangle D^2$ *there exists an eigenfunction* V *which can be analytically continued into* $\mathbf{R}^2/(D^1 \cup D^2)$ *and* $V - U^i = e^{ikr}/r^{1/2}F + o(r^{-1}), \quad r \to \infty$.
The main properties of ES.
1) *the nonanalytic points of the contours* $\partial D^{1,2}$ *are the same;*
2) *if* $D^{1,2}$ are ES, *then* $D^1 \cap D^2 \neq \emptyset$.

THEOREM 1.2.2. (Uniqueness (Lax and Phyllips, 1967)). *For any external excitation, the contour ∂D is uniquely determined by the* FFF $F(\varphi, k)$, $\varphi \in [0, 2\pi]$, $k \in (k_1, k_2) \subset \mathbf{R}^+$.

THEOREM 1.2.3. (Uniqueness (Colton and Sleeman, 1983; Eremin and Zakharov, 1987)). *Assume that a scatterer D is contained in a circle of radius R and let N be the number of zeros of $J_m(x)$, $x \in (0, kR]$, $m \leq N$. Then D is uniquely determined*
1) *by the* FFP $F(\varphi, t_m)$, $\varphi \in [0, 2\pi]$, $t_1, t_2, \ldots, t_n$, $n \leq N$, *where t_m are the directions of the incident plane wave* $U^i = e^{ikt}$;
2) *by the* FFP $F(\varphi, M_n^0)$, $\varphi \in [0, 2\pi]$, $M_n^0 \in \mathbf{R}^2/\overline{K}_R$, *where M_n^0 is a singular point of the source* $U_m^i = H_0^{(1)}(kR_{MM_n^0})$, $n \leq N$.

These theorems make it possible to formulate the multifrequency and multiposition sounding criteria that guarantee the uniqueness of a solution of the ISP.

1.3. Methods for reconstructing the boundary of a scatterer

Today the methods of solution of the ISP reconstructing the boundary of a scatterer that do not require the solution of the DP are believed to be the most perspective ones. In this group of methods, we consider the method introduced in (Kirsch and Kress, 1977). Its principal idea consists in using scattering data to construct a near field and to determine zero level of the total field corresponding to the Dirichlet boundary condition. The problem is reduced to the minimization of the Tikhonov regularization functional (Tikhonov and Arsenin, 1979). It is

necessary to use also supplementary information concerning the number of incidence angles of plane waves (Kirsch et al., 1988).
We will describe another approach basing upon the results given in (Colton and Monk, 1985) when we consider the ISP for a penetrable scatterer.

2. DIFFRACTION PROBLEM FOR AN IMPERFECTLY CONDUCTING OBSTACLE

2.1. Direct problem
The mathematical formulation of the DP in this case is

$$\Delta U + k^2 U = 0 \quad \text{in } D_e,$$

$$\partial U(P) - \chi U(P) = 0, \quad P \in \partial D, \tag{2.1.1}$$

$$\frac{\partial U^s}{\partial r} - ikU^s = o(r^{-1/2}), \quad r \to \infty.$$

If we assume that Im $\chi \geq 0$, $\chi \in C(\partial D)$ then problem (2.1.1) has an unique solution.

THEOREM 2.1.1. *For any $k > 0$ $F(\varphi, \chi)$ is an entire function of the independent variable φ. The* FFP *is also an analytic function of the parameter χ.*

We shall consider the operator T of the DP: $T\{U^i, D, \chi\} = F$.
Let us state one important property of the operator $T\{U^i\}$ which holds under the assumption that D and χ are given.

THEOREM 2.1.2. (Angell et al., 1989). *Let U^i be the set of all entire solutions of the Helmholtz equation in $\mathbf{R}^2$ and the variable impedance* Im $\chi > 0$. *Then* $\overline{R(T)} = \mathbf{L}_2[0, 2\pi]$.

2.2. Inverse problem
We shall now focus on the existence of the inverse operator T^{-1},

$$T^{-1}\{U^i, F\} = \{D, \chi\}, \quad F \in R(T).$$

Definition of ES . For given k and U^i, two scatterers are equivalent, $\{D^1, \chi_1\} \backsim \{D^2, \chi_2\}$, if $F_1 = F_2$, but either

$$\text{mes}(D^1 \bigtriangleup D^2) \neq 0 \text{ or } \chi_1 \not\equiv \chi_2.$$

It is possible to construct an example showing the existence of ES for the imperfectly conducting obstacles by applying the idea suggested in (Sveshnikov et al., 1979). In this case the situation is more complicated than for a perfectly conducting scatterer because in a bounded domain an unlimited number of ES can exist.

THEOREM 2.2.1. (Uniqueness (Colton and Kress, 1983)). *Assume that the shape of an obstacle D and an external excitation U^i are given. Then the surface non-vanishing impedance χ: $\mathrm{Im}\ \chi > 0$ is uniquely determined by the* FFP $F(\varphi, k)$, $\varphi \in [0, 2\pi]$ *at a fixed frequency.*

THEOREM 2.2.2. (Uniqueness (Ramm, 1986; Eremin and Zakharov, 1987)). *A scatterer $\{D, \chi\}$ is uniquely determined by the following* ID:
1) *by the* FFP $F(\varphi, k)$, *for any* U^i, $\varphi \in [0, 2\pi]$, $k \in (k_1, k_2) \subset \mathbf{R}^+$;
2) *by the* FFP $F(\varphi, t)$, $\varphi \in [0, 2\pi]$, $\{\bar{t}\} = \Omega$, *if k is fixed;*
3) *by the* FFP $F(\varphi, M_n^0), \varphi \in [0, 2\pi]$, $\{M_n^0\} = d \subset D_e$, *if k is fixed.*

The results given above make it possible to formulate the multifrequency and multiposition sounding criteria which are sufficient for the uniqueness of a solution of the ISP.

3. SCATTERING BY A PENETRABLE OBSTACLE

3.1. Direct problem

Let D denot a bounded connected obstacle in $\mathbf{R}^2$ with the boundary $\partial D \in C^{(2,\alpha)}$. The scattering problem now consists in solving the following boundary-value problem (BVP):

$$\begin{gathered} \Delta U_e + k_e^2 U_e = 0 \quad \text{in } D_e, \\ p \,\mathrm{div}\, (1/p \nabla U_i) + k_i^2 U_i = 0 \quad \text{in } D, \\ [U] = 0, \quad [1/p \partial U] = 0 \quad \text{on } \partial D, \\ U^s = e^{ikr}/r^{1/2} F(\varphi) + o(r^{-1}), \quad r \to \infty. \end{gathered} \tag{3.1.1}$$

Here, $U = U^s + U^i$, U^i is an external excitation, U_i is the total field inside the obstacle, $[\cdot]$ is the jump of the corresponding quantity in crossing ∂D, $k_i^2 = \omega^2 \varepsilon(x) \mu(x)$, $\mathrm{Im}\ \varepsilon, \mu > 0$, $\varepsilon, \mu \in C(D)$, $p(M)$ equals either $\varepsilon/\varepsilon_e$ or μ/μ_e, depending on the polarization, ∂U is the normal derivative at ∂D. Under these conditions there exists an unique solution of BVP (3.1.1).
The operator T of the DP transforms $\{U^i, D, \varepsilon, \mu\}$ into the FFP F

$$T\{U^i, D, \varepsilon, \mu\} = F.$$

Solution of the ISP for the full ID is rather complicated. Therefore we shall consider two reduced sets of ID which are of practical interest.
A. Assume that the values $\mu = \mu_e$ and $\varepsilon = \text{const}$ are given and the aim is to reconstruct the contour D of the obstacle.
B. Assume that ∂D, and $\mu = \mu_e$ are given and the aim is to determine $\varepsilon(x)$.
The determination of the range $R(T)$ of the operator T the is essentially based on the following internal BVP:

$$\Delta V + k_e^2 V = 0 \quad \text{in } D,$$

$$p \,\mathrm{div}\, (1/p \nabla W) + k_e^2 p W = 0 \quad \text{in } D, \tag{3.1.2}$$

$$W = V, \qquad 1/\varepsilon\ \partial W = 1/\varepsilon_e\ \partial V \quad \text{on } \partial D.$$

The existence of a solution of problem (3.1.2) is a necessary condition for an invisible scatterer (IS) corresponding to ID $\{D, \varepsilon\}$ to exist (Eremin, 1988).
If this is the case and a pair $\{V; W\}$ is a nontrivial solution of the BVP (3.1.2) then the scatterer $\{D, \varepsilon\}$ is a nonradiating object. Therefore, this situation is an analogous completely to the case where a nonradiating current exists for a perfectly conducting obstacle.

Definition 3.1.1. If a nontrivial solution of the BVP (3.1.2) exists, then the quantity k_e^2 is called an eigenvalue of the IS (EVIS) and the function V is called an eigenfunction of the IS (EFIS).

THEOREM 3.1.1. *Assume that k_e^2 is not an* EVIS, *then* $\overline{R(T_{\mathbf{A}})} = \mathbf{L}_2[0, 2\pi]$. *If k_e^2 is an* EVIS, *then*
1) *if none* EFIS $V \notin G$ *(1.1.2), then* $\overline{R(T_{\mathbf{A}})} = \mathbf{L}_2[0, 2\pi]$;
2) *if at least one* EFIS $V \in G$, *then* $R^{\perp}(T_{\mathbf{A}}) = \text{span}\ \{g\}$, *where the span* $\{g\}$ *is the linear* span *of the Gerglotz kernel* (Colton and Monk, 1987).

Remark 3.1.1. A detailed analysis of the spectrum of the BVP (3.1.1) is given in (Eremin, 1988), where it is shown that if $\text{Im} k_i \not\equiv 0$, then (3.1.1) has only the trivial solution. The spectrum of the BVP is discrete with no finite points of accumulation.
In the case **B**, where $\text{Im} k_i \not\equiv 0$, the following theorem can be proved.

THEOREM 3.1.2. *If U^i is the set of all entire functions in* $\mathbf{R}^2$ *or* $U^i = e^{ikt}$, $m \in \mathbf{N}$, $|t_m| = 1$, $\overline{\{t_m\}} = \Omega$, *the unit circle, or* $U^i = H_0^{(1)}(k_e R_{MM_n^0})$, $\overline{\{M_n^0\}} = d \subset D_e$, *then* $\overline{R(T_{\mathbf{B}})} = \mathbf{L}_2[0, 2\pi]$.

3.2 Inverse problem

We shall define ES taking into account the fact that μ is known for the sets **A** and **B**.

Definition 3.2.1. Scatterers $\{D^{1,2}, \varepsilon_{1,2}\}$ are said to be ES $\{D^1, \varepsilon_1\} \backsim \{D^2, \varepsilon_2\}$, if $F_1 = F_2$, but either $\text{mes}(D^1 \triangle D^2) \neq 0$ or $|\varepsilon - \varepsilon_2| \neq 0$.
The main properties of ES.

THEOREM 3.2.1. (Eremin and Sveshnikov, 1987). *Assume that there exist two* ES $\{D^1, \varepsilon_1\} \backsim \{D^2, \varepsilon_2\}$. *Then*
1. *the nonanalytic points of the contours* $\partial D^{1,2}$ *are identical;*
2. *if* $D^1 \cap D^2 \neq 0$ *and* $\text{Im}\ \varepsilon_1 \equiv 0$, *then* $\text{Im}\ \varepsilon_2 \equiv 0$;
3. *if* $D^1 \cap D^2 = 0$, *then* $\{D^1, \varepsilon_1\}$ *and* $\{D^2, \varepsilon_2\}$ *are not* ES.

An example of the existence of ES is given in (Eremin and Sveshnikov, 1987). Same as in the case of an imperfectly conducting scatterer, in a bounded region there exist infinitely many ES.
We shall study properties of the inverse operator in the case **A**:

$$T_{\mathbf{A}}^{-1}\{U^i, \mu, \varepsilon\} = D, \; F \in R(T_{\mathbf{A}}).$$

THEOREM 3.2.2. (Uniqueness (Eremin and Sveshnikov, 1987)). *In the case* **A** *the contour* ∂D *is uniquely determined by the* ID
1) *by the* FFP $F(\varphi, k_e)$, *for any* U^i, $\varphi \in [0, 2\pi]$, $k_e \in \underline{(k_1, k_2)} \subset R^+$;
2) *by the* FFP $F(\varphi, t_m)$, $\varphi \in [0, 2\pi]$, $m \in \mathbf{N}, |t_m| = 1$ *and* $\overline{\{t_m\}} = \Omega$ *if* k_e *is given* ;
3) *by the* FFP $F(\varphi, M_n^0)$, $\varphi \in [0, 2\pi]$, $n \in \mathbf{N}; \overline{\{M_n^0\}} = d \subset D_e$.

Now we shall consider the structure of ES in the case **B**.

THEOREM 3.2.3. (Necesstiy and sufficient conditions of the existence of ES (Eremin and Sveshnikov, 1987)). *A necessary condition of the existence of* ES $\{D, \varepsilon_1\} \backsim \{D, \varepsilon_2\}$ *is the existence of a nontrivial solution* $\{V_1; V_2\}$ *of the BVP:*

$$p_j \text{div}\,(1/p_j \nabla V_j) + k_e^2 p_j V_j = 0 \quad \text{in } D, \quad j = 1, 2;$$

$$V_1 = V_2, \qquad 1/\varepsilon_1\, \partial V_1 = 1/\varepsilon_2\, \partial V_2 \quad \text{on } \partial D. \tag{3.2.1}$$

Assume that there exists a nontrivial solution $\{V_1; V_2\}$ of (3.2.1). Then $V_{1,2}$ are fields of ES only when they satisfy in $D^{1,2}$ the integral equation

$$V_{1,2}(q) - k_e^2 \int_D (1 - p_{1,2}(t))\psi_e(q,t) V_{1,2}(t)\, d\tau_t = U^i(q), \quad q \in D,$$

where $\psi_e = i/4 H_0^{(1)}(k_e R_{qt})$.

Remark 3.2.1. The set of eigenvalues k_e^2 of BVP (3.2.1) is discrete and has no finite points of accumulation. Therefore in the case **B** we can formulate an uniqueness theorem similar to Theorem 3.2.3.
The results obtained make it possible to formulate the multifrequency and multiposition sounding criteria.

3.3. Methods of reconstruction of a penetrable obstacle
In recent years the most perspective approach to solution of the ISP is the one that takes into consideration the orthogonal complement to the range of the DP operator and does not require the solution of the DP. This approach has been developed in (Colton and Monk, 1985, 1987).
The main steps for reconstructing the boundary of an obstacle (the case **A**) are the following:
Step 1. The solution of the BVP

$$\Delta V + k_e^2 V = 0 \quad \text{in } D,$$

$$p \text{ div}\,(1/p \nabla W) + k_e^2 p W = 0,$$

$$W = V - H_0^{(1)}(k_e R_{0P}), \quad P \in \partial D$$

$$1/\varepsilon\, \partial W = 1/\varepsilon_e \{\partial V - \partial H_0^{(1)}(k_e R_{0P})\}$$

is constructed in the form of the Gerglotz wave functions

$$V(\varphi) = \int_0^{2\pi} g(\theta) \exp\big(- ik_e\rho(\theta)\big) \cos\big(\theta - \varphi\big)\, d\theta,$$

$$W(\varphi) = \int_0^{2\pi} h(\theta) \exp\big(- ik_e\rho(\theta)\big) \cos\big(\theta - \varphi\big)\, d\theta.$$

Step 2. Assume that the contour ∂D is starlike with respect to the origin $\rho^0(\varphi)$. Then the contour ∂D is determined as a solution of the minization problem for the functional

$$\xi(F, M, N) = \min_{\rho \in U} \left\{ \sum_{n=1}^{N} \int_0^{2\pi} \left| F(\theta, \alpha_n) g^*(\theta)\, d\theta + e^{1/4i} \sqrt{2/\pi k_e} \right|^2 \right.$$

$$+ \int_0^{2\pi} \left| W\big(\rho(\varphi), \varphi\big) - V\big(\rho(\varphi), \varphi\big) + H_0^{(1)}\big(k_e\rho(\varphi)\big) \right|^2 d\varphi$$

$$\left. + \int_0^{2\pi} \left| 1/\varepsilon \partial W(\rho, \varphi) - 1/\varepsilon_e \Big\{ \partial V(\rho, \varphi) + \partial H_0^{(1)}\big(k_e\rho(\varphi)\big) \Big\} \right|^2 d\varphi \right\}.$$

THEOREM 3.3.1. *Assume that k_e^2 is not an* EVIS *and ∂D is such that $\rho^0(\varphi)$ and $F \in R(T_{\mathbf{A}})$, then for all N* $\lim_{M \to \infty} \xi(F, M, N) = 0$ (Colton and Monk, 1987).

4. IDENTIFICATION OF A DIELECTRIC BODY LOCATED UNDERGROUND

The identification problem is a part of a more general recognition problem. Its specific feature is that it must be solved for a compact class of scatterers. This makes it possible to restrict the input data with respect to the information which is sufficient for the uniqueness of the ISP. We shall consider the identification problem for the homogeneous dielectric bodies of revolution located in an absorbing halfspace. Let $\mathbf{R}^3$ be cut by the plane ($z = 0$) into two halfspaces $D_{0,1}$. The halfspace D_0 has the characteristics ε_0, μ_0 whill D_1 has the characteristics ε_1, μ_1. We assume that in D_1 there is a local scatterer D_i with the characteristics ε_1, μ_1. We consider the amplitude characteristics of a scattered field when the incident plane wave propagates along the spin axes. By means of a numerical experiment the most informative parameters have been determined which enable us to solve the identification problem for a fixed number of scatterers. As a result it was found that the amplitude-frequency parameter of an anomalous field (AF) is the most informative parameter. It was shown that there exist specific features (SF) of dielectric scatterers which are determined by the positions of extreme of the amplitude-frequency characteristics of the AF. Moreover, the SF do not depend on the half-space electromagnetic characteristics ε_1, μ_1 even in the presence of frequency dependence (Fig.1). Neither do they depend on the depth of the positions of scatterers and on the measurements of the height of the AF not on the point of observation and the AF components (Fig. 2). The SF are entirely determined by the shape and dielectric permitivity of the scatterer (Eremin and Orlov, 1988).

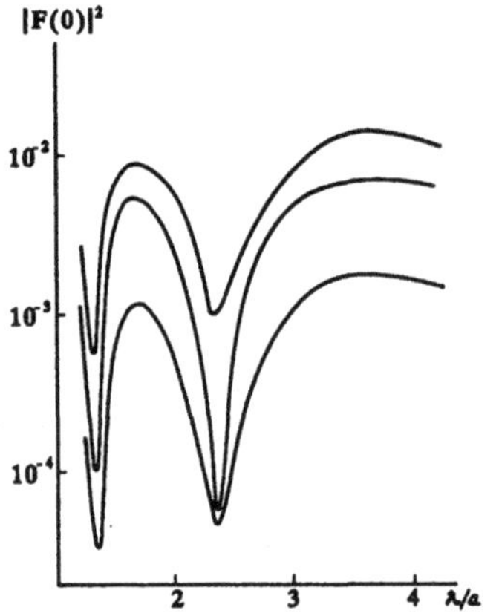

Fig. 1

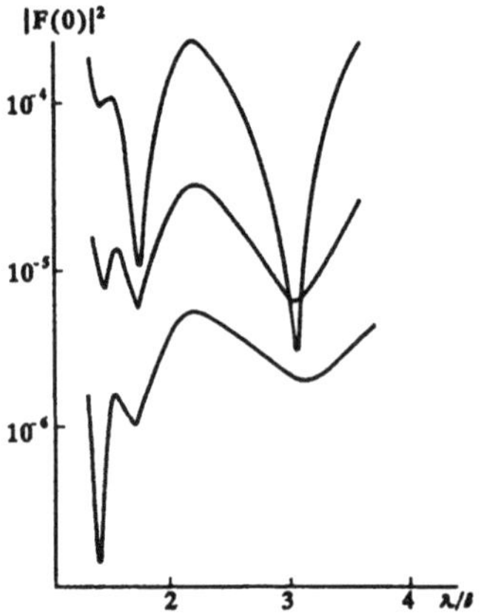

Fig. 2

5. SYNTHESIS OF THE VARYING CONDUCTIVITY

In a number of practical problems of radiophysics and applied electrodynamics there arises the problem of constructing electromagnetic screens and other more complicated devices with desired scattering properties. When solving these problems it is necessary to study synthesis problems for thin conducting screens. In many cases such a screen is a nonclosed surface of revolution excited by a plane electromagnetic wave propagating along the axes of revolution. Our aim is to create a distribution of conductivity on the screen surface which provides the minimum of back scattering (BS) at the frequency band

$$\sigma = \lim_{R \to \infty} 4\pi R^2 |E|^2 / |E^0|^2. \tag{5.1}$$

The level of BS given at the frequency band is not sufficient to reconstruct the distribution of conductivity on the surface. Therefore, in this case it is difficult to avoid the solution of the DP. Moreover, the application of the algorithm of the DP solution allows taking into account physical and constructive limitations for impedance. Therefore the solution can be represented as

$$\min \sigma = 4\pi / k_e^2 \langle Z,\ J \rangle^2, \tag{5.2}$$

where J is the current at the screen surface, Z is the FFP of J. To solve the DP we use the integral equation for the electric field

$$(\eta + G)J = e^0. \tag{5.3}$$

The distribution of surface conductivity along the meridian of the screen is taken in the form

$$\eta(t) = \sum_{m=1}^{N} \eta_m X_m(t),$$

where η_m are constants, and $\{X_m(t)\}$ is a basis at the meridian.
The solution of the minimization problem (5.2) is essentially simplified by the fact that the gradient of the functional (5.2) has an explicit form. Its components are

$$\frac{\partial \sigma}{\partial \eta_m} = 8\pi/k_e^2 \ \mathrm{Re}\ \{\langle Z, b_m\rangle \langle Z, J\rangle^*\}, \tag{5.4}$$

b_m: $(\eta + G)\ b_m = -X_m J$, $m = 1, \ldots, N$ (Eremin and Sveshnikov, 1983).
If the integral equation is solved by an iterative scheme then the iteration must be repeated $N + 1$ times (once for the determination of J and N times for the calculation of the gradient components). To avoid this procedure we solve the conjugate system (Eremin and Sveshnikov, 1988). Then the calculation of the gradient components is carried out by the following scheme:

$$\frac{\partial \sigma}{\partial \eta_m} = -8\pi/k_e^2 \mathrm{Re}\ \{\langle a, X_m J\rangle \langle Z, J\rangle^*\}, \quad m = 1, \ldots, N,$$

a: $(\eta + G)^T\ a = Z$.
In this case it is enough to use the iterative scheme only twice per one gradient step.

REFERENCES

Angell, T.S., Colton, D., and Kress, R. (1989). Far field patterns and inverse scattering problems for imperfectly conducting obstacles. *Math. Proc. Camb. Phil. Soc.* **106**, 553–569.

Colton, D. (1984). The inverse scattering problem for time-harmonic acoustic waves. *SIAM Reviw.* **24**, 324–350.

Colton, D. and Kress, R. (1987). *Integral Equation Methods in Scattering Theory.* John Wiley, New York.

Colton, D. and Monk, P. (1985). A novel method for solving the inverse scattering problem for time harmonic acoustic waves in the resonance region. *SIAM J. Appl. Math.* **45**, 1039–1053.

Colton, D. and Monk, P. (1987). The inverse scattering problem for time-harmonic acoustic waves in a penetrable medium. *Quart. J. Mech. Appl. Math.* **40**, 189–212.

Colton, D. and Sleeman, B. (1983). Uniqueness theorems for the inverse problem of acoustic scattering. *IMA J. Appl. Math.* **31**, 253–259.

Eremin, Ju.A. (1988). The problem of existence of an invisible scatterer in the diffraction theory. *Different Equat.* **24**, 684–687 (in Russian).

Eremin, Ju.A. and Orlov, N.V. (1988). Discrete sources method for the problems of diffraction on the body of revolution located in a dissipative halfspace. *Radiotechnica and Electronica* **32**, 2506–2514 (in Russian).

Eremin, Ju.A. and Sveshnikov, A.G. (1983). Provement of the nonorthogonal series method and the investigation of some inverse diffraction problems. *Zh. Vychisl. Mat. i Mat. Fiz.* **23**, 734–742 (in Russian).

Eremin, Ju.A. and Sveshnikov, A.G. (1987). About the existence of equivalent scatterers in the inverse diffraction problems. *Dokl. Akad. Nauk SSSR* **297**, 1095–1099.

Eremin, Ju.A. and Zakharov, E.V. (1987). About some direct and inverse diffraction theory problems In: *Integral Equation Methods in Scattering Theory.* Ed.by D. Colton and R. Kress. Mir, Moscow, pp. 290–309.

Eremin, Ju.A. and Sveshnikov, A.G. (1988). Conjugate equations using at the auxiliary sources method. *Zh. Vychisl. Mat. i Mat. Fiz.* **28**, 734–742 (in Russian).

Kirsch, A. and Kress, R. (1977). On integral equation of the first kind in inverse acoustic scattering In: *Inverse Problems.* Birkhauser, Basel, pp. 93–102.

Kirsch, A., Kress, R., Monk, P., and Zinn, A. (1988). Two methods for solving the inverse acoustic scattering problem. *Inverse Probl.* **4**, 749–770.

Lax, P. and Phyllips, R. (1967). *Scattering Theory.* Academic Press, New York.

Ramm, A.G. (1986). *Scattering by Obstacles.* Reidel, Dordrecht.

Sveshnikov, A.G., Eremin, Ju.A., and Chiviljov, A.V. (1979). Investigation of the uniqueness of a solution of an inverse diffraction problem. *Differentsial'nye Uravneniya* **15**, 2205–2209.

Tikhonov, A.N. and Arsenin, V.Ja. (1979). *Methods for Solution of Ill-Posed Problems.* Nauka, Moscow.

Ill-Posed Problems in Natural Sciences, pp. 515 – 524
A. Tikhonov (Ed.)
1992 VSP/TVP

THE CONSTRUCTION OF NUMERICAL ALGORITHMS FOR ILL-POSED PROBLEMS WITH RANDOM ERRORS IN INITIAL DATA

A.M. FEDOTOV
Computing Center, Siberian Branch of the Russia Academy of Sciences, Academgorodok, 660036 Krasnoyarsk, Russia

ABSTRACT

This paper is devoted to the theory of constructing in optimal approximate solution for the ill-posed problems with random errors in initial data and to the estimating of the accuracy of an arbitrary admissible method of solution. Linear and non-linear problems, the problem of estimating the value of a linear functional by the solution of the Main Problem for a linear operator by the measurements of a finite set of linear functionals of the initial data, the computational stability of algorithms are considered.

1. THE MAIN PROBLEM

The majority of ill-posed problems can be described in abstract form by the operator equation of the first kind

$$Ax = y, \tag{1}$$

where x is an unknown, y is a given elements of separable linear vector spaces X and Y, respectively, and A is a specified continuous one-to-one mapping with a dense range, acting from X into Y. Let us consider a problem of calculating the value of some linear operator G: $X \to Z$ on the solution of operator equation (1) as the *Main Problem*. For convenience X, Y, and Z are considered to be Hilbert spaces.

The results represented here are generalized to an arbitrary enough separable space X, the reflexive Banach space Y and the space Z of continuous functions with a compact storage medium.

The main results of the theory of constructing the approximate numerical algorithms are given in the book (Fedotov, 1990).

2. MEASUREMENTS

Let the exact equation (1) have a unique solution x_t, and initial data for the problem solution be obtained from the measurements of the exact value of initial data y_t, i.e., we know their approximate value

$$\tilde{y} = Py_t + \sigma u, \qquad y_t = Ax_t, \tag{2}$$

where $P: Y \to \widetilde{Y}$ is a linear operator of measurements, and an element $u \in \widetilde{Y}$ simulates errors in the specification of initial data, which accompany measurements, $\sigma > 0$ is a parameter, characterizing the noise power in the specification of initial data. As a model of errors in the initial data specification we consider realizations of some weak random value ξ with values in the space $\widetilde{Y}$ with a zero average and finite second moment

$$\mathbf{E}_\xi[u,\xi]_{\widetilde{Y}} = 0, \qquad \mathbf{E}_\xi[u,\xi][v,\xi]_{\widetilde{Y}} = [Ru,v]_{\widetilde{Y}}, \qquad \|R\|_{\widetilde{Y}} = 1$$

for every u, $v \in \widetilde{Y}$, where R is a correlative operator, $\mathbf{E}_\xi$ is a mathematical expectation on distribution μ_ξ of the random value ξ, $[\cdot,\cdot]_{\widetilde{Y}}$ is the inner product in $\widetilde{Y}$. Only the finite set of linear functionals of the value y_t is usually accessible as measurements thus the weak random value is the most adequate model of errors, which accompany measurements (Fedotov, 1990).

3. SOLVING PROCEDURES

Our approach to the construction of approximate solutions of ill-posed problems with random errors in initial data is based on the idea suggested by M.M. Lavrent'ev which consists in the substitution of the initial operator A by some operator, the choice of which is based on a priori specified criteria of the approximate solution quality. This approach is close to the approach, suggested by A. Wald (Wald, 1950) for constructing the theory of statistical solving functions. A separate set D of all mappings, acting from the measurement space $\widetilde{Y}$ into the solution space Z, suitable for the construction of approximate solutions of the Main Problem by initial data (2).

Definition 1. An arbitrary continuous mapping d: $\widetilde{Y} \to Z$ is called solving procedure (solving method) for the construction of the Main Problem solution (the approximate calculation of the value of an operator G on the solution of equation (1) with initial data (2)).
Note that the suggested definition of the solving procedure for the construction of the Main Problem approximate solution has the connection neither with equation (1) and operator G, nor with the model of errors in the specification of the initial data. In order to obtain such connection and to compare various solving procedures, one should take into account, that here the approximate solution of the Main Problem is discussed. The significant characteristic of an arbitrary solving procedure is its error, which is determined from the quality requirements of approximate solutions, which establish such connections. The quality of an arbitrary solving procedure is characterized by the following function:

$$\mathfrak{R}(x,d,A,G\mu_\xi) = \mathbf{E}_\xi\|Gx - d(Ax+\xi)\|_Z^2. \tag{3}$$

We call function (3) risk function of the solving procedure d according to the mathematical statistics terminology.
Constructing an approximate solution is aimed at choosing the solving procedure whose risk function (considered as a function of the argument x) is as small as

possible. Note that the problem of constructing approximate solutions thus formulated, becomes "well-posed" in the sense of the continuous dependence of the risk function on its arguments (see (Fedotov, 1990)). Unfortunately, the choice of the best solving procedure meets a number of unsurmountable difficulties, connected with ordering the risk functions, since on the set of all risk functions considered as functon of an argument x, one can introduce only the relationship of the partial order and in the majority of cases it's not clear, which function should be considered as the less one.

4. A PRIORI INFORMATION

The possible solution can be found in the introduction of an ordering on the set of all risk functions, using a priori information. In the majority of cases a priori information is specified as an unknown exact solution belonging to some subset of the space X ($x \in W \subset \mathrm{Dom}\,(A)$), for example, the most frequent W is a compact nested subset. Taking into account of the fact that the exact solution belongs to some limited subset we represent a priori set in the form:

$$W = \bigcup_{\rho>0} \rho W_1 = \bigcup_{\rho>0} W_\rho,$$

where W_1 is some absolute convex limited set, which is called generating,

$$W_\rho = \left\{x \in X\colon\ \frac{1}{\rho}x \in W_1\right\}.$$

Note that a form of the optimal solving procedure essentially depends on the form of the generating set.
Using the correctness set (a priori set) $W \subset X$, we introduce a basic restriction on the Main Problem, the error model in the specification of the initial data being:

$$A[W] \subset \mathrm{Ran}\,(R^{1/2}) \tag{4}$$

and having the essential influence of errors on data (see (Fedotov, 1990)).

5. THE ERROR FUNCTIONAL

The order relationship can be chosen, prescribing on the whole set of risk functions, considered as functions of an argument $x \in W$, some functional of the mean value type — the error functional of the solving procedure

$$\Omega(d, A, G, \mu_\xi, W) \equiv \Omega(\mathfrak{R}(\cdot, d, A, G, \mu_\xi)).$$

From our point of view the order relationship, specified by the functional of the mean value type, corresponds mostly to the intuitive representation of the best solving procedure. As an example of the error functional consider the following:

$$\Omega_\rho(d) = \sup\left\{\mathfrak{R}(x, d, A, G, \mu_\xi)\colon x \in W_\rho\right\}. \tag{5}$$

6. THE OPTIMAL SOLVING PROCEDURES

Let Ω be some functional of the mean value type (the error functional), defined on the whole set of risk functions of the Main Problem solving procedures. Then we can introduce the notions of optimal and optimal by the order solving procedures with respect to this functional.

Definition 2. A solving procedure d_0 (if it exists) is called optimal, with respect to the error functional Ω, if

$$\Omega(d_0) \equiv \Omega\big(\mathfrak{R}(\cdot, d_0, A, G, \mu_\xi)\big) = \inf_{d \in D} \Omega(d).$$

By analogy a solving procedure optimal by the order is defined as the procedure d_{00} for which

$$\Omega(d_{00}) \leq \text{const} \inf_{d \in D} \Omega(d).$$

Solutions, optimal with respect to functional (5), are called minimax.
The problem of constructing approximate solutions (the choice of solving procedures) for ill-posed problems thus stated has an advantage namely, that with the general enough assumptions concerning the Main Problem and the chosen functionals of mean value type, the risk functions and error functional are continuous functions of their arguments (see (Fedotov, 1990)). This allows us to construct stable numerical algorithms for the solution of the ill-posed problems.
An optimal continuous solving procedure does not always exist. Therefore, by constructing an algorithm of an ill-posed problem approximate solution we mean the construction of a sequence of continuous solving procedures, close to the optimal in the sense of error. This approach allows us to consider the problem of constructing the approximate solutions of the Main Problem as well-posed (see (Fedotov, 1990)). Note that for the linear equation and for solving procedures of the linear class such sequence should be constructed within the class of linear operators of Hilbert–Schmidt type.

7. LINEAR PROBLEM

There is no necessity to dwell upon the construction of the optimal solving procedures with respect to some criterion. The theory of their construction is given in detail in the book (Fedotov, 1990). As an example, consider the problem with a linear operator A.

THEOREM 1. *The optimal linear solving procedure is optimal by order with respect to any functional of the error with constant* 3, *i.e.*,

$$\inf\big\{\Omega(L)\colon L \in \mathfrak{L}(\widetilde{Y}, Z)\big\} \leq 3 \inf\big\{\Omega(d)\colon d \in D\big\}.$$

Condition (4) allows us to consider an information operator of the Main Problem

$$F = (PA)^* R^{-1} PA\colon X \to X.$$

The optimal linear solving procedure has a form

$$L_{\text{opt}} \equiv L(\alpha, B_\alpha) = GB_\alpha(\alpha E + FB_\alpha)^{-1}(PA)^* R^{-1}, \tag{6}$$

where $\alpha = \sigma^2/\rho^2$, B_α: $X \to X$ is the solution of some extreme problem, which depends on the set W_1, specifying a priori information, and on the error functional: $B_\alpha = B_\alpha^*$, $\text{tr}\,(B_\alpha) \le 1$,

$$B_\alpha = \arg\max \{\gamma(B) \colon B \in T_\Omega(W_1)\}, \tag{7}$$

where $\gamma(B) = \text{tr}\,(GB(\alpha E + FB)^{-1}G^*)$, $T_\Omega(W_1)$ is a subset of the set $T(W_1)$, depending on the form of functional Ω, for example, for the error (5) functional we put $T_\Omega(W_1) = T(W_1)$. The set of operators $T(W_1)$ is the absolute convex shell $\mathfrak{A}(\{S_x\}_{x \in W_1})$ of operators of the unit rank $S_x u = [u, x]x$. The error of the optimal linear solving procedure

$$\Omega(L_{\text{opt}}) = \max \{\gamma(B) \colon B \in T_\Omega(W_1)\} = \gamma_\Omega.$$

Note that solving procedures of type (6) have stability, i.e., if $\{B_n\}_{n=1}^\infty \subset T_\Omega(W_1)$ is a sequence of operators, which maximize the function $\gamma(B)$, then the sequence of solving procedures $\{L(\alpha, B_n)\}_{n=1}^\infty$ minimizes the corresponding error functional, and

$$\Omega(L(\alpha, B_n)) - \Omega(L_{\text{opt}}) = \gamma_\Omega - \gamma(B_n).$$

Solution (7) of the extreme problem is unique, if the information problem operator F is the operator of full rank.

8. REGULARIZATION

It is easy to see that solving procedure (6) coincides in form with the solving procedure, obtained from Euler equation for Tikhonov functional, therefore, there arises a question of the possibility of its application for the construction of regularizing algorithms.

THEOREM 2. *For every $B \in T(W_1)$ and any dependence $\alpha = \alpha(\sigma)$ such that $\alpha(\sigma) \to 0$ with $\sigma \to 0$, $\alpha(\sigma) \ge \text{const}\,\sigma^2$ the family of solving procedures $L(\alpha, B)$ is reqularizing, i.e.,*

$$\lim_{\sigma \to 0} \mathfrak{R}\Big(x, L(\alpha(\sigma), B), A, G, \mu_\sigma\Big) = 0.$$

The parameter α can be chosen a priori from the conditions of the theorem or by a statistical version of the error method from the equation

$$E_M(\alpha) = \frac{1}{M} \sum_{m=1}^{M} [PAx_\alpha - \tilde{y}, \varphi_m]^2 = \lambda\sigma^2,$$

where $\{\varphi_m\}$ is the sequence of eigenvectors of the correlation operator R, $\lambda \in (1, 2]$. The best results of computer experiments have been obtained with $\lambda = 1.7$.

9. COMPUTATIONAL STABILITY

The main computational difficulties, arising in the course of solution, especially with small noise levels in the specification of the initial data, are mainly connected with the rounding-off errors in the inversion of ill-posed matrices. In many cases, such as the problem of analytical continuation, even the usage of regularization doesn't come to the rescue. The traditional solution is to use a singular expansion. In the approach proposed however it's possible to avoid this complicated procedure, when we use some gradient method for the solution of extreme problem (7).
Consider the method of quickest gradient, which leads to the following iterative procedure of the problem (7) solution:

1) $B_0 = S_{x_0}, \quad x_0 = \arg\max\{\gamma(S_x)\colon x \in W_1\}$;

2) $x_{n+1} = \arg\max\left\{\|(\alpha E + FB_n)^{-1}x\|\colon x \in W_1\right\}$;

3) $a_{n+1} = \arg\max\left\{\gamma((1-a)B_n + aS_{x_{n+1}})\colon a \in [0,1]\right\}$;

4) $B_{n+1} = (1 - a_{n+1})B_n + a_{n+1}S_{x_{n+1}}$.

Therefore, at each step the extreme operator is a finite convex combination of the unit rank operators, i.e., we have a natural finite dimensionalization of the problem. Thus, by stopping the iterative process of the function $\gamma(B)$ maximization as the mantissa is exhausted, we obtain a numerically stable process of the Main Problem solution. This process will be represented in detail below by the example of an analytical continuation problem.

10. NONLINEAR CASE

Consider equation (1) with a Gateaux-differentiable operator A. We will define the information operator for this problem

$$F(x) = \big(P\partial A(x)\big)^* R^{-1} P\big(\partial A(x)\big), \tag{8}$$

where $\partial A(x)$ is Gateaux derivative of operator A at the point x. Let the unknown exact solution of the problem be represented in the form

$$x = x_0 + \varepsilon x_\varepsilon,$$

where ε is a small enough number. Then due to the continuity of risk function and of error functional the linear solving procedure of the form

$$L(x_0) = GB_\alpha\big(\alpha E + F(x_0)B_\alpha\big)^{-1}(P\partial A))^* R^{-1}$$

gives a solution close to the optimal. Here B_α is the solution of extreme problem (7) with information operator (8).

THEOREM 3. *If the Gateaux derivative of an operator A is Lipschitz-continuous, then the optimal solving procedure d_{opt} exists and for small enough $\varepsilon \le \varepsilon_0$*

1) $\mathbf{E}_\xi \|d_{\text{opt}}(\tilde{y}) - L(x_0)\tilde{y}\|_Z \le C_1\varepsilon$;

2) $\Omega(L(x_0)) - \Omega(d_{\text{opt}}) \le C_2\varepsilon^2$.

3) *The iterative process of gradient type*

$$\begin{aligned} x_{n+1} &= x_n + K(x_n)(\tilde{y} - PA(x_n)),\\ K(x) &= B_\alpha\big(\alpha E + F(x)B_\alpha\big)^{-1}\big(P\partial A(x)\big)^* R^{-1} \end{aligned}$$

converges to the optimal solution with the estimate of the convergence velocity

$$\mathbf{E}_\xi \big\|d_{\text{opt}}(\tilde{y}) - Gx_n(\tilde{y})\big\|^2 \le C_3 q^{2n},$$

where $q < \min\{\alpha\|(\alpha E + F(x)B_\alpha)^{-1}\|,\ \|B_\alpha(\alpha E + F(x)B_\alpha)^{-1}\|\} \le 1$.

Note that, unfortunately, the optimal solving procedure is not always a continuous mapping.
The iterative solving procedure constructed is regularized with any $B \in T(W_1)$ and $\alpha = \alpha(\sigma) \ge \text{const}\,\sigma^2$, $\alpha(\sigma) \to 0$ the parameter α can be chosen from the statistical error equation

$$E_M(\alpha) = \lambda\sigma^2 + \nu(A, G, P), \tag{9}$$

where $\nu(A, G, P)$ is the measure of the problem incompatibility.

11. ALGORITHMIC UNIQUENESS

One of the important questions, which arises while solving inverse problems, is the question of the uniqueness of the Main Problem solution. A sufficient condition for the uniqueness of the Main Problem solution is the nongeneration of the information operator, but in the majority of practically important cases this condition is not satisfied because, as a rule, the space $\tilde{Y}$ is finite-dimensional (the Euclidean space E_M). For a finite-dimensional space $\tilde{Y}$ in the linear case the necessary and sufficient condition is the condition $\text{Ker}\,(F) \cap W_\rho \subset \text{Ker}\,(G) \cap W_\rho$ which is difficult to check.

THEOREM 4. *The following conditions are eguivalent:*

1) *The solution of the Main Problem is unique;*

2) $\lim\limits_{\sigma \to 0} \Re\Big(x,\ d(\alpha(\sigma), B_{\alpha(\sigma)}),\ A,\ G,\ \mu_\sigma\Big) = 0;$

3) $\lim\limits_{\sigma \to 0} d\big(\alpha(\sigma), B_{\alpha(\sigma)}\big)PAx = Gx;$

where $d(\alpha(\sigma), B_{\alpha(\sigma)})$ *is the optimal solving procedure.*

This theorem gives a fundamental possibility for algorithmic check of the uniqueness conditions: the sufficiency of the information, contained in initial data, for the uniqueness of the problem solution. For the practical use Theorem 4 can be reformulated in the following form:

THEOREM 5. *Let* $\tilde{Y} = E_M$ *and* $Z = E_N$ *be finite-dimensional euclidean spaces,* $R = E$, $d_{MN}(\alpha)$ *is the optimal solving procedure. Then for every* $x \in W_\rho$, *the following conditions are eguivalent:*

1) *The solution of the Main Problem is unique;*

2) $\lim_{M\to\infty} \mathfrak{R}(x, d_{MN}(\alpha), A, G, \mu_{\sigma M}) = 0;$

3) $\lim_{M\to\infty} d_{MN}(\alpha(M))P_M Ax = Gx.$

The theorem remains true, if the parameter α is chosen from statistical error equation (9).
As an example, we cite a problem of restoring interfaces with the acoustic or electromagnetic sounding. The problem is solved in small, i.e., when the surface is known up to some small component. For the purely reflecting surface of interface the convergency is observed with one disturbance source. For the surface with impedance condition it is observed with three sources; with one source the divergence is observed: with two sources the convergence is present in some cases, in other cases it is absent.

12. LINEAR FUNCTIONALS

Consider the problem of estimating the linear functional value by the solution of equation (1) with a linear operator A by measurements of the finite set of linear functionals of the initial data: Z is a one-dimensional Euclidean space, $G(x) = [g, x]$, $\widetilde{Y} = E_M$ is a finite-dimensional Euclidean space. For convenience consider only error functional (5).

THEOREM 6. *For the problem of estimating the linear functional value by measurements of a finite set of linear functionals the optimal solving procedure exists in class D and*

1) *if $\sigma = 0$, then the linear procedure exists among optimal solving procedures;*

2) *if $\sigma > 0$, then the optimal linear solving procedure is optimal by order with constant not greater than 2.*

The optimal linear solving procedure has a form

$$L_{\text{opt}} \equiv L(\alpha, x_0) = G(x_0)/(\alpha + [Fx_0, x_0])\, R^{-\frac{1}{2}} F^{\frac{1}{2}} x_0, \tag{10}$$

where $x_0 \in W_1$ is the solution of the following extreme problem

$$x_0 = \arg\max \{\omega(x) \colon x \in W_1\}, \tag{11}$$

$$\omega(x) = G^2(x)/(\alpha + [Fx, x]). \tag{12}$$

Unfortunately, the solution of the extreme problem (11) is not unique, i.e., there exist several optimal linear solving prosedures. The possibility to use numerical methods for the solution of problem (11) follows from the fact, that each local maximum of the function (12) is global both for any maximizing sequence $\{x_n\}$ for function (12) $L(\alpha, x_n) \to L_{\text{opt}}$ in E_M^* and as the error (5) functional.
Let $\{e_m\}$ be an arbitrary basis in the space E_M; g_w be a vector g projection on the subspace X_W, generated by $\text{Lin}(W)$; $\{q_m\}_{m=1}^{M} \subset X^*$ be the set of functionals of the form $q_m(x) = [PAx, e_m]_E = [q_m, x]_X$.

THEOREM 7. *For the problem of estimation of the linear functional value by measurements of the finite set of linear functionals for* $\alpha = \alpha(\sigma) \geq \text{const}\,\sigma^2$ *and* $\alpha(\sigma) \to 0$ *with* $\sigma \to 0$

$$\lim_{\sigma \to 0} \Omega_\rho\Big(L(\alpha(\sigma), x_0),\, A,\, g,\, \mu_\sigma,\, W_\rho\Big) = \rho^2 \|g_W^\perp\|^2,$$

where $g_W^\perp$ *is the orthogonal complement to the vector* g_W *projection on the subspace* X_M*, generated by* $\text{Lin}\,(\{q_m\})$.

THEOREM 8. *For the infinite set of functionals* $\{q_m\}_{m=1}^{\infty}$ *such that* $g_W \in C_0\,\text{Lin}\,(\{q_m\})$ *and* $R_M = E$*, for every* $x \in W_\rho$*, with* $\alpha = \sigma^2/\rho^2$ *the following statement is true*

$$\lim_{M \to \infty} \Re\big(x,\, L_M(\alpha, x_0),\, A,\, g,\, \mu_M\big) = 0.$$

Note that the theorem remains true while choosing the parameter α from the statistical error equation.

13. COMPUTATIONAL STABILITY OF AN ESTIMATION ALGORITHM

Consider the computational stability of an algorithm within the conditions of Theorem 8. Function (12) in this case is

$$\omega_M(x) = g^2(x)/\big(\alpha + \sum_{m=1}^{M} q_m^2(x)\big), \tag{13}$$

and its maximum is achieved on the finite-dimensional subspace X_M, generated by $\text{Lin}\,(\{q_m\}_{m=1}^{M})$. Let $x_{0M} \in X_M$ be a maximum of function (13), then the estimate takes the form

$$\hat{g}(x) = \sum_{m=1}^{M} a_m \tilde{y}_m, \quad a_m = g(x_{0M})/\big(\alpha + \sum_{m=1}^{M} q_m^2(x_{0M})\big) q_m(x_{0M}).$$

The calculation errors occur when calculating the sum and functionals, included into it. One can check calculation errors, choosing "the most informative" subbasis in X_M, in which function (13) is maximized with the desired accuracy, or one can solve the extreme problem approximately, which is equivalent. Besides, as the check measure the accuracy of the difference $\|g_W\|^2 - \|g_M\|^2 \equiv \|g_W\|^2 \sum_{m=1}^{M} \hat{a}_m q_m = \|g_W^\perp\|^2$ calculation could serve, the calculation of the latter value for the majority of problems is simple enough and is practically reduced to the definition of mantissa length, from which the calculations begin.

14. ANALYTICAL CONTINUATION

A problem of the analytical continuation from discrete sets is formulated as follows: let $f(z)$ be a function of a complex variable, analytical in the domain D of the complex plane, from class $H_2(D)$ which is known on the discrete set of points $U_N = \{z_n\}_{n=1}^N \subset D$. We should define the value of the function $f(z)$ at a point $z_0 \in D$, which does not belong to the set U_N. The value of function f at the point $z_n \in D$, $n = 0, \dots, N$ can be represented in the form:

$$q_n(f) = f(z_n) = \frac{1}{2\pi i} \int\limits_{\partial D} \frac{f(\zeta)}{\zeta - z_n} d\zeta = \langle q_n, f \rangle.$$

The optimal minimax solving procedure always exists on the set $W_\rho = \{f \in H_2(D)\colon \|f\| \leq \rho\}$ and can be found analytically (without solving extreme problem (11))

$$L_{N,\mathrm{opt}} = \left(\sigma^2/\rho^2 E + Q\right)^{-1} Q_0. \tag{14}$$

If points $\{z_n\}_{n=0}^N$ belong to the circle of unit radius, then the components of the matrix Q and the vector Q_0 are respectively

$$Q_{n,m} = 1/(1 - z_n \bar{z}_m), \qquad Q_{0,n} = 1/(1 - z_n \bar{z}_0).$$

Thus, with the absence of errors we obtain the formula, based on Blashke products. In spite of the regularization, calculations by formula (14) are not stable. The construction of the stable algorithm is possible either by the singular expansion of the matrix Q, seeking the definition of the most informative basis for the approximation of the functional q_0 by the family of functionals $\{q_m\}_{m=1}^N$, or by the approximate solution of the extreme problem (12), taking account the length of mantissa.

REFERENCES

Fedotov, A.M (1990). *Ill-Posed Problems with Random Error in the Data.* Nauka, Novosibirsk (in Russian).

Wald, A. (1950). *Statistical Decision Function.* John Wiley, New York.

Ill-Posed Problems in Natural Sciences, pp. 525 – 534
A. Tikhonov (Ed.)
1992 VSP/TVP

A UNIFIED APPROACH TO THE CREATION OF KNOWLEDGE-BASED SYSTEMS FOR IDENTIFICATION OF SOME CLASSES OF NONSTATIONARY PROCESSES

V.Ya. GALKIN and V.A. KARPENKO
Moscow State University, Department of Computational Mathematics and Cybernetics, Statistical Simulation Laboratory, Len. Gory, 119899 Moscow, Russia

ABSTRACT

Problems of creation of expert systems in the field of stochastic analysis of nonstationary systems are considered. It is shown how knowledge-based systems can be used to solve the identification problems for nonstationary processes, that arise in various applications. The architecture of such intellectual systems is discussed. Special attention is paid to the algorithms of identification of nonstationary processes on the basis of a priori information, which are the source of the knowledge base of expert systems in the applied field under consideration.

1. INTRODUCTION

One of the principal problems in many natural sciences and applications is the identification problem for time-varying processes and systems. By such an identification problem we mean the possibility of determine required parameters of such evolutionary systems on the basis of given data and some supplementary information about their characteristics. These problems appear in such applied fields as mechanics, acoustics, speech processing, medicine, radio- and hydrolocation, etc.
A large number of different methods and algorithms for identification of characteristics of time-varying systems were suggested during last decades (Maople, 1987; Ahmed, 1988). Based on these mathematical methods, many interesting applied program packages were created to be used both in research and in different applications for the solution of specific practical problems.
However, all these methods are, as a rule, based on the assumption of stationarity of the process under study, whereas in practice this assumption is in many cases true only within bounded time intervals or is not valid at all. So the problem of creation of adaptive algorithms adjusted to the analysis of some classes of nonstationary processes, different from each other by the types of a priori information, is very urgent. But even within the chosen class the degree of uncertainty (Franks, 1969) is rather high to choose the analysis methods and parameters of the system

under investigation. These facts explain the importance of the multivariant analysis of nonstationary systems. This, together with the applied orientation of the problems mentioned above, calls for using expert systems (ES) in the subject field (SF) of stochastic analysis of evolutionary systems (SAES).
In § 2 of the present paper we justify the possibility and reasonability of the ES construction in the field of SAES. Then we analyze specific features of ES in the field of statistics and time series analysis, providing the possibility to take into account the available a priori information.
In § 3 and 4 we discuss principles of knowledge base construction of such expert systems and present the original procedures, which constitute their foundation.

2. EXPERT SYSTEMS AND ANALYSIS OF NONSTATIONARY STOCHASTIC SYSTEMS

The most significant achievement in the field of Artificial Intelligence during recent years was the development of powerful computer systems, the so-called "expert" systems. The base of such systems is constituted by programs for representation of real knowledge from specific fields and its application to the solution of problems. Same as a human expert, these systems use symbolic logic and heuristics-empirical rules for seeking a solution. Same as human experts, they can make mistakes. However, expert systems have the ability to learn from their own mistakes. At the same time, in contrast to the human expertise, such an artificial expertise is noncontradictory, easy transferred, documented and specified, and also have more robust and reproducible results. As for the ES cost, such systems are rather cheap in operation, whereas human experts, especially the high-skilled ones, cost a pretty penny.
It is clear, however, that ES development costs are much lager, than the creation of usual applied program packages. Thus, before proceeding to the creation of such system it is necessary to answer the questions about possibility and justification of its development (Waterman, 1986).
For this purpose, let us consider shortly the specific features of the SAES procedures. First of all, these procedures are multistage ones and have repetition character. Each of the analysis stages supposes the necessity of the choice of methods and parameters of processing the system. In the case of nonstationary systems analysis there are two kinds of such repetition: on the one hand, because the process parameters vary in time, it is necessary to analyze this process on the set of the consequent homogeneity intervals (Galkin and Karpenko, 1991) characterized by the property that these parameters are constant or change in accordance with a known rule; on the other hand, for the quality improvement of obtained estimates, it is advisable to play through different processing scripts for investigated system. In all the cases mentioned above the decision can not be made on the basis of versatile and completely justified mathematical criteria, though in some cases such criteria are used essentially. It makes necessary to utilize the heuristic approaches to the solution of identification problems for the processes under study. At the same time the formulations of such problems are similar and quite clear, their solution requires intellectual skills, and the solution methods of different classes of problem may be described by experts. In view of what was said above, it is

possible to create the ES in the applied field of SAES (Waterman, 1986).

The justification of the ES development in this SF consists in the fact that expert's experience may and should be utilized in many applications by users, having only superficial knowledge in statistics and time series analysis area (Waterman, 1986). Besides, such experience may be lost.

3. ES ARCHITECTURE FOR THE ANALYSIS OF NON–STATIONARY STOCHASTIC SYSTEMS

As the majority of expert systems, ES in the field of SAES should include the following components:

1) a data base;

2) a knowledge base;

3) a means of explanation and inference.

The foundation of the first element consists in the use of models of stationary and nonstationary processes, possible forms of the given a priori information, etc. The core of each ES is its knowledge base. For the expert systems in the considered SF it should include the methods of classical and modern time series analysis in the form of libraries or sub-packages of applied programs (Marple, 1987), algorithms for the data class identification (stationarity, garmonicity, etc.) (Bendat and Piersol, 1986), and specific algorithms for the nonstationary processes analysis. The latter component of knowledge base will be discussed in details in the §4.

The majority of ES in the considered SF should be rule based, as it seems to us, because the procedures of evolutionary system analysis, as mentioned above, have multistage character, and within bounds of each stage the algorithmic way of problem solution is possible, and we can guess beforehand the branching points (Waterman, 1986).

Let us note also that the adaptation of methods of time series analysis to the different applications is restrained by the fact that there are many users with rather low qualification in this mathematical field. Frequently they use existing analysis methods improperly and this leads to unpredictable results. So it would be highly desirable, that ES in the considered SF simplify the usage of statistical methods in different applications for the account of intellectualization of statistical processing procedures. For supporting such a regime the explanation and inference means should be constructed accordingly. It should include backward chains demonstration, give the opportunity to alter the analysis script when user is unsatisfied by its results (Matsuyama, 1989), and to repeat the analysis procedures in accordance with different scripts. Rather frequently ES are able to offer software means for the justification of usage of particular methods and strategies, based on the expert knowledge of professionals. When designing such expert systems it is reasonable to envisage repeated reference to choose only those facts, which have been changed during previous stage (Mettrey, 1991).

4. KNOWLEDGE BASE FEATURES OF EXPERT SYSTEMS FOR THE ANALYSIS OF NONSTATIONARY PROCESSES

One of the main features of expert systems for the analysis of nonstationary systems is the necessity to include the specific algorithms in the knowledge base, taking into account the nonstationarity type and allowing to obtain robust estimates of the required parameters. At present time, as a rule, such accountment of nonstationarity is produced by adaptation of processing parameters for the analyzed process empirically on the base of researcher's experience separately in every particular problem.
However in some cases one can propose unified algorithms for nonstationary systems analysis, based on the a priori information about their characteristics. It is difficult to choose a particular approach for the analysis of such systems because the bounds of different nonstationary classes are rather fuzzy. This confirms that it is desirable to use utilizing of different heuristic rules for the solution of identification problem on different classes of nonstationary systems.

A. Interpretation conditions of spectral estimation results of some nonstationary systems
In this section we shall consider an approach to the analysis of one class of nonstationary systems, represented in the form

$$u(t) = \sum_{j=1}^{N} B_j(t) \exp\{i\varphi_j(t, \overline{\theta}_j)\} + n(t), \tag{1}$$

where N is the number of quasiharmonical components, $\overline{\theta}_j$, $j = 1, \ldots, N$, are vectors of parameters, $\varphi_j(t, \overline{\theta}_j)$ are known phase functions, $B_j(t)$ are unknown amplitude terms, $n(t)$ is the noise component.
The analysis of such systems is based on two main assumptions: a) the amplitude and frequency functions $\omega_j(t, \overline{\theta}_j) = \varphi'_j(t, \overline{\theta}_j)$ are slowly varying in time functions; b) the components $u(t)$ either depend on the common vector of parameters or the phase functions are equal for different indices, but $\overline{\theta}_j \neq \overline{\theta}_m$ if $j \neq m$.
The identification problem consists in estimation of the number of components N, the parameters $\overline{\theta}_j$ and the form of functions $B_j(t)$ by the given realization. Its solution is based on the averaging method. It implies approximation of (1) on the set of consequent intervals $I_k = [t_k - h_k/2,\ t_k + h_k/2]$, $k = 1, 2, \ldots$, by the polyharmonical functions

$$\tilde{u}(t_k; \tau) = \sum_{j=1}^{N} B_j(t_k) \exp\{i[\varphi_j(t_k, \overline{\theta}_j) + \varphi'_j(t_k, \overline{\theta}_j)\tau]\}, \tag{2}$$

where $B_j(t_k)$ are supposed to be constants when $\tau \in [-h_k/2, h_k/2]$. Further, more using methods of spectral analysis we can obtain the sequence of estimates $B_j(t_k)$ and $\varphi'_j(t_k, \overline{\theta}_j)$. This sequence permits to estimate $\overline{\theta}_j$. Let us note that the choice of spectral estimation methods may be performed by charge the expert system.

The main problem, appearing when the above approach is used, is connected with the estimation of the lengths of I_k. Usually it is solved empirically, and that leads to the unrobustness of the obtained estimates. We propose a knowledge based approach to the solution of this problem, which provides interpretation of the spectral estimation results on the basis of available a priori information about the analyzed process. We shall suppose that this information is given in the form of regions Ω_j of possible values of $\overline{\theta}_j$.

Let us note, however, that frequently a priori information is given in the form of Table 1.

In this table the column P means the confidence level of the following event: $\overline{\theta}_j$ belongs to Ω_j, Q means confidence level of the opposite event. Then the interpretation conditions, defining restrictions on admissible values h_k of the lengths of stationarity intervals, are also obtained with a certain confidence level. This level may be estimated on the basis of fuzzy set theory tools (Kandel and Schneider, 1989).

To obtain reliable estimates of the required parameters by the spectrums, it is necessary to choose the spectral processing parameters on the basis of interpretation conditions, including the following ones:

1) compatibility of accepted polyharmonical model (2) with analyzed process (1);
2) resolution of the spectral peak, used for parameters estimation of each component, from remained spectral peaks.

Compatibility conditions may be written in the form (Galkin et al., 1989)

$$W_j(t_k, \overline{\theta}_j, h_k) < C/h_k,$$

where

$$W_j(t_k, \overline{\theta}_j, h_k) = \max_{t \in I_k}\{f_j(t, \overline{\theta}_j)\} - \min_{t \in I_k}\{f_j(t, \overline{\theta}_j)\},$$

and $f_j(t_k, \overline{\theta}_j)$ is the frequency of the j-th component of polyharmonical model (2). The value of C is defined only by the type of the used spectral estimation method (SEM).

Resolution conditions may be represented in the form (Galkin et al., 1989) $F_l(t_k) > K/h_k$, where

$$F_l(t_k) = \min_{\substack{1 \le j \le N \\ j \ne l}} \min_{\substack{\overline{\theta}_j \in \Omega_j \\ \overline{\theta}_l \in \Omega_l}} |f_l(t_k, \overline{\theta}_l) - f_j(t_k, \overline{\theta}_j)|,$$

and K, as C, is also defined by the SEM type.

Interpretation conditions allow to obtain those regions of parameters localization, where it is possible to get reliable estimates of required parameters.

Interpretation region is defined by the relationship (Galkin et al., 1989)

$$\Omega_{in}(t_k, h_k) = \Omega_c(t_k, h_k) \cap \Omega_r(t_k, h_k),$$

where

$$\Omega_r(t_k, h_k) = \prod_{l=1}^{N} \Omega_{rl}(t_k, h_k), \quad \Omega_c(t_k, h_k) = \prod_{j-1}^{N} \Omega_{cj}(t_k, h_k),$$

$$\Omega_{rl}(t_k, h_k) = \{\bar{\theta}_l \in \Omega_l\colon\ F_l(t_k) > K/h_k\}, \quad l = 1, \ldots, N,$$

$$\Omega_{cj}(t_k, h_k) = \{\bar{\theta}_j \in \Omega_j\colon\ W_j(t_k, \bar{\theta}_j, h_k) < C/h_k.$$

Let us note, that compatibility conditions and resolution conditions are contradictory from the point of view of the choice h_k: when h_k increases, the measure $\Omega_r(t_k, h_k)$ also increases, but at the same time the measure $\Omega_c(t_k, h_k)$ decreases. One of the ways to expand the interpretation region is to calculate spectra for several h_{km} for each k. Then the resulting interpretation region may be written in the form

$$\Omega_{in}(t_k) = \bigcup_m \Omega_{in}(t_k, h_{km}).$$

In conclusion of this section we observe that the approximating model may differ from polyharmonical model (2). Instead of the decomposition of the phase functions in the Tailor series up to the second term, it is possible to use the polynomial approximation containing an arbitrary number of terms. Such approach allows to extend significantly the interpretation region and to improve the quality of obtained estimates of parameters (Galkin et al., 1990). Under such approach the identification problem is solved on the basis of dynamic spectral transforms of the second or higher orders (Gerlach et al., 1989), but not on the basis of traditional periodogram method (Brillindger, 1975).

B. Stationarity and homogeneity conditions for some classes of nonstationary stochastic systems

In contrast to the stationary case, in the analysis of nonstationary processes it is not possible to offer universal approach, similar to the periodogram method, which would be indifferent to the analyzed process type. So it is necessary to solve the identification problem of system's characteristics within the bounds of some model. Such model may be either one of wellknown models (autoregression (AR), autoregression-moving average (ARMA), polyharmonal), or their generalizations (AR model with time-varying parameters (Grenier, 1983), quasiharmonical model with polynomial phases (Gerlach et al., 1989), etc.). Models with time varying coefficients describe some nonstationary systems, so their spectral characteristics also will be time dependent. This fact shows the necessity of introduction of the notion of "evolutionary spectra". In (Loynes, 1968) the requirements on spectral function were given (nonnegativity, convexity, fulfillment of Wiener–Khinchin theorem (Brillindger, 1975), etc.). However in the nonstationary case, apparently, there is no spectral function satisfying all these requirements, so that spectrum notion should be chosen which is most convenient in the considering problem (Loynes, 1968).

We shall consider the notion of "evolutionary" spectra $F(t, \omega)$ of the process $X(t)$ introduced in (Priestley, 1981) which satisfies the nonnegativity property. Thus, $F(t, \omega)$ is written in the form

$$R(t, t) = \int_T dF(t, \omega), \tag{3}$$

where $R(s,t) = E[X(s)\ X^*(t)]$, T is a set: $T \subseteq \mathbf{R}$. We shall obtain a uniquely defined estimate of "evolutionary" spectrum on the class of generalized oscillatory processes (Karpenko, 1990), belonging to the Karhunen class (Rao, 1985)

$$X(t) = \int_T A(t,\omega)\exp\{i\omega t\}dZ(\omega), \tag{4}$$

where $Z(\omega)$ is the orthogonal process, $A(t,\omega)$ is a Borel function on $\mathbf{L}^2(T,\cdot)$, $T = \mathbf{R}$ in the continuous case, $T = [-\pi,\pi]$ in the discrete one, and the functions $A(t,\omega)$ satisfy an a priori relationship in the form

$$A(t+\tau,\omega) = \sum_{j=1}^{p} a_j(t)A(t+\tau-j,\omega) + c(t,\omega), \tag{5a}$$

$$c(t,\omega) = \sum_{l=-p_1}^{p_1} c_l(t)\exp\{-il\omega\}, \tag{5b}$$

where p is the markovian order of the considered process, p_1 is a positive number, $a_j(t)$, $j = 1,\dots,p$, $c_l(t)$, $l = -p_1,\dots,p_1$ are smooth enough functions of t.
In the identification of models (5a)–(5b) it is convinient to represent $a_j(t)$ and $c_l(t)$ in the form of finite weighted sums of orthogonal functions (Grenier, 1983) (polynomials, complex exponents, etc.). Then the problem arises of found a sequence of intervals I_k, such that the analyzed process may be considered to be close to the model process in some metric within the bounds of these intervals.
To simplify further considerations and to avoid bulky computations we shall examine this problem when $c(t,\omega) = 0$ and assume that the model process may be written in the form

$$X_M(t) = \int_T B(t,\omega)\exp\{i\omega t\}dZ(\omega),$$

$$B(t+\tau,\omega) = \sum_{j=1}^{p} b_j(t)B(t+\tau-j,\omega),$$

$$b_j(t) = \sum_{k=0}^{q_j} b_{jk}f_k(t), \quad j = 1,\dots,p, \tag{6}$$

where q_j are fixed natural numbers, b_{jk} are unknown coefficients of the decomposition of (6) by orthogonal functions $f_k(t)$.
Let the a priori information about $a(t)$ is given in the form

$$\varepsilon_{01} \le a_j(t_q - h_q/2) \le \varepsilon_{02}, \qquad \varepsilon_k^{(1)} \le |a_j^{(k)}(t)| \le \varepsilon_k^{(2)} \tag{7}$$
$$k = 0,\dots,q_j, \qquad t \in [t_q - h_q/2,\ t_q + h_q/2],$$

where $\varepsilon_{0i}, \varepsilon_k^{(i)}$, $i = 1,2$, are known numbers: either $\varepsilon_k^{(1)} > 0$, $0 < \varepsilon_k^{(2)} \le 1 - \delta_1$ and $0 < \varepsilon_{0i} \le 1 - \delta_2$, or $\varepsilon_k^{(1)} > 1 + \delta_1$ and $\varepsilon_{0i} > 1 + \delta_2$, where δ_1, δ_2 are known positive numbers, $a_j^{(k)}(t) = d^k a_j(t)/dt^k$, t_q, $q = 1, 2, \ldots$, are some points in the time axis, h_q is the length of the q-th homogenity interval which should be found. Then h_q is obtained by the condition (Galkin and Karpenko, 1991)

$$h_q = 2\Big[\max_{t>t_q}\{t\colon \mu|\Psi_1(t) - \Psi_2(t)|^2 \le \Delta\} - t_q\Big],$$

where

$$\Psi_1(t) = \varepsilon_{01} + \sum_{k=0}^{q_j} \varepsilon_k^{(1)} f_k(t), \qquad \Psi_2(t) = \varepsilon_{02} + \sum_{k=0}^{q_j} \varepsilon_k^{(2)} f_k(t),$$

μ depends only on ε_{0i}, $\varepsilon_k^{(i)}$, and Δ is defined by the requirements on the compatibility accuracy.

As in the item 4.A, restrictions (7) may be given with some confidence level, which defines according by some confidence level in the homogenity conditions. The presence of fuzzy given information in the source data is an unargued reason for the use of ES in the field of stochastic analysis of evolutionary systems (Waterman, 1986).

Let us note that spectrum (3) is defined not uniquely because decomposition (4) by $A(t,\omega)$ functions is not unique. To single out an unique spectrum we propose to use the Maximum Entropy (ME) principle. This criterion is chosen because the ME filter minimizes the lose and spoiling of information when estimating spectrum (Jaynes, 1982). The ME spectrum is obtained as a result of the solution of a conditional extremum problem for the entropy functional under restrictions defined by the autocorrelation function points. It is written in the form (Karpenko, 1991b)

$$S(t,\omega) = \left|\sum_{\theta=1}^{r} \lambda_\theta \sum_{j=1}^{p} u_{\theta-j} a_j(t) \exp\{i\omega\theta\}\right|^{-2},$$

where λ_θ are the Lagrange terms, r is the number of the autocorrelation function points estimated by the data on the current homogeneity interval, $u_{\theta-j}$ are defined by the coefficients $a_j(t)$ and $c_l(t)$, estimated on the preceding homogeneity interval.

5. CONCLUSION

In the present paper we have justified the expediency of the creation of expert systems for the analysis of some classes of nonstationary stochastic systems. In the §3 architecture of such ES was considered and their features have been discussed. The central point of our discussion is the most complex and significant part of such systems, i.e., their knowledge base. The essential element of the KB is the adaptive algorithms of the analysis of nonstationary processes. In the §4 we proposed some approaches to the creation of spectral and correlation estimation algorithms for some classes of nonstationary stochastic systems. Special attention was paid to the algorithms for the solution of identification problems for nonstationary systems. This was done because this question is rathen complicated and not yet enough investigated as well as because it is important for the solution of some applied problems.

Table 1. Setting of the a priori information in the fuzzy form.

Number	Parameter localization region	P	Q	Uncertainty
1	Ω_1	p_1	q_1	$1 - p_1 - q_1$
2	Ω_2	p_2	q_2	$1 - p_2 - q_2$
...	...	...	...	...
N	Ω_N	p_N	q_N	$1 - p_N - q_N$

REFERENCES

Ahmed, N.V. (1988). *Elements of Finite-Dimensional Systems and Control Theory*. Jonh Wiley, New York.

Bendat, J.S. and Piersol, A.G. (1986). *Random Data: Analysis and Measurement Procedures*. Second Ed. Wiley-Interscience, New York.

Brillindger, D.R. (1975). *Time Series: Data Analysis and Theory*. Holt, Rinehart, and Winston, New York.

Franks, L.E. (1969). *Signal Theory*. Prentice-Hall, Engelwood Clifts, New Jersey.

Galkin, V.Ya. et al. (1989). To proof of utilizing of spectral methods in identification problem of the signals of one class. In: *Actual Questions of Applied Mathematics*. Ed. by A.N. Tikhonov, A.A. Samarsky, Moscow Univ. Press, Moscow, pp. 9–17 (in Russian).

Galkin, V.Ya. et al. (1990). Dynamic spectral transforms in the identification problem of parameters of one class signals. In: *Mathematical Models and Optimization of Calculating Algorithms*. Ed. by A.N. Tikhonov and A.A. Samarsky, Moscow Univ. Press, Moscow (in Russian).

Galkin, V.Ya. and Karpenko, V.A. (1991). Correlative analysis of nonstationary processes on the basis of a priori information. In: *Mathematical Simulation and Solution of Mathematical Physics Inverse Problems*. Ed. by A.N. Tikhonov and A.A. Samarsky, Moscow Univ. Press, Moscow (in Russian).

Gerlach, A.A. et. al. (1989). Dynamic spectral transforms: properties of the canted spectral transforms. *J. Acoust. Soc. Am.* **86(1)**, 141–149.

Grenier, Y. (1983). Time-dependent ARMA modelling of nonstationary signals. *IEEE Trans. on ASSP.* **31**, 899–911.

Jaynes, E.T. (1982). On the rationale of maximum entropy methods. *IEEE Proc.* **70**, 33–50.

Kandel, A. and Schneider, M. (1989). Fuzzy sets and their applications to artificial intelligence. *Advances in Computers.* **28**, 69–103.

Karpenko, V.A. (1990). Linear prediction for the general oscillatory random processes. In: *Mathematical Models and Optimization of Calculating algorithms*, Ed. by A.N. Tikhonov, A.A. Samarsky, Moscow Univ. Press, Moscow (in Russian).

Karpenko, V.A. (1991). Spectral estimation of some nonstationary random processes on the basis of Maximum Entropy Method. In: *Proceedings of IV-th All-Union Conference on Perspective Methods of Experiment Analysis and Planning*. Petrozavodsk, pp. 16–17.

Loynes, R.M. (1968). On the concept of the spectrum for nonstationary processes. *J. Roy. Stat. Soc.* **30B**, 1–30.

Marple, S.L., Jr. (1987). *Digital Spectral Analysis with Applications.* Prentice-Hall, Englewood Cliffs, New Jersey.

Matsuyama, T. (1989). Expert systems for image processing: knowledge-based composition of image analysis processes. *Computer Vision, Graphics and Image Processing.* **48**, 22–49.

Mettrey, W. (1991). A comparative evaluation of expert system tools. *Computer.* **2**, 19–31.

Priestley, M.B. (1981). *Spectral Analysis and Time Series.* Academic Press, New York.

Rao, M.M. (1985). Harmonizable, Cramer and Karhunen classes of processes. In: *Handbook of Statistics,* Vol. 5. Ed. by M.M. Rao, North–Holland/Elsevier, Amsterdam, pp. 279–310.

Waterman, D.A. (1986). *A Guide to Expert Systems.* Addison-Wesley, New York.

Ill-Posed Problems in Natural Sciences, pp. 535 - 542
A. Tikhonov (Ed.)
1992 VSP/TVP

INVERSE PROBLEMS IN VIBRATIONAL SPECTROSCOPY

I.V. KOCHIKOV, G.M. KURAMSHINA, Yu.A. PENTIN, and A.G. YAGOLA
Moscow State University, Len. Gory, 119899 Moscow, Russia

ABSTRACT

The correctness of statements of various inverse problems related to processing the experimental data in vibrational spectroscopy is discussed. Stable numerical methods of determining parameters of the force fields of multi-atomic molecules based on Tikhonov regularization method are proposed.

1. INTRODUCTION

The main purpose of solving the inverse problems in vibrational spectroscopy is to determine parameters of the molecular force field using experimental data. The concept on force field arises from the attempt to consider a molecule as a mechanical system of nuclei while all the interactions due to the electrons are included into the effective potential function. This function (as well as masses of nuclei) determines the most important properties of a molecule.

Let us introduce the generalized coordinates of nuclei $\{q_1, \ldots, q_n\}$, where $n = 3N - 6$ (N is the number of atoms in the molecule), then the force field is characterized by potential function $U(q_1, \ldots, q_n)$.

One of the most important characteristics of any molecule, its equilibrium geometrical configuration, satisfies (provided it exists) the relations

$$\left.\frac{\partial U}{\partial q_i}\right|_{eq} = 0, \quad (i = 1, 2, \ldots, n).$$

We can redefine the coordinates $q_1, \ldots, q_n$ in such a way that $q_1 = q_2 = \ldots = q_n = 0$ in the equilibrium configuration, in this case in a neighbourhood of this configuration the following expansion is valid:

$$U(q_1, \ldots, q_n) = U_0 + \frac{1}{2}\sum_{ij} f_{ij} q_i q_j + o(\|q\|^3), \tag{1}$$

where U_0 is a constant and the force constants

$$f_{ij} = \left.\frac{\partial^2 U}{\partial q_i \partial q_j}\right|_{eq}$$

constitute a positive-definite matrix F. Elements of the matrix determine all the molecular characteristics pertinent to small deformations (including vibrations). Since the general problem of determining potential function $U(q_1, \dots, q_n)$ seems far from being solvable, the inverse vibrational problems are usually aimed at finding the matrix elements f_{ij}, assuming the equilibrium configuration to be a priori known (as is usually the case).

Mathematically the concept of force field can be obtained through the adiabatic theory of perturbations with the use of a small parameter related to the ratio of the mass of electrons to the mass of nuclei. It can be shown that in a certain approximation the nuclei can be treated as particles moving in the force field determined by (1).

During the last 30 years numerous attempts to compute force fields directly from the Schrödinger's equation were undertaken. The use of very restricting assumptions was (and still is) necessary to make the computations technically possible, and so far we have no hope of getting the results compatible with experimental data within the accuracy of measurements. We nevertheless expect the further rapid progress in these computations (widely known as ab initio methods) and later in this paper we shall propose the way of utilization even the not so accurate results of ab initio computations.

Determination of the force constants of molecules from experimental data can be used in two general directions:

(a) to check the validity of various model assumptions frequently used by spectroscopists to simplify the potential function calculation;

(b) to predict the vibrational properties of molecules (including molecules which do not exist but which can be synthesized).

The latter possibility is based on the fundamental properties of the force field (e.g. its transferrability in the series of molecules containing similar fragments, etc.) and will be discussed later in more detail.

2. EXPERIMENTAL DATA. MAIN EQUATIONS

2.1. *Vibrational spectra*

The vibrational molecular spectra provide the most convenient and useful information concerning force fields. The main role is played by the infrared absorption and Raman spectra which correspond to the normal vibrations of molecules. Quantum and classical treatments of the vibrational problem both lead to the eigenvalue equation

$$(GF)L = L\Lambda \tag{2}$$

connecting the force constants matrix F with the frequencies of normal vibrations $\omega_1, \dots, \omega_n$ constituting the diagonal matrix Λ:

$$\Lambda = \operatorname{diag}\{\omega_1^2, \dots, \omega_n^2\}.$$

In equation (2) G is the $n \times n$-matrix of the kinetic energy in the momentum representation and can be calculated using only masses and equilibrium configuration of nuclei (which we assume to be known, possibly with some errors). Matrix

L characterizes the shape of each normal vibration (i.e., relative amplitudes of vibrations speaking in classical terms).
Equation (2) is so for the main source of data to determine the force constants. It is evident, however, that (with the exception of diatomic molecules) $n(n+1)/2$ parameters of F cannot be found from n frequencies $\omega_1, \ldots, \omega_n$.
This leads, to the attempts on the one hand, to use certain model assumptions concerning the structure of matrix F, and, on the other hand, to attract additional experimental data.

2.2. *Isotopic species*
Within the approximation considered, the force field of a molecule does not depend on the masses of atoms, unlike matrix G and vibrational frequencies. This fact enables simultaneous treatment of several isotopic species of the molecule which yields instead of (2) the system of equations

$$(G_i F) L_i = L_i \Lambda_i, \tag{3}$$

where subscript i stands for various isotopic species of the molecule. The introduction of isotopes allows usually to obtain a limited number of independent equations in (3) and so leaving the inverse problem underdefined.

2.3. *Coriolis constants*
The additional information can be extracted from the Coriolis constants which characterize the vibrational-rotational interaction in a molecule. These constants are determined experimentally from the splitting of degenerate vibrational levels (these are present in the molecules of medium and high symmetry). Coriolis constants constitute matrix $\vec{\xi}$ with vector elements satisfying the equation

$$\vec{\xi} = \frac{1}{M^2} L^* \vec{A} \mathcal{M} \vec{A}^* L, \tag{4}$$

where $\mathcal{M} = \mathrm{diag}\{m_1, \ldots, m_N\}$, $m_1, \ldots, m_N$ are masses of nuclei, $M = m_1 + m_2 + \ldots + m_N$ and matrix $\vec{A}$ can be determined, the same as matrix G, using the equilibrium configuration and masses of nuclei. This matrix proved to be useful in analyzing vibrations because of the relation

$$\overrightarrow{\Delta r_i} = \sum_{k=1}^{n} \vec{A}_{ik} q_k,$$

where $\{q_k\}$ are the generalized coordinates of nuclei, and Δr_i $(i = 1, 2, \ldots, N)$ correspond to the Cartesian components of nuclei displacements from the equilibrium position.

2.4. *Centrifugal distortion*
Yet another result of the rotation of a molecule is the distortion of its equilibrium configuration due to the centrifugal forces. This leads also to splitting the degenerate vibrational levels and the transitions between them appear as the result of

it. The splitting is characterized by the centrifugal distortion constants

$$\tau_{\alpha\beta\gamma\delta} = -\frac{\sum_{ij} J^0_{\alpha\beta i} J^0_{\gamma\delta j} F^{-1}_{ij}}{2 I^0_\alpha I^0_\beta I^0_\gamma I^0_\delta}, \tag{5}$$

where I^0, J^0 can be determined beforehand (the same as matrices G and $\vec{A}$).

2.5. *Amplitudes of vibrations.*
Experimental methods of gas electron diffraction can be used to determine the mean square amplitudes of vibrations for interatomic distances. These amplitudes (squared) correspond to the diagonal elements of the matrix

$$D = HL\Delta L^* H^*, \tag{6}$$

where H is completely determined by the same data as matrices $\vec{A}$ and G, and Δ is a diagonal matrix with the elements given by quantum thermodynamics:

$$\Delta_i = \frac{\hbar}{2\omega_i} \tanh^{-1} \frac{\hbar\omega_i}{KT} \quad (i = 1, \dots, n),$$

where $\hbar$, K are Planck and Boltzmann constants, T denotes the absolute temperature.
It is not so easy to obtain the experimental information mentioned in Sections 2.2–2.5 but they are very useful because they need no introducing any additional unknown parameters of any kind (unlike, e.g., the intensities of infrared spectra).

3. FORMULATION OF THE PROBLEM

Equations (2)–(6) with respect to matrix F can be written as a single operator equation

$$AF = \mathcal{L}, \tag{7}$$

where $\mathcal{L}$ represents the set of experimental data available (frequencies ω, constants $\vec{\xi}$, D, τ, etc.) and A is an operator putting into correspondence to any matrix F the set of values determined by equations (2)–(6). The equilibrium configuration of the molecule is considered to be given.
To make some mathematical statements we shall consider matrices F as elements of Euclidean space with the norm

$$\|F\| = \left(\sum_{i,j=1}^{n} f_{ij}^2 \right)^{1/2}$$

The set of data $\mathcal{L}$ (which is a finite set of numbers) will be considered as an element of the finite-dimensional space with the weighted norm

$$\|\mathcal{L}\| = \left(\sum_{i=1}^{m} l_i^2 p_i \right)^{1/2},$$

where l_i $(i = 1, \dots, m)$ are the components of $\mathcal{L}$, and $p_1, p_2, \dots, p_m$ are some positive weights. These weights can be used to take into account for the different degree of accuracy of various kinds of experimental data.

Operator A, as it is clearly seen from (2)–(6), is nonlinear and includes some not very evident dependencies. For instance, matrix elements of $\overrightarrow{\xi}$ in (4) are related to matrix L which is constituted by eigenvectors of problem (2) and so being connected with matrix F. Nevertheless, it can be shown that A is continuous and even differentiable.

Analysis of operator equation (7) shows that (depending on the experimental data available) it can have finite or infinite set of solutions or to have no solutions at all. The latter case can result from the roughness of the mathematical model which does not take into account some essential points, e.g., unharmonicity of vibrations. Incompatibility of (7) can be caused also by inevitable errors in the experimental data.

The errors can appear in operator A because the parameters of equilibrium configuration present in (2)–(6) are usually measured in experiment. The errors in right-hand side and operator, however small, can cause under certain conditions finite (and large) variations of the solution (the common example is related to three isotopic species of methane, see (Koptev and Pentin, 1977)).

All the properties of equation (7) mentioned above make it clear that this inverse problem is ill-posed, and regularization technique should be applied to solve it in a stable way. The most methods designed to determine the force constants have so far used explicitly or implicitly formulated restrictions posed upon the solution to be found. Sometimes this enabled the researchers to obtain reasonable solutions using the "common sense reqularization". The common sense being subjective, it has caused certain problems in comparing the results of different authors.

The essence of Tikhonov's regularization method lies in the formulation of principle of choosing some solution from the set of possible solutions (this set consists of all the solutions compatible with the experimental data within the range of experimental errors). Kochikov (Kochikov et al., 1981; 1985) have proposed to use one of the variants of such a principle – the concept of normal solution – to take explicitly into account the poorly formalized features of the force constants matrix.

Now we are able to give a mathematical formulation of the problem. Let equation (7) with the exact data (A and $\mathcal{L}$) have a unique normal (with respect to a certain a priori specified matrix F^0) solution $\bar{F}$ and, instead of A, $\mathcal{L}$ we are given some approximations A_h, $\mathcal{L}_\delta$ such that

$$\|A_h F - AF\| \leq \varphi(h, F), \qquad \|\mathcal{L}_\delta - \mathcal{L}\| \leq \delta,$$

where h, δ are known, and $\varphi(h, F)$ is a continuous function satisfying the condition $\sup_{\|f\|<C} \varphi(h, F) \to 0$ as $h \to 0$ for any finite C.

It is required to construct approximations $F_{h\delta}$ to the normal solution $\bar{F}$ such that $F_{h\delta} \to \bar{F}$ as h, $\delta \to 0$. The algorithms satisfying these conditions are called Tikhonov regularizing (Tikhonov, 1963).

It should be noted that if problem (7) with the exactly given data has a unique solution, it coincides with $\bar{F}$. If equation (7) has no solutions, it has to be formulated as the problem of finding a pseudo-solution (that is the matrix F minimizing

$\|AF - \mathcal{L}\|$). A regularizing algorithm in this case must provide approximations converging to the normal pseudo-solution of (7).

The last case to be discussed is a relatively unprobable one when the problem (7) has more than one normal solution (or pseudo-solution). The approximations $F_{h\delta}$ converge in this case to the set of such solutions (in the sense of β-convergence).

The concept of normal solution leads to the problem of determining matrix F which is compatible with the experimental data and is the nearest one to a certain matrix F^0. We suggest using F^0 for several purposes:

(a) to find the solutions of (7) satisfying (though not strictly) certain model assumptions. For instance, we can require that some elements of F are small with respect to others;

(b) to take into consideration the quantum-mechanical (ab initio) computations which do not reproduce the experimental data but nevertheless reveal some physical essence of interactions between the atoms;

(c) to find the "transferrable" solutions of (7), i.e., the sets of force constants which do not change (or change insignificantly) in different molecules including similar fragments; see § 5 for a more detailed discussion;

(d) to eliminate the difficulties connected with the usage of redundant coordinate sets when more than $n = 3N - 6$ coordinates are introduced. The usage of redundant coordinates is often justified by convenience in taking into account symmetry of molecules and in interpretation of results.

4. REGULARIZING ALGORITHMS

The paper (Kochikov et al., 1987) (which we recommend to those interested in details) has proposed some regularizing algorithms which are efficiently computable. Two such algorithms will be formulated below.

The first one is based on the minimization of Tikhonov functional

$$M^\alpha[F] = \|A_h F - \mathcal{L}_\delta\|^2 + \alpha\|F - F^0\|^2, \tag{8}$$

where regularizing parameter $\alpha > 0$ is determined from a version of generalized discrepancy principle. For nonlinear problems this method was proposed by Goncharskii (Goncharskii et al., 1972) and developed later by Leonov (Leonov, 1982). The modification of this principle used in the present work is valid for the special estimate of operator A_h errors, namely $\varphi(h, F) = h\|A_h F\|$.

The estimate of this kind defines the relative error of operator A_h $(h < 1)$ and can be derived for the most commonly used equations (2) and (3). In this case parameter α can be found from the equation

$$\rho(\alpha) \equiv \|A_h F^\alpha - \mathcal{L}_\delta\| - \frac{1}{1-h}\Big[\hat{\mu} + k\big(\delta + h\|\mathcal{L}_\delta\|\big)\Big], \tag{9}$$

where F^α is any extremal of (8), $k > 1$ is a constant of purely theoretical significance, and

$$\hat{\mu} = \inf_F \left\{\|A_h F - \mathcal{L}_\delta\| + \delta + h\|A_h F\|\right\}$$

is introduced to take into account the possible incompatibility of (7) with exact data (otherwise it should be omitted). Since $\rho(\alpha)$ can be not continuous, the generalized solution of (9) should be used.

The extremal of (8) with parameter α defined by (9) is an approximation for $F_{h\delta}$.

The specific kind of $\varphi(h, F)$ estimate is restricting for the formulations of inverse vibrational problems including more exotic data (see (4)–(6)). The following algorithm is somewhat more complicated as regards computations, but it implies no restrictions on $\varphi(h, F)$ besides those stated in § 3.

Let us find $\arg\min\limits_{F\in\mathcal{F}} \|F - F^0\|$, where

$$\mathcal{F} = \{F\colon \|A_h F - \mathcal{L}_\delta\| \le \delta + \varphi(h, F) + \hat{\mu}\}.$$

This is an optimization problem with nonlinear restrictions. It can be proved that the set $\mathcal{F}$ is never empty, and the solution of this problem can serve as a regularized approximation to the normal solution (pseudo-solution) of (7).

The algorithms discussed above served as a basis to compile a package of programs for processing spectroscopic data by a computer (the latest version to appear was specially adapted to IBM PC). This package allows to take into account molecular symmetry by introducing symmetry coordinates which, when possible, leads to much higher efficiency. Some examples together with package description can be found in (Kochikov et al., 1984; 1985b; Kochikov and Kuramshina, 1985).

5. SOME ADDITIONAL FEATURES

Here we describe one of the most interesting extensions of the program package. We have already mentioned the problem of transferrability of force constants which means that the constants related to the similar fragments in different molecules are expected to be nearly the same.

The first and most evident way to obtain a transferrable set of force constants is to use matrix F^0 composed of the elements (already computed) for the simpler molecules with similar fragments.

The more promising though complicated way is to treat several molecules simultaneously. In this generalized procedure of computation the restrictions should be imposed that certain force constants in different molecules must be equal. The list of such restrictions is provided by a user.

In fact several equations of kind (7) are being solved paralelly using several sets of experimental data and several different matrices are obtained, the only links between these equations being represented by the equality requirements.

Mathematically the problem is formulated as an operator equation with linear restrictions which are taken into account by the projection procedure. Otherwise the algorithm is quite similar to that described in § 4. Transfer to symmetry coordinates (which can be different for different molecules) causes numerous but purely technical difficulties.

REFERENCES

Goncharskii, A.V., Leonov, A.S., and Yagola, A.G. (1972). A regularizing algorithm for ill-posed problems with approximately specified operator. *USSR Computer Math. and Math. Phys.* **12**, 1592–1594 (in Russian).

Kochikov, I.V. and Kuramshina, G.M. (1985). A package of programs for calculating force fields of polyatomic molecules using the Tikhonov regularization method. *Vestnik Moscow Univ. Ser. 2.Chemistry.* **26**, 354–358 (in Russian).

Kochikov, I.V., Kuramshina, G.M., Pentin, Yu.A., and Yagola, A.G. (1981). Regularizing algorithm for solving the inverse vibrational problem. *Dokl. Akad. Nauk SSSR.* **261**, 1104–1106 (in Russian).

Kochikov, I.V., Kuramshina, G.M., Pentin, Yu.A., and Yagola, A.G. (1985a). Calculation of force fields of polyatomic molecules using the Tikhonov regularization method. *Dokl. Akad. Nauk SSSR.* **283**, 850–854 (in Russian).

Kochikov, I.V., Kuramshina, G.M., and Yagola, A.G. (1987). Stable numerical methods of solving certain inverse problems of vibrational spectroscopy. *USSR Comput. Math. and Math. Phys.* **27**, 33–40 (in Russian).

Kochikov, I.V., Yagola, A.G., Kuramshina, G.M., Kovba, V.M., and Pentin, Yu.A. (1984). Force fields and mean amplitudes of vibration of chromium, molibdenum and tungsten oxotetrafluorides. *J. Mol. Str.* **106**, 355–360.

Kochikov, I.V., Yagola, A.G., Kuramshina, G.M., Kovba, V.M., and Pentin, Yu.A. (1985b). Calculation of force fields of chromium, molybdenum and tungsten hexafluorides and dioxodifluorides by means of the Tikhonov regularization method. *Spectrochim. Acta. Ser. A.* **41**, 185–189.

Koptev, G.S. and Pentin, Yu.A. (1977). *Calculation of the Vibrations of Molecules.* Moscow Univ. Press, Moscow (in Russian).

Leonov, A.S. (1982). The usage of the generalized discrepancy to solve ill-posed extremal problems. *Dokl. Akad. Nauk SSSR.* **262**, 1306–1311 (in Russian).

Tikhonov, A.N. (1963). On the solution of ill-posed problems and regularization method. *Dokl. Akad. Nauk SSSR.* **151**, 501–504 (in Russian).

Ill-Posed Problems in Natural Sciences, pp. 543 – 552
A. Tikhonov (Ed.)
1992 VSP/TVP

INVERSE PROBLEMS IN ELECTROENCEPHALOGRAPHY AND THEIR NUMERICAL SOLVING

Yu.M. KOPTELOV and E.V. ZAKHAROV
Institute for Physico-Technological Research, Moscow State University, Len. Gory, 119899 Moscow, Russia

ABSTRACT

Statements of the inverse electroencephalography problems and stable algorithms for their numerical solution are considered. These problems are of particular concern in the noninvasive diagnostics of functional human brain condition both in normals and in cases of variety pathology. For the diagnostics, use is made of electroencephalography (EEG) data registrated on the head surface. Two classes of inverse problems are investigated: the first class involves the primary bioelectrical sources field parameters reconstruction due to EEG data (EEG structure filtering problems) and the second class deals with the problems of primary sources localization within the brain. The first class is considered in quasi-static approximation with using inhomogeneous conductive medium. The spherically symmetric head model is used in the transition to the linear operational equation of the first kind and, to solve it, the regularization methods are applied. To settle the inverse problems of the other class, the sequence of homogeneous and inhomogeneous head models is used and the solution is seeked over a finiteparametric set of equivalent current sources. Results of numerical simulations of experiment and mathematical treatment of EEG data are shown for the brain pathology cases in comparison with tomographic information.

1. INTRODUCTION

The brain is of vital importance for human organizm functioning. However, while other human organs had well studied, the brain still remains terra incognita. This is due to the fact that the direct physiological methods of human brain investigation present much difficulty and their application commonly requires complex neurosurgery operations.

Therefore noninvasive (nondestructive) methods of investigation are of particular significance. These methods include computer tomography and also registration of physical brain fields, namely, electric field (electroencephalography method), magnetic field (magnitoencefalography method), infra red radiation (thermoencephaloscopy method) (see (Nunez, 1981; Weinberg et al., 1985; Shevelev et al., 1989)). Among these methods, the electroencephalography one has the widest acceptance. This technique involves registration of electric brain potentials on the head surface (or eleclroencephalogram, EEG).

To pose an EEG interpretation problem, one has to consider two classes of inverse ill-posed problems. The first class includes problems of reconstructing primary bioelectric brain sources field parameters by using EEG data (the EEG structure filtering problems) and the second deals with localizing the primary bioelectric sources within the brain. Both these classes will be considered in what follows below.

2. EEG STRUCTURE FILTERING

2.1. The difficulty of the physiological interpretation and analysis of EEG is due to a number of passive shells (soft and solid brain shells, cerebro-spinal fluid, skull layers, scalp) surrounding the brain containing the primary bioelectric EEG sources, i.e., neurons and glyacytes. These shells distort the initial potential distribution pattern obtained due to primary sources. These potentials are of interest since they give the main diagnostic information about different brain structure states. These distortions are caused by both damp effect of shells and influence of secondary sources potential induced on the shells' boundary with different conductivities.

To describe mathematically the electric field of the brain, one uses the Maxwell system of equations. Taking into account that the frequency of bioelectrical sources oscillation is not very high (not greater than 1000 Hz) the inductive and capacity brain properties are not significant, and, finally, the brain size is not large we can neglect the influence of the limited field spreading rate towards the registration points on the scalp. We may also ignore the displacement currents. Then the Maxwell system of equations reduces to the system of equations of electromagnetic field of static currents (Plonsy, 1969). The field value is defined by this system of equations at any time moment independently of the field value in the previous period.

Let consider a spherically symmetric head model consisting of k regions: a sphere of radius R_k with electric conductivity λ_k modeling the brain (the electrically active domain containing the primary bioelectrical sources) and $k-1$ concentric sphere layers G_1 with electric conductivities λ_i $i = 1, 2, ..., k-1$, modeling the basic inhomogeneous regions of the brain (scalp, skull layers, cerebrospinal fluid and others). Each of the regions G_i, $i = 1, 2, ..., k-1$, is located between concentric spheres Γ_i and Γ_{i+1} of radii R_i and R_{i+1}, respectively, $R_{i+1} < R_i$. Out side a sphere Γ_1 of radius R_1 we have $\lambda = 0$.

We denote the brain region without primary sources by G_k. This region is restricted by the spheres Γ_k and Γ_{k+1} of radii R_k and R_{k+1}, respectively, $R_{k+1} = R_k - h_\varepsilon$, where h_ε is the distance between the brain surface Γ_k and the nearest primary source, $h_\varepsilon > 0$. Then the brain potential u of electric field in the region without sources $G = \bigcup_{i=1}^{k} G_i$ satisfies the Laplace equation $\Delta u = 0$ and can be represented in terms of the spherical coordinates $\{r, \theta, \varphi\}$ as follows (Tikhonov and

Samarsky, 1977):

$$u(r,\theta,\varphi)=\sum_{0}^{\infty}\left(a_{lm}^{(i)}r^{-l-1}+b_{lm}^{(i)}r^{l}\right)Y_{lm}(\theta,\varphi), \quad (1)$$

$$R_{i+1}<r<R_i, \quad \varnothing\le\theta\le\pi, \quad 0\le\varphi<2\pi, \quad i=1,2,...,k,$$

where

$$Y_{lm}(\theta,\varphi)=P_{lm}\begin{cases}P_l^m(\cos\theta)\cos m\varphi, & m=0,1,...,l,\\ P_l^{|m|}(\cos\theta)\sin|m|\varphi, & m=-1,-2,...,-l,\end{cases}$$

are spherical functions, $P_l^m(\xi)$ are attached Lagrangian functions,

$$P_{lm}=\left[\frac{(2l+1)(l-|m|)!}{2\pi(1+\delta_{0m})(l+|m|)!}\right]^{1/2}, \quad \delta_{0m}=\begin{cases}1, & m=0,\\ 0, & m\ne0,\end{cases}$$

P_{lm} are normalization factors, $\sum_0^\infty\equiv\sum_{l=0}^\infty\sum_{m=-l}^{l}$. In this case a sequence of potential continuity and normal current component continuity the function u satisfies the following conditions

$$\left.\frac{\partial u}{\partial r}\right|_{\Gamma_1}=0, \quad [u]_{\Gamma_i}=0, \quad \left[\lambda\frac{\partial u}{\partial r}\right]_{\Gamma_i}=0, \quad i=2,3,...,k, \quad (2)$$

where $[\xi]_{\Gamma_i}$ denotes the jump in ξ which occurs when passing through the surface Γ_i from outside.

The potential (1) within the region G_k may be divided into two components:

$$u(r,\theta,\varphi)=v_{pr}(r,\theta,\varphi)+u_{\sec}^{(k)}(r,\theta,\varphi),$$

where

$$v_{pr}(r,\theta,\varphi)=\sum_{0}^{\infty}a_{lm}^{(k)}r^{-l-1}Y_{lm}(\theta,\varphi), \quad r>R_{k+1} \quad (3)$$

is the potential of the sources located inside a sphere Γ_{k+1} of radius R_{k+1} (it is primary EEG sources potential in the homogenious unbounded medium with electric conductivity λ_k), and

$$u_{\sec}^{(k)}(r,\theta,\varphi)=\sum_{0}^{\infty}b_{lm}^{(k)}r^{l}Y_{lm}(\theta,\varphi), \quad r<R_k,$$

is the potential of the secondary sources, induced on the boundary of inhomogeneities, spheres Γ_1, Γ_2, ..., Γ_k. Let us denote the part of the sphere where EEG is recording as S. Since is practice one measures not the absolute value of the potential but the potential difference due to some lead system, let us introduce a lead operator A_0 in $\mathbf{L}_2(S)$:

$$A_0w\equiv w(\theta,\varphi)-w(\theta_0,\varphi_0), \quad (\theta,\varphi),\ (\theta_0,\varphi_0)\in S.$$

Let now

$$A_0\left[u|_S\right]=f(\theta,\varphi), \quad (4)$$

where $f(\theta, \varphi)$ is the brain electric field potential (EEG) measured on S by means of the lead system with the operator A_0. Then the primary sources potential reconstructing problem can be formulated as follows: given a function $f(\theta, \varphi)$ on S, it is required to determine a function $v_{pr}(r, \theta, \varphi)$ provided that (1)–(4) hold. Let us denote the primary sources potential on the sphere Γ_0 of radius r_0 by $v(\theta, \varphi)$, $v(\theta, \varphi) \equiv v_{pr}(r_0, \theta, \varphi)$, $R_{k+1} < r_0 < R_k$. Let $v(\theta, \varphi) \in V$, $V = \{v \in \mathbf{L}_2(\Gamma_0) : v = \sum_1^\infty c_{lm} Y_{lm}(\theta, \varphi),\ \Delta_\theta v \in \mathbf{L}_2(\Gamma_0)\}$ (here Δ_θ is the Laplace–Beltrami spherical operator), $f(\theta, \varphi) \in \mathbf{L}_2(S)$. Then the correspondence between the functions v and f can be represented by the following operator equation of the first kind (Zakharov and Koptelov, 1987; 1990):

$$Av = f, \tag{5}$$

where A is the continuous linear one-to-one operator which maps the functions $v(\theta, \varphi)$ from the space V,

$$v(\theta, \varphi) = \sum_1^\infty c_{lm} Y_{lm}(\theta, \varphi), \quad c_{lm} = (v, Y_{lm})_{\mathbf{L}_2(\Gamma_0)},$$

into functions from $\mathbf{L}_2(S)$ according to the following rule:

$$Av = \sum_1^\infty q_l c_{lm} \Big[Y_{lm}(\theta, \varphi) - Y_{lm}(\theta_0, \varphi_0) \Big], \tag{6}$$

$$(\theta, \varphi), (\theta_0, \varphi_0) \in S,$$

where $q_l = (2l+1)/l(B_l^{(k)})_1^{-1}(r_0/R_1)^{l+1}$, $l = 1, 2, ...$, $(B_l^k)_1$ is the first element from the 2×1-matrix $B_l^{(k)}$, $B_l^{(k)} = D_l^{(k)} D_l^{(k-1)} ... D_l^{(2)} D_l^{(1)}$, $D_l^{(1)} = (1, (l+1)/(lR^{2l+1}))^T$ is a 2×1-matrix, $D_l^{(i)}, i = 2, 3, ..., k$, are 2×2-matrices,

$$D_l^{(i)} = \begin{pmatrix} \frac{l}{2l+1}\left(1 + \frac{l+1}{l}\frac{\lambda_{i-1}}{\lambda_i}\right) & \frac{l}{2l+1}\left(1 - \frac{\lambda_{i-1}}{\lambda_i}\right) R_i^{2l+1} \\ \frac{l+1}{2l+1}\left(1 - \frac{\lambda_{i-1}}{\lambda_i}\right) R_i^{-2l-1} & \frac{l+1}{2l+1}\left(1 + \frac{l}{l+1}\frac{\lambda_{i-1}}{\lambda_i}\right) \end{pmatrix}$$

. **2.2.** Since the inverse operator A^{-1} is not continuous, the problem of finding v satisfying equation (5) with a given f may be considered as an ill-posed problem (Tikhonov and Arsenin, 1986). Let the EEG-potential be measuared on S with an error δ, $\|f_\delta - \bar{f}\|_{\mathbf{L}_2(S)} \leq \delta$, $\bar{f} = A\bar{v}$, and let $\bar{v}$ be an exact solution of equation (5) corresponding to the exact right-hand side vector $\bar{f}$. Assume that, instead of the exact operator A, we know its approximation A_h,

$$\|A_h - A\| = \sup_{v \neq 0} \Big[\|A_h v - Av\|_{\mathbf{L}_2(S)} / \|v\|_V \Big] \leq h, \quad h \geq 0,$$

where $\|v\|_V = \|\Delta_\theta v\|_{\mathbf{L}_2(\Gamma_0)} = \left[\sum_1^\infty l^2 (l+1)^2 c_{lm}^2 \right]^{1/2}$, $c_{lm} = (v, Y_{lm})_{\mathbf{L}_2(\Gamma_0)}$.

Then the function

$$v_L^\alpha(\theta,\varphi) = \sum_1^L c_{lm}^\alpha Y_{lm}(\theta,\varphi) \tag{7}$$

can be treated as an approximate solution of (5) with the approximate initial data $\{A_h,\ f_\delta,\ \eta\}$, $\eta = (\delta, h)$. This function provides minimum to the smoothing functional (Tikhonov and Arsenin, 1986)

$$M^\alpha\Big[v,\ f_\delta,\ A_h\Big] = \|A_h v - f_\delta\|^2 + \alpha\|v\|^2 \tag{8}$$

over the set $V_L = \{v \in V\colon\ v = \sum_1^L c_{lm}Y_{lm}(\theta,\varphi)\}$. The regularization parameter $\alpha = \alpha(\eta)$ is selected according to the principle of generalized discrepancy (Tikhonov et al., 1983) as a root of the equation

$$\|A_h v_L^\alpha - f_\delta\|^2 - (\delta + h\|v_L^\alpha\|)^2 = 0.$$

Then $v_L^\alpha \to \bar{v}$ as $L \to \infty$, $\delta, h \to 0$ (Tanana, 1981; Tikhonov et al., 1983).

2.3. Let the area S be divided into N parts of equal size ΔS with centres at $(\theta_j,\ \varphi_j), j = 1, 2, ..., N$. Denote the differences between the potential values at the points (θ_j, φ_j) and (θ_0, φ_0) by $f_j \equiv f_\delta(\theta_j, \varphi_j)$, $j = 1, 2, ..., N$ (one moment EEG values). Then the expression for the potential of the primary sources corresponding to the one-moment EEG vector $F = (f_1, f_2, ..., f_N)^T$ can be represented as

$$\tilde{v}_L^\alpha(\theta,\varphi) = \sum_1^L (q_l^h)^{-1}\tilde{c}_{lm}^\alpha Y_{lm}(\theta,\varphi), \tag{9}$$

where $\tilde{c}_{lm}^\alpha$, $l = 1, 2, ..., L$, $m = 0, \pm 1, ..., \pm l$, are the components of the vector $\tilde{C}^\alpha$ of dimension $M = \sum_{l=1}^L (2l+1)$ which is the solution of the linear vector equation $(Y^TY + \alpha Q)\tilde{C}^\alpha = Y^TF$ implied by the minimality conditions for the functional (8), Y is an $N \times M$-matrix with elements $(Y_{lm}(\theta_j, \varphi_j) - Y_{lm}(\theta_0, \varphi_0))$, Q is the diagonal $M \times M$-matrix of operator A_h with elements $Q_l = [l(l+1)/q_l^h]^2 E_l$, E_l is the unit $l \times l$-matrix. In this case the regularization parameter $\alpha = \alpha(\eta)$ is taken to be a root of the equation

$$(Y\tilde{C}^\alpha - F)^T(Y\tilde{C}^\alpha - F) - \Big\{\delta(\Delta S)^{-1/2} + h(\Delta S)^{-1/2}\big[(\tilde{C}^\alpha)^T Q\tilde{C}^\alpha\big]^{1/2}\Big\}^2 = 0.$$

The expression for the potential on the head surface Γ_1 deduced from (9) by taking into consideration expression (6) is

$$v_{S,L}^\alpha(\theta,\varphi) = \sum_1^L \tilde{c}_{lm}^\alpha Y_{lm}(\theta,\varphi). \tag{10}$$

Expression (10) can be used for interpolating the potential between the measuarements points $(\theta_j, \varphi_j) \in S$ and extrapolating it on $\Gamma_1 \backslash S$.

2.4. If the approximate solution of equation (5) (i.e., the primary sources potential) is available in the form (7) or (9), then it is possible to obtain also an expression for the components of the tensity vector of the electric field of primary sources, $\mathbf{E} = \{E_r, E_\theta, E_\varphi\}$. To this end, let us use the expression $\mathbf{E} = -\mathrm{grad}\, v$ in a spherical coordinate system (Stratton, 1941):

$$\mathbf{E} = -\frac{\partial v}{\partial r}\mathbf{i}_r - r^{-1}\frac{\partial v}{\partial \theta}\mathbf{i}_\theta - (r\sin\theta)^{-1}\frac{\partial v}{\partial \varphi}\mathbf{i}_\varphi, \tag{11}$$

where $\mathbf{i}_r, \mathbf{i}_\theta, \mathbf{i}_\varphi$ are the unit coordinate vectors. If we substitute (9) into (11) and make use of (3), we obtain the following expressions for components of the tensity vector of the electric field of primary sources at points (θ, φ) of a sphere Γ_0 of radius r_0 :

$$\widetilde{E}_r^{\alpha,L}(\theta,\varphi) = \frac{1}{r_0}\sum_1^L (l+1)\tilde{c}_{lm}^{\alpha}/q_l^h Y_{lm}(\theta,\varphi),$$

$$\widetilde{E}_\theta^{\alpha,L}(\theta,\varphi) = -\frac{1}{r_0}\sum_1^L \tilde{c}_{lm}^{\alpha}/q_l^h \frac{\partial Y_{lm}(\theta,\varphi)}{\partial\theta},$$

$$\widetilde{E}_\varphi^{\alpha,L}(\theta,\varphi) = -(r_0\sin\theta)^{-1}\sum_1^L \tilde{c}_{lm}^{\alpha}\frac{\partial Y_{lm}(\theta,\varphi)}{\partial\varphi}.$$

In the same way we can derive the expressions for the components of the tensity vector of electric field on the head surface Γ_1 :

$$\widetilde{E}_{s,r}^{\alpha,L}(\theta,\varphi) = \frac{1}{R_1}\sum_1^L (l+1)\tilde{c}_{lm}^{\alpha} Y_{lm}(\theta,\varphi),$$

$$\widetilde{E}_{s,\theta}^{\alpha,L}(\theta,\varphi) = -\frac{1}{R_1}\sum_1^L \tilde{c}_{lm}^{\alpha}\frac{\partial Y_{lm}(\theta,\varphi)}{\partial\theta},$$

$$\widetilde{E}_{s,\varphi}^{\alpha,L}(\theta,\varphi) = -(R_1\sin\theta)^{-1}\sum_1^L \tilde{c}_{lm}^{\alpha}\frac{\partial Y_{lm}(\theta,\varphi)}{\partial\theta}.$$

Since the relation $\lambda\mathbf{E} = \mathbf{j}$ holds true in the conductive medium, the expressions for the components of the current density vector both at points of the sphere Γ_0 and on the head surface Γ_1 are obtained from the expressions for the corresponding components of the tensity vector by multipliing them by the medium electric conductivity value.

3. EQUIVALENT EEG SOURCES LOCALIZATION

3.1. In contrast with the problems considered above, the problem of finding the sources in the brain for the electric potential measured on the head surface has not, generally, a unique solutions. Therefore, to solve this problem, a finite-parametric set of equivalent current sources is commonly employed (Nunez, 1981). The current dipole is a simple equivalent source for the brain being electrically neutral as a whole. Such dipole is generally defined by six independent time-varying parameters (3 coordinates of the centre, x_d, y_d, z_d, and 3 orthogonal projections of the vector moment, M_X, M_Y, M_Z, characterizing the intensity and the orientation of the dipole current source). Although the current dipole model is simple enough, it provides a reasonable approximation of the electrical activity of limited brain regions and is widely used to estimate the equivalent sources of scalp recorded potentials (Nunez, 1981).

The estimation of the equivalent dipole parameters from the experimental data by the method of least squares reduces to the solution of the following extremal value problem

$$\mathbf{p}^* = \arg \min_{\mathbf{p} \in G_\mathbf{p}} \Phi(\mathbf{p}), \tag{12}$$

where $\mathbf{p} = \{x_d, y_d, z_d, M_X, M_Y, M_Z\}$ is the dipole parameter vector, $G_\mathbf{p} = \{\mathbf{p}: x_d^2 + y_d^2 + z_d^2 \leq R_{K+1}^2, M_X^2 + M_Y^2 + M_Z^2 \leq C_M^2\}$ is the set of feasible values of the dipole parameters, C_M is the maximum value of the dipole moment,

$$\Phi(\mathbf{p}) = \sum_{j=1}^{N} \left(\eta_j(\mathbf{p}) - f_j\right)^2 \tag{13}$$

is the square discrepancy functional, f_j, $j = 1, 2, ..., N$, are the potential values measured simultaneously at points (θ_j, φ_j) on the head surface (one-moment EEG values), $\eta_j(p) = A_0[u_d(\theta_j, \varphi_j, \mathbf{p})], u_d(\theta_j, \varphi_j, \rho)$ is the potential value at a point (θ_j, φ_j) on the surface Γ_1 of the multilayer spherical head model (see §1) of the current dipole with moment $\mathbf{m} = (M_X, M_Y, M_Z)$ located at a point (x_d, y_d, z_d) (Koptelov, 1988; Zakharov and Koptelov, 1990).

3.2. The total computing time required for solving the inverse problem (12) is proportional the time required for of solving the direct problem and consists in calculating the dipole potential on the head surface with arbitrary dipole parameters $\mathbf{p} \in G_\mathbf{p}$. Using the multilayer inhomogeneous head model causes great losses of machine time. It gives no possibility to process EEG-data in the quasi-real regime.

To reduce the amount of computation let us replace the multilayer inhomogeneous head model with an equivalent homogeneous head model which basically consists of homogeneous sphere of radius $R_{eq} = R_K + H_{eq}$, $H_{eq} > 0$, and electric conductivity λ_K surrounding unbounded medium of the same conductivity. The internal sphere of radius R_K models the brain and the surrounding spherical layer of thickness H_{eq} simulates the external shells of the brain.

Since the absolute value of the dipole potential depends linearly upon the absolute value of the dipole moment H_{eq} can be found from the condition that the normalized potential values on the surfaces of the multilayer (the sphere Γ_1 of radius R_1) equal approximately to those of the homogeneous (the sphere of radius R_{eq}) head models: $c_1 v_{pr}(R_{eq}, \theta, \varphi) = c_2 v_s(\theta, \varphi)$. Taking into account (3), (9) and (10), we obtain $W_l(H_{eq}) \equiv (2l+1)/l \times (B_l^{(K)})_1^{-1}[(R_K + H_{eq})/R_1]^{l+1} = c_2/c_1 = \text{const}$, $l = 1, 2, ..., L$. Since the constants c_1 and c_2 are unknown, we are led to the following variational problem:

$$H_{eq} = \arg\min \sum_{l=1}^{L} \left[W_l(H) - \frac{1}{L} \sum_{i=1}^{L} W_i(H)\right]^2. \tag{14}$$

After H_{eq} is determined from (14), the functions $\eta_j(p)$ involved in (13) in the case of the nomogeneous head model are representable as $\eta_j(\mathbf{p}) = A_0[u_d^0(\theta_j, \varphi_j, \mathbf{p})]$, where

$$U_d^0(\theta, \varphi, \mathbf{p}) = \frac{1}{4\pi\lambda_k} \frac{(x - x_d)M_X + (y - y_d)M_Y + (z - z_d)M_Z}{\left[(x - x_d)^2 + (y - y_d)^2 + (z - z_d)^2\right]^{3/2}}$$

is the potential at a point (x, y, z), $x = R_{eq} \sin\theta \cos\varphi$, $y = R_{eq} \sin\theta \sin\varphi$, $z = R_{eq} \cos\theta$, on the sphere of radius $R_{eq} = R_K + H_{eq}$ in a homogeneous medium of electric conductivity λ_K of the current dipole with parameters $\mathbf{p} = \{x_d, y_d, z_d, M_X, M_Y, M_Z\}$.

The computer experiment has shown that the standard error of determination of the coordinates of the dipole sources provided by replacing the multilayer head model by an equivalent homogeneous model for $\mathbf{p} \in G_\mathbf{p}$ is 0.5 cm which equals the same standard error in the case of a multilayer model under the assumption that the initial data have 5

Thus, applying an equivalent homogeneous head model decreases the amount of computation required for solving the inverse problem of determination of the coordinates of the dipole sources within the brain with no essential loss of the solution accuracy. Thus, the algoritms considered above can readily be used on modern personal computers for studying the brain functioning in quasi-real time.

4. RESULTS OF EEG-DATA PROCESSING

As an example, results of the EEG-data processing for a patient diagnosed as having epilepsy are shown in Fig. 1 and Fig. 2. These results were obtained by the algorithms considered above.

Fig. 1 shows the potential isoline map (Fig. 1,a) and the current density vector distribution map (Fig. 1,b) on the head surface obtained for some time moment during brain epileptical discharge activity. Positive potential isolines are shown by solid lines, negative potential isolines are depicted by broken lines. The dotted lines present zero potential isolines and the circles mark the projections of lead electrodes placed on the head surface according to the 10–20 international system (see (Rusinov, 1973)). The isolines have a 0,00002 V step. The arrows denote the current density vectors. The X-axis is directed from the occipital domain towards the frontal one and the Y-axis is directed from the right ear towards the left one. The maps presented in Fig. 1 are typical of one current dipole placed within the brain with a concrete orientation to the head surface.

Fig. 2 shows two orthogonal projections (a) of the head–side sight and (b) the back sight, with equivalent dipole sources represented by arrows. These sources were calculated for six different time moments during epileptical brain activity. The arrow lengths characterize the intensities and its direction indicates the orientation of the corresponding dipole source. The interval between the points on the coordinate axis is 1 cm. The atlas of the scale brain cutting was used to introduce the equivalent dipole sources coordinates in a specific brain structure. The comparison of the results of equivalent dipole localization with the data obtained by using the X-ray computer thomography or in the course of the neurosurgery

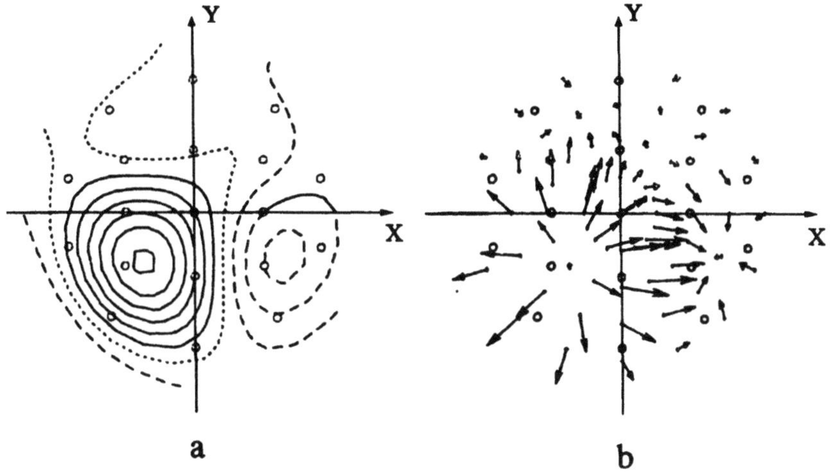

Fig. 1

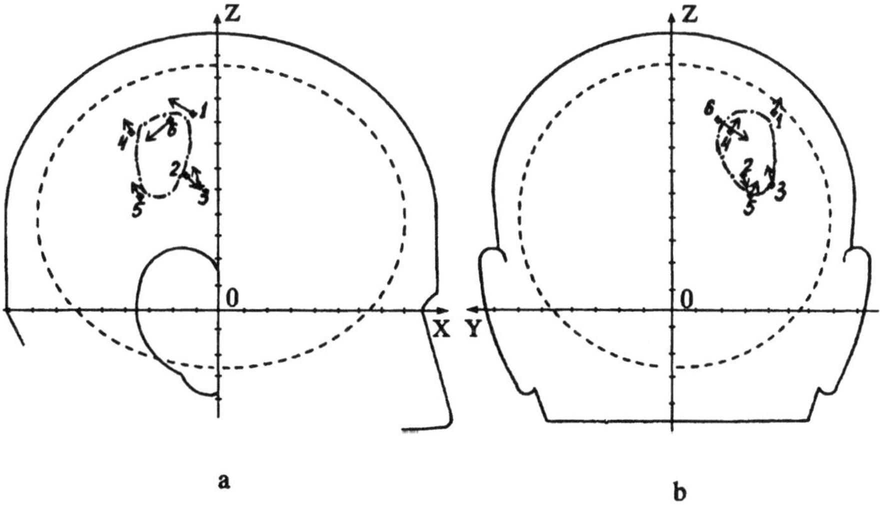

Fig. 2

tumorectomy operation proves that the equivalent dipole sources are located in the perifocal zone of tumor (the boundary of the tumor is shown in Fig. 2 by broken-dotted line). It is in agreement with the known facts about the genesis of pathology brain discharge activity in the perifocal zone of disorder region (Rusinov, 1973). To conclude, we note that application of the algorithms considered above is very promising for the analysis of spike and rhythmical activity, evoked potentials and other kinds of electrical brain activity both in normals and in the cases of pathology.

REFERENCES

Koptelov, Yu.M. (1988). *Research and Numerical Solution of Some Inverse Problems in Electroencephalography.* Candidate Thesis. Moscow Univ. Press, Moscow (in Russian).

Nunez, P.L. (1981). *Electric Fields of the Brain.* Oxford Univ. Press, New York.

Plonsy, R. and Flemind, D.G. (1969). *Bioelectric Phenomena.* McGraw-Hill, New York.

Rusinov, V.S. (Ed.) (1973). *Clinical Electroencephalography.* Medicina, Moscow (in Russian).

Shevelev, I.A., Kuznetsova, G.D., Tsicalov, E.N., Gorbach, A.M., Budko, K.P., and Sharaev G.F. (1989). *Thermoencephaloscopy.* Nauka, Moscow (in Russian).

Stratton, J.A. (1941). *Electromagnetic Theory.* McGraw-Hill, New York.

Tanana, V.P. (1981). *Methods for Solving Operator Equations.* Nauka, Moscow (in Russian).

Tikhonov, A.N. and Arsenin, V. Ya. (1986). *Methods for Solving Ill-Posed Problems.* Nauka, Moscow (in Russian).

Tikhonov, A.N., Goncharsky, A.V., Stepanov, V.V., and Yagola, A.G. (1983). *Regularizing Algorithms and Priori Information.* Nauka, Moscow (in Russian).

Tikhonov, A.N. and Samarsky, A.A. (1977). *Equations of Mathematical Physics.* Nauka, Moscow (in Russian).

Weinberg, H. et al (Eds.) (1985). *Biomagnetism: Applications and Theory.* Pergamon Press, New York.

Zakharov, E.V. and Koptelov, Yu.M. (1987). On solving one problem on mathematical treatment of electroencephalographic data. *Dokl. Akad. Nauk SSSR,* **292**, 576–581 (in Russian)..

Zakharov, E.V. and Koptelov, Yu. M. (1990). Statement and numerical solving inverse problems in electroencephalography In: *Methods of Mathematical Simulation and Computer Diagnostics.* Ed. by A.N. Tikhonov and A.A. Samarsky. Moscow Univ. Press, Moscow, pp. 18–28 (in Russian)..

Ill-Posed Problems in Natural Sciences, pp. 553 – 562
A. Tikhonov (Ed.)
1992 VSP/TVP

A NUMERICAL METHOD OF CRACK DETERMINATION BY THE BOUNDARY INTEGRAL EQUATION METHOD

N. NISHIMURA
Division of Global Environment Engineering, Kyoto University, Kyoto 606, Japan

ABSTRACT

This paper presents a method of determining the unknown shape of a crack in a domain via a numerical analysis based on the boundary integral equation method. We consider a bounded domain Ω in R^3. This domain is known to contain a piece of smooth surface S, or a crack, in its interior, but the shape and the location of S are unknown. To determine S we carry out an experiment using a physical quantity u governed by Helmholtz' equation. Namely, we prescribe on $\Gamma := \partial\Omega$ several Dirichlet data for u and measure the corresponding Neumann data or vice versa. From these data and the homogeneous Neumann condition for u on S we determine S in a manner that the error of a certain integral equation on Γ is minimised. This minimisation is carried out numerically with the help of the boundary integral equation method and a nonlinear programming technique. A three-dimensional numerical example is presented to demonstrate the applicability of the proposed method. Finally a few comments on the convergence of the method and on analogous formulations in other inverse problems are made.

1. INTRODUCTION

Let D be a bounded domain in R^3 which has a smooth boundary $\Gamma := \partial\Omega$. This domain Ω is known to contain a smooth piece of non-self-intersecting surface S,or a *crack*, in its interior. The shape and the location of S, however, are unknown. We now wish to determine S from experimental data obtained in boundary measurements. To be specific we consider an experiment using a physical quantity u governed by Helmholtz' equation

$$(\Delta + k^2)u = 0 \quad \text{in } \Omega\backslash\overline{S}, \tag{1}$$

where $k \geq 0$ is the wave number. The physics of the problem requires that u satisfies

$$\left(\frac{\partial u}{\partial n}\right)^{\pm} = 0 \quad \text{on } S, \tag{2}$$

$$\lim_{x(\in S)\to x_0(\in\partial S)} \varphi(x) = 0, \tag{3}$$

where $\varphi(x) := u^+(x) - u^-(x)$ and the superposed + (−) indicates the limit on S from the positive (negative) side, with the positive side of S indicating the one into which the unit normal vector n to S points. We now carry out measurements which mathematically amount to prescribing Dirichlet data on Γ and measuring the corresponding Neumann data, or vice versa. As a result of these boundary measurements, we know several pairs of both Dirichlet and Neumann data on Γ denoted respectively by $\mathring{u}^I$ and $\mathring{v}^I$ for $I = 1 \sim M$, where M is the number of experiments. From these data we attempt to reconstruct S. Problems of this type in the Laplace case ($k = 0$) have been considered by engineers as well as by mathematicians; the reader is referred to (Kubo, 1988) and (Nishimura and Kobayashi, 1991) for some of these efforts.Also an analogous problem of determining the shape of an ordinary (non-flat) scatterer from far field patterns is considered in (Colton and Monk, 1987) and (Kress, 1989), among others. In this paper we shall extend the two-dimensional numerical method to solve the crack determination problem discussed in (Nishimura, 1991) to three-dimensional with the help of the so called boundary integral equation method (BIEM).

2. DIRECT PROBLEM

We first discuss the direct problem, where the shape and the location of S are known. The statement of the problem is the following: *Solve* (1) *subject to* (2), (3) *and a boundary condition of either Dirichlet or Neumann type on* Γ. As is well known, the solution to this problem has the following integral representation:

$$u(x) = \int_\Gamma G(x-y)\frac{\partial u(y)}{\partial n}\, dS_y - \int_\Gamma \frac{\partial G(x-y)}{\partial n_y} u(y)\, dS_y$$
$$+ \int_S \frac{\partial G(x-y)}{\partial n_y}\varphi(y)\, dS_y,$$

where $G(x-y)$ is the fundamental solution of Helmholtz' equation given by

$$G(x-y) = \frac{e^{ik|x-y|}}{4\pi|x-y|}, \quad k \geq 0.$$

This representation, however, includes unknown quantities, i.e., the unknown data on Γ (either u or $\partial u/\partial n$) and φ on S. They are obtained as the solution to the following system of 'integral' equations:

$$0 = \frac{u(x)}{2} - \int_\Gamma G(x-y)\frac{\partial u(y)}{\partial n} dS_y + \int_\Gamma \frac{\partial G(x-y)}{\partial n_y} u(y) dS_y$$
$$- \int_S \frac{\partial G(x-y)}{\partial n_y}\varphi(y) dS_y, \quad x \in \Gamma \tag{4}$$

$$0 = \frac{\partial}{\partial n_x}\left(\int_\Gamma G(x-y)\frac{\partial u(y)}{\partial n} dS_y - \int_\Gamma \frac{\partial G(x-y)}{\partial n_y} u(y) dS_y\right)$$

$$+\fint_S \frac{\partial}{\partial n_x}\frac{\partial}{\partial n_y}G(x-y)\varphi(y)\,dS_y, \quad x \in S \tag{5}$$

where the integration sign with a superimposed = indicates an integration in the sense of the finite part. This system of equations has a unique solution unless k coincides with one of the eigenfrequencies of the homogeneous problem associated with the given boundary value problem. In engineering applications one usually discretises these equations with the help of certain shape functions and collocation, and then solves the resulting algebraic equations numerically. It has been shown, however,that the use of C^1 elements for φ is necessary in order to obtain a good numerical solution of (5) with collocation (see (Nishimura and Kobayashi, 1991)). Satisfying this requirement for the shape function in three-dimensional problems is a considerable difficulty because it is not easy to generate C^1 shape functions on a surface. This is incontrast to the two-dimensional case where one may use spline functions for example. We therefore choose to use a variational approach in the discretisation of (5), because this approach allows the use of C^0 elements. In addition to this relaxed requirements for the shape functions, the variational approach has an advantage of reducing the hypersingular integral in (5) to an ordinary one. This reduction of the singularity is sometimes called "regularisation", although it is not related to Tikhonov's regularisation.

To show the variational equation in the present application, we consider a simplified version of (5) given by

$$\frac{\partial g(x)}{\partial n} = \fint_S \frac{\partial}{\partial n_x}\frac{\partial}{\partial n_y}G(x-y)\varphi(y)\,dS_y, \quad x \in S \tag{6}$$

where g is a known function defined in Ω. A variational equation associated with (6) is written as (Hamdi, 1981)

$$\int_{S_0} e_{PQ}\frac{\partial\psi(x)}{\partial s_P}\frac{\partial x_p(x)}{\partial s_Q}\,ds_{1_x}ds_{2_x}\int_{S_0} G(x-y)e_{KL}\frac{\partial\varphi(y)}{\partial s_K}\frac{\partial x_p(y)}{\partial s_L}\,ds_{1_y}ds_{2_y}$$

$$-k^2\int_S \psi(x)n_i(x)dS_x\int_S G(x-y)n_i(y)\varphi(y)dS_y = -\int_S \psi g_{,j}n_j dS, \tag{7}$$

where ψ is a test function which vanishes on ∂S, e_{PQ} is the two-dimensional permutation symbol, $s_{1,2}$ are curvilinear coordinates by which a fixed domain $S_0 \subset R^2$ is mapped to S and $_{,i} := \partial/\partial x_i$. We henceforth use the summation convention to repeated lower case (upper case) indices which run from 1 to 3 (1 to 2). The solution φ to (7) and (3) is unique. Notice that the kernel functions in this variational equation is integrable on S in the ordinary sense. Therefore the discretisation of this equation with Galerkin's method and a finite element interpolation for φ is straightforward. See (Nishimura and Kobayashi, 1991) for further details in the Laplace case.

A practical way of obtaining a numerical solution to the system given by (4) and (5) is to use collocation to (4) and the variational method to (5). In this case the function g on the left hand side of (6) is unknown: it depends linearly on the discretised unknown quantities on Γ, whose coefficients are written in terms of integrals with bounded integrands (see (5)). Therefore the discretisation of this system is done without difficulty.

3. INVERSE PROBLEM

We next discuss the inverse problem mentioned in the introduction. This problem is now formulated as follows: *Given M pairs $(\overset{\circ}{u}^I, \overset{\circ}{v}^I)$, $(I = 1 \sim M)$ on Γ, find a crack S in a way that there exists a function $u^I(x)$ $(x \in \Omega \setminus \overline{S}, I = 1 \sim M)$ which satisfies* (1) $\sim$ (3) *and*

$$u^I = \overset{\circ}{u}^I, \tag{8}$$

$$\frac{\partial u^I}{\partial n} = \overset{\circ}{v}^I \qquad \text{on } \Gamma \quad (I = 1 \sim M) \tag{9}$$

To the best of the knowledge of this author, there are only a few known theoretical results concerning this problem. For example, (Friedman and Vogelius, 1989) showed that 2 experiments ($M = 2$) are sufficient for the uniqueness of the solution (assumed to exist) to this problem in the two-dimensional Laplace case ($k = 0$). It seems that their results have not been extended to the general three-dimensional Helmholtz case where k is nonzero (see (Nishimura, 1991) for more comments on this subject.). In this paper, however, we shall not get into the uniqueness issue any further because we are more interested in actually constructing the most plausible crack.

In general our inverse problem may not have a solution. Guided by (4) and (5), however, we consider the following minimisation problem: *Minimise$_S$ $J(S)$ subject to $S \subset \Omega$*, where the cost function $J(S)$ is defined as follows: Given an arbitrary S, we solve (see (5), (8) and (9))

$$\frac{\partial g^I(x)}{\partial n} = \oint_S \frac{\partial}{\partial n_x}\frac{\partial}{\partial n_y} G(x-y)\varphi^I(y)\, dS_y, \quad x \in S, \tag{10}$$

for the unknown φ^I on S with the help of (7),where

$$g^I(x) = \int_\Gamma \frac{\partial G(x-y)}{\partial n_y} \overset{\circ}{u}^I(y) dS_y - \int_\Gamma G(x-y)\overset{\circ}{v}^I(y) dS_y \quad \text{in } \Omega,$$

which is a known function because both $\overset{\circ}{u}^I$ and $\overset{\circ}{v}^I$ are known on Γ via measurements. We now define $J(S)$ (see (4)) by

$$J(S) = \frac{1}{2}\sum_I \int_\Gamma \left| \frac{\overset{\circ}{u}^I(x)}{2} - \int_\Gamma G(x-y)\overset{\circ}{v}^I(y) dS_y \right.$$

$$\left. + \int_\Gamma \frac{\partial G(x-y)}{\partial n_y} \overset{\circ}{u}^I(y) dS_y - \int_S \frac{\partial G(x-y)}{\partial n_y}\varphi^I(y) dS_y \right|^2 dS.$$

The most plausible crack S is then obtained as the minimiser of $J(S)$. In the next section we present a numerical method for solving this minimisation problem.

4. NUMERICAL METHODS

4.1. *Minimisation*

In practice one may introduce simplifying assumptions on the shape of S and describe S by some shape parameters. For example, the assumption that S is planar and elliptic enables us to identify S by 8 parameters, as we shall see. Since the cost $J(S)$ is then a function of these parameters, we can minimise $J(S)$ by using nonlinear programming techniques. We here choose to use Powell's variable metric method, which belongs to the family of quasi-Newton methods. This nonlinear programming method needs derivatives of $J(S)$ with respect to the shape parameters, which are computed as

$$\overset{\triangledown}{J}(S) = \sum_I \mathrm{Re}\Big\{ \int_S \Big[e_{jip} e_{stp} \overset{\triangledown}{x}_j n_t \frac{\partial \varphi^I}{\partial x_s} - n_i \overset{\triangledown}{\varphi}^I \Big] dS \int_\Gamma G\overline{F^I} dS + k^2 \int_S \varphi^I \overset{\triangledown}{x}_i n_i \, dS \int_\Gamma G\overline{F^I} dS \Big\}, \qquad (11)$$

where the superposed triangle signifies one of such derivatives,e_{ijk} indicates the permutation symbol, stands for the complex conjugate and

$$F^I(y) = \frac{\overset{\circ}{u}^I(x)}{2} - \int_\Gamma G(x-y)\overset{\circ}{v}^I(y)\, dS_y + \int_\Gamma \frac{\partial G(x-y)}{\partial n_y}\overset{\circ}{u}^I(y)\, dS_y - \int_S \frac{\partial G(x-y)}{\partial n_y}\varphi^I(y)\, dS_y.$$

All the quantities included in (11) except for $\overset{\triangledown}{\varphi}^I$ are known from the boundary data $(\overset{\circ}{u}^I, \overset{\circ}{v}^I)$ and the solution to (10). Also, a variational equation given by

$$\int_{S_0} e_{PQ}\frac{\partial\psi}{\partial s_P}(x)\frac{\partial x_p}{\partial s_Q}(x)\, ds_{1_x} ds_{2_x} \int_{S_0} G(x-y) e_{KL} \frac{\partial \overset{\triangledown}{\varphi}^I}{\partial s_K}(y)\frac{\partial x_p}{\partial s_L}(y)\, ds_{1_y} ds_{2_y}$$

$$- k^2 \int_S \psi(x) n_i(x)\, dS_x \int_S G(x-y)\overset{\triangledown}{\varphi}^I(y) n_i(y)\, dS_y$$

$$= \int_{S_0} e_{PQ}\frac{\partial\psi}{\partial s_P}(x)\frac{\partial x_p}{\partial s_Q}(x) e_{pij} g_i^I(x)\overset{\triangledown}{x}_j(x)\, ds_1 ds_2$$

$$+ k^2\int_S \psi(x) g^I(x) n_i(x)\overset{\triangledown}{x}_i(x)\, dS$$

$$-\int_{S_0} e_{PQ}\frac{\partial\psi}{\partial s_P}(x)\frac{\partial \overset{\triangledown}{x}_p}{\partial s_Q}(x)\,ds_{1_x}ds_{2_x}\int_{S_0} G(x-y)e_{KL}\frac{\partial\varphi^I}{\partial s_K}(y)\frac{\partial x_p}{\partial s_L}(y)\,ds_{1_y}ds_{2_y}$$

$$-\int_{S_0} e_{PQ}\frac{\partial\psi}{\partial s_P}(x)\frac{\partial x_p}{\partial s_Q}(x)\,ds_{1_x}ds_{2_x}$$

$$\times\int_{S_0} G_{,k}(x-y)(\overset{\triangledown}{x}_k(x)-\overset{\triangledown}{x}_k(y))e_{KL}\frac{\partial\varphi^I}{\partial s_K}(y)\frac{\partial x_p}{\partial s_L}(y)\,ds_{1_y}ds_{2_y}$$

$$-\int_{S_0} e_{PQ}\frac{\partial\psi}{\partial s_P}(x)\frac{\partial x_p}{\partial s_Q}(x)\,ds_{1_x}ds_{2_x}\int_{S_0} G(x-y)e_{KL}\frac{\partial\varphi^I}{\partial s_K}(y)\frac{\partial \overset{\triangledown}{x}_p}{\partial s_L}(y)\,ds_{1_y}ds_{2_y}$$

$$+k^2\int_{S_0}\psi(x)e_{ikp}e_{KL}\frac{\partial\overset{\triangledown}{x}_k}{\partial s_K}(x)\frac{\partial x_p}{\partial s_L}(x)\,ds_{1_x}ds_{2_x}\int_{S} G(x-y)\varphi^I(y)n_i(y)\,dS_y$$

$$+k^2\int_{S}\psi(x)n_i(x)\,dS_x\int_{S} G_{,k}(x-y)(\overset{\triangledown}{x}_k(x)-\overset{\triangledown}{x}_k(y))\varphi^I(y)n_i(y)\,dS_y$$

$$+k^2\int_{S}\psi(x)n_i(x)\,dS_x\int_{S_0} G(x-y)\varphi^I(y)e_{ikp}e_{KL}\frac{\partial\overset{\triangledown}{x}_k}{\partial s_K}(y)\frac{\partial x_p}{\partial s_L}(y)\,ds_{1_y}ds_{2_y} \qquad (12)$$

for $\overset{\triangledown}{\varphi}^I$ is obtained as one takes the "triangle derivative" of both sides of (10). This equation can be solved numerically for $\overset{\triangledown}{\varphi}^I$ in a manner similar to the method of solving (7) if φ^I is known.

4.2. Computation of integrals

The iterative nature of nonlinear programming suggests that it is important to find fast ways to compute various integrals included in the proposed method. In the numerical examples to be shown in this paper we restrict our attention to planar cracks. Also, we use the ordinary three-node triangular finite elements (piecewise linear) for the discretisation of φ^I. In this case the "inner" integrals, included in the double integrals in (7) after the discretisation, allow the following expressions:

$$\int_{E_S} G(x-y)\,dS_y, \qquad (13)$$

$$\int_{E_S} G(x-y)\Phi(y)\,dS_y, \qquad x\in S, \qquad (14)$$

where $\Phi(x)$ is a piecewise linear shape function defined on a plane element E_S on S. To compute these integrals economically one splits G into

$$G(x-y) = G(x-y)|_{k=0} + \frac{i}{2\pi r} e^{ikr/2} \sin\left(\frac{kr}{2}\right), \quad r := |x-y|,$$

and decomposes the integrals in (13) and (14) accordingly. The integrals including $G|_{k=0}$ are evaluated analytically and the remaining integrals are computed numerically. We can thus avoid costly numerical integrations of a singular function $G(x-y)|_{k=0}$. The "outer"' integrals in (7) are then evaluated numerically. Notice that the algebraic equations obtained from (7) and (12) with Galerkin's method are identical. We therefore store the triangular-decomposed matrix equation for φ^I so that we can reuse it later in the computation of $\overset{\nabla}{\varphi}{}^I$.

4.3. *Elliptical crack*

In the special case of a planar elliptical crack we use 8 parameters t_i $(i = 1, \ldots, 8)$ to describe S as in the following formulae:

$$x_i = aQ_{iJ}A_{JK}\zeta_K + d_i, \quad x \in S, \ |\zeta| \le 1.$$

where $a = \sqrt{t_1^2 + t_2^2 + t_3^2}$, $d_i = t_{i+3}\,(i = 1 \sim 3)$ and

$$aQ = \begin{bmatrix} t_3 + \dfrac{t_2^2}{a+t_3} & -\dfrac{t_1 t_2}{a+t_3} & t_1 \\ -\dfrac{t_1 t_2}{a+t_3} & t_3 + \dfrac{t_1^2}{a+t_3} & t_2 \\ -t_1 & -t_2 & t_3 \end{bmatrix}, \quad A = \begin{bmatrix} 1+t_7 & t_8 \\ t_8 & 1-t_7 \end{bmatrix}, \quad t_7^2 + t_8^2 \le 1.$$

The matrix Q is seen to be orthogonal with its third column coinciding with the unit normal vector n to S. This matrix is designed so that the parametrisation does not introduce discontinuities to $\overset{\nabla}{\varphi}{}^I$ except when $t_3 = -a$ (This discontinuity is unavoidable because there exists no continuous unit tangential vector field on a unit sphere.). The variational equations (7) and (12) now simplify to

$$\int_{S'} \frac{\partial \psi}{\partial \xi_J}(\xi)\, d\xi_1 d\xi_2 \int_{S'} G(x(\xi) - x(\eta)) \frac{\partial \varphi^I}{\partial \eta_J}(x(\eta))\, d\eta_1 d\eta_2$$

$$-(ka)^2 \int_{S'} \psi(\xi)\, d\xi_1 d\xi_2 \int_{S'} G(x(\xi) - x(\eta))\varphi^I(x(\eta))\, d\eta_1 d\eta_2$$

$$= -\int_{S'} \psi(\xi) g^I_{,i}(x(\xi)) n_i(x(\xi))\, d\xi_1 d\xi_2,$$

and

$$\int_{S'} \frac{\partial \psi}{\partial \xi_J}(\xi)\, d\xi_1 d\xi_2 \int_{S'} G(x(\xi) - x(\eta)) \frac{\partial \overset{\nabla}{\varphi}{}^I}{\partial \eta_J}(x(\eta))\, d\eta_1 d\eta_2$$

$$-(ka)^2 \int_{S'} \psi(\xi)\, d\xi_1 d\xi_2 \int_{S'} G(x(\xi)-x(\eta))\overset{\triangledown}{\varphi}{}^I(x(\eta))\, d\eta_1 d\eta_2$$

$$= \frac{1}{a}\Bigg\{ \int_{S'} e_{JK}\frac{\partial \psi}{\partial \xi_J}(\xi) Q_{pK} e_{pij} g^I_{,i}(x(\xi))\overset{\triangledown}{x}_j(x(\xi))\, d\xi_1 d\xi_2$$

$$+k^2 a \int_{S'} \psi(\xi) g^I(x(\xi)) n_i(x(\xi))\overset{\triangledown}{x}_i(x(\xi))\, d\xi_1 d\xi_2$$

$$+\left(\overset{\triangledown}{a} + a\frac{L_{JJ}}{2}\right)\int_{S'} \psi(\xi) g^I_{,i}(x(\xi))\, n_i\, d\xi_1 d\xi_2 \Bigg\}$$

$$+k^2 a\left(\overset{\triangledown}{a} + a\frac{L_{JJ}}{2}\right)\int_{S'} \psi(\xi)\, d\xi_1 d\xi_2 \int_{S'} G(x(\xi)-x(\eta))\varphi^I(x(\eta))\, d\eta_1 d\eta_2$$

$$+L'_{JK}\int_{S'} \frac{\partial \psi}{\partial \xi_J}(\xi)\, d\xi_1 d\xi_2 \int_{S'} G(x(\xi)-x(\eta))\frac{\partial \varphi^I}{\partial \eta_K}(x(\eta))\, d\eta_1 d\eta_2$$

respectively, where $L_{JK} := \overset{\triangledown}{A}_{JL}A^{-1}_{LK}$, $\Omega_{iJ} := Q_{ji}\overset{\triangledown}{Q}_{jJ}$, $L'_{JK} := \frac{1}{2}(L_{JK} + L_{KJ} - L_{PP}\delta_{JK})$, $S' = \{\eta | \eta_I = A_{IJ}\zeta_J,\ |\zeta| \le 1\}$ and the points ξ and η on S' are referred to a rectangular coordinate system on S'.The reader should bear in mind that the quantities L and Ω depend on the particular choice for t_i in the definition of $\triangledown := \partial/\partial t_i$.
Finally we note that t_7 and t_8 are absent, that A_{JK} reduces to the identity and that L vanishes when S is circular. In the Laplace case (Nishimura and Kobayashi, 1991) pointed out that the circular crack assumption keeps the matrix equation for φ^I unchanged throughout the whole minimisation process. Unfortunately, however, this property does not hold any longer for a nonzero k.

5. NUMERICAL EXAMPLE

We consider a cubic domain Ω containing a circular crack S as shown in Figure 1. The edge length of Ω is l. The Dirichlet data given on the boundary are $\overset{\circ}{u}{}^I = e^{ikx_I}$ $(I = 1 \sim 3)$, and the corresponding Neumann data $\overset{\circ}{v}{}^I$ are prepared numerically rather than experimentally via

$$\overset{\circ}{v}{}^I = \frac{\partial e^{ikx_I}}{\partial n} + (1+\mathrm{RND})\left(\frac{\partial u^I}{\partial n} - \frac{\partial e^{ikx_I}}{\partial n}\right),$$

where $\partial u^I/\partial n$ is the Neumann data computed with BIEM (boundary integral equation method) using the true geometry of S and RND is a random number.

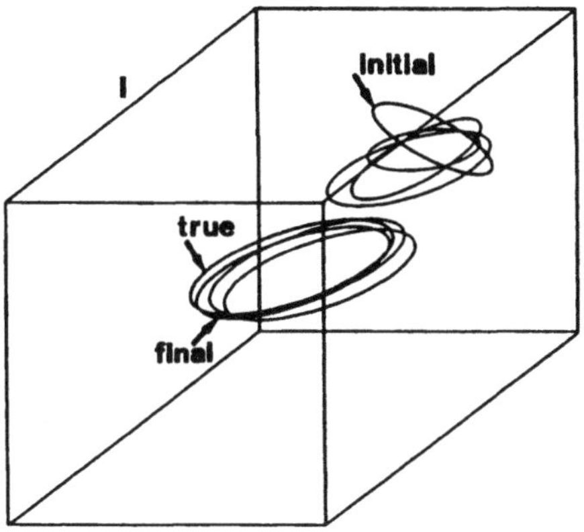

Fig. 1

We here set RND to be a uniform random number in the interval of $[-1/2, 1/2]$. The wave number is chosen to be $kl = 1.2$. Galerkin's method, together with (7), is used to solve the three-dimensional hypersingular integral equation in (10). The exterior boundary is divided into 216 piecewise constant rectangular elements of the same size and the crack S is discretised with piecewise linear triangular elements having 21 DOF in total. An analysis starting from the initial guess shown in Figure 1 converged to the crack which is sufficiently close to the true one as indicated in the same figure. Figure 1 also includes intermediate crack configurations after every (approximately) 5 iteration steps.

6. CONCLUDING REMARKS

1. During the minimisation of $J(S)$ one is sometimes trapped by local minima of $J(S)$. This is especially true when the wave number is large and the initial guess is far away from the true crack. Our experience shows that this problem of local minima can be avoided as one uses a small k. However, the convergence seems to be faster with a bigger k when the initial guess is not very far from the true crack.
2. The method developed in this paper is applicable to analogous problems for the wave equation in time domain as well. Indeed, such attempts have already been made both in 2D and 3D problems. One may also apply similar methods to elasticity problems.

REFERENCES

Colton, D. and Monk, P. (1987). The numerical solution of the three-dimensional inverse scattering problem for time harmonic acoustic waves. *SIAM J. Sci. Stat. Comput.* **8**, 278–291.

Friedman, A. and Vogelius, M. (1989). Determining cracks by boundary measurements. *Indiana Univ. Math. J.* **38**, 527–556.

Hamdi, M.A. (1981). Une formulation variationnelle par équations intégrales pour la résolution de l'équation de Helmholtz avec des conditions au limites mixtes. *C. R. Acad. Sci. Paris, (Sér. II)* **292**, 17–21.

Kress, R. (1989). *Linear Integral Equations.* Springer, Verlag, Berlin.

Kubo, S. (1988). Inverse problems related to the mechanics and fracture of solids and structures. *JSME Int. J.* **31**, 157–166.

Nishimura, N. (1991). Regularised integral equations for crack shape determination problems. In: *ICM-90 Satellite Conference Proceedings,* Springer-Verlag, Tokyo (to appear).

Nishimura, N. and Kobayashi, S. (1991). A boundary integral equation method for an inverse problem related to crack detection (to appear).

Ill-Posed Problems in Natural Sciences, pp. 563 – 567
A. Tikhonov (Ed.)
1992 VSP/TVP

REGULARIZATION METHOD IN NONSTATIONARY INVERSE SCATTERING PROBLEM

L. NIZHNIK and R. ROMANENKO
Institute of Mathematics, Ukrainian Academy of Sciences, Repin str. 3, 252601 Kiev, Ukraine

ABSTRACT

The results on the two-dimensional inverse scattering problem for the hyperbolic Dirac system of equations are formulated.
The inverse scattering problem for the hyperbolic equations (the problem of reconstruction of the coefficients of equations from the scattering data) is an ill-posed problem in the natural spaces. Estimates of the conditional stability for this problem are established.
The algorithm of the solution of the inverse scattering problem, with a two-parameter set of integral equations as the basic part, admits the Tikhonov regularization.
This permits us to construct the stable algorithms for approximate solution of the inverse scattering problem from approximately and incompletely known initial data.

1. TWO-DIMENSIONAL NONSTATIONARY INVERSE SCATTERING PROBLEM FOR THE DIRAC SYSTEM

We shall consider the two-dimensional nonstationary hyperbolic Dirac system of equations of the form

$$\frac{\partial \psi(x,t)}{\partial t} = \sigma \frac{\partial \psi(x,t)}{\partial x} + U(x,t)\psi(x,t) \tag{1.1}$$

with the matrix coefficients

$$\sigma = \begin{pmatrix} 1 & 0 \\ 0 & -1 \end{pmatrix}, \qquad U(x,t) = \begin{pmatrix} 0 & u_1(x,t) \\ u_2(x,t) & 0 \end{pmatrix}$$

on the whole axis $-\infty < x < +\infty$ under the assumption that the coefficients $u_1(x,t)$, $u_2(x,t)$ are complex-valued measurable in x, t functions which satisfy the a priori condition

$$u_1,\ u_2 \in \mathbf{L}_2(E^2). \tag{1.2}$$

The admissible solution of the system (1.1), that is the solution with the property $\psi(\cdot\,,t) \in \mathbf{L}_2$, has the following asymptotic behaviour at infinity

$$\psi_1(x,t) = a_1(t+x) + o(1), \quad x \to +\infty,$$

$$\psi_2(x,t) = a_2(t-x) + o(1), \quad x \to -\infty,$$
$$\psi_1(x,t) = b_1(t+x) + o(1), \quad x \to -\infty,$$
$$\psi_2(x,t) = b_2(t-x) + o(1), \quad x \to +\infty,$$

with a_1, a_2, b_1, $b_2 \in \mathbf{L}_2$.
The vector-functions $a(\xi) = (a_1(\xi), a_2(\xi))$, $b(\xi) = (b_1(\xi), b_2(\xi))$ define the profile of incident and scattered waves respectively.
The operator S transforming the incident waves into scattered waves is called the scattering operator

$$b(\xi) = Sa(\xi).$$

If the coefficients $u_i(x,t)$, $i = 1,2$, satisfy condition (1.2), then in the space $\mathbf{L}_2(-\infty, +\infty; E^2)$ the operator S^{-1} exists and

$$S = I + F, \qquad S^{-1} = I + G,$$

where $F = \{F_{ij}\}_{i,j=1,2}$, $G = \{G_{ij}\}_{i,j=1,2}$ are the matrix Hilbert–Schmidt integral operators with the matrix kernels $F(\xi,\eta)$, $G(\xi,\eta)$.
One of the pairs of operators (F_{12}, G_{21}), (F_{21}, G_{12}) will be called the scattering data for system (1.1).
The two-dimensional Nonstationary Inverse Scattering Problem (NISP) consists in finding the coefficients $u_1(x,t), u_2(x,t)$ of the Dirac sytem (1.1) when the scattering operator S or the scattering data, say (F_{12}, G_{21}), are given. Under assumption (1.2) NISP is uniquely solvable.
The algorithm of reconstructing the coefficients of equations (1.1) from the initial data follows from the formulas

$$\mathcal{T}_x S \mathcal{T}_{-x} = \left(I + K_-(x)\right)^{-1}\left(I + K_+(x)\right), \tag{1.3}$$

$$U(x,t) \doteq \sigma K_-(x;t,t)\sigma - K_-(x;t,t) = K_+(x;t,t) - \sigma K_+(x;t,t)\sigma, \tag{1.4}$$

where $\mathcal{T}_x = \operatorname{diag}\{T_x, T_{-x}\}$, $T_x f(t) = f(t+x)$ are the matrix and the usual shift operators; $K_+(x)$, $K_-(x)$ are the Volterra operators with variable upper and lower limits of integration.
Finding the kernels $K_\pm(x;t,t)$ of the operators $K_\pm(x)$ is reduced to solving the fundamental equations for NISP which follow from (1.3) and are the equations of Gel'fand–Levitan–Marchenko type. The coefficients $u_i(x,t)$, $i = 1,2$, can be defined in terms of these kernels by (1.4).
In the theory of multidimensional inverse problems the results on the direct and inverse scattering problems for two-dimensional nonstationary Dirac system (1.1) have been obtained by (Nizhnik, 1973; 1988; 1991).

2. ESTIMATES OF CONDITIONAL STABILITY FOR NISP

In the natural, from the practical point of view, spaces NISP is an ill-posed problem. The question of its conditional stability arises.
The conditional stability of NISP means that for $\forall \varepsilon > 0$

$$\|U^{(1)} - U^{(2)}\|_{\Phi_1} \le \varepsilon,$$

if

$$\|S^{(1)} - S^{(2)}\|_{\Phi_2} \leq \delta(\varepsilon), \quad U^{(i)} \in \mathcal{M} \subset \Phi_1, \quad i = 1, 2,$$

where Φ_1 is the space of solutions, Φ_2 is the space of initial data, $\mathcal{M}$ is the set of correctness for NISP.

The choice of the spaces depends on the requirements of applications.

As Φ_1, Φ_2 we have considered the following spaces of functions $f(x,y)$ depending on two variables: $\mathbf{L}_p(E^2)$, $p \geq 1$; $W_p(E^2)$, $p = 1, 2$, which is the space of functions whose Fourier transforms belong to $\mathbf{L}_p(E^2)$; $B_\mu^m(E^2)$ which is the weight space with the norm

$$\|f\|_{B_\mu^m(E^2)} = \underset{x,y,\alpha_1+\alpha_2 \leq m}{\text{vrai sup}} (1 + x^2 + y^2)^\mu \left| \frac{\partial^{|\alpha|} f(x,y)}{\partial x^{\alpha_1} \partial y^{\alpha_2}} \right|.$$

As Φ_2 the following "mixed" spaces of functions depending on two variables have been also considered: the spaces $\mathbf{L}_p W_q$, $W_q \mathbf{L}_p$ with the norm defined in the order of variables, and the spaces $\mathbf{L}_p(W_q)$, $W_q(\mathbf{L}_p)$ with the norm, defined in the inverse order. These spaces permit to consider the cases when the profile or the harmonics composition of the incident and scattered waves are measured.

We have chosen the sphere of the radius R in the space $B_\mu^m(E^2)$ as the set of correctness $\mathcal{M}_m$ for NISP.

The following estimates of the conditional stability have been established by (Nizhnik and Romanenko, 1981).

THEOREM 1. *If the coefficients $U^{(1)}(x,t)$, $U^{(2)}(x,t)$ of two Dirac systems* (1.1) *belong to the set of correctness $\mathcal{M}_m$, then there exists a constant $\mathcal{A}$ such that the following estimates hold*

$$\|U^{(1)} - U^{(2)}\|_{\Phi_1} \leq \mathcal{A} \|S^{(1)} - S^{(2)}\|_{\Phi_2}, \tag{2.1}$$

$$\|U^{(1)} - U^{(2)}\|_{\Phi_1} \leq \mathcal{A}(\|F_{12}^{(1)} - F_{12}^{(2)}\|_{\Phi_2} + \|G_{21}^{(1)} - G_{21}^{(2)}\|_{\Phi_2}), \tag{2.2}$$

where for $m = 0$ as the pair (Φ_1, Φ_2) of spaces one of the pairs $(\mathbf{L}_p, \mathbf{L}_p)$, $(\mathbf{L}_2, W_2)$, (B_μ^0, B_μ^0) can be chosen and for $m = 2$ as the pair (Φ_1, Φ_2) one of the pairs (C, W_1), $(\mathbf{L}_\infty, \mathbf{L}_\infty(W_1))$ can be considered.

In practice we can not use estimates (2.1), (2.2) if the initial data are known not only approximately but also incompletely in the sense, that they are not given for all the values of the arguments.

We shall say that a defined for all $x \in R^n$ function $f(x)$ from the functional space Φ is given approximately with the accuracy δ and with the degree of incompleteness N, if this function is known on the sphere $S_N = \{x\colon |x| \leq N\}$ of the radius N with the accuracy δ on Φ-metric, i.e., if the (a) function f_{apr} defined on S_N exists such that

$$\left\|\theta_N(f - f_{\text{apr}})\right\|_\Phi \leq \delta,$$

where θ_N is the operator of multiplying by the characteristic function of the sphere S_N.

For the first time the estimates of the conditional stability of the inverse scattering problem for the Sturm–Liouville equation have been established by (Lundina and Marchenko, 1969).

The following estimates of accuracy of the reconstruction of the coefficients of Dirac system (1.1) from incompletely known initial data have been obtained by (Romanenko, 1985).

THEOREM 2. *If the coefficients $U^{(i)}(x,t)$, $i = 1,2$, of two Dirac systems (1.1) belong to the set of correctness $\mathcal{M}_m$, then defined by this condition constant $\mathcal{A}$ exists such that, for any $N > 1$,*

$$\|U^{(1)} - U^{(2)}\|_{\Phi_1} \leq \mathcal{A}\Big(\|\theta_N(S^{(1)} - S^{(2)})\|_{\Phi_2} + N^{-r}\Big), \quad r > 0,$$

where $r = 2(\mu - 1/p)$ for $\Phi_1 = \Phi_2 = \mathbf{L}_p$ and $r = m_{-1}$ in the case $\Phi_1 = \mathbf{L}_2$, $\Phi_2 = W_2$;

$$\|U^{(1)} - U^{(2)}\|_{\Phi_1} \leq \mathcal{A}\Big(\|\theta_N(F_{12}^{(1)} - F_{12}^{(2)})\|_{\Phi_2}$$

$$+\|\theta_N(G_{21}^{(1)} - G_{21}^{(2)})\|_{\Phi_2} + N^{-r}\Big), \quad r > 0,$$

where $r = m - 1$ when $\Phi_1 = \mathbf{L}_2$, $\Phi_2 = W_2$ and $r = m_{-2}$ for $\Phi_1 = C$, $\Phi_2 = W_1$.

3. REGULARIZATION METHOD IN NONSTATIONARY INVERSE SCATTERING PROBLEM

One of the most wides pread methods of constructing the regularizated solution of the linear equation $Ax = h$ is expressed by the formula $x_\alpha = (A^*A + \alpha I)^{-1}A^*h$, where α is the parameter of regularization and A^* is the adjoint to A operator. The defined by this formula regularization will be called the standard regularization.
The solution of the fundamental equations of NISP, which is constructed from the approximately given initial data, can be not unique. Therefore it is necessary to construct regularization in the problem of factorization (1.3) of the Fredholm operator by the Volterra multipliers.
The following results have been obtained by (Nizhnik and Rozanov, 1982).
Standartly regularized solutions of the fundamental equation of the problem of factorization give the regularization in the problem of the normal factorization of the Fredholm operator.
The usual algorithm of NISP, whose fundamental equations being constructed with the help of the approximately known initial data, are solved by the method of standard regularization, gives the regularization in NISP.

REFERENCES

Lundina, D. and Marchenko, V. (1969). Correction of inequalities characterizing stability of inverse scattering problem. *Math. USSR Sbornik.* **78**, 475–484 (in Russian).

Nizhnik, L. (1973). *Nonstationary Inverse Scattering Problem.* Naukova Dumka, Kiev (in Russian).

Nizhnik, L. (1988). The inverse scattering problems for the hyperbolic equations and their application to non linear integrable systems. *Reports Math. Phys.* **26**, 261–283.

Nizhnik, L. (1991). *Inverse Scattering Problem for Hyperbolic Equations.* Naukova Dumka, Kiev (in Russian).

Nizhnik, L. and Romanenko, R. (1981). Conditional stability of nonstationary inverse scattering problem. *Ukrainian Math. J.* **33**, 694–696.

Nizhnik, L. and Rozanov, R. (1982). Method of regularization in factorization problem. *USSR Comput. Math. Math. Phis.* **22**, 540–547 (in Russian).

Romanenko, R. (1985). Conditional stability of nonstationary inverse scattering problem with incompletely known scattering data. In: *Direct and Inverse Problems of Spectral Theory of Differential Operators.* Math. Inst. Press, Ukrainian Academy Sci., Kiev, pp. 119–123 (in Russian).

Ill-Posed Problems in Natural Sciences, pp. 568 – 583
A. Tikhonov (Ed.)
1992 VSP/TVP

NUMERICAL METHODS FOR INVERSE SCATTERING PROBLEMS IN TWO-DIMENSIONAL SCALAR FIELD

K. ONISHI
Department of Mathematics II, Science University of Tokyo, Wakamiya-cho 26, Shinjuku-ku, Tokyo 162, Japan

ABSTRACT

Methods for the numerical solution of inverse scattering problems for shape identification from the scattering amplitude are presented. The field is governed by two-dimensional Helmholtz equation in exterior domain subject to plane incident waves. The scatterer is assumed either soft or hard infinitely long cylindrical obstacle. Some numerical results of shape identification problems as well as direct problems corresponding to the inverse problems are demonstrated.

1. INTRODUCTION

Boundary integral equation method is suited to the scattering problem of homogeneous media, for the method can not only reduce the exterior boundary value problem with the domain extending to infinity to a corresponding integral equation on the compact surface of the domain, but also reduce the dimensionality of the problem by one. The Galerkin-collocation method using lower-order finite element interpolation functions has been applied to the numerical solution of the boundary integral equation by the name of Boundary Element Method.

The integral equation method for inverse scattering problems has been investigated by several authors. Among them, (Landsberga and Borovikova, 1982; Colton and Kress, 1983; Kirsch and Kress, 1987a, b) presented formulations and some numerical examples for acoustic and electromagnetic waves. (Hirose, 1987) discussed some problems for elastic waves. To cope with the difficulty of ill-posedness in the inverse problems, methods of regularization are presented in (Groetsch, 1984; Morozov, 1984, and Hofmann, 1986) as well.

In this report the author would like to summarize some recent works concerning the numerical solution to inverse scattering problem being done in his office. The problem consists of finding the shape of the smooth obstacle based on the observation of scattering amplitude (far-field pattern) of acoustic waves of low frequencies. If the far-field pattern in all directions is known exactly, the problem will be reduced to a mathematical minimization problem. If the far-field pattern is known only in a finite number of directions, we shall also consider the restoration of the pattern in the rest of directions as one of the moment problems. We shall also consider a special case when some Fourier components of the far-field pattern are given. Before embarking the solution of inverse problems, one is required to be able to resolve the direct problems satisfactorily. Numerical examples in direct scattering problem are presented, including scattering from circular cylindrical obstacles, numerical demonstration of the Fraunhofer diffraction. A simple numerical example in inverse problem is presented for the shape identification based on the exact knowledge of far-field pattern corresponding to a soft scatterer.

2. INVERSE SCATTERING PROBLEMS

We shall first describe direct scattering problem rather than we start the problem statements with the description of the inverse problem. Let Ω be a simply connected bounded domain enclosed by the boundary Γ in two-dimensional Euclidean space with the coordinates $x = (x_1, x_2)$. We may think of Ω as the cross section of an infinitely long cylindrical scatterer. We assume that Γ is smooth of the class $\mathbf{C}^2$. By Ω^e we shall denote the exterior domain to Ω.
The scattering to be considered in this paper is described by the spatial form $u(x)$ of the scalar wave

$$u(x) = u^i(x) + u^s(x), \quad x \in \overline{\Omega}^e, \tag{1}$$

where u^i denotes the incident plane wave of the form

$$u^i = e^{ik(d_1 x_1 + d_2 x_2)}, \tag{2}$$

with the wave number k, the incident direction (d_1, d_2), and u^s the scattered wave. The scattered wave has to satisfy the Helmholtz equation

$$(\Delta + k^2)u^s(x) = 0 \quad \text{in} \quad \Omega^e. \tag{3}$$

The Sommerfeld radiation condition in two-dimensional case, which inhibits the reflection of waves from the infinity, is imposed as

$$\frac{\partial u^s}{\partial r} - iku^s = o\Big(\frac{1}{\sqrt{r}}\Big) \quad \text{as} \quad r = |x| \to \infty, \tag{4}$$

uniformly with respect to the angle $\theta = \arctan(x_2/x_1)$.
If Ω is a soft obstacle, the boundary condition is given in terms of the total wave u by the homogeneous Dirichlet condition

$$u = 0 \quad \text{on} \quad \Gamma. \tag{5}$$

If Ω is a hard obstacle, the corresponding boundary condition is the homogeneous Neumann condition

$$\frac{\partial u}{\partial \nu} = 0 \quad \text{on} \quad \Gamma, \tag{6}$$

where ν denotes the exterior unit normal to Γ. The direct problem is to find u^s in $\mathbf{C}^2(\Omega^e) \cap \mathbf{C}(\Omega^e \cup \Gamma)$ satisfying (1)–(6) for given u^i.

Both the exterior Dirichlet and Neumann problems of the Helmholtz equation with the radiation condition are uniquely solvable. The solution u^s has the integral representation

$$u^s(\xi) = \int_\Gamma \left\{ u^s(x) \frac{\partial G}{\partial \nu(x)}(x,\xi) - \frac{\partial u^s}{\partial \nu}(x) G(x,\xi) \right\} d\Gamma(x), \quad \xi \in \Omega^e \tag{7}$$

with the fundamental solution G to the operator $\Delta + k^2$

$$(\Delta + k^2) G(x,\xi) = -\delta(\xi), \tag{8}$$

$$G(x,\xi) = \frac{i}{4} H_0^{(1)}(k|x-\xi|), \tag{9}$$

where $H_0(1)$ is the Hankel function of the first kind of order zero satisfying the radiation condition (4). According to (Kress, 1989), we know that the scattered wave has an asymptotic expansion of the form

$$u^s(\xi) = \sum_{n=-\infty}^{+\infty} a_n H_n^{(1)}(kr) e^{in\theta} \quad \text{as} \quad r \to \infty, \tag{10}$$

with $\xi = (r,\theta)$ and the constants a_n.
The scattered wave has the asymptotic form

$$u^s(\xi) = \frac{e^{ikr}}{\sqrt{r}} \left\{ f(\theta) + O\left(\frac{1}{r}\right) \right\} \quad \text{as} \quad r \to \infty, \tag{11}$$

where $f(\theta)$ is the scattering amplitude or far-field pattern of the scattered wave. The scattering cross section is defined by $\sigma(\theta) = 4|f(\theta)|^2/k$. We notice that the far-field pattern can be expanded in the Fourier series

$$f(\theta) = \sum_{n=-\infty}^{+\infty} b_n e^{in\theta}, \tag{12}$$

with

$$b_n = a_n \sqrt{\frac{2}{\pi k}} e^{-\frac{2n+1}{4}\pi i}. \tag{13}$$

This shows that $f(\theta) = 0$ implies $u^s = 0$; one-to-one correspondence between solutions (10) to the Helmholtz equations satisfying the radiation condition and their far-field patterns $f(\theta)$.
In terms of the total wave, the representation (7) has the alternative form

$$u(\xi) = u^i(\xi) + \int_\Gamma \left(u \frac{\partial G}{\partial \nu} - \frac{\partial u}{\partial \nu} G \right) d\Gamma, \quad \xi \in \Omega^e. \tag{14}$$

In the limit as the point ξ tends to a point on the boundary, we can see

$$\frac{\Theta(\xi)}{2\pi} u(\xi) = u^i(\xi) + \int_\Gamma \left(u \frac{\partial G}{\partial \nu} - \frac{\partial u}{\partial \nu} G \right) d\Gamma, \quad \xi \in \Gamma, \tag{15}$$

where $\Theta(\xi)$ denotes the exterior planar angle to Ω at the boundary point ξ. For a smooth Γ, we have $\Theta(\xi) = \pi$. The equation (15) is the boundary integral equation corresponding to the exterior problem of the Helmholtz equation (3).

We now describe the inverse problem. We first assume that there is only one scatterer Ω from *a priori* knowledge. Consider a Jordan curve S in Ω^e, which contains Ω in its interior. We assume that S is sufficiently smooth. We may think of S as an observatory contour, along which the waves are measured. Suppose that the total wave u_∞ is obtained at every point on S. The inverse scattering problem is to find shape and location of the scatterer Ω from given u^∞ on S for known one or more incident waves u^i.

The inverse problem in a usual sense is to find the shape of the scatterer Ω from given far-field pattern $f(\theta)$ for one or more incident plane waves.

Suppose that the far-field patterns can be known exactly. The uniqueness of the solution to the conventional inverse scattering problem is given by a theorem due to Schiffer (see, e.g., (Colton and Kress, 1983), Theorem 6.10), stating that Ω is uniquely determined by a knowledge of $f(\theta)$ on some line element of the unit circle ($\theta_0 \leqslant \theta \leqslant \theta_1$) for one incident wave u_i and for any interval of positive wave numbers k ($0 < k_0 \leqslant k \leqslant k_1$).

We notice that the far-field pattern $f(z)$ regarded as a complex-valued function of the complex numbers z is an entire function (see (Colton and Kress, 1983)). From the knowledge of f on a line element of the unit circle, we can determine f on the entire circle by analytic continuation. Moreover, it is sufficient to know f only on a set of infinitely many number of points on a unit circle, which has at least an accumulating point, in order to determine f on the entire unit circle by the unicity theorem.

For the question of uniqueness of the inverse problem, we can also refer to (Kirsch and Kress, 1987b), mentioning that Ω is determined by $f(\theta)$ for some finite number of u_i of the same wave number k with different incoming directions.

The far-field pattern $f(\theta)$ depends continuously on the shape of Ω, provided the surface is of $\mathbf{C}^2$-class. However, we notice that Ω does not depend continuously on $f(\theta)$. Moreover, in practice the far-field pattern is only determined from measurements inherently subject to a certain amount of experimental error. The available far-field pattern may not belong to the class of functions (entire functions of exponential type) corresponding to the direct scattering problem. In this case, no solution exists to our inverse scattering problem. We must also suffer from the lack of information in the observation. The far-field pattern is usually measured at a few selected points. To restore the far-field pattern on the entire unit circle, the Legendre polynomial interpolation may be applied as numerical continuation of analytic functions (see (John, 1967)). We shall apply the Backus–Gilbert method for the restoration in the next section.

3. THE NUMERICAL OPTIMIZATION METHOD

We shall confine the geometry of Γ to a star-like curve: $r = \phi(\theta)$, $0 \leqslant \theta \leqslant 2\pi$ with the periodic real positive single-valued function ϕ of $\mathbf{C}^2$-class with the period 2π. We assume that for a given $\epsilon > 0$, there exists $N(\epsilon)$ such that ϕ is uniformly approximated by an N-parameter function ψ as $\left|\phi(0) - \psi(\theta; c_1, c_2, \ldots, c_N)\right| \leqslant \epsilon$.

To this purpose, we may think of the Fourier series approximation of ϕ, as the Fourier series of such functions converge uniformly.

3.1. *Exact data*

We assume that, for a given incident wave u_i, the exact far-field pattern $f(\theta)$ are known on an entire unit circle. To make our problem specific, let Ω be a soft obstacle. Corresponding boundary condition on the surface Γ of Ω is the Dirichlet condition (5). For a hard obstacle, the method can be developed in a similar way. Our inverse scattering problem is to find Γ from the knowledge of $f(\theta)$.
The method of solution is iterative and two-fold (solution of direct problem plus mathematical minimization using homotopy method) nonlinear parameter optimization. It will be safe to first assume that an initial guess $\Gamma(0)$ to Γ is known. For $n = 0, 1, 2, \ldots$, we consider the following process:
(i) Solve the boundary integral equation of the first kind

$$Gq^{(n)}(\xi) = \int_{\Gamma(n)} q^{(n)}(x)G(x,\xi)d\Gamma(x) = u^i(\xi), \quad \xi \in \Gamma(n) \tag{16}$$

for unknown $q^{(n)}(x)$, $x \in \Gamma(n)$.
(ii) Calculate corresponding far-field pattern $f^{(n)}(\theta)$, $0 \leqslant \theta < 2\pi$ using (1), (11) and (14).
If $f^{(n)}(\theta)$ coincides with the given $f(\theta)$, then $\Gamma^{(n)}$ is a required solution of our inverse scattering problem. Otherwise, $\Gamma^{(n)}$ must be modified to $\Gamma^{(n+1)}$ so that $f^{(n)}(\theta)$ approaches nearer to the given $f(\theta)$ in some sense and the above process (i) and (ii) is continued for next n.
The integral operator G is deduced from the single-layer acoustic potential. We notice that the equation (16) is uniquely solvable in $\mathbf{C}(\Gamma^{(n)})$ if and only if the wave number k is not an interior Dirichlet eigenvalue (Colton and Kress, 1983). Since the equation is a Fredholm integral equation of the first kind, the numerical solution is likely to stain by computational errors. To circumvent the difficulty, we shall consider the operator G in $\mathbf{L}^2$ and apply the Tikhonov regularization: Minimize the functional

$$\left\| \int_{\Gamma(n)} q^{(n)} G d\Gamma - u^i \right\|_{\mathbf{L}^2(\Gamma(n))} + \alpha \| q^{(n)} \|_{\mathbf{L}^2(\Gamma(n))} \tag{17}$$

with respect to $q^{(n)}$ with the regularization parameter α. Let the minimum be denoted by $q\alpha^{(n)}$. It is given as the solution of the linear equation

$$(G^*G + \alpha I)q_\alpha^{(n)} = G^*u^i, \tag{18}$$

where G^*: $\mathbf{L}^2(\Gamma(n)) \to \mathbf{L}^2(\Gamma(n))$ is the adjoint of G.
We shall have some remarks. If we consider a hard obstacle, the corresponding integral operator will be associated with the double-layer acoustic potential and it is a compact operator: $C(\Gamma^{(n)}) \to C(\Gamma^{(n)})$. From the Riesz–Fredholm theory we can see that the equation is uniquely solvable in $C(\Gamma^{(n)})$ if and only if k is not an

interior Dirichlet eigenvalue. If $\Gamma^{(n)}$ has corners, the integral operator fails to be compact in general. Under an assumption on the geometry of corners, which may not hinder engineering applications, (Kleinman and Wendland, 1977) presented the Fredholm theory for the integral equation of the second kind.
The algorithm is not complete, unless some way of improvement from $\Gamma^{(n)}$ to $\Gamma^{(n+1)}$ is specified. Our plan is the reduction to the nonlinear minimization problem

$$\min_{c_j^{(n)}} \int_0^{2\pi} | f^{(n)}(\theta) - f(\theta)|^2 \, d\theta. \tag{19}$$

This implies that N parameters $c_j^{(n)}$ are searched for so that the integral in (15) is to be minimized. The minimization process can be realized by one of the methods in mathematical optimization. Unless $\Gamma^{(0)}$ is sufficiently close to Γ, the Newton-like methods may not work out well. We shall employ robust methods, e.g., Box's complex algorithm, for constrained minimization for this purpose (Huester and Mize, 1973).
However, due to our numerical experience the Box's algorithm applied to the inverse problems under consideration often converges to several local minima, depending on the initialy assumed shape of the scatterer. To circumvent the drawback and to make sure of the convergence to a minimum corresponding to the exact shape, we introduce a far-field pattern $g(\theta)$ of the unit circle subject to the same incident wave u^i and we consider the homotopy

$$h(\theta, t) = t f(\theta) + (1 - t) g(\theta), \quad 0 \leqslant t \leqslant 1. \tag{20}$$

From the linearity of the set of all far-field patterns of the form (12), we know that $h(\theta, t)$ corresponds to a scattered wave for all such t. Instead of (19), we shall consider

$$\min_{c_j^{(n)}(t)} \int_0^{2\pi} |h^{(n)}(\theta, t) - h(\theta, t)|^2 \, d\theta. \tag{21}$$

Starting with $t = 0$, we determine $c_j^{(n)}(t)$ as we increase t up to $t = 1$. When $t = 1$, the estimated shape with $c_j^{(n)}(t)$ is considered to correspond the given far-field pattern $f(\theta)$. The method is called homotopy continuation method (Kojima, 1981).

3.2. *Exact but lack of information*

In a practical situation, we have to cope with the lack of observations. We shall assume that the scattered wave is observed only at M distinct locations $\xi_j (j = 1, 2, \ldots, M)$. Let $f(\theta_j)$ be the corresponding values of the far-field pattern. We assume that $f(\theta_j)$ are exact. To restore the analytic function $f(\theta)$ for all θ, we consider the following Cauchy integral representation on the unit circle $C\colon |\zeta| = 1$.

$$\frac{1}{\pi i} (\mathcal{P}) \int_C \frac{f(\zeta)}{\zeta - z} \, d\zeta = f(z), \qquad |z| = 1 \tag{22}$$

in the sense of Cauchy's principal values. This leads us to the linear moment problem with M equations

$$\frac{1}{\pi i}(\mathcal{P})\int\limits_C \frac{f(\zeta)}{\zeta - z_j}\,d\zeta = f(\theta_j), \qquad e^{i\theta_j} = z_j/|z_j|. \tag{23}$$

The problem can be resolved by the Backus–Gilbert method (see Colton and Kress, 1983; Section 7.1) in the form

$$f_M(\theta) = \sum_{j=1}^{M} f(\theta_j)a_j(z), \qquad z = e^{i\theta}, \tag{24}$$

where the functions $a_j(z)$ are given by the expressions

$$a = (a_1, a_2, \ldots, a_M)^T = \frac{1}{g^T S^{-1} g} S^{-1} g, \tag{25}$$

with

$$S_{jk}(z) = \frac{1}{\pi i}(\mathcal{P})\int\limits_C (\zeta - z)^2 \frac{1}{(\zeta - z_j)(\zeta - z_k)}\,d\zeta = z_k + z_j - 2z, \tag{26}$$

and

$$g_k = \frac{1}{\pi i}(\mathcal{P})\int\limits_C \frac{1}{\zeta - z_k}\,d\zeta = 1. \tag{27}$$

The algorithm in this case is reduces to the nonlinear minimization problem

$$\min_{c_j^{(n)}} \int\limits_0^{2\pi} \left|f^{(n)}(\theta) - f_M(\theta)\right|^2 d\theta. \tag{28}$$

3.3. *Fourier components*

Instead of the far-field pattern $f(\theta)$, we shall assume that some Fourier components of the far-field pattern are given. To make our problem specific, let the lower M components c_j in the Fourier expansion

$$f(\theta) = \sum_{j=-\infty}^{+\infty} c_j e^{ij\theta}, \qquad c_j = \frac{1}{2\pi}\int\limits_0^{2\pi} f(\theta)e^{-ij\theta}\,d\theta \tag{29}$$

are given. The restoration of $f(\theta)$ can be reduced to the inverse Fourier transform and the minimization in (28) is considered again. Alternatively in an application of the Backus–Gilbert method, we can see

$$S_{jk} = \begin{cases} \frac{1}{i\pi(j+k)}\left(\theta - \pi + \frac{i}{j+k}\right), & j+k \neq 0, \\ \frac{1}{6\pi}(3\theta^2 - 6\pi\theta + 4\pi^2), & j+k = 0, \end{cases} \tag{30}$$

$$g_k = \delta_{k_0}. \tag{31}$$

4. EXAMPLES

We shall present two examples from the direct scattering problem.

4.1. *Direct problems*

As an incident wave, we consider the plane wave

$$u^i = e^{ikx_1} \tag{32}$$

of the unit amplitude. The exact solution of the direct problem for a circular scatterer with the radius a is presented by (King and Wu, 1959) for $ka = 3.1$.

We solve the problem numerically by the boundary element collocation method using linear finite element interpolation functions. Fig. 1 shows the mesh used in the approximation.

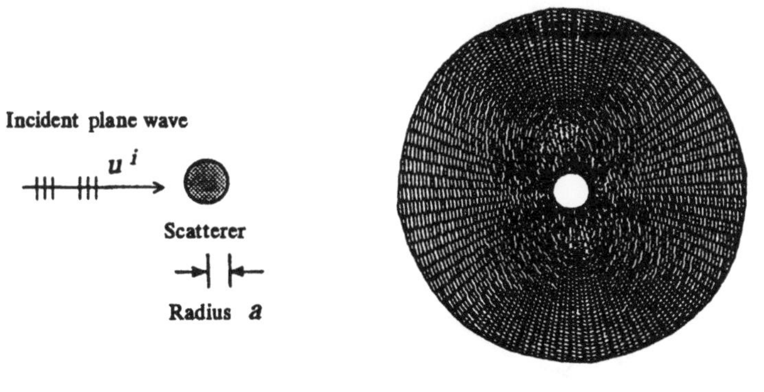

Fig. 1. Scattering by a circular cylinder and the mesh

Suppose for the moment that Ω is a soft obstacle. Fig. 2 shows the calculated contours for the wave components; real, imaginary, absolute and argument of the total wave u.

We shall assess the sufficient magnitude of r in the asymptotic form (11) when we want $f(\theta)$ from calculated scattered wave $u^s(x)$. To this purpose, we calculated far-field patterns for different radii $r = 10, 10^2, 10^3, 10^4$ as the higher order term decreases of the same order of $1/r$. We know from Fig. 3 that $r = 10^3$ is sufficient for our present purpose.

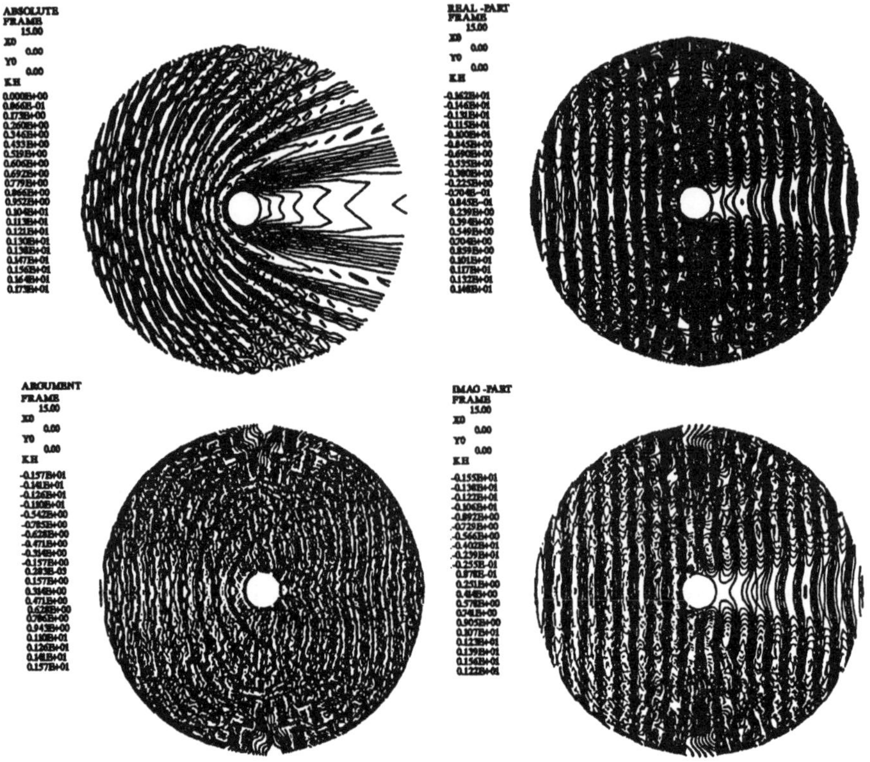

Fig. 2. Calculated waves for a soft circular obstacle ($ka = 3.1$)

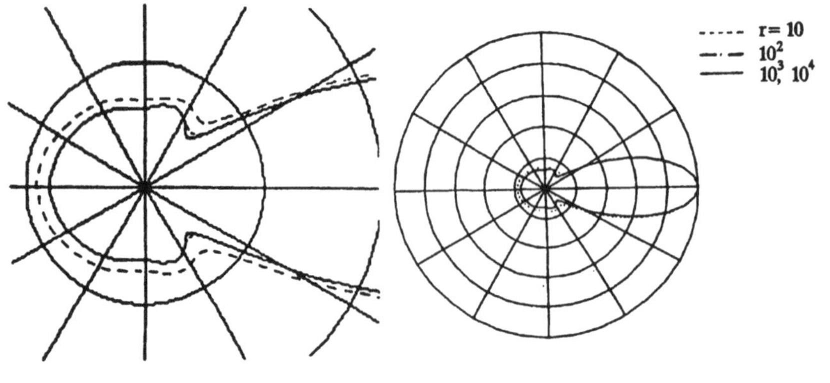

Fig. 3. Convergence of scattering cross sections for a unit circular cylinder as $r \to \infty$

Since the equation (15) reduces in this case to the Fredholm integral equation of the first kind, the Tikhonov regularization is applied with the regularization parameter $\alpha = 0.1$ which was determined intuitively from our previous experiment. Fig. 4 shows the effect of regularization. We can see smoother variations in boundary fluxes.

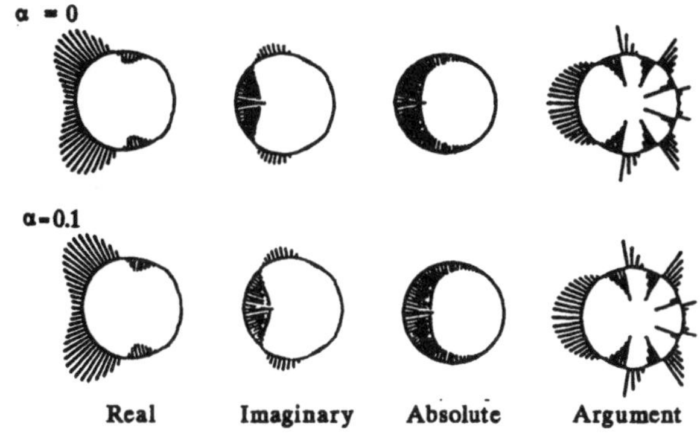

Fig. 4. Boundary fluxes

The exterior Dirichlet problem of the Helmholtz equation (3) is uniquely solvable. However, the boundary integral equation (15) is not uniquely solvable for the interior eigenvalues ka satisfying

$$J_n(ka) = 0\,(n = 0, 1, 2, \ldots) \tag{33}$$

with the Bessel functions of order n with a zero $ka = 2.40483\ldots$ $(n = 0)$. The direct search for corresponding zero in numerical computation showed that $ka = 2.42349$ with the $\log|$ *determinant of the coefficient matrix* $| = -122.934$. Fig. 5 shows calculated boundary fluxes corresponding to the interior numerical eigenvalue. The calculation was done using single precision arithmatics.

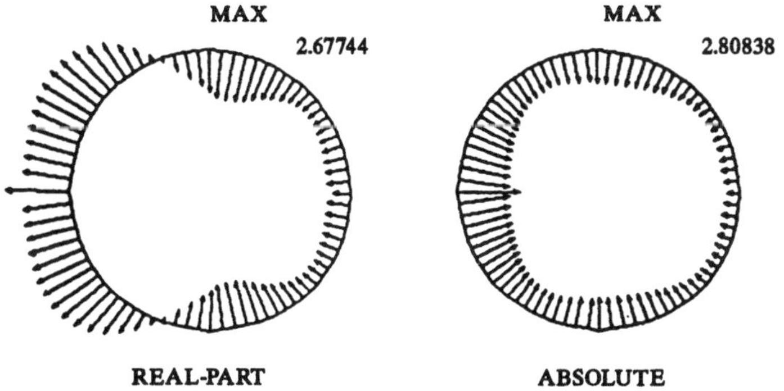

Fig. 5. Calculated fluxes with an interior resonant solution ($ka = 2.42349$)

Suppose that Ω is now a hard obstacle. Fig. 6 shows the calculated contours of the wave components.

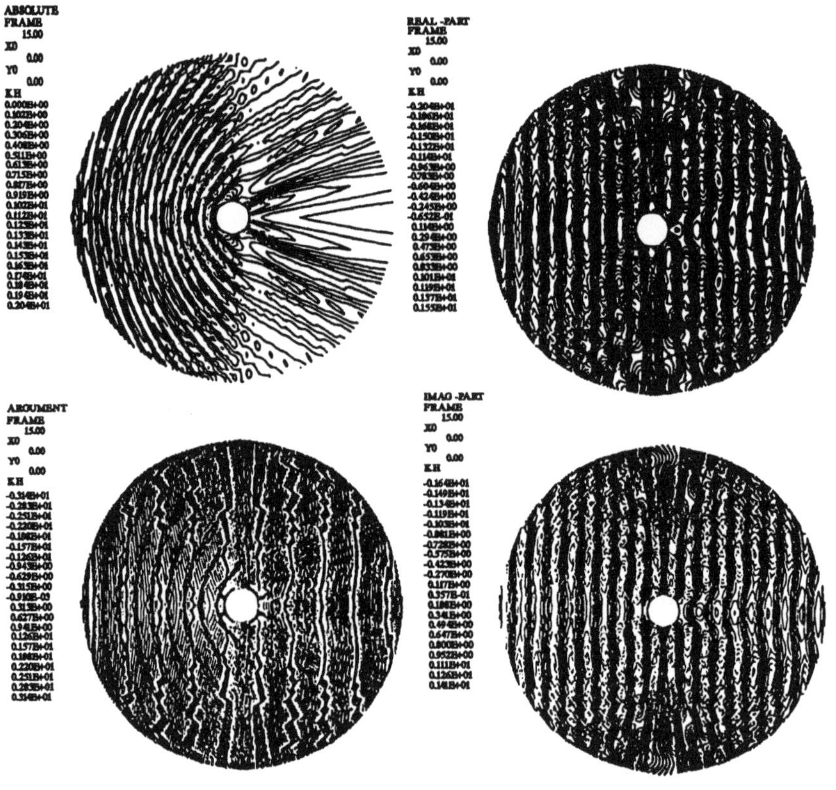

Fig. 6. Calculated waves for a hard obstacle ($ka = 3.1$)

The exterior Neumann problem of the Helmholtz equation is uniquely solvable. However, the integral equation (15) is not uniquely solvable for the interior eigenvalues ka satisfying (33). The direct search for corresponding zero in numerical computation shows that $ka = 2.40889$ with the $\log|$ *determinant of the coefficient matrix* $| = -24.3440$. However, fairly good results can be obtained even for the interior numerical eigenvalue.

4.2. *Diffraction*

We consider diffraction of the plane wave (32) through a slit as shown in Fig. 7. The width δ of the slit and the wave length λ are chosen as $\delta = \lambda = 1$. The wave number $k = 2\pi$.

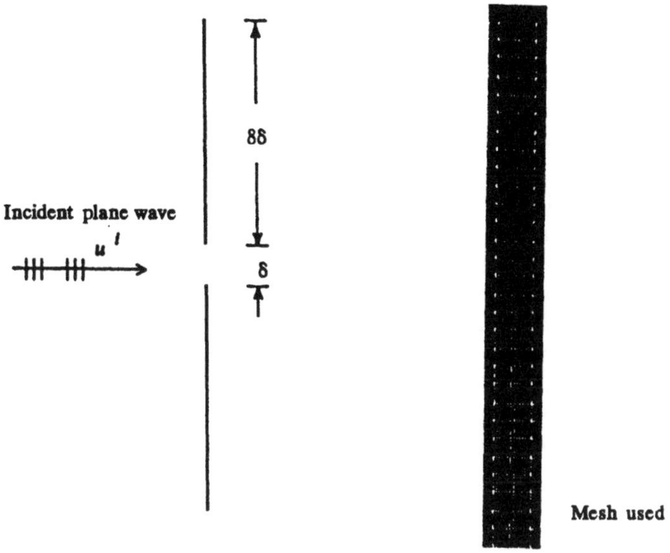

Fig. 7. Diffraction of a plane wave through a slit

Fig. 8 shows calculated results. The Fraunhofer diffraction can be seen from the contours of the absolute values of total wave.

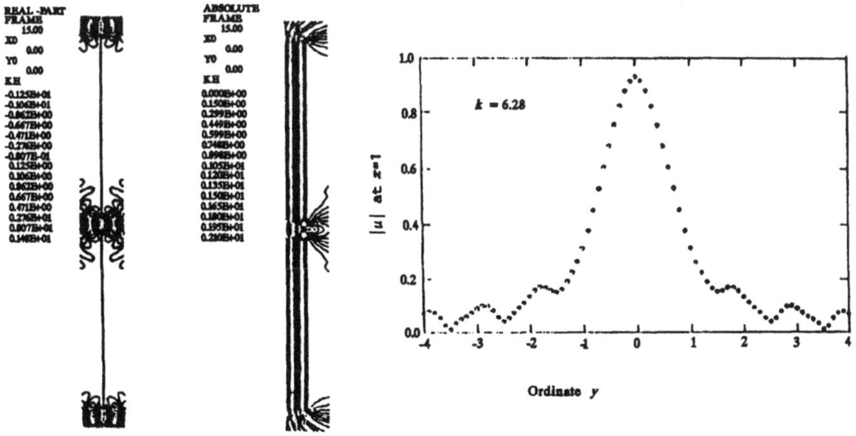

Fig. 8. Calculated diffraction by a slit of soft obstacle

4.3. *Inverse problems*

We shall present an inverse problem consisting of finding the shape of an obstacle. We took the complex-valued far field pattern $f(\theta)$. As the class of functions approximating the shape of cylindrical cross section we assumed

$$\psi(\theta; c_1, c_2, \dots, c_N) = r_0 + a_1 \cos\theta + b_1 \sin\theta \tag{34}$$

with unknown parameters r_0, a_1, b_1 to be determined.
We shall consider the case when the far-field pattern $f(\theta)$ corresponding to a soft scattering obstacle is known exactly in all directions $0 \leqslant \theta < 2\pi$. As we mentioned in the last section, no one knows whether the shape of the obstacle can be uniquely determined mathematically unless far-field patterns are given for all wave numbers k contained in some interval $0 \leqslant k_1 \leqslant k \leqslant k_2$. In what follows, we shall study numerically how much the shape can be identified based on the exact knowledge of one far-field pattern corresponding to the plane wave (32) with $k = 3.1$.
Box's complex algorithm: Prior to the solution of the inverse problem, we calculated values of the exact far-field pattern $f(\theta_j)$ for $\theta_j = 2\pi j/90$ $(j = 1, 2, \dots, 90)$ corresponding to a cylinder with the cross section of the unit circle $r_0 = 1$, $a_1 = b_1 = 0$ in the expression (34). The shape was approximated by an equilateral polygon with 40 vertices. The object function to be minimized is

$$\sqrt{\sum_{j=1}^{90} \left| f^{(n)}(\theta_j) - f(\theta_j) \right|^2}. \tag{35}$$

We took $(\alpha, \beta, \gamma, \delta) = (1.3, 10^{-6}, 5, 10^{-5})$ for the parameters involved in the Box's complex algorithm. The set of constraints we set are $0 < \psi$, $0 < r_0 \leqslant 2$, $-0.5 \leqslant a_1$, $b_1 \leqslant 0.5$. Starting with the initial values $r_0(0) = 1.5$, $a_1(0) = 0.2$, $b_1(0) = -0.2$, the minimization process converged unduly with the minimum value 2.406, resulting in the shape $r_0(^{100}) = 0.555$, $a_1(^{100}) = -0.5002$, $b_1(^{100}) = -0.00181$. Process of the estimation is shown in Fig. 9. The experiment failed to identify the shape of the scatterer.

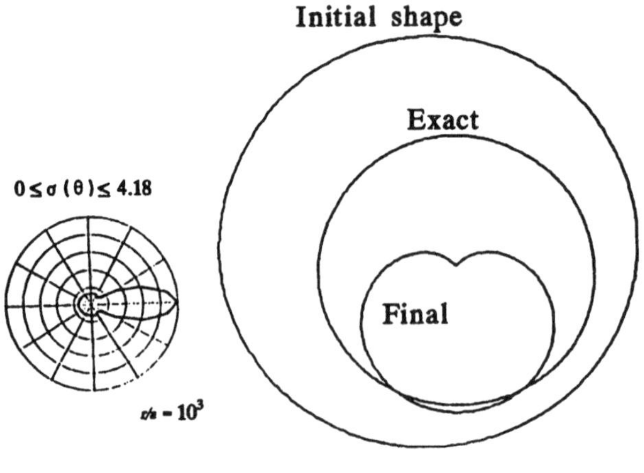

Fig. 9. Process of shape identification

To examine the reason for the failure, we depicted the surface of the objective function for $0.1 \leqslant r_0 \leqslant 2$, $-0.5 \leqslant a_1 = b_1 \leqslant 0.5$ as shown in Fig. 10. We can observe that the objective function is not unimodal, has several local minima in particuler in the vicinity of the global minimum.

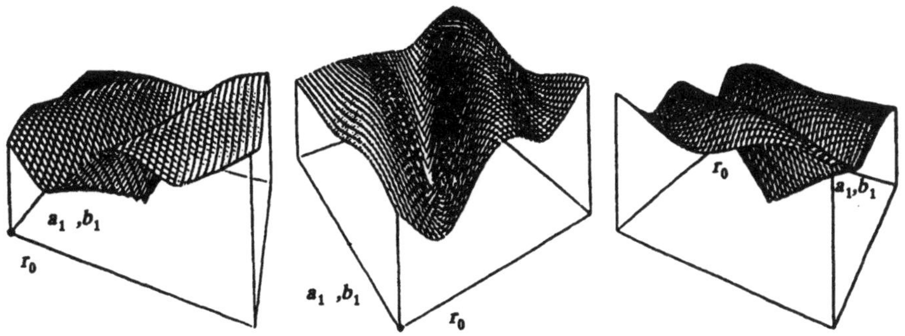

Fig. 10. Surface of the objective function

Homotopy continuation method. We applied the homotopy to make sure the convergence to the global minimum, independently on the initial shape assumed. For a numerical experiment, we considered a cardiac cylindrical scatterer $r_0 = 0.53840$, $a_1 = b_1 = -0.33333$ as the target. Fig. 11 shows the homotopy of far-field patterns and process of identification, starting with the initially assumed unit circular cylinder converging toward the exact shape.

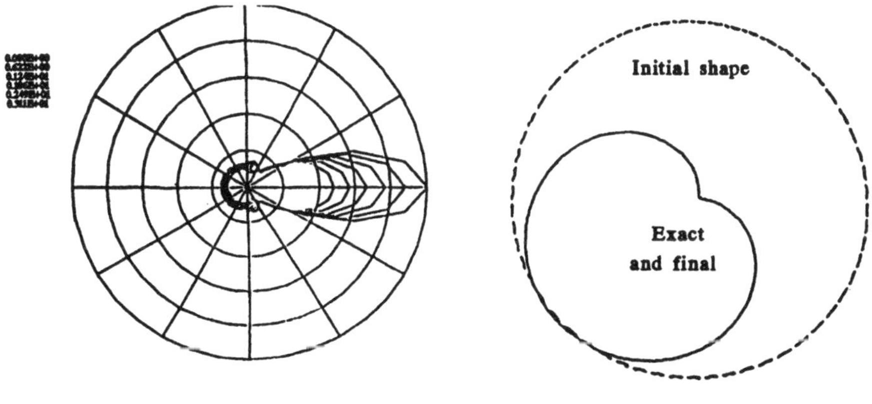

Fig. 11. Identification using homotopy method

The numerical process of convergence is summarized in Table 1. The residual is the value of objective function (35). The final residual corresponding to the homotopy $t = 1$ is small. However, residuals corresponding to the intermediate values of t are large, possibly because far-field patterns $h(\theta, t)$ with $0 < t < 1$ do not correspond the shape that can be represented by the expression (34)

Table 1. Homotopy process.

Homotopy No.	# of iterations	r_0	a_1	b_1	Residual
0	–	1	0	0	–
1	32	0.96451	−0.03139	−0.17009	0.17408
2	31	0.92578	−0.06606	−0.03724	0.32051
3	31	0.88193	−0.10451	−0.06314	0.43542
4	41	0.83947	−0.13915	−0.09488	0.51607
5	29	0.75543	−0.20861	−0.13200	0.52556
6	46	0.72182	−0.22172	−0.16664	0.50036
7	47	0.66874	−0.25776	−0.20926	0.42621
8	27	0.62215	−0.28703	−0.25094	0.31384
9	40	0.58133	−0.30905	−0.29302	0.17010
10	121	0.53872	−0.33295	−0.33330	0.00089
Exact		0.53840	−0.33333	−0.33333	

N.B. $r_0 + a_1 \cos\theta + b_1 \sin\theta$ with the constraints $0 < r_0 \leqslant 2$, $-0.5 \leqslant a_1, b_1 \leqslant 0.5$.
Box's complex algorithm with $\alpha = 1.3$, $\beta = 10^{-4}$, $\gamma = 5$, $\delta = 10^{-4}$.

5. CONCLUDING REMARKS

We presented some numerical examples in direct and inverse scattering problems. Based on the numerical technique we have together with the formulation proposed in this report, we are going futher to develop computational method for the determination of the shape of obstacles.

Acknowledgements
Computer implementation is much due to Mrs. Y. Ohura and Mrs. K. Kobayashi.

REFERENCES

Colton, D. and Kress, R. (1983). *Integral Equation Methods in Scattering Theory.* John Wiley, New York.

Groetsch, C. (1984). *The Theory of Tikhonov Regularization for Fredholm Equations of the First Kind.* (Ser. Research Notes in Mathematics, **105**) Pitman Advanced Publishing Program, Boston.

Hirose, S. (1987). *A Study on Direct and Inverse Scattering Problems of Elastic Waves.* Doctor Thesis. Kyoto Univ. Press, Kyoto.

Hofmann, B. (1986). *Regularization for Applied Inverse and Ill-Posed problems.* (Ser. Teubner-Texte zur Mathematik, **85**) Teubner Verlagsgesellschaft, Leipzig.

John, F. (1967). *Lectures on Advanced Numerical Analysis.* Gordon and Breach, Science Publishers, New York.

King, R.W.P. and Wu, T.T. (1959). *The Scattering and Diffractions of Waves.* Harvard Univ. Press, Cambridge, Massachusetts.

Kirsch, A. and Kress, R. (1987a). A numerical method for an inverse scattering problem. In: *Inverse and Ill-Posed Problems.* Ed. by H.W. Engl and C.W. Groetsch. Academic Press, Boston, pp. 279–289.

Kirsch, A. and Kress, R. (1987b). An optimization method in the inverse acoustic scattering. In: *Boundary Elements IX, Vol. 3, Fluid Flow and Potential Appl.* Ed. by C.A. Brebbia, W.L. Wendland and G. Kuhn. Springer-Verlag, Berlin, pp. 3–18.

Klienman, R.E. and Wendland, W.L. (1977). On Neumann's method for the exterior Neumann problem for the Helmholtz equation *J. Math. Anal. Appl.* **57**, 170–202.

Kojima, M. (1981). *Complementarity and Fixed Points.* (Ser. Mathematical Programming, **9**) Sangyou-Tosho, Tokyo (in Japanese).

Kress, R. (1989). *Linear Integral Equations.* (Ser. Applied Mathematical Sciences, **82**) Springer-Verlag, Berlin.

Kuester, J.L. and Mize, J.H. (1973). *Optimization Techniques.* McGraw-Hill, New York.

Landsberga, P.L. and Borovikova, V.A. (1982). *Numerical Methods of Diffraction Theory.* Mir, Moscow (in Russian).

Morozov, V.A. (1984). *Methods for Solving Incorrectly Posed Problems.* Springer-Verlag, New York.

Ill-Posed Problems in Natural Sciences, pp. 584 – 594
A. Tikhonov (Ed.)
1992 VSP/TVP

ON THE HYPERSINGULAR FIRST KIND INTEGRAL EQUATIONS FOR PROBLEMS OF ELECTROMAGNETIC SCATTERING FROM SCREEN SURFACES

E.V. ZAKHAROV and I.V. HALEJEVA
Moscow State University, Len. Gory, 119899 Moscow, Russia

ABSTRACT

One-dimensional and surface hypersingular integral equations for two- and three-dimensional problems of electromagnetic scattering from perfectly conducting screens are considered. Conditions of existence, uniqueness and regularity of solutions are investigated. Properties of hypersingular operators and their inverse operators acting in some spaces of Hölder continuous functions are studied. Efficient numerical methods for solving these equations are developed.

1. INTRODUCTION

A number of diffraction problems describing electromagnetic scattering from perfectly conducting screen surfaces are of great importance for applied electrodynamics, radio location, antenna designing, etc. Recently a numerical study of such scattering problems was successfully conducted using hypersingular integral equations (HIE's) on an nonclosed curve in the two-dimensional case or on nonclosed surfaces in the three-dimensional case. This method made it possible to develop and implement a number of computational algorithms (see e.g., (Zakharov and Pimenov, 1982; Davydov et al., 1984)). Therefore the study of the HIE's arising in diffraction theory is of major importance. The purpose of this paper is twofold: (i) to investigate properties of one-dimensional and surface hypersingular integral operators (HIO's) and (ii) to prove the convergence of the traditional numerical scheme for one-dimensional HIE and to suggest an efficient numerical method for surface HIE.

In §2 we consider the one-dimensional HIE of the two-dimensional Neumann problem for the Helmholts equation that describes the electromagnetic scattering of TM-waves from infinitely long cylindrical screens. The uniqueness and the existence of the solution are proved. The investigation of the HIO and its inverse operator when they are bounded and unbounded is also presented.

The numerical method of piecewise constant approximation and collocation for solving one-dimensional HIE is described in §3. A uniform estimate of convergence of the approximate solution to the exact one is obtained.

The surface HIE of the three-dimensional problem of electromagnetic scattering from perfectly conducting screens is considered in §4. The solution of this HIE in the space of Hölder uniformly continuous tangential vector fields on any inner part of an open surface is an ill-posed problem because both the direct and the left inverse operators are unbounded. It is proved that an HIO is bounded in the space of Hölder continuous tangential fields with Hölder continuous surface divergence on any interior part of an unclosed surface. In the particular case of unclosed surfaces of revolution the above mentioned space proves to be the correctness set as inverse operator is bounded (§5).
Finally, in §6 an efficient numerical technique for solving surface HIE's is described. This technique is based on piecewise-linear approximation and collocation. The generalized minimized residual (GMRES) method permits to solve surface HIE's for electrically large surfaces in many cases where direct methods are not available because of large sizes of the matrices. A necessary condition for the convergence of GMRES method is obtained.

2. THE ONE-DIMENSIONAL HYPERSINGULAR INTEGRAL EQUATION

The two-dimensional Neumann problem for the Helmholts equation

$$\Delta u + k^2 u = 0, \quad r \in \mathbf{R}^2 \backslash \overline{\Gamma} \tag{1}$$

with the boundary condition

$$\frac{\partial u}{\partial n} = \frac{\partial u^0}{\partial n}, \quad r \in \Gamma \tag{2}$$

on the unclosed contour $\Gamma \in C^{2,\mu}$, $0 < \mu \leq 1$, combined with the Sommerfeld radiation condition and the edge condition is reduced to the solution of the first-kind integral equation (see (Zakharov and Pimenov, 1982))

$$\int_{-1}^{1} \frac{\varphi(t)}{(x-t)^2}\, dt + \int_{-1}^{1} \mathcal{K}(x,t)\varphi(t)\, dt = f(x), \tag{3}$$

where

$$\mathcal{K}(x,t) = \frac{\mathcal{K}_0(x,t)}{|x-t|^{\lambda}}, \qquad \mathcal{K}_0(x,t) \in C^{0,\alpha},$$

$$f(x) \in C^{0,\alpha}, \quad \alpha < 1 - \lambda < \mu.$$

In the operator form equation (3) can be written as

$$T\varphi = (T_0 + K)\varphi = f. \tag{4}$$

The principal part of the operator T is an integral operator with strong singularity of the kernel, namely, it is a one-dimensional hypersingular operator with the integral understood as Hadamard finite part integral (see (Zakharov et al., 1989))

Let us consider some properties of a one-dimensional HIO T. At first introduce the notation for some spaces and auxiliary operators. We shall consider the following Banach spaces of functions:

$$C_*^{0,\alpha} := \{\varphi(t) | \varphi = \varphi^*(t)(1-t^2)^{-1/2}, \quad \varphi^* \in C^{0,\alpha}\},$$
$$C_*^{1,\alpha} := \{\varphi(t) | \varphi = \varphi^*(t)(1-t^2)^{1/2}, \quad \varphi' \in C_*^{0,\alpha}, \quad \varphi^* \in C^{0,\alpha}\}.$$

Let us consider the one-dimensional singular integral operator with index $\varkappa (\varkappa = -1, 1)$

$$(L_\varkappa \varphi)(x) := \int_{-1}^{1} \frac{\varphi(t)}{t-x}\, dt.$$

From the properties of this operator $L_\varkappa$ (see (Muskhelishvili, 1968)) it follows in particular that the operator L_1: $C_*^{0,\alpha} \to C_*^{0,\beta}$ is bounded and has the bounded right operator L_1^*: $C_*^{0,\alpha} \to C_*^{0,\beta}$; the operator L_{-1}: $C^{0,\alpha} \to C^{0,\beta}$ is bounded and has the bounded left operator L_{-1}^*: $C^{0,\alpha} \to C^{0,\beta}$, where

$$\beta = \min\{\alpha - \varepsilon, 1/2\} \tag{5}$$

and ε is any positive constant. Introduce the following notation for the differentiation operator:

$$(Df)(x) := \frac{df}{dx},$$

and the right inverse of the operator D:

$$(D_{x_0}^{-1} f)(x) := \int_{x_0}^{x} f(t)\, dt.$$

Using these notation and the properties of Hadamard finite part integral (see (Zakharov et al., 1989)) we can represent the one-dimensional HIO T and its inverse in the form

$$T_0 = DL_{-1} = L_1 D, \tag{6}$$
$$T_0^{-1} = L_{-1}^* D_{x_0}^{-1} = D_{-1}^{-1} L_1^*. \tag{7}$$

For HIE (4) the following existence theorem is true.

THEOREM 1. *If the homogeneous equation corresponding to* (4) *has only the trivial solution, then for any right-hand side $f \in C^{0,\alpha}$ equation* (4) *has a solution $\varphi \in C_*^{1,\beta}$, where β is defined by* (5).

Proof. By inverting the principal hypersingular part of the operator T in (4), we obtain the following equation:

$$(E + T_0^{-1} K)\varphi = T_0^{-1} f, \tag{8}$$

where E is the unit operator. From (7) it follows that $T_0^{-1} f \in C_*^{1,\beta}$ for any function $f \in C^{0,\alpha}$. Since the space $C_*^{1,\beta}$ is compactly imbedded in $C^{0,\alpha}$, the operator K: $C^{0,\alpha} \to C^{0,\alpha}$ is compact and the operator T_0^{-1}: $C^{0,\alpha} \to C_*^{1,\beta}$ is bounded, we have that the operator $T_0^{-1}K$: $C_*^{1,\beta} \to C_*^{1,\beta}$ is compact. Therefore equation (8) is a Fredholm equation of the second kind in the Banach space $C_*^{1,\beta}$. By the Fredholm alternative, the uniqueness of the solution of the HIE implies its existence.

Remark. Note that the uniqueness of the solution of the HIE follows from the uniqueness theorem for solutions of the corresponding boundary-value problem (1),(2) (see (Hönl et al., 1961)).
From representation (6) and Theorem 1 we see that the HIO T is bounded in the pair of spaces T: $C_*^{1,\beta} \to C^{0,\beta}$ and has the bounded inverse operator T^{-1}: $C^{0,\beta} \to C_*^{1,\beta}$. In the wider space of Hölder uniformly continuous functions $C^{0,\beta}$ in the pair of operators (T, T^{-1}) the direct operator T is unbounded whereas its inverse T^{-1} is compact. Such "one-sided" improperness suggests solving by approximate equation (4) by a numerical method based on piecewise constant approximation of the unknown function and referred to as the method of piecewise constant approximation and collocation.

3. NUMERICAL METHOD FOR SOLUTION OF ONE-DIMENSIONAL HIE

Consider the equispaced partition of the segment $[-1, 1]$ by the system of points $\{x_i\}$, $x_i = -1 + h(i-1)$, $i = 1, \dots, N+1$ into partial segments $[x_i,\ x_{i+1}]$ of length $h = 2/N$ and choose the system of nodes $x_{0i} = x_i + h/2$, $i = 1, \dots, N$. Let us take the characteristic functions of the partial segments $\lambda_i, i = 1, \dots, N$, as the system of basis local functions. Denote by H^N the linear span of this system and introduce on this space the norm of the space L_∞

$$\|f\|_{H^N} = \|f\|_{L_\infty} = \operatorname{ess}\sup_{[-1,1]} |f|.$$

Let Q_N be the operator of piecewise constant approximation of function f by its values at the points $\{x_{0i}\}$:

$$(Q_N f)(x) := \sum_{i=1}^{N} f(x_{0i})\lambda_i(x) = f^N(x).$$

Then the method of piecewise constant approximation and collocation of the solution of equation (4) leads us to the following equation in the space H^N:

$$Q_N(T_0 + K)Q_N\widetilde{\varphi} = f^N \tag{9}$$

and we obtain the following theorem.

THEOREM 2. *There exists a number N_0 such that equation (9) is uniquely solvable for any $N \geq N_0$ if equation (4) has a unique solution $\varphi(t)$. The approximate solution $\widetilde{\varphi}(t)$ of (9) converges to the exact solution of (4) as $N \to \infty$. Furthermore, we have the uniform estimate*

$$\|\widetilde{\varphi}^N(t) - \varphi(t)\|_{L_\infty} \leq Ch^\beta, \tag{10}$$

where h is the length of the partial segment, C is a constant not depending on of h, and β is defined by (5).

Proof. In Theorem 1 equation (4) is reduced to the equation

$$(E + A)\varphi = F, \tag{11}$$

where $F = T_0^{-1}f$, $A = T_0^{-1}K$, which is the Fredholm equation of the second kind in the space L_∞ since the operator A: $L_\infty \to L_\infty$ is compact. There exists the inverse operator of the principal part of the discrete operator in equation (9) $T_{0N} = Q_N T_0$ (see (Belotsercovskii and Lifanov, 1985)) and the following estimate holds:

$$\|Q_N T_0^{-1} - T_{0N}^{-1} Q_N\|_{C^{0,\alpha} \to L_\infty} \leq C h^\beta. \tag{12}$$

Inverting the principal part of the operator T_{0N} in (9) and taking into account estimate (12) we obtain the following equation in the space H^N:

$$(E + A_N)\tilde{\varphi}^N = \widetilde{F}^N, \tag{13}$$

where

$$\|F^N - \widetilde{F}^N\|_{L_\infty} \leq C h^\beta, \qquad A_N = T_{0N}^{-1} Q_N K.$$

Using the compactness of the operator A: $L_\infty \to C^{0,\beta}$ and estimate (12) one can easily show that

$$\|A_N - A\|_{L_\infty} \leq C h^\beta.$$

This inequality shows that for a sufficiently large N equation (13) is uniquely solvable because equation (11) has a unique solution; for the operator $E + A_N$, the inverse operator $(E + A_N)^{-1}$ exists and is bounded uniformly in N, and for the solutions of equations (4),(9) estimate (10) is valid.

4. SURFACE HYPERSINGULAR INTEGRAL EQUATION

The three-dimensional problem of diffraction of the time-harmonic electromagnetic field $\{E, H\}$ from the open perfectly conducting surface $S \in C^{2,\alpha}$ with edge $C_0 \in C^{2,\alpha}$ which is placed in a homogeneous isotropic medium is described mathematically by the boundary value problem for Maxwell equations

$$\nabla \times H = ikE, \qquad \nabla \times E = -ikH, \qquad r \in \mathbf{R}^3 \backslash \overline{S} \tag{14}$$

with the boundary condition

$$n \times E|_\pm = -n \times E^0, \quad r \in S, \tag{15}$$

the radiation conditions and the edge condition (see (Hönl et al., 1961)). This problem has been reduced to the solution of the equation (see (Davydov et al. 1984))

$$\widehat{N} J = -in \times [n \times E^0], \tag{16}$$

where the operator $\widehat{N}$ is defined by

$$(\widehat{N}J)(r) = n \times \left[\frac{n}{k} \times \left\{ k^2 \int_S \Phi J \, dS' + \int_S (J \cdot \nabla) \nabla \Phi \, dS' \right\} \right]. \tag{17}$$

Here, n denotes the unit normal vector, its direction determines the positive (+) side of the surface , k is the wave number

$$\mathrm{Im}(k) \leq 0, \tag{18}$$

E^0 is the initial electric field, and

$$\Phi(|r-r'|) = \frac{\exp(-ik|r-r'|)}{4\pi|r-r'|}$$

is the fundamental solution of the Helmholts equation in the three-dimensional space. The unknown function $J = J_+ + J_-$ has the physical meaning of the surface current induced on both sides of the surface. The second integral in the right-hand side of (17) is hypersingular and is considered as an Hadamard finite part integral (see (Zakharov et al., 1989)).

Let us partition the surface S into two parts: the inner part $S_\delta = S \backslash U_\delta$ and the boundary part $S_* = S \cap U_\delta$, where U_δ is the cylindrical neighborhood of the contour C_0 of radius δ. Let ν, τ, n be the orthogonal basis (on the curve C_0, ν denotes the unit normal vector to C_0 which is perpendicular to the surface normal and τ denotes the unit vector tangent to C_0). By φ_δ (φ_*) we denote the contraction of the function φ on $S_\delta(S_*)$. Define the weight function w which is equal to 1 in S_δ and is proportional to the square root of the distance from the curve C_0 in S_* and introduce the space of scalar functions

$$C_*^{0,\alpha}(S) := \{\varphi\colon S \to \mathbf{C}^3 | \varphi_\delta \in C^{0,\alpha},\ \varphi_* = \varphi^* w^{-1},\ \varphi^* \in C^{0,\alpha}\}.$$

Furthermore, introduce the spaces

$$N_{l,m}^\alpha(S) := \{J\colon S \to \mathbf{C}^3 | J \times n = 0,\ J_\delta \in C^{0,\alpha},$$

$$J_* = w^l J_\nu^* \nu + w^m J_\tau^* \tau;\ J_\nu^*, J_\tau^* \tau \in C^{0,\alpha}\}, \quad l,m = -1,0,1$$

of tangential Hölder uniformly continuous fields on any inner part of S and their subspaces of tangential fields possessing the Hölder continuous surface divergence on any inner part of S:

$$M_{l,m}^\alpha(S) := \{J \in N_{l,m}^\alpha | \mathrm{Div}\ J \in C_*^{0,\alpha}\}.$$

Consider also the space $N^{\alpha\perp}$ defined by

$$N^{\alpha\perp}(S) := \{g\colon S\, \mathbf{C}^3 | gn = 0,\ n \times g \in N^\alpha\},$$

and the space $M^{\alpha\perp}$ defined in a similar way.

The vector functions

$$E = \frac{1}{ik}\nabla \times \nabla \times A, \qquad H = \nabla \times A \tag{19}$$

expressed via the vector potential

$$A(r) = \int_S \Phi(|r-r'|)J(r')\,dS' \tag{20}$$

with the tangential density $J \in M_{1,-1}^\alpha$ solve the boundary-value problem (14), (15) provided J is a solution of HIE (16) (see (Zakharov et al., 1989)). The uniqueness of the solution of the HIE follows from the uniqueness theorem for the solution of the corresponding boundary problem (see (Hönl et al., 1961)). The following theorem is valid.

THEOREM 3. *The hypersingular operator $\widehat{N}$ is bounded in the pair of spaces $\widehat{N}$: $M^{\beta}_{1,-1}(S) \to M^{\beta\perp}_{0,-1}(S)$ and unbounded in the wider spaces $\widehat{N}$: $N^{\beta}_{1,-1}(S) \to N^{\beta\perp}_{0,-1}(S)$.*

Proof. The regularity properties of the vector potential (20) and its derivatives for the case of a closed surface (see (Colton and Kress, 1983)) are obviously extended to any inner part of the open surface $\overline{S}_\delta \subset S$. The study of the surface potential behavior in a sufficiently small neighborhood of a point of the contour C_0 is based on the analysis of one-dimensional single-layer potential and its derivatives. Since $\widehat{N}J = \frac{1}{k} n \times [n \times \text{rot rot } A]|_{\pm}$ and $n \times$ rot rot $A|_{\pm} \in N^{\beta}_{0,-1}$, it follows that the operator $\widehat{N}$: $M^{\beta}_{1,-1} \to N^{\beta\perp}_{0,-1}$ is bounded. For any $J \in M^{\beta}_{1,-1}$ the surface divergence Div $[n \times \widehat{N}J]$ exists and is given by the formula Div $[n \times \widehat{N}J] = nk$ rot $A|_{\pm} \in C^{0,\beta}_{*}$. Therefore the operator $\widehat{N}$: $M^{\beta}_{1,-1}(S) \to M^{\beta\perp}_{0,-1}(S)$ is bounded. Since the following representation for $\widehat{N}$ is valid (see (Zakharov et al., 1989))

$$(\widehat{N}J)(r) = n \times \left[\frac{n}{k} \times \left\{k^2 \int\limits_S \Phi J \, ds' + \int\limits_S \nabla\Phi \text{Div } J \, ds'\right\}\right],$$

the operator $\widehat{N}$: $N^{\beta}_{1,-1}(S) \to N^{\beta\perp}_{0,-1}(S)$ is unbounded.

5. SYSTEMS OF EQUATIONS FOR SURFACES OF REVOLUTION

In the particular case of open surfaces of revolution, the vector equation (16) breaks up into a number of systems of two one-dimensional equations for the Fourier coefficients J_n of the current (see (Davydov et al., 1983))

$$R_n J_n = (R_{n0} + M)J_n = f_n. \tag{21}$$

The principal part R_{n0} of the matrix operator R_n is given by

$$R_{n0} = \begin{bmatrix} T_0 & \text{in } L \\ \text{in } L & (n^2 - k^2\rho^2)S \end{bmatrix},$$

where T_0 is a hypersingular operator, L is a singular operator, S is an operator with logarithmic singularity in the kernel, and ρ is the distance from the point on the surface to the axis of revolution. From the properties of operators T_0, L, S it follows that the operator R_n: $C^{1,\beta}_{*} \times C^{0,\beta}_{*} \to C^{0,\beta} \times C^{1,\beta}$ is bounded. The existence theorem for the matrix equation (21) follows from the uniqueness of the solution as in the one-dimensional case.

THEOREM 4. *If $\rho \neq 0$ and the homogeneous equation corresponding to (21) has only trivial solution in the space $C^{1,\beta}_{*} \times C^{0,\beta}_{*}$, then equation (21) has a solution in that space.*

Proof. By inverting the principal part of the operator R_n in equation (21), we obtain the Fredholm equation of the second kind in the Banach space $C^{1,\beta}_{*} \times C^{0,\beta}_{*}$:

$$(E + R_{n0}^{-1}M)J_n = R_{n0}^{-1}f_n$$

which is equivalent to (21). To complete the proof, it suffices to apply the Fredholm alternative.

Remark. Note that the condition $\rho \neq 0$ ensures the existence of the inverse operator R_{n0}^{-1}. It appears here due to our proof technique which is based on the inversion of the principal part of the operator and does not seem to be essential. The inverse of the operator R_n exists and is a bounded operator: R_n^{-1}: $C^{0,\beta} \times C^{1,\beta} \to C_*^{1,\beta} \times C_*^{0,\beta}$. If we consider equation (21) in wider spaces of Hölder functions, then both the operator R_n: $C^{0,\beta} \times C_*^{0,\beta} \to C^{0,\beta} \times C^{0,\beta}$ and its inverse R_n^{-1}: $C^{0,\beta} \times C^{0,\beta} \to C^{0,\beta} \times C_*^{0,\beta}$ are unbounded due to the hypersingular components $(R_n)_{11}$, $(R_n^{-1})_{22}$. In (Davis and Mittra, 1977), numerical results of calculation of the first Fourier coefficient of the current $J_1 = J_{\tau 1}\tau + J_{\varphi_1}\varphi$ induced on a finite circular cylinder are presented which are based on piecewise constant approximation of the unknown function. The calculation of J_{τ_1} is stable whereas that of J_{φ_1} is unstable, the latter component having fast oscillating behavior. The lack of stability of the calculation of J_{φ_1} is due to the fact that $(R_1^{-1})_{22}$ is unbounded.

6. NUMERICAL METHOD FOR SOLUTION OF SURFACE HIE

We can reformulate the results of §5 in terms of the operator $\widehat{N}$: in the particular case of open surfaces of revolution the operator $\widehat{N}$: $M_{1,-1}^{\beta} \to M_{0,-1}^{\beta\perp}$ has a bounded inverse: $\widehat{N}^{-1}$: $M_{0,-1}^{\beta\perp} \to M_{1,-1}^{\beta}$ whereas both operators $\widehat{N}$ and $\widehat{N}^{-1}$ are unbounded in the wider spaces of Hölder tangential fields ($N_{1,-1}^{\beta} \to N_{0,-1}^{\beta\perp}$).
The use of piecewise constant functions as a basis approximation system leads to instability of calculation of the unknown function $J(r)$ because both the operator and its inverse are unbounded. Let us consider a numerical method which requires no special treatment of the edge condition, preserves the simplicity of calculation of matrix elements as in the case of piecewise constant approximation and collocation and ensures the stability of calculation.
Let us describe a method for numerical solution of equation (16) in the case where the surface S is a rectangle with sides a_x, a_y situated on the coordinate plane XOY. The method is based on piecewise linear approximation of the current density. Let us partition the rectangle S into the partial rectangles $S_{m,n} = [x_{m-1}, x_m] \times [y_{n-1}, y_n]$, $m = 1, \ldots, N_x$, $n = 1, \ldots, N_y$, where

$$r_{\alpha,\beta} = (x_\alpha, y_\beta), \qquad x_\alpha = \alpha h_x, \qquad y_\beta = \beta h_y,$$

$$h_x = a_x/Nx, \qquad h_y = a_y/N_y.$$

Choose the set of rooftop functions (see (Glisson and Wilton, 1980)) as a basis system of finite elements

$$f_x^{m,n}(x,y) = \begin{cases} 1 - |x - x_m|/h_x, & r \in S_{m,n} \cup S_{m+1,n}, \\ 0, & r \notin S_{m,n} \cup S_{m+1,n}, \end{cases}$$

$$m = 1, \ldots, N_x - 1, \qquad n = 1, \ldots, N_y;$$

$$f_y^{m,n}(x,y) = \begin{cases} 1 - |y - y_n|/h_y, & r \in S_{m,n} \cup S_{m,n+1}, \\ 0, & r \notin S_{m,n} \cup S_{m,n+1}, \end{cases}$$

$$m = 1, \dots, N_x, \qquad n = 1, \dots, N_y - 1.$$

We shall seek for the cartesian components of the unknown function J^N in the form

$$J_x^N = \sum_{n=1}^{N_y} \sum_{m=1}^{N_x-1} J_x^{m,n} f_x^{m,n}(r),$$

$$J_y^N = \sum_{n=1}^{N_y-1} \sum_{m=1}^{N_x} J_y^{m,n} f_y^{m,n}(r). \tag{22}$$

By substituting functions (22) in equation (16) and using Stokes theorem for hypersingular operators, we reduce the second integral in (17) to

$$\int_S (J^N \cdot \nabla) \frac{\partial \Phi}{\partial \eta} \, dx' \, dy' = \frac{\partial}{\partial \eta} \int_S \operatorname{Div} J^N \Phi \, dx' \, dy', \quad \eta = x, y. \tag{23}$$

Using the collocation method with the points of collocation $r_{i,j,-1/2}$ for the x – component of equation (16) and $r_{i-1/2,j}$ for the y – component, replacing the derivatives in (23) by the central difference derivatives we obtain the system of equations

$$\begin{aligned} &\sum_{m=1}^{N_x-1} \sum_{n=1}^{N_y} J_x^{m,n} \Big\{ k^2 I_{t,l} + \frac{I_{t+1,l} - 2I_{t,l} + I_{t-1,l}}{h_x^2} \Big\} \\ &+ \sum_{m=1}^{N_x} \sum_{n=1}^{N_y-1} J_y^{m,n} \frac{I_{t+1,l} - I_{t,l} - I_{t+1,l-1} + I_{t,l-1}}{h_x h_y} \\ &= -ikE_x^0(r_{i,j-1/2}), \quad i = 1, \dots, N_x - 1; \quad j = 1, \dots, N_y, \\ &\sum_{m=1}^{N_x-1} \sum_{n=1}^{N_y} J_x^{m,n} \frac{I_{t,l+1} - I_{t,l} - I_{t+1,l+1} + I_{t-1,l}}{h_x h_y} \\ &+ \sum_{m=1}^{N_x} \sum_{n=1}^{N_y-1} J_y^{m,n} \Big\{ k^2 I_{t,l} + \frac{I_{t,l+1} - 2I_{t,l} + I_{t,l-1}}{h_y^2} \Big\} \\ &= -ikE_y^0(r_{i-1/2,j}), \quad i = 1, \dots, N_x, \quad j = 1, \dots, N_y - 1, \end{aligned} \tag{24}$$

where

$$I_{t,l} = \int_{S_{1/2,1/2}} \Phi(|r' - r_{t,l}|) \, dx' \, dy', \qquad t = i - m, \quad l = j - n.$$

Remark. The use of the system of rooftop functions is convenient for the surfaces with a rectangular parameter domain. In the case of a surface of an arbitrary form one can consider the more general system of piecewise linear basis functions defined on a pair of adjacent triangles built on the base of triangulation of the surface (see (Rao et al., 1982)).

The numerical solution of the HIE of electromagnetic scattering problems for electrically large surfaces leads to linear algebraic systems of equations of large orders. Direct methods are not applicable to the solution of such systems if their matrices are of general form. The generalized minimized residual (GMRES) method (see (Eremin and Orlov, 1990)) has been used for numerical solution of linear system of equations (24). The necessary condition for the GMRES method to converge is fulfilled due to dissipation of the hypersingular operator $\widehat{N}$.

THEOREM 5. *For any function $J \in M^{\alpha}_{1,-1}(S)$, $J \not\equiv 0$ the following statement holds:*

$$\operatorname{Im}\langle \widehat{N}J, J\rangle = \operatorname{Im}\int_S (\widehat{N}J \cdot \overline{J})\, ds > 0.$$

Proof. Let us take an arbitrary function $J \in M^{\alpha}_{1,-1}$, $J \not\equiv 0$ as the density of vector potential (20) and let us construct vector functions (19). Let the surface S be surrounded by a smooth closed surface S_ε. We apply the Gauss' theorem to the function $\overline{E} \times H$ in the domain bounded by the surface S_ε and the sphere of radius R. Using the radiation condition at infinity, we obtain, after taking the real part,

$$\operatorname{Re}\int_{S_\varepsilon} (n, \overline{E}, H)\, ds = -\operatorname{Im}(k)\int_{\mathbf{R}^3} |H|^2 + |E|^2\, dx$$
$$+ \int_{|x|=1} |F_E|^2\, d\sigma,$$

where F_E is the far-field pattern of E. Letting S_ε tend to S, and taking into account the edge condition and the relations $\widehat{N}J = \mathrm{i}n \times [n \times E]|_{\pm}$ and $n \times (H_+ - H_-) = J$ we obtain the representation

$$\operatorname{Im}\,\langle \widehat{N}J, J\rangle = -\operatorname{Im}\,(k)\int_{\mathbf{R}^3} |H|^2 + |E|^2\, dx + \int_{|x|=1} |F_E|^2\, d\sigma.$$

Since, by the uniqueness of the solution of diffraction problem (14), (15), the condition $|F_E| \equiv 0$ implies the equality $J \equiv 0$ (see (Colton and Kress, 1983)) and k satisfies inequality (18), the latter representation completes the proof.
The computational experiments have shown that the GMRES method is more effective for the solution of HIE's than other iterative methods (the minimized residual method, the conjugate gradient method).

REFERENCES

Belotsercovskii, S.M. and Lifanov, I.K. (1985). *Numerical Methods in Singular Integral Equations*. Nauka, Moscow (in Russian).

Colton, D. and Kress, R. (1983). *Integral Equation Methods in Scattering Theory*. Jonh Wiley, New York.

Davis, W.A. and Mittra, D.R. (1977). A new approach to the thin scatterer problem using the hybrid equations. *IEEE Trans. Ant. Prop.* **AP-25**, 402–406.

Davydov, A.G., Zakharov, E.V., and Pimenov, Yu.V. (1983). Numerical solution method for problems of electromagnetic scattering by open surfaces of revolution. *Dokl. Akad. Nauk SSSR* **269**, 329–332 (in Russian).

Davydov, A.G., Zakharov, E.V., and Pimenov, Yu.V. (1984). Numerical solution method for problems of electromagnetic scattering by open surfaces of arbitrary shape. *Dokl. Akad. Nauk SSSR* **276**, 96–100 (in Russian).

Eryomin, Yu.A. and Orlov, N.V. (1990). Quasi-solution of problems of acoustic scattering by thin space screens. *Matem. Modelirovanie.* **2**, 103–110 (in Russian).

Glisson, A.W. and Wilton, D.R. (1980). Simple and efficient numerical methods for problems of electromagnetic radiation and scattering from surfaces. *IEEE Trans. Ant. Prop.* **AP-30**, 409–418.

Hönl, H., Maue, A.W., and Westpfahl, K. (1961). Theorie der beugung In: *Handbuch der Phisik,* **25/1**, Ed. by S. Flugge. Sprinder-Verlag, Berlin

Muskhelishvily, N.I. (1968). *Singular Integral Equations.* Nauka, Moscow (in Russian).

Rao, S.M., Wilton, D.R., and Glisson, A.W. (1982). Electromagnetic scattering by surfaces of arbitrary shape. *IEEE Trans. Ant. Prop.* **AP-30**, 409–418.

Zakharov, E.V. and Pimenov, Yu.V. (1982). *Numerical Analysis of Radio Waves Diffraction.* Radio i Svyaz, Moscow (in Russian).

Zakharov, E.V., Davydov, A.G. and Halejeva, I.V. (1989). Integral equations with Hadamard finite part integrals in diffraction problems. In: *Actual Questions of Applied Mathematics,* Ed. by A.N. Tichonov, and A.A. Samarskii, Moscow Univ. Press, Moscow, pp. 118–127 (in Russian).

LIST OF CONTRIBUTORS

NATIONAL ORGANIZING COMMITTEE

Acknowledgement

It is our please to express our gratitude to the editorial group of TVP Sci. Publ. that has prepared these proceedings for publication. Of those people whose contribution to this work was of particular importance we would like to mention V.I. Khokhlov, M.I. Tikhomirova, and N.V. Tretyakov.

The Editors

Bei Fragen zur Produktsicherheit wenden Sie sich bitte an:
If you have any questions regarding product safety,
please contact:

Walter de Gruyter GmbH
Genthiner Straße 13
10785 Berlin
productsafety@degruyterbrill.com